FUNDAMENTALS OF COGNITIVE NEUROSCIENCE

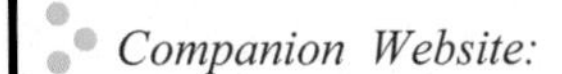

Companion Website:

http://textbooks.elsevier.com/web/Manuals.aspx?isbn=9780128038130

Fundamentals of Cognitive Neuroscience, 2e

Nicole M. Gage, Bernard J. Baars, Editors

Available Resources:

- Downloadable complete lecture slides for each chapter
- Student Study Guide for each chapter
- Instructor Test Bank and Key for Multiple Choice and Short Answer Questions for each chapter
- Sample Syllabi for Semester and Quarter Terms

FUNDAMENTALS OF COGNITIVE NEUROSCIENCE

A BEGINNER'S GUIDE

SECOND EDITION

Nicole M. Gage
Bernard J. Baars

Academic Press is an imprint of Elsevier
125 London Wall, London EC2Y 5AS, United Kingdom
525 B Street, Suite 1800, San Diego, CA 92101-4495, United States
50 Hampshire Street, 5th Floor, Cambridge, MA 02139, United States
The Boulevard, Langford Lane, Kidlington, Oxford OX5 1GB, United Kingdom

Library of Congress Cataloging-in-Publication Data
A catalog record for this book is available from the Library of Congress

British Library Cataloguing-in-Publication Data
A catalogue record for this book is available from the British Library

ISBN: 978-0-12-803813-0

For information on all Academic Press publications visit our website at https://www.elsevier.com/books-and-journals

Publisher: Nikki Levy
Acquisition Editor: Joslyn Chaiprasert-Paguio
Editorial Project Manager: Timothy Bennett
Production Project Manager: Anusha Sambamoorthy
Designer: Victoria Pearson

Typeset by TNQ Books and Journals

Contents

4. The Art of Seeing

5. Sound, Speech, and Music Perception

8. Attention and Consciousness

9. Decisions, Goals, and Actions

10. Humans Are Social Beings

11. Feelings

12. Sleep and Levels of Consciousness

13. Disorders of Consciousness

14. Growing Up

Preface

We invite you to join us in exploring the mind and brain in the second edition of *Fundamentals of Cognitive Neuroscience: A Beginner's Guide*. What brain processes are at work as we watch a beautiful sunset, listen to a favorite song, or remember the face and name of a new friend? In this edition, we explore the internal universe of the human brain. We look at the structures and functions of the brain to investigate which brain structures are at work as we form our perceptions, make and retrieve memories, and develop our social selves.

Woven through these experiences are the conscious and unconscious threads of our mind as we learn, experience, explore, and make our way through our daily lives. Join us as we see how brain imaging techniques are revolutionizing the way we can "see" into the human brain to unlock the mysteries of how the human mind works.

Our second edition text takes a unique thematic approach, guiding students along a clear path to understand the latest findings whether or not they have a background in neuroscience. It includes case studies and everyday examples designed to help students understand the more challenging aspects of the material.

New chapters in this second edition include a chapter exploring the brain bases of sleep, dreaming, and levels of consciousness from full wakefulness to deep sleep to coma. The neural bases of sleep disorders such as insomnia, sleepwalking, and narcolepsy are provided along with scientific and practical ideas about human sleep—where we spend a full third of our lives.

Another new chapter, *Disorders of Consciousness*, explores the mystery of what happens following brain damage that produces long-term unconsciousness and vegetative states. We provide a discussion of end-of-life decisions and how brain imaging techniques are providing new ways to investigate brain function in an otherwise nonresponsive individual.

A third new chapter, *Feelings*, explores the complex ballet of neuromodulators dancing through the brain, arousing us, helping us calm down, and enabling us to understand the emotional cues of those around us. We end the chapter with a discussion of mood disorders, ranging from depression to posttraumatic stress disorder to bipolar disorder.

Key Features of our Second Edition:

- Provides a complete introduction to mind-brain science, written to be highly accessible to undergraduates with limited neuroscience training
- Richly illustrated with carefully selected color graphics to enhance understanding
- Full update of all chapters from the first edition with new scientific findings as well as ongoing mysteries
- New chapters on *Sleep and Levels of Consciousness, Disorders of Consciousness, and Feelings*
- New text boxes on topics such as
  - #YourBrainOnTwitter...Your brain on Social Media
  - When the War Never Ends...-Posttraumatic Stress Disorder

- Go ahead and sleep in…or why it might be a bad idea to sign up for that 8 a.m. class
- The NFL versus Neuroscience: the CTE Controversy
- Visual Illusions and how they work
- The Neural Language of Music
- Eight Problems for the Mirror Neuron Theory of Action Understanding
- End-of-life decisions in Disorders of Consciousness
- Infantile Amnesia…Why I cannot remember my first birthday
- The Connectogram…A new way to Visualize the Brain
- And many more…

* Enhanced pedagogy highlights key concepts for the student and aids in teaching—chapter outlines, study questions, glossary, and image collection are also available on the student's companion website
* Ancillary support saves instructor's time and facilitates learning—test questions, image collection, and lecture slides available on the instructor's manual website
* Enhanced pedagogy highlights key concepts for the student and aids in teaching. Chapter outlines, study questions, glossary, and image collection are also available on the student's companion website.

We have many people to thank for their guidance, assistance, and support throughout the process of preparing this new book. We thank our editors, April Farr and Joslyn Chaiprasert-Paguio, for their support and guidance throughout this process. Their enthusiasm and friendship were an extra benefit for us! Timothy Bennett, our editorial project manager, guided us through the complex process of transforming our written words into a printed book, and we thank him for his patience and wisdom during the process.

Nicole Gage wishes to thank Greg Hickok and David Poeppel for their mentoring and friendship as she began her journey to understand the mind and brain. She thanks her family and friends for their patience while we created this textbook. She in particular wants to thank her husband Kim for his insight and love.

Bernard Baars also owes a debt of gratitude to Gerald M. Edelman and many colleagues at the Neurosciences Institute in San Diego for a first-rate education in the biology of brains—human, virtual, and squid.

Nicole M. Gage and Bernard J. Baars

POSTSCRIPT

Praise for our first edition!

Awards

Fundamentals of Cognitive Neuroscience is a comprehensive and easy-to-follow guide to cognitive neuroscience. Winner of a ***2013 Most Promising New Textbook Award*** from the Text and Academic Authors Association, this book was written by two leading experts in the field to be highly accessible to undergraduates with limited neuroscience training. It covers all aspects of the field—the neural framework, sight, sound, consciousness, learning/memory, problem-solving, speech, executive control, emotions, socialization and development—in a student-friendly format with extensive pedagogy and ancillaries to aid both the student and professor.

Reviews

This introductory textbook on cognitive neuroscience is a welcome addition to the field...The book won the 2013 Most Promising New Textbook Award in the life sciences from the Text and Academic Authors Association, an award that recognizes excellence in first-year edition textbooks and learning materials.

The Quarterly Review of Biology, June 2014

This introductory text offers a comprehensive and easy-to- follow guide to cognitive neuroscience. Chapters cover all aspects of the field…in a student-friendly format with extensive pedagogy and ancillaries to aid both the student and professor. Throughout the text, case studies and everyday examples are used to help students understand the more challenging aspects of the material. Written by two leading experts in the field, the text takes a unique thematic approach, guiding students along a clear path to understand the latest findings whether or not they have a background in neuroscience.

Doody.com, April 24, 2013

Fundamentals of Cognitive Neuroscience: A Beginner's Guide should be widely used as the required text in focused cognitive neuroscience courses taught at the undergraduate level. Additionally, the information it contains will likely be of use to those professors teaching a variety of psychology and biology elective courses and should be consulted for applicable reading material due to its clarity and style.

MedicalScienceBooks.com (2012)

This introductory text offers a comprehensive and easy-to-follow guide to cognitive neuroscience. Chapters cover all aspects of the field - the neural framework, sight, sound, consciousness, learning/memory, problem solving, speech, executive control, emotions, socialization and development - in a student-friendly format with extensive pedagogy and ancillaries to aid both the student and professor. Throughout the text, case studies and everyday examples are used to help students understand the more challenging aspects of the material. Written by two leading experts in the field, the text takes a unique thematic approach, guiding students along a clear path to understand the latest findings whether or not they have a background in neuroscience.

MedicalScienceBooks.com

There was a lot of ambiguity in the previous texts that I used for my undergraduate cognitive neuroscience course. There was so much waffling between positions and opinions in the other books that it was hard for beginning students to get a handle on basic concepts. Overall the student feedback was quite poor. I think the new Baars/Gage *Fundamentals of Cognitive Neuroscience* book is much more straight forward and to the point with the concepts.

Michael S. Cannizzaro, Ph.D., CCC-SLP
University of Vermont

CHAPTER

1

A Framework for Mind and Brain

THE NEURAL SYMPHONY OF THE MIND AND BRAIN

Consciousness is somehow a by-product of the simultaneous, high frequency firing of neurons in different parts of the brain. It's the meshing of these frequencies that generates consciousness, just as tones from individual instruments produce the rich, complex, and seamless sounds of a symphony orchestra. ***Francis Crick, British Scientist.***

A stylized interpretation of the neurons of the brain. *Source: Public domain.*

1. INTRODUCTION TO THE STUDY OF MIND AND BRAIN

In this chapter, we invite you to join us in exploring the mind-brain. What brain processes are at work as we watch a beautiful sunset, listen to a favorite song, or remember the face and name of a new friend? Join us as we see how brain imaging techniques are revolutionizing the way we can "see" into the human brain to unlock the mysteries of just how the human mind works.

Fundamentals of Cognitive Neuroscience
https://doi.org/10.1016/B978-0-12-803813-0.00001-5

FIGURE 1.1 The Thinker, Rodin, 1902. *Source: Public domain.*

Cognitive neuroscience is the combined study of the mind and the brain. The brain is said to be the most complex structure in the known universe. It can be changed by drinking a cup of cappuccino or by listening to a favorite song. Some neuronal events happen over a thousandth of a second, while others take decades.

This book is written to simplify the facts. We are in a revolution in our understanding of the human mind and brain. The ability to record from the living brain has brought out new facts, raised new ideas, and stirred new questions. Just 15 years ago, we might not have seen a link between cognition and genes, brain molecules, or the mathematics of complex networks. Today, those are hot topics. Some traditional ideas are returning, like consciousness, unconscious beliefs and goals, mental imagery, voluntary control, intuitions, emotions, and the executive self. Many unsolved puzzles remain, but we can see progress (Fig. 1.1).

1.1 Where Do We Begin?

Just as exploring the universe in outer space requires us to consider what level of analysis we will use—*will we investigate the universe at the level of galaxies? stars systems? stars? … or the molecules that form their basis?*—studying the human brain requires us to determine our level of investigative analysis. In Chapter 2, The Brain, we explore the *structure, cellular bases, functional processing, pathways,* and *rhythms of the brain.* To do this, we use five aspects of neuroscience investigation: *neuroanatomy, neurophysiology, functional neuroanatomy, neuroconnectivity,* and *neurodynamics.* These disciplines allow us to separately examine the diverse aspects of the brain as we explore the internal universe inside our heads: our mind-brain (Figs. 1.2 and 1.3).

In Chapter 3, Observing the Brain, we explore the ways to observe the living brain along the same aspects of investigation introduced in Chapter 2. Imaging the living, acting, thinking brain has provided new insights into how our minds work. In Chapter 3, we provide a tutorial of how these methods work and how they relate to human cognition.

FIGURE 1.2 The large spiral galaxy NGC 1232 taken Sep. 21, 1998 during a period of good observing conditions. *Source: http://www.eso.org/public/images/eso9845d/ Produced by the European Southern Observatory, free to use under license of the Creative Commons Attribution 4.0 International License, available on commons.wikimedia.org.*

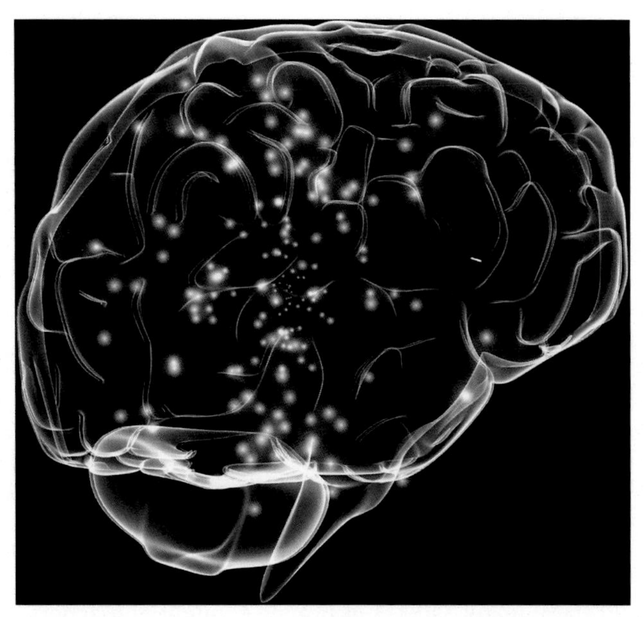

FIGURE 1.3 An abstract image of the brain. *Source: Brain image created by the Defense Advanced Research Projects Agency (DARPA) Systems-Based Neurotechnology for Emerging Therapies (SUBNETS) program. Courtesy of Massachusetts General Hospital and Draper Labs. Public domain figure provided by the US federal government.*

2. CONSCIOUSNESS, CORTICAL CORE, CONNECTIVITY, AND CONSISTENCY

Several themes tend to recur when investigating the brain. A current focus of many scientists' attention is the study of human consciousness in the brain: *the neural correlates of consciousness*. Consciousness themes occur in virtually every chapter of the book but are specifically highlighted in Chapter 8, Attention and Consciousness, and Chapter 13, Disorders of Consciousness.

The *connectivity pathways* of the brain include fiber tracts that form the highways, streets, and small roads that provide the bases for neural communication traffic flow. These pathways provide a key linkage for the billions of nerve cells—neurons—that are the mighty working units of the brain. The *dynamics* of these billions of cells together provide the rhythms of the brain that hum together across varying frequencies that together provide the instantiation of the symphony of neural activity that underlies our mind-brain.

A key feature of the brain is its *consistency*: despite the overwhelmingly complex sights, sounds, textures, aromas, and tastes of our daily experience, the brain extracts consistent elements that form the basis for *that is a chair* or *this smells like chocolate*. This consistency allows us to extract important patterns so that we know that *a dog is a dog* despite being very near to us, and therefore relatively large, or far away, and therefore relatively small, or in an abstract painting, and therefore relatively unrealistic, or in a cartoon, and therefore relatively simplified, or simply as a word on a page, and therefore relatively abstract.

3. CONSCIOUSNESS

The scientific study of human consciousness is once again front-and-center in the field of cognitive neuroscience. Swept away from the foreground of the scientific study of the mind during periods of behavioral and psychophysical investigations, human consciousness—as a valid and scientifically viable topic of research—has landed squarely in the center of contemporary neuroscience. Leaders in the field that opened the doors to the current study of human consciousness include Francis Crick, Christoph Koch, Gerald Edelman, Bernard Baars, Stan Dehaene, Guilio Tononi, and many many more.

3.1 A Useful Framework for Investigating Conscious and Unconscious Brain Processing

Human sensation, perception, language, cognition, attention, goals, memory, actions, and emotions form a multidimensioned discipline of study, each with their core bases in terms of phenomena and each with a distributed—yet highly interactive—brain network subserving their neural bases.

To help understand these many aspects of human thought and action, we provide a functional framework as a simplified way to organize the broad study of the human mind (Fig. 1.4).

The functional framework provides an overview of how sensory information from the world enters our mind-brain to form the input to our mental processes (left side of Fig. 1.4). Chapter 4, The Art of Seeing, and Chapter 5, Sound, Speech, and Music Perception, describe in detail how the sights and sounds of our everyday world combine to form our visual and auditory

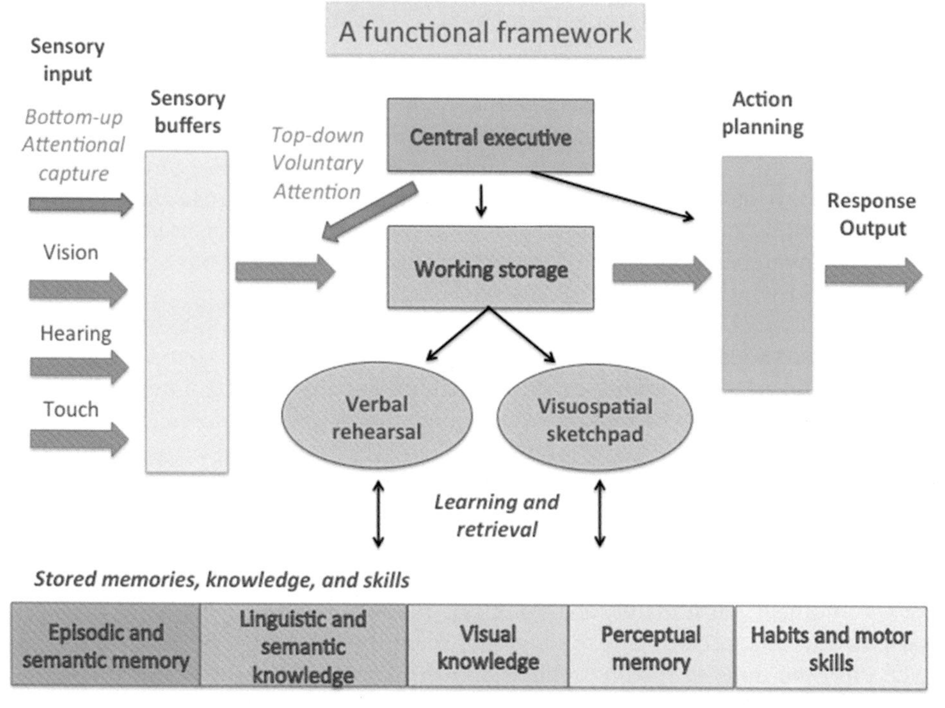

FIGURE 1.4 A functional framework for human cognition. Sensory inputs (vision, hearing, touch, left side of figure) enter the brain system through the sensory pathways. The executive and attentional processes governed by frontal lobe regions serve to focus attention, hold information temporarily in working storage (center of figure), and prepare it for long-term storage (green boxes at bottom of figure). Action plans and response outputs are shown (right side of figure). Note that green-shaded boxes grow lighter in color as the processes and information they represent are mediated by more and more unconscious processing. *Source: Baars & Gage.*

knowledge and ultimately our conceptual representation. That conceptual knowledge is held in networks and brain regions that form our long-term memory cores and language knowledge (bottom of Fig. 1.4) that together provide the basis for decision making and goal setting processes for action planning (right side of Fig. 1.4). Chapter 6, Language and Thought, and Chapter 7, Learning and Remembering, discuss the neural bases of these processes in detail. Chapter 14, Growing Up, describes their development as we grow from infants to adults.

In the center of Fig. 1.4—and at the core of the neural processes that guide our behavior—are *executive functions* that allow us to focus our attention on *this* and not on *that*, to keep something in mind, and to act on goals. Chapter 9, Decisions, Goals, and Actions, provides an overview of the many brain areas that together form our *central executive* that allows us to work through the daily and life-long problems we must solve to make our way through life.

Not shown on the functional framework but key to the very essence of being human are the social and emotional processes that form our personalities, guide our behavior, and lead us through the human interactions that are essential to our survival as happy living creatures. The brain processes that underlie our social and emotional selves are presented in Chapter 10, Humans Are Social Beings, and Chapter 11, Feelings.

3.2 Conscious and Unconscious Processes Together Form the Bases of Our Mental Processes

As Bernard Baars has written, "*Consciousness is the water in which we swim. Like fish in the ocean, we can't jump out to see how it looks from the outside. As soon as we lose consciousness, we can no longer see anything.*" We are well aware of our conscious selves. We can *report* on what is occurring right now as we experience an event or taste a food. Our knowledge of what is occurring moment by moment of our conscious life—*the contents of our consciousness*—forms the base of our awareness. Fluid, flexible, ever-changing contents of our consciousness seem to be a vast sea in which we swim.

Careful experimentation has time and time again proven this feeling to be false. In fact, the contents of our consciousness are quite *limited*. At any given time, only so much information, so many senses, feelings, and thoughts, can share the mind-space of our consciousness. More and more evidence is providing support for the notion that it is our *unconscious processing* that forms the overwhelming majority of brain functions. Our vast store of memories, language knowledge, automaticities, and procedural learning combine to form the largely unconscious *but accessible if recalled* storage of what we have learned and experienced throughout our life. Shown in the green boxes at the base of Fig. 1.4, this information is retrievable, in large part, and forms the bulk of the iceberg of our knowledge store (Fig. 1.5). Conscious contents, at any given moment, form merely the tip of this massive iceberg.

Conscious and unconscious threads together form the thoughts and ideas and actions on any given day. We may be reading a book and thinking consciously about the topic of the book while keeping an eye on the time so we are not late for work. At the same time that these conscious processes enter and recede from the contents of our consciousness, largely unconscious processes are also at work outside our attention or our influence. If we are skilled readers, the process of *reading the book*—decoding the shapes of the print to form letters, decoding the strings of letters for form words, parsing the sequence of words to form sentences and so on—is not something we are usually conscious of. Similarly, we may be *walking around the room* while reading the book—the act of walking is an overlearned process and so the individual motor movements and balance required are largely automatic. Together the conscious, task-based activities of our lives combine with their unconscious and automatic elements to help us achieve our goals, large and small.

3.3 Global States: Waking, Sleeping, and Dreaming

For healthy individuals, our typical conscious state includes a balance of full *wakefulness* and *awareness* (Fig. 1.6). As we move through the three global states of *waking, sleeping, and dreaming*, our level of awareness (contents of our consciousness, y-axis on Fig. 1.6) is coupled with our level of wakefulness (x-axis on Fig. 1.6). As we begin to feel sleepy, then, both the levels of awareness and wakefulness drop until we reach deep sleep where we are neither aware nor awake (Laureys, 2005). We will discuss the *waking conscious state* in more detail in Chapter 8. We will discuss the *sleeping and dreaming states* in Chapter 12, Sleep and Levels of Consciousness. *Disorders of Consciousness*, including coma and vegetative states, are described in Chapter 13.

FIGURE 1.5 Most brain processes are not conscious: the limited capacity of the contents of consciousness at any given moment are represented by the "tip of the iceberg." The vast store of largely unconscious knowledge and representations is not available; however, much of it is retrievable from stable knowledge stores. *Source: http://keywordsuggest.org/gallery/246650.html.*

3.4 The Theater of the Mind and Its Stage

Several aspects of the contents of consciousness are important to note: they are limited in capacity, controlled by voluntary attention, and directed by executive functions. Baars developed a "theater of consciousness" analogy to help understand the many "players" on the stage of consciousness (Fig. 1.7).

According to this analogy, the entire theater—stage, audience, players, and backstage areas—form the basis of conscious and unconscious brain processes (Fig. 1.8). The *theater stage* represents *working memory*. A spotlight on the stage represents voluntary attention. *Only the stage contents illuminated by the attentional spotlight* are conscious. The rest of the theater represents the vast unconscious store of knowledge and memories that can enter the

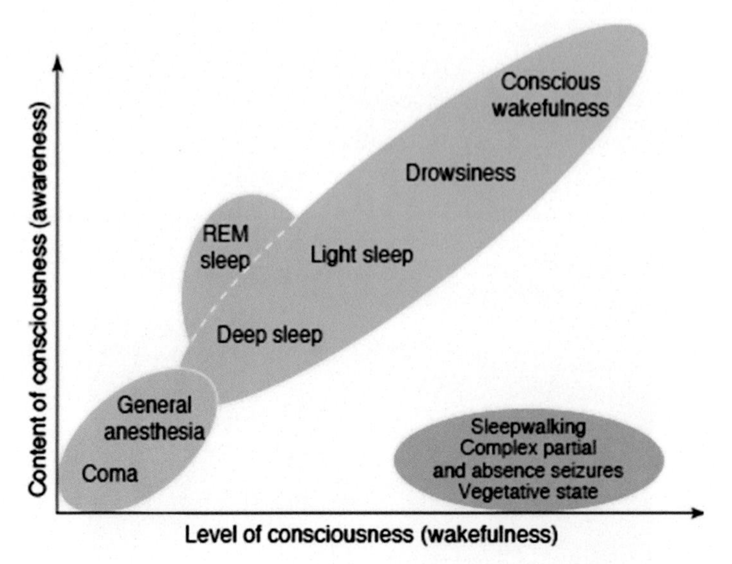

FIGURE 1.6 A simplified version of the global states waking, sleeping, and dreaming showing levels of the two major components of consciousness: awareness and wakefulness. On the vertical y-axis, levels of awareness are shown beginning at the bottom of the axis with no level of awareness, as in a coma or when under anesthesia (*shown in pink*). Moving up the y-axis, level of awareness increases as one moves through heavy sleep, light sleep, and into drowsiness and finally to full conscious wakefulness (*shown in blue*). On the horizontal x-axis, levels of wakefulness are shown beginning with coma and moving similarly through stages of sleep to conscious wakefulness. According to this simplified version, levels of awareness and wakefulness are roughly similar as one moves through the stages from coma to deep sleep to wakefulness, represented by the *pink and blue shapes* forming a diagonal from the bottom left corner, where no awareness or wakefulness are observed in coma, to the upper right, where both levels of awareness and wakefulness are high during conscious wakefulness. For more on this topic, see Chapter 13. *Source: Laureys, 2005.*

FIGURE 1.7 The United Palace, Manhattan, New York. *Source: Commons.wikimedia.org, Author: Beyond my Ken, open source permission under the terms of GNU Free Documentation License.*

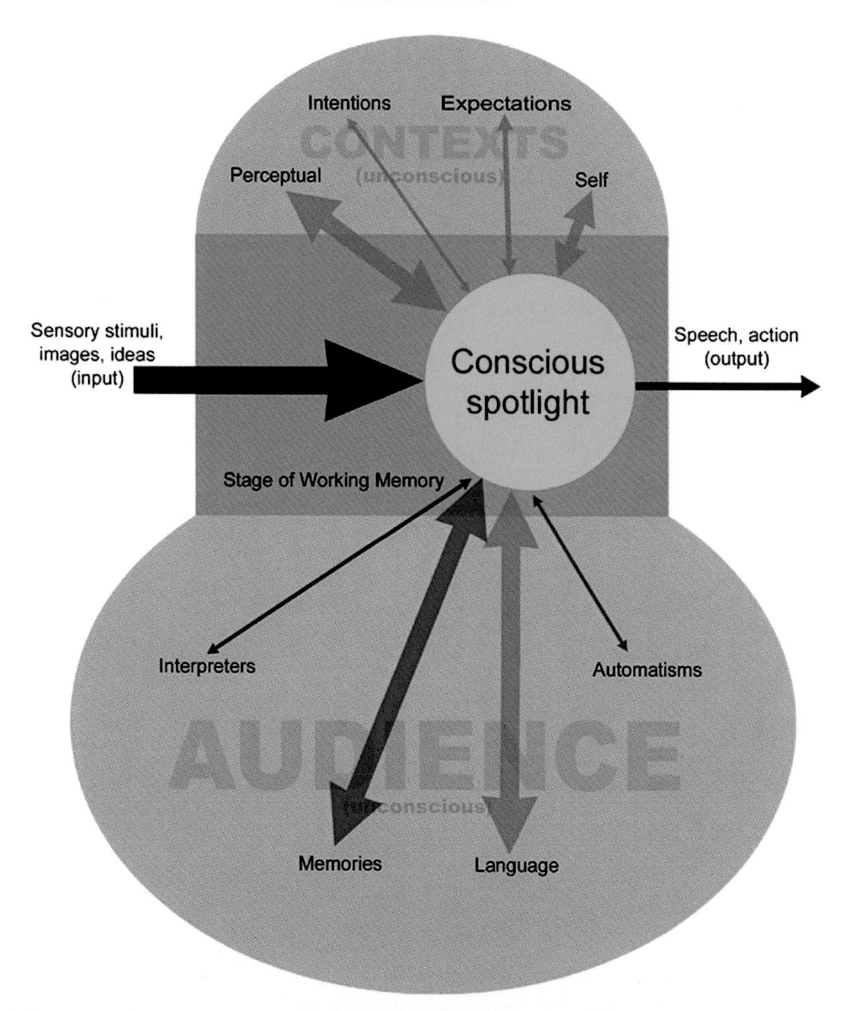

FIGURE 1.8 The theater analogy for the global workspace theory. According to this analogy, the entire theater—stage, audience, players, and backstage areas—form the basis of conscious and unconscious brain processes. The theater stage (*shown in teal*) represents working memory. A spotlight (*shown in turquoise*) on the stage represents voluntary attention. Only the stage contents illuminated by the attentional spotlight are conscious. The rest of the theater (*shown in purple*) represents the vast unconscious store of knowledge and memories that can enter the contents of consciousness once they are on the stage *and* under the spotlight. A key point here is that the spotlight of attention on the stage is very limited in capacity: it represents just a small portion of the stage (working memory), which in turn represents a small portion of the vast theater (unconscious knowledge and processes). *Source: Baars.*

contents of consciousness once they are on the stage *and* under the spotlight. A key point here is that the spotlight of attention on the stage is very limited in capacity: it represents just a small portion of the stage (working memory), which in turn represents a small portion of the vast theater (unconscious knowledge and processes).

4. THE CORTICAL CORE

A generally held belief, supported by mounting evidence, is that the *cortical core* supports human consciousness: the mighty thalamus located deep in the heart of the brain connects to nearly every region in the cortex and together they form the cortical core, the central machinery of the brain.

The two thalami are nestled into the center of the brain and form a massive influence over the cortex and the brain in general (Fig. 1.9, top). As we will discuss in more detail in later chapters, the thalamo-cortical core (Fig. 1.9, bottom), made up of the thalami and the cortex,

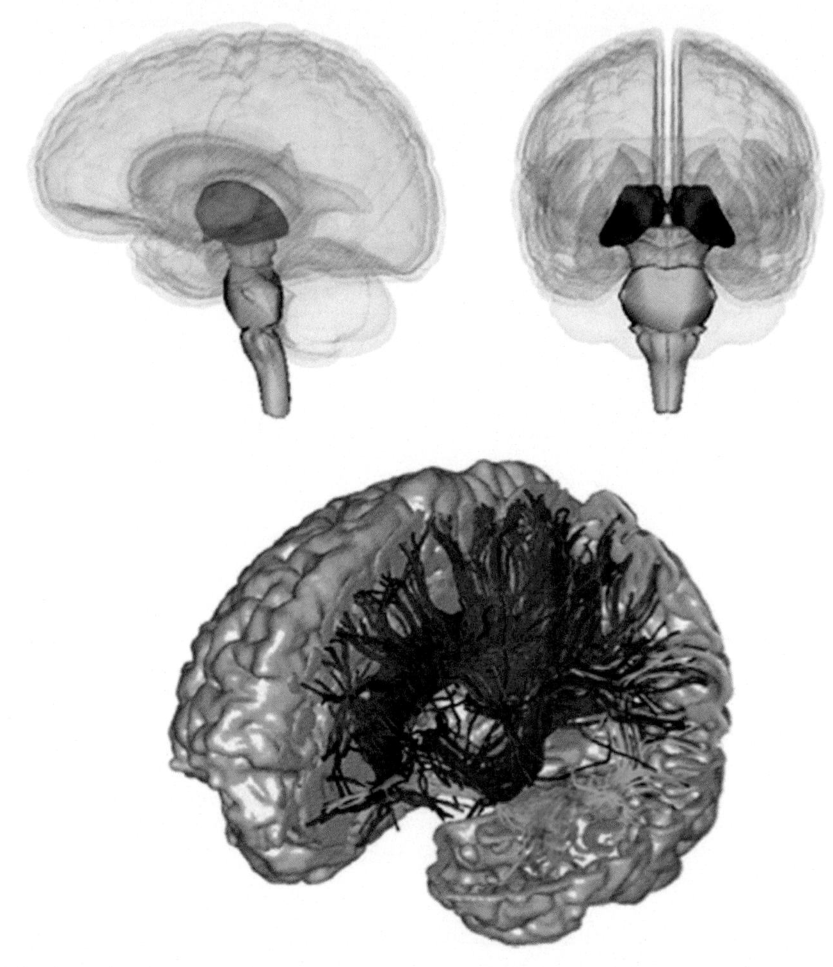

FIGURE 1.9 Top: the thalamus (*shown in red*) is located in the middle of the brain in the subcortical region. Bottom: the thalamus is highly interconnected with the cortex, with fibers extending to the frontal, parietal, temporal, and occipital lobes (*shown in red, blue, and green* in this diffusion tensor study). *Source: Left: Images are from Anatomography maintained by Life Science Databases (LSDB). Life Science Databases (LSDB)* ⓘ *Anatomography. Right: Izhikevich, E. M, & Edelman, G. M. 2007.*

can almost be thought of as a single massive hub that provides the lightning fast connectivity needed for waking cognition and the modulatory influences required for shifting the brain through the stages of sleep and the states of consciousness (Izhikevich & Edelman, 2007). Together they form the target for anesthesia and, when damaged, the basis for sustained unconsciousness and vegetative states.

5. CONNECTIVITY

The massive fiber tracts of the brain form the connective highways streets and small roads that provide the bases for neural communication traffic flow (Fig. 1.10). These tracts, large and small, provide the pathways for the billions of neurons to propagate the flow of information throughout the brain. Populations of neurons *oscillate* at similar rates to form synchronized frequency bands, such as alpha, beta, and gamma (Fig. 1.11) that carry information across the hemispheres and along the front-back regions of the cortex.

The massively interconnected cortex, subcortex, thalamus, and brainstem are difficult to visualize. While images such as those presented in Fig. 1.10 and 1.11 convey the complexity of the fiber tracts and pathways in the brain, they lack the specificity to address the question "Does *this brain area* connect directly with *this brain area*? The Human Connectome Project of

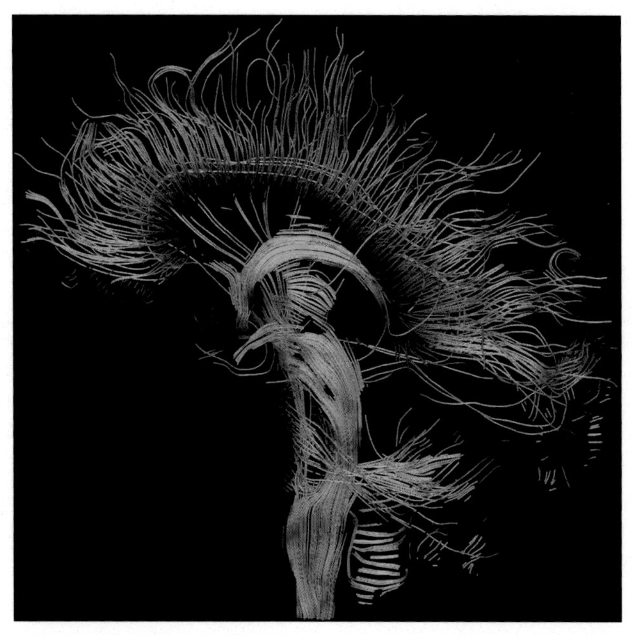

FIGURE 1.10 Visualization of a diffusion tensor imaging (DTI) measurement of a human brain. Reconstructed fiber tracts course through the midsagittal plane. The front of the head is at the left of the figure, the back of the head is at the right of the figure. The brainstem and spinal cord are shown in purple in the center of the figure. *Source: Thomas Schultz, commons.wikimedia.org, open source, licensed through GNU Free Documentation License, Creative Commons Attribution-Share Alike 3.0 Unported License.*

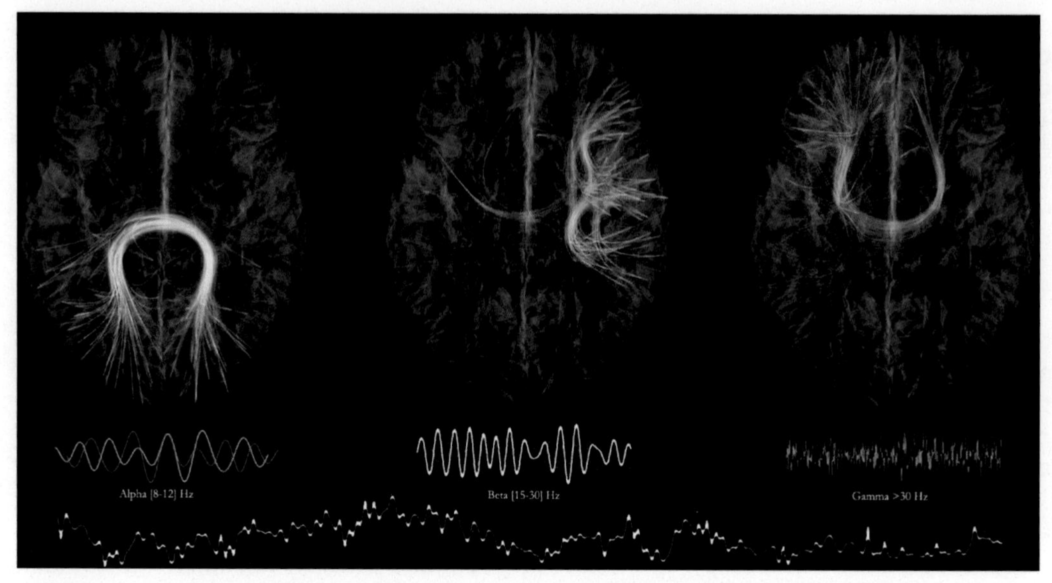

FIGURE 1.11 Neural oscillations in the alpha, beta, and gamma bands. *Source: Sebastien Dery McGill University.*

the National Institutes of Health (https://www.neuroscienceblueprint.nih.gov/connectome/) has developed brain analysis (Irimia, Chambers, Torgerson, & Van Horn, 2012; Fig. 1.12) and visualization methods (Fig. 1.13) to provide a more intuitive method for understanding whole-brain connectivity patterns. See Chapter 2, The Brain, for more on how to "read" the *connectograms* like the one shown in Fig. 1.13.

6. CONSISTENCY

Conscious perception seems to have a *consistency constraint* in the brain. Simple shapes are always perceived (consciously) as internally consistent, three-dimensional objects rather than the two-dimensional, partially occluding drawings they actually are, because that is, after all, the world we live in (Fig. 1.14).

When the brain receives any ambiguous or even cluttered sensory input, we tend to impose consistent interpretations, at least consciously. In recent decades it has become clear that unconsciously the brain can consider many interpretations of many inputs, some of them inconsistent with each other. The examples are rife. Our brain's ability to extract consistency in an inconsistent-to-lifelike image underlies our love for and understanding of abstract art forms, such as Impressionism (Fig. 1.15).

7. SUMMARY

Together and separately, the themes of consciousness/unconsciousness, cortical core, connectivity, and consistency will appear and reappear throughout this book as they are key

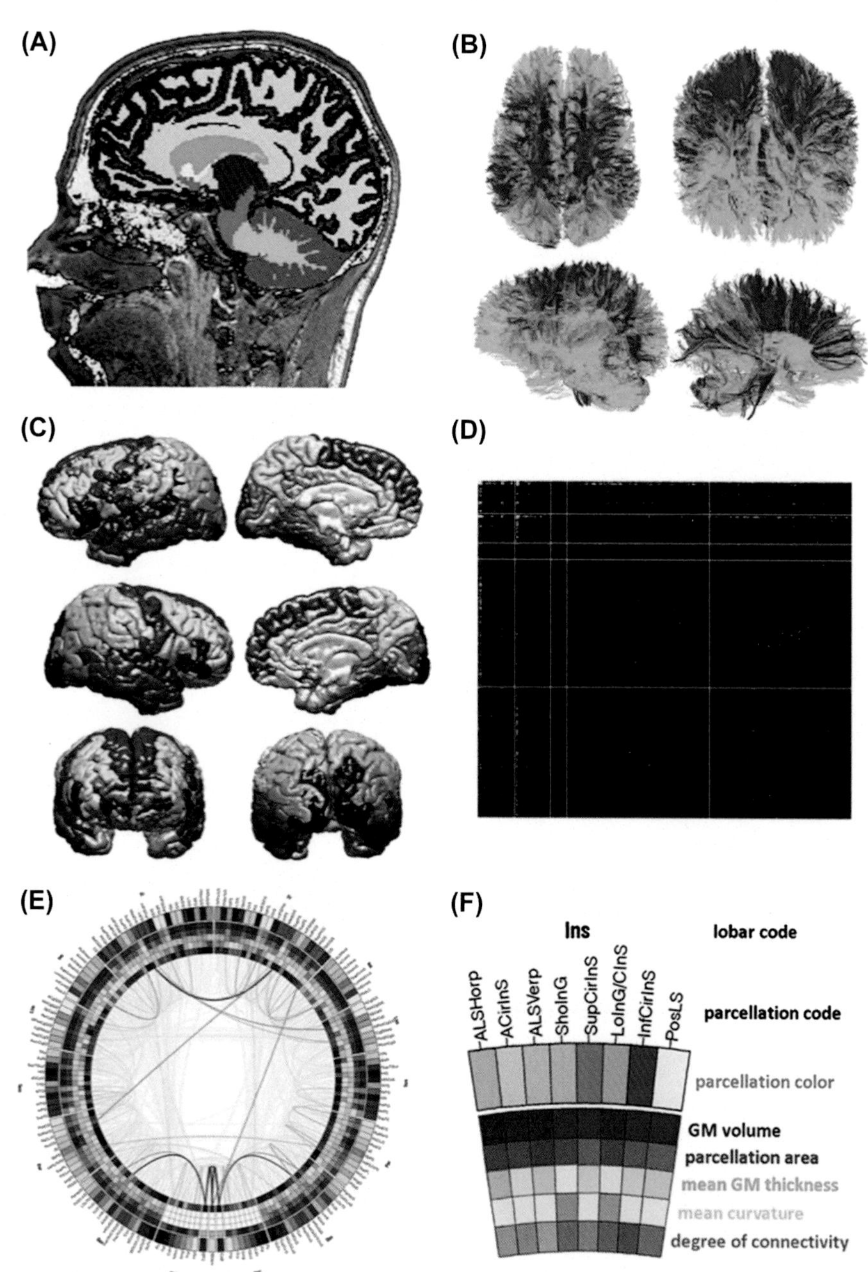

FIGURE 1.12 Constructing the connectogram. (A) Brain areas are segmented, results are shown for a sample subject. (B) Results of DTI tractography (for white matter reconstruction) are shown for a sample subject. (C) Reconstructed cortical surface is shown for a sample subject. (D) Example of a connectivity matrix computed for a sample subject. (E) A sample connectogram created for a single subject (see Figure 1.13 for details). (F) Legend of the representations shown in (E). The legend describing the rings of the Connectogram, which show different attributes of the corresponding cortical regions. From outermost to innermost, the rings represent the grey matter volume, surface area, cortical thickness, curvature, and degree of connectivity. *Source: Irimia, Chambers, Torgerson, & Van Horn, J. D., 2012.*

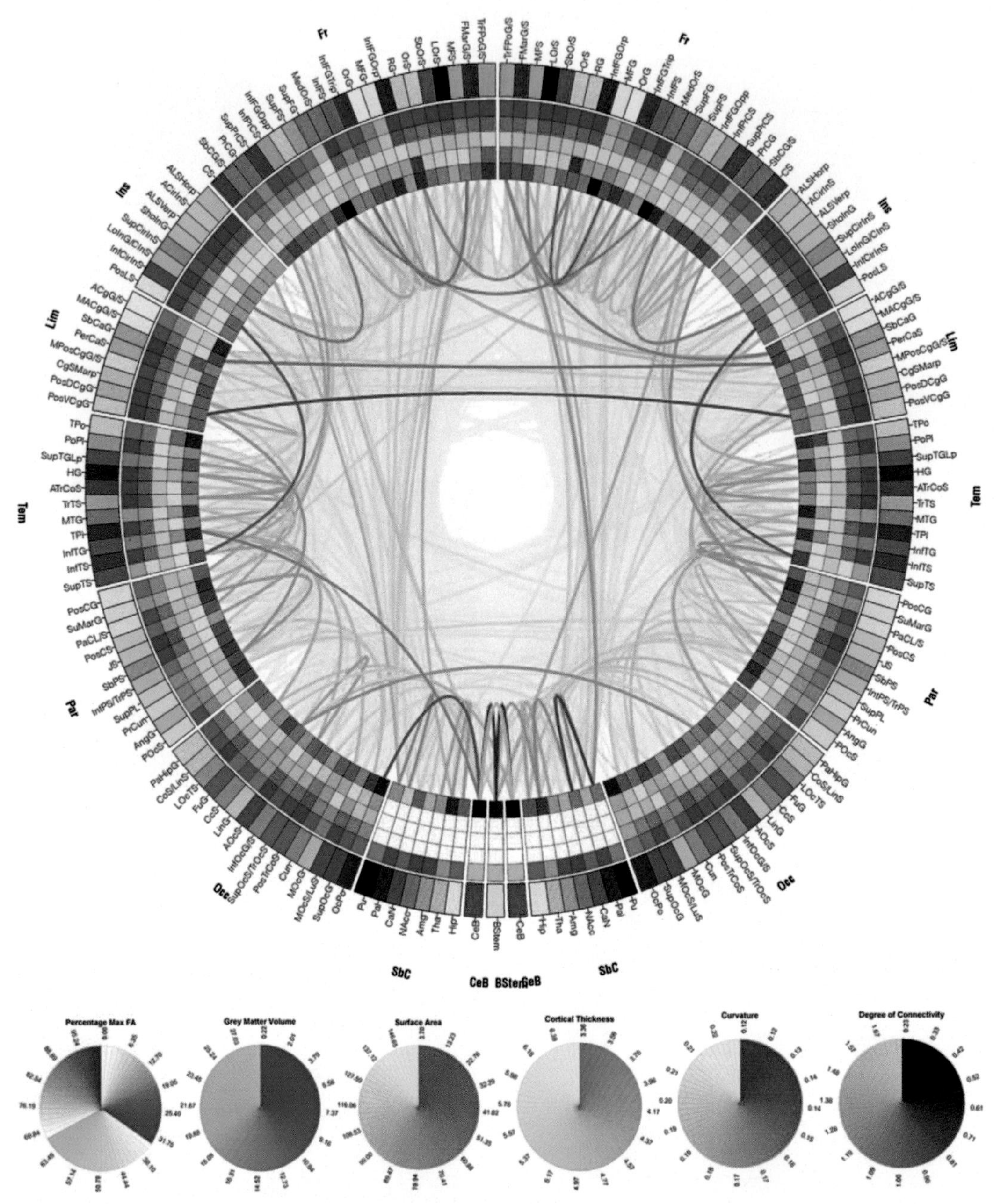

FIGURE 1.13 The Connectogram. The circular Connectogram, is a graphic way to show brain connectivity. The left half of the circle depicts the left hemisphere and the right half depicts the right hemisphere. The lobes and subregions of each hemisphere are shown and labeled, beginning at the top of the circle, as frontal lobe (fr), insular cortex (Ins), limbic lobe (Lim), temporal lobe (Tem), parietal lobe (Par), occipital lobe (Occ), subcortical structures (SbC), and cerebellum. The brainstem (BStem) is shown at the bottom of the circle, between the two hemispheres. Within the lobes, each cortical area is labeled with an abbreviation and an assigned color. Note the five rings that are inside the Conneectogram, just next to the color-coded cortical regions. The legend for these rings is shown in at the bottom of the figure. In the center of the Connectogram, lines connect regions that are structurally connected. The relative density (number of fibers) of these connections is reflected in the opacity of the lines, with the more opaque lines reflecting less densely connected regions and the less opaque lines reflecting more densely connected regions. The color of the line is color coded to reflect the density of fibers within that connection. *Source: Irimia, Chambers, Torgerson, & Van Horn, J. D., 2012.*

FIGURE 1.14 Consistency in visual processing: basic shapes are easily perceived as three-dimensional due to their color shading, despite the fact that they are actually two-dimensional and partially occluding one another. *Source: Author Mysid, Elisabethd; available at https://commons.wikimedia.org/wiki/File:Basic_shapes.svg, Licensed under the Creative Commons Attribution-Share aLike 2.5 Generic.*

FIGURE 1.15 Woman with a Parasol - Madame Monet and her Son, Claude Monet, 1874 Impressionism as an example of visual consistency: although the clouds in the sky are just daubs of paint and Madame. Monet's face is largely veiled; our perception of this painting is typically clear and consistent with how a more realistic image of Madame Monet would be interpreted due to our visual experience. *Source: Public domain.*

aspects of the mind and brain. The trading relationships between conscious and unconscious processing; the massive hub of the cortical core and its nightly descent into deep sleep; the lightning speed of the fiber tracts and their synchronized slowing as we transition from wake to sleep; and the strength of consistency coupled with the agility of learning, mental flexibility, and adaptation are just a few of the fascinating aspects of the almost limitless brain processes that occur within our heads.

CHAPTER

2

The Brain

THE BRAIN

The human brain has 100 billion neurons, each neuron connected to 10,000 other neurons. Sitting on your shoulders is the most complicated object in the known universe. ***Michio Kaku, American Physicist.***

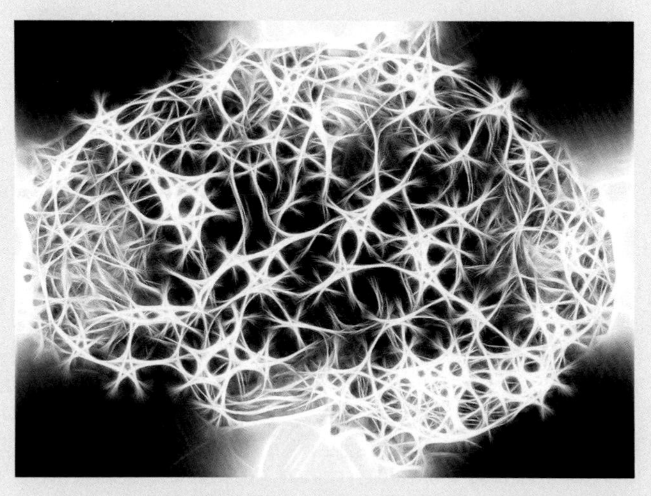

An abstract view of the brain with neurons. *Source: Public domain.*

1. INTRODUCTION

Ultimately we are each a single person: one human being with one mind and one brain. As we explore the structures and functions of the human brain in this chapter, we will see that the brain is far from a unitary structure, although we do label it as such: *the brain*. Its complexity is staggering. Formed of two major hemispheres, each with four cortical lobes, along with midbrain, brainstem, and cerebellum, the brain does not look like a unitary structure at all. Its many structures and regions combine with its vast connectivity to form a complex universe that drives human thought, actions, desire, and cognition.

Fundamentals of Cognitive Neuroscience
https://doi.org/10.1016/B978-0-12-803813-0.00002-7

FIGURE 2.1 The planet Earth as seen from space. *Source: Public domain.*

Perhaps it is not surprising that our brain contains so many discrete regions and structures when we consider our mind. It turns out that we humans are not of one mind at all. Scattered throughout human conversations are phrases like "I'm of two minds about this …", "I need to leave but I don't want to leave …", and "I feel I ought to tell him but I don't want to …". We humans are full of conflict. It is represented in classical psychological theory describing the human *id*, *ego*, and *super ego*: levels of human consciousness and behavior that are divided sharply despite representing our personality as a whole (Freud, 1923/1961). So it seems our minds are not so unitary at all, but full of levels of conflicting thought and emotion, and internal dilemmas. Why, then, should our brain be thought of as unitary?

In this chapter, we will explore the many regions of the brain and their relative functions. A central issue in understanding the brain is determining our *level of analysis*. Just as looking at the planet Earth from a distant view from space provides us with a global view of the continents and oceans of the Earth (Fig. 2.1), looking at the human brain as a whole provides information about its major structures and connections (Fig. 2.2). Yet just as a view of the Earth as a whole informs of the entirety of the planet, it provides little information about the depth and breadth of the Florida Everglades, the topography of the Grand Canyon, or the density of the forests of Yosemite National Park. To understand these subregions, we need to "zoom" our analysis in to a more detailed view. So it is with the brain: to understand the totality of brain function, we need to be able to first understand some of its constituent parts such as the cortex, the thalamus, and the brainstem. And then we need to "zoom" in to different subregions to understand them in more depth.

In this chapter, we will describe the key major structures of the brain (Fig. 2.2) and then delve into a bit more detail about the tiny but major working unit of the brain, the neuron. These are the disciplines of *neuroanatomy* and *neurophysiology*. Next, we will see how the many brain areas contribute to our functional selves: ranging from emotional processing to social cognition, to

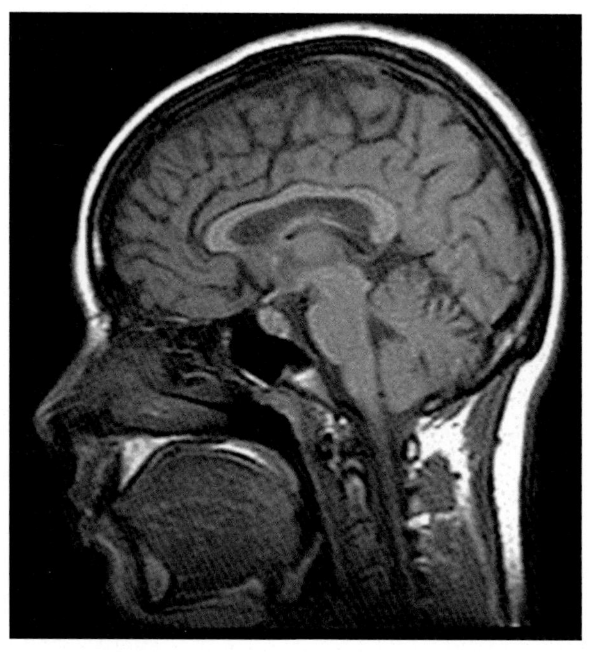

FIGURE 2.2 A side (sagittal) view of the human brain. *Source: Open access. Patrick J. Lynch, medical illustrator; C. Carl Jaffe, MD, cardiologist. http://creativecommons.org/licenses/by/2.5/.*

language and thought, and to attentional processes and future planning. These aspects of the brain are studied in *functional neuroanatomy* investigations. Then we will describe the many connective pathways that form the *street map* of the brain. Investigations into the *neuroconnectivity* of the brain provide important information about how information travels throughout the brain, brainstem, and spinal cord. Finally we will look at the "traffic" on those streets as we explore the brain's *dynamic processes*. *Neurodynamics* is an exciting, relatively new field of study. These aspects of the brain form the separate disciplines of neuroanatomy, neurophysiology, functional neuroanatomy, neuroconnectivity, and neurodynamics, which combine to shed light on the structure and function of the mind/brain.

2. BRAIN STRUCTURE—NEUROANATOMY

The brain sits at the top of the central nervous system (Fig. 2.3), nestled under the protective armor of the skull, and highly interconnected with the body through brainstem regions and the massive spinal cord. The vast peripheral nervous system that extends from the brain to regions throughout the body and back again sends and receives neural signals that provide information spanning from pain, pressure, and touch impulses, movement and balance, to the senses of vision and audition, and the chemical senses of smell and taste.

2.1 The Cortex

The *cortex* or *neocortex* (the terms are used interchangeably) is the seat of most of human cognition, from sensory processing of visual, auditory, somatosensory, smell, and taste inputs, to internal processes such as decision making and planning for the future. The cortex

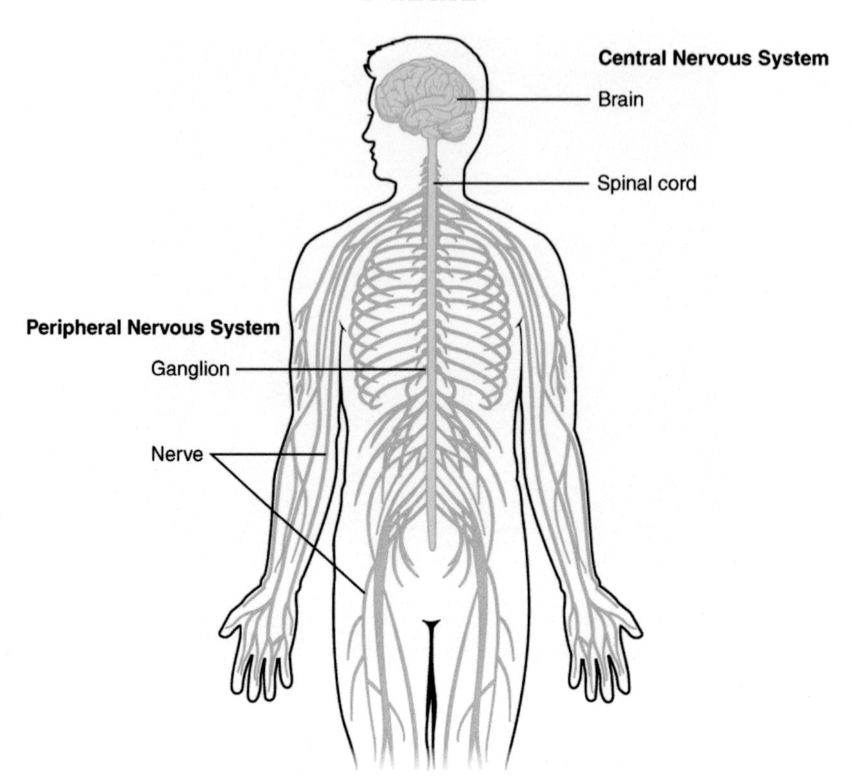

FIGURE 2.3 The central nervous system includes the brain and the spinal cord, while the peripheral nervous system includes the nerves throughout the body. *Source: Public domain. Version 8.25 from the Textbook OpenStax Anatomy and Physiology Published May 18, 2016 https://cnx.org/contents/FPtK1zmh@8.25:fEI3C8Ot@10/Preface.*

is subdivided into regions that are meaningful when we explore the functional roles of the brain later in this chapter and throughout this book. They have signature roles in the many functions that make up brain processing.

2.1.1 Planes of the Brain

Classically, the brain has been "sliced up" using three planes, and these planes are described in similar ways whether the slicing is actual slicing of the brain during, for example, a post mortem examination, or if it is virtual slicing using magnetic resonance images of the brain. The three planes are shown in Fig. 2.4: slicing sideways across the brain—so that you can see the left and right hemispheres—is called an *axial* slice. Sometimes an axial slice is called a *horizontal* slice because it is a horizontal cut through the brain. A second way to slice through the brain is called a *sagittal* slice. This is a slice that cuts down through the brain beginning in one hemisphere, continuing on until the middle of the brain is met: this is called a *mid-sagittal* slice, and continuing still further until the second hemisphere is shown. Think of this slicing as beginning at one ear and continuing through the brain towards the middle of the head and onto the other ear. The third type of brain slice is the *coronal* slice. Think about a slice that begins at the ears but this time the slices will continue forward towards the front of the head or backward towards the back of the head. A coronal slice will show both hemispheres, like the axial slice. These three plane terms—axial, sagittal, and coronal—will be used throughout this book.

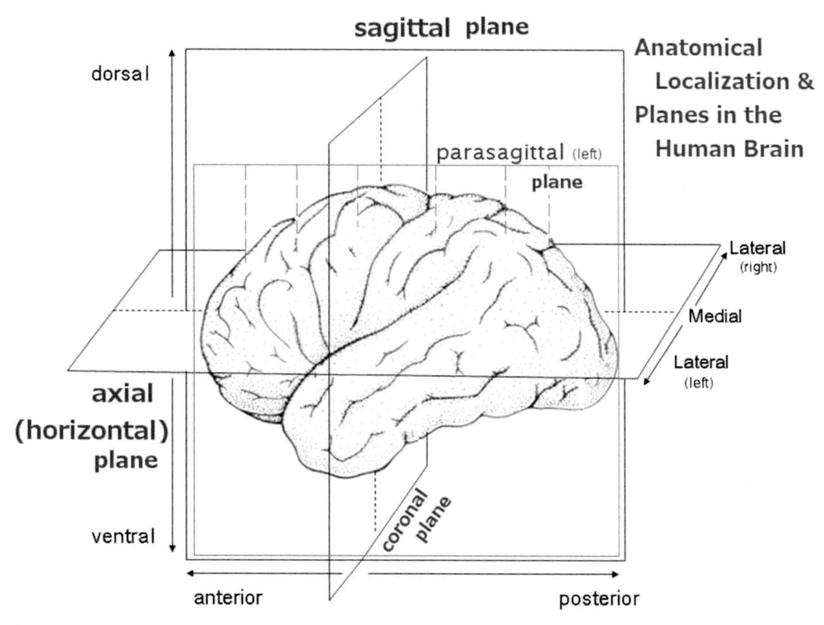

FIGURE 2.4 The planes of the brain. For investigative and clinical purposes, the brain is "sliced" along one of three major planes: *axial, sagittal*, and *coronal*. The axial or horizontal plane slices the brain horizontally from the top of the brain to the bottom, the sagittal plane slices the brain vertically from the left side to the right side, and the coronal plane slices the brain vertically from the front of the brain to the back. Other key brain-related terms are *dorsal*, indicating the top of the brain, and *ventral*, indicating the bottom of the brain. The front of the brain is the *anterior* and the back of the brain is the *posterior*. The middle part of the brain is termed the *medial* section and the left and right outside edges are called *lateral*. These terms are frequently used to denote where a region is in the brain, for example the dorsolateral prefrontal cortex or the ventromedial prefrontal cortex. *Source: Open access. Diagram with English annotations for planes and localisations in the human brain. JonRichfield 2014. Creative Commons Attribution-Share Alike 4.0 International license.*

Another set of terms for the brain describes where you are in the brain from the front of the brain, the *anterior* of the brain, to the back of the brain, the *posterior* of the brain (Fig. 2.4). A second set of terms describes where you are in the brain from the top of the brain, the *dorsal* or *superior* part of the brain, to the bottom of the brain, the *ventral* or *inferior* part of the brain. Finally, a third set of terms describes where you are in the brain from the outside edge of the brain, the *lateral* surface, to the middle of the brain, the *medial* surface located between the hemispheres within the longitudinal fissure.

These terms are used to name brain regions, such as the dorsal lateral prefrontal cortex (PFC) or the ventral medial PFC. Once learned, these terms can help you identify where in the brain a specific region is located. They can also help you identify the relative locations of brain areas: for example, the temporal lobe is posterior to the frontal lobe and inferior to the parietal lobe.

2.1.2 Cortical Anatomy

The cortex is the outer layer of the brain. *Cortex* comes from the Latin word for bark, and indeed the cortex has a look of the bark of a tree with all of its fine ridges and grooves. The

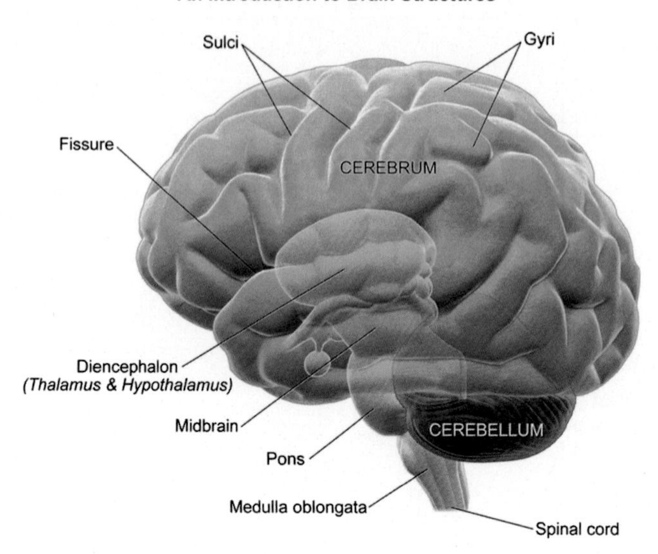

FIGURE 2.5 The many small grooves of the cortex are called sulci: each *sulcus* is a tiny valley that contains more of the *grey matter* that is visible on the surface of the cortex. A large sulcus is called a *fissure*. The many ridges of the cortex are called *gyri*: each *gyrus* is a small bump on the surface of the cortex, also made up of grey matter. *Source: Open access. Blausen.com staff (2014). "Medical gallery of Blausen Medical 2014." Wikiversity Journal of Medicine 1 (2): 10. https://doi.org/10.15347/wjm/2014.010. ISSN 2002-4436.*

cortex is made up of *grey matter*: the cell bodies of the billions of nerve cells—*neurons*—that form the key element of the cortex. The many small grooves of the cortex are called sulci: each *sulcus* is a tiny valley that contains more of the *grey matter* that is visible on the surface of the cortex (Fig. 2.5). The many ridges of the cortex are called *gyri*: each *gyrus* is a small bump on the surface of the cortex, also made up of grey matter.

There are two cerebral hemispheres: a right and a left hemisphere that are completely separate except for the massive connectivity bridge called the *corpus callosum* (Fig. 2.6). The corpus callosum is a major landmark as you observe the brain from various angles and views (Fig. 2.7). Although in Fig. 2.7 the corpus callosum is shaded in red for easy identification, more realistic images of the corpus callosum show that it is very light in color: this is because it is made up of *white matter*, which is formed by the *myelinated* axons that extend from neurons. The *myelin sheath* around the axon is made up of a white fatlike substance that serves to increase the signal transmission along the axon.

Within each hemisphere, there are four cortical lobes: *frontal, parietal, temporal, and occipital* (Fig. 2.8). While the four lobes are all part of the cortex, each lobe differs in shape and size, and each has a signature role in human cognition, as we will discuss later in this chapter and throughout this book.

Other key landmarks are three giant sulci: the Sylvian fissure (sometimes referred to as the Lateral Fissure) extending from the frontal lobe back between the parietal and temporal lobes; the central sulcus, which separates the frontal and parietal lobes (Fig. 2.9); and the longitudinal fissure (Fig. 2.6), which runs between the two hemispheres.

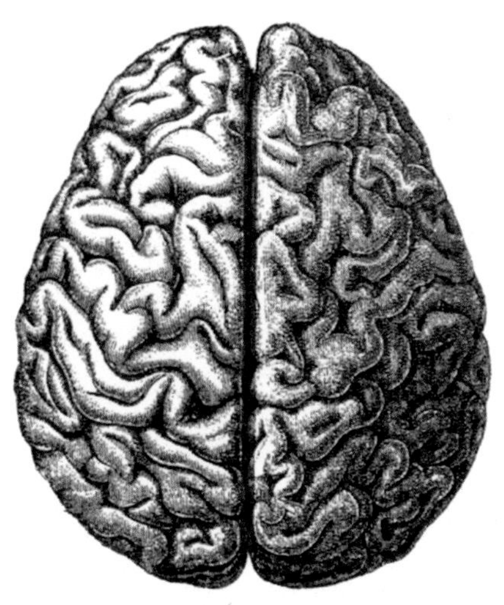

FIGURE 2.6 The brain seen from above showing two cerebral hemispheres separated by the large longitudinal fissure (shown in *red*). The two hemispheres are linked by the corpus callosum. *Source: Public domain. Wikimedia Commons, the free media repository.*

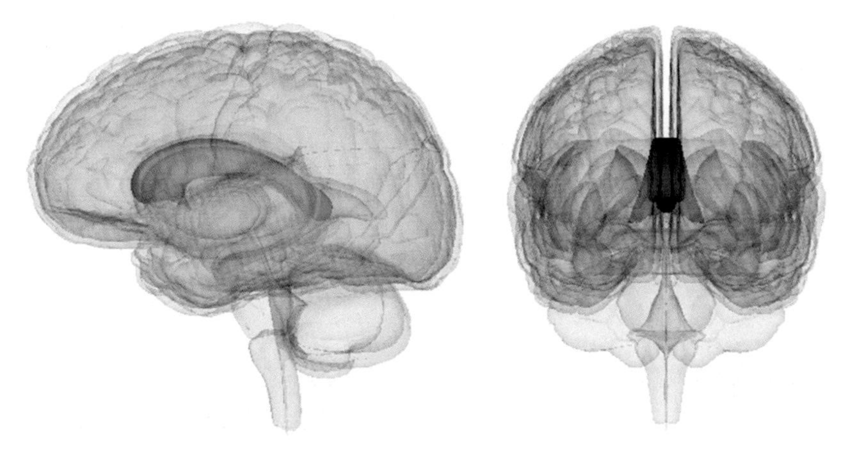

FIGURE 2.7 The massive corpus callosum (shown in *red*) is a white matter tract that links the two cerebral hemispheres. Left: the corpus callosum seen when looking through the cortex in the left hemisphere; right: the corpus callosum seen when looking through the brain from the front. *Source: Open access. Images are from Anatomography maintained by Life Science Databases (LSDB).*

2.2 The Subcortex

Beneath the cortex, the subcortical region contains many bodies or nuclei that perform important functions ranging from control of movement to emotional processing. Chief among these structures is the mighty *thalamus*. Located deep in the center of the brain, the two thalami (one in the left hemisphere, one in the right) are highly interconnected with all parts of the cortex (Fig. 2.10A). Together, they form the *thalamocortical system*. In fact,

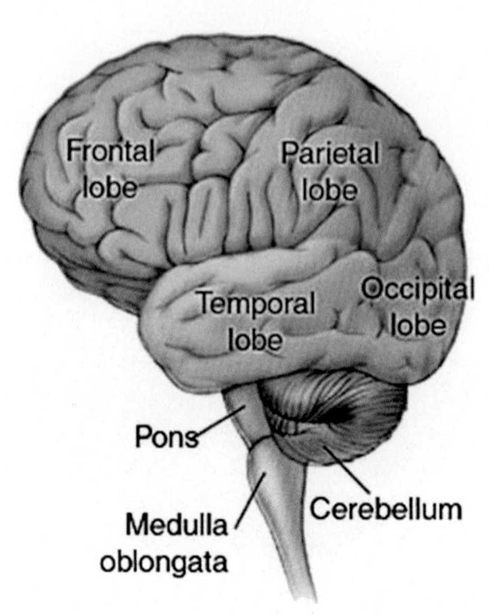

FIGURE 2.8 The four lobes of the cerebrum: frontal (shown in *pink*), parietal (shown in *purple*), temporal (shown in *green*), and occipital (shown in *orange*). *Source: Public domain. https://www.wpclipart.com/medical/anatomy/brain/brain_2/brain_anatomy.png.html.*

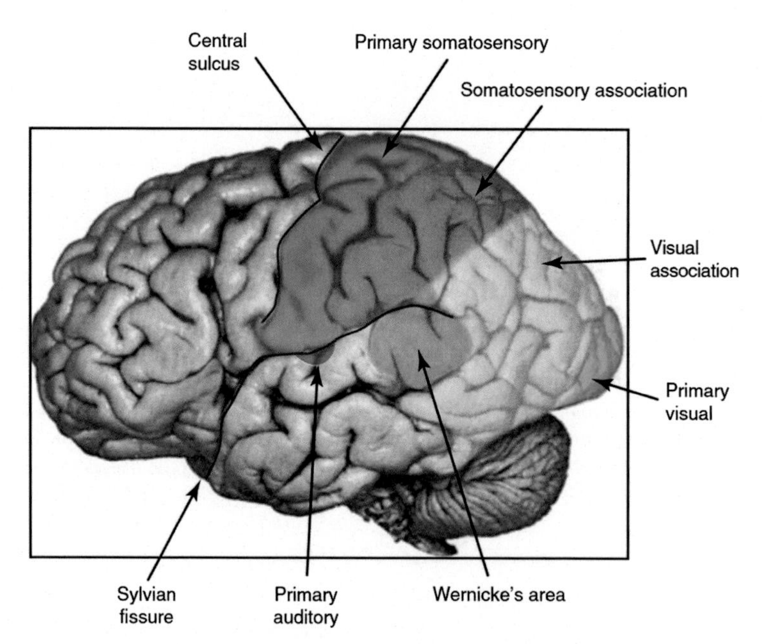

FIGURE 2.9 Brain landmarks: the massive Sylvian fissure runs from the ventral portion of the frontal lobe to the posterior of the brain, separating the parietal and temporal lobes. The central sulcus is a large sulcus separating the frontal lobe from the parietal lobe. *Source: Fig. 22.6, Fundamentals of Sensory Systems, SH Hendry & SS Hsiao, in LR Squire, D Berg, FE Bloom, S du Lac, A Ghosh, & NC Spitzer (Eds),* Fundamental Neuroscience, *4th edition, (pp. 499–511). San Diego: Academic Press.*

the thalamus and cortex are so highly interconnected that they can be thought of as a single massive brain system. The thalamic connections with the four lobes of the cortex are highly structured, with complex connective patterns between the thalamus and each cortical lobe as well as many other subcortical bodies and regions (Fig. 2.10B).

Basal ganglia are a group of subcortical nuclei or bodies that together are primarily engaged in influencing motor (movement) control along with the motor cortex and the spinal cord (Fig. 2.11A). Damage to the basal ganglia leads to motor dysfunction despite an intact motor cortex and spinal cord. The nuclei of the basal ganglia are located posterior to the frontal lobe and include the *putamen*, the *caudate nucleus*, the *globus pallidus*, and the *nucleus accumbens* (Fig. 2.11B). These nuclei play key roles in motor as well as emotional and cognitive functions of the brain.

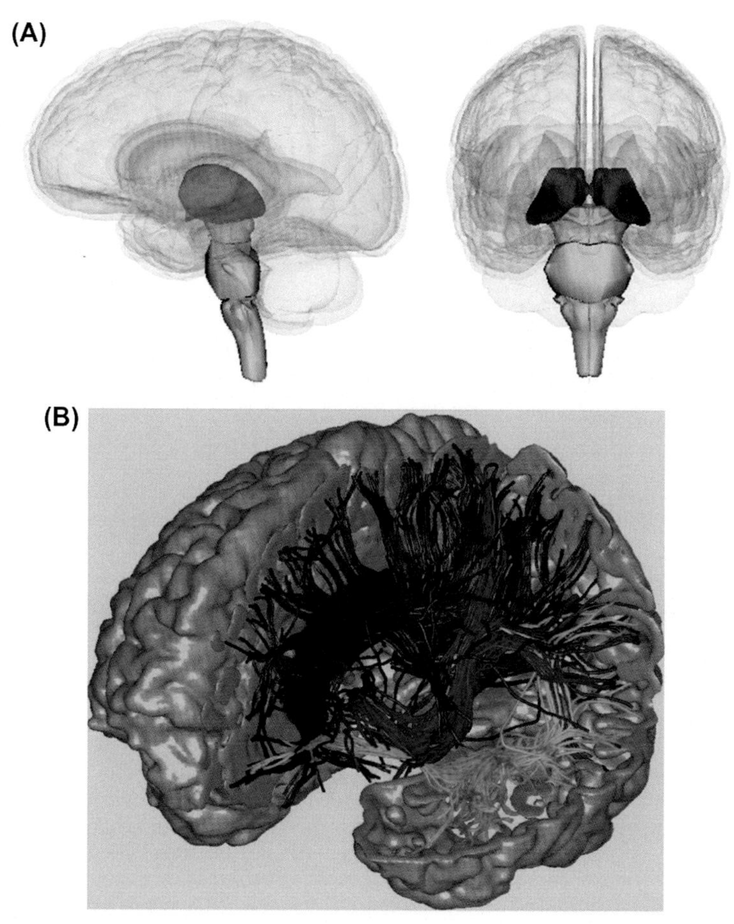

FIGURE 2.10 (A) The thalamus (shown in *red*) is located in the middle of the brain in the subcortical region. (B) The thalamus is highly interconnected with the cortex, with fibers extending to the frontal, parietal, temporal, and occipital lobes (shown in *red*, *blue*, and *green* in this diffusion tensor study). *Source: Left: Images are from Anatomography maintained by Life Science Databases(LSDB). Life Science Databases (LSDB)* Ⓒ *Anatomography. Right: Izhikevich, E. M, & Edelman, G. M. 2007.*

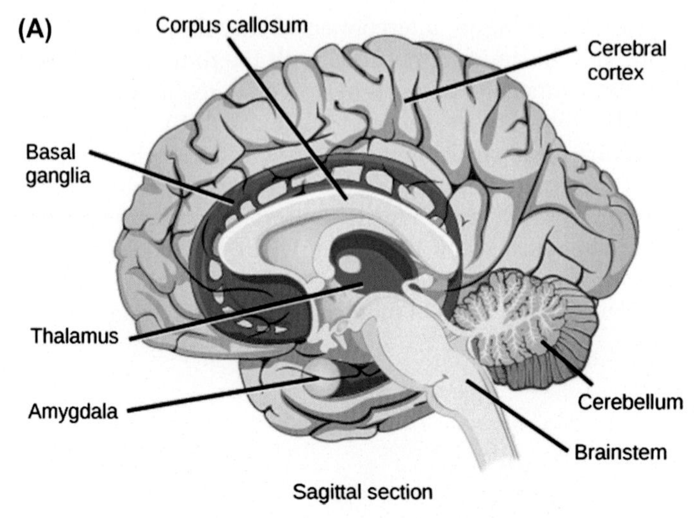

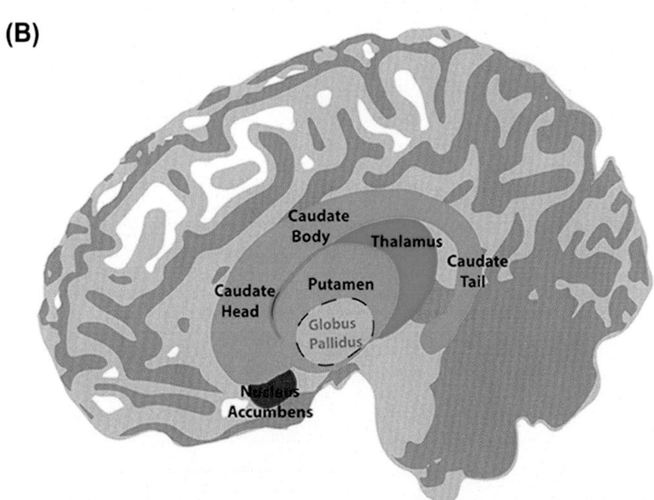

FIGURE 2.11 (A) *Basal ganglia* are a group of subcortical nuclei or bodies that together are primarily engaged in influencing motor (movement) control along with the motor cortex and the spinal cord. *The amygdala* and other regions in the subcortical region primarily function as our emotional processors. The tiny amygdala gets its name from the Latin word for almond: the amygdala is indeed an almond-shaped small body located in the center of the brain below the thalamus. (B) The nuclei of the basal ganglia are located posterior to the frontal lobe and include the *putamen*, the *caudate nucleus*, the *globus pallidus*, and the *nucleus accumbens*. These nuclei play key roles in motor as well as emotional and cognitive functions of the brain. *Source: A. Open Access. Courtesy of https://courses.candelalearning.com/biologymajors/chapter/chapter35-the-nervous-system/. Creative Commons Attribution. B. Open Access. Lim S-J, Fiez JA and Holt LL (2014) How may the basal ganglia contribute to auditory categorization and speech perception? Front. Neurosci. 8:230. https://doi.org/10.3389/fnins.2014.00230. http://journal.frontiersin.org/article/10.3389/fnins.2014.00230/full.*

The amygdala and other regions in the subcortical region primarily function as our emotional processors. The tiny amygdala gets its name from the Latin word for almond: the amygdala is indeed an almond-shaped small body located in the center of the brain below the thalamus (Fig. 2.11A). Highly interconnected with other subcortical regions and the cortex, the amygdala is part of the emotional system in the brain frequently referred to as the *limbic system*. Other brain areas that form part of the limbic system include the olfactory bulb, which provides information regarding the sense of smell; the hippocampus, which is critical for memory processing; and the cingulate gyrus. Together regions in the limbic system provide information to the cortex about emotionally relevant information in our sensory world, such as an angry face or voice. They are also key for the encoding and retrieval of emotional memories.

2.3 The Cerebellum

Located at the most posterior portion of the brain behind the temporal and occipital lobes, the cerebellum (shown in Figs. 2.5, 2.8, and 2.11A) has a variety of roles in human movement, learning, and cognition. While the cerebellum does not initiate motor commands—these come via the cortical motor pathways—the cerebellum plays a key role in modifying those commands. This results in the cerebellum controlling aspects of movement such as the maintenance of balance and posture and the coordination of complex movements.

The cerebellum also plays a role in motor learning. As you learn to drive a car, surf a wave, or shoot a basketball, the circuitry in the cerebellum is providing adaptive movement information to guide the motor learning process. Although there is less evidence about the cerebellum's role in cognition, converging evidence is providing support for a role for the cerebellum in language processing, which may be related to the complex movements entailed in forming human speech.

2.4 The Brainstem

Connecting the brain to the spinal cord and the rest of the body, the *brainstem* includes three key regions that are critical for human life: the *midbrain*, the *pons*, and the *medulla* (Fig. 2.12). This hub connects the spinal cord, with its many motor and sensory impulses sending key signals to the brain, to the brain itself. Passing through the brainstem region are major nerves that control movement, sensation, breathing control, and other core aspects of the human body. In the midbrain, the *inferior colliculus* transmits auditory signals. The *superior colliculus*, just above, transmits visual signals. The *reticular formation* will be discussed often in this book: this area within the midbrain transmits signals throughout the brain and is implicated in arousal and in human consciousness.

The pons, the middle section of the brainstem, comes from the Latin for *bridge*. Nestled between the midbrain and the medulla, the pons transmits signals from the cortex and subcortex to the medulla and the cerebellum as well as to the thalamus. The pons carries signals that control basic functions such as sleeping, equilibrium, and posture. The medulla, situated below the pons, has similar basic functions but is mostly in control of involuntary functions such as breathing, heart rate, and blood pressure. Damage to the brainstem can cause

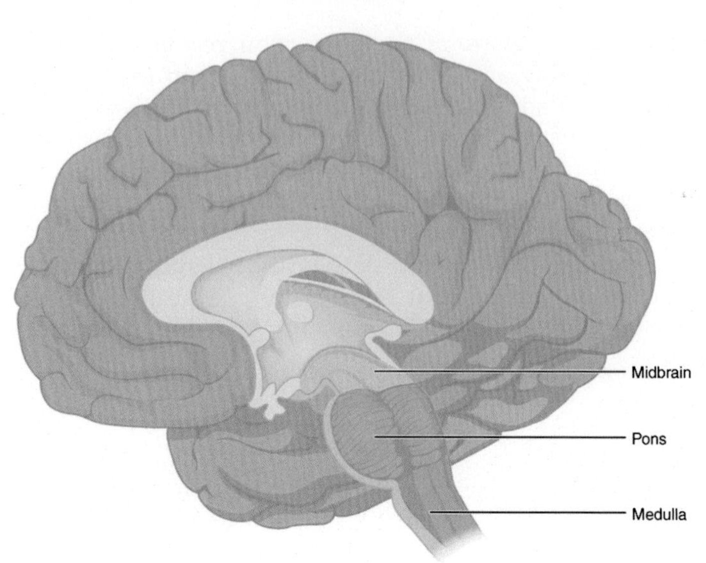

FIGURE 2.12 The brainstem. Connecting the brain to the spinal cord and the rest of the body, the *brainstem* includes three key regions that are critical for human life: the *midbrain* (shown in *green*), the *pons* (shown in *pink*), and the *medulla* (shown in *purple*). This hub connects the spinal cord, with its many motor and sensory impulses sending key signals to the brain, to the brain itself. Passing through the brainstem region are major nerves that control movement, sensation, breathing control, and other core aspects of the human body. *Source: Open Access. Version 8.25 from the Textbook OpenStax Anatomy and Physiology Published May 18, 2016. By OpenStax - https://cnx.org/contents/FPtK1zmh@8.25:fEI3C8Ot@10/Preface, CC BY 4.0.*

widespread impairments ranging from vision and hearing loss, weakness, paralysis, and even death. A key point to remember is that brainstem damage can cause paralysis and loss of movement control; however, if the cortex is unharmed, human awareness, cognition, and thought may be unaffected despite the obvious physical limitations due to brainstem damage. This condition is referred to as the *Locked-in Syndrome* because the patient cannot speak to verbalize their condition (see Chapter 13, Disorders of Consciousness).

Throughout the cortex, the subcortical regions, the cerebellum, and the brainstem, there are billions of *neurons*: the microscopic worker bees of the brain. Although there are hundreds of types of neurons, they share a classic structure, which we will refer to as the "idealized neuron." Let us turn now to a description of how neurons communicate throughout these disparate regions of the brain that, together as an ensemble, form the intricate language of the brain and its connections.

3. BRAIN CELLS—NEUROPHYSIOLOGY

Neurophysiology is the study of the cells in the brain. While there are many types of cells in the brain performing vital functions, we will be discussing the *nerve cell* or the *neuron* in this chapter.

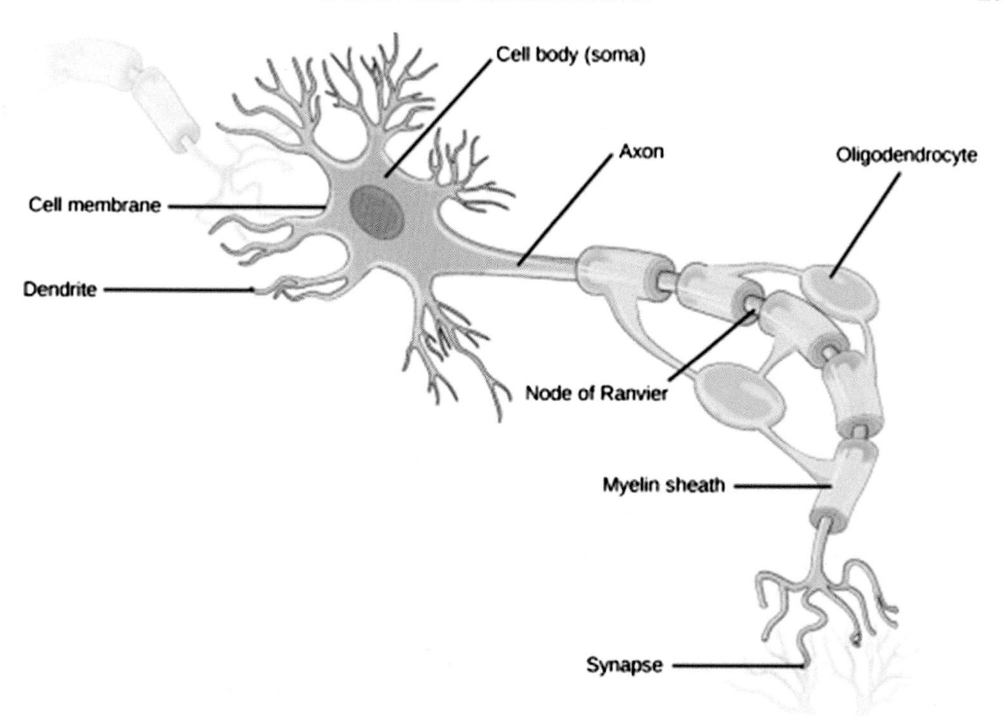

FIGURE 2.13 An idealized neuron. Although there are more than 200 types of neurons in the brain, most share some key features. *Dendrites* (left side of figure) are treelike structures around the *cell body or soma* of the neuron. The many dendrites receive incoming signals from other neurons. This is the *input* to the neuron. Extending from the cell body is the *axon*: axons vary in length and can extend long distances along a hemisphere or across the two hemispheres. Covering many axons is a white fatlike substance: this is the myelin sheath that forms an insulating layer over the axon that increases the electrical conduction rate. When many axons are together in a bundle, the myelinated coverings form a light-colored fiber tract that is referred to as *white matter*. Note the Nodes of Ranvier in the myelin sheath: these are gaps in the myelin sheath that serve an important function. The gaps in the myelin sheath that exposes the axon aid in action potential generation. *Source: Public domain. Courtesy of https://courses.candelalearning.com/biologymajors/chapter/chapter35-the-nervous-system/. Creative Commons Attribution.*

3.1 The Structure of an Idealized Neuron

An "idealized" neuron is shown in Fig. 2.13. While there are hundreds of types of neurons in the brain, they share key features. *Dendrites* are treelike structures around the *cell body or soma* of the neuron. The many dendrites receive incoming signals from other neurons. This is the *input* to the neuron. Extending from the cell body is the *axon*: axons vary in length and can extend long distances along a hemisphere or across the two hemispheres. Covering many axons is a white fatlike substance: this is the myelin sheath, which forms an insulating layer over the axon that increases the electrical conduction rate. When many axons are together in a bundle, the myelinated coverings form a light-colored fiber tract that is referred to as *white matter*. In Fig. 2.13, note the Nodes of Ranvier in the myelin sheath: these are gaps in the myelin sheath that serve an important function. The gaps in the myelin sheath that exposes the axon aid in action potential generation are discussed in the following section.

At the end of the axon, synaptic terminals form the communication method for the *output* of the neuron: *the synapse*. Neurons do not directly connect with one another. Rather, the impulses propagate down to the axon to the *synaptic terminals* where they cross the *synaptic cleft*: a separation between transmitting neuron and the receptors. Although there are many complex ways in which brain cells interact, the *action potential* forms the key communication method for a neuron to communicate with another neuron or group of neurons.

3.2 Action Potentials

Neurons communicate with each other through a combination of electric and chemical processes. The cell membrane of a neuron has a *resting potential*—the voltage of the neuron—that is approximately −70 mV. The chemical aspect of the communication comes via the three basic ions that exist on the inside and outside of the cell membrane: potassium (K^+), calcium (Ca^{2+}), and sodium (Na^+) ions. When the neuron is not firing—i.e., when it is at rest—the *resting potential* of the neuron has potassium ions (K^+) that are more highly concentrated on the inside of the cell while calcium (Ca^{2+}) and sodium (Na^+) ions are more concentrated on the outside (Fig. 2.14).

When incoming input to the neuron increases to a given threshold—the *threshold potential*—the cell membrane will *depolarize*, lowering the voltage within the membrane. Ion channels open, the concentration of K^+, Ca^{2+}, and Na^+ ions within and without the membrane shifts, and an *action potential* occurs. The depolarization spreads down the axon towards the axon terminals where the synapse will occur (Fig. 2.15). The action potential or firing of the neuron is frequently termed a *spike* or an *impulse*. This action potential is propagated to the next neuron(s) via the synapse, thus forming the basic communication method for neurons. Note that the action potential is an *all-or-nothing event*: the neuron either *fires* or it does *not fire*.

3.3 Connectivity Basics

Connectivity patterns within the cortex and throughout the brain are complex and not easily simplified. There are a few basic working assumptions, however, that we will use to describe a general way in which brain connectivity occurs.

First, building upon the description of the action potential, above, neurons work by adding graded voltage inputs until the total membrane voltage on the target neuron goes past a threshold value. If it does, an all-or-nothing spike fires down the output branch, the axon. We will refer to these neurons as *integrate-and-fire* neurons.

Next, connections between neurons are either *excitatory* or *inhibitory*. Thus they can either cause the neuron they are synapsing with to fire (excitatory connection) or not to fire (inhibitory connection).

Third, neurons can form one-way pathways, such as from the optic nerve of the eye to the visual region of the thalamus (the lateral geniculate nucleus). However, one-way pathways are actually quite rare. More likely, neurons run in two directions, forming *two-directional pathways* and networks in which activity at Point A triggers activity at Point B and vice versa. This is often referred to as *reentrant connectivity* (Edelman, 1989).

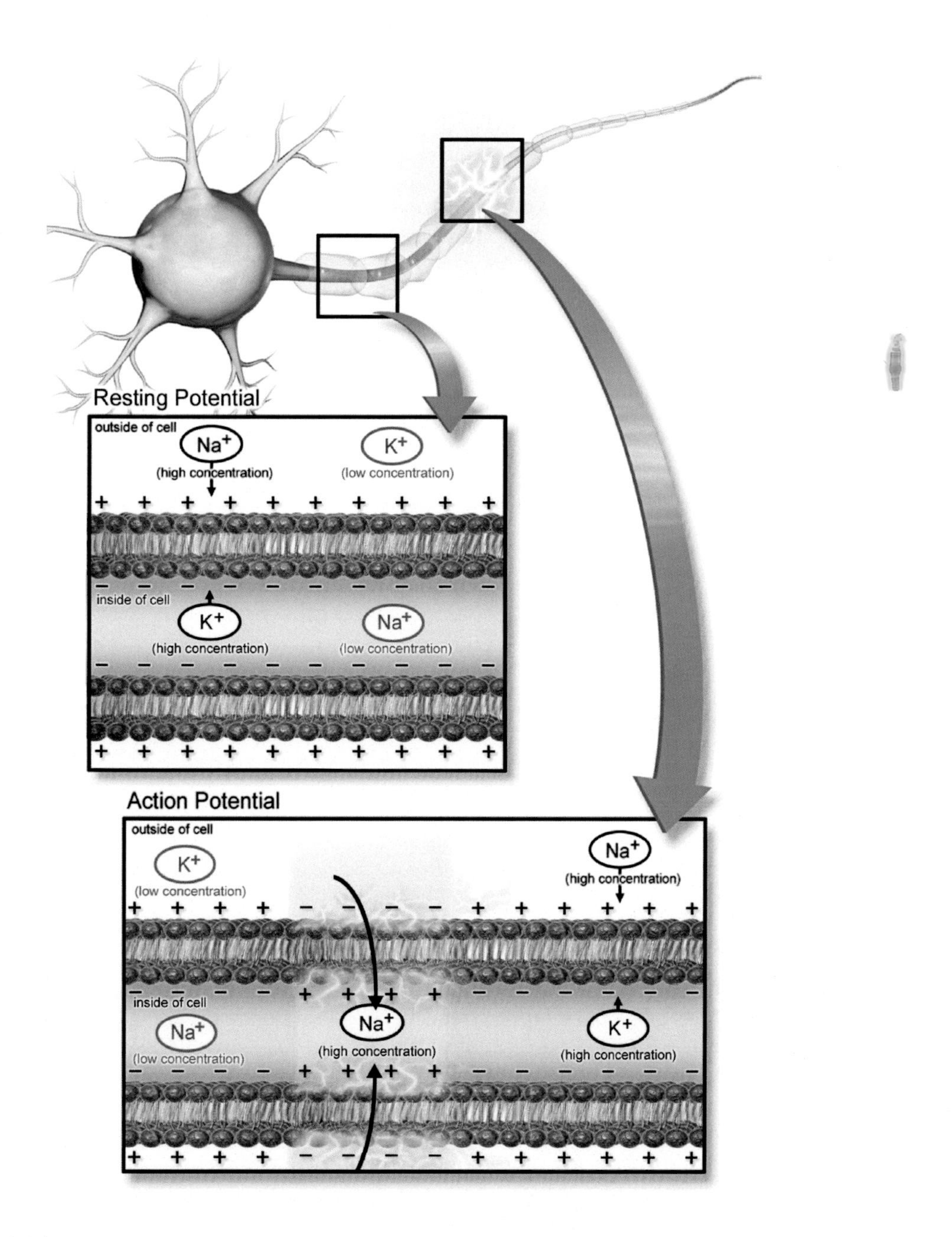

FIGURE 2.14 The resting and action potentials. Neurons communicate with each other through a combination of electric and chemical processes. The cell membrane of a neuron has a *resting potential*—the voltage of the neuron—that is approximately −70 mV. The chemical aspect of the communication comes via the three basic ions that exist on the inside and outside of the cell membrane: potassium ions (K^+), calcium (Ca^{2+}), and sodium (Na^+). When the neuron is not firing—i.e., when it is at rest—the *resting potential* (top insert box) of the neuron has potassium ions (K^+) that are more highly concentrated on the inside of the cell while calcium (Ca^{2+}) and sodium (Na^+) ions are more concentrated on the outside. When incoming input to the neuron increases to a given threshold—the *threshold potential*—the cell membrane will *depolarize*, lowering the voltage within the membrane. Ion channels open, the concentration of K^+, Ca^{2+}, and Na^+ ions within and without the membrane shifts, and an *action potential* occurs (bottom insert box). *Source: Public domain. Courtesy of https://courses.candelalearning.com/biologymajors/chapter/chapter35-the-nervous-system/. Creative Commons Attribution.*

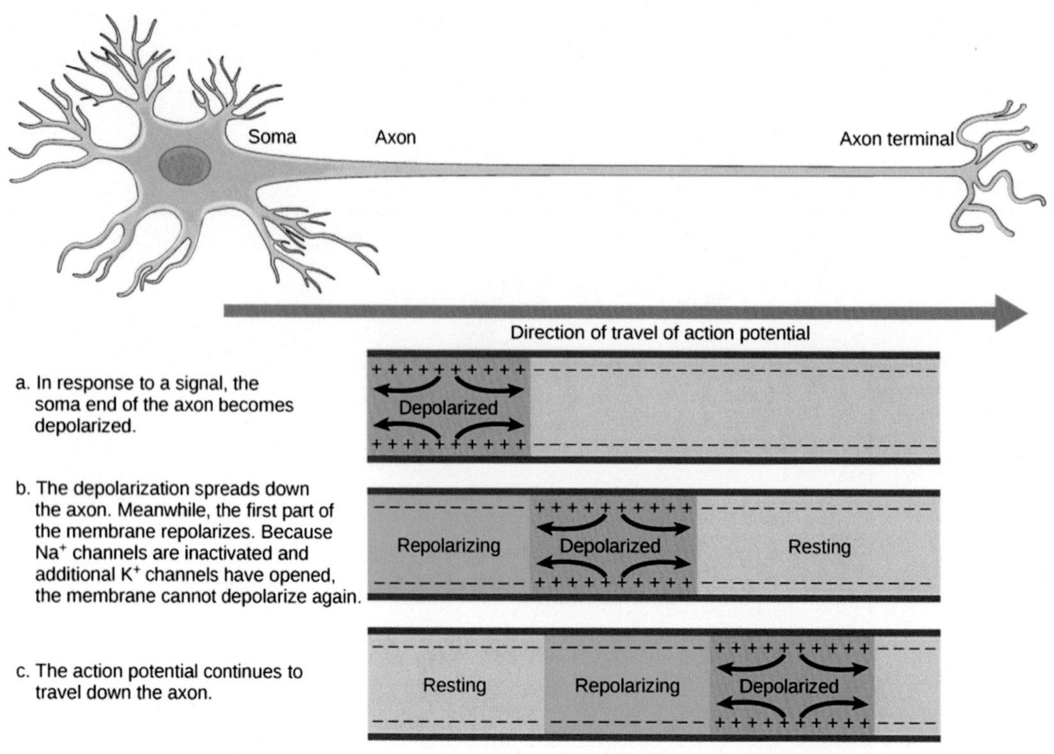

FIGURE 2.15 Action potential propagation. When the action potential occurs, the depolarization spreads down the axon towards the axon terminals where the synapse will occur. This action potential is propagated to the next neuron(s) via the synapse, thus forming the basic communication method for neurons. Note that the action potential is an *all-or-nothing event*: the neuron either fires or it does not. *Source: Public domain. Courtesy of /. Creative Commons Attribution.*

3.4 Brodmann Areas

And finally, the cerebral cortex is a massive six-layer array of neurons, with an estimated 10 billion cells and trillions of synaptic connections. Neurons are organized into arrays, which form maps that can reflect the perceptual world around us, as we will see in Chapter 4, The Art of Seeing. We mentioned at the beginning of this section that there are many, many types of neurons in the brain: more than 200. The study of the arrangement of neurons in the brain—typically under a microscope—provides key information about cortical regions in the brain. This type of study provides information about the cytoarchitecture—the cell architecture—of neurons in the brain. The German anatomist Korbinian Brodmann defined and numbered cortical regions according to their cytoarchitectural structure. These regions, called the Brodmann areas, are still in wide use today in describing discrete cortical subregions (Fig. 2.16, Brodmann regions open source).

While the neuron forms the basic "worker bee" of the brain, the more complete story behind how the brain functions comes from the *dynamical aspects* of their communication across neural arrays and networks. We will discuss these in more depth later in the chapter. First, let us take a look at the basic functions of the brain.

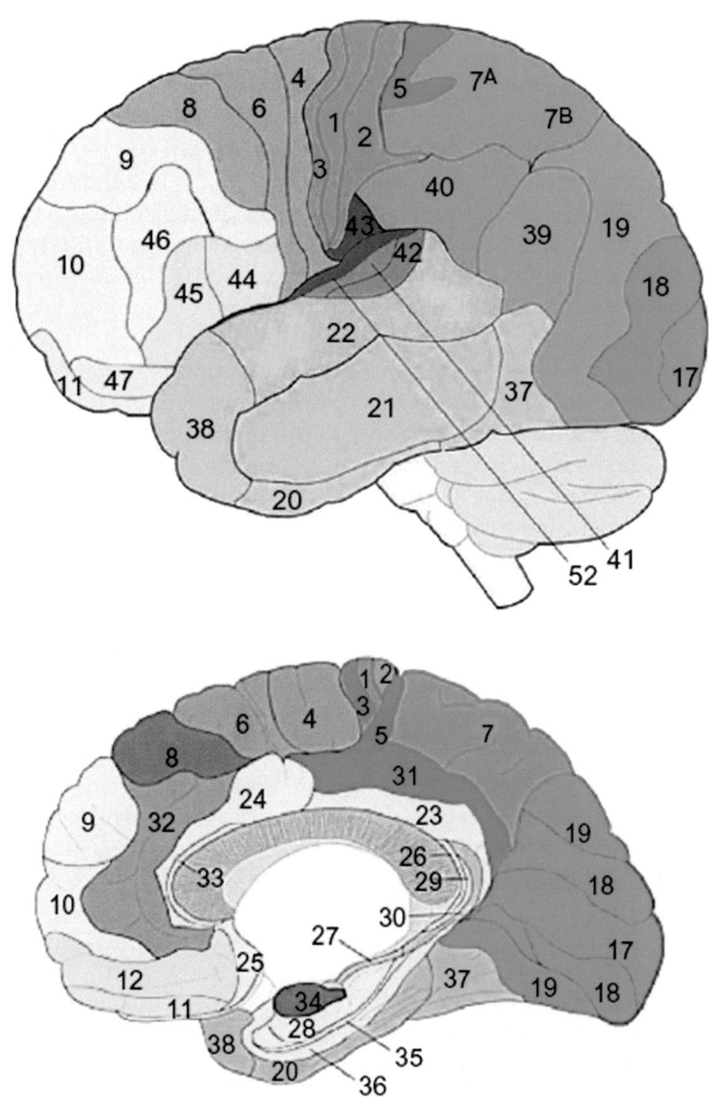

FIGURE 2.16 Brodmann areas. The German anatomist Korbinian Brodmann defined and numbered cortical regions according to their cytoarchitectural structure. These regions, called the Brodmann areas, are still in wide use today in describing discrete cortical subregions. *Source: Baars & Fu, with permission.*

4. BRAIN FUNCTION—FUNCTIONAL NEUROANATOMY

4.1 Right Brain—Left Brain

As we have discussed, there are two cerebral hemispheres linked only by the massive fiber structures that course through the corpus callosum (Fig. 2.7). While the two hemispheres are very similar in structure, they are not identical: there are proportional differences in various regions and especially around the Sylvian Fissure. These anatomical differences have led to investigations of whether the two hemispheres perform differing cognitive functions: are some cognitive functions lateralized to a single hemisphere? Or do both hemispheres participate in all cognitive functions?

Perhaps the most well-understood lateralization of function is human language: for most humans regardless of race or natural language, spoken language function is lateralized to the left hemisphere. We will discuss this in far more depth in Chapter 6, Language and Thought. Briefly, we have learned through cases of brain damage that spoken language can be selectively disrupted if that damage is in the left frontal region of the brain. Other, more subtle lateralization of cognitive functions appears to exist in the two hemispheres, leading to many investigations into "right brain—left brain" cognition. However, as a general rule, both hemispheres participate in cognitive functions and it is likely that the hemispheres play similar but slightly differing roles in these functions rather than performing as specialized for cognitive processes such as attention, memory, and language.

4.2 The "Front-Back" Division

We have mentioned one way to divide the human brain—the left and right cerebral hemispheres form one way to separate regions of the brain. Another basic way to divide the brain's functions is to look at the "front" of the brain and the "back" of the brain. The front-back division in the brain distinguishes the brain regions where most of the incoming sensory information arrives from the periphery (the "back" of the brain) from the brain regions where most of the outgoing motor acts or movements are generated (the "front" of the brain).

Take a look back at Fig. 2.9: the frontal lobe is separated from the remaining three lobes by the Central Sulcus at the top of the brain and the Sylvian fissure in the middle of the brain as seen from the side or lateral view. While the "back" of the brain has many functions it perform, one key set of functions is that it forms the primary "landing place" for sensory inputs that come from the eyes, the ears, and the nerves of the body. These senses, vision, audition, and somatosensory, are the three key cortical senses. Rounding out the five senses are the chemical senses of taste and smell, which have noncortical landing places.

4.2.1 Sensory and Motor Functions

Sensory information comes to the cortex from pathways from the eyes, ears, and body to regions of cortex specialized for processing that sensory information (Fig. 2.17). The brain areas that correspond to these specialized regions are called *primary sensory areas* because they are typically the first or main "landing spot" for those pathways. Thus there is a *Primary Visual Cortex, Primary Auditory Cortex,* and *Primary Somatosensory Cortex* located in each

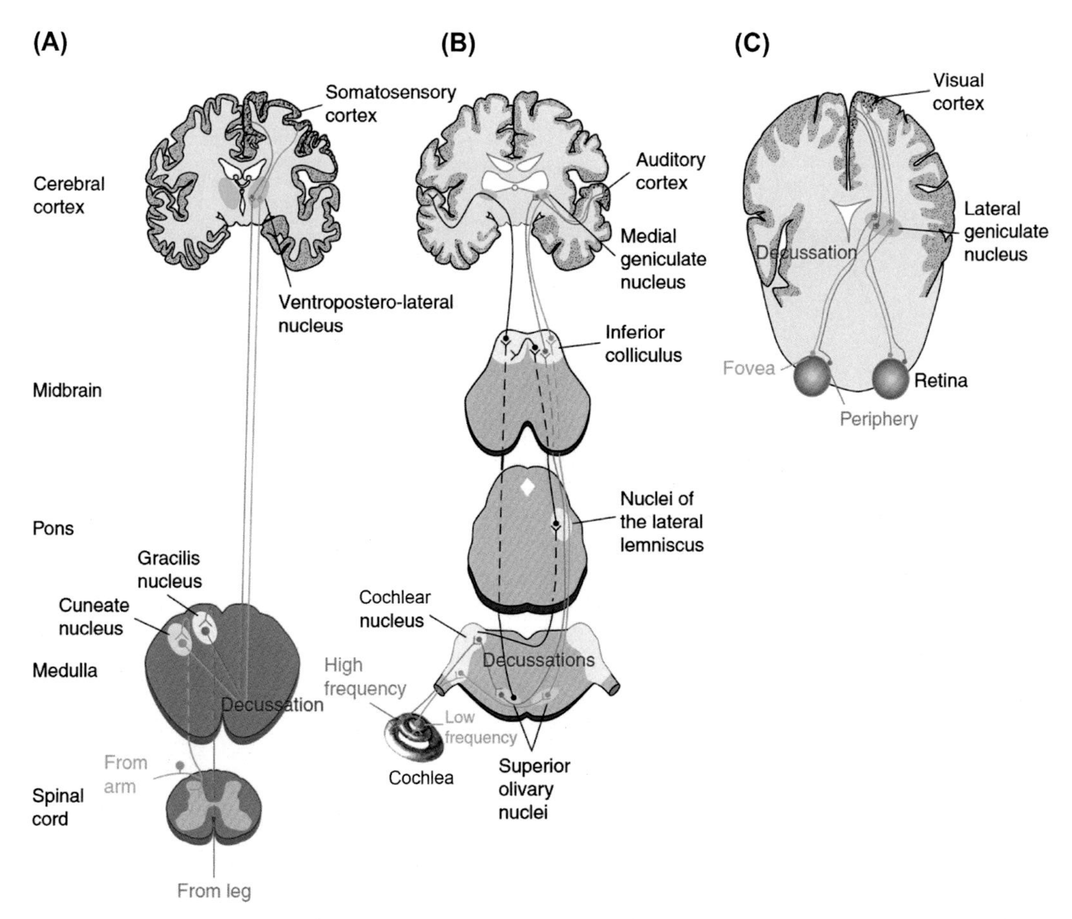

FIGURE 2.17 Sensory information comes to the cortex from pathways from the body, ears, and eyes to regions of cortex specialized for processing that sensory information. The somatosensory pathways from the body to the somatosensory cortex in the parietal lobe are shown in (A). These pathways carry pain, pressure, and touch signals. The auditory pathways from the ears to the auditory cortex in the temporal lobe are shown in (B). These pathways carry sound signals. The visual pathways from the eyes to the visual cortex are shown in (C). These pathways carry visual signals. The brain areas that correspond to these specialized regions are called *primary sensory areas* because they are typically the first or main "landing spot" for those pathways. Thus there is a *Primary Somatosensory Cortex*, a *Primary Auditory Cortex*, and a *Primary Visual Cortex* located in each hemisphere: therefore there are actually two Primacy Somatosensory Cortices, two Primary Auditory Cortices, and two Primary Visual Cortices. *Source: Fig. 22.5, Fundamentals of Sensory Systems, SH Hendry & SS Hsiao, in LR Squire, D Berg, FE Bloom, S du Lac, A Ghosh, & NC Spitzer (Eds),* Fundamental Neuroscience, *4th edition, (pp. 499–511). San Diego: Academic Press.*

hemisphere: therefore there are actually two Primacy Visual Cortices, two Primary Auditory Cortices, and two Primary Somatosensory Cortices.

Motor information goes from the motor cortex to the body via the brainstem and spinal cord. As in the case of the sensory systems, there is a *Primary Motor Cortex*. In this case, the Primary Motor Cortex is generally the *last* cortical region before the signals are sent on

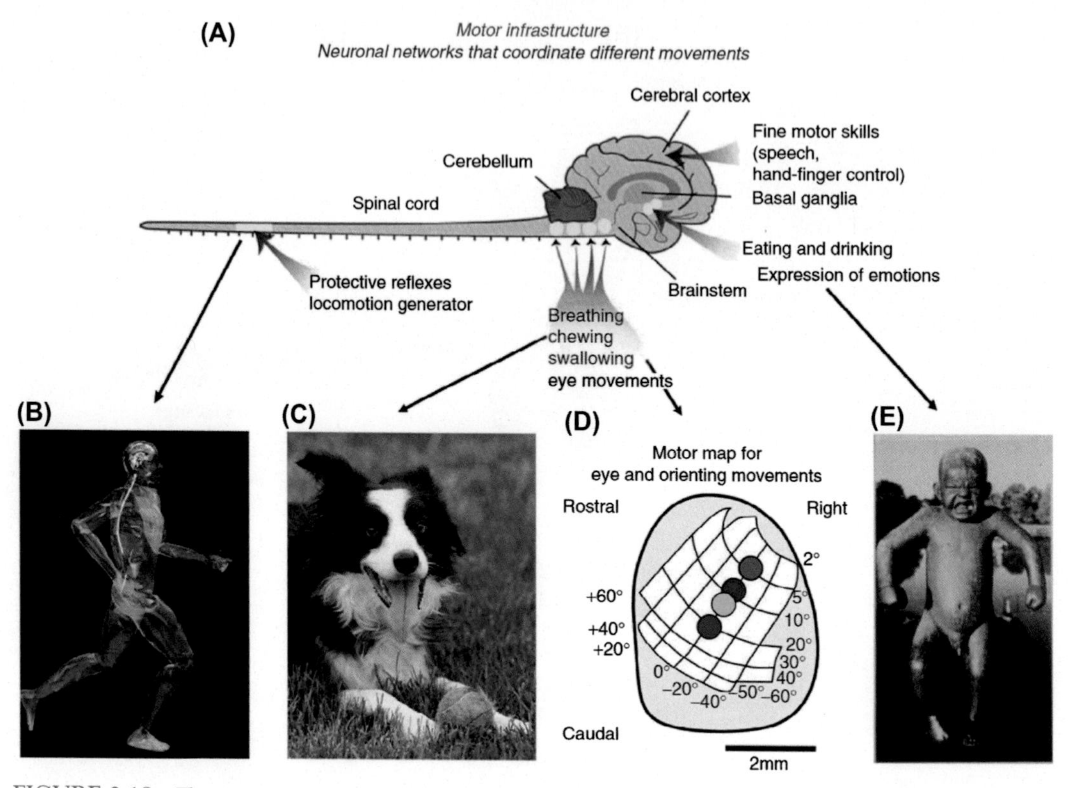

FIGURE 2.18 The motor system is complex, with many layers of circuitry that control fine motor skills, such as speech, and more automatic functions, such as breathing and swallowing, and even more basic motor functions, such as reflexes. The neural circuitry of the motor system reflects these disparate functions. (A) The neural circuitry of the motor system reflects these disparate functions, with functions ranging from the spinal cord (left of figure), to the cerebellum (center of figure), brainstem (to the right of the cerebellum), and to the cerebral cortex (top right of figure). (B) The spinal cord contains the network for moving or locomotion. (C) The brainstem contains networks that control basic and largely automatic movements such as breathing, swallowing, and eye movements. (D) A motor "map" in the brainstem for controlling eye movements. (E) More complex movements such as those underlying the expression of emotion (as shown in this sculpture of an angry child by Vigeland), are controlled largely in the cerebral cortex. *Source: Fig. 27.5, Fundamentals of Motor Systems, S Grillner, in LR Squire, D Berg, FE Bloom, S du Lac, A Ghosh, & NC Spitzer (Eds),* Fundamental Neuroscience, *fourth edition, (pp. 599–611). San Diego: Academic Press.*

to the body. The motor system is complex, however, with many layers of circuitry that control fine-motor skills, such as speech, and more automatic functions, such as breathing and swallowing, and even more basic motor functions, such as reflexes. The neural circuitry of the motor system reflects these disparate functions (Fig. 2.18).

4.3 The Cerebral Lobes

While the two hemispheres and the front-back division of the brain provide a basic sense of how the brain's functions are organized, the four cerebral lobes are perhaps the clearest way to understand the way sensory and cognitive processes are organized in the cortex. Underlying the functional roles that are organized across the four lobes are the vast white matter networks that connect them and the key subcortical regions that form the massive hub of the thalamocortical network.

4.3.1 *The Frontal Lobe*

As we discussed in the "front-back division" section, the frontal lobe is the site for motor planning and motor output. The motor areas are located just anterior to the Central Sulcus (Fig. 2.9). The motor cortex is structured with a "map" of the body: this is referred to as the *homunculus* ("little man") as it contains all parts of the body. The representation of differing body parts in the motor cortex is not proportional to those body parts; however, some areas of the body—such as the face, mouth, and hands—have a proportionately larger representation in motor cortex than other body areas such as the back and trunk of the body (Fig. 2.19). The fine dexterity of the mouth and speech–generating apparatus and the hands and fingers reflects the high proportion of neurons governing their movements from the motor cortex. The relatively low dexterity and fineness of motion control in the back and trunk of the body also reflects the lower proportion of neurons governing their movements.

Note that the motor cortex is closely situated to the somatosensory areas that are located just posterior to the Central Sulcus. In fact, there is a homunculus in this region as well, with a similar disproportionate representation of the face, hands, and fingers for the sensations of touch, pressure, and pain. This makes some logical sense as you consider that the senses of pain, pressure, and touch are tightly coupled with motor movements and reactions.

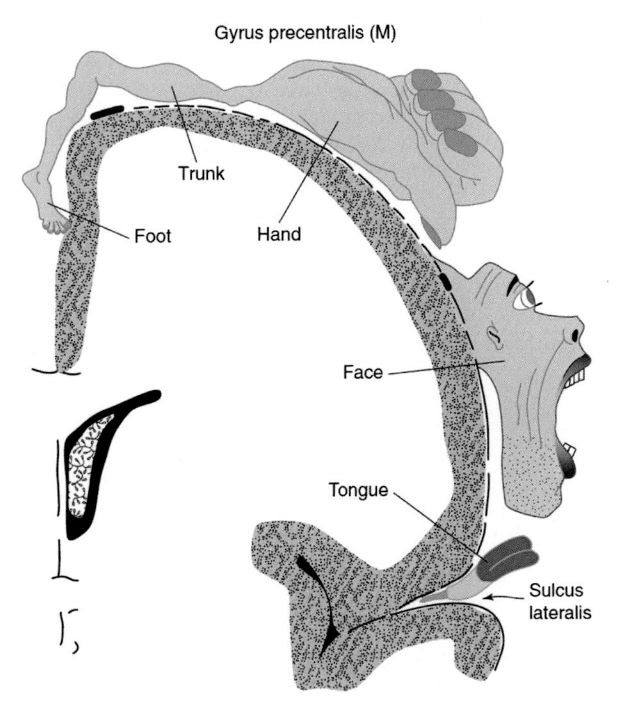

FIGURE 2.19 The motor homunculus. The motor cortex is structured with a "map" of the body: this is referred to as the *homunculus* ("little man") as it contains all parts of the body. The representation of differing body parts in the motor cortex is not proportional to those body parts; however, some areas of the body—such as the face, mouth, and hands—have a proportionately larger representation in motor cortex than other body areas such as the back and trunk of the body. *Source: Fig. 27.6, Fundamentals of Motor Systems, S Grillner, in LR Squire, D Berg, FE Bloom, S du Lac, A Ghosh, & NC Spitzer (Eds),* Fundamental Neuroscience, *4th edition, (pp. 599–611). San Diego: Academic Press.*

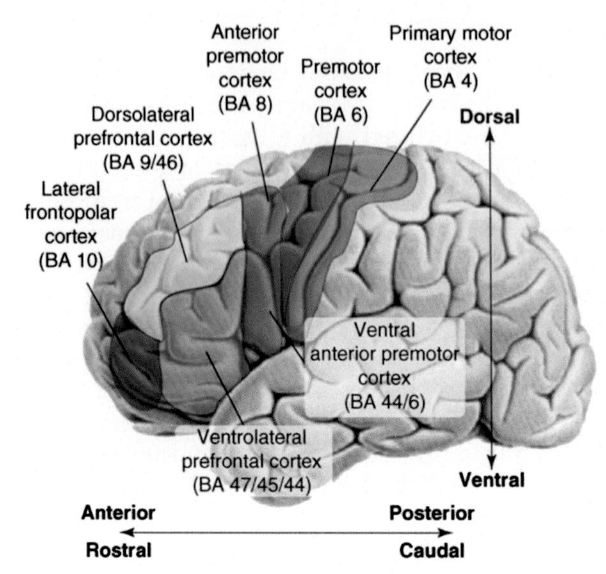

FIGURE 2.20 The functional organization of the frontal lobe. The frontal lobe is home to Motor Cortex, which controls movement. However, much of the frontal lobe is devoted to many key areas of human cognition. The *prefrontal cortex (PFC)*—that is, the area in front region of the frontal lobe—is the nonmotor area of the frontal lobe. In some ways, the PFC is the most cognitive region of the brain. The PFC is highly interconnected with the thalamus, other subcortical regions, as well as with the other three lobes of the cerebral cortex. Note: BA refers to approximate Brodmann area locations. *Source: Fig. 1, Badre, D. 2008 Cognitive control, hierarchy, and the rostro-caudal organization of the frontal lobes. TICS 12(5), 193–200, with permission.*

Returning to the functional organization of the frontal lobe, many other key cognitive functions are located in the massive frontal lobes. The Prefrontal Cortex—PFC—that is, the area in the front region of the frontal lobe—is the nonmotor area of the frontal lobe. In some ways, the PFC is the most cognitive region of the brain. It is here that our "*executive*" functions are located: those processes that allow us to plan for the future, make decisions, and focus our attention on one thing and not another. Like an executive in a large firm, the PFC does not do all of the cognitive "work" of the brain but rather controls it and synthesizes it. The PFC is highly interconnected with the thalamus, other subcortical regions, as well as with the other three lobes of the cerebral cortex (Fig. 2.20).

The PFC is critical for many key human processes such as

- initiating activities
- planning
- holding critical information ready for use (an aspect of working memory)
- changing mental set from one line of thinking to another, mental flexibility
- monitoring the effectiveness of one's actions
- detecting and resolving conflicting plans for action
- inhibiting plans and actions that are ineffective or self-defeating

This brief list—and the many other cognitive processes that the PFC regulates—demonstrates how critical this part of the brain is to human cognition. Other key functions of the mighty frontal lobe include emotional and personality processing that is important for social cognition. Human expressive speech systems are also located in the frontal lobe in a region called Broca's area, named after Paul Pierre Broca (see Chapter 6, for more on this region).

4.3.2 *The Parietal Lobe*

Just posterior to the frontal lobe is the parietal lobe. The anterior region of the parietal lobe, just posterior to the Central Sulcus, is home to the somatosensory cortex (Fig. 2.9). As mentioned previously, the somatosensory cortex has a body "map"—homunculus—similar to the one in the motor cortex (Fig. 2.19) that has a disproportionate representation of the body, similar to the motor homunculus. In the case of the somatosensory cortex, the expanded neural representation corresponds to finer senses of touch, pressure, and pain in regions such as the hands, mouth, and face.

The location of the somatosensory cortex in the parietal lobe is just the beginning of this lobe's complex role on human brain function. The parietal lobe is a key region of the visual "where" pathway (see Chapter 4, for more on this). This pathway is instrumental in object location, among many other visual processes. Another important function of the parietal lobe is the integration of multiple maps of body space. These maps reflect several key regions for mapping motor activity with the visual system, for example. Think about approaching a coffee cup on a nearby table. Your visual system is mapping where the cup is, while your body maps are preparing the neural code for the action of reaching your hand out to grasp the cup. Let us imagine that you were a distance away from that coffee cup: other body maps would integrate information from the visual system with your body's motor system so that you could guide your movements to walk to the table, reach for the cup, grasp the cup, and so on. These body maps provide us with the idea of where we are in space: multiple maps can guide our hands vis a vis our mouth, for example, while we are eating, while others can guide us in other complex body mapping situations such as learning to skate board or surf. While these body maps seem to be tightly related to the visual system in helping to guide our movements, other sensory inputs are also actively involved.

The inferior portion of the parietal lobe—the inferior parietal lobe or IPL—is thought to be a multisensory region where information from the somatosensory, visual, and auditory systems is combined or integrated. The specific nature of the IPL and the parietal lobe's role in cognition is still being elaborated.

4.3.3 *The Temporal Lobe*

The temporal lobe is nestled posterior to the frontal lobe and inferior to the parietal lobe. The temporal lobe is home to the auditory system for sound processing: Primary Auditory Cortex is tucked into the upper bank of the temporal lobe within the Sylvian fissure (Fig. 2.9), thus is it not visible from a lateral view of the cortex. Just posterior to the auditory cortex is a region known as Wernicke's area, named after Carl Wernicke who studied speech perception pathways in the brain (see Chapter 6, for more on this). The middle sections of the lateral aspect of the temporal lobe are key regions for conceptual knowledge, with many theories of conceptual representation in the brain identifying these areas as important for conceptual knowledge storage. The temporal lobe is a key region of the visual "what" pathway (see Chapter 4, for more on this). This pathway is instrumental in object and face recognition, among many other visual processes.

4.3.4 *The Medial Temporal Lobe*

The medial section of the temporal lobe plays such a key role in human cognition that it is singled out with its own functional description. Located in the heart of the brain in the medial region of the temporal lobe (Fig. 2.21), the function and the anatomy of the

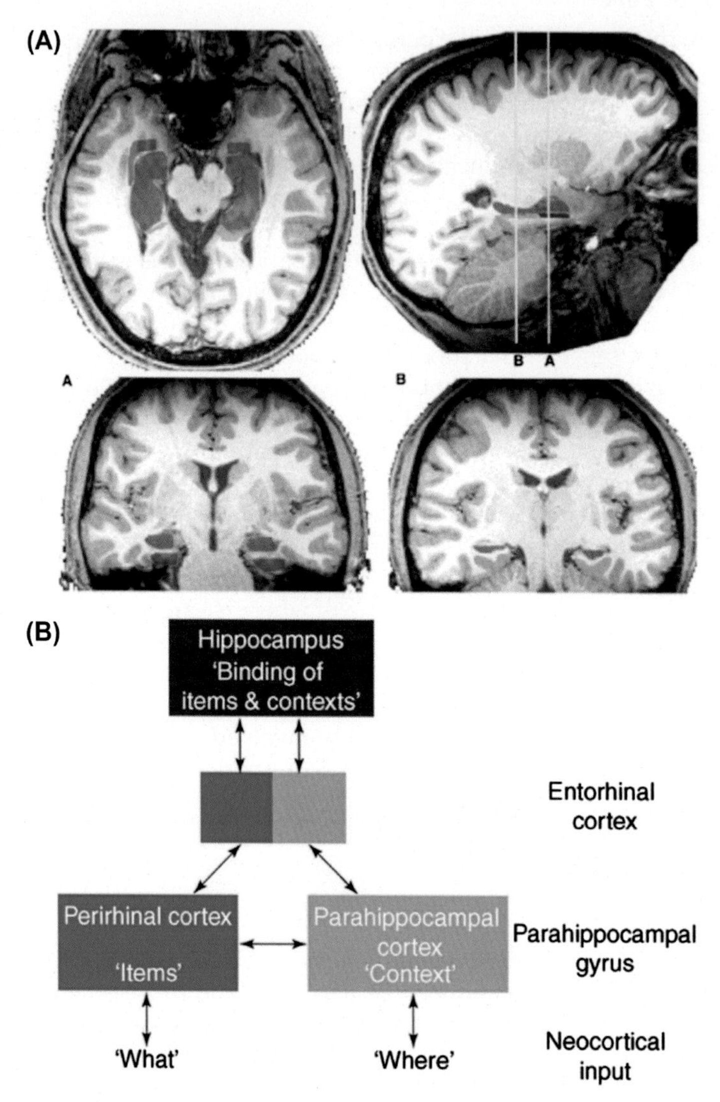

FIGURE 2.21 The medial temporal lobe. (A) The anatomy of the medial temporal lobe (MTL) is shown on axial (left top), sagittal (top right), and coronal (bottom central) slices of the brain. The approximate locations of the hippocampus are shown in *red*, the perirhinal cortex is shown in *blue*, and the parahippocampal cortex is shown in *green*. Note that these brain structures tucked into the MTL are overlapping due to their close proximity (B). The hypothesized connectivity and functional roles of the regions within the MTL in memory processing according to the authors' '"binding of item and context"' (BIC) model: the *arrows* between the hippocampus (*red rectangle*) extending through the entorhinal cortex to the perirhinal cortex (in *blue*) and parahippocampal gyrus (in *green*) reflect bidirectional connections for encoding "'item-related aspects'" such as what is an item (*blue rectangle*) and "'context-related aspects"' such as where that item is located (*green rectangle*). *Source: Diana, Yonelinas, and Ranganath (2007). Medial Temporal Lobe Trends in Cognitive Sciences 11(9), 379–386. Fig. 3, with permission.*

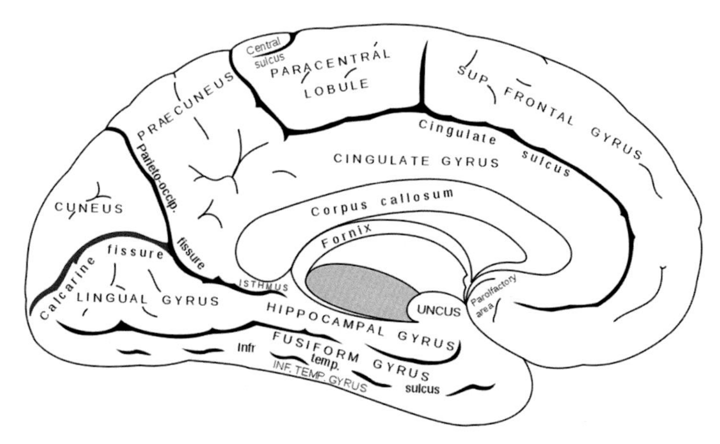

FIGURE 2.22 The Calcarine fissure. Much of visual cortex is tucked into the *Calcarine sulcus* (or fissure) that is located on the inside of the occipital lobe between the hemispheres (shown in *red* in the posterior region of the brain). *Source: Public domain. https://en.wikipedia.org/wiki/Calcarine_sulcus#/.*

medial temporal lobe (MTL) differs strikingly from the rest of the temporal lobe and is further justification to provide a separate description of its role in the brain. The MTL is home to the hippocampus and related regions that are associated with memory functions, including the perirhinal, parahippocampal, and entorhinal cortical regions (Fig. 2.21). As we will see in more detail in Chapter 7, Learning and Remembering, the MTL is key to memory formation and memory storage.

4.3.5 The Occipital Lobe

The occipital lobe, posterior to both the parietal and temporal lobes, is home to Primary and Association Visual Cortex (Fig. 2.9). Much of visual cortex is tucked into the *Calcarine sulcus* (or fissure) that is located on the inside of the occipital lobe between the hemispheres (Fig. 2.22). Thus, the occipital lobe contains the complex, multifaceted neural ensembles that form the basis for human vision and visual perception. As we will see in Chapter 4, there are many subregions within visual cortex in the occipital lobe that play central roles in our sense of the sights in the world around us.

5. BRAIN PATHWAYS: NEUROCONNECTIVITY

The brain has massive interconnectivity, the heart of which is the *thalamocortical hub* connecting virtually every region of cortex with the thalamus and other cortical and subcortical regions (Fig. 2.10) (Box 2.1).

Several major fiber pathways—white matter tracts—course through the brain. Three major ones are the *arcuate fasciculus*, the *corpus callosum*, and the *internal capsule/corona radiata*; however there are many more.

BOX 2.1

THE CONNECTOGRAM

Is there an easy way to "see" the brain's massive, multidimensional interconnectivity? This has been a challenge for the field of neuroscience but recent progress has revolutionized the way we visual the brain and its connections. Enter the *Connectogram* (Fig. 2.23).

Visualizing Brain Interconnectivity

Brain regions are hugely interconnected: for example, the thalamocortical circuitry connects almost every region of the cortex to the mighty thalamus and back. A central goal in understanding brain connectivity is to provide ways to visualize these highly interconnected pathways that course throughout the brain. Many methods for displaying them use three-dimensional graphical programs and a user interface. These methods do not lend themselves to an easy-to-use two-dimensional website, figure, or printout of the brain's connectivity.

The Connectogram was developed as a graphical representation of brain connectivity data studied using *connectomics*, the discipline for mapping and interpreting fiber connections in the brain. An early version was devised by Andrei Irimia and Jack Van Horn and colleagues (Irimia, Chambers, Torgerson, & Van Horn, 2012). Today there are many versions of connectograms, all building around a similar theme.

Interpreting A Connectogram

Although not an easy graph to understand at first sight, the connectogram provides many levels of analysis in a single graph. Let us step through those levels of analysis.

Take a look at Fig. 2.23: the circle reflects the entire brain (but dominated by the cortex). The left side of the circle reflects the left hemisphere, and the right side reflects the right hemisphere. On the outside edge of the circle, the lobes and brain regions are listed beginning with the left and right frontal lobes on the left and right upper sides of the circle. Moving down the outside of the circle, you will see Insula cortex, Limbic lobe, and down through the brain regions until you arrive at the brainstem at the center bottom of the connectogram. Subregions within each lobe and brain region are detailed along the outside edge of the connectogram. Each subregion has been given a color coding. Their abbreviations and fully detailed region names can be found on Wikipedia (https://en.wikipedia.org/wiki/Connectogram).

Concentric Rings Provide More Structural Information

Next, there are five rings that form concentric circles inside the connectogram. The legend for these rings is shown in Fig. 2.23B, and briefly refers to structural attributes of each region such as gray matter volume and surface area. The innermost ring is labeled as "degree of connectivity," with the darker colors reflecting more fibers (connections) initiating or terminating in that brain region. Take a look at the upper left section of the connectogram in Fig. 2.23 and search for a

BOX 2.1 *(cont'd)*

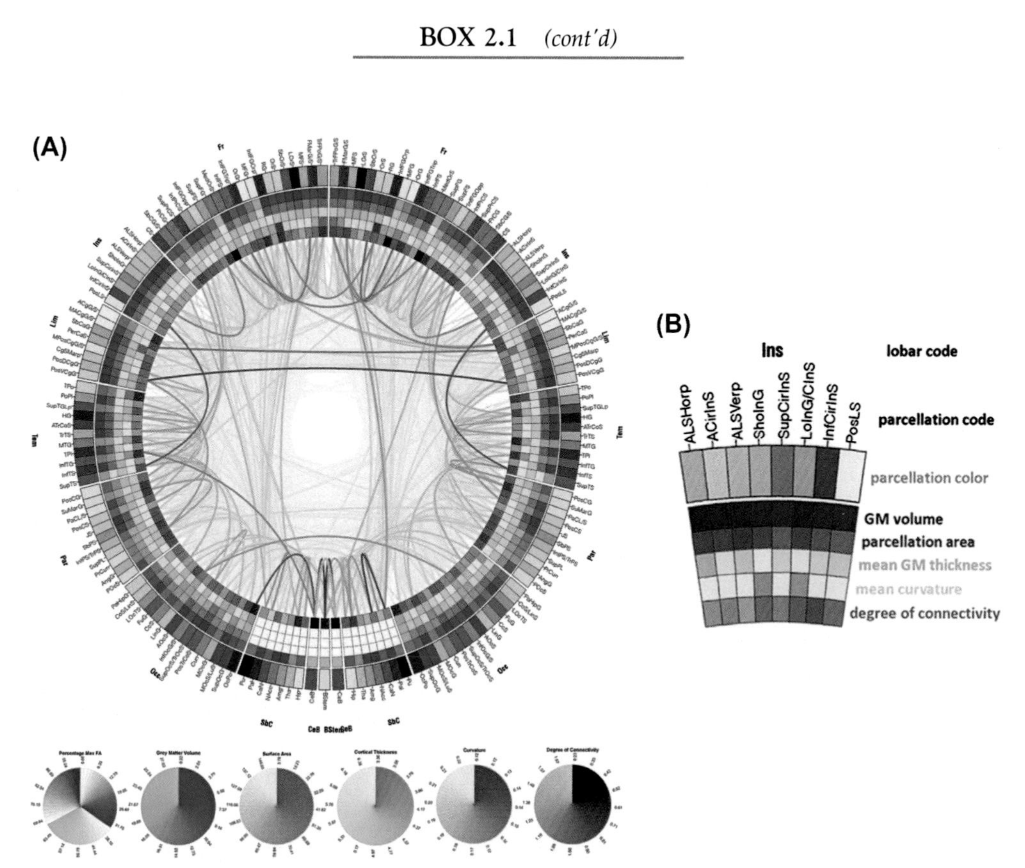

FIGURE 2.23 The Connectogram. (A) The circular Connectogram is a graphic way to show brain connectivity. The left half of the circle depicts the left hemisphere and the right half depicts the right hemisphere. The lobes and subregions of each hemisphere are shown and labeled, beginning at the top of the circle, as frontal lobe (fr), insular cortex (Ins), limbic lobe (Lim), temporal lobe (Tem), parietal lobe (Par), occipital lobe (Occ), subcortical structures (SbC), and cerebellum. The brainstem (BStem) is shown at the bottom of the circle, between the two hemispheres. Within the lobes, each cortical area is labeled with an abbreviation and an assigned color. Note the five rings that are inside the Conneectogram, just next to the color-coded cortical regions. The legend for these rings is shown in B. In the center of the Connectogram, lines connect regions that are structurally connected. The relative density (number of fibers) of these connections is reflected in the opacity of the lines, with the more opaque lines reflecting less densely connected regions and the less opaque lines reflecting more densely connected regions. The color of the line is color coded to reflect the density of fibers within that connection. (B) The legend describing the rings of the Connectogram, which show different attributes of the corresponding cortical regions. From outermost to innermost, the rings represent the grey matter volume, surface area, cortical thickness, curvature, and degree of connectivity. *Source: Open access. John Darrell Van Horn - PLoS One, http://www.plosone.org/article/info%3Adoi%2F10.1371%2Fjournal.pone.0037454. A connectogram showing the average connections and cortical measures of 110 normal, right-handed males, aged 25–36.*

Continued

BOX 2.1 *(cont'd)*

very dark brown, almost black, inner ring. You will find one labeled as SupFG: this is the Superior Frontal Gyrus in the frontal lobe that has many fibers beginning and ending in its region as compared to the rest of the brain. At the bottom of the connectogram, you will see similar dark boxes in the innermost ring at the BStem, the brainstem.

Connectivity Shown With Lines in the Center of the Connectogram

Now take a look at the center of the connectogram. This is a key region for beginning to visualize brain connectivity. You will see many lines that extend from one brain region to another that vary sharply in their opacity. The *opacity* of the line reflects the *density of the fibers* in that connection, with darker or less opaque lines referring to dense connections between those brain regions. The color of the lines reflects the *fractional anisotropy* of the connection, with blue associated with lower values, green with mid values, and red with higher values. Fractional anisotropy is a measure of the directional dependence of the connection, which is based on the diffusion tensor imaging technique, which provides the data for the color-coding of these lines (see Chapter 3, Observing the Brain).

Comparing Group Data to Individuals With Brain Damage

One interesting use of the connectogram is to visualize individual brain data as compared to group data of healthy individuals. The connectogram shown in Fig. 2.23 reflects data from a group of healthy young men between the ages of 25–35 years. How is the connectogram changed when evaluating someone with brain damage?

The Famous Case of Phineas Gage

You may have read about the case of Phineas Gage, a 25-year old railroad construction supervisor. In 1848, an accident caused a tamping iron to shoot through his skull and brain in the left frontal area (Fig. 2.24A). His cognitive and personality changes following the accident were well documented and led to our understanding of the role of the left ventral PFC in human cognition. What happened to the massively interconnected frontal lobe when this terrible brain injury occurred? Irimia, Van Horn, and colleagues were curious about this question and using the CT scan of Gage's brain, developed a connectogram just for him (Fig. 2.24B, Van Horn et al., 2012). In this connectogram, the color of the lines in the center of the figure that reflect the connectivity of the brain regions is coded so that gray-scale lines reflect completely severed connections between those brain regions while tan-scale lines reflect partially severed connections.

The Future of the Connectogram

The connectogram showed in Fig. 2.23 was developed from data from a sample of men aged 25–35 years and thus represents brain connectivity in healthy young men. This raises many interesting questions for future research. Does a connectogram differ for healthy young women? For children?

BOX 2.1 (cont'd)

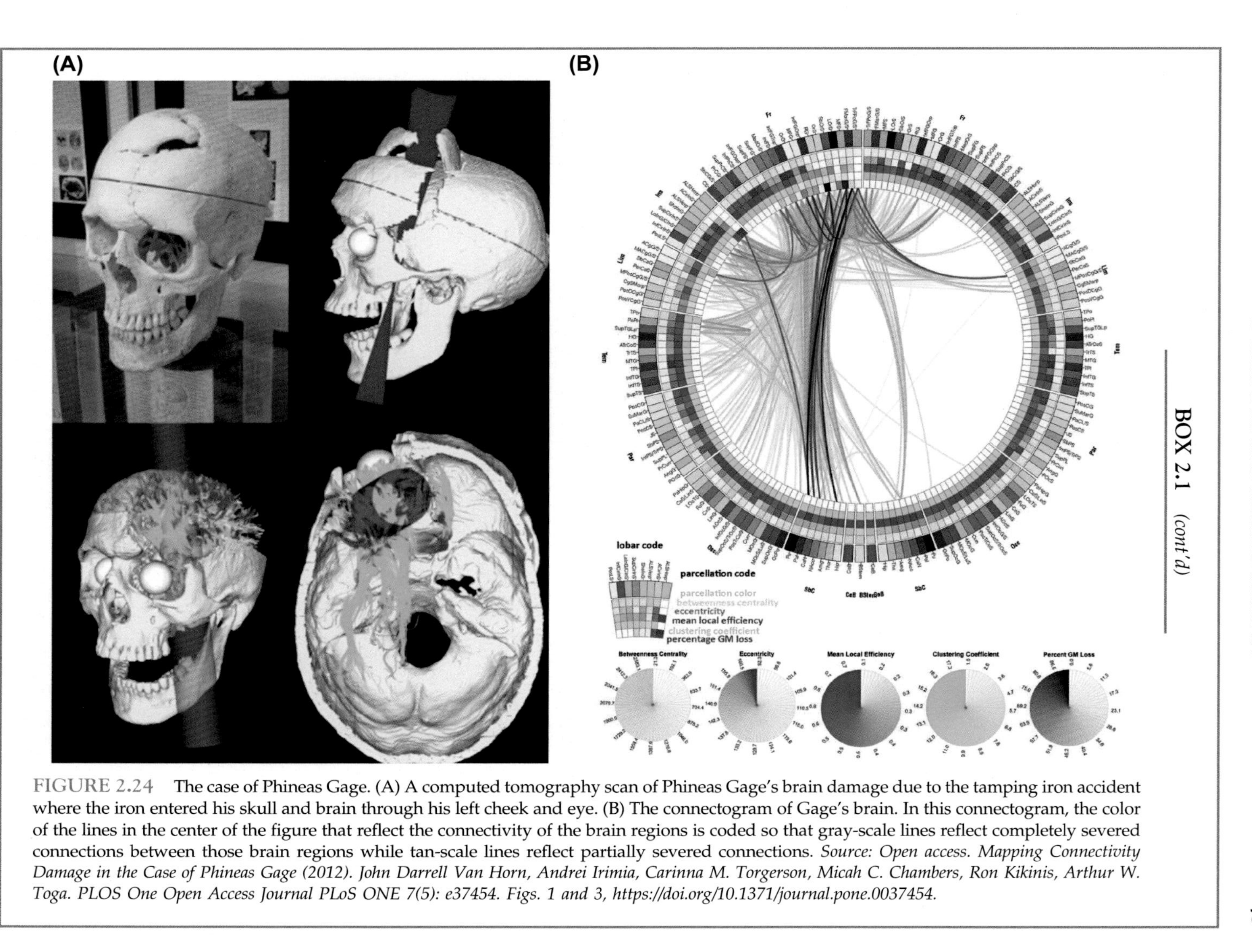

FIGURE 2.24 The case of Phineas Gage. (A) A computed tomography scan of Phineas Gage's brain damage due to the tamping iron accident where the iron entered his skull and brain through his left cheek and eye. (B) The connectogram of Gage's brain. In this connectogram, the color of the lines in the center of the figure that reflect the connectivity of the brain regions is coded so that gray-scale lines reflect completely severed connections between those brain regions while tan-scale lines reflect partially severed connections. *Source: Open access. Mapping Connectivity Damage in the Case of Phineas Gage (2012). John Darrell Van Horn, Andrei Irimia, Carinna M. Torgerson, Micah C. Chambers, Ron Kikinis, Arthur W. Toga. PLOS One Open Access Journal PLoS ONE 7(5): e37454. Figs. 1 and 3, https://doi.org/10.1371/journal.pone.0037454.*

Continued

BOX 2.1 *(cont'd)*

And how does it change in normal aging and in disorders and diseases such as autism, Alzheimer disease, and schizophrenia? Because the connectogram can be developed for a single individual, as well as in a group study, there are likely many clinical applications that will be very useful for individuals with traumatic brain injury or stroke. Perhaps the connectogram will provide information for recovery from these disorders and diseases—or at least provide information about which pathways have been affected so that clinical treatment and therapy can be individually adapted for each person.

References

(1) Irimia, A., Chambers, M. C., Torgerson, C. M., & Van Horn, J. D. (2012). Circular representation of human cortical networks for subject and population-level connectomic visualization. *NeuroImage, 60*(2), 1340–1351. http://doi.org/10.1016/j.neuroimage.2012.01.107

(2) Van Horn, J. D., Irimia, A., Torgerson, C. M., Chambers, M. C., Kikinis, R., Toga, A. W. (2012). Mapping connectivity damage in the case of Phineas Gage. *PLoS One, 7*(5), 1–23.

5.1 The Arcuate Fasciculus

The arcuate fasciculus is a major anterior/posterior tract (Fig. 2.25). The arcuate fasciculus is a white matter bundle that contains both long and short fibers that connect the frontal, parietal, and temporal lobes (Catani & Thiebaut de Schotten, 2008). The arcuate plays a key role in the left hemisphere in language processing (Fig. 2.26) and in the right hemisphere in visuospatial processing and some aspects of language processing (Fig. 2.27), such as prosody and semantics (Catani & Thiebaut de Schotten, 2008).

5.2 The Corpus Callosum

The corpus callosum contains millions of axon fibers sending neural transmissions across the two hemispheres. Take a look at Fig. 2.28, and you will get a sense of the mighty mass of connectivity that forms the linkage between the two hemispheres of the brain. Most of the callosal connections across the hemispheres are connecting similar regions in the opposite hemisphere: right hemisphere visual cortex callosal fibers travel across to left hemisphere visual cortex, and vice versa. Similarly, PFC fibers connect similar regions in the left and right frontal lobe.

5.3 The Internal Capsule/Corona Radiate

A third major white matter tract is the internal capsule/corona radiata (Fig. 2.29). These pathways contain fibers that ascend from the thalamus up to the cortex. They also contain descending fibers from the frontal and parietal lobes down to subcortical bodies such as

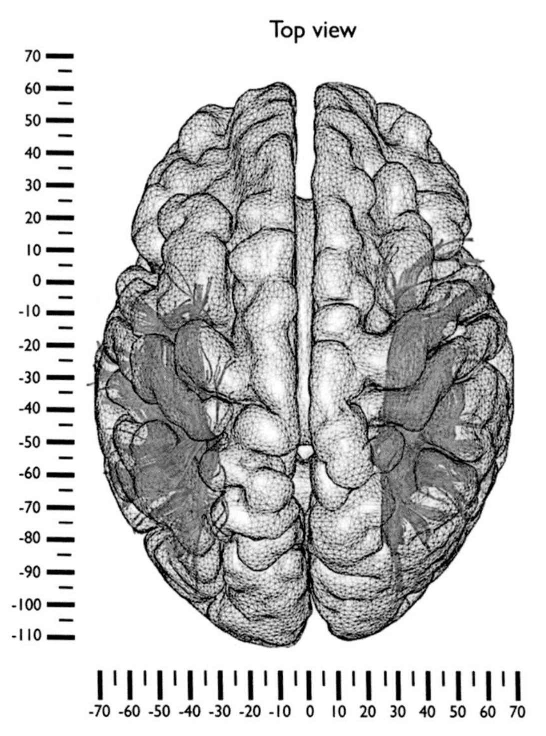

FIGURE 2.25 Arcuate fasciculus. The large fiber tract arcuate fasciculus, which connects the frontal, parietal, and temporal lobes, viewed from the top of the brain. The left hemisphere is shown on the left side of the figure, the right hemisphere is shown on the right. *Source: Catani & Thiebaut de Schotten 2008 Cortex 44, 1105–1132. Fig. 1, with permission.*

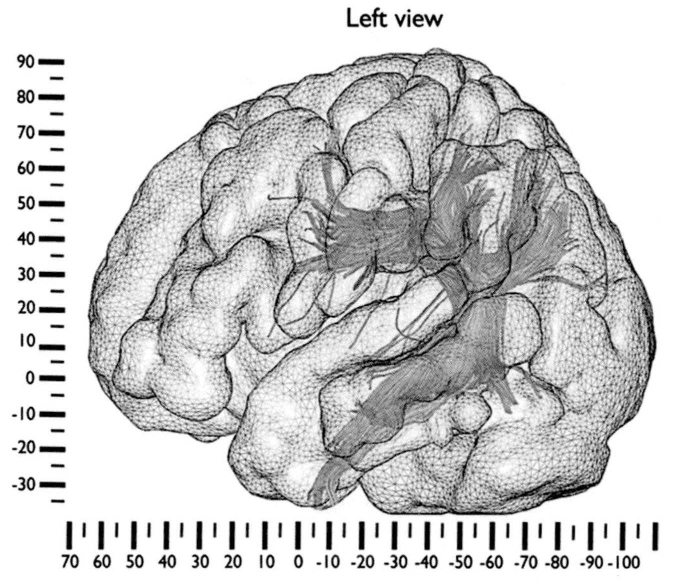

FIGURE 2.26 Arcuate fasciculus in the left hemisphere. The arcuate plays a key role in the left hemisphere in language processing. *Source: Catani & Thiebaut de Schotten 2008 Cortex 44, 1105–1132. Fig. 1, with permission.*

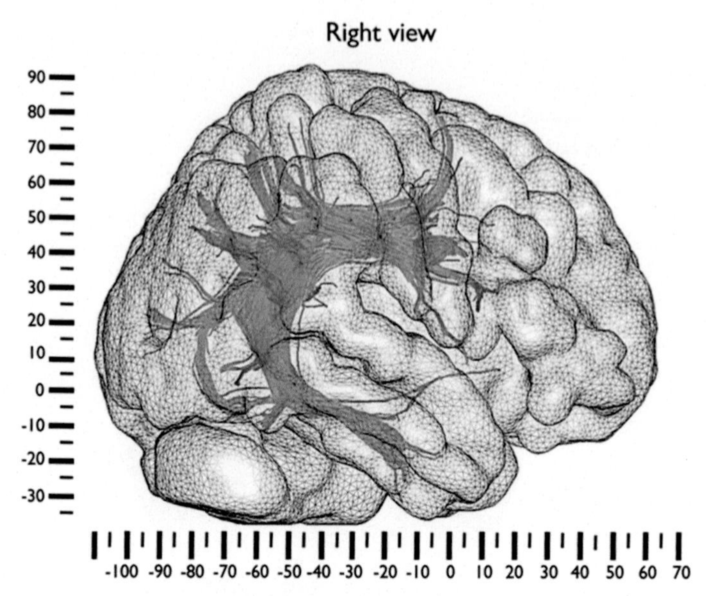

FIGURE 2.27 Arcuate fasciculus in the right hemisphere. The arcuate plays a key role in the right hemisphere in visuospatial processing and some aspects of language processing. *Source: Catani & Thiebaut de Schotten 2008 Cortex 44, 1105–1132. Fig. 1, with permission.*

the basal ganglia, and also to the brainstem and spinal cord. The internal capsule/corona radiate is the anatomic linkage that supports cognitive and perceptual and motor systems in the cortex.

In addition to these three massive white matter tracts, there are other white matter bundles, large and small, connecting cortical areas and subcortical bodies. Together these pathways form a multidirectional connectivity system that links the brain's lobes and hemispheres to each and to the body via the spinal cord.

Despite this massive array of pathways, it is important to note that not every region in the brain is connected to every other region. Think of a large state highway system: there are large superhighways that course long distances between major cities, smaller highways connecting smaller cities to larger ones, and still smaller routes and streets connecting areas of cities to other areas and suburban regions. Typically, although there is usually a major superhighway connecting the large metropolitan regions in a state, there are also other options for travel that may be less rapid but provide more options for stopping off at various subroutes along the way. Pathways in the brain are laid out in a similar manner: there are major white matter tracts that course throughout the brain and across the hemispheres combined with smaller white matter bundles that provide more localized connectivity.

6. BRAIN DYNAMICS—BRAIN RHYTHMS AND OSCILLATIONS

We mentioned earlier in the chapter that most neural pathways are bidirectional. We have described some of the major anterior-posterior and left-right pathways above, as well as the

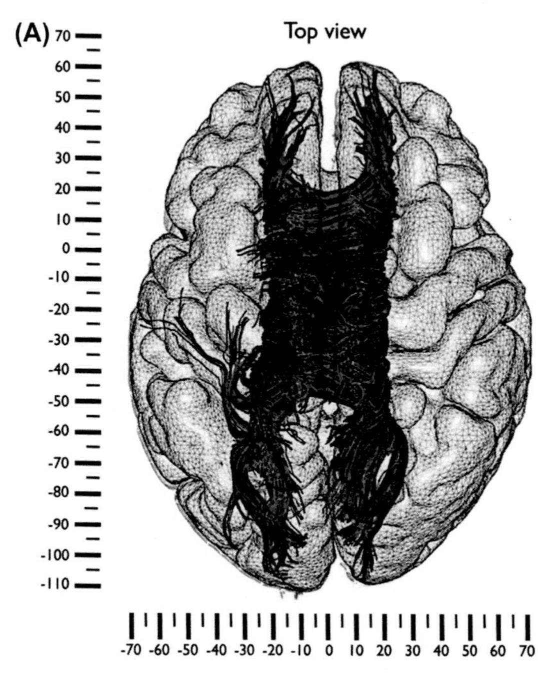

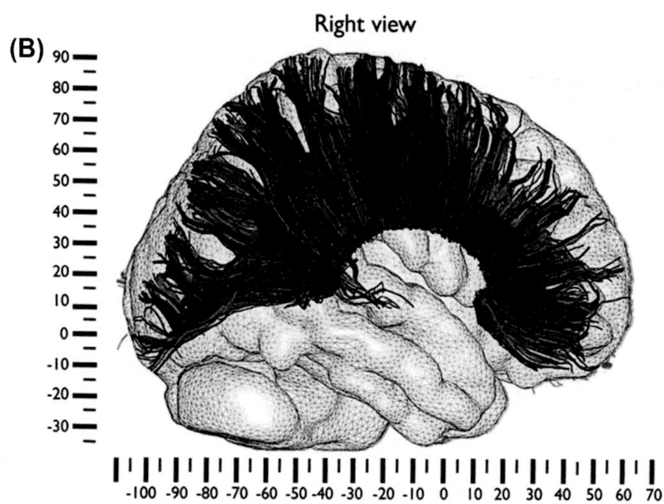

FIGURE 2.28 Corpus callosum. The fiber tracts of the massive corpus callosum are shown from the top of the brain (A), right side of the brain (B), and left side of the brain (C). *Source: Catani & Thiebaut de Schotten 2008 Cortex 44, 1105–1132. Fig. 6, with permission.*

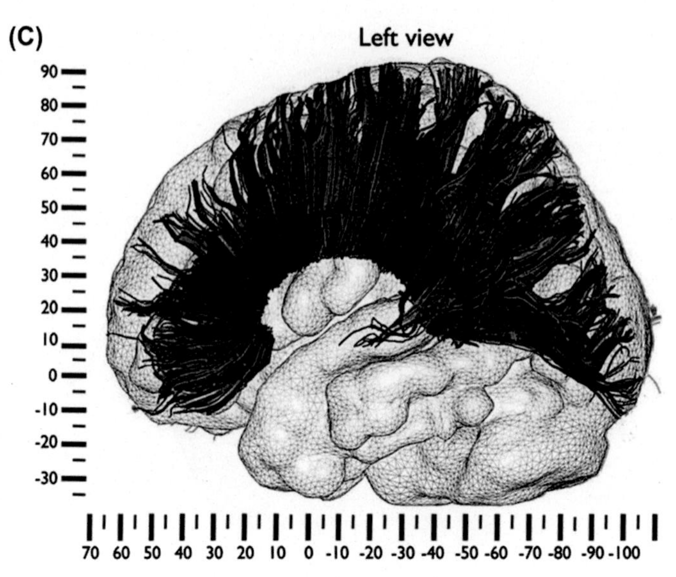

FIGURE 2.28 cont'd

massive thalamocortical hub formed by the thalamus and the cerebral cortex. A key aspect of brain function is elucidating how these many bidirectional pathways form our mind/brain, with an integrated sense of awareness of our world around us, our inner thoughts, and our very consciousness.

Wolf Singer and his colleagues have shed new light on these processes in their investigations into neural synchrony in cortical networks (for example, see Uhlhaas et al., 2009 for a review). Singer and his colleagues describe the cortex as a system comprised of many distinct, distributed networks that process information in parallel through these bidirectional pathways that are called "re-entrant." Since these many networks do not converge on a single hub or higher order region in the brain, a central question is how this distributed system gives rise to our conscious experience. One hypothesis for how this happens in the brain is the *synchrony of the oscillations* produced by these many networks across brain regions. The *temporal dynamics* of the oscillations across the brain are thought to provide the neural bases of neural synchrony in the brain that leads to our perceptions, thoughts, and coherent experience with the world around us, as well as with our internal thoughts and consciousness. More specifically, the action potentials of groups of neurons phase-lock with the wider scale oscillations to form a method to bind or link neural activity across wide ranging cortical regions.

7. PUTTING IT ALL TOGETHER

We have discussed five central aspects of brain study: *neuroanatomy* for studying the structure of the brain; *neurophysiology* for studying the cellular aspects of the brain; *functional*

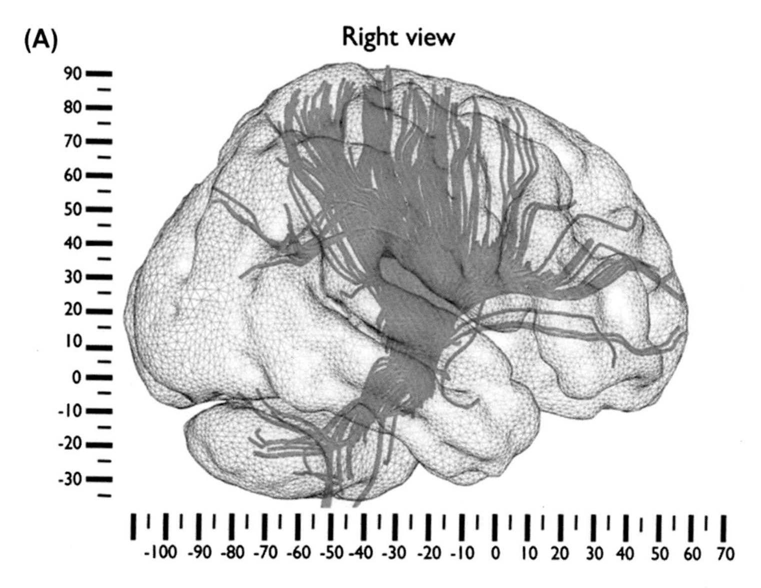

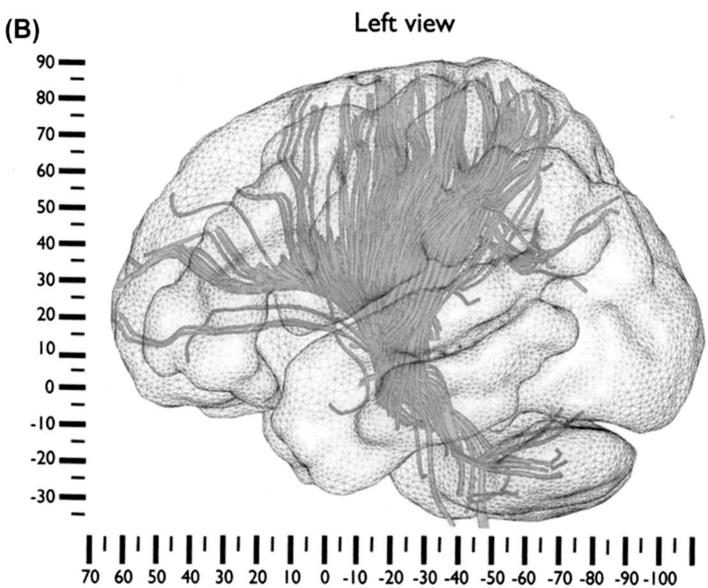

FIGURE 2.29 The internal capsule/corona radiata. The fiber tracts of the internal capsule/corona radiata are shown from the right side (A) and left side of the brain (B). *Source: Catani & Thiebaut de Schotten 2008 Cortex 44, 1105–1132. Fig. 9, with permission.*

neuroanatomy for studying the perceptual, motor, and cognitive processes of the brain; *neuroconnectivity* for studying brain pathways; and neurodynamics to study the rhythms and oscillations of the brain. Together these disciplines enable us to understand both small-scale and wide ranging aspects of the human brain. In the following chapter, we will investigate how brain visualization systems and methods allow us to further the study of these five key aspects of the brain and how they correspond to human behavior.

8. STUDY QUESTIONS

1. What does "level of analysis" refer to when investigating the human brain?
2. What is the difference between *neuroanatomy* and *functional neuroanatomy*?
3. Name the three planes of the brain and describe how they "slice" the brain.
4. What are some major landmarks of the brain?
5. What is an action potential? How (when) does it occur?
6. What role does the PFC play in human cognition?
7. What is a *homunculus*?
8. Describe three major white matter (fiber) tracts in the brain.
9. What is a connectogram and what information does it provide about the brain?

CHAPTER

3

Observing the Brain

OBSERVING THE VAST UNIVERSE OF THE HUMAN BRAIN

As long as our brain is a mystery, the universe, the reflection of the structure of the brain will also be a mystery. ***Santiago Ramón y Cajal, Spanish Neuroscientist***

A typical barred spiral in space. *Source: NASA public domain figure.*

1. INTRODUCTION

The ability to more directly observe the living brain has created a scientific turning point, much like Galileo's first glimpse of the moons of Jupiter. Humans have studied the sky for many centuries, but when glass lenses and telescopes were invented, the pace of discovery took off. But just as Galileo's telescope required constant improvement, our "brain scopes" have their limits. We should know their limits as well as their capabilities.

Fundamentals of Cognitive Neuroscience
https://doi.org/10.1016/B978-0-12-803813-0.00003-9

A perfect brain observer would keep track of tens of billions of neurons many times per second. The perfect observer should then be able to track the shifting interplay between groups of neurons, with tens of thousands of signals traveling back and forth. By analogy, a perfect spy satellite would see every human being on Earth as well as all the things we say to one another.

Such a perfect observer does not exist. We are more like distant space explorers beginning to observe a new planet. We pick up signals without knowing exactly where they come from, whether they are made by living creatures, what languages they speak, or even whether the signals mean anything.

We know that the brain has major pathways and maplike arrays of neurons and that single spikes (action potentials) as well as waves (oscillations) can travel between brain maps. Brain oscillations vary between 0.5 and 120 Hz (i.e., up to about 120 cycles per second), with occasional faster events. If we add in the neurotransmitters and neuromodulators that shape signal transmission, the potential number of signals is enormous (Fig. 3.1).

Neurons have *electrical*, *magnetic*, *chemical*, and *anatomical* properties. Each of these can be measured. As you know from Chapter 2, The Brain, every neuron can send a fast electrical signal down its axon. We can record this activity in the cell or in the surrounding fluid. The brain also has a rich fuel supply, and when a specific region of the brain is working harder, it needs extra fuel. Those facts give us all our brain measurement techniques: neurons, and networks of neurons, generate electrical and magnetic signals. Metabolic processes can be picked up using methods like positron emission tomography (PET) and functional magnetic resonance imaging

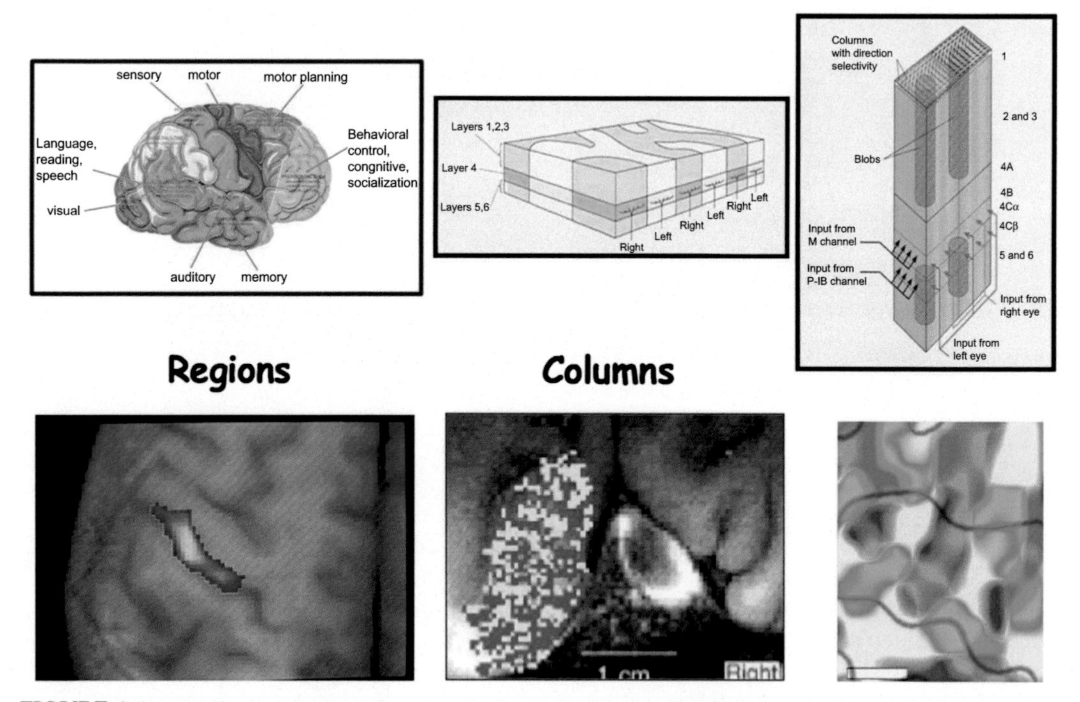

FIGURE 3.1 Levels of analysis in observing the brain. Techniques for observing the human brain range from functional magnetic resonance imaging methods for observing brain regions (left), to microscopic methods for observing neural columns in the cortex (center), to microscopic methods for observing within-column structures (right). *Source: Bandettini (2012).*

(fMRI). The anatomical shape of brain structures can be detected by computerized axial tomography (CAT) scans and magnetic resonance imaging (MRI). As a result, we can now see functional brain activities for speech, action control, motivation, sensory perception, and more.

1.1 Basics

Brain imaging has been a breakthrough technology for cognitive neuroscience, building on decades of cognitive psychology, behavioral conditioning, psychophysics, and brain science. Before imaging techniques matured, our knowledge came from animal studies and the haphazard injuries incurred by human beings. But brain injuries are extremely imprecise, and to locate the damage, neurologists often had to rely on postmortem examination of patients' brains—as in the case of Broca's and Wernicke's patients. The brain can often compensate for damage, so lesions change over time as cells die and adaptation occurs. Therefore, postmortem examinations do not necessarily reflect the injury at the time of diagnosis. Animal studies depend on presumed homologies—similarities across species—that were often not convincing. No other animals have language and other human specializations. It was therefore very difficult to understand how brain functions in animals mapped onto human cognition.

Today, we have perhaps a dozen techniques that are rapidly becoming more precise. Medical needs often drive this expensive technology because it applies to many organs in the body. As a result, we now have ways to study the distribution of billions of neurochemical receptors in the brain, the thickness of cortex, the great highway system of white fiber bundles, and, most important for cognitive neuroscience, the *functional* activity of the brain—the basis of its adaptive capacities. New advances are allowing scientists to investigate not only specific brain regions but also the dynamic pattern of connectivity between them.

1.2 Accuracy in Space and Time

Fig. 3.2 shows today's methods and their accuracy in space and time. Techniques like fMRI, which records brain changes like blood oxygenation, have good spatial resolution and relatively poor time resolution. fMRI has a response time of about 6 s because the fMRI signal (called the blood oxygen level–dependent [BOLD] signal) reflects a flow of oxygen-rich blood traveling to "hot spots" that are doing extra work. The changes in blood flow take several seconds to catch up with the neuronal activity. fMRI is therefore too slow for tracking single neurons and populations in "real time."

We do not have a complete census of all the neurons in the brain the way a human society might conduct a census of the whole population. We are always sampling from a *very large set* of active neurons. For that reason we cannot be sure that we know every single cell type in the brain, down to the smallest level. Brain anatomists are constantly discovering new specialized neurons in some local neighborhood. For example, the light receptors that adjust our body to sunlight and darkness were only discovered in recent years.

fMRI has very good spatial specificity compared to electroencephalography (EEG) and magnetoencephalography (MEG), which use electrical and magnetic signals, respectively. Thus, fMRI is used to localize brain functions. But EEG and MEG have excellent temporal resolution—almost instantaneous—and relatively poor spatial precision. They can track neuron population

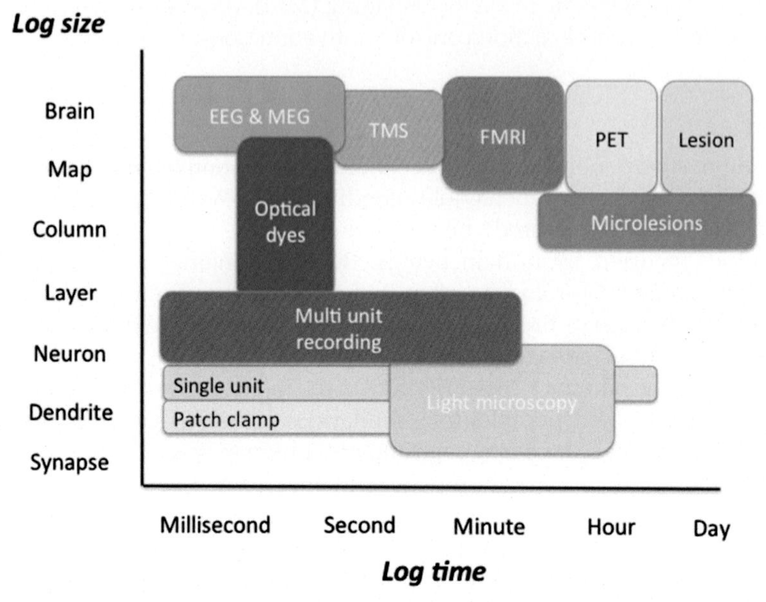

FIGURE 3.2 How good are current methods? Pros and cons of imaging techniques: differing imaging modalities have different resolutions. While some approaches have a very high temporal (time-based) resolution but a low spatial (space-based) resolution (such as EEG), other modalities have an opposite relation (such as light microscopy). *Source: Gage with permission.*

activity as quickly as tens and hundreds of milliseconds, but it is hard to know *which* set of neurons is doing it. fMRI is sometimes used in combination with EEG to obtain the best temporal *and* spatial precision. Combined measures may give us the best of both worlds.

1.3 A Brain in a Shoebox: Coordinates

When sailors learned to draw imaginary lines of latitude and longitude to place a coordinate system around the earth, it became possible to specify any place with a precise "x" and "y" number. The earth's coordinate system became a major advance in navigation.

The shape of the brain is more complex than the earth's, but the strategy of placing it in a three-dimensional space is the same. Scientists can specify a precise location in the brain by placing it in a virtual shoebox so each point in the brain has a unique address in three orthogonal dimensions. Each point can be specified as x, y, and z. The best example is the Talairach Coordinate System illustrated in Fig. 3.3.

Different people have different brains. Just as our heads have distinctive shapes, so do the organs inside. We begin to lose neurons after birth, so infants have more neurons than older children, teenagers, or adults. On the other hand, brain myelination (the wrapping of long axons by white protective cells) keeps growing until about age 30. Serious illnesses, growth and aging, learning and exercise, brain injury, and even malnutrition can add or subtract tissue. The brain keeps changing.

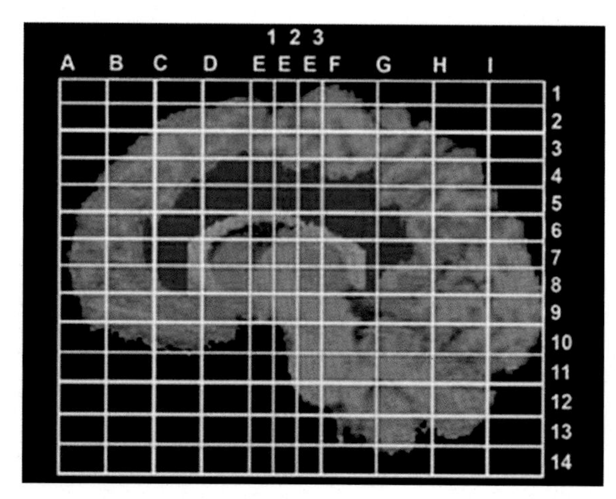

FIGURE 3.3 A coordinate system for the brain. The brain is typically imaged in three dimensions: axial, sagittal, and coronal. Putting the brain together into a three-dimensional image allows the coordinates to be determined across these three dimensions. This is a midsagittal slice of the brain (front of the head is to the left of the image) with the Talairach coordinates superimposed on the brain image. *Source: Woodward, Kaloupek, Schaer, Martinez, and Eliez (2008).*

Individual brains therefore need individual images. MRI and CAT scans are used to take a snapshot of the three-dimensional brain at any particular moment. Figs. 3.4 and 3.5 show the smallest unit imaged using MRI: a "voxel." The actual size of a voxel varies depending upon factors such as the resolution of the MRI scanner, the size of the brain being scanned, and the brain region being scanned. A typical voxel for a T1-weighted scan is about one cubic millimeter (mm^3). If it is from the cortex, a single voxel may contain tens of thousands of neurons. Fig. 3.6 shows a brain navigation program with a screenshot of the standard coordinate system used in most MRI research.

Brain surgeons need to know where to remove disease-causing tissue and which regions need to be left untouched. Structural images from MRI and CAT scans give us a three-dimensional map of the brain. But maps are also essential for understanding functional measures, such as EEG, MEG, fMRI, and deep brain recording. Typically, measures of brain structure and function are combined.

2. OBSERVING BRAIN STRUCTURE: NEUROANATOMY

MRI is the predominant anatomical imaging device used in hospitals, clinics, and research laboratories today. Its emergence only a few decades ago revolutionized both clinical and research brain investigations. MRI scanners contain strong magnets that range between 0.5 and 7 T (5000–70,000 G) and beyond. MRI utilizes the magnetic properties of the tiny cells—in the body and in the brain—to render sharp, precise images. A hydrogen nucleus (a single proton) in a cell is present in water, which in turn is present in all body tissues. MRI scanners make use of the magnetic properties in protons to align them using the strong magnetic field that is present in the MRI scanner (Fig. 3.7). The protons are rotated by the radio waves in the magnetic field of the scanner and detected by coils designed to collect information across the three dimensions of the area of the body or brain being scanned.

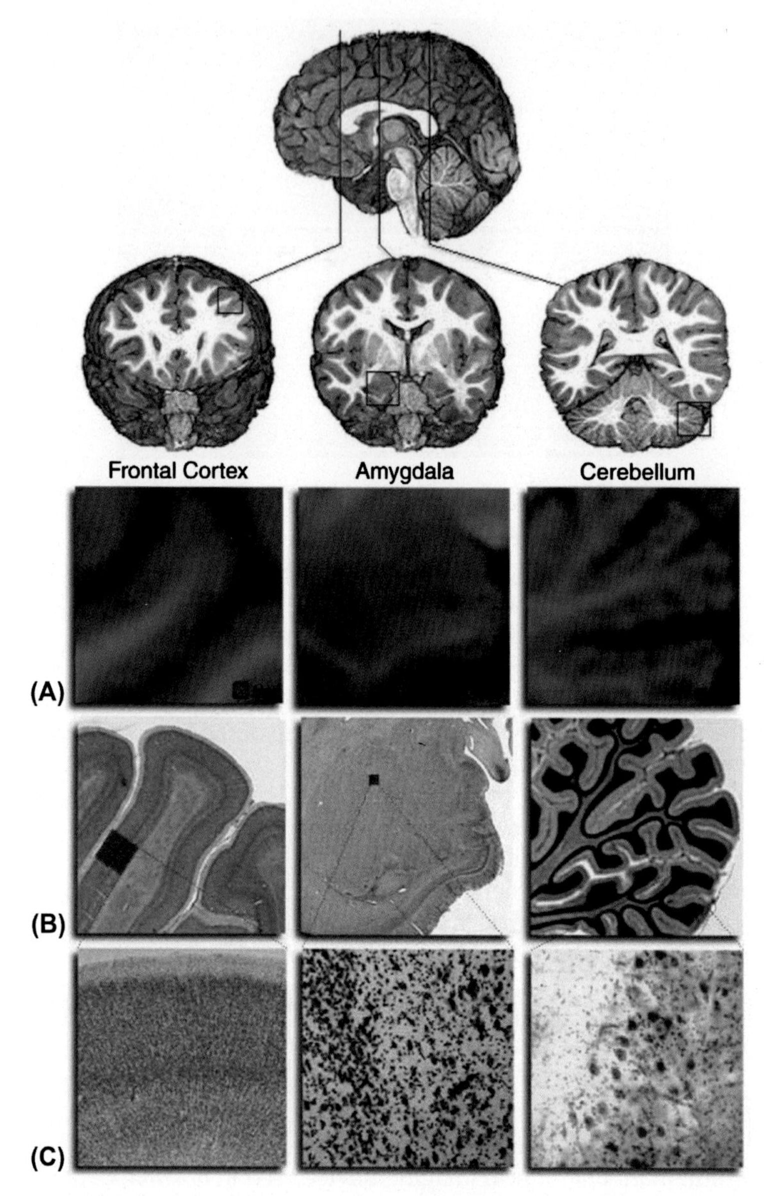

FIGURE 3.4 The smallest unit imaged using magnetic resonance imaging (MRI): a "voxel." The actual size of a voxel varies depending upon factors such as the resolution of the MRI scanner, the size of the brain being scanned, and the brain region being scanned. A typical voxel for a T1-weighted scan is about 1 mm^3. If it is from the cortex, a single voxel may contain tens of thousands of neurons. (A) An example of a single voxel (~1 mm^3, shown in red) recorded from a high-resolution MRI scan from a region in the frontal cortex (left panel), from the amygdala in the subcortical region (center panel), and from the cerebellum (right panel). Each region selected (cortical, subcortical, and cerebellar) has differing cell densities and types. (B) Postmortem brain tissue from the same brain regions shown in (A), showing the resolution of ~1 mm^3 in postmortem tissue versus MRI image: the differences in cell distribution and morphology across these brain areas are recognizably similar across MRI scanning and postmortem tissue preparation. (C) At higher magnification, the differences in the distribution of cell bodies in the frontal cortex (left), amygdala (center), and cerebellum (right) can be seen. *Source: Schumann and Nordahl (2011).*

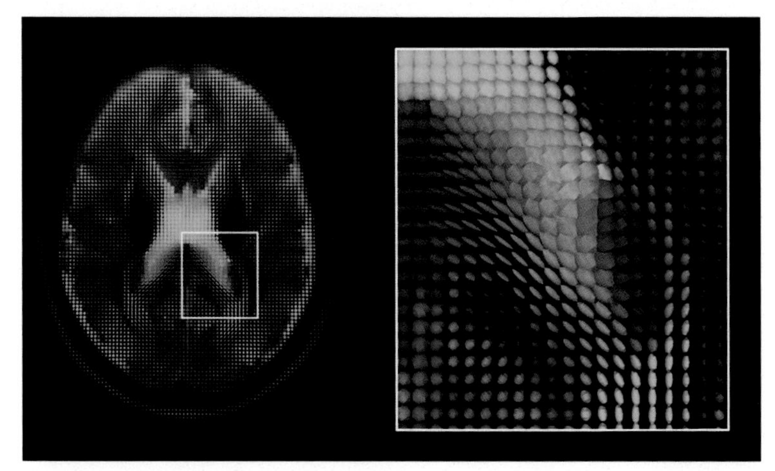

FIGURE 3.5 Voxels in the tensor field in a diffusion tensor imaging analysis. Each colored shape represents an individual voxel color coded according to the tensor field analysis method. Left: axial slice of the brain showing the left and right hemispheres on the left and right side of the figure with the top of the head at the top of the figure and the back of the head at the bottom of the figure. Right: close up view of the voxels showing their orientation. *Source: public domain: https://en.wikipedia.org/wiki/File:DiffusionMRI_glyphs.png.*

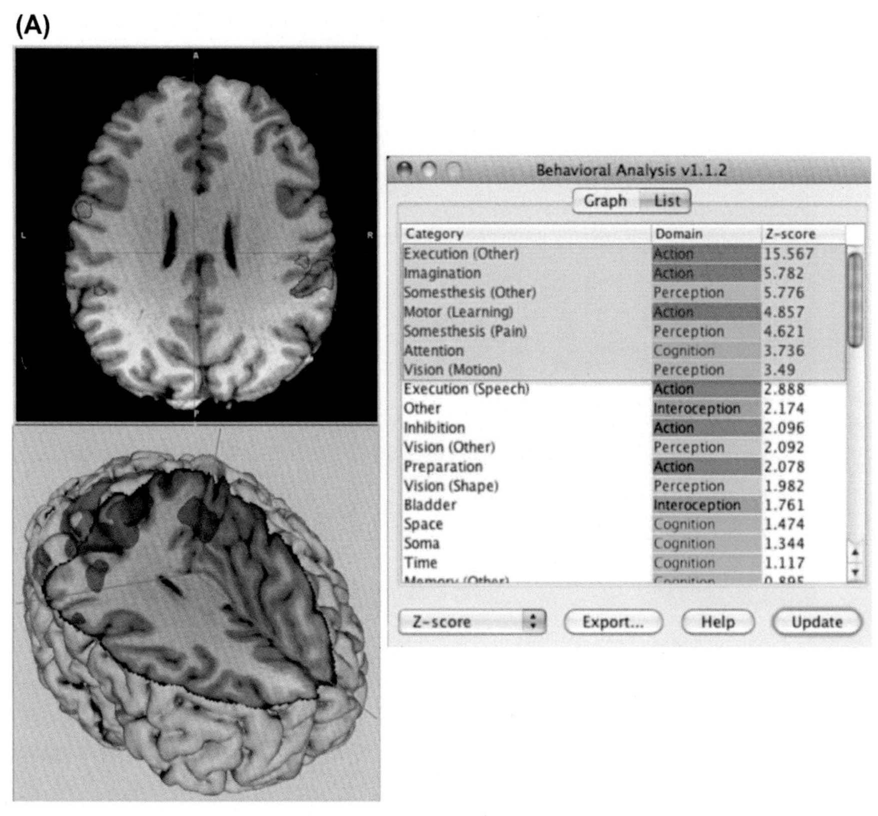

Category	Domain	Z-score
Execution (Other)	Action	15.567
Imagination	Action	5.782
Somesthesis (Other)	Perception	5.776
Motor (Learning)	Action	4.857
Somesthesis (Pain)	Perception	4.621
Attention	Cognition	3.736
Vision (Motion)	Perception	3.49
Execution (Speech)	Action	2.888
Other	Interoception	2.174
Inhibition	Action	2.096
Vision (Other)	Perception	2.092
Preparation	Action	2.078
Vision (Shape)	Perception	1.982
Bladder	Interoception	1.761
Space	Cognition	1.474
Soma	Cognition	1.344
Time	Cognition	1.117

FIGURE 3.6 Brain navigation software allows the user to translate precise locations into conventional brain locations to compare results across patients. The standard orientation of the brain in the x, y, and z planes developed by Talairach and Tournoux (1988) is used for analysis. (A) Upper left: an axial image of the brain with the left hemisphere shown on the left side. Red-shaded areas show regions of functional magnetic resonance imaging activation. Lower left: a surface rendering shows a three-dimensional view of the brain. Right: regions of interest are color coded in the analysis software package. (B) Activations (in teal) are overlaid on axial (top), coronal (lower left), and sagittal (lower right) images of the gray matter in the brain using the Talairach Daemon software. *Source: Lancaster, Laird, Eickhoff, Martinez, Fox, and Fox (2012).*

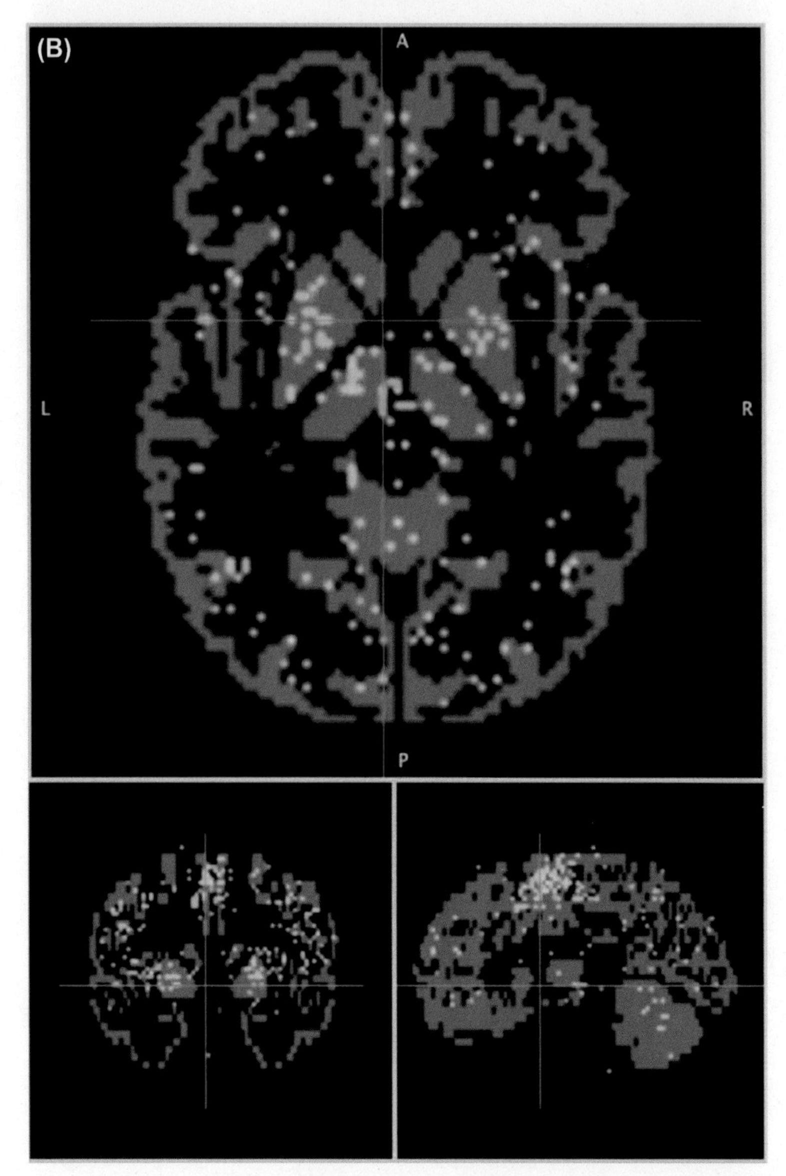

FIGURE 3.6 cont'd

The result is a detailed "picture" of soft tissue, bones, cartilage, etc., in whatever is being scanned. For our purposes, the region being scanned is the brain. Fig. 3.8 shows an axial (left) and a midsagittal (right) "slice" of a brain inside the skull. On the axial slice, you can clearly see the sulci and gyri of the cortex around the edges of the slice, the space between the two hemispheres along the centerline, along with the ventricles containing cerebral fluid. On the midsagittal slice, you can see brain structures that lie in the center of brain between the two hemispheres. The corpus callosum—the massive fiber tract that forms the link between the

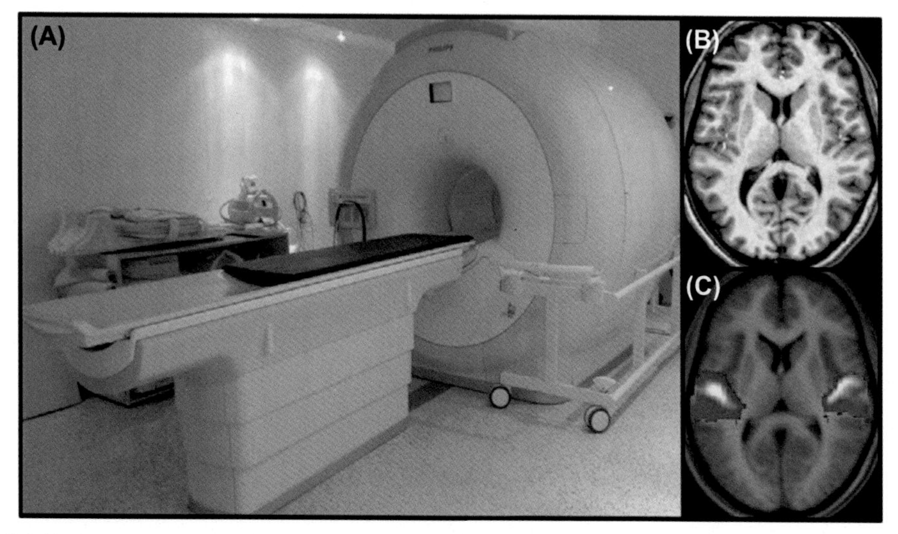

FIGURE 3.7 (A) A magnetic resonance image (MRI) scanner. The patient/subject lies on the elevated bed, which is rolled into the bore of the scanner. A large magnet of 1.5–5.0 T provides the method for collecting anatomical images. (B) An MRI image in the axial plane, with the forehead shown at the top of the image, the left and right sides of the image reflecting left and right hemispheres, and the back of the head shown at the bottom of the image. (C) A functional MRI image that combines the MRI image with functional, hemodynamic changes measured during experimental paradigms. *Source: Hall & Susi, in Celesia & Hickok, 2015.*

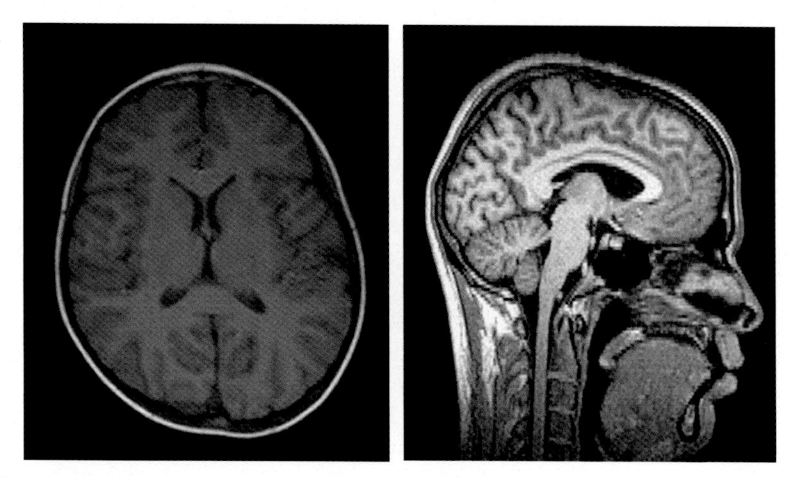

FIGURE 3.8 Magnetic resonance images. Left: An axial plane image, with the forehead shown at the top of the image, the left and right sides of the image reflecting left and right hemispheres, and the back of the head shown at the bottom of the image. Right: A midsagittal plane image, with the forehead shown on the right side of the image, the back of the head shown at the left side of the image, the top of the head shown at the top of the image, and neck and spinal cord shown at the bottom of the image. *Source: public domain.*

two hemispheres—appears in white, with the brainstem below it extending to the spine, and the cerebellum at the back of the head, below the cortex. You can easily discern with the naked eye that brain areas must have differing types of structures as you compare the look of the cerebellum with that of the cortex and the brainstem.

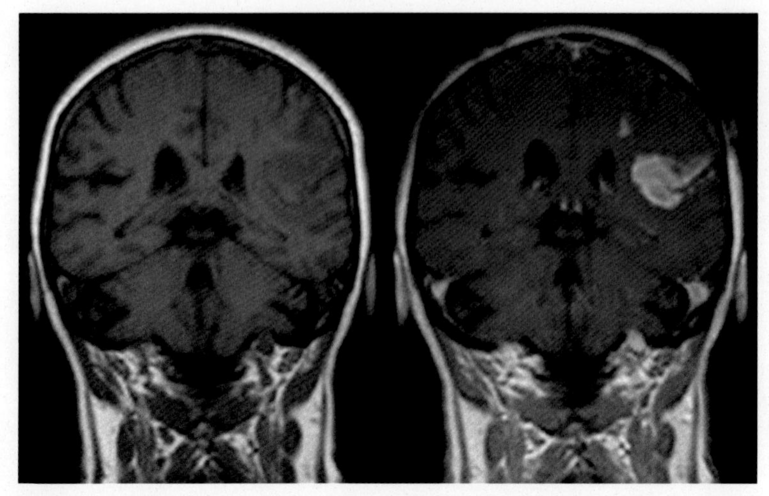

FIGURE 3.9 Magnetic resonance images. Left: A coronal plane image, with the top of the head shown at the top of the image, the left and right hemispheres shown at the left and right sides of the image, and the neck and spinal cord shown at the bottom of the image. Right: A similar coronal image, this one shows evidence for brain damage or tumor on the right side (white mass). *Source: public domain.*

Structural MRI's provide fine-grained details about brain structures that have fueled research into how the human brain functions. They have also provided medical professionals with precise diagnostic tools to visual brain injuries, tumors, and stroke (Fig. 3.9).

3. OBSERVING BRAIN CELLS: NEUROPHYSIOLOGY

Even with fast-improving imaging techniques, the most *direct* evidence about the living brain still comes from *intracranial* (within the cranium) electrical recordings. One reason is that the electrical voltages in the brain are much greater than on the scalp—on the order of *millivolts* (*thousandths of a volt*) rather than *microvolts*. Surface EEG is filtered through a watery medium of brain tissue, skin, bone, and muscle. When you frown, the muscles above your eyes contract, and thin layers of muscle across the scalp stretch and adjust. Even eye movements have large effects on the scalp-recorded EEG. Thus surface EEG recordings mix many electrical sources, as well as being filtered through layers of tissue. About 99.9% of the signal strength is lost.

3.1 Recording Neuron Activity From Inside the Brain

Direct brain recording therefore has a great advantage. The biggest drawback is that it requires invasive surgery. In humans it is never done without medical justification. However, many animal studies use direct brain recording, and these provide much of the basic evidence.

3.1.1 Recording From Single and Clusters of Neurons

Single neurons have electrical and magnetic properties, similar to electrical batteries. We can measure the charge left in a battery and the amount of work it can do for us. We can also measure its magnetic field, as well as its chemistry and overall structure. Hubel and Wiesel (1962) recorded single feature–sensitive cells in the visual cortex of the cat—an achievement for which they received a Nobel Prize in 1981. More recent work has focused on recording from single neurons, clusters of neurons, and grids within the cortex using electrodes and grids of differing sizes (Figs. 3.10 and 3.11). Like every method, electrical recording of neuronal firing has its limitations, but it continues to be a major source of information.

Modern unit recording devices provide researchers with fine-grained tools to understand *where* in the brain they are recording neural activity and *when* that activity is taking place (Figs. 3.12–3.15).

3.2 Recording From Single and Clusters of Neurons in Humans

Invasive in-brain recording is only done for humans when it is deemed medically necessary. One frequent use of invasive recording employs chronic (relatively long term) electrocorticography (ECoG) recording. Wilder Penfield and his colleagues pioneered ECoG in humans in the 1950s (Fig. 3.16). Epileptics with uncontrolled seizures can benefit from surgical removal of seizure-causing scars in the cortex. ECoG recordings can show where such "epileptogenic foci" are located (Fig. 3.17). In addition, the surgeons need to know which areas to avoid damaging because they are vital for perception and language. ECoG exploratory studies are relatively safe and practical. Probing the cortical surface is generally pain-free because the cortex does not have pain receptors. The scientific benefits have been very important. ECoG studies in conscious humans are helping to uncover the neural basis of language, conscious perception, and voluntary control (Fig. 3.18).

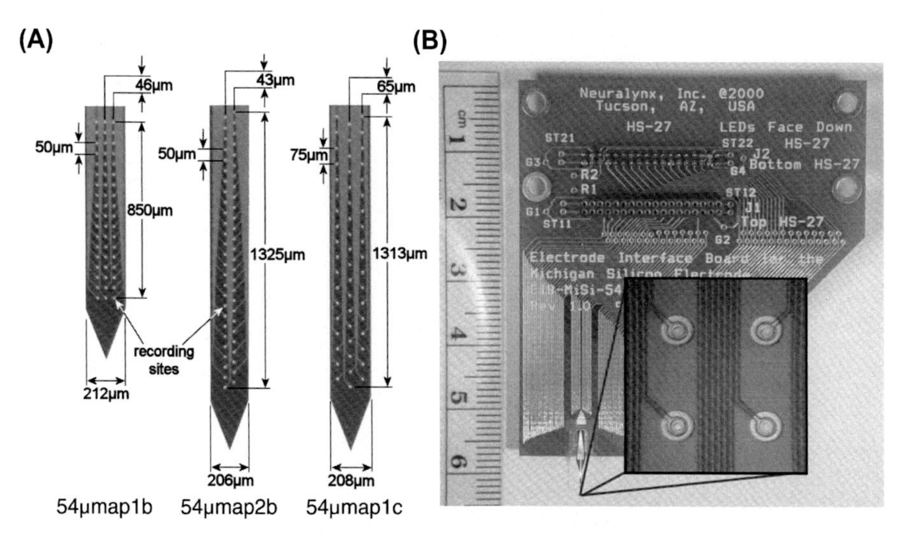

FIGURE 3.10 (A) In addition to using electrodes for single and cluster unit (neuron) recording, polytrodes can be used that can record high-density simultaneous recordings from multiple (>50) sites along the polytrode. Left panel shows three polytrode designs that have differing sizes and can record from sites spaced at differing intervals along two or three vertical columns. (B) A polytrode (lower left) can also be bonded to a head-stage board, providing highly detailed recording across multiple sites in three columns. The enlarged square to the right of the polytrode shows a close-up of the recording sites. *Source: Blanche, Spacek, Hetke, and Swindale (2005).*

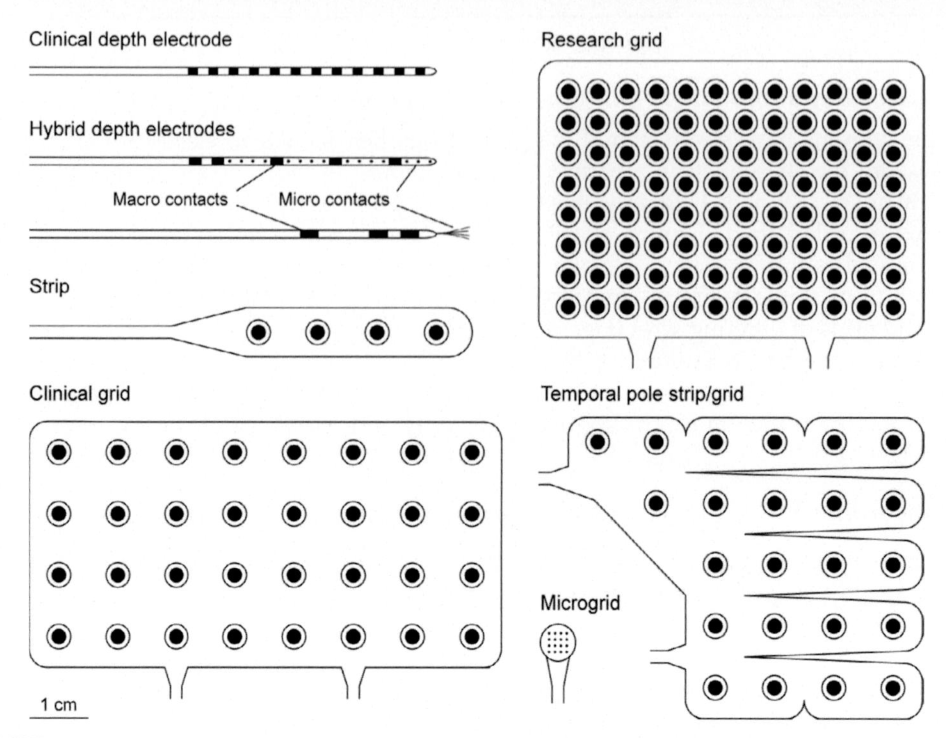

FIGURE 3.11 Ways to record within the brain: depth electrodes can vary in size and shape, clinical and research grids can be used to cover larger brain areas and can be modified to adjust to the brain's curving shape, such as at the temporal pole. *Source: Nourski & Howard, III, in Celesia & Hickok, 2015.*

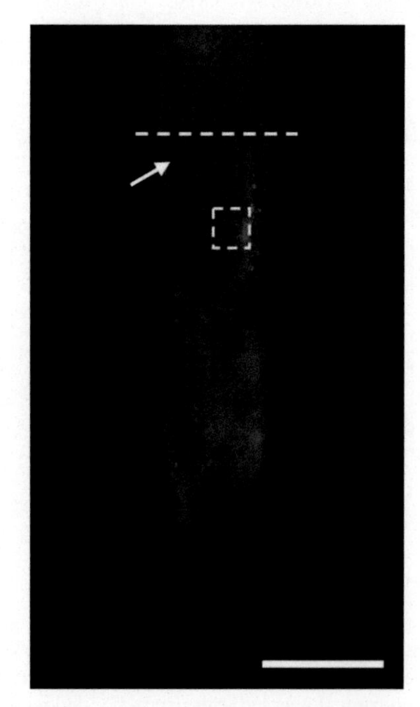

FIGURE 3.12 Single and cluster unit recording: knowing *where* the electrode is recording in precise terms. A rat brain section is stained with fluorescent dil deposited by the polytrode whose outline can be seen faintly in this image. *Source: Blanche et al. (2005).*

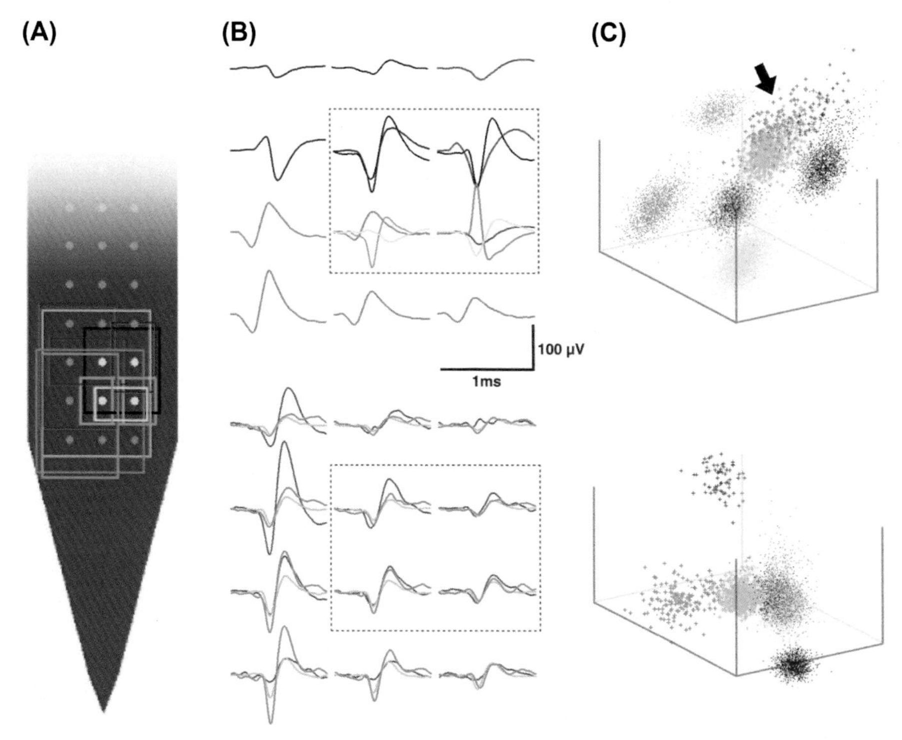

FIGURE 3.13 An example of how a polytrode can help isolate neuron responses in an improved manner as compared with single or cluster electrode or tetrode recordings. (A) Modern polytrodes can simultaneously record from multiple sites along their length and width, providing improved resolution of neuronal activity as compared with single or cluster unit electrodes or tetrodes. Here the subrecording sites of a polytrode are color-coded to show their neuron recording coverage of about 10 neurons. Each neuron's spiking activity is color-coded on the polytrode and in the waveforms shown in (B). (B) Spike waveforms of neuron firing are shown using the color codes from the polytrode in (A). (C) Cluster plots of neuronal firing from the 10 neurons. Top plot shows the clusters as if each region of the polytrode was a single electrode or tetrode. Note that some of the responses from the green, pink, purple, and orange electrode sites/neurons could not be isolated from each other in this plot using the single electrode analysis. Bottom plot shows the color-coded polytrode/neuron firing responses. Using the polytrode, the neuronal responses from the green, pink, purple, and orange color-coded neurons were better isolated. *Source: Blanche et al. (2005).*

4. OBSERVING BRAIN FUNCTION: FUNCTIONAL NEUROANATOMY

The primary techniques using in brain research to understand brain function are fMRI and PET. Although each provides a colorful "picture" of brain activity, neither is a direct measure and they differ sharply in just what they measure. fMRI measures *changes in blood flow*—hemodynamic—response and PET measures *metabolic activity*. While PET scans used to play a central role in brain research, they now take a much smaller role due to their invasive nature. fMRI, a relatively safe and noninvasive method, has taken center stage.

4.1 Functional Neuroimaging Using Functional Magnetic Resonance Imaging: A Blood Oxygen Level–Dependent New World

Currently, the most popular method is fMRI (Figs. 3.19 and 3.20) and especially the kind that measures the oxygen level of the local blood circulation (called BOLD).

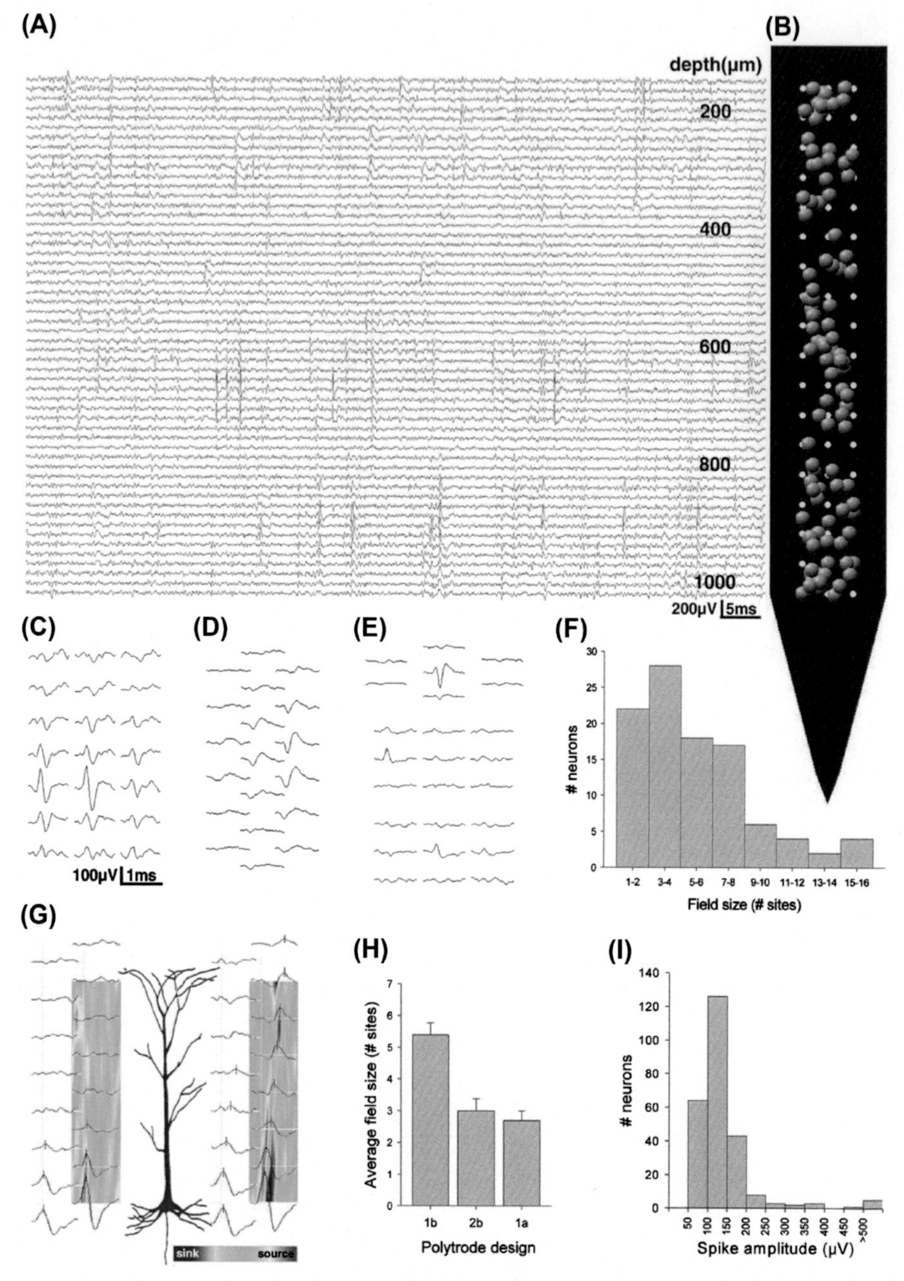

FIGURE 3.14 The relative site of neuron ensembles being recorded along the length (depth) of the polytrode. (A) Recording depths can be determined and recorded data can be assessed accordingly. (B) Cartoon of cells being recorded in locations on the polytrode. (C), (D), (E) Individual neural spikes (action potentials). (F) A histogram of the spike data. (G) A putative five-layer pyramidal cell with two distinct spatial distributions. (H) A histogram showing a comparison of the average number of recording sites per neuron for different site configurations. (I) The distribution of peak-channel spike amplitudes from multiple penetrations. *Source: Blanche et al. (2005).*

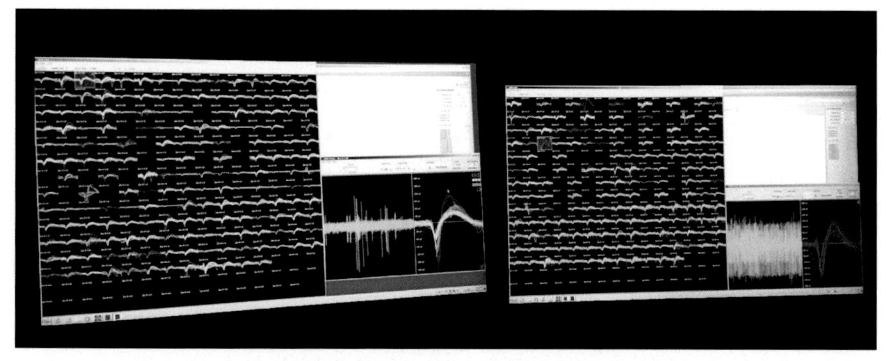

FIGURE 3.15 Current modern single unit recording methods allow for the simultaneous viewing of spike activity from 384 electrodes using Ripple Grapevine software. *Source: Mitz et al. (2017).*

FIGURE 3.16 Wilder Penfield (1891–1976). Wilder Penfield was a Canadian neurosurgeon who expanded brain surgery's techniques by mapping the brain surface intraoperatively. Penfield and colleagues devised open-brain neurosurgery for untreatable epilepsy in the 1950s. His techniques revolutionized neurosurgical approaches. *Source: wikipedia.*

When neurons fire, they consume oxygen and glucose and secrete metabolic waste products. An active brain region consumes its local blood oxygen supply, and as oxygen is consumed, we can see a small drop in the BOLD signal. In a fraction of a second, the loss of regional oxygen triggers a new influx of oxygen-rich blood to that region producing a recovery of the signal. However, as the compensatory mechanism overshoots, flooding more oxygenated blood into the area than is needed, the signal rises high above the baseline. Finally, as unused oxygen-rich blood flushes out of the region, we can see a drop in the BOLD signal back to the baseline (Fig. 3.21).

Thus, as the oxygen content of blood produces changes, we can measure neural activation indirectly. The BOLD signal comes about 6 s after the onset of neuronal firing. The relationship between neural activation and the BOLD fMRI signal is shown in Fig. 3.22. Note that these studies frequently use a block design, where a certain task is cycled on and off and the BOLD signal is measured across the ON and OFF blocks. Other experimental methods are also used to refine the contrasts between experimental conditions and their corresponding functional brain responses (Fig. 3.23).

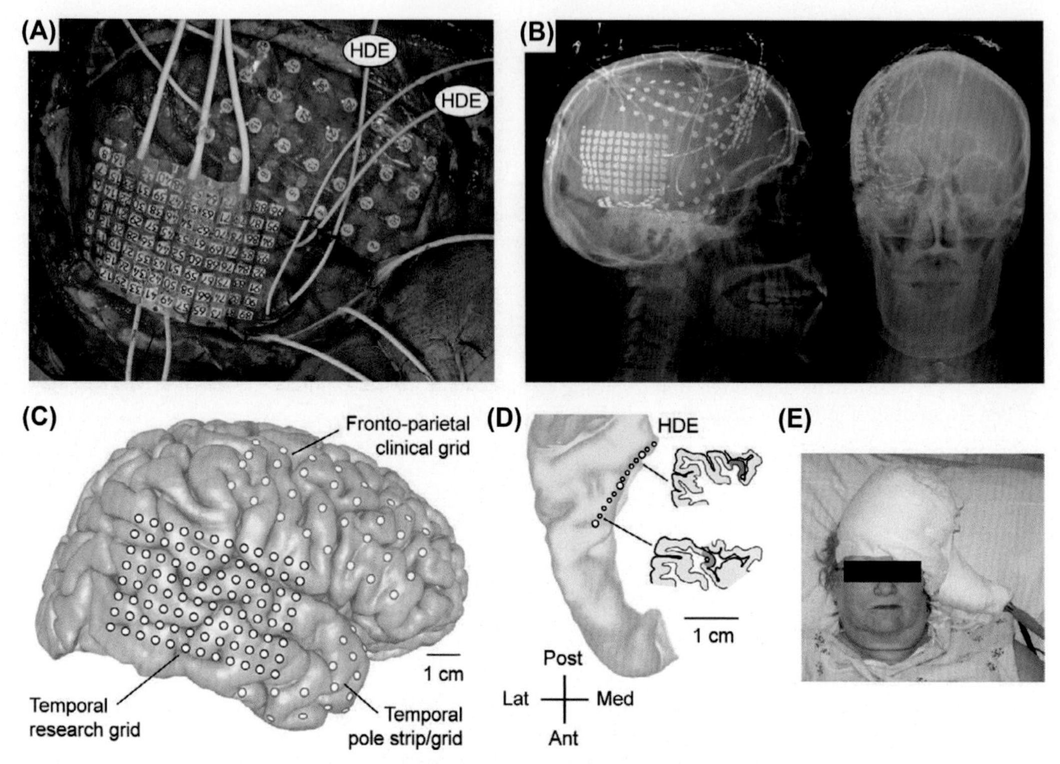

FIGURE 3.17 Electrocorticography records from a grid using depth electrodes. In cases when humans need to have an invasive within brain recording of neural activity, such as in localizing the source(s) of epileptic "spikes" that lead to seizures, (A) a grid array is surgically inserted, including hybrid depth electrodes (HDE), (B) postimplantation X-rays show the location of the implanted grad, (C) the grid is projected onto a magnetic resonance image of the patient's brain, (D) an anatomic reconstruction of the depth electrodes in the specific brain region targeted, and (E) the patient is monitored in hospital during the procedure. *Source: Nourski & Howard, III, in Celesia & Hickok, 2015.*

Analysis of fMRI data includes many steps and levels (Fig. 3.24). While fMRI results reflect *the change in brain response* (*hemodynamics*) between two or more experimental conditions and are thus an indirect measure, new experimental and analytic methods have provided ways to refine both the experimental procedure and data analysis, which in turn has provided better ways for interpreting brain response using fMRI (Figs. 3.25 and 3.26). For example, many early fMRI studies employed a design contrasting a *Task condition* with a *Rest condition*. In the Task condition, the subject may observe a stimulus, press a button, detect a sound, etc. In the Rest condition, the subject is asked to relax with eyes closed. During data analysis, typically the hemodynamic response for Rest would be *subtracted from* the response for the Task condition providing *that brain activity that corresponded to the given task*.

Yet there is a hidden assumption in these studies. If you are asked just to lie still and "rest," what will you do? Will your mind be a blank screen? A great deal of evidence shows that people just go back to their everyday thoughts, images, and feelings. You might be thinking about something that happened the day before or that you need to do later or you might just daydream a bit. You are likely to have inner speech, visual imagery, episodic memories

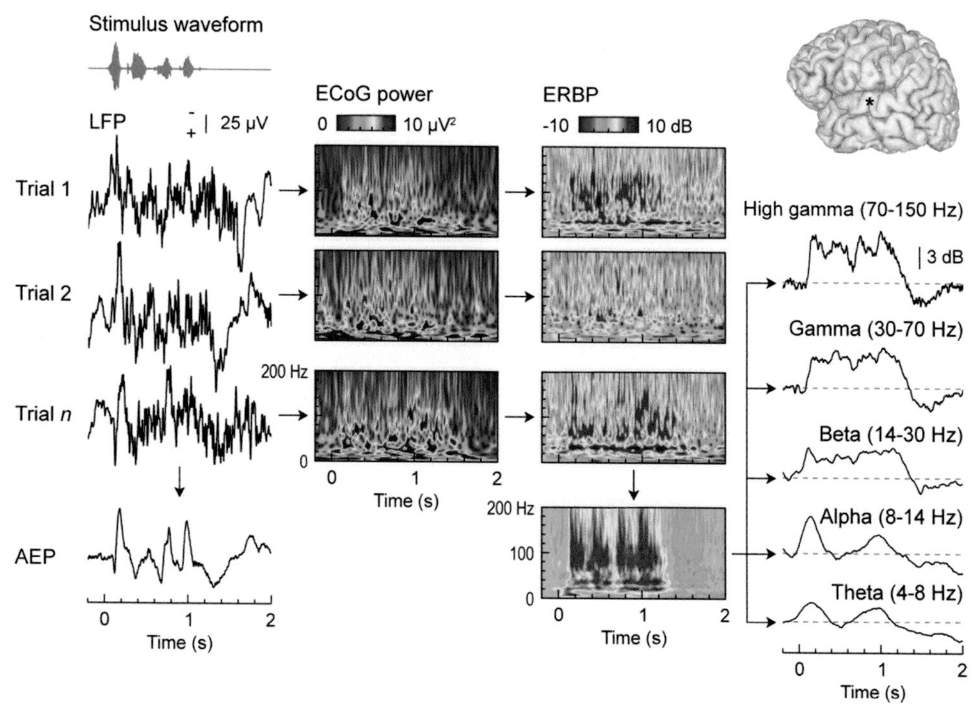

FIGURE 3.18 Electrocorticography (ECoG) recording of sentences. If a patient having depth recordings for medically motivated reasons consents, sometimes research experiments are conducted using the ECoG electrodes. Here is an example of evoked and induced auditory cortical activity using a grid implanted in the auditory region of the temporal lobe. Left panel: Auditory evoked responses to a sentence "Black cars cannot park" (stimulus waveform is shown at the top). Blue panels next to the auditory evoked response show the ECoG power spectrum of the response. The event-related band power (ERBP) is shown in green panels and is decomposed (right panel) into frequency bands ranging from theta (bottom) through alpha and beta, to gamma (top). *AEP*, auditory evoked potential; *LFP*, local field potential. *Source: Nourski & Howard, III, in Celesia & Hickok, 2015.*

coming to mind, and so forth. For the brain, that is not "rest." Instead, the experimental comparison is really between two active states of the brain. One is driven by experimental task demands, while the other reflects our own thoughts, hopes, feelings, images, inner speech, and the like. In some ways, *spontaneous activity* may tell us more about the natural conditions of human cognitive activity than specific experimental tasks. Both are important.

Brain researchers Fox and Raichle asked the question: what is occurring during the Rest condition? What are we actually subtracting from the Task condition (or any other type of experimental condition employed in an fMRI paradigm)? And more importantly, from a broader, neuroscientific perspective: what is occurring in the brain during a nontask situation? Fox and Raichle began an effort to understand the resting state during fMRI experiments by introducing new analysis approaches to understanding the resting state brain activity called *the default network*. Analysis of resting state activation in the brain has provided important new data for understanding brain connectivity that is naturally and intrinsically occurring "in the background" of our mental functions (Fox & Raichle, 2007; Fig. 3.27).

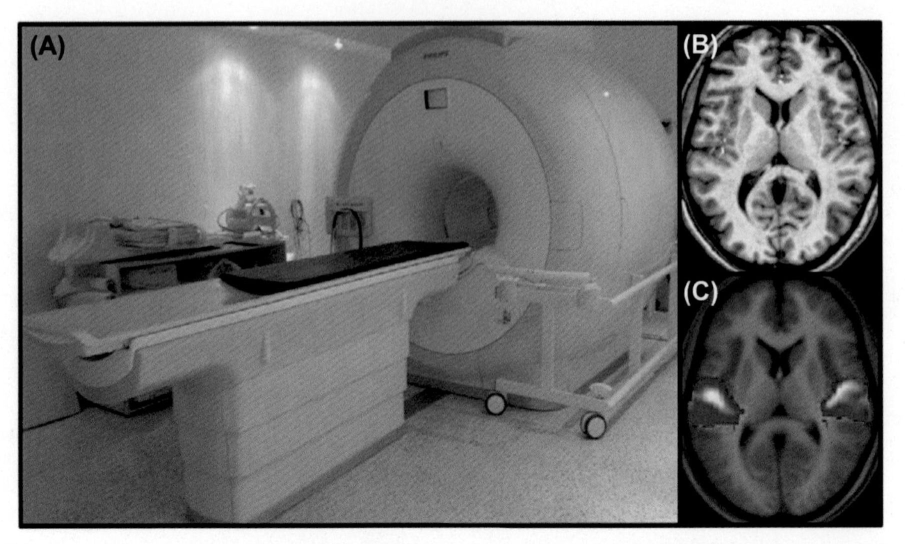

FIGURE 3.19 (A) A magnetic resonance image (MRI) machine. The patient/subject lies on the elevated bed and is rolled into the bore of the machine where a strong magnet (typically 1.5–5 T) is constantly at field strength. (B) An MRI image in the axial plane, with the top of the figure reflecting the front of the head, the sides of the figure reflecting the left and right sides of the head/brain, and the bottom of the figure reflecting the back of the head. (C) A functional MRI image with hemodynamic response shown in red/orange/yellow in left and right temporal lobes. *Source: Nourski & Howard, III, in Celesia & Hickok, 2015.*

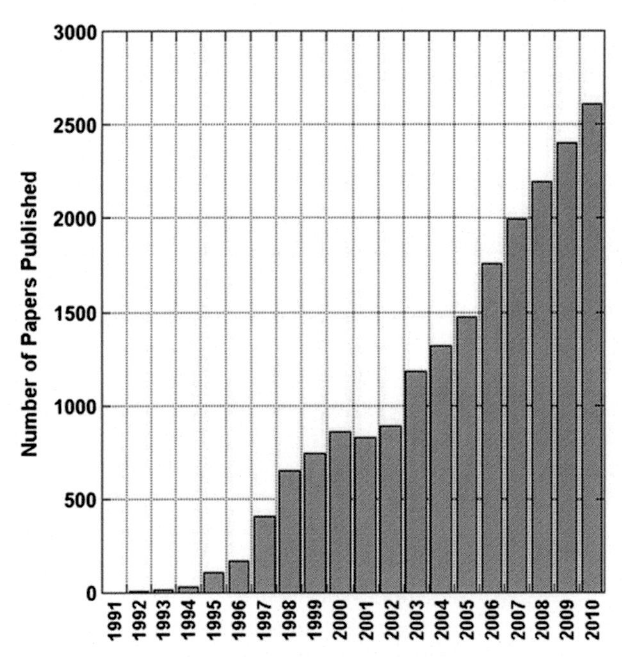

FIGURE 3.20 The number of scientific papers published each year, which contain functional magnetic resonance imaging experiments, has virtually exploded in recent years. *Source: Bandettini (2012).*

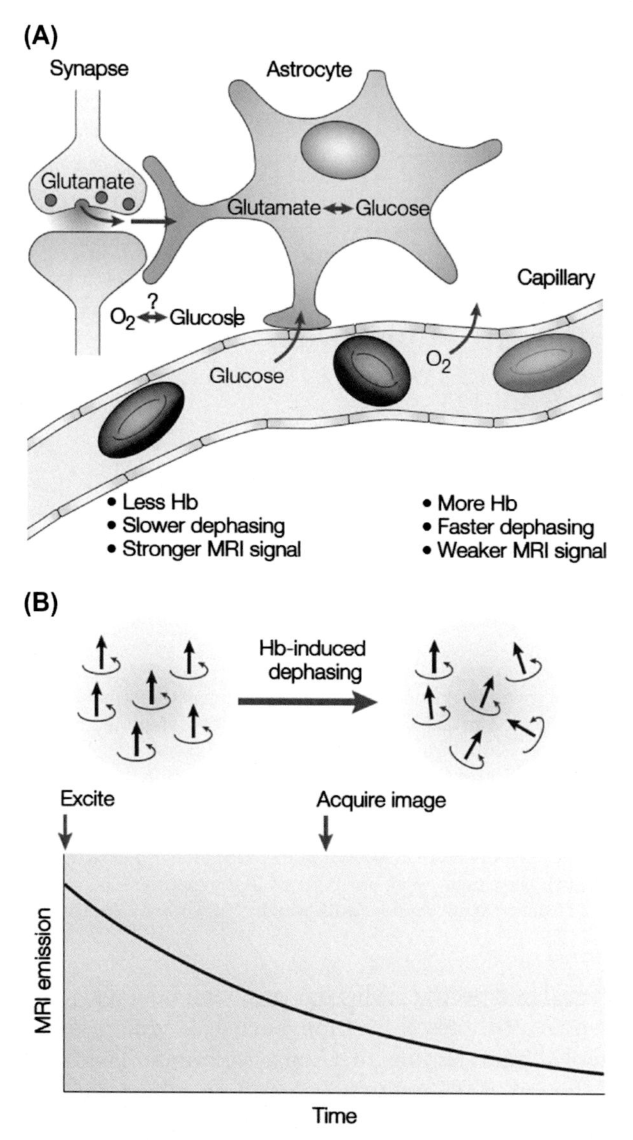

FIGURE 3.21 The blood oxygen level–dependent (BOLD) signal. (A) BOLD functional magnetic resonance imaging (MRI) signal comes from hydrogen atoms, which are abundant in water molecules of the brain. Hemoglobin (Hb) can be deoxy- or oxyhemoglobin depending on local blood oxygenation levels. (B) Changes in the oxygenation levels of Hb are detectable in the MRI due to their different magnetic properties. An *increase* in the concentration of deoxyhemoglobin causes a *decrease in image intensity* and a *decrease* in deoxyhemoglobin causes an *increase in image intensity*. The changes in blood oxygen level due to transient increases in neuronal activity produces a three-phased response: (1) an initial small decrease in image intensity below baseline during the initial period of oxygen consumption, (2) a large increase above baseline (an oversupply of oxygenated blood), and (3) a decrease back to below the baseline after the oversupply of oxygenated blood has diminished. *Source: Heeger and Ress (2002).*

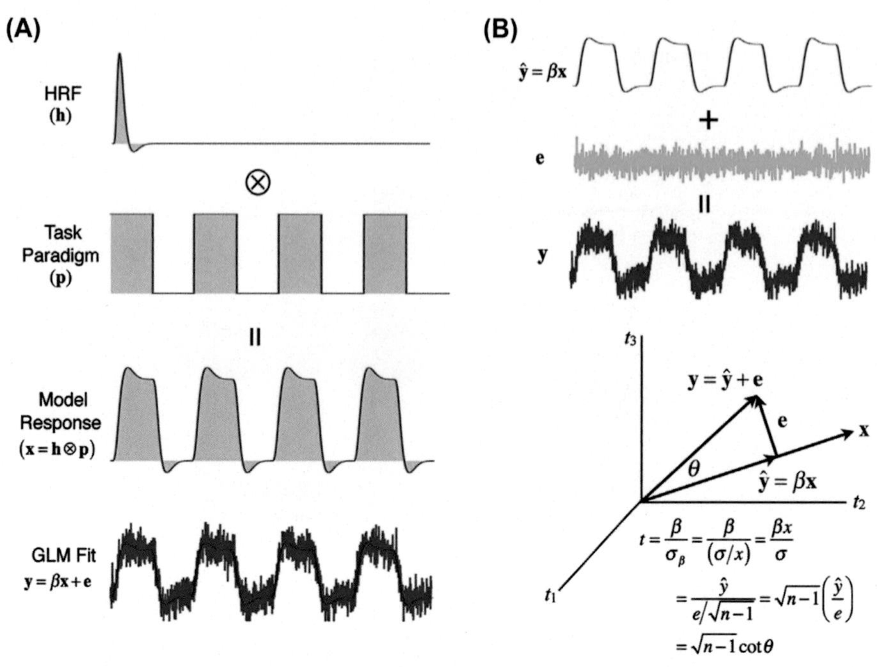

FIGURE 3.22 Functional magnetic resonance imaging (fMRI) shows the activity of the brain, using a blocked design, where a stimulus is presented in blocks separated by a resting state, the blood oxygen level–dependent (BOLD) signal cycles on and off as neural activity changes. A general linear model (GLM) is used in the statistical analysis of the fMRI data. (A) A model for the BOLD response (x) is computed by convolution of the hemodynamic response function (HRF, shown as (h) at top of left panel) and the blocked task paradigm is shown as (p). The modeled response is shown along with x regressor in the GLM fit (shown at bottom of left panel). (B) The GLM decomposes the signal y into two components: the model estimate shown in the dark blue in the right panel, and the residual error shown in green in the right panel. The closest approximation of y in the space spanned by x is computed as a vector in n-dimensional space (bottom of right panel). *Source: Buchbinder, in Masdeu and Gonzalez (2016).*

From a medical/clinical perspective, fMRI has provided new ways to noninvasively investigate important functional attributes of the brain such as language laterality. During surgical planning that may involve brain regions that subserve human language function, neurosurgeons frequently call for an fMRI study to determine where that individual's language regions are located in the brain. While for most individuals, expressive (spoken) language is lateralized to the left hemisphere, this is not the case for everyone. Also, important receptive (heard speech) language processes are mediated *bilaterally* in left and right cortical regions. Thus, presurgical knowledge of the location of both lateralized and bilateral language function provides the neurosurgeon with the ability to avoid those regions during surgery (Fig. 3.28).

fMRI can be combined with a measure for measuring fiber pathways (diffusion tensor imaging, DTI; see Section 5) to provide the neurosurgical team with key information about both functional activity and association pathways in sensory, cognitive, and motor regions in brain areas targeted for surgery (Figs. 3.29 and 3.30).

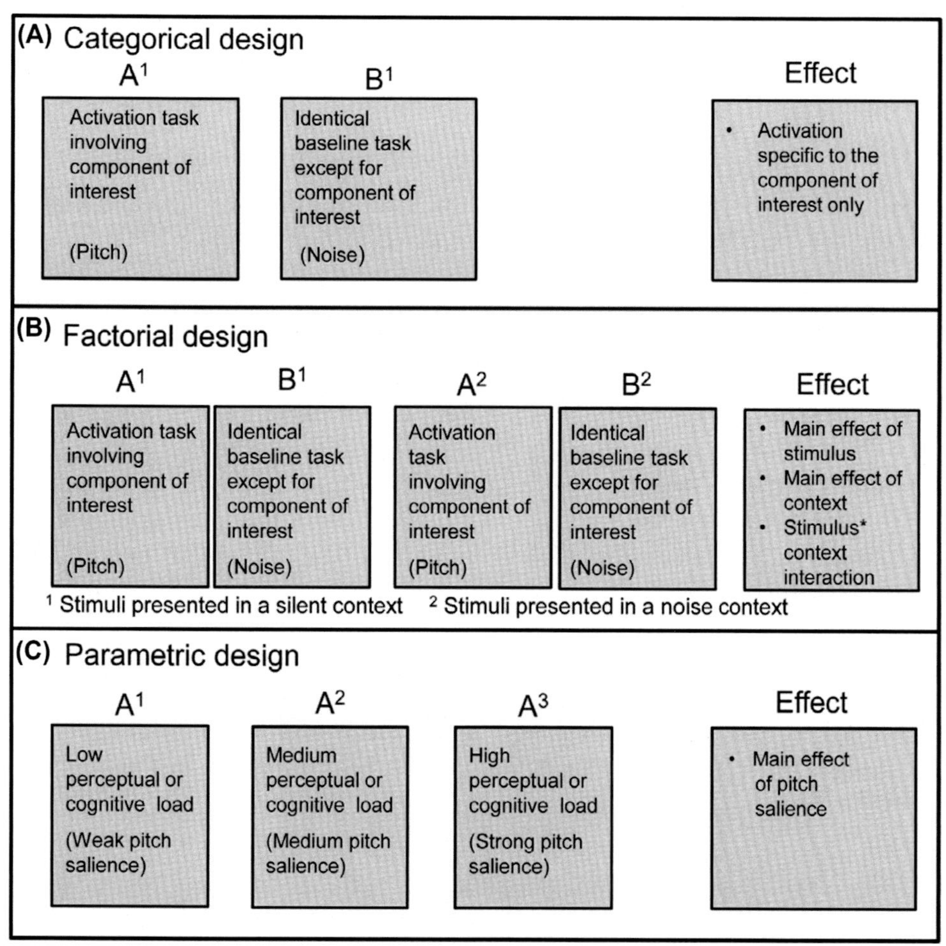

FIGURE 3.23 Functional magnetic resonance imaging experimental designs. Three major experimental designs are shown: (A) Categorical design includes a simple subtraction of one task or experimental condition from another. Frequently the subtracted task is a *Rest* condition or nontask condition, but this is not always the case. The *effect* of the task or condition is thus the *difference*, or subtracted response, from the selected task or condition. (B) A 2 × 2 factorial design includes two factors, *the signal* (for example, pitch and noise), and two factors, *the context* (stimulus* context interaction [stimulus context in this example include a stimulus presented in silent context (1) or in a noise context (2) see footnotes 1, 2]). Activations associated—the *main effects*—with either the signal or the context can be separately determined as the *interactions* between the two factors. (C) A parametric design includes levels of factors A^1 (weak pitch salience), A^2 (medium pitch salience), A^3 (strong pitch salience) as levels of a *pitch* condition. The *effect* in this example is the *main effect of pitch salience. Source: Hall & Susi, in Gelesia & Hickok, 2015.*

4.2 Measuring Metabolic Brain Activity: Positron Emission Tomography

PET was developed much earlier than MRI or fMRI, and it provides a measure of metabolic brain activity. PET is used less often for research today because it is very expensive, requiring a cyclotron. It also requires subjects to be injected with a radioactive tracer. For nonmedical

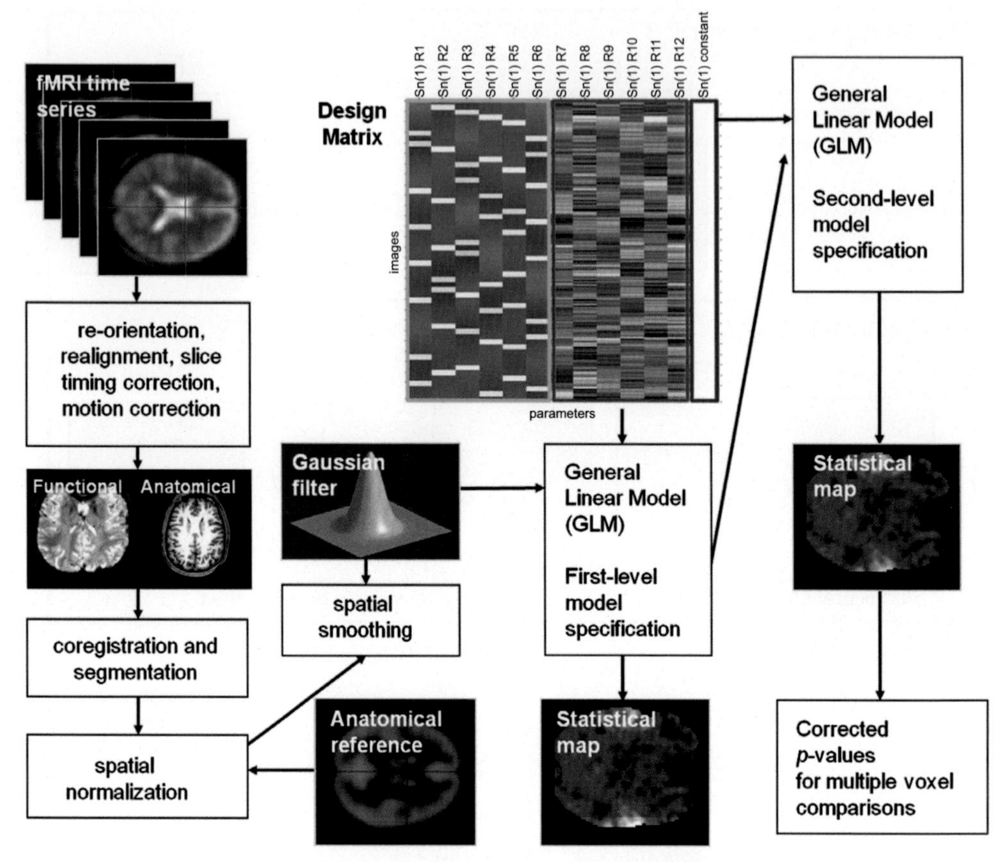

FIGURE 3.24 Statistical parametric mapping includes many preprocessing and statistical analysis steps. (Upper left) Functional magnetic resonance imaging (fMRI) time series images are reoriented and realigned with timing and motion correction. Anatomical and functional responses are coregistered and segmented, and then spatially normalized (bottom left). Spatial smoothing and anatomical references are added and then the general linear model (GLM) is used for model specifications (center). Second-level GLM is then statistically mapped (bottom right). *Source: Hall & Susi, in Gelesia & Hickok, 2015.*

investigations, MRI and fMRI have largely taken over the research field. Medically, PET is still important because it can aid in the diagnosis and treatment options for neurodegenerative diseases and disorders, infection, epilepsy, psychiatric disorders, and tumors.

PET works by emitting radiopharmaceuticals labeled with *positron-emitting radioisotopes* (Figs. 3.31 and 3.32). The positron collides with an electron in the brain tissue and during this process it converts mass to energy in the form of *two photons*. The PET camera detects these photons and converts them to an electrical signal, which is measured. Commonly, PET is used to measure glucose consumption, which provides information about metabolic activity within brain (and body) regions.

Results of a PET scan provide key information about relatively active metabolic regions in the brain, with *warm colors* (red, orange, yellow) showing higher levels of metabolic activity

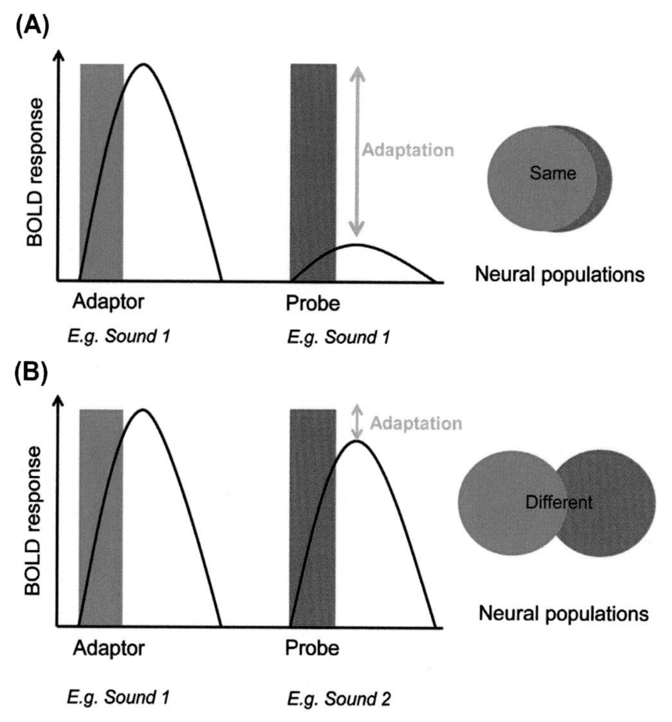

FIGURE 3.25 Another functional magnetic resonance imaging (fMRI) experimental design uses *adaptation* rather than a categorical, factorial, or parametric approach. (A) The underlying assumption is that *the same neural population* underlies processing of an *Adaptor stimulus* and a *Probe stimulus*, and thus the response to the Probe will be reduced as compared to the Adaptor. (B) If, however, the Adaptor and Probe are processed using *differing neural populations*, the response to the Probe will not be reduced. For example, if you wanted to understand if a region within auditory cortex responded to speech and to music with the same underlying neural networks, you might use an adaptation experimental approach to see if a Probe stimulus (musical note) showed a reduced response versus an Adaptor stimulus (spoken sound). *Source: Hall & Susi, in Gelesia & Hickok, 2015.*

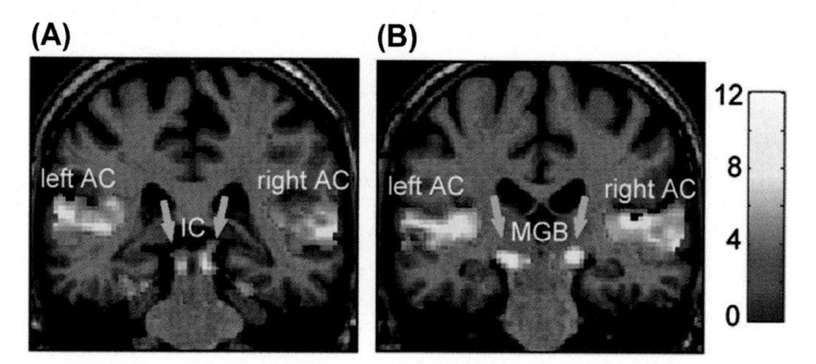

FIGURE 3.26 An example of group-averaged (over subjects) functional magnetic resonance imaging (fMRI) activation. Auditory stimuli produced fMRI responses in left auditory cortex (left AC), right AC, and in the subcortical regions of (A) the inferior colliculus (IC) and (B) the medial geniculate body (MGB) in the thalamus. (A) and (B) show two representative "slices" of the brain to reveal the subcortical activation. *Source: Hall & Susi, in Gelesia & Hickok, 2015.*

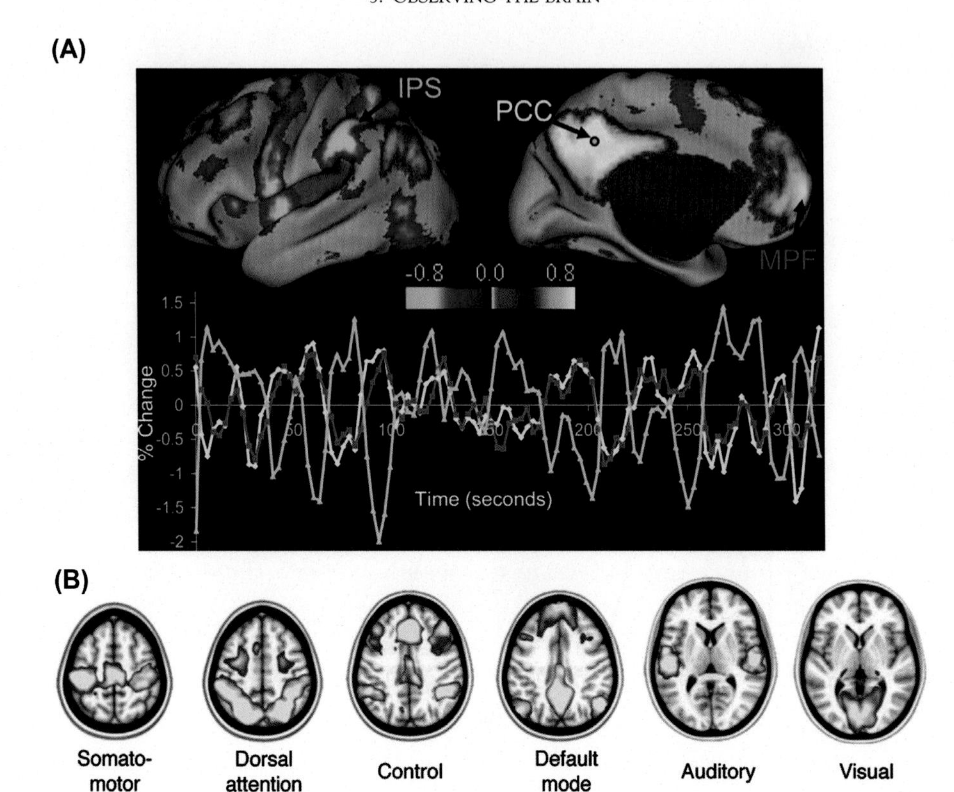

FIGURE 3.27 (A) (Top left panel) An example of the *resting state default network* showing functional magnetic resonance imaging responses across experimental conditions. A seed voxel was placed in the posterior cingulate cortex/precuneus (PCC, small green disc in top right panel). The time course of this seed voxel was correlated with all other brain voxels and shown on a partially inflated cortex (bottom panel). Red–yellow areas indicate positive correlation with the seed voxel (coactivations) and blue–turquoise areas indicate negative correlation. Correlated regions included the default network including the PCC, medial prefrontal (MPF) cortex, and lateral parietal cortex (not labeled). Anticorrelated regions included intraparietal sulcus (IPS), frontal eye fields, and middle temporal regions (not labeled). (B) Examples of six resting state networks: somatomotor, dorsal attention, executive control, default mode, auditory, and visual. Red–yellow areas include regions positively correlated with the seed voxel in the PCC. *Source: Buchbinder, in Masdeu & Gonzalez, 2016.*

and *cool colors* (blue, green) showing lower levels (Fig. 3.33). PET scans are useful in determining lower-than-normal metabolic activity levels in patients with neurodegenerative disorders, such as Alzheimer's disease (Fig. 3.34).

4.3 Measuring Electrical and Magnetic Fields: Electroencephalography and Magnetoencephalography Recording

The brain has billions of neurons lined up more or less in parallel. The long pyramidal neurons of the cortex mostly point their axons inward to the thalamus, and later many axons join the great hubs and highways of the inner white matter. However, pyramidal cells have input dendrites that are mostly spread out horizontally, following the curve of the cortex.

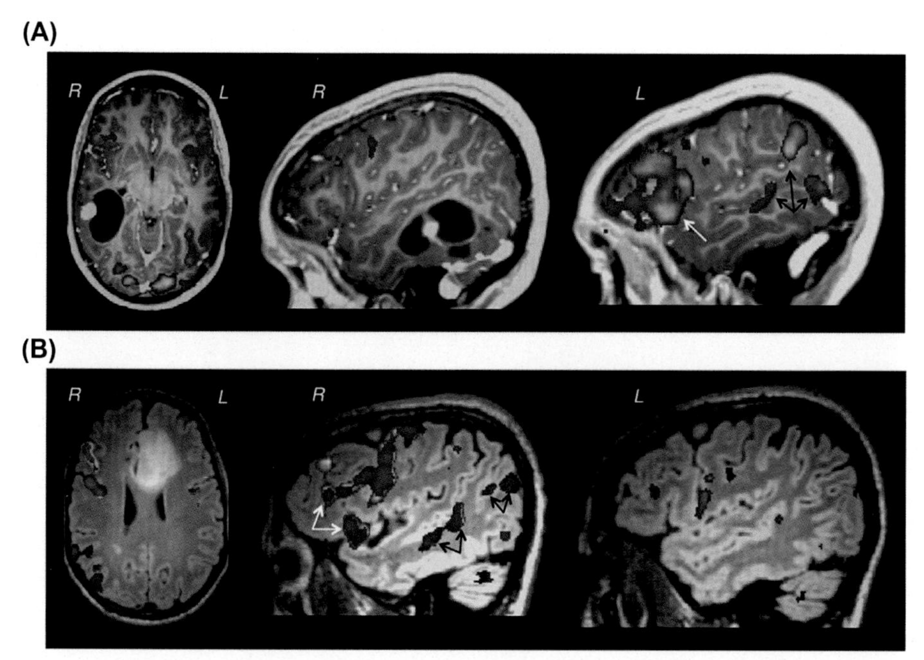

FIGURE 3.28 Functional magnetic resonance imaging (fMRI) can be utilized in clinical situations such as mapping language regions in surgical patients. (A) A 15-year-old left-handed girl with a brain tumor (white mass and dark region in the right hemisphere) underwent an fMRI language mapping experiment using tasks designed to tap cortical language regions. Responses are shown in the top panel in the axial (left) and sagittal (center and right) planes. Language activations (shown in red–yellow) were predominantly in the left (nonsurgically targeted) hemisphere. Her language function was completely intact after she underwent surgery to resect the right hemisphere tumor. (B) A 34-year-old left-handed man was also targeted for brain tumor surgery (white mass in the frontal left hemisphere). In the same planes and shading as in (A), fMRI responses for him showed right hemisphere language activations. His language function was completely intact after he underwent surgery to resect the left hemisphere tumor. *Source: Buchbinder, in Masdeu & Gonzalez, 2016.*

When pyramidal neurons fire a spike down the axons, their dendrites also change. Those flat and bushy horizontal "arbors" of dendrites are picked up as EEG activity at the scalp because their electrical fields happen to point outward.

All streams of electrons generate both electrical and magnetic fields, which radiate at right angles to each other. Therefore, we can also pick up the brain's magnetic field—at least those components that conveniently point outward from the scalp. Most the field strength in the brain is not measurable from the outside. We just take advantage of radiating fields that can be picked up.

4.3.1 Electroencephalography: The Electrical Fields of the Brain

EEG is a relatively direct measure of the brain's electrical activity. But with tens of billions of cortical neurons firing about 10 Hz, we have several trillion electrical events per second. EEG is relatively inexpensive and is in wide use in medical and clinical locations and in research laboratories.

In a typical EEG recording, electrodes or electrode caps are placed on the head and electrical activity that reaches the scalp from the brain is measured (Fig. 3.35). The temporal

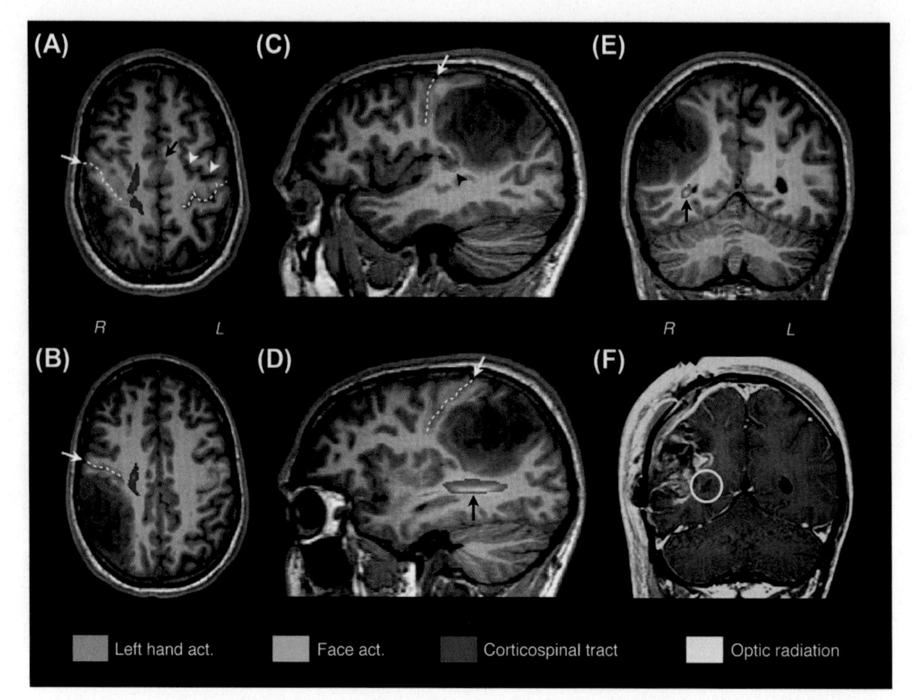

FIGURE 3.29 Functional magnetic resonance imaging (fMRI) can be combined with diffusion tensor imaging (DTI) to determine white matter pathways in addition to hemodynamic response. Results of combined fMRI and DTI mapping are shown on (A) an axial slice with the right hemisphere shown on the left side of the image according to radiological convention, (B) a second axial slice presented as in (A), (C) a sagittal slice of a lateral view of the right hemisphere, (D) a second sagittal slice as in (C), (E) a coronal slice with the right hemisphere shown on the left side per convention, and (F) a coronal slice showing the brain following the tumor resection. In this case, a 21-year-old man had a massive tumor (gray-shaded region in the right hemisphere, apparent on each magnetic resonance image). Left hand activation (shown in orange) showed responses in the right and left region abutting the central sulcus in somatosensory cortex (shown as a *white dotted line* on (A–D)). Face activation (shown in green on (A–D)) showed activation in similar regions. Premotor and motor responses are highlighted with white arrows (shown on (A–D)) and supplemental motor cortex responses are highlighted with *black arrows* (shown on (A) and (D)). DTI revealed the corticospinal tract (showed in purple on (A) and (B)) that courses from the Rolandic region and is near the tumor to be resected. The tumor resection disrupted the optic radiation (*yellow circle* shown in F), which resulted in a postoperative reduction of sight in the patient's left visual field. The ability to map both functional responses and white matter tracts allowed for surgical planning that reduced postoperative cognitive and somatosensory loss of function after the removal of the very large tumor. *Source: Buchbinder, in Masdeu & Gonzalez, 2016.*

resolution (*when is neural activity occurring*) of EEG is very high, with submillisecond (ms) precision. The spatial resolution (*where is the source of that neural activity*) of EEG is low; however, with the advent of larger and larger electrode arrays, this is improving.

Data from EEG recordings can be analyzed in many ways, depending on the clinical or research question being asked. In this section, we will focus on a temporal analysis: what is occurring *over each sampled time point, averaged across electrodes*. This analysis relates well to understanding cortical function. In Section 6.1, we will discuss a spectral analysis: *what is occurring over each electrode in the power spectrum, averaged over time*. This analysis relates well to understanding cortical dynamics. It is important to note that a single EEG recording

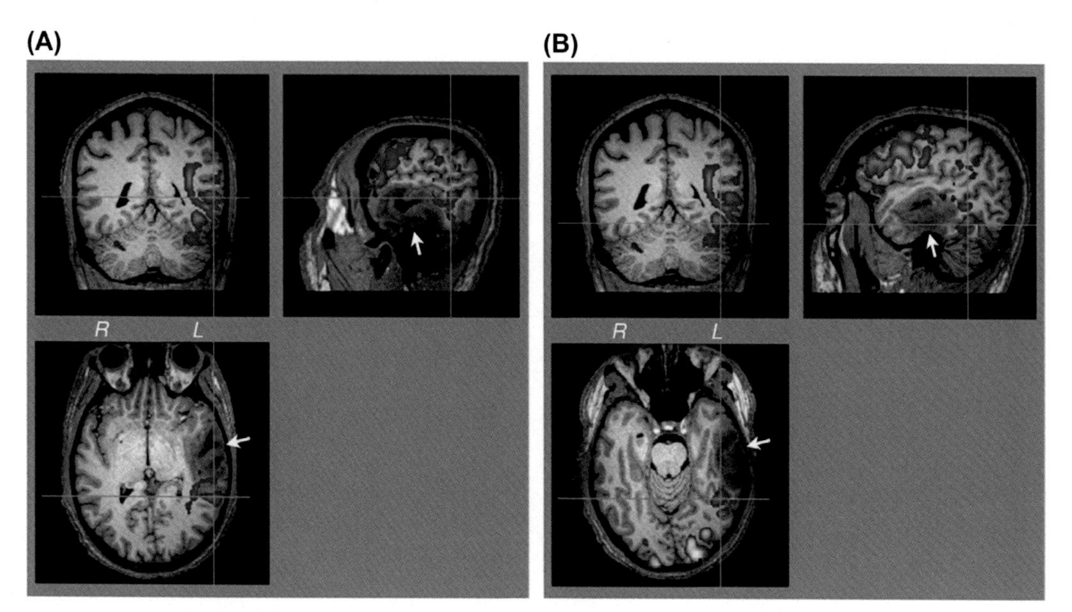

FIGURE 3.30 Presurgical clinical language mapping combines functional magnetic resonance imaging (fMRI) and diffusion tensor imaging (DTI) techniques to maximize the ability to localize language regions. A 30-year-old male with a left hemisphere tumor underwent language mapping. (A) *White arrows* reflect the putative (hypothesized) Wernicke's area, a region for processing speech sounds. (B) *White arrows* reflect the putative visual word form area for reading words. The results of the combined fMRI and DTI responses alerted the neurosurgeon to the proximity of the tumor to be resected to brain regions specialized for speech sound and reading processing. *Source: Buchbinder, in Masdeu & Gonzalez, 2016.*

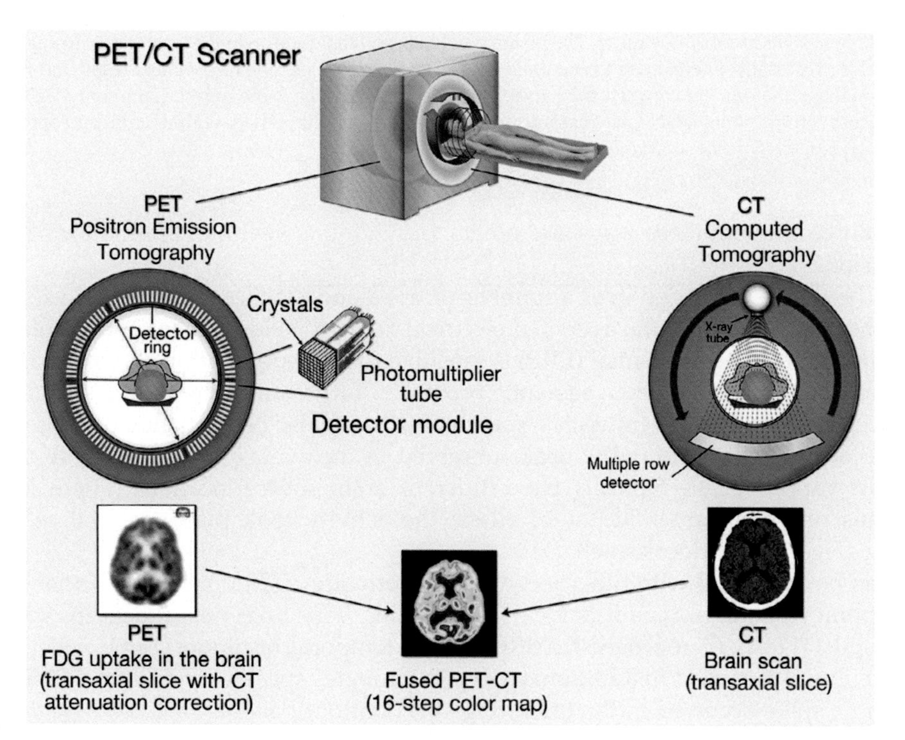

FIGURE 3.31 An example of a combined positron emission tomography (PET) and computed tomography (CT) scanner. The PET scanner (left side of figure) provides molecular imaging of metabolic activity and the CT scanner (right side of figure) provides anatomical imaging. The PET-obtained metabolic color map data is coregistered on the CT-obtained brain images. *Source: Lameka, Farwell, & Ichise, in Masdeu & Gonzalez, 2016.*

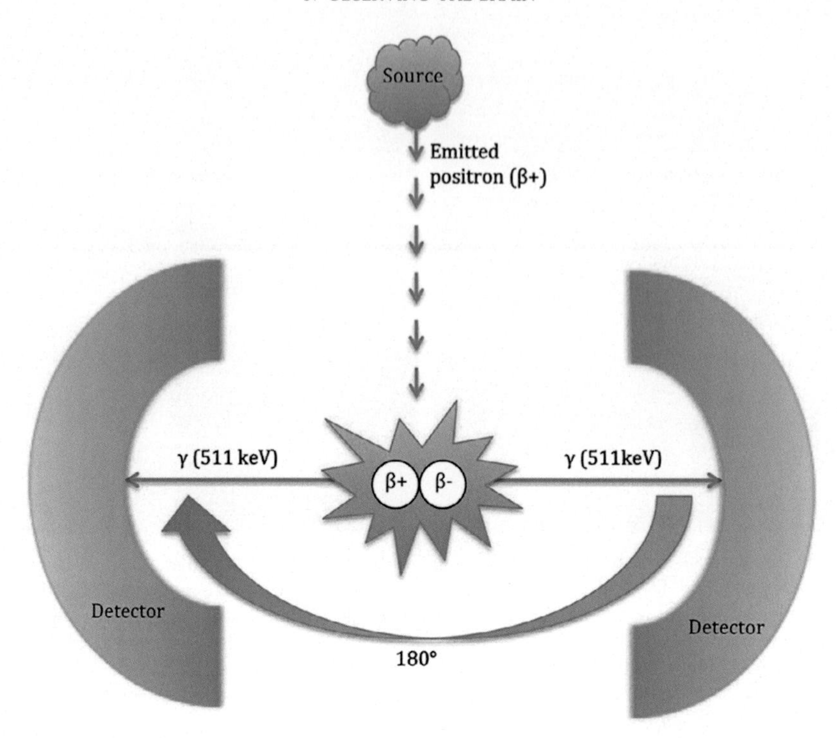

FIGURE 3.32 Positron emission tomography (PET) works by emitting radiopharmaceuticals labeled with *positron-emitting radioisotopes* (top of figure). The positron collides with an electron in the brain tissue (center of figure) and during this process it converts mass to energy in the form of *two photons*. The PET camera (right and left sides of figure) detects these photons and converts them to an electrical signal, which is measured. Commonly, PET is used to measure glucose consumption, which provides information about metabolic activity within brain (and body) regions. *Source: Lameka, Farwell, & Ichise, in Masdeu & Gonzalez, 2016.*

can be analyzed in both these ways and others, providing a wealth of brain information in a single session.

When the EEG is averaged over a number of trials and "locked" to a specific zero point, like the onset of a stimulus, the averaged electrical activity yields elegant and regular waveforms. This event-related potential (ERP) is sensitive to large population activity that characterizes visual, auditory, and even semantic processes. ERPs can be contrasted for individual events, such as stimulus type (music vs. speech). ERPs can also be evaluated over time within a given recording by assessing the peaks observed in the averaged response (Fig. 3.36, top panel). The various peaks typically have different brain source locations (Fig. 3.36, center panel), thus providing an indicator of where the activity took place, as well as when it took place (Fig. 3.36, bottom panel).

EEG can be combined with other techniques to provide additional information. EEG has high temporal resolution, thus it is ideal for detecting very brief neural activity such as an epileptic spike (Fig. 3.37, top panel). fMRI has low temporal resolution but excellent spatial resolution, thus when used in addition, to EEG, it provides specific information about where that epileptic spike occurred in the brain (Fig. 3.37, bottom panel).

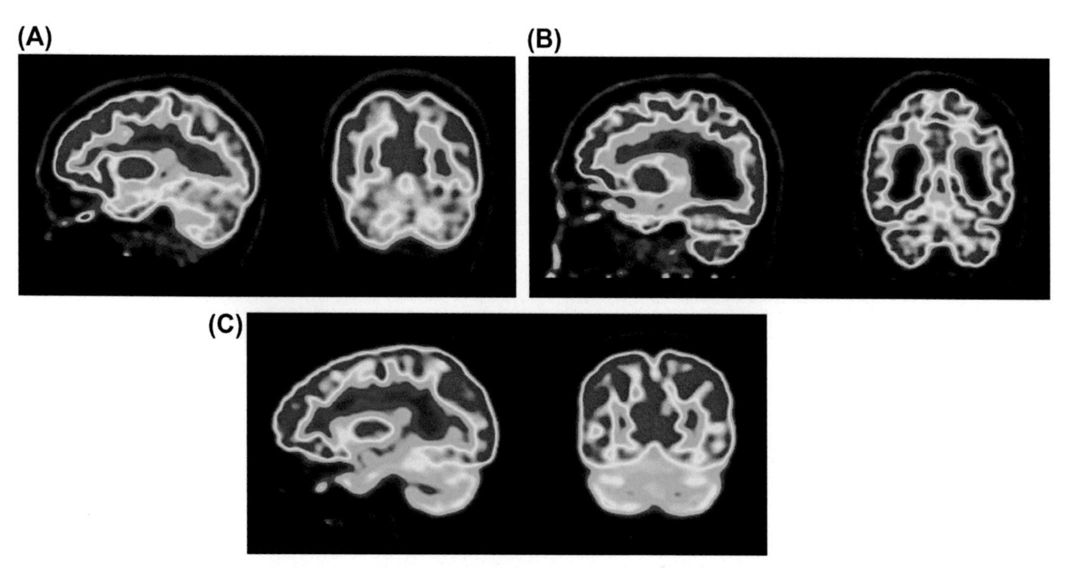

FIGURE 3.33 (A) F-fluordeoxyglucose (FDG) positron emission tomography (PET) results for a normal adult are shown in a midsagittal plane (left) and a coronal plane (right). *Red-yellow shading* shows regions with high metabolic activation. (B) The same measures for an adult with cerebral atrophy. Although his anatomical images showed no clear difference from the normal case shown in (A), PET scanning revealed clear difference in metabolic activity. In this case, reduced metabolic activation versus the normal adult (shown in A) is apparent throughout the cortex but especially in the frontal lobe shown on the left side of the midsagittal plane and at the top of the coronal plane. (C) Same measures for a normal 6-year-old girl who shows similar cortical metabolic activation but much reduced metabolic activation in the cerebellum at the bottom posterior section of the brain. The reduced cerebellum metabolic activity shown in this child is a normal stage of development. *Source: Lameka, Farwell, & Ichise, in Masdeu & Gonzalez, 2016.*

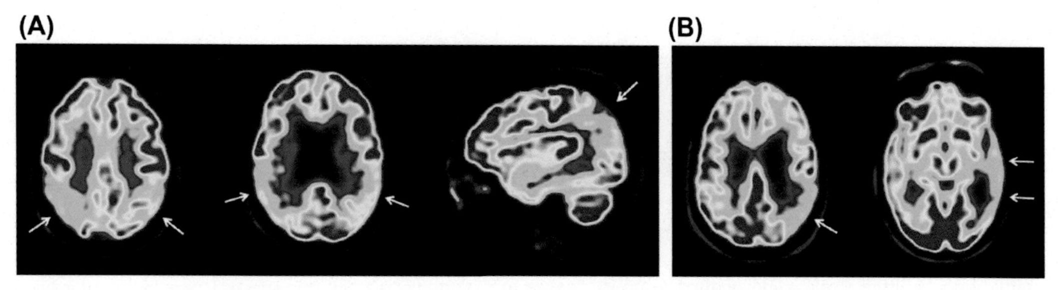

FIGURE 3.34 (A) F-fluordeoxyglucose (FDG) positron emission tomography (PET) results for a patient with Alzheimer's disease are shown on axial (left and center) and sagittal (right) planes. According to radiologic convention, the left hemisphere is shown on the right side of the image and the right hemisphere is shown on the left side of the image. *Red-yellow shading* shows regions with high metabolic activation. Measures show reduced metabolic activation in temporal and parietal lobes (shown with *white arrows*). (B) A second patient with Alzheimer's disease is shown. In this case, reduced metabolic activation is shown in the left hemisphere (right side of the image) in parietal and temporal regions. Asymmetric diminished metabolism is often observed in cases with Alzheimer's disease. *Source: Lameka, Farwell, & Ichise, in Masdeu & Gonzalez, 2016.*

(A) From low-density to high-density montages

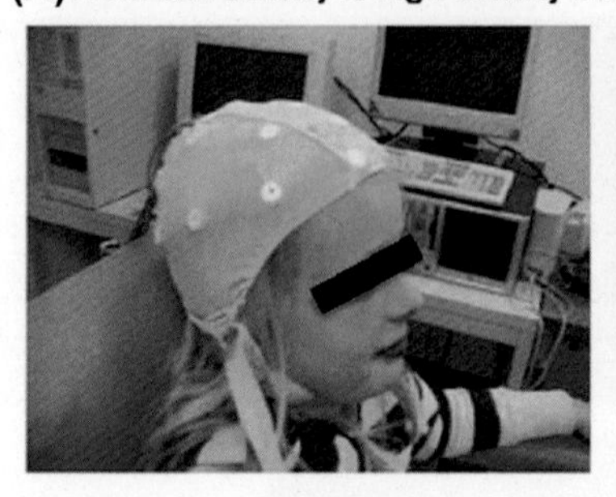

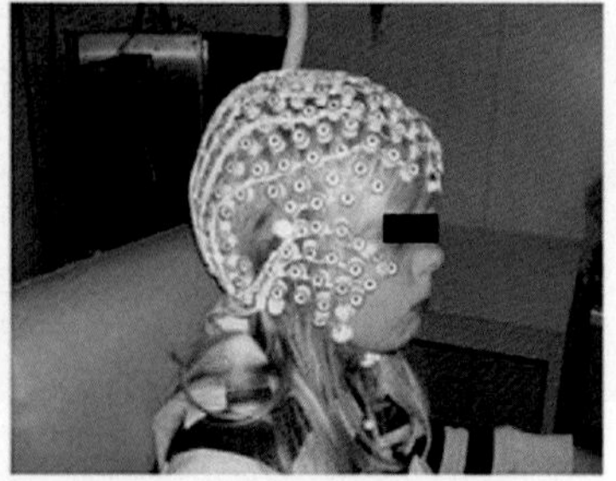

(B) From voltage waveforms to topographic representations

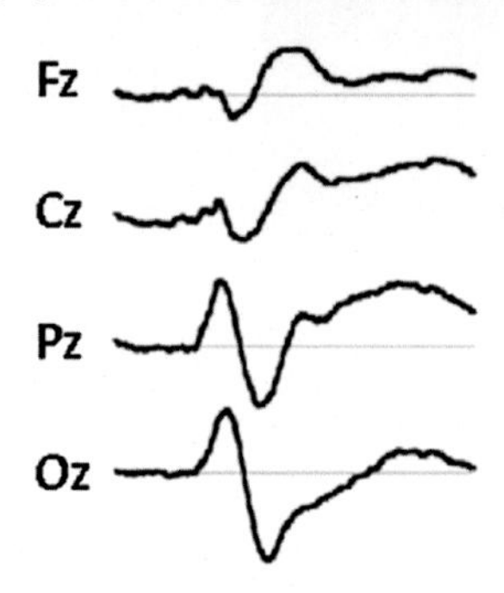

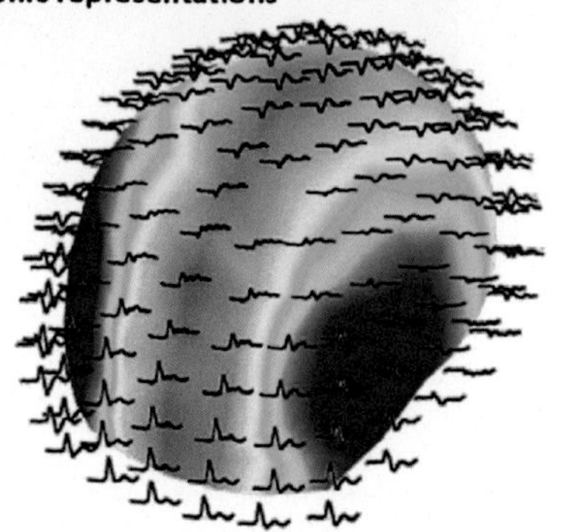

(C) From equivalent current dipole to distributed source models

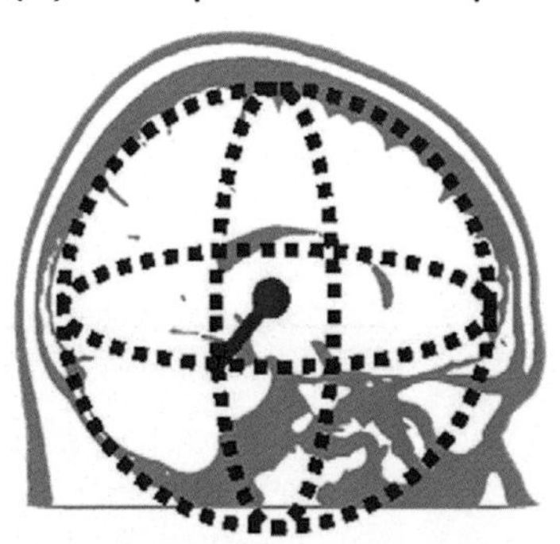

FIGURE 3.35 (A) Electroencephalographic (EEG) recordings can use low- to high-density electrode montages. (B) Results from individual electrodes are averaged over trials and form a traditional evoked-response wave with negative and positive peaks. The evoked-response waves and their corresponding spectral power of recordings (shown in *blue*, *green*, *yellow*, and *orange shading*) can be plotted for each electrode in the montage. (C) Data are analyzed using equivalent current dipole or distributed source models to provide the localized sources of the evoked response for each electrode. *Source: Michel and Murray (2012).*

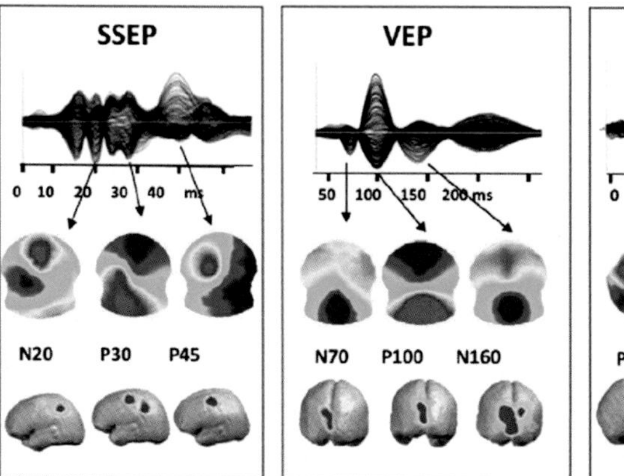

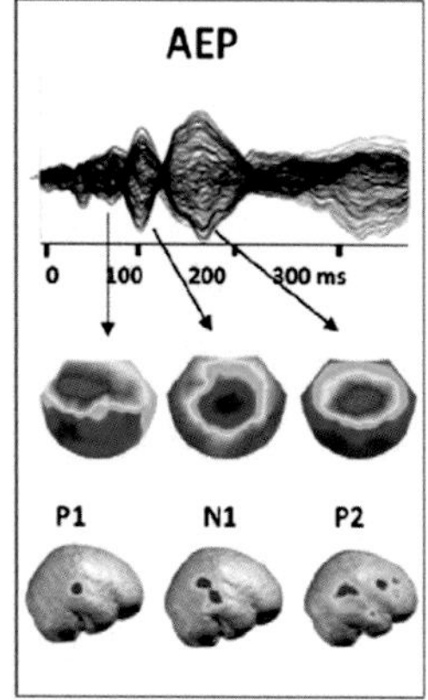

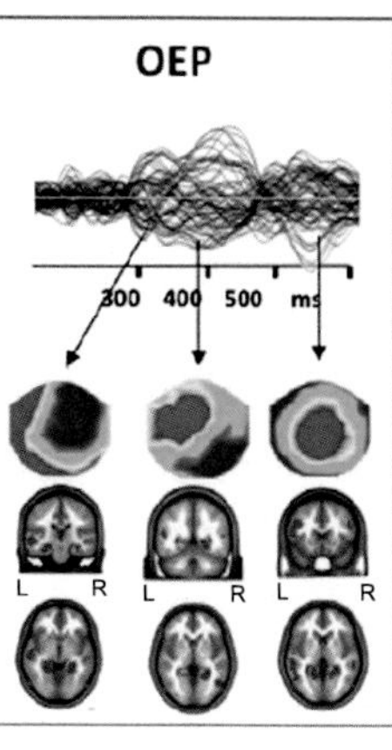

FIGURE 3.36 Examples of electroencephalography event-related evoked responses and their corresponding source images in the brain. Evoked response components are named for their negative- or positive-going peak (e.g., N or P) and the time in which they peak following stimulus onset (in milliseconds, e.g., N20 or P45). In each panel, the top figure shows the averaged event-related response, the center figure shows the scalp potential maps of the peaked components, and the bottom figure shows the corresponding location in the brain. *AEP*, auditory evoked potential; *OEP*, olfactory evoked potential; *SSEP*, somatosensory evoked potential; *VEP*, visual evoked potential. *Source: Michel and Murray (2012).*

4.3.2 Magnetoencephalography: Magnetic Fields of the Brain

MEG measures the magnetic field produced by electrical activity in the brain (Fig. 3.38). Its spatial resolution is now approaching a few millimeters, while its temporal resolution is in milliseconds; thus it provides source localization ability that far exceeds that is available using EEG. MEG is expensive, however, and requires liquid helium to maintain its superconducting equipment. There are only a relative handful of MEG laboratories in the world compared to the tens and hundreds of thousands EEG and MRI sites.

Because of the physics of magnetism, MEG is highly sensitive to dendritic flow at right angles to the walls of the sulci (the cortical folds). MEG results are superimposed upon a structural image of the brain. MEG uses a process called *magnetic source imaging* to coregister the magnetic sources of brain activity onto anatomical pictures provided by MRI. MEG has the advantage of being entirely silent and noninvasive. Thus, MEG is attractive for use with children and vulnerable people. MEG is easy for young children to tolerate as long as they can stay relatively still.

Similar to EEG, event-related MEG can provide information sensory, motor, and cognitive brain processes using the peaks in the magnetoencephalogram and their corresponding source locations (Figs. 3.39 and 3.40). Enhanced spatial resolution of MEG combined with its submillisecond temporal resolution provides experimenters with the capability to assess the *time course* of neural activity *across brain regions* during the performance of a given task (Fig. 3.41).

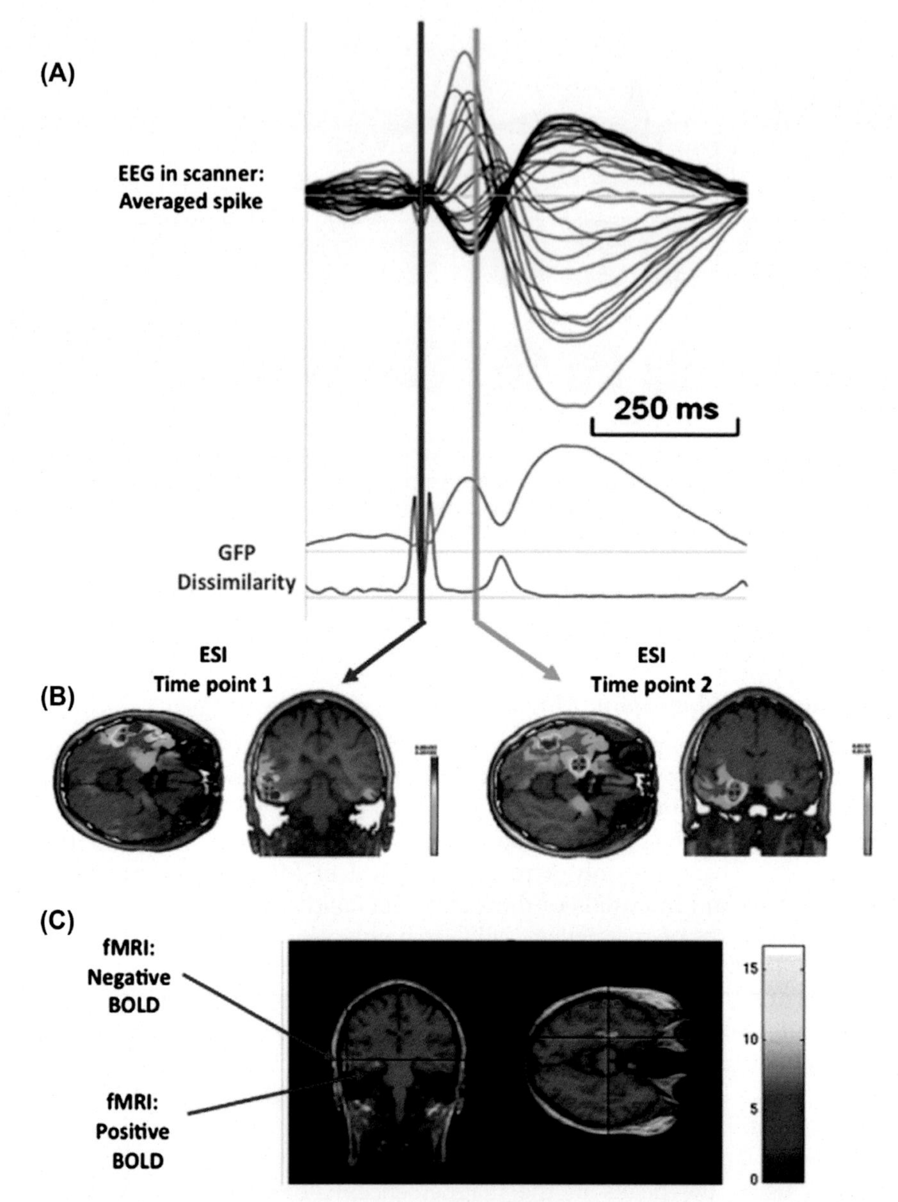

FIGURE 3.37 Combined electroencephalography (EEG) recordings and source imaging and functional magnetic resonance imaging (fMRI) aided in detecting and localizing epileptic spikes. (A) and (B) Long-term EEG recording revealed epileptic spikes shown at the top of the figure in (A), which were localized in the brain using source imaging (B); however, none were detected using an fMRI paradigm for this patient. (C) Combining the two methods provided sources of the EEG-recorded spikes onto the fMRI blood oxygen level–dependent (BOLD) dataset, revealing one source in the lateral temporal lobe (fMRI: Negative BOLD) and one in the mesial temporal lobe (fMRI: Positive BOLD). *Source: Michel and Murray (2012).*

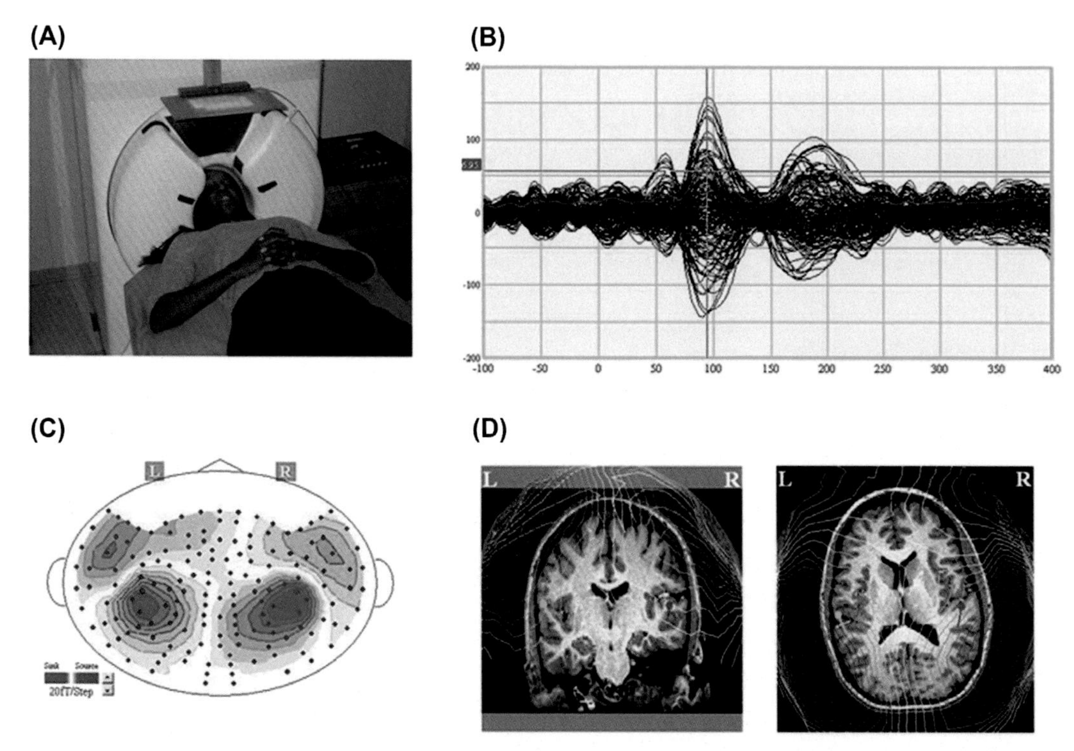

FIGURE 3.38 Magnetoencephalography (MEG) using a whole-head biomagnetometer records (A) the tiny magnetic waves that emanate from the skull due to neural activity at the population level. Each detector outputs the time-varying magnetic field at the scalp. These are averaged over detectors and plotted as an evoked response with time along the x-axis (in ms) and amplitude of the response along the y-axis. The large peak in the center of the plot is the M100, occurring ~100 ms poststimulus onset. (B) The contour plot shows the direction and amplitude of the response at the M100 peak, with the source shown in red and the sink shown in blue on a scale of 20 femtoTesla (fT)/step. (C). (D) MEG source models can be coregistered on an individual subject's MRI. *Source: Poeppel & Hickok, in Celesia & Hickok, 2015.*

4.4 Zapping the Brain: Transcranial Magnetic Stimulation

As opposed to *recording neural activity* using functional techniques such as fMRI, PET, EEG, and MEG, another method to investigate the brain's function comes from *stimulating the brain* and observing the outcomes.

It is now possible to simulate brain lesions in healthy subjects. Without cutting a person's brain, we can alter the brain's level of activity locally. Brief magnetic pulses over the scalp either inhibit or excite a region of cortex. For example, if you stimulate the hand area of the motor cortex, the subject's hand will move and twist. Applying an inhibitory pulse over the same area will cause subjects to have difficulty moving their hands. This is called *transcranial magnetic stimulation* (TMS) or, as one leading researcher called it, "zapping the brain" (Cowey & Walsh, 2001) (Figs. 3.42 and 3.43).

TMS appears to be generally safe. By applying TMS, we can test causal hypotheses about the contribution of specific brain regions to cognitive processes. Since the TMS works at the millisecond scale, it is also possible to study how rapid waves of processing develop. Recent

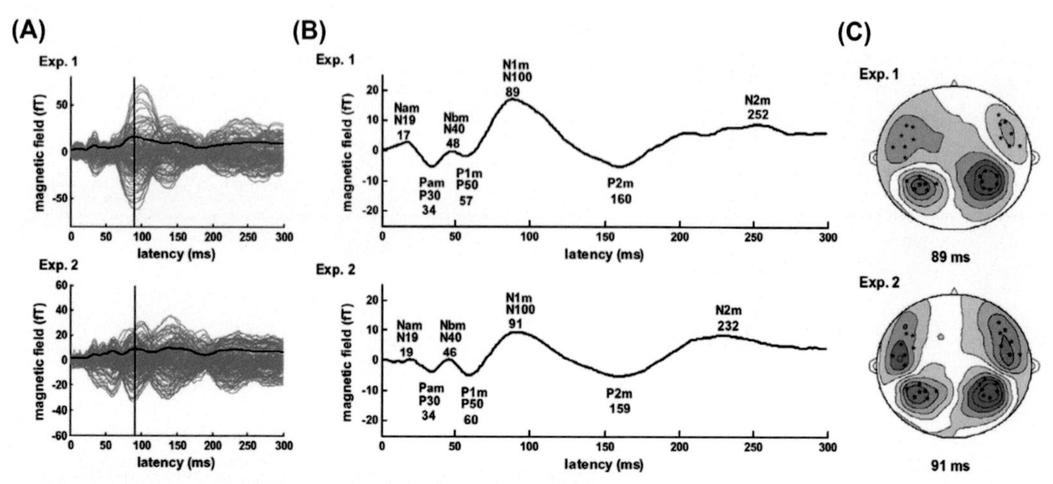

FIGURE 3.39 Magnetoencephalography (MEG) auditory evoked response to simple click sounds shows the canonic topography of the M100 component. Two experiments (1—upper plot and 2—lower plot) using clicks show similar responses in the auditory evoked magnetic field (A). The timing of the latency peaks are very similar across the two experiments, with the M100 peaking for Experiment 1 at 89 ms and for Experiment 2 at 91 ms (B). (C) Sources in the contour map are also highly similar across the two experiments, with sources shown in red for the left and right hemispheres, demonstrating that the M100 is a consistent and reliably generated component for investigating auditory cortical processing. *Source: Poeppel & Hickok, in Celesia & Hickok, 2015.*

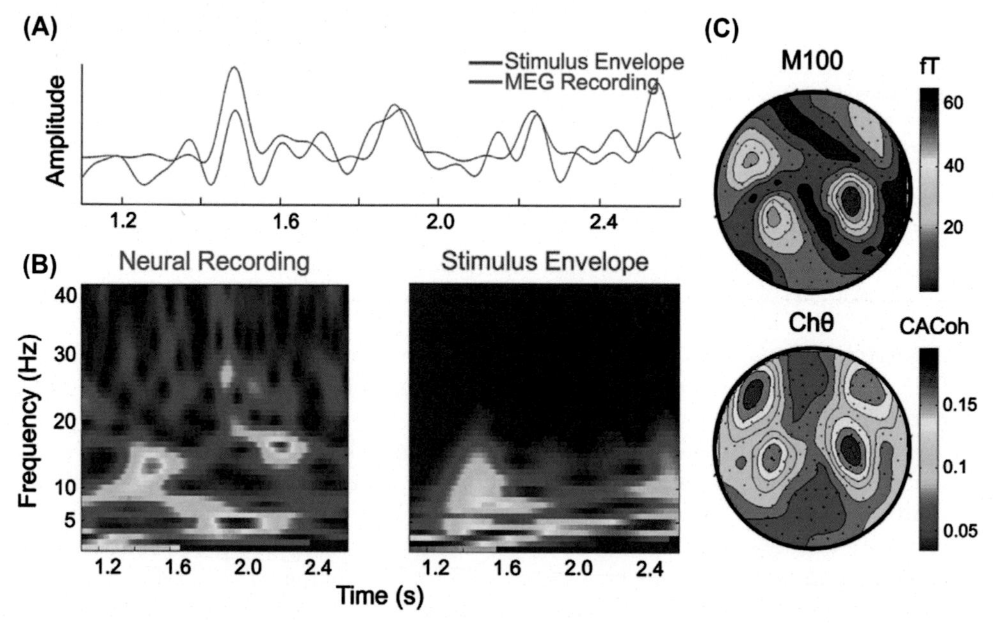

FIGURE 3.40 The high temporal resolution provided by magnetoencephalography (MEG) recording reflects fine-grained features in sound stimuli. (A) The amplitude of the MEG response (shown in blue) closely matches the envelope of the spoken sentence stimulus (shown in red). (B) Time-frequency analysis shows a similar pattern of neural recording reflecting the stimulus envelope. The time-frequency analysis is used to calculate the cerebroacoustic coherence (CACoh) to assess how closely the neural response reflects the envelope of the given stimulus. A legend for CACoh is shown at the right side of the figure, with blue showing none of very little coherence and warmer colors (yellow, orange, red) showing higher levels of coherence. (C) The top circle shows the contour map for the M100 in response to a 1000 Hz tone for a single subject. The bottom circle shows a different analysis: in this case, the average CACoh of the auditory response for one condition across subjects is shown. *Source: Poeppel & Hickok, in Celesia & Hickok, 2015.*

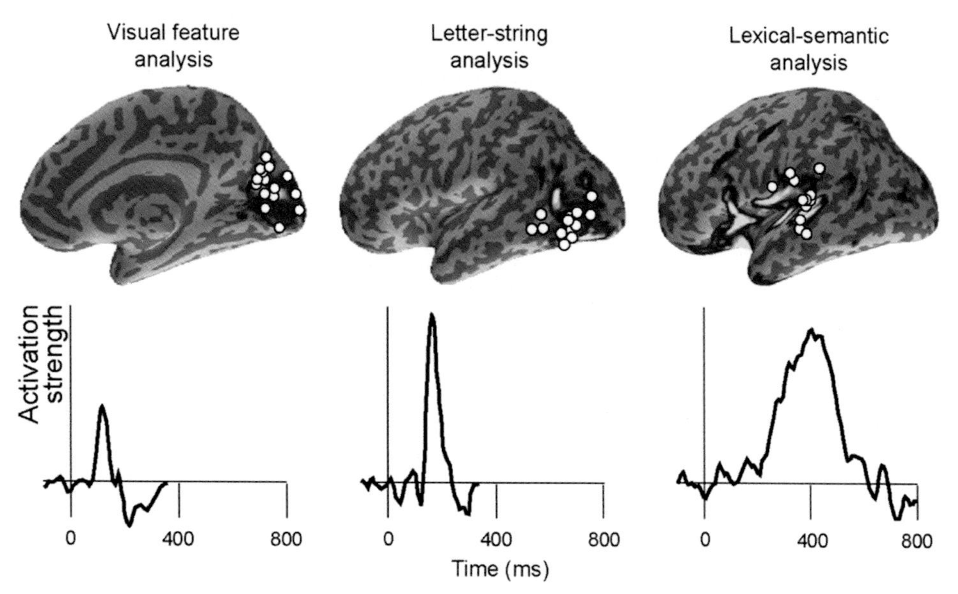

FIGURE 3.41 The high temporal resolution provided by magnetoencephalography (MEG) recording along with the ability to localize sources in the brain allows investigators to look across time in an experiment to see where activation patterns are observed. In this case, the subject's task was to silently read. Evoked responses were measured and plotted (bottom of the figure). Peaks and their sources were plotted on the inflated brain for early visual feature analysis of the written material (left), the subsequent letter-string analysis of word recognition processes (center), and the lexical-semantic networks as the written material was decoded at the sentence level (right). Sources of the activity moved from visual cortical regions for the visual feature, expanding into temporal lobe regions for letter-strong analysis, and expanded into a large lexical-semantic processing region in frontal and temporal lobes for the lexical-semantic analysis. *Source: Hari and Salmelin (2012).*

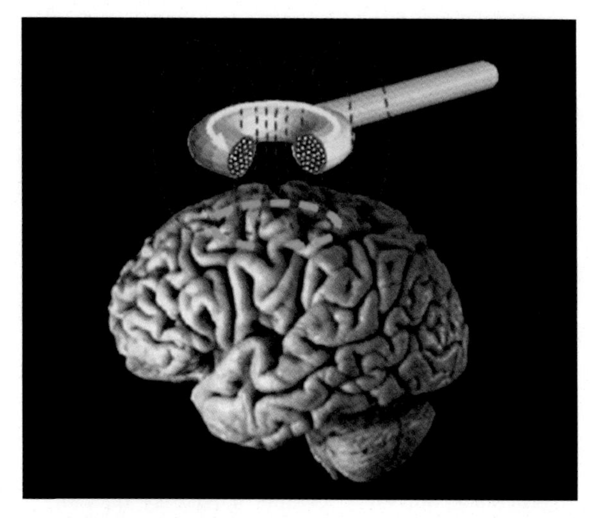

FIGURE 3.42 Transcranial magnetic stimulation (TMS) provides brief magnetic pulses over the scalp, which either inhibit or excite a region of cortex. TMS provides a relatively safe way to change brain function for brief periods in an experiment. *Source: Eric Wassermann, M.D.*

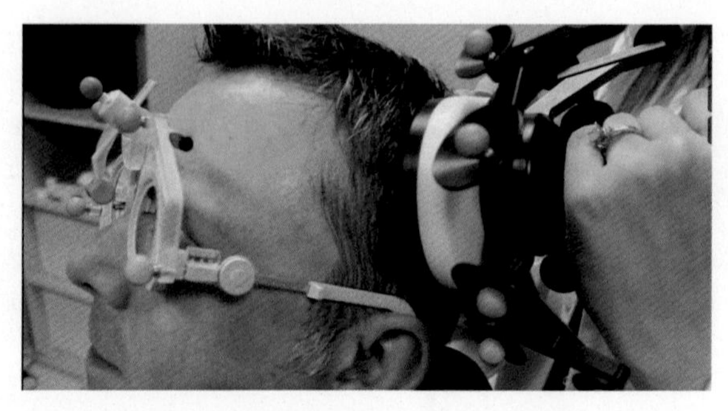

FIGURE 3.43 An example of a transcranial magnetic stimulation (TMS) experiment, with the TMS stimulator applied to the surface of the head. *Source: public domain.*

TMS studies emphasize that magnetic pulses rarely have simple, local effects. Rather, like other kinds of brain stimulation, magnetic pulses often trigger widespread activities, depending on the subject's expectations and ongoing goals.

5. OBSERVING BRAIN PATHWAYS: NEUROCONNECTIVITY

White matter fiber tracts are the vast internal highway system of the cortex. We can visualize these fiber tracts using an MRI method called *DTI*. DTI uses water flow along the axons surrounded by white myelin to measure the relative direction of white matter tracts (Figs. 3.44–3.46). DTI helps us to understand brain connectivity patterns in the healthy brain, as well as investigate these patterns in individuals with brain diseases that affect white matter, such as multiple sclerosis.

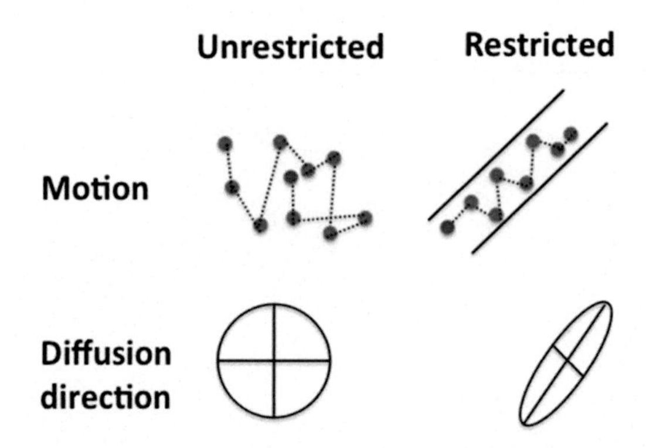

FIGURE 3.44 Diffusion magnetic resonance imaging movement of water can generally be described by Brownian motion. If the water molecule is in a relatively open substance, such as gray matter, movement of the molecules is relatively unrestricted and can occur in all directions (isotropy). If, however, the water molecules are in a relatively restricted substance, such as within an axon, movement of the molecules is relatively restricted, which leads to a preferential diffusion direction (anisotropy). *Source: Gage with permission.*

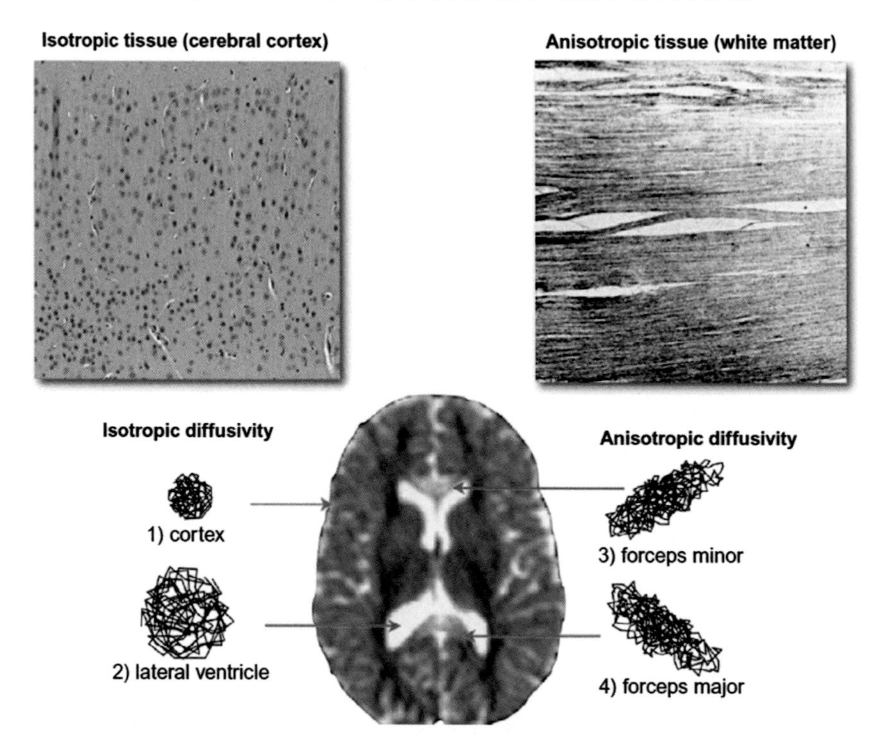

FIGURE 3.45 A schematic representation of the diffusion of water molecules in isotropic (shown in pink on the left) and anisotropic (shown in green on the right) tissue. In isotropic tissue, such as the gray matter of the cortex and in the ventricles, the physical properties are identical in all directions. This allows the water molecules to be reduced equally along all orientations. In anisotropic tissue, such as white matter in the forceps, axonal membranes and myelin sheath allow displacement of water molecules along the direction of the axons (rather than perpendicular to them). *Source: Maffei, Soria, Prats-Galino, & Catani, in Celesia & Hickok, 2015.*

By far the largest fiber bundle in the brain is the corpus callosum, which connects the two hemispheres, but there are many other fiber bundles or tracts that connect regions within the hemispheres. A DTI analysis of a section of the corpus callosum is shown in Fig. 3.47.

When white matter tract cross, such as the massive interhemispheric corpus callosum and the large posterior/anterior running arcuate fasciculus, DTI can map the two fiber bundles using fractional anisotropy (FA) (Fig. 3.48).

In this way, separate sections of large white matter tracts, such as the arcuate fasciculus, and separate white matter tracts can be mapped in the brain, providing key data for understanding the neural pathways that underlie human cognition (Fig. 3.49).

6. OBSERVING BRAIN DYNAMICS: BRAIN RHYTHMS AND OSCILLATIONS

The rhythms of our lives: you may have a sense that your daily life is full of rhythms that can change suddenly or continue unchanged depending on your situation. The college years

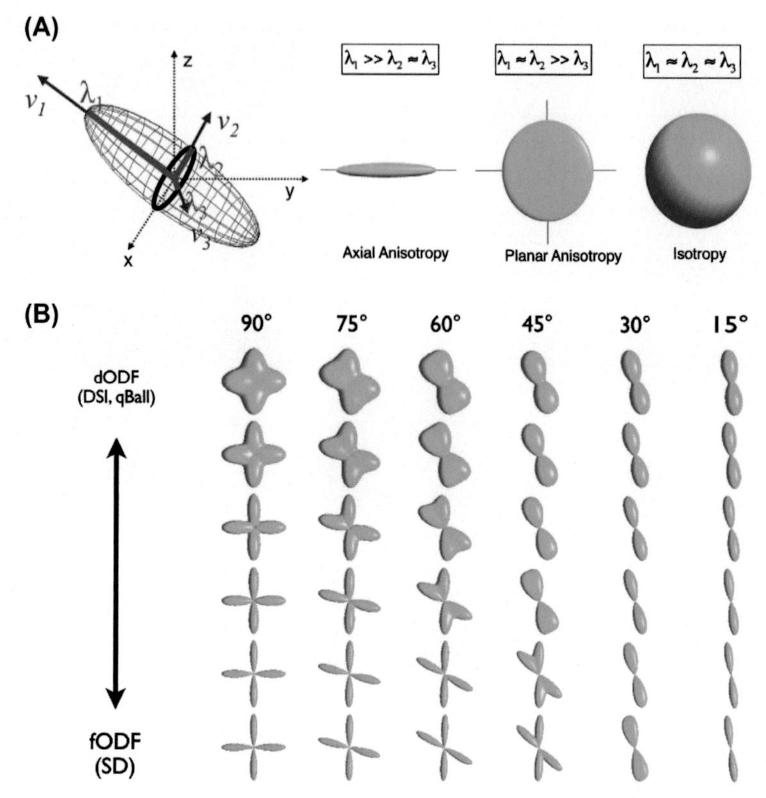

FIGURE 3.46 This figure provides a way to visualize the diffusion tensor as a diffusion ellipsoid. (A) The size and shape of the tensor has three configurations: axial anisotropy, which is typically found in voxels containing parallel fibers; planar anisotropy, which is typically found in voxels containing groups of crossing or diverging fibers; and isotropy, which is typically found in voxels containing gray matter. (B) The orientation distribution function profiles according to different diffusion model methods. *Source: Maffei, Soria, Prats-Galino, & Catani, in Celesia & Hickok, 2015.*

are busy, exciting, emotional, and highly social times in our lives: fast and suddenly changing rhythms occur. Old age is quieter, less changing, more serene: rhythmic patterns seem to be more established and maintained.

The brain is full of rhythms that provide the push–pull for human cognition, from concentrating attention on an individual sentence in a book to becoming globally sleepy. Brain rhythms help modulate brain activity on a daily basis as we move from the waking state to the sleep and dreaming states and back. We will discuss these brain "waves" in more detail in Chapter 8, Attention and Consciousness, and Chapter 12, Sleep and Levels of Consciousness. Briefly, *neural oscillations* reflect the rhythmic activity of populations of neurons in the brain. These oscillations can be recorded from the surface of the brain using EEG and recording electrodes (Fig. 3.50A). The recorded signal can then be analyzed in the power domain for each frequency (Fig. 3.50B). The signal can also be analyzed in the

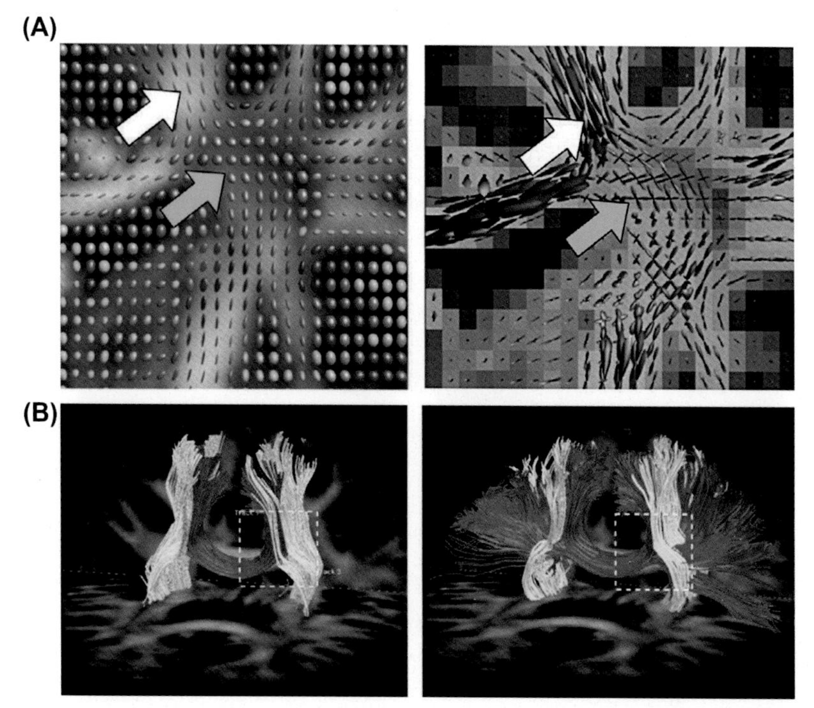

FIGURE 3.47 A comparison of DTI tractography models: tensor-based and spherical deconvolution. (A) The white matter can be visualized based on the tensor model (left panel) or the spherical deconvolution model (right panel). The *white arrow* is indicating the corpus callosum, which is one fiber population. Both models are adept at showing the visualization of this large fiber tract. However, in regions with more than one population of crossing fibers (*cyan arrows*), the spherical deconvolution model better separates different fiber components and their orientations. (B) Diffusion tensor-based model (left) and spherical deconvolution model (right) tractography results show differing results. Left: tensor-based model reconstructs only the most central part of the corpus callosum (shown in red). Right: spherical deconvolution model shows several streamlines of the corpus callosum (in red) crossing the streamlines of the corticospinal tract (in yellow). *Source: Maffei, Soria, Prats-Galino, & Catani, in Celesia & Hickok, 2015.*

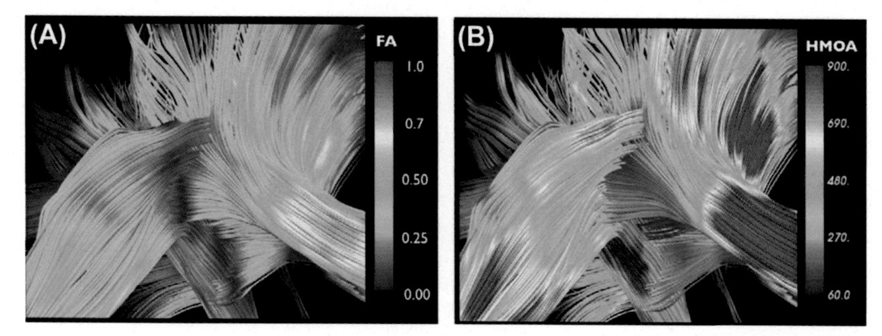

FIGURE 3.48 A comparison of two anisotropy techniques for visualizing crossing fibers. (A) The arcuate fasciculus and the lateral projections of the corpus callosum at the level of the corona radiate are shown. Left panel shows a fractional anisotropy (FA) analysis that does not distinguish the crossing fibers (shown in mostly blue and green). (B) A hindrance-modulated orientational anisotropy (HMOA) index provides more distinct diffusion characteristics for the two crossing fiber tracts (shown in red, yellow, green, and blue). *Source: Maffei, Soria, Prats-Galino, & Catani, in Celesia & Hickok, 2015.*

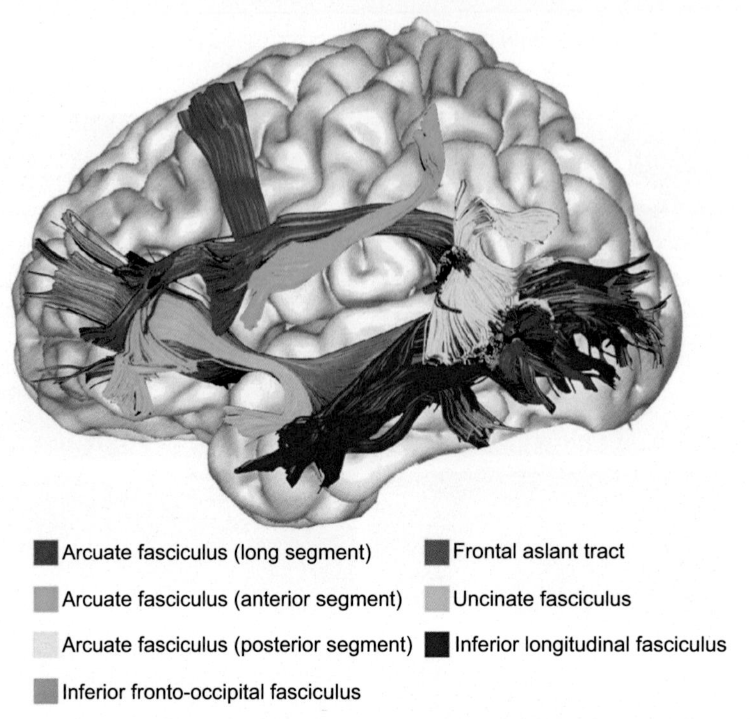

FIGURE 3.49 Techniques for visualizing fiber tracts in the brain. Techniques described for Figs. 3.47 and 3.48 combine to provide a tractography reconstruction of the major association pathways in the brain for auditory and language processing. They include the arcuate fasciculus (long segment, shown in red), the arcuate fasciculus (anterior segment, shown in green), the arcuate fasciculus (posterior segment, shown in yellow), the inferior fronto-occipital fasciculus (shown in orange), the frontal aslant tract (shown in purple), the uncinate fasciculus (shown in turquoise), and the inferior longitudinal fasciculus (shown in blue). *Source: Maffei, Soria, Prats-Galino, & Catani, in Celesia & Hickok, 2015.*

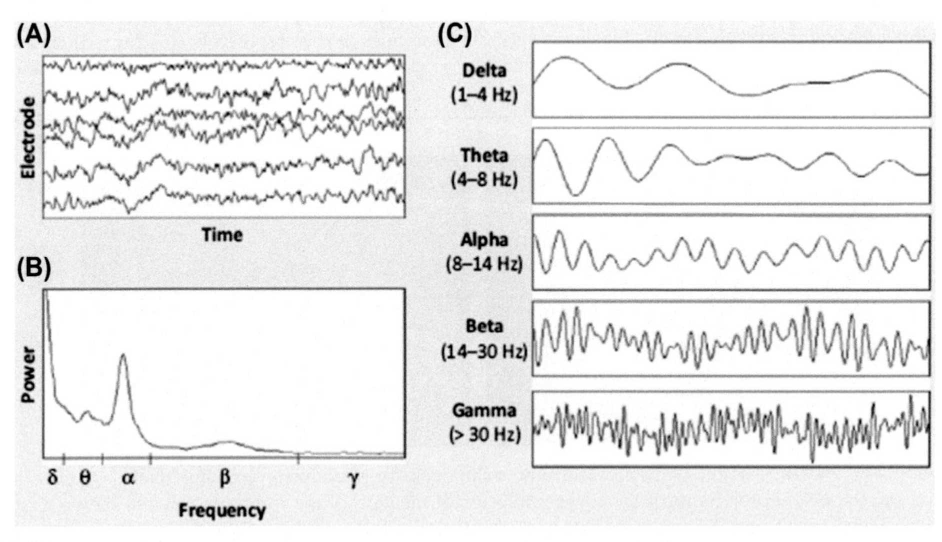

FIGURE 3.50 *Neural oscillations* reflect the rhythmic activity of populations of neurons in the brain. These oscillations can be recorded from the surface of the brain using electroencephalography and recording electrodes (A). The recorded signal can then be analyzed in the power domain for each frequency (B). The signal can also be analyzed in the temporal (time) domain for each frequency (C). Characteristic neural oscillation frequency bands are delta (1–4 cycles per second, Hz), theta (4–8 Hz), alpha (8–14 Hz), beta (14–30 Hz), and gamma (>30 Hz). *Source: Clayton, Yeung, and Kadosh (2015).*

temporal (time) domain for each frequency (Fig. 3.50C). Characteristic neural oscillation frequency bands are delta (1–4 cycles per second, Hz), theta (4–8 Hz), alpha (8–14 Hz), beta (14–30 Hz), and gamma (>30 Hz). Resting state EEG typically shows these bands with a prominent waveform in the alpha range (Fig. 3.51).

Singer and colleagues have investigated the role of the synchrony of neural oscillations in human cognition. According to Singer and colleagues, neural synchrony in cortical networks provides a general mechanism for the coordination of distributed neural networks across brain regions (Uhlhaas et al., 2009). Neural synchrony provides the neural underpinnings of human perception, thought, and even consciousness. Further, neural synchrony is abnormal in major brain disorders such as schizophrenia and autism.

Excitatory and inhibitory neural networks are hypothesized to communicate through dynamic relays that become coupled through patterns of oscillations (Fig. 3.52). Changes in the strength of synchronization across neural networks reflect sensory processes such as detecting brightness contrasts in a visual stimulus (Fig. 3.53).

This synchronization is carried in very large scale throughout long-ranging brain networks such as those initiated in the prefrontal cortex, which are thought to sustain cognitive functions such as sustained attention (Fig. 3.54).

In this way, neural communication flows throughout the brain as voluntary attention and executive control functions mediate cognition while attentional capture and emotional awareness functions compete for conscious entry. The ebb and flow of neural oscillations throughout the day and during our lifetime provide the flexibility to provide the neural resources needed as we adapt to our ever-changing environments.

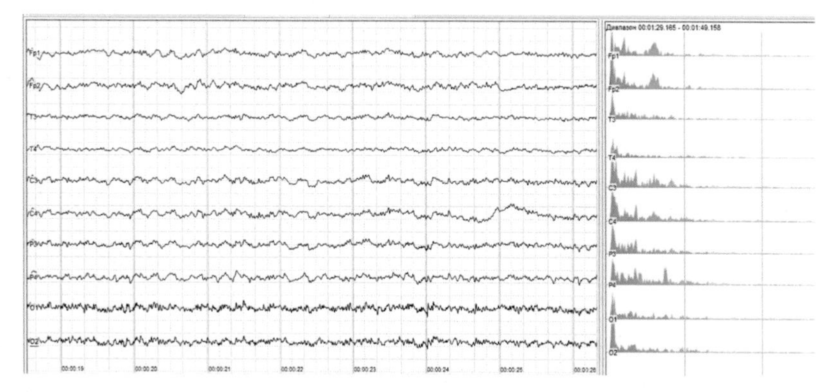

FIGURE 3.51 Resting state electroencephalography (EEG) typically shows EEG traces (left) for each recording electrode and their corresponding frequency bands (right). This example includes a prominent waveform in the alpha range (shown in purple). *Source: Andrii Cherninskyi.*

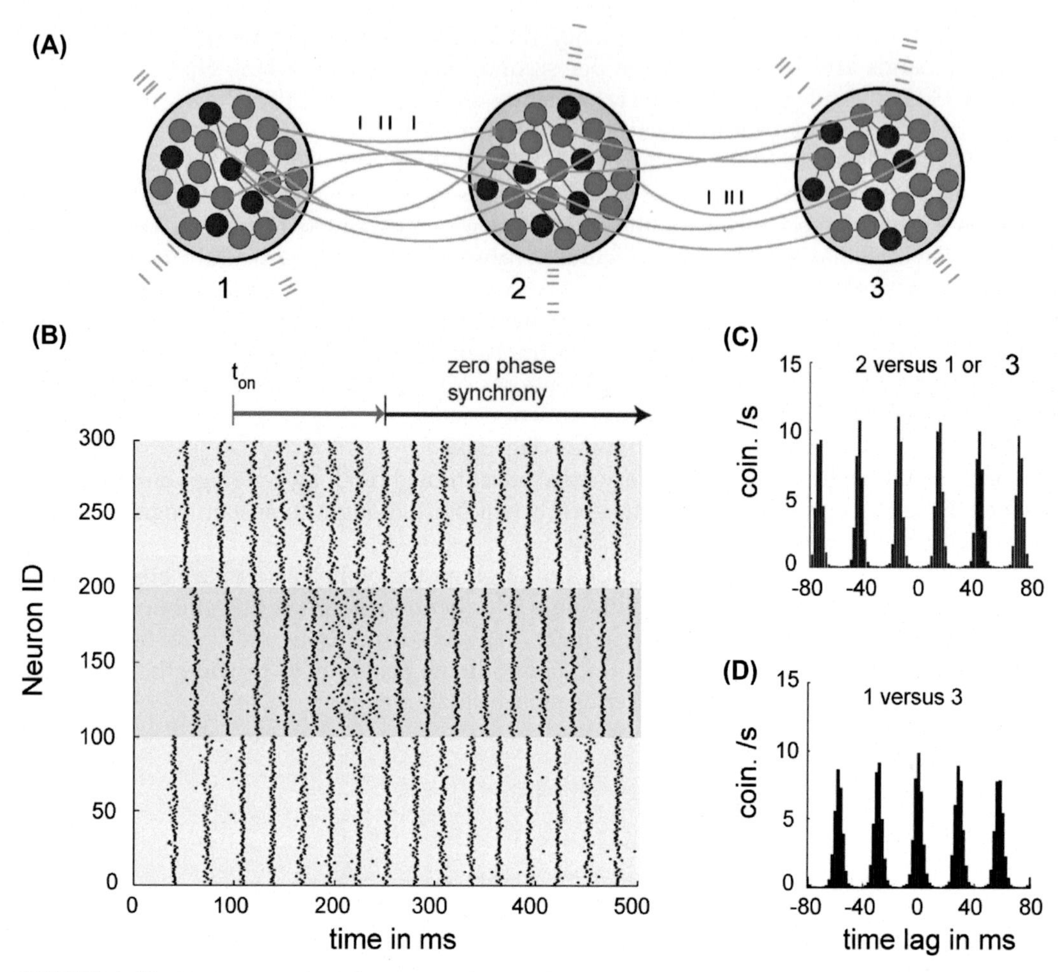

FIGURE 3.52 Excitatory and inhibitory neural networks are hypothesized to communicate through dynamic relays that become coupled through patterns of oscillations. (A) Three networks (1, 2, and 3) are each composed of 20 inhibitory (red) and 80 excitatory (blue) neurons. In this demonstration, external input, with a Poisson distribution (*gray spikes*) is used to control the activity level of each population in the three networks. (B) Proposed activity of the three networks proposed in (A). A raster plot shows the activity (firing) of each of the 100 neurons in population 1 (bottom of the plot in yellow), the 100 neurons shown in population 2 (middle of the plot in green), and the 100 neurons shown in population 3 (top of the plot in yellow). *Blue-colored spikes* show excitatory responses and *red-colored spikes* show inhibitory responses. The coupling between the networks is shown with the *gray and red lines* above the raster plot. Beginning around 100 ms, t_{on} indicates the beginning of the coupling of the three networks. The coupling begins to become phase locked at around 90 ms later and lasts for about 130 ms. Note that population 2 (shown in green) is not as phase locked as populations 1 and 3: this may indicate that while population 2 *relays* phase coupling information, it is populations 1 and 3 that form the tightest coupling. (C) The average cross-correlation between populations 1 or 3 and the relaying population 2. (D) Average cross-correlation between the outer populations 1 and 3. Uhlhaas et al. (2009).

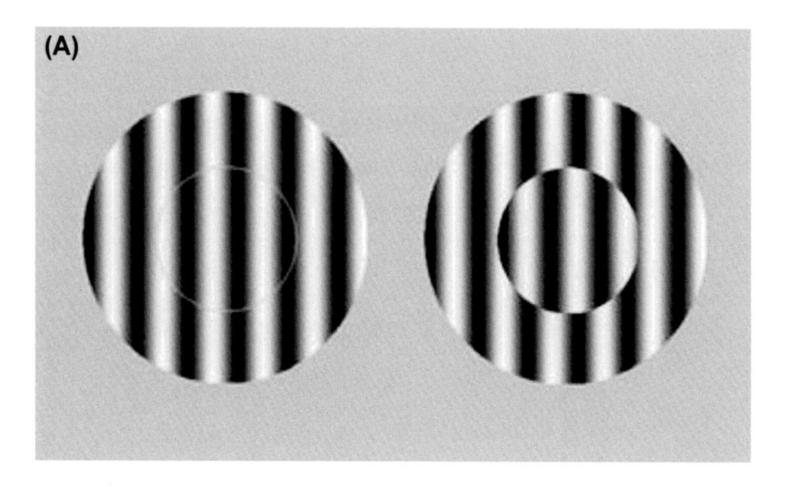

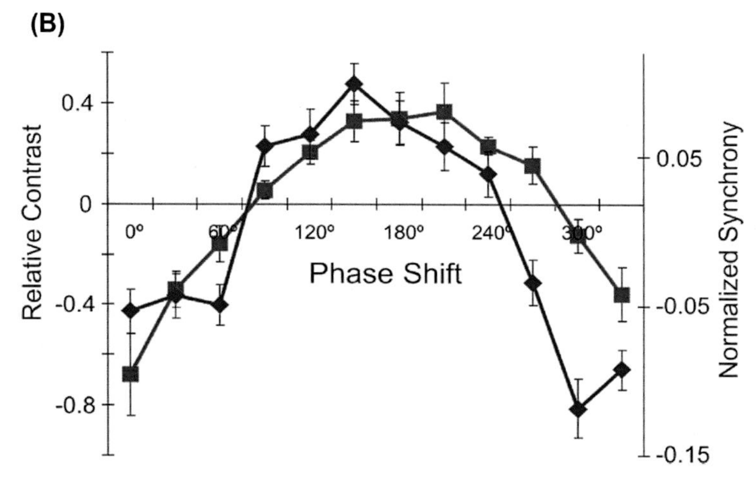

FIGURE 3.53 Neural synchrony and the perception of brightness. (A) Visual stimuli on the left and right have the identical center images; however, they look different because of their context. The stimulus on the right is typically reported to have a higher contrast where the center is offset from the surround. (B) *Red squares and lines* show a human psychophysical judgments of the perceived contrasts as a function of the phase offset of the center versus the surround in stimuli such as those presented above in (A). *Blue diamonds and lines* are data from animal studies showing the strength of neural synchronization in visual regions where phase and orientation features of visual stimuli are encoded. Together, results show that perception of contrasts in visual stimuli reflects level of synchronization in the neural networks tuned to their encoding. *Source: Uhlhaas et al. (2009).*

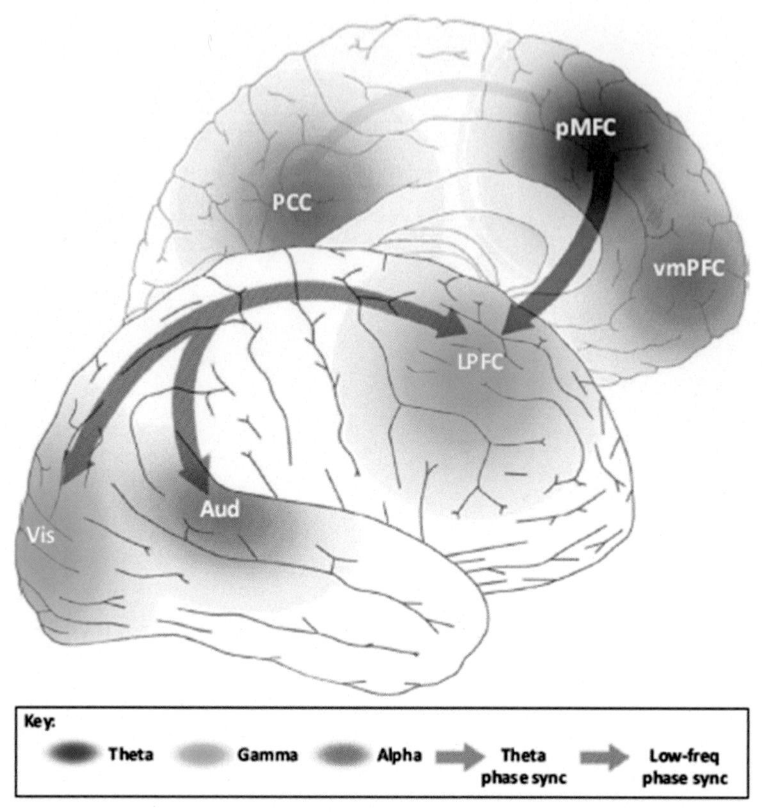

FIGURE 3.54 According to the model proposed by Clayton and colleagues, *monitoring of attention* is supported by theta oscillations in posterior medial frontal cortex (pMFC, shown in purple). The neural communication between the pMFC and lateral prefrontal cortex (LPFC) is aided by phase synchronization in the theta frequency band (shown in the *purple arrow*). Next, neural communication between the frontal lobe LPFC and the posterior sensorimotor regions is supported by phase synchronization in the lower-frequency bands (theta, alpha, shown in *gray arrows*). Sustaining attention during a cognitive task requires the excitation of task-relevant areas in the brain with the inhibition of task-irrelevant areas. Clayton et al. (2015) proposes that this is accomplished through neural oscillations from the frontal lobe to posterior sensory cortex. Thus the influence of the frontal cortex over posterior regions provided by the low-frequency bands of oscillations described above allows frontal processes to promote high-frequency gamma band oscillations in a task-relevant area of the brain (in this case, visual cortex shown in green) and lower-frequency alpha band oscillations in a task-irrelevant area of the brain (in this case, auditory cortex shown in orange). The influence of the frontal cortex over visual cortex allows the continuous activation of a task-relevant activity (in this example, a visual task with visual cortex activation) and suppression of task-irrelevant activity (in this example, auditory cortex with suppression of auditory processing). *Source: Clayton et al. (2015).*

7. SUMMARY

In Chapter 2, we discussed five central aspects of brain study: *neuroanatomy* for studying the structure of the brain; *neurophysiology* for studying the cellular aspects of the brain; *functional neuroanatomy* for studying the perceptual, motor, and cognitive processes of the brain; *neuroconnectivity* for studying brain pathways; and *neurodynamics* to study the rhythms and oscillations of the brain. Together these disciplines enable us to understand both small scale and wide ranging aspects of the human brain.

In this chapter, we have discussed the imaging techniques used to investigate these five central aspects of the brain. Each method and their brain targets, from structural MRI images to functional studies to white matter pathways, provide new ways to visualize brain system's processes. Their combined use, such as EEG and fMRI to determining when and where an epileptic seizure occurs, has revolutionized both clinical and research neuroscientific investigations.

8. STUDY QUESTIONS

1. What is a voxel and what does it measure in the brain?
2. Discuss the difference between structural MRI and functional MRI? How are each used?
3. What is a single unit recording? What is the "unit" it records?
4. What is the BOLD response and how does it reflect brain activity?
5. How does EEG differ from MEG?
6. What is the role of neural oscillations in brain dynamics?

CHAPTER

4

The Art of Seeing

ART OF SEEING

Whilst part of what we perceive comes through our senses from the object before us, another part (and it may be the larger part) always comes … out of our own mind. ***William James (1890, V2, p 103), American Psychologist and Philosopher.***

Morning on the Siene, Claude Monet, 1898. *Source: Public domain.*

1. INTRODUCTION

Human vision is a dynamic system that *constructs* our visual environment, beginning with the images projected to the back of our eyes—the retinas—projecting to multiple brain regions within the visual cortex. Why do we say "construct"? Because unlike a camera, which provides a veridical image of whatever scene is captured by the lens, the visual system does

Fundamentals of Cognitive Neuroscience
https://doi.org/10.1016/B978-0-12-803813-0.00004-0

not actually produce a true picture of the world in our view. We will see as we progress through this chapter that the visual system is highly interactive and dynamic in representing the world around us—and in many cases the actual physical images projected onto the retina are not part of our visual perception at all! We are literally blind to them.

However, before we discuss this aspect of the visual system, let us work through the neural machinery that forms the human visual system. Next, we will discuss the constructive nature of human vision, from visual illusions to depth perception to color constancy—and how this is accomplished at various stages and places in the visual system. Lastly, we will look at the effects of visual awareness and unawareness in the study of visual consciousness (Box 4.1).

BOX 4.1

VISUAL ILLUSIONS: SHEDDING LIGHT ON VISUAL PROCESSING MECHANISMS

Everyone likes visual illusions—or most everyone. And no one seems to like them more than vision scientists. Although looking at cool visual illusions is fun, they play a serious role in vision science. The difference between what is actually presented in a book or on a website and your perception of it tells us a lot about how the visual system works. Notice that we do not call them "optical illusions." That would mean they were due to the eyes—but visual illusions are coming from your brain.

Illusory motion

Illusory motion—perceiving movement from a still or "static" figure—is a very popular visual illusion; they are fun to look at! On the serious side, however, illusory motion illusions are playing a key role in visual science studies of eye movements, luminance gradients, and motion detection systems. There is no general agreement as to precisely how they happen, but much progress has been made lately through carefully designed experiments. More on that later. But for now, take a look at the visual illusion shown in Fig. 4.1. Focus in on the center of the figure for about 10 s, and then begin moving your eyes around the perimeter. Does it appear to be moving, shimmering? Follow the outermost groove, and watch it change from a groove to a hump as your eyes follow it around the perimeter of the figure.

We think this perception of motion is related to the Pinna effect, first described by Dr. Baingio Pinna, an Italian vision scientist. His first description of this was described in an article published in the Italian language in 1990 (2), but more recent work has been published in English and presents a full discussion of the effect (3). There are many variants on this effect, but one general notion is that when you fixate on the center of the figure, your vision system is also encoding the elements in the figure that extend past the fixation point, in the periphery. The illusory perception of motion can be affected by a myriad of experimental and illusion parameters ranging from the luminance of the elements in the illusion, head and eye movement, and so on.

BOX 4.1 *(cont'd)*

FIGURE 4.1 The moving spiral. Focus in on the center of the figure for about 10 s, and then begin moving your eyes around the perimeter. Does it appear to be moving, shimmering? Follow the outermost groove, and watch it change from a groove to a hump as your eyes follow it around the perimeter of the figure. *Source: Optical Illusions and Pictures, http://www.123opticalillusions.com/pages/opticalillusions14.php, with permission.*

But What About the Rotating Snakes?

One of our favorite illusory motion illusions comes to us from a master illusion maker and vision scientist, Dr. Akiyoshi Kitaoka (4). Rotating Snakes is a complex figure with circular patterns seeming to rotate in opposite directions despite being a static figure (Fig. 4.2). This particular illusion is a variant of the Fraser–Wilcox illusion (5) where illusory motion is perceived along repeated blocks of colored regions within a figure. In these early illusions, viewers reported motion along differing directions (e.g., clockwise vs. counter clockwise). In Rotating Snakes and other similar illusions prepared by Kitaoka, the direction of motion perceived can be controlled by varying factors in the illusion such as luminance and color contrast. However, while color enhances the effects, they are present—and consistent in the direction of motion perceived—in a gray-tone version as well. According to Kitaoka (6), the motion effects are probably due to the shading gradients across small sections of the figure that vary in luminance, contrast, and color. Most vision scientists would agree that eye movements also play a role, as does the neural adaptation in the retina and motion perception systems in the visual cortex.

Continued

BOX 4.1 *(cont'd)*

FIGURE 4.2 The Rotating Snakes. This is a very popular illusory motion illusion created by Kitaoka as a variant of the Fraser–Wilcox illusion. What do you see? Which way are the spirals spinning? The direction of the motion illusions moves from black to blue, white, yellow, back to black. Can you gaze at the figure and see how these directions across the colors appear to be in motion? We highly suggest that you check out Kitaoka's website in Ref. (4) to see his latest and greatest illusions and their explanations. *Source: http://www.ritsumei.ac.jp/~akitaoka/index-e.html.*

Have you wondered how the visual system could be so easily tricked into perceiving motion when there clearly is none in a fixed 2-D image? It turns out that this is a central part of human vision—demonstrated using simple visual illusions such as the ones presented here—and fundamental to how we interpret our complex visual world.

References

(1) Optical Illusions and Pictures, http://www.123opticalillusions.com/pages/opticalillusions14.php.

(2) Pinna, B. (1990). *Il Dubbio sull'Apparire.* Padua, Upsel Editore.

(3) Pinna, B., & Brelstaff, G. J. (2000). A new visual illusion of relative motion. *Vision Research, 40,* 2091–2096.

(4) Akiyoshi's illusion pages, http://www.ritsumei.ac.jp/~akitaoka/index-e.html.

(5) Fraser, A., & Wilcox, K.J. (1979). Perception of illusory movement. *Nature, 281,* 565–566.

(6) Kitaoka, A., (2014). Color-dependent motion illusions in stationary images and their phenomenal dimorphism. *Perception, 43*(9), 914–925.

2. FUNCTIONAL ORGANIZATION OF THE VISUAL SYSTEM

When the light coming from an object reaches our eyes, it triggers a cascade of neural events as this visual pattern is converted into neural impulses that travel up the visual system from one brain area to the next. Through a series of neural processes, the activity of neurons in numerous brain areas leads to the visual experience and recognition of the object and its many component features. Although the visual system is hierarchical, with feedforward connections beginning at the retina and extending throughout the visual cortex, it is important to keep in mind that this dynamic system also has many feedback and parallel connections. We will discuss that in greater detail later in the chapter.

For now, let us trace the sequence of events that occur to understand how the brain processes visual information at each stage of visual perception. This will help us understand how different visual areas of the brain contribute to visual perception.

2.1 The Retina

There are two types of photoreceptors in the retina: *cones* and *rods* (Fig. 4.3). Cones are color selective, less sensitive to dim light than rods, and important for detailed color vision in daylight. Each cone contains one of the three kinds of *photopigments*, specialized proteins that are sensitive to different wavelengths of light. These wavelengths roughly correspond to our ability to distinguish red, green, and blue. When light strikes a photopigment molecule, the light energy is absorbed and the molecule then changes shape in a way that modifies the flow of electrical current in that photoreceptor neuron. Cones are densely packed into the *fovea*, the central part of the retina that we use to look directly at objects to perceive their fine details. In the periphery, cones are more spread out and scattered, which is why objects in the periphery appear blurrier and their colors are less vivid.

Rods contain a different photopigment that is much more sensitive to low levels of light. Rods are important for *night vision*. We rely on seeing with our rods once our eyes have adapted to the darkness (*dark adaptation*). Curiously, there are no rods in the fovea, only cones, and the proportion of rods increases in the periphery. This is why you may have noticed when gazing at the night sky that a very faint star may be easier to see if you look slightly off to one side.

There are far more rods in the retina than cones, with roughly 120 million rods distributed throughout the retina except for the fovea, and 6–7 million cones that are concentrated in that fovea.

The signals from photoreceptors are processed by a collection of intermediary neurons, *bipolar cells*, *horizontal cells*, and *amacrine cells*, before they reach the *ganglion cells*, the final processing stage in the retina before the signals leave the eye. The actual cell bodies of ganglion cells are located in the retina, but these cells have long axons that leave the retina at the *blind spot* and form the *optic nerve*. Each ganglion cell receives excitatory inputs from a collection of rods and cones; this distillation of information forms a *receptive field*. Ganglion cells at the fovea receive information from only a small number of cones, whereas ganglion cells in the periphery receive inputs from many rods (sometimes thousands). With so many rods providing converging input to a single ganglion cell, if any one of these rods is activated by photons of light, this may activate the ganglion cell, which increases the likelihood of being able to detect dim, scattered light. However, this increase in sensitivity to dim light is

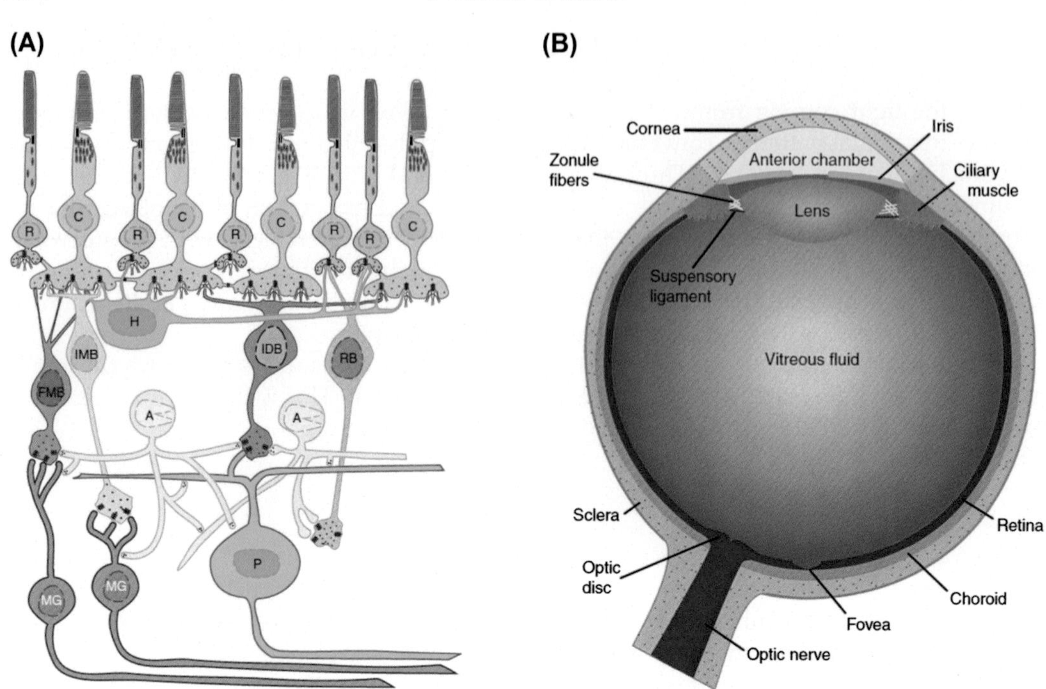

FIGURE 4.3 The eye. (A) There are two types of photoreceptors in the retina: *cones* and *rods*. Cones (labeled C) are color selective, less sensitive to dim light than rods (labeled R), and important for detailed color vision in daylight. Other cells in the retina include the horizontal cell (labeled H), the flat midget bipolar (labeled FMB), invaginating midget bipolar (labeled IMB), invaginating diffuse bipolar (labeled IDB), rod bipolar (labeled RB), amacrine cell (labeled A), parasol cell (labeled P), and midget ganglion cell (labeled MG). (B) Cones are densely packed into the *fovea*, the central part of the retina that we use to look directly at objects to perceive their fine details. In the periphery, cones are more spread out and scattered, which is why objects in the periphery appear blurrier, and their colors are less vivid. *Source: Reid and Usrey in Squire et al., 2013.*

achieved at the cost of poorer resolution; rods provide not only more sensitivity but also a "blurrier" picture than the sharp daytime image provided by cone vision.

Retinal ganglion cells receive both excitatory and inhibitory inputs from bipolar neurons, and the spatial pattern of these inputs determines the cell's *receptive field* (Fig. 4.4A). A neuron's receptive field refers to the portion of the visual field that can activate or strongly inhibit the response of that cell. Retinal ganglion neurons have center-surround receptive fields. For example, a cell with an *on-center off-surround* receptive field will respond strongly if a spot of light is presented at the center of the receptive field. As that spot of light is enlarged, responses will increase up to the point where light begins to spread beyond the boundaries of the on-center region. After that, the response of the ganglion cell starts to decline as the spot of light gets bigger and stimulates more and more of the surrounding off-region. Similarly, a cell with an off-center on-surround receptive field will respond best to a dark spot presented in the center of the receptive field.

How can the behavior of retinal ganglion cells be understood? A key concept is that of *lateral inhibition* (Kuffler, 1953). Lateral inhibition means that the activity of a neuron may be inhibited by inputs coming from neurons that respond to neighboring regions of the visual field. For example, the retinal ganglion cell in Fig. 4.4B receives excitatory inputs from cells

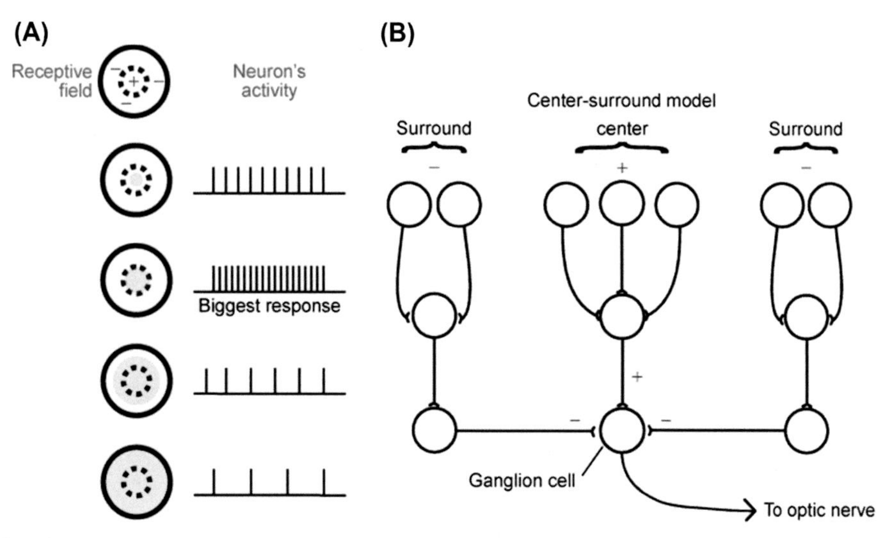

FIGURE 4.4 Center-surround receptive fields. (A) Schematic example of a center-surround cell's response to different-sized patches of light. Notice that the biggest spiking response (shown by the lines on the right) occurs for the intermediate-sized center light patch. The spot of light has to be just the right size to get the maximum response out of that particular neuron. (B) A model of how a center-surround receptive field might be achieved by the collaboration and competition between different connective neurons in the retina. *Source: Frank Tong, with permission.*

corresponding to the on-center region and inhibitory inputs from the off-center region. The strengths of these excitatory and inhibitory inputs are usually balanced, so if uniform light is presented across both on- and off-regions, the neuron will not respond to uniform illumination.

Lateral inhibition is important for enhancing the neural representation of *edges*, regions of an image where the light intensity sharply changes. These sudden changes indicate the presence of possible contours, features, shapes, or objects in any visual scene, whereas uniform parts of a picture are not particularly informative or interesting. Fig. 4.5 shows a picture of a fox in original form and after using a computer to filter out just the edges (right picture) so that the regions in black show where ganglion cells would respond most strongly to the

FIGURE 4.5 The edges hold most information. An example of how most of the information in the picture comes from the edges of objects. Figure on the left is the original, on the right is the information from the edges only—taken from the image using a computer algorithm. *Source: Frank Tong, with permission.*

image. Lateral inhibition also leads to more efficient neural representation because only the neurons corresponding to the edge of a stimulus will fire strongly; other neurons with receptive fields that lie in a uniform region do not. Because the firing of neurons takes a lot of *metabolic energy*, this is much more efficient. This is an example of *efficient neural coding*; only a small number of neurons need to be active at any time to represent a particular visual stimulus.

Lateral inhibition also helps to ensure that the brain responds in a similar way to an object or a visual scene on a cloudy day and on a sunny day. Changes in the absolute level of brightness will not affect the pattern of activity on the retina very much at all; it is the relative brightness of objects that matters most. An example of this is you see a friend wearing a red shirt. The *absolute level of brightness* of that shirt when you see your friend outside your house on a sunny day versus inside your house in a sheltered room will differ, but this will not affect the pattern of activity on the retina. On the other hand, the *relative brightness* of the shirt compared with other nearby objects or the background scene will make a difference on the retinal activity. Finally, lateral inhibition at multiple levels of visual processing, including the retina, lateral geniculate nucleus (LGN), and visual cortex, may lead to interesting visual illusions such as the Hermann grid illusion (Fig. 4.6). We will discuss more about this type of illusion in Section 3.

2.2 Lateral Geniculate Nucleus

From the eye, retinal ganglion cells send their axons to a structure in the thalamus called the *LGN*. Specifically, the left half of each retina projects to the left LGN; the right half of the retina projects to the right LGN. For this to happen, the inputs from the nasal portion of each retina must cross at the *optic chiasm* to project to the opposite LGN (Fig. 4.7). The result is that

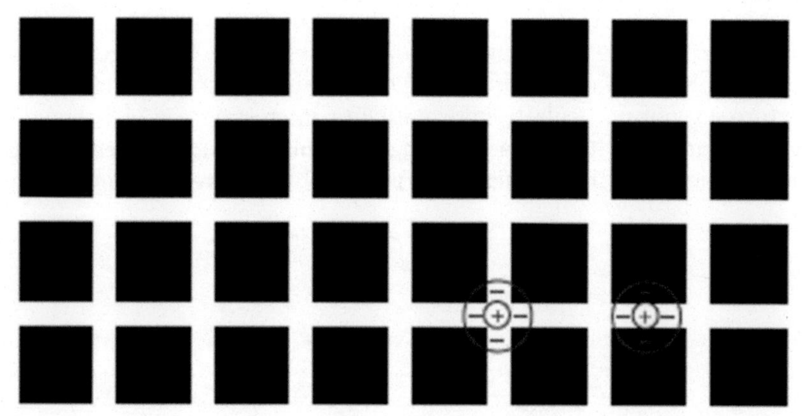

FIGURE 4.6 Hermann grid illusion. Take a careful look at the collection of black squares in the figure. Do you notice anything unusual? Do you have the impression of seeing small dark circles in between the black squares in the periphery? Do not be alarmed, this is completely normal. This is a great example of receptive fields with lateral inhibition at work (Herman, 1870). In the rightmost matrix of squares, some possible receptive fields are shown. A receptive field that falls between the corners of four dark squares will have more of its inhibitory surround stimulated by the white parts of the grid than a receptive field that lies between just two of the dark squares. As a result, neurons with receptive fields positioned between four dark squares will fire more weakly, leading to the impression of small dark patches at these cross points. At the fovea, receptive fields are much smaller so the illusion is only seen in the periphery. *Source: Frank Tong, with permission.*

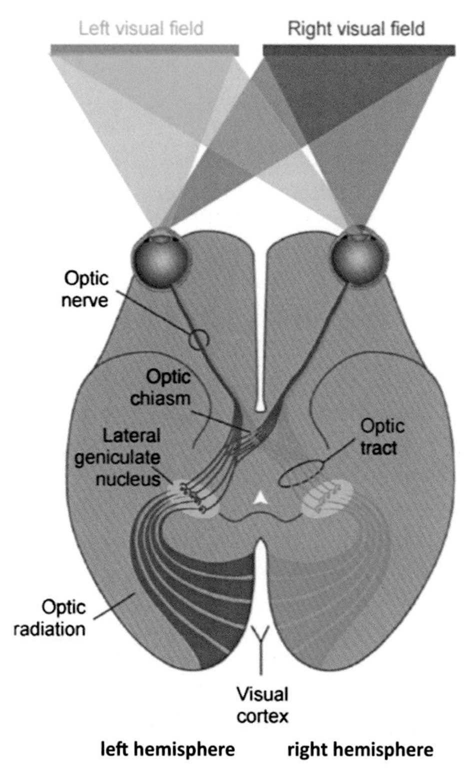

FIGURE 4.7 The visual pathways from retina to cortex. Schematic illustration showing the visual pathways from the retina in the eyes to the primary visual cortex at the back of the brain. You can see here that the neural information from the nasal or inner sides of the eyes crosses over at the optic chiasm, to be processed in the contralateral side of the brain. The left visual field, in blue, is processed by the right visual cortex (also blue). The LGN, displayed in green, relays the visual information to the primary visual areas of the cortex. *Source: Reid & Usrey in Squire et al., 2013.*

the left LGN receives input from the right visual field, and the right LGN receives input from the left visual field, so each LGN serves to represent the *contralateral* (i.e., opposite) visual field. Note that the inputs from each eye go to separate monocular layers of the LGN, so signals from the two eyes remain separate until they reach the *primary visual cortex,* where these signals are combined.

Receptive fields in the LGN share the same shape and basic properties of the retinal ganglion cells with center-surround receptive fields. The thalamus is often considered a way station for signals finally to reach the cerebral cortex, where the neurons start to respond in very different ways.

2.3 Primary—or Striate—Visual Cortex

From the LGN, neurons send their signals to the primary visual cortex, sometimes called *V1* because this region is the first cortical visual area. V1 is frequently referred to as "striate" cortex because of its distinguishing striped—or striate—appearance. About 90% of the outputs from the retina project first to the LGN and then onward to V1. The left LGN projects to V1 in the left hemisphere; the right LGN projects to right V1 (see Fig. 4.7). In V1, the spatial layout of inputs from the retina is still preserved. Left V1 receives an orderly set of inputs from the left half of both retinas via the thalamus. The foveal (central) part of the visual field is represented in the posterior part of the occipital lobe near the occipital pole, and the more peripheral parts of the visual field are represented more anteriorly. Left V1 therefore contains a *retinotopic map* of the entire right visual field, whereas right V1 contains a map of the entire left visual field. This retinotopic organization is very prevalent in early visual areas (V1 through V4), where neurons have small receptive fields, but it becomes weaker and less orderly in higher visual areas outside the occipital lobe.

Neurons in V1 are sensitive to a whole host of visual features not seen in the LGN. One of the most important visual features is *orientation* (Hubel & Wiesel, 1962; 1968). Some V1 neurons respond best to vertical lines, some to 20-degree tilted lines, and still others to horizontal lines and so forth. Fig. 4.8 shows an example of a model for V1 orientation selectivity. If a V1 neuron receives excitatory input from three LGN neurons with aligned center-surround receptive fields, then the V1 neuron will respond best to a matching oriented line. For example, if a vertical bar is presented, the neuron shown in the figure will respond at its strongest because the entire excitatory region will be stimulated, whereas the inhibitory surround will not be stimulated. If the bar is tilted somewhat away from vertical, the neuron will respond more weakly because part of the inhibitory surround will now be stimulated and part of the excitatory center will not. Finally, if a horizontal bar is presented, the neuron may not fire at all because equal proportions of the center and the surround regions will be receiving stimulation, leading to a balance in the strength of incoming excitatory and inhibitory inputs. This configuration of center-surround receptive fields can explain the orientation selectivity of V1 neurons.

V1 neurons are also sensitive to many other *visual features* besides orientation (Hubel & Wiesel, 1998). Some neurons respond best to a particular *direction of motion*, such as upward motion, leftward motion, or downward motion. Other neurons respond best to particular colors or color differences (e.g., red vs. green, yellow vs. blue), although some basic types of color-sensitive neurons can also be found in the retina and LGN. Finally, some neurons respond best to particular *binocular disparities* (Barlow et al., 1967; Cumming, 2002), which refer to the degree of alignment between images in the two eyes. Small displacements between images in the two eyes are what allow us to perceive stereodepth when we look at objects with both eyes open. (Close one eye, extend your arms to full length, and try to quickly bring your two index fingers to meet each other. Now try this again with both eyes open. If you have normal binocular vision, this should be much easier to do with both eyes open because you can better judge the distance of your two fingers.)

So how will V1 neurons respond to the outline of the house shown in Fig. 4.9? A V1 neuron that is tuned to 45-degree tilted lines and has its receptive field in the position along the roof may fire strongly to the angled roof. A V1 neuron that responds best to vertical will help signal the presence of the vertical wall, and a horizontal neuron will respond to the ceiling or floor of the house. In this sense, it can be readily seen that V1 neurons do much more than respond to simple spots of light, as the LGN does. V1 neurons provide a *neural representation* of the orientation of visual features that comprise the contours and shapes of objects. Fig. 4.9 provides a summary of the hierarchy of visual processing. From the LGN, V1, and V4 to the ventral temporal cortex, you can see that neurons gradually respond to more complex stimuli from one area to the next.

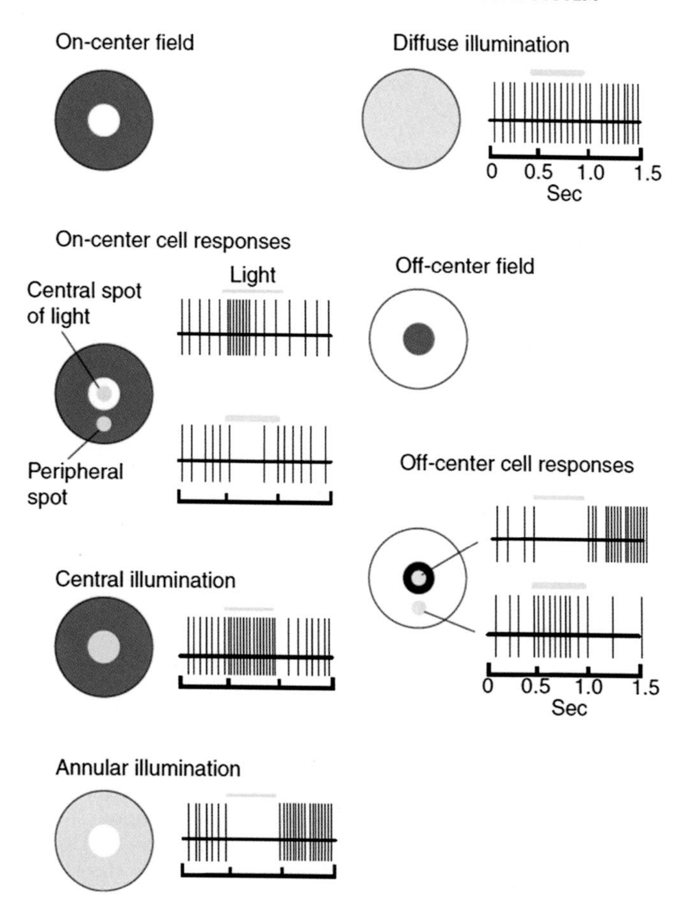

FIGURE 4.8 Orientation selectivity in V1. An example of how a collection of center-surround receptive fields could lead to orientation selectivity in V1 neurons. The overlapping circles on the left show center-surround receptive fields. When the bar of light lies vertically, it triggers all the on-centers (1) of each receptive field, whereas when its orientation changes (the *bar tilts*) fewer centers and more surrounds are activated resulting in a smaller neural response (fewer *spike bars* in the graph). Hence the vertical bar gives a larger neural response; when the stimulus is oriented, the magnitude of the V1 response is reduced. This constitutes orientation selectivity—only observed in the cortex. *Source: Reid and Usrey in Squire et al., 2013.*

To summarize, V1 is important for analyzing the visual features at a fine level of detail. These neurons have small receptive fields that are sensitive to orientation, color, motion, or binocular disparity. After visual signals are analyzed in V1, they are sent to higher visual areas for further processing.

2.3.1 *A Subcortical Route to Visual Cortex*

We have just described in detail the main pathway for information coming from the retina to the LGN to visual cortex. But, there is another pathway that we will introduce here and then discuss in more detail as we look at conscious and nonconscious perception later in this chapter. While about 90% of the outputs from the retina project to the LGN and onto V1, a small

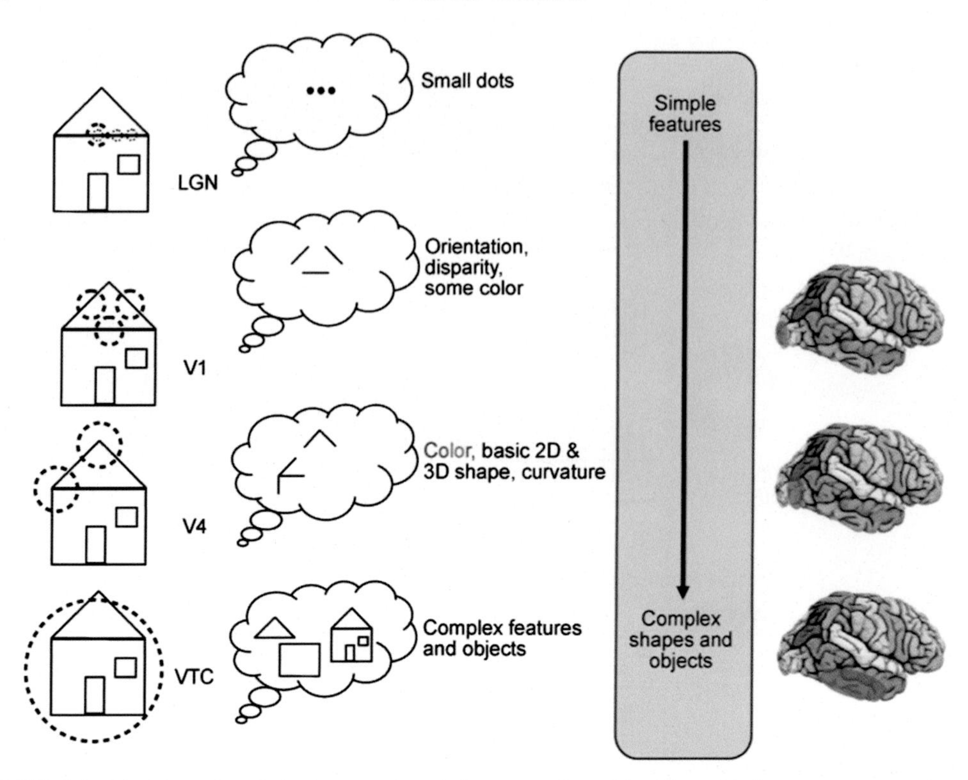

FIGURE 4.9 The hierarchy of visual processing. A demonstration of the hierarchical response properties of the visual system to simple and complex stimuli. The leftmost column shows our house stimulus and what receptive fields of each visual area we would see in the balloons. Not only do the receptive field sizes increase in each visual area but also the complexity of the shapes they respond to. The rightmost column shows an estimate of where each area is in the brain. You can see that early visual areas respond to simple features, and, as we move along the processing stream, areas respond to more complex shapes and objects. This is a well-established theme of the visual system. *Source: Frank Tong, with permission.*

percentage of information coming from the retina follows a different pathway to the cortex. This pathway, shown in Fig. 4.10, projects through recently discovered (2005–2006) middle layers of the LGN onto the subcortical nuclei called the superior colliculus (SC) and the pulvinar. This pathway continues to visual cortex; however, it does not project to area V1, instead, it projects to extrastriate visual cortex. This "old" pathway—from a brain evolutionary perspective—responds to certain types of visual stimuli and is connected to the subcortical regions that form part of the limbic region for emotional processing. Recent studies have linked this secondary pathway to the processing of emotional stimuli by patients who are otherwise blind (Tamietto & de Gelder, 2010). Curious? We will discuss this further later in the chapter.

2.4 Extrastriate Visual Areas—Outside V1

V1 sends *feedforward* signals to many higher visual areas, including areas such as V2, V3, V4, and *motion-sensitive area middle temporal* (*MT*), to name a few (Fig. 4.11) (Felleman & Van

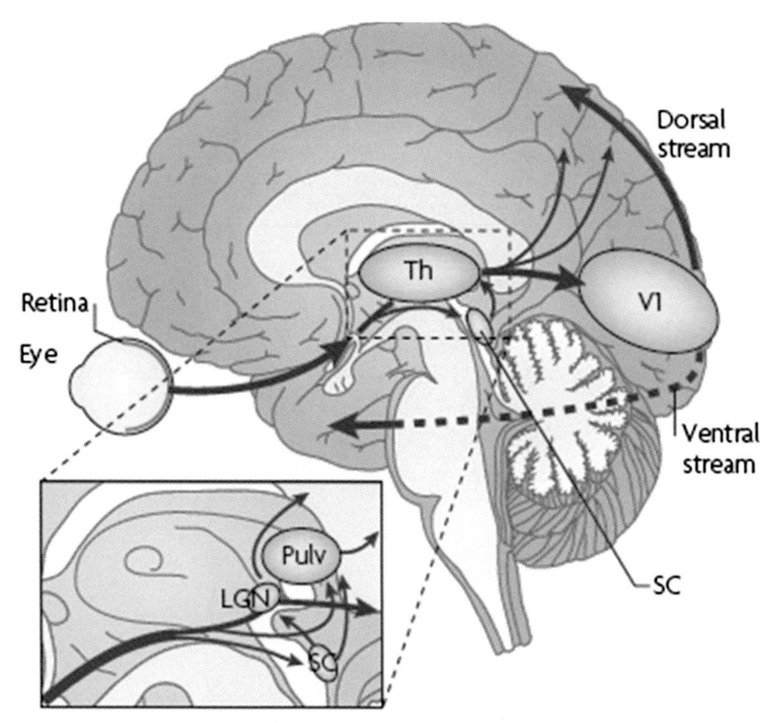

FIGURE 4.10 Cortical and subcortical pathways for vision. The primary visual pathway—from the retina to the lateral geniculate nucleus (LGN) in the thalamus (Th) and then onto V1 in the cortex—is shown by the thick black arrows. From V1, visual information reaches the extrastriate cortex through dorsal stream and ventral stream pathways. A small proportion of fibers from the retina projects through middle layers of the LGN and onto the superior colliculus (SC) and the pulvinar (Pulv). From these subcortical nuclei, projections then extend to extrastriate vision cortex, bypassing area V1. Another pathway extends from the LGN to the SC and then onto the dorsal stream in extrastriate visual cortex, again bypassing area V1 (shown in the thin arrows). Although these projections are from a small proportion of the total fibers that project from the retina, they nevertheless form a candidate visual pathway that may explain how visual information can reach the extrastriate cortex without the observer being aware of seeing that information due to the bypassing of area V1. *Source: Tamietto and de Gelder (2010).*

Essen, 1991). *Area V4* is known to be especially important for the *perception of color* (Zeki, 1977), and some neurons in this area respond well to more complex features or combinations of features (Pasupathy & Connor, 2002). For example, some V4 neurons are sensitive to curvature or to two lines that meet at a specific angle. These neurons might signal the presence of a curving contour or a corner. From our example of the house, a V4 neuron might respond best to the meeting of the two lines forming the point of the roof or to another corner of the house.

How, then, are these various bits and parts of the house, as represented by simple line orientations and corners, eventually represented as an entire object? Area V4 sends many outputs to higher visual areas in the *ventral visual pathway*, which is important for object recognition (Ungerleider & Mishkin, 1982). The anterior part of the ventral visual pathway consists of the ventral temporal cortex, which is especially important for object recognition.

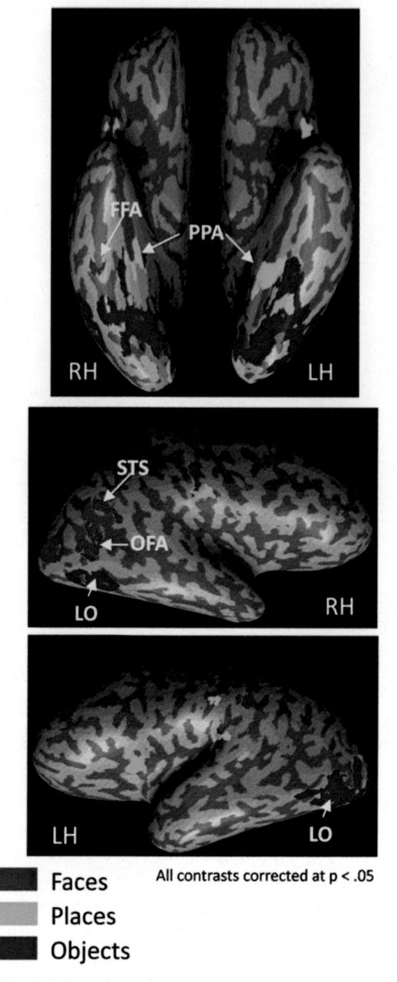

FIGURE 4.11 Visual areas of the brain. Functionally defined visual areas of the human brain, shown on a reconstruction from anatomical brain scans. Fusiform face area (FFA, shown in red), parahippocampal place area (PPA, shown in green), and the lateral occipital cortex (LOC, shown in blue). *Source: Modified from Scherf, Luna, Minshew, and Behrmann (2010).*

2.5 Area Middle Temporal

The middle-temporal area, or what is commonly called area MT (see Fig. 4.11), is important for motion perception. Almost all of the neurons in area MT are direction selective, meaning that they respond selectively to a certain range of motion directions and do not respond to directions beyond that range (Albright, 1984; Zeki, 1974). Moreover, some of these neurons respond well to patterns of motion (Albright, 1992), meaning that these neurons can integrate many different motion directions and calculate what the overall direction of an object might be. As we will see, the activity in this region seems to be closely related to motion perception and, when activity in this region is disrupted, motion perception may be severely impaired.

2.6 Connectivity Patterns in Visual Cortex: Feedforward, Feedback, and Parallel Flows

There are three types of connections in the visual system: *feedforward* flows of information coming from lower (earlier) visual areas (such as from V1 to V2), *feedback* flows from higher visual areas to lower ones (such as from V4 to V2), and *parallel* connections across visual regions. These parallel flows are sometimes referred to as *horizontal* flows because they exchange information across visual regions at the same level. These connections are shown in Fig. 4.12A: where feedforward connections are shown in pink, feedback connections are shown in blue, and parallel or horizontal connections are shown in yellow.

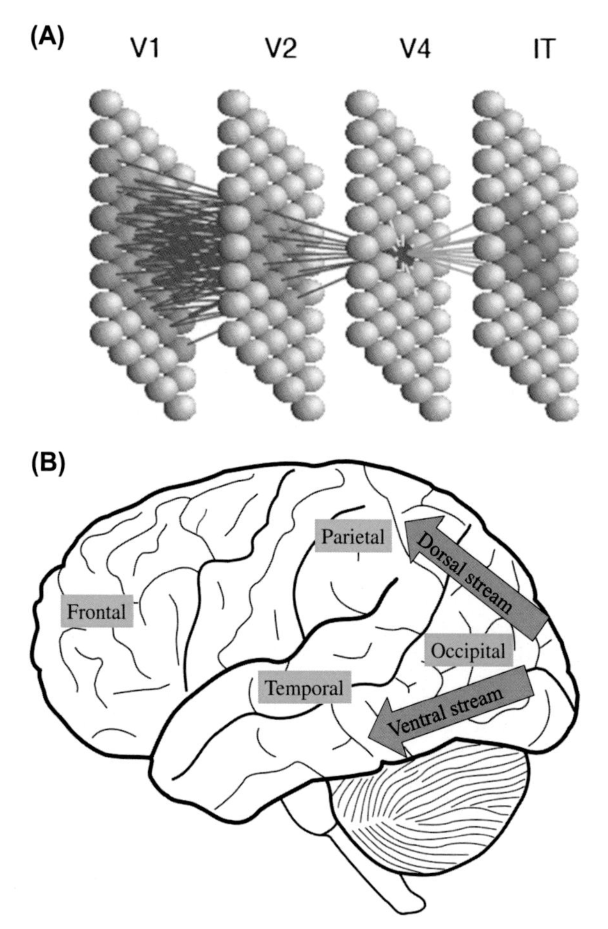

FIGURE 4.12 Anatomical organization of the visual system. (A) An example of how feedforward (shown in pink), feedback (shown in blue), and parallel (shown in yellow) flows of information are theorized to be established in the visual cortex. Note the preponderance of *feedforward flows* in area V1 and V2, the focus of *parallel flows* within area V4, and the large amount of *feedback* flowing from inferior temporal (IT). (B) Beyond extrastriate cortex: visual information flows in streams along two major pathways in the cortex. The dorsal stream projects from visual areas toward the parietal lobe and is described as a "where" stream, providing spatial information about seen objects and scenes. The dorsal stream includes dorsal cortical areas such as middle temporal and parietal cortex. The ventral stream projects from visual areas toward the temporal lobe and is described as a "what" stream, providing object identity information. The ventral stream incudes areas such as V4, lateral occipital complex, and IT. *Source: (A) Lamme and Roelfsema (2000); (B) Gage, with permission.*

2.7 The Ventral and Dorsal Pathways: Knowing What and Where

The projections from V1 to higher areas in the cortex can be roughly divided according to two major parallel pathways: a *ventral pathway* leading from V1 to the *temporal lobe* that is important for representing "*what*" objects are and a *dorsal pathway* leading from V1 to the *parietal lobe* that is important for representing "*where*" objects are located (Fig. 4.12B). This distinction between the ventral and dorsal pathways, sometimes referred to as the *what* and *where* pathways, respectively, is an important organizational principle of the visual system proposed by Ungerleider and Mishkin (1982).

In the ventral pathway, many signals from V1 travel to ventral extrastriate areas V2, V3, and V4 and onward to many areas of the temporal lobe. The ventral or "what" pathway is important for processing information about the color, shape, and identity of visual objects, processing which emphasizes the stable, *invariant* properties of objects. For example, the ventral pathway is less concerned about the exact size, orientation, and position of, say, a coffee mug sitting on a table; instead, its goal is to be able to identify such an object anywhere in the visual field and to be able to tell it apart from other similar objects (e.g., cups, bowls, teapots).

In the dorsal pathway, signals from V1 travel to dorsal extrastriate areas, such as area MT and V3A, which then send major projections to many regions of the parietal lobe. The dorsal pathway is important for representing the *locations* of objects so the visual system can guide actions toward those objects (Goodale & Humphrey, 1998). Consider what is involved in reaching for any object, such as that coffee mug; this type of vision requires detailed information about the precise location, size, and orientation of the object. Without such detailed information, you might reach toward the wrong location, your grasp might not match the size of the handle of the mug, or your hand might not be angled properly for gripping the handle. Areas MT and V3A are important for processing visual motion and stereodepth, while specific regions in the parietal lobe are specialized for guiding eye movements or hand movements to specific locations in visual space.

While this dorsal–ventral pathway distinction is useful for grouping areas of the brain and understanding how much of the information flows back and forth between visual areas, it should not be taken as an absolute distinction. There is plenty of cross talk between the two pathways. Also, the parietal and temporal lobes send projections to some common regions in the *prefrontal cortex*, where information from each pathway can also be reunited.

2.8 Areas Involved in Object Recognition

Human neuroimaging studies have revealed many brain areas involved in processing objects. These object-sensitive areas, which lie just anterior to early visual areas V1–V4, respond more strongly to coherent shapes and objects, as compared with scrambled, meaningless stimuli. In this section, we will focus on three such brain areas (see Fig. 4.11). The *lateral occipital complex* (LOC) lies on the lateral surface of the occipital lobe, just posterior to area MT. Because this region is strongly involved in object recognition, we will consider it as part of the ventral pathway, even though its position is quite dorsal when compared with other object areas. The *fusiform face area (FFA)* lies on the fusiform gyrus, on the ventral surface of the posterior temporal lobe. The *parahippocampal place area* (*PPA*) lies on the parahippocampal gyrus, which lies just medial to the fusiform gyrus on the ventral surface of the temporal lobe.

2.8.1 The Lateral Occipital Complex

The LOC seems to have a general role in object recognition and responds strongly to a variety of shapes and objects (Malach et al., 1995). Fig. 4.13 shows an example of the neural activity in LOC compared with V1. As the picture becomes more and more scrambled, V1 continues to respond, and even gives a larger response, whereas activity in LOC declines. This shows that LOC prefers intact shapes and objects more than scrambled visual features.

2.8.2 The Fusiform Face Area

Human neuroimaging studies have shown that a region in the fusiform gyrus, called the FFA, responds more strongly to faces than to just about any other category of objects (Kanwisher et al., 1997). This region responds more to human, animal, and cartoon faces than to a variety of nonface stimuli, including hands, bodies, eyes shown alone, back views of heads, flowers, buildings, and inanimate objects (Kanwisher et al., 1997; McCarthy et al., 1997; Schwarzlose et al., 2005; Tong et al., 2000). In a recent study, researchers tried scanning the brains of monkeys to see if they might also have a face-selective area in the ventral temporal cortex, and it turns out that they do too (Tsao et al., 2006). The researchers then recorded the activity of single neurons in this face area and discovered that 97% of the neurons in this region responded more to faces than to other kinds of objects. Moreover, these neurons were very good at telling apart different identities of faces but poor at telling apart different identities of objects, suggesting they may have an important role in recognizing and telling apart different faces. As we will see, this region seems to be important for the conscious perception of faces (Fig. 4.11B–D).

2.8.3 The Parahippocampal Place Area

The PPA is another strongly category-selective region that responds best to houses, landmarks, and indoor and outdoor scenes (Epstein & Kanwisher, 1998). In comparison, this brain area responds more weakly to other types of stimuli, such as faces, bodies, or inanimate objects. Because this region responds to very different stimuli than the FFA, many studies have taken advantage of the different response properties of the FFA and PPA to study the neural correlates of visual awareness, as we will see later.

3. CONSTRUCTIVE PERCEPTION

We have discussed that human vision is a dynamic, interactive process with neural pathways that extend from the eyes through subcortical pathways to vision cortex at the back of the brain. Along the way, there are many feedforward, feedback, and parallel pathways that

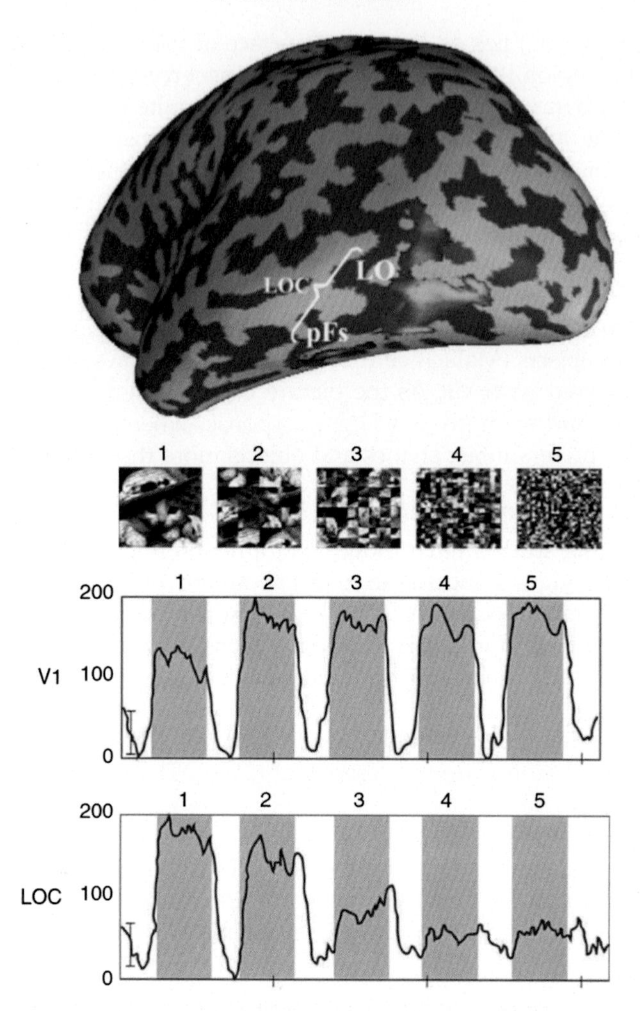

FIGURE 4.13 Neural response from low- and high-level areas. The response of primary visual cortex (V1) and lateral occipital complex (LOC) to a picture of a kitten at different coherencies. As the picture is scrambled, V1 continues to respond—in fact, it actually increases its response once the image is scrambled. This demonstrates that it is not the image of the kitten, which is driving responses in V1 but local patches of luminance and orientation. Conversely, the activity in lateral occipital cortex shows a large response to the kitten, but, as the picture is scrambled, the activity in LOC drops down dramatically. This demonstrates that unlike V1, the activity in LOC is in response to the kitten. *Source: Bell et al., in Squire et al., 2013.*

are connecting at lightning speed as you look at the world around you. And, not just the visual system is at play here—your memories, your knowledge, your experience, your emotions all play a role in just what you are perceiving at any given moment.

A fascinating aspect of vision is just how much we perceive that is not actually in the visual stimulus we are gazing at. And, conversely, how much we miss of what is right in front of our eyes. Let us break this aspect of visual perception down into some separate phenomena, which, together, form our visual experience.

Our world is a busy place, with even simple visual scenes that have complexity to them: some items in the foreground are blocking items farther away, some items are in bright sunshine and some are in shade, and some items are far away and look small while others are close by and look large. However, despite these various aspects of the visual landscape, we need to be able to navigate through it with some consistency. Here are some of the properties of the visual system that aid us in this task.

3.1 Perceptual Filling-In: The Blind Spot

If you drive a car, then you probably know what a blind spot is: for the driver, it is the area next to or behind the car that cannot be seen in the side- or rear-view mirrors. Our eyes also have a *blind spot* at the back of the retina where the axons of the retinal ganglion cells meet to form the optic nerve as it exits the eye (see Fig. 4.3B). There are no photoreceptors in the blind spot, so we are blind in that area.

It is easy to demonstrate the blind spot. Look at the diagram in Fig. 4.14A. Close your left eye, look directly at the cross with your right eye, and move the textbook close to your nose. Then move it slowly away from your face, while keeping your eye fixed on the cross. At the right distance, which should be around 12 in. (30 cm) away from the page, you should notice the dot vanish. As the image of the dot on your retina moves into the blind spot, it disappears!

Hopefully this demonstration worked on you, and perhaps you are thinking, "Wait a minute! If there is a blind spot in my vision all the time, then why don't I see a hole in my vision when I look around at things with one eye covered?" In this case, the brain does something remarkable: it actually *fills in* the perception of the blind spot. The brain uses the visual information from around the blind spot to infer what should be in the blind spot, and it constructs awareness of what it "thinks" should be there. Filling in at the blind spot is an example of *constructive perception* or *perceptual filling in.*

Another way to demonstrate the filling in can be seen by viewing Fig. 4.14B. Move the page around until the gap between the lines is in the blind spot. When each group of lines is abutting the blind spot, what you will see are continuous long lines. The red dot should be replaced by one continuous set of lines. The brain fills in the path of the lines so you do not see the red dot in the gap anymore. This is a case of perceptual filling in; the brain actively creates the perception of a complete object from the separate segments that lie on either side of the blind spot.

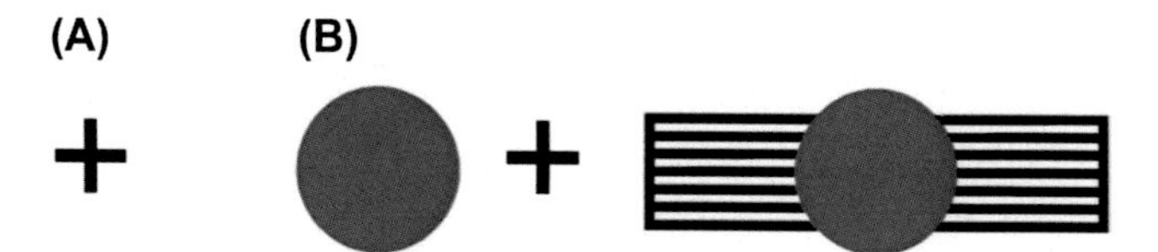

FIGURE 4.14 Demonstrations of the blind spot. (A) Close your left eye, look directly at the cross with your right eye, move the page up close to your nose, and then move it slowly away from your face, while keeping your eye fixed on the cross. At the right distance, which should be around 12 in. (30 cm) away from the page, you should notice the red dot vanish. As the image of the dot on your retina moves into the blind spot, which has no photoreceptors, it disappears! (B) Likewise, notice how the black stripes now fill in; they become joined, and the red dot vanishes. *Source: Frank Tong, with permission.*

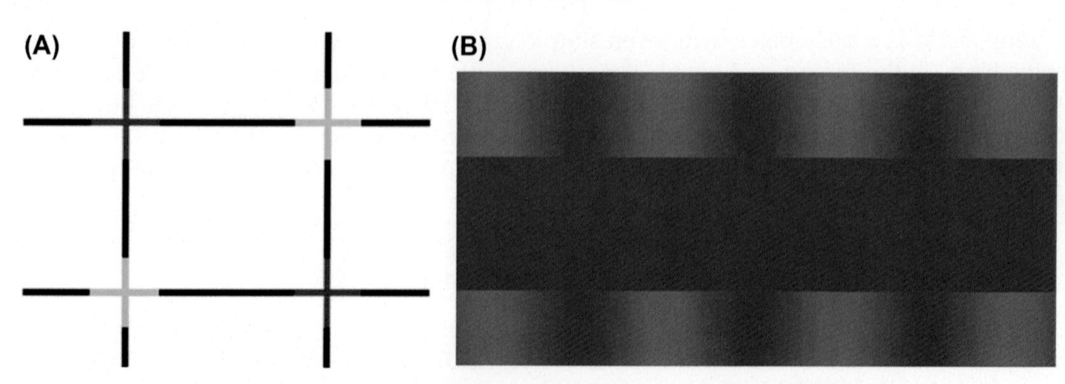

FIGURE 4.15 Demonstrations of perceptual filling in. (A) In these two examples the background is white—it has no color. However, you might notice that the red and the blue tend to fill in, coloring the white background. (B) The light patches and dark patches in the top and bottom panels tend to give the impression of light and dark sectors along the center strip, even though the center strip is a uniform gray. This illusion works much better when moving. *Source: Frank Tong, with permission.*

Perceptual filling in not only happens in the blind spot, but it also occurs in other parts of the visual field. Notice how the area between the colored lines in Fig. 4.15A somehow appears colored. However, careful inspection reveals that this background area is actually just white—no color at all. This illusion is called *neon color spreading,* and the experience of color here seems to come from constructive filling-in mechanisms at work in the visual cortex. Recent brain imaging studies have found that activity in V1 is greater in corresponding regions where subjects perceive neon color spreading (Sasaki & Watanabe, 2004). This suggests that neurons as early as V1 may be important for perceptual filling in and our experience of color, even illusory colors.

During another type of filling in known as *visual phantoms*, our visual system can also fill in gaps between two patterns. In Fig. 4.15B, you may have the impression of dark bands continuing across the blank gap, even though there is no stimulus being presented there at all. Using moving stimuli can further enhance the visual phantom illusion. When we experience this type of filling in, cortical areas V1 and V2 respond as if a real pattern were being presented in the gap (Meng et al., 2005).

3.2 Apparent Motion

The brain can fill in not only color and patterns but also motion. If you flash a spot of light in one location, then follow it by a similar flash at a different location at the appropriate time, observers experience the illusion of movement—*apparent motion*. This kind of trick is used all the time in overhead shop signs, and it is the basis for why movies look so smooth and real. If you look at the actual film running through the projector, you will find that there is a series of stills that are flashed onto the screen rapidly, one after another: the experience of smooth motion is simply an illusion. Studies have found that the perceived path of apparent motion is represented by neural activity in V1 (Jancke et al., 2004; Muckli et al., 2005).

3.3 Illusory Contours

Your visual perception system is continually trying to make sense of the complex visual scenes that are presented to it. You can imagine that your visual brain has a lot of experience in decoding the visual scene—it has been at it since you were born. The concept of *illusory contours* (ICs) is an example of that learning. ICs are images that you perceive in a way that is *not actually present* in the image itself. See Fig. 4.16 for an example—pictures are better than words here!

Look at the three black partial circles in Fig. 4.16: most people see a white triangle occluding three black circles. And most people also "see" a white edge to the triangle. However, there is no line present in this image, no actual triangle. Your brain has interpreted this image as representing a two-layer image of a triangle on top of circles. This is a simple example of a rather complex process: the visual perceptual system decodes the physical (sensory) stimulus that is projected to the retina and comes up with its own "answer" as to what is being seen—it *constructs* the triangle over circles. Several models have been presented to describe the brain mechanisms that subserve ICs: a central view is that IC perception involves feedforward, feedback, and parallel connections in decoding the visual scene (Murray & Herrmann, 2013). Thus, your experience with seeing actual triangles that are in front of circles—or something similar—is brought to bear on your visual interpretation of figures such as the one in Fig. 4.16 with the help of feedback and parallel processing flows.

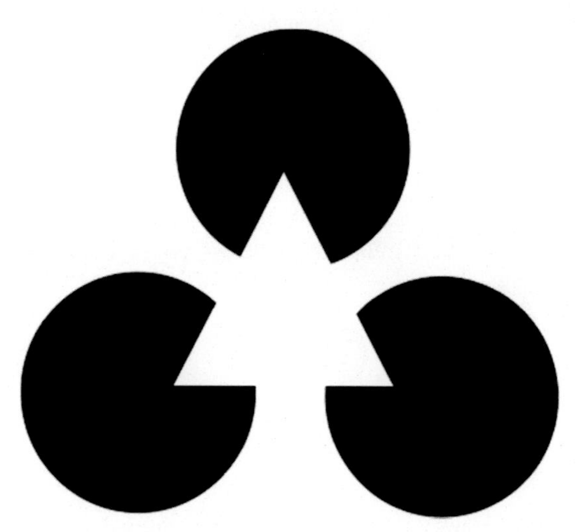

FIGURE 4.16 Illusory contours. An example of a figure, which induces illusory contours for most observers. Based on a lot of experience living and seeing in a 3-D world, when most observers look at this figure, they "see" a white triangle in front of—or occluding—three black circles. There is no actual triangle in the figure at all—merely three partial circles, yet we see it as a triangle and even see the "white lines" of the outside edges of the triangle, despite the fact that they are not there. This illusion goes away, by the way, if the proportion and relative sizes of the partial circles are changed. *Source: Gage with permission.*

3.4 Gestalt Principles

The phenomenon of ICs in constructive perception relates to the Gestalt principles of visual perception. The *Gestalt* psychologists proposed that perception could not be understood by simply studying the basic elements of perception (Wertheimer, 1912; Koffka, 1935). The German word *Gestalt* is difficult to translate directly, but it expresses the idea that *the whole is greater than the sum of the parts*. These psychologists proposed the *Gestalt laws of perceptual grouping*, such as the laws of *similarity, proximity, good continuation, common fate*, and so forth (Fig. 4.17). These laws suggest that elements that are more similar in color or shape are more likely to be perceived as a group. Likewise, if a set of elements is arranged in a way that they are more closely spaced in rows or columns, it will determine whether they are perceived as rows or columns.

Why is perceptual grouping so important? It helps us to perceive which features belong to a possible object, and it helps us to distinguish an object from the background. For example, imagine seeing a golden retriever lying in tall grass. Grouping by similarity may help us see that the golden dog is separate from the green grass, and the neighboring golden wagging tail might be recognized as part of the dog, even though only the end of the tail can be seen. These perceptual grouping processes aid in object recognition—which we will discuss further in Section 4.

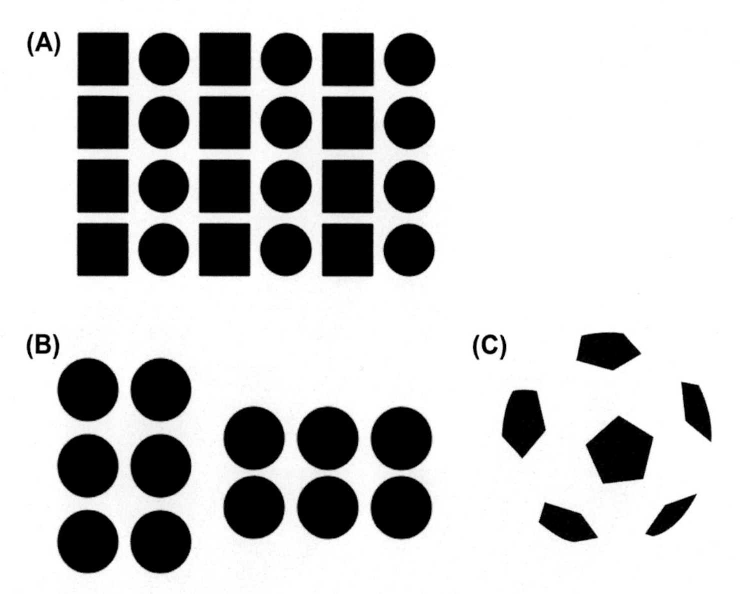

FIGURE 4.17 Gestalt principles or laws. (A) Law of similarity: The square dots are typically seen as grouped with the other square dots because of their similar shape. The same is true of the round dots; they are also seen as forming a group due to their similarity. (B) Law of proximity: Here, even though all the objects are circles, they are perceived as two groups due to their relative spatial location—or proximity—to each other. (C) Law of closure: This is similar to an illusory contour. When an object is incomplete or a space is not completely enclosed, we nevertheless "see" the object as complete if enough of the shape is present for us to "close" the separate shapes to form one well-known object, such as this soccer ball. *Source: Gage with permission.*

3.5 2-D to 3-D Mapping

Early mankind—think cavemen here—experienced their world in three-dimensions (3-D). Modern man and woman experience many visual scenes and objects in two-dimensional (2-D) form: books, photographs, web pages, newspapers, movies, smart phones, games, navigation devices, art, etc., you name it, and we can find a 2-D version of our 3-D world. How does our visual system construct a 3-D view of a 2-D image?

First, let us explore how our visual system determines the details of a 3-D scene. We live in a 3-D world. We were born into this world, we have had all of our experience in this world, and our visual system has learned to perceive the world as 3-D. There are many cues in an image or scene that let the visual system help to construct a 3-D "picture": *relative size* cues inform us if an object is close or far away, vertical lines that converge in the distance provides *perspective*, which give a sense of depth or distance, and the partial *occlusion* of an object by another object gives the sense that it is in the background. Relative sizes, perspective, and occlusion are some of the cues that form the bases for depth perception that provides us with a 3-D view of the world. Other key cues are *lighting, shading,* and *texture gradients* across an object or scene.

Most of us use two eyes when we view the world—the distance between the eyes (*binocular disparity*) and the slightly differing view they form aid us in 3-D perception. Binocular disparity, or *stereopsis,* helps our visual system use the slightly disparate views from each eye to triangulate the image, providing important depth perception cues. However, we use depth perception cues such as relative size and occlusion, even when we cover one eye and view the world with just one eye—*monocular* vision. A key aspect of the depth perception we sense using monocular vision is our lifetime of experience viewing the world with binocular vision. This experience aids us in detecting more subtle depth clues when using just one eye.

This vast experience using visual cues about depth in a scene is used when we view any 2-D scene or object: your visual brain will "see" the 2-D scene in 3-D. In fact, it is very difficult *not to see* 2-D scenes in anything but 3-D! Take a look at the simple objects shown in Fig. 4.18. It is virtually impossible for you not to interpret these figures as 3-D, with the circles on the left seen as convex and the ones on the right as concave. Why do we see them this way? The

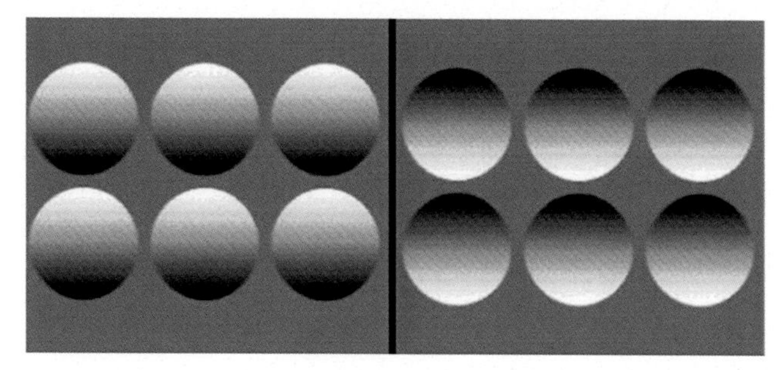

FIGURE 4.18 An example of 3-D cues at work when perceiving a 2-D object: Despite the fact that this figure is shown to you in 2-D, you will likely "see" it in 3-D. The cues at work here to guide your visual system to perceive the circles as being convex (left side of figure) or concave (right side of figure) come from the texture and gradient shading. These cues simulate the actual convex and concave circles that we have experienced seeing in the natural world, aiding us in seeing them as 3-D in this 2-D figure. *Source: Public Domain.*

shadows and play of light follow the "rules" of how they would play out on actual objects that had convex or concave aspects to them. And thus, that is how your visual system interprets these 2-D images.

3.6 Color Perception

On a bright and sunny day, as you glance around you, your strong impression is that the entire visual scene is shown in vibrant color. Not so. Only a small portion (~1/20th) of the photoreceptor cells—the cones—encodes color information (see Fig. 4.3). How is it that we have the sense that the entire scene is in color?

The eyes make rapid voluntary or involuntary movements—*saccades*—back and forth across the visual scene. The involuntary saccades include some momentary eye fixations, but in general the eyes are in constant motion (Fig. 4.19). The cones in the fovea are encoding fine-grained information in the visual scene—and also detecting color—as the eyes saccade rapidly across the visual field. Since the eyes are continually doing the rapid involuntary saccadic movements, the entire visual field is encoded by the cones in the fovea of the retina. For this reason, the color information across the entire scene is encoded. Result? We *see* the entire visual scene in color.

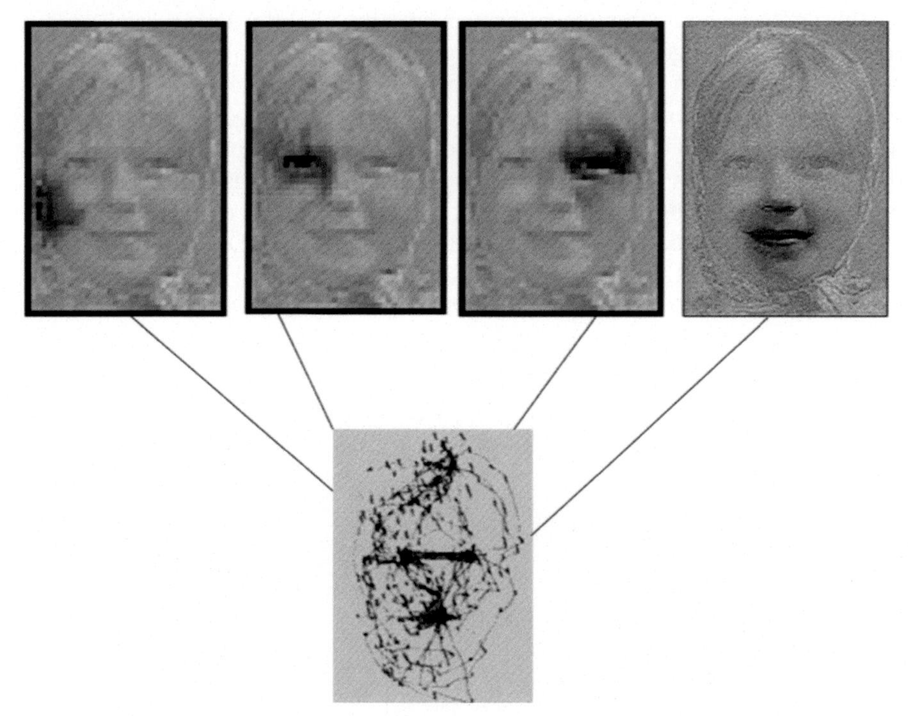

FIGURE 4.19 Color perception across the entire visual field. Although only the central—or foveal—part of the retina contains cones that encode color and fine detail in the visual field, we nevertheless have the perception that the entire field of our view is in color. This occurs, at least partly, because our eyes are always in motion using a combination of voluntary and involuntary eye movements across the visual field: saccades. The information obtained from these near-constant eye movements combines to form a percept of a fully colored scene. *Source: Modified from M. Dubin with permission.*

3.7 Color Constancy

Color constancy is when the perceived color of an item or object remains constant even though the amount of illumination differs starkly. Take, for example, the red T-shirt your friend is wearing to the beach. The perceived color of red that you "see" when you look at your friend will remain constant despite the physical change in its illumination that occurs when he is (1) in his apartment when you knock on the door; (2) in your car on the way to the beach; (3) at the beach in the bright sunlight; and (4) in a restaurant when you are having lunch. The actual physical properties of that T-shirt—as a complex spectrum of light reflects on it in brightly lit, partially lit, and electrically lit areas—do not change, but the reflections emanating from that T-shirt and that your eyes are encoding differ vastly. Here is a simple example: take a look at the squares labeled A and B shown in Fig. 4.20A. Do they look like they are the same color? Most people see square A as much darker than square B, but in fact they are exactly the same color. This is due to color constancy as they lie out of or in the shadow formed by the green cylinder. Not convinced? Look at Fig. 4.20B. We have added a gray rectangle that is the same color as squares A and B.

How Does This Occur?

Almost everything that we see in our world is visible to us due to the surface reflections of that object from the source of light: the sun, an electrical light, starlight, etc. The *changing*

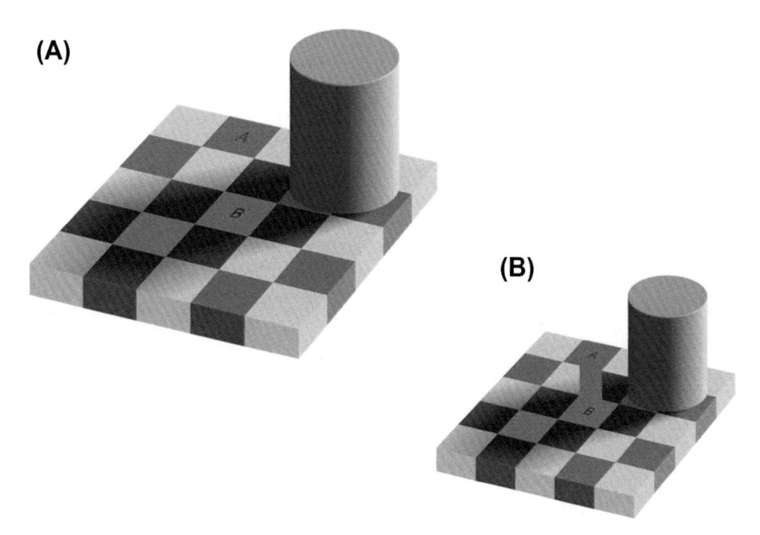

FIGURE 4.20 Color constancy. Color constancy is when the perceived color of an object remains constant even though the amount of illumination differs starkly. Illusions shed light on how much difference in the actual color of an object can be perceived as the same or different depending on the perceived relative illumination in objects nearby. (A) In this figure, squares A and B appear to be very different shades of gray. This illusion is aided by the perceived shadow extending from the green cylinder. Our visual experience tells us to take into account this shadow. However, squares A and B are *exactly* the same shade of gray. (B) To prove it to you, we added a patch of the same color of gray, extending between squares A and B. *Source: File created by Adrian Pingstone, based on the original created by Edward H. Adelson.*

illumination that occurs across viewing conditions—as the source of light changes or the intensity of the light varies—poses challenge for our color perception. The variation in the illumination is thought to be a key factor in how neural mechanisms create a sense of color constancy: neural mechanisms have been proposed that include some that adapt to these changes (respond invariantly to them) and others that are responsive to the many other cues to color in the visual scene (see the discussion in Section 2.1 on lateral inhibition).

Although we are all quite familiar with the concept of color constancy once it is described to us, many aspects of color constancy have made it difficult to investigate experimentally (Foster, 2011). First, it is difficult to really know exactly what color one observer is "*seeing*" when they report that a T-shirt, for example, is red. These color judgments are highly subjective across individuals and difficult to control experimentally. Second, just what kind of experiment is reasonable for investigating color constancy? Should they use "real" stimuli—individuals observing objects in a natural setting—or artificial laboratory-based settings? Third, how can the neural mechanisms that subserve color constancy be studied? Early single-cell recordings of neurons in area V4 in monkey (Zeki, 1983) implicated that area as the brain region for color constancy. However, since those early studies, it has become apparent that color constancy is complex and is not locatable within a single focal region within the visual system. At least four areas within the visual system have been proposed as forming the neural basis for color constancy: the retina (specifically, the cones), LGN, some regions within V1, and area V4. The existence of multiple visual regions forming the neural bases for color constancy and the difficulties in investigating color constancy experimentally, detailed earlier, combine to make the investigation of the neural mechanisms that underlie color constancy a work in progress.

3.8 Crowding

Have you ever entered a store that was full of colorful items covering every available space on every wall, shelf, and fixture? You may have had the impression that you "could not see the forest for the trees." *Crowding* is the difficulty in recognizing individual objects in clusters of other objects or natural scenes. Note that we stated "recognizing"—despite being in a crowded part of your visual field, individual features of that crowded scene are detected. The breakdown comes when they are to be identified as individual objects. More on that in a minute. For now, let us describe how crowding begins to happen in the retina.

Do you recall that we discussed the fact that although our impression is that we see color information across our entire visual field view, we actually only encode color information with the cones that are in the fovea of the retina (see Section 3.6)? It is the almost constant saccades of the eyes that provide us with the color details across the visual scene; the rods in the retina periphery do not encode color or the fine-grained details in the visual scene. The same concept is true for our sense of seeing the entire visual scene in *high resolution*; again, this is due to the saccades that allow the entire scene to come under the encoding mechanisms of the cones in the fovea. When we try to recognize objects in the periphery—despite the advantages of the saccades on providing higher resolution of the visual scene—we have difficulty in recognizing objects when there are several of them and if they are overlapping. This is *crowding*. Look at Fig. 4.21. Focus on the bull's eye in the upper center of the scene. It

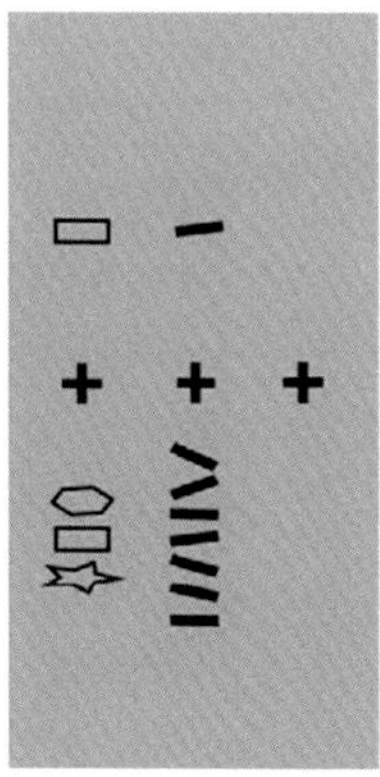

FIGURE 4.21 Crowding. Crowding is the difficulty in recognizing individual objects when they are clustered in the periphery of a scene. Focus on the bull's eye in the top center of the figure. You may find that it is difficult to recognize the child on the left as an individual object due to the clustering of the signs nearby. Alternatively, you may find that it is far easier to recognize the child on the right due to the lack of nearby objects. It is important to note that crowding occurs in the peripheral regions of a scene, which is why we test it when you are focusing on the center of this scene. It is also important to note that it is *object recognition* that is affected by clustering in the periphery, not *object detection*. Another way to state that is that the step of *detecting an object's or objects' features* and combining them to form a recognizable object that is impaired due to clustering. *Source: Whitney and Levi (2011).*

is difficult to recognize the small child on the left side of the road because of the signs that are close by; this is crowding (Whitney & Levi, 2011). It is relatively easy to recognize the small child on the right side of the road because there are far fewer objects there.

The neural mechanisms that underlie the phenomenon of crowding have yet to be uncovered; however, they are thought to be in the visual cortex as opposed to the retina or LGN—although each of these stages of visual processing is a key to ultimate successful object recognition. A general view is that when a visual object is undergoing an object recognition process in the visual system, individual features of that object are combined to form the recognition of the object. The phenomenon of crowding occurs, it is theorized, when the *combination of features' step* of processing fails due to the close proximity of other objects; importantly, this occurs when the objects are in the periphery of vision, not in central or foveal vision (Pelli, Palomares & Majaj, 2004). Thus it is *the clustering of objects within a given space* that produces the phenomenon of crowding, not the actual size of the objects.

3.9 Multistable Perception

Have you ever been in the dark, perhaps lying in bed, staring at some strange shape across the room? At one moment it might look like a person, then like some strange static animal, then like a marble statue. After racking your brain to identify this mysterious object, you finally hit the light switch, and, lo and behold, it is only the dim light catching your jacket on the chair. In this example, when vision was difficult and *ambiguous*, perception did something creative and useful: it faltered or alternated between different interpretations of what you were seeing, providing you with each potential "solution" of just what you were trying to see in the dim room. This is an example of *multistable perception*: The jacket on the chair (the

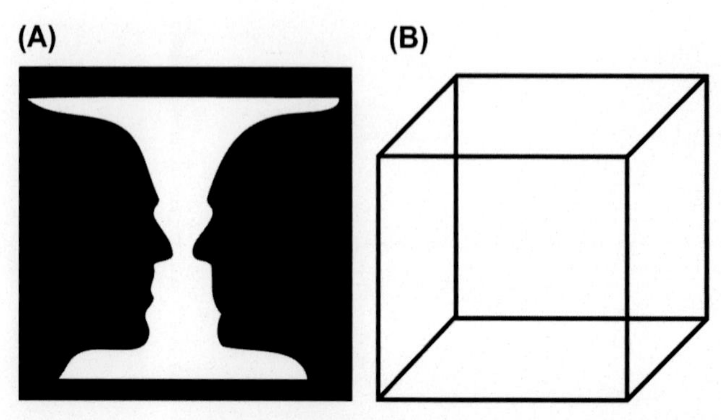

FIGURE 4.22 Bistable figures. (A) After looking at this figure for a while you will notice that there are two interpretations. One is a central vase; the second one silhouettes of two faces looking in at each other. This image is bistable; while you look at it, your perception will alternate between the vase and the faces. (B) This wireframe cube, typically known as the Necker cube, has two equally likely spatial interpretations. Perception tends to alternate between the configuration of the closest side projecting upward and the closest side projecting downward. Like the vase and silhouettes aforementioned, this is bistable. This bistability allows a dissociation of low level stimulation and awareness. The physical pattern does not change, but your awareness of it does! *Source: Public Domain.*

physical stimulus) did not change, but your perception of it did! The phenomenon of multistable perception is a valuable tool since it enables scientists to study changes in visual awareness independent of any changes in the visual stimulus.

There are many examples of *multistable patterns* or ambiguous figures that scientists can use to investigate these neural correlates of consciousness. Patterns that primarily have only two primary interpretations are called *bistable patterns* (Fig. 4.22). Try to see if you can perceive both interpretations of each ambiguous figure, the face and the vase and the two views of the Necker cube. Bistable perception can still occur if you keep your eyes fixed on a single point in the middle of the stimulus. In this case, the image on your retinas—and thus the stimulus-driven activity—is pretty much constant over time, while your perception fluctuates. By looking for brain areas that show activity changes correlated with perceptual changes, scientists can identify the neural correlates of consciousness. We will discuss this topic in more depth later in this chapter!

Another way to separate physical stimulation and perceptual awareness is to get people to do a *visual detection* task. Here, a subject has to detect and say when he or she sees a particular pattern. The researcher makes the pattern harder and harder to see, so on different instances the pattern might be easy to see, whereas at other times it is almost impossible, and sometimes it will not even be there. Because this task gets difficult, people will get it wrong sometimes. There will be occasions when someone reports seeing the pattern when it was not even presented and other times when he or she will miss the pattern and reports seeing nothing. When this kind of experiment is done in the functional magnetic resonance imaging (fMRI) scanner, we can see the pattern of brain activity corresponding to the visual pattern. Also, when no pattern is displayed and subjects report that they do not see the pattern (*true negative*), we do not see that type of brain activity for the pattern. However, what do you think happens when someone gets it wrong? This is the case when no pattern is actually presented, but the subject thinks he or she saw something and so reports, "yes, the pattern was there" (*false positive*). Interestingly, activity in areas V1, V2, and V3 closely follows what *we think we saw*. In other words, on trials where a faint stimulus is presented but the subject fails to detect it, activity is much weaker in these areas (Ress & Heeger, 2003). However, if no stimulus is presented, but the subject mistakenly thinks that a faint stimulus was presented, it turns out that activity is greater in V1, V2, and V3 on these trials.

This is another example of how the brain's activity may closely reflect the *phenomenal experience* of seeing something, even when nothing was actually presented. This is interesting because it demonstrates that it does not matter as much just what physical pattern is presented to a person's eyes; what does matter is what is happening in the brain!

3.10 Perceived Motion

Look back at visual illusions shown in Figs. 4.1 and 4.2. Recall that you have a sense of motion despite the fact that these images are 2-D and stationary. How does this occur? One of the ways that the visual system detects motion begins in the retina; this is called *first-order motion perception*. *Change in luminance* across areas in the visual field is detected as *motion* by neurons on the retina. These changes in luminance can be studied experimentally and are a key reason for the use of visual illusions in the scientific study of human vision. Studying just what changes in luminance affect the perception of motion has provided vision scientists with a vast amount of information about first-order motion processes in the retina, as well as second-order motion processes in the visual cortex. Of course, we experience perceptive motion every time we view a movie, or a store sign that has lights turning on in rapid succession; we "see" these visual objects as having motion.

4. OBJECT (AND FACE) RECOGNITION

4.1 A Two-Step Process for Object Recognition

We touched on the subject of object recognition in our descriptions of the *FFAs and the PPAs* and in the concept of *crowding*, aforementioned. A central theory about how visual objects are recognized is that it consists of a two-step process. Individual *features* of an object are detected first; let us say you are looking at a dog: you detect fur, a long nose, a wagging tail, and four legs. Next, these features are combined using neural mechanisms that are still being discovered. This *combination of features* that forms the recognition process for an object has been called integration, binding, segmentation, or grouping (Treisman & Gelade, 1980). You can imagine that these features or parts of an object can vary in complexity depending on the complexity of the object being viewed. For example, while most dogs share some features such as four legs, fur, and a tail, dogs vary widely in their size, color, shape, and other features. Even recognizing a relatively simple object such as a dog is complicated. Other factors in addition to the complexity of the object to be recognized include the visual crowding of potential objects that we discussed earlier. Attention places a key role in object recognition as well.

And then there is the issue of faces and places.

4.2 Faces and Places

Two main theories exist about how the visual system manages to process and recognize special objects: faces and places. The brain could either do it in a *modular* fashion, with distinct modules for processing faces, or the processing could be done in a *distributed* way across multiple areas of the ventral temporal cortex. According to the modular view, object perception is broken down into neural modules, specific areas of the brain that specialize in processing a particular object category. Research suggests that the FFA might be a *specialized module* for the processing and recognition of upright faces (Fig. 4.11) (Kanwisher et al., 1997; Kanwisher & Yovel, 2006; Tsao et al., 2006). In addition, an area that responds to the presentation of places (e.g., a house, a park) seems also to be a specialized module (Epstein & Kanwisher, 1998). This area has become known as the PPA. This trend for modular representation of objects does not span every object; in fact, it is primarily observed only for faces and places. For example, there is not a banana or shoe area in the human brain.

An interesting twist on the modular hypothesis is the *expertise hypothesis*, which proposes that the so-called FFA is actually specialized for expert object recognition (Gauthier et al., 2000). We probably spend more time looking at faces than at any other object (especially if you include watching faces on TV and in movies and videos). Just stop a moment and think about how much information you can get from all the subtle changes in someone's face when you are talking to him or her, and you soon realize that you are indeed a face expert. It has been proposed that the FFA is responsible for the recognition process of any object we are "experts" at. Research shows that, while looking at pictures of birds, bird experts show somewhat stronger activity in the FFA than the nonbird experts do (Gauthier et al., 2000). Whether the FFA is specific for face recognition or any object of expertise, both cases involved a specialized structure that is distinct from regions of the ventral temporal cortex involved in object processing. Look at the two faces in Fig. 4.23. Do you notice anything strange? Now turn the book upside down and look again. Now do you see it? This is an example of the face inversion effect. When faces are upside down, we are really bad at identifying them. This is an example of how specialized we are at face perception.

The other hypothesis is that the brain processes faces and objects in a distributed way across multiple areas in the ventral pathway. A study by Haxby et al. (2001) demonstrated that regions outside the FFA still show differential responses to faces, compared with other types of stimuli. Even if the response difference between faces and different objects is quite small in many of these areas, there is enough information across all of these regions outside the FFA to tell the difference between faces and objects. However, neuropsychological evidence of double dissociations between face recognition and object recognition are difficult to explain according to this account. One possible resolution is that the activity of highly face-selective neurons in regions such as the FFA may be important for telling apart subtle differences between individual faces but that more distributed activity patterns outside these highly face-selective regions are enough for telling apart basic differences between faces and objects and perhaps more obvious differences between faces (e.g., male vs. female, young vs. old).

FIGURE 4.23 The face inversion effect. Demonstration of how bad we are at recognizing upside down faces. Look at the two pictures of former president Barack Obama. Do you notice anything strange? Now turn the page upside down and look again. Now you should see that one of the pictures has been severely distorted. This effect, called the face inversion effect, demonstrates just how specialized our visual system is for processing upright faces. *Source: Public Domain.*

5. HOW THE VISUAL SYSTEM CREATES YOUR PERCEPTION

We have briefly described three types of information flows in the visual processing stream: feedforward flows from lower visual areas to higher ones, feedback flows from higher visual areas to lower ones, and parallel or horizontal flows across visual areas at the same level (Fig. 4.12A). Together, these flows of information have been described as *reentrant processing*, and they play a key role in many visual functions (Edelman, 1989). Although no consensus has been found for describing just what functions the flows perform in visual perception, their complex interactions are the subject of much study, and together these flows provide an intuitive explanation of how our dynamic and highly interactive visual system works.

As you gaze at the scene around you, information from the retina is being sent to the cortex via LGN. Arriving at V1, this flow of information continues in a feedforward flow through the many visual areas and separates into the dorsal and ventral streams that we have discussed. The dorsal stream, or the *where stream*, provides spatial information about the objects being viewed. The ventral stream, or the *what stream*, provides object recognition information. This flow of information is quite rapid—information arrives in higher visual areas such as V4 and MT in a fraction of a second.

Closely following the initial feedforward flows, feedback and parallel flows combine to form neural networks within and across visual areas for processing the visual scene. Neurons in V1 dynamically change their tuning properties with the arrival of new information from these flows. Consider this situation: you have just walked into a room and are looking at the people in the room. Your face perception mechanisms are at work here—and you are

FIGURE 4.24 Disambiguating a natural scene can be a challenge to the visual system. Here are some examples of the need for both feedforward as well as feedback and parallel flows of information when perceiving a complex scene. *Source: http://www.wherecoolthingshappen.com/22-stunning-optical-illusion-photographs/.*

beginning to recognize some faces. But, as you continue to gaze at those faces, you are also beginning to recognize facets of their faces beyond their identity—*who is it?*—such as their facial expressions. You can *almost feel* your visual system tuning in to more and more details in the scene as you continue to look about. The combined feedforward, feedback, and parallel flows of visual information also contribute to discerning aspects of the complex visual scene that may be ambiguous, such as in multistable perception. Together, they provide the neural machinery that underlies the dynamic and complex human vision system. See Fig. 4.24 for some vision-perception-challenging scenes!

6. VISUAL CONSCIOUSNESS

With all of the information available in any crowded visual scene, it seems intuitively correct that we can only be fully aware of a subset of the visual information at any given time. In other words, only a portion of the information in the visual scene *gains access to our consciousness*. This is the subset of information that we can report on, we can describe to another person. This is also the subset of information that we can remember later. Conscious access has limitations: we are intuitively aware that only some of the visual scene is being fully processed and is active within our conscious mind. Other facts of the visual scene may be preconscious, available to be accessed by conscious processes such as attention. Other facets of the visual scene may be hidden from conscious access. More on that later.

Different cortical visual areas seem to play different roles in our conscious visual experience. An emerging view is that many of the same brain areas and neurons involved in processing specific kinds of visual stimuli, such as orientation, motion, faces, or objects, are also involved in representing these types of stimuli in consciousness. Many neurons are more active when a person is conscious of seeing a stimulus than when the stimulus is shown but fails to reach consciousness, both within visual areas and across the cortex (Fig. 4.25). So far, there does not seem to be any single area in the brain that is solely responsible for

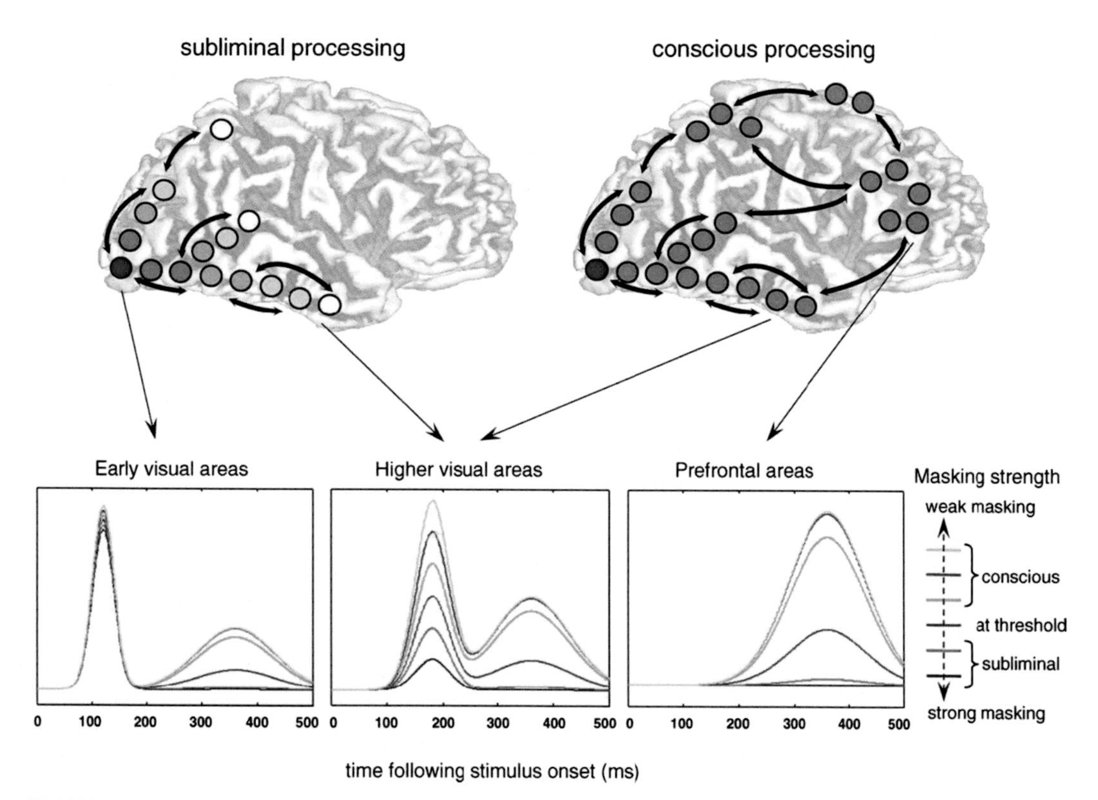

FIGURE 4.25 Different cortical areas are activated during our conscious visual experience. The brain shown on the left side of the figure represents an example of a cortical activation pattern in response to an object that is presented subliminally—that is, either too faintly, too quickly, or *masked* in another manner to enter our visual awareness. The proposed activation pattern is limited to the visual cortex. The brain shown on the right side of the figure represents an example of a cortical activation pattern to objects presented that may be consciously perceived. In this case, there is far more cortical activation across far-reaching regions in the cortex, including the frontal lobe. Below the brain figures are modeled time courses for the activation as a function of the *masking strength*—that is, how difficult it is to consciously perceive a visual stimulus due to experimental manipulations of the viewing parameters. The blue and green lines show proposed cortical activation time courses when a stimulus is presented with *strong masking*, such that it is only perceived on a subliminal way. The teal, pink, and yellow lines show proposed cortical activation time courses when a stimulus is presented with *weak masking*, such that it is readily consciously perceived. Note that it is only in these conditions that prefrontal activations are proposed. *Source: Del Cul et al. (2007).*

consciousness. Instead, many brain areas seem to work together to achieve this remarkable feat. An emerging view is that our conscious experience may reflect the distributed pattern of brain activity involving many visual areas, a kind of dialog between neurons in early visual areas, including V1, and high-level areas such as those in the ventral temporal cortex and the parietal lobe (Del Cul, Baillet, & Dehaene, 2007).

How are conscious and unconscious aspects of visual perception studied? Early studies involved patients with damage to visual regions in the brain. Studies of what these patients were aware of, that is, conscious of, and what they were unaware of formed the early basis for investigating visual areas that are key to visual consciousness. One such focus of investigation was the phenomenon of *blindsight* (Stoerig & Cowey, 1997). This refers to the ability to have some visual perception preserved despite the damage to V1. The key aspect of blindsight as studied in humans is that the patients *are not aware* that they have some visual processing intact. These studies have shed light on the key role area V1 plays in visual consciousness. Studies of animals and humans with brain damage—*lesion studies*—have provided a wealth of information about which visual areas are required to be intact for visual consciousness to arise. A second key avenue of research is using experiments conducted in a laboratory with normally sighted humans, with carefully devised studies to determine what aspects of stimulus presentation allow a given object to reach conscious access in a healthy brain.

6.1 Brain Areas Necessary for Visual Awareness: Lesion Studies

Studies of people or animals with selective brain damage are important for understanding what brain areas may be necessary for certain kinds of visual awareness—awareness of color, motion, faces, objects, or the capacity to be aware of seeing anything at all! Brain lesions may be performed experimentally in animal studies, or they may be investigated in humans who have suffered from unfortunate injury to certain parts of the brain, which may result from strokes, tumors, trauma, or neurodegenerative diseases. By studying these patients, it may be possible to understand the neural causes of their impairment, which may inform scientists about the brain function and eventually lead to new ways to help treat such impairments.

6.1.1 Consequences of Damage to Early Visual Areas

Different visual deficits can result from neural damage at different levels of the visual processing hierarchy. Damage to the retina or optic nerve of one eye can result in monocular blindness—the loss of sight from one eye. Damage to the LGN, the optic radiations that travel to V1, or V1 itself, can lead to loss of vision in the contralateral visual field (see Fig. 4.7). Damage to a small part of V1 can lead to a clearly defined scotoma, a region of the visual field where perception is lost. Damage to a larger region of V1—or all of it—leads to clinical blindness; although the patient's retinas and subcortical pathways are intact, loss of V1 causes the lack of awareness of visual perception.

6.1.1.1 PRIMARY VISUAL CORTEX AND BLINDSIGHT

Do lesions to V1 lead to a complete loss of visual function? It is difficult to ask an animal if it is conscious of something or not, but a recent study of monkeys with unilateral V1 lesions suggests that they might not be aware of what they see. In this study, monkeys were able to report the locations of objects in their "blind" hemifield, accurately, if they were forced to make a choice between two options. However, if they were given the choice of reporting whether an object was presented or not and an object was sometimes presented in the good hemifield, in the blind hemifield, or not at all, they would fail to report objects presented in the blind hemifield (Cowey & Stoerig, 1995; Stoerig et al., 2002). It is as if these objects were not "seen."

Interestingly, humans with V1 lesions may show similar above-chance performance, even though they report a lack of visual experience in their blind hemifield. Patient DB had tumor surgery that required the removal of his right V1. Careful testing revealed that he was able to perform visual tasks at above-chance levels, despite his reports of lacking any visual impressions, a phenomenon known as *blindsight* (Weiskrantz et al., 1974). These findings suggest that there can be dissociations between visual processing in the brain and a person's subjective awareness. Sufficient information was reaching DB's visual system to allow him to make forced-choice discriminations, but this information was not sufficient to support awareness (Weiskrantz, 1986).

The investigation of patients with blindsight reveals that although V1 is necessary for visual consciousness, information from the eyes gains access to the cortical visual system, despite the damaged V1. How does this happen? Studies in monkey and a recent study with a human with a rare pattern of V1 damage combine to shed light on a potential newly found pathway to visual cortex (see Tamietto & de Gelder, 2010) that bypasses V1 (see Section 2.3.1). Clever studies using fMRI in monkey reveal that despite lesions to V1, when visual stimuli were presented, higher visual areas (V2, V4, MT) were activated—as long as the LGN was intact. When the activity of the LGN was temporarily blocked, the higher visual areas were no longer activated (Schmid et al., 2010). These scientists report a subcortical pathway that extends from the retina to middle layers of the LGN, onto the SC, and eventually onto higher visual cortex, bypassing V1.

A case study of a patient who had two successive strokes has led to further new information about a potential subcortical vision pathway that bypasses area V1 (de Gelder et al., 2008). In this case, the patient had a large stroke that destroyed V1 in one hemisphere and then shortly thereafter he suffered a second stroke that destroyed V1 in the other hemisphere. This rare occurrence caused the patient to be clinically blind. While de Gelder and her colleagues were investigating the potential for any spared visual processes in this patient, they discovered that the patient was able to navigate successfully down a long corridor that was strewn with many differing barriers. During testing, the patient is seen carefully avoiding even the smallest of barriers (for a video of this, go to https://www.youtube.com/watch?v=nFJvXNGJsws). Importantly, the patient was not aware that he was doing this and reported that he was completely blind and saw no barriers. Based on these findings and the previous findings from Schmid et al., de Gelder et al. proposed a subcortical visual pathway that is shown in Fig. 4.10. A small proportion of projections from the retina extend through the middle layers of the LGN and onto the SC and other subcortical nuclei. Along the way, this pathway makes contact with the pulvinar and regions in limbic region known for processing emotion, thus this newly described pathway may also explain previous findings of the role of emotional faces that are presented at subliminal (preconscious) stimulus parameters.

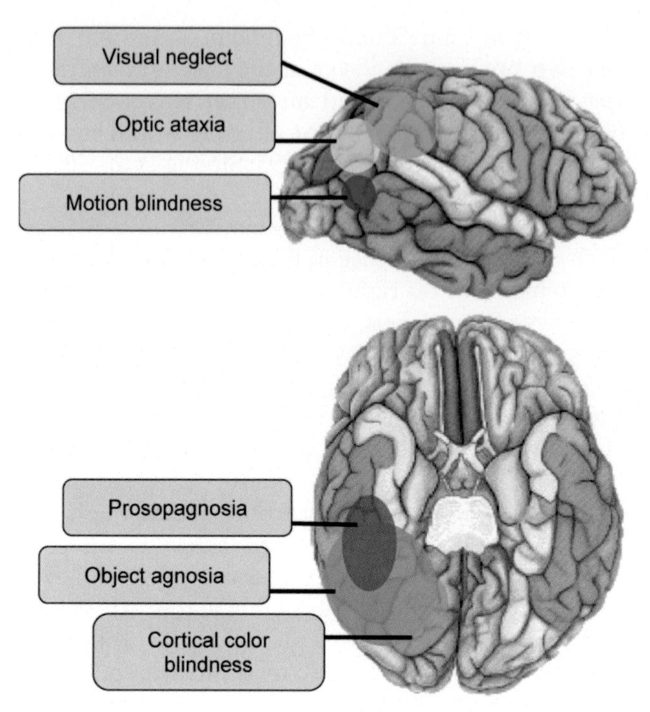

FIGURE 4.26 Visual deficits and brain areas. The areas of the brain in which damage can result in the associated visual deficit. Here, the areas are only shown on one hemisphere, although for some deficits such as motion blindness, damage to both hemispheres is required. *Source: Frank Tong, with permission.*

6.1.2 Extrastriate Lesions—Damage Outside Area V1

What happens when extrastriate areas are damaged? Fig. 4.26 shows a schematic map of visual areas and corresponding deficits.

6.1.2.1 MOTION BLINDNESS

Perhaps, because we are so sensitive to seeing motion, it is very rare for brain damage to lead to a complete loss of motion perception. However, there is a striking example of one patient who can no longer perceive motion after suffering strokes leading to large bilateral lesions that encompassed area MT and extensive surrounding areas. For this patient, the world appeared to be a series of still snapshots, like living in a strobe-lit world. Simple tasks such as crossing the street became dangerous because she could not tell how fast the cars were approaching (Fig. 4.27A). Even pouring a cup of coffee became a challenge because she could not tell how fast the liquid was rising and the cup would overflow.

Other studies have found that smaller lesions just to area MT usually lead to more moderate deficits in the ability to perceive motion, and both patients and animals may also show considerable recovery over time (Pasternak & Merigan, 1994; Plant et al., 1993). So it seems that area MT is very important for motion perception, but other visual areas can contribute to motion perception even when MT is damaged.

FIGURE 4.27 Color and motion blindness. (A) Damage to motion area MT in both hemispheres can lead to a loss of motion perception, *akinotopsia*. Patients describe seeing multiple still frames instead of smooth motion. This can make simple tasks such as crossing the road challenging and dangerous. (B) Damage to color areas in only one hemisphere of the cortex can result in a loss of color perception to one side of visual space. Cortical color blindness is called *achromatopsia*. *Source: Frank Tong, with permission.*

6.1.2.2 CORTICAL COLOR BLINDNESS

Damage to ventral area V4 can lead to cortical color blindness or what is sometimes called *achromatopsia* (Bouvier & Engel, 2006; Meadows, 1974a). Patients report that the world appears to be drained of color, almost like shades of gray, perhaps like the illustration in Fig. 4.27B. These patients can still perceive the boundaries between colors but have difficulty with identifying the colors themselves. Achromatopsia is typically associated with lesions that include area V4 and possibly regions just anterior to area V4. Damage to one hemisphere can even lead to selective loss of color perception in the contralateral visual field.

6.1.2.3 DAMAGE TO VENTRAL OBJECT AREAS—VISUAL AGNOSIAS

Patients with *visual agnosia* have difficulties with recognizing objects because of impairments in basic perceptual processing or higher-level recognition processes. Such patients can still recognize objects by using other senses such as touch, hearing, or smell, so the loss of function is strictly visual. The word *agnosia* can be translated from Greek as meaning "to lack knowledge of," so visual agnosia implies a loss of visual knowledge.

Here, we will discuss three types of visual agnosia: apperceptive agnosia, associative agnosia, and prosopagnosia. Patients with *apperceptive agnosia* can still detect the appearance of visually presented items, but they have difficulty perceiving their shape and cannot recognize or name them. These patients usually fail at shape-copying tests and may have difficulty copying very simple shapes, such as a circle, a square, or perhaps even a single tilted line. Carbon monoxide poisoning is a frequent cause of apperceptive agnosia, since this can lead to profuse damage throughout the occipital lobe.

Remarkably, some apperceptive agnosics show evidence of unconscious visual processing of visual features that they cannot consciously perceive. Goodale et al. (1991) tested an apperceptive agnosic patient, DF, who had difficulty in reporting the orientation of simple lines or real objects. When asked to report the orientation of a narrow slot cut into the face of a drum, she was unable to report the angle of the slot and made many errors. However, when asked to post a card through the slot, she could do so with remarkable accuracy. Surprisingly, when she was asked just to hold the letter by her side and rotate it to match the angle of the slot, her performance was poor once again, and she reported that the slot seemed "less clear" than when she was allowed to post the card. What can account for this behavioral dissociation between DF's ability to report the angle of the slot and to perform a visually guided action?

Patient DF provides strong evidence to suggest that there are separate pathways for processing "what" an object is and "where" it is with respect to performing a visually guided action. According to Goodale and Milner, patient DF has damage to the ventral visual pathway but intact processing in the dorsal pathway, a claim that has recently been supported by brain imaging studies (James et al., 2003). They propose that the dorsal system is not only responsible for processing "where" objects are but also "how" actions can be performed toward a particular object, such as pointing or reaching for that object. Apparently, visual processing in the dorsal system is not accessible to consciousness—the patient cannot report the orientation of the slot—yet the dorsal system can guide the right action.

Associative agnosia refers to the inability to recognize objects, despite the apparently intact perception of the object. For example, when asked to copy a simple picture, patients with associative agnosia manage to do a reasonable job, especially if given enough time. Dr. Oliver Sacks described a patient who, as a result of brain injury, "mistook his wife for a hat." The patient had great difficulty in identifying objects, even though his vision was otherwise normal, and he could describe the features of what he saw. When presented with a rose from Dr. Sacks's lapel, the patient described it as "a convoluted red form with a linear green attachment," but only on smelling it did he realize that it was a rose. When his wife came to meet him at the doctor's office, he accidentally reached for her head when he wanted to retrieve his hat from the coat rack (Sacks, 1985). Associative agnosia usually results from damage to the ventral temporal cortex.

Although most patients with visual agnosia will have difficulty with recognizing both faces and objects, there are some remarkable exceptions that have been reported. Patients with *prosopagnosia* are still able to recognize objects well but have great difficulty in recognizing or telling apart faces (Bodamer, 1947; Meadows, 1974b). Deficits can be severe; some prosopagnosic patients can no longer recognize close family members or friends and, instead, must rely on other cues such as the person's voice or clothing to recognize that person. Some patients can no longer recognize their own face in photos or even in the mirror.

What type of brain damage leads to prosopagnosia? Prosopagnosia can result from bilateral damage around the regions of the lateral occipital cortex, the inferior temporal cortex, and the fusiform gyrus (Bouvier & Engel, 2006; Meadows, 1974b). In some cases, unilateral damage to the right hemisphere may lead to this impairment. Because lesions are usually quite large and might damage fiber tracts leading to a critical brain region, it is difficult to identify a precise site. Nonetheless, the brain lesion sites associated with prosopagnosia appear to encompass the FFA and extend much more posteriorly.

6.1.2.4 DAMAGE TO DORSAL PARIETAL AREAS

Damage to the posterior parietal lobe (or superior temporal gyrus) can lead to a striking global modulation of visual awareness called *neglect*, in which a patient completely ignores or does not respond to objects in the contralateral hemifield (Driver & Mattingley, 1998).

Patients with right parietal damage may ignore the left half of the visual field, eat just half of the food on their plate, or apply makeup to just half of their face. They may also ignore sounds or touches coming from their left.

This syndrome can resemble a disorder of visual perception. However, neglect happens in the absence of damage to the visual system and can involve multimodal deficits, including motor and tactile deficits. Moreover, when patients are instructed to attend to the neglected field, they can sometimes respond to these stimuli (Posner, 1980). So this syndrome is more a deficit of *attention* than an inability to perceive stimuli. Without specific cuing, patients find it difficult to perceive objects in their neglected field.

6.2 Brain Areas Necessary for Visual Awareness: Experimental Studies

One major technique for investigating visual awareness in humans was first designed by the English scientist Charles Wheatstone in 1838 (Dehaene, 2014). Wheatstone noticed that when two very different images were presented at the same time to the two eyes, the brain could not *fuse* the disparate images into a single one, and the observer experienced the perception of first one of the images and then the second one. It is now referred to as *binocular rivalry*, and we describe it in more detail below. A second major way to investigate visual awareness involves the studies of *inattentional blindness*; it has a far more modern origin—1999—and can be accomplished in very natural settings. Described by Dan Simons and Christopher Chabris (Simons & Chabris, 1999), inattentional blindness sheds light on just how much of a natural scene we can be unaware of when our attention is focused elsewhere. These two experimental approaches provide evidence of visual awareness or consciousness across a wide spectrum of visual processes; binocular rivalry sheds light on which of the two competing stimuli *will gain access* to our consciousness at any given moment in a highly controlled laboratory setting whereas inattentional blindness reveals how much of a visual scene *does not gain* conscious access in a complex natural setting.

6.2.1 Binocular Rivalry Experiments

One of the most powerful and best-studied examples of multistable—in this case, bistable—perception is a phenomenon called *binocular rivalry*. When two very different patterns are shown, one to each eye, the brain cannot fuse them together as it would normally because they are so different. What then happens is quite striking: Awareness of one pattern lasts for a few seconds, and then the other pattern seems magically to appear and wipe away the previously visible pattern. It is like the two patterns are fighting it out in the brain for your perceptual awareness! Fig. 4.28 shows an fMRI experiment approach to understanding binocular rivalry in humans using an image of a face (shown in green) and an image of a house (shown in red). The images are sufficiently different so they cannot be "fused" into a single image—they will be seen by the subjects as alternatively a face or a house. Brain areas activated by either the face or house were then examined.

What happens in the brain during binocular rivalry? Tong, Nakayama, Vaughan, and Kanwisher (1998) tackled this problem by focusing on two category-selective areas in the ventral temporal lobes: the FFA and the PPA. They used red-green filter glasses to present a face to one eye and a house to the other eye while measuring fMRI activity from these brain

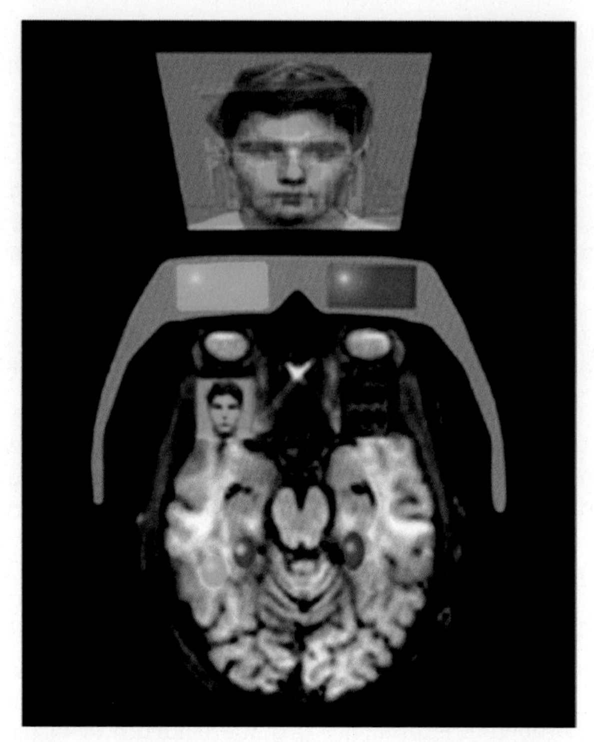

FIGURE 4.28 Binocular rivalry. Schematic of how Tong et al. (1998) used red-green glasses to attain binocular rivalry in the functional magnetic resonance imaging scanner and the fusiform face area and parahippocampal place area where they found activity that correlated with awareness. *Source: Frank Tong, with permission.*

areas (Fig. 4.29). In this study, participants might first perceive the house, then flip to the face and then back again—as is typical of binocular rivalry (Fig. 4.29A). Remarkably, the FFA was active only when subjects reported that they saw the face. Likewise, the PPA was active only when the participants reported that they saw the picture of the house (Fig. 4.29B). Next, the researchers tested physical alternations between the two pictures, switching one picture on while switching the other off. The resulting stimulus-driven responses in the FFA and PPA were about the same strength as those measured during binocular rivalry and, surprisingly, not stronger. It seems that the activity in these brain areas closely mirrors what the observer perceives during rivalry and does not reflect the temporarily suppressed stimulus that is still activating the retina.

6.2.2 *Attentional Blindness Experiments—Or Did You See the Gorilla?*

You may have the feeling that looking at a complex visual scene demands a certain amount of attention to be fully perceived—and you would be correct. Our voluntary and involuntary attentional processes interact with our visual processing and help to define

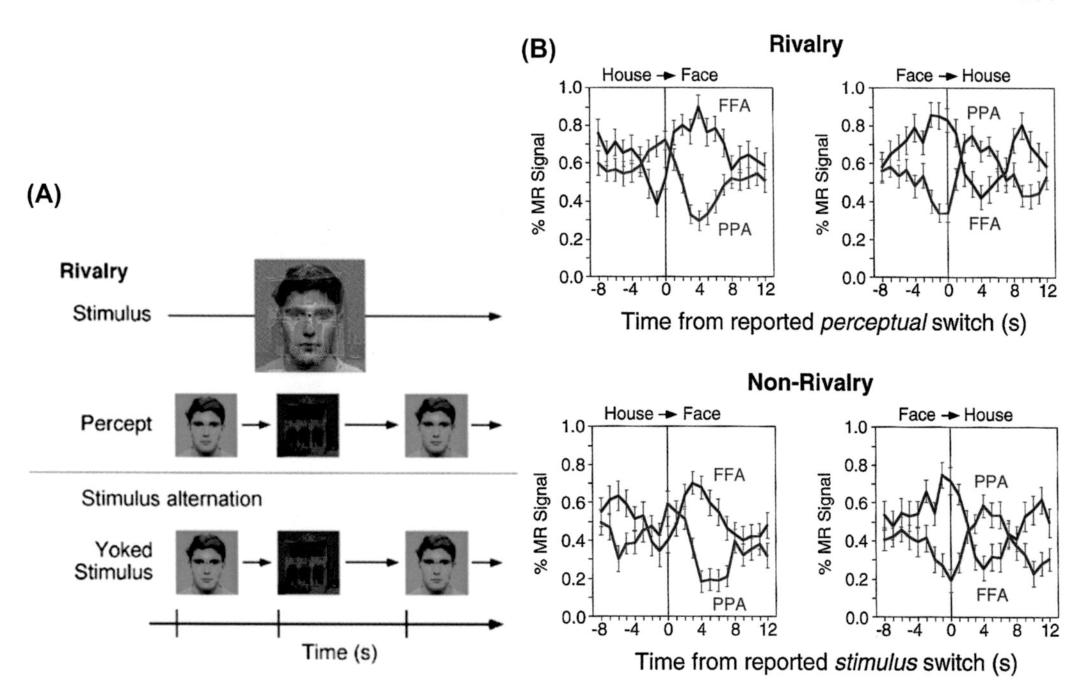

FIGURE 4.29 The stimuli and data from Tong et al. (1998). (A) Left panel shows the binocular rivalry condition. Subjects experienced first the face, then the house, then the face, etc. Lower part of the panel (Yoked Stimulus) shows a control condition with no binocular rivalry; the images were switched on and off the screen. (B) Top panel shows the fluctuations in activity in both the fusiform face area (FFA) (blue) and parahippocampal place area (PPA) (red) during binocular rivalry. When each image became perceptually dominant, activity in the corresponding area increased. Lower panel shows the neural response in the same areas to the control condition with no binocular rivalry. The alternations in activity in both conditions were around the same size. *Source: Tong et al. (1998).*

what aspects of a visual scene gain access to our consciousness at any given moment. Very briefly, we can think of attention as having two basic sources (but see Chapter 8, Attention and Consciousness, for a full description): *Bottom-up attentional capture* is a process when an element in our visual world gains sudden access to our awareness. Think about a friend suddenly leaping in front of you. You *will* pay attention to this new stimulus—it has "captured" your attention! Bottom-up refers to an item or items in the visual scene that are capturing your attention. *Top-down voluntary attention* is a process where you direct or guide your attention to a part of the visual scene that you want to explore more fully. Top-down refers to your executive processes from the frontal lobe, which are guiding your visual apparatus (your eyes) to a certain region or area.

The phenomena of bottom-up and top-down attentional processing have intrigued Dan Simons and his colleagues, who have perfected the study of the role of attention on visual awareness using a clever and very funny set of experiments. Check out the many videos

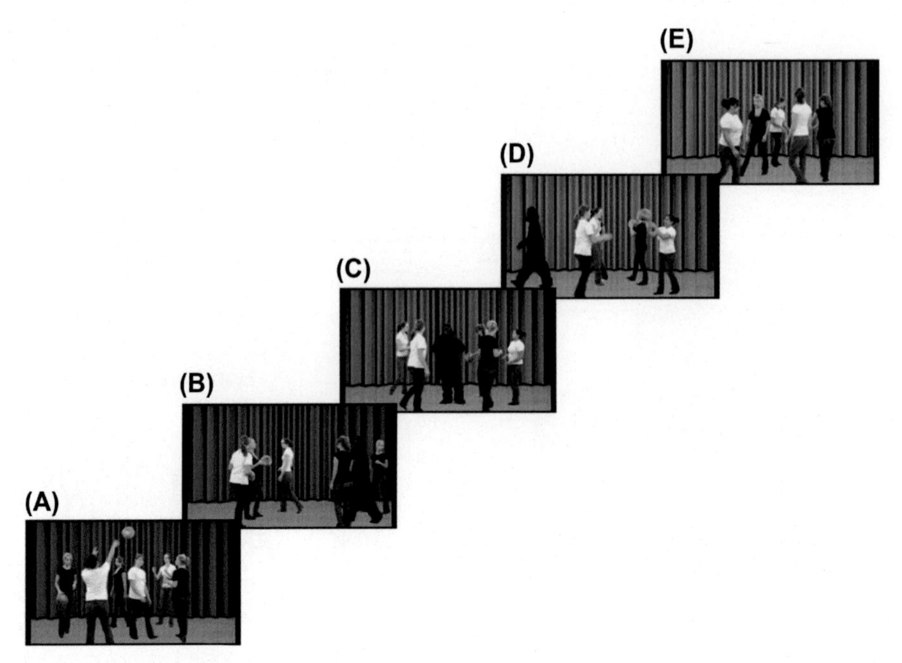

FIGURE 4.30 The gorilla in our midst. The observer's task is to count how many ball passes the white or black team make during the video. When given this task—which demands the attentional following of the ball across the entire stage—observers fail to notice major events that are hypothesized to "capture attention" such as a person in a gorilla suit walking across the stage or the curtains changing color in the background. This effect, termed inattentional blindness, shows how much is shown right before us that nevertheless is never consciously perceived. Here are five frames from Simon's video depicting the gorilla event, the curtain change, and the change in the number of players wearing black. Frame A shows the scene before the appearance of the gorilla. Frame B shows the gorilla entering the scene and one of the players in black backing out of the scene, both on the right side of the image. Frame C shows the gorilla in the center of the scene thumping its chest. Frame D shows the gorilla exiting the scene on the left. Frame E shows the final appearance of the curtain and black team after the gorilla left the scene. *Source: Simons (2010).*

on these experiments by searching for *Simons* and *gorilla*. Trust us; you will be amused and intrigued. Simons' studies have shed new light on how attentional demands affect awareness of even major aspects of a visual scene. These studies have shown that they can insert a visual element to a scene that would be thought to completely capture one's attention through the bottom-up attentional process mechanisms; however, they actually *do not* capture attention at all, if the proper attentional demands are added to the equation. These experiments include inserting a person in a gorilla suit into the middle of a video: with the proper task assignment, most people completely fail to "see" the gorilla (Fig. 4.30). What is more, even when a person is aware of the "gorilla in our midst" videos that have reached a wide audience, they can still miss major facets in the visual scene, if properly distracted by task demands. Simons' research includes inattentional blindness experiments along with experiments to investigate our blindness to even large change in a scene, called *change blindness*.

The bottom line is that although we have traced the visual pathways from the retina to the LGN to V1 and higher visual areas, it is clear that *some aspects* of the visual scene can enter the visual cortex even if one is clinically blind and that *other aspects* of the visual scene are *not seen* by otherwise completely normal-sighted people.

7. SUMMARY

The visual system has a strongly hierarchical structure, with projections that extend from the retina to the LGN to V1 and higher cortical areas that have been well mapped in human and animal studies. Details of just what features in visual images are decoded in distinct visual areas have also been clearly mapped out in literally thousands of studies. And yet, despite this seemingly orderly progression of visual information to the cortex, with increasingly complex visual decoding mechanisms, we find on further investigation that the visual system is hugely adaptive and constructive—and easily fooled by well-made visual illusions. Also, this well-structured system is prone to huge failures in detecting major events such as a man in a gorilla suit walking across a stage—and it is unaware of some of its visual processing abilities as demonstrated by investigations of patients with blindsight. These are aspects of visual perception that have long captured our imagination and our attention as we study this dynamic and complex sensory system.

8. STUDY QUESTIONS

1. Visual input from your left visual field is projected to which area in the cortex?
   a. Area V1 in the right hemisphere
   b. Area V1 in the left hemisphere
   c. Area MT
   d. Area V1 in both the right and left hemispheres
2. How do receptive fields differ in area V1 versus V4?
3. What are the two visual processing streams and how do they differ? What brain areas do they involve?
4. How is visual consciousness studied experimentally?
5. What does the "gorilla in our midst" experiment tell us about human vision?

CHAPTER

5

Sound, Speech, and Music Perception

MUSIC'S VOICE

Where words fail, music speaks. ***Hans Christian Anderson, Danish Author.***

Musician with acoustic guitar. *Source: Public domain.*

1. INTRODUCTION

This chapter provides an overview of how we hear and understand sounds—from simple sounds, to complex speech, to symphonies. We begin with basic information about how we

Fundamentals of Cognitive Neuroscience
https://doi.org/10.1016/B978-0-12-803813-0.00005-2

process sounds: from the ear, through the ascending auditory pathways, to the auditory cortex. Next, we discuss specific types of sound processing such as speech and music perception. Unlike the visual system, where visual objects and scenes are frequently stationary, the hallmark of the auditory system is *time*. Sounds have a temporal dimension as they flow from the environment in the form of complex vibrations as they reach our ears. The exquisite temporal resolution of the auditory system—submillisecond—defines, decodes, and deciphers these complex and overlapping sets of vibrations to bring us meaningful sound, speech, and music.

1.1 A Model for Sound Processing

Our environment is frequently quite noisy, with many types of sounds reaching our ears at the same time. Think about a large college classroom before a lecture begins: there are the sounds of students' voices, chairs scraping, doors opening and closing, backpacks being unzipped, books being dropped onto desktops. All of these sounds hit our ears at the same time, and yet we have little difficulty in perceiving them as separate events or auditory "objects." This process is called *auditory scene analysis*, and it forms the basis for understanding how the auditory system decodes a complex listening environment (Bregman, 1990). We will discuss how the auditory system decodes this type of scene. We begin with a functional framework with which to understand auditory system processes and how they interact with other brain systems.

1.1.1 A Working Framework for Sound Perception

In Chapter 1, A Framework for Mind and Brain, we discussed a framework for understanding brain processing. The same general concepts hold for auditory processing: sensory (sound) inputs enter the system, and there is a very brief storage (echoic memory) for these inputs (Fig. 5.1). Selective attention allows the system to direct its attention to a subset of the inputs for further processing. At this stage, there are complex interactions between the new inputs and existing memory and experiences, as well as with other sensory systems. The ultimate goal or "action" to be performed is important as well, and it will affect how information is encoded and stored. It is important to note that this model for auditory processing is not a one-way process, with sounds being decoded, understood, and then stored into long-term memory. There are interactions that occur throughout the encoding of sounds, both within the auditory system itself and across other sensory, cognitive, memory, and motor systems. The anatomy and connectivity of the auditory system reflects this complexity, with multiple stages of processing and neural pathways, including the ascending pathways from the ear to the brain, descending pathways that carry information back to the peripheral system, and many parallel pathways within brain regions and across the two hemispheres.

1.1.2 Limited and Large Capacity

Brain processes have both limited and large capacity aspects: this is the case for the auditory system. There are some specific limitations in decoding sound inputs. For example, if you present speech through headphones, it is easy to attend to each word uttered. However, if two different speech streams are presented to the two ears, it becomes a very difficult task to try to attend to each stream. In fact, we selectively listen to just one stream

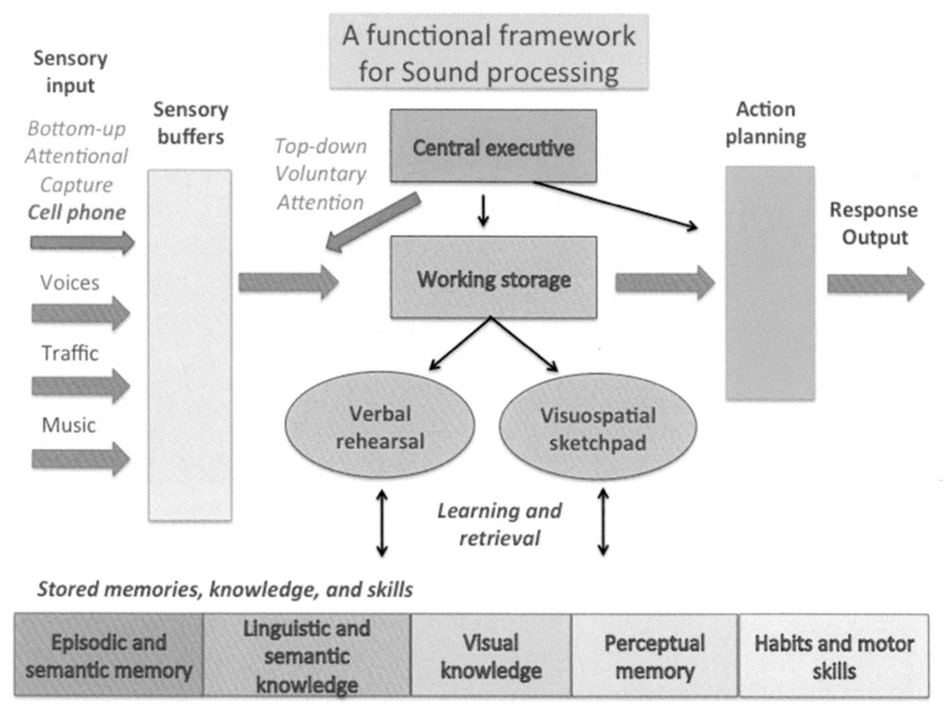

FIGURE 5.1 A functional framework for auditory processing, adapted from the general functional framework presented in Chapter 1. Sensory inputs, such as the sound of someone's voice or a cell phone ring, enter the system (see *green arrows* on the left side of the figure). There are early influences from bottom-up and top-down attentional processes (*orange arrows*). These inputs make contact with the working storage, long-term knowledge, and action systems. It is important to keep in mind that the processes underlying auditory function are highly interactive, with feedforward, feedback, and integrative processes. *Source: Gage with permission.*

or the other (Broadbent, 1982). Thus there are some limits to the capacity for decoding complex sounds entering the auditory system and a role for central executive function in directing attention selectively to some of the sounds in a complex listening environment. On the other hand, our capacity for learning new sounds or auditory objects (such as spoken words) continues throughout life and appears to be virtually unlimited in capacity. In fact, an average adult's vocabulary is estimated at more than 100,000 words. The same is true for recognizing new melodies and the voices of new friends and acquaintances. Therefore while some capacity limits exist in attending to sounds during perception and encoding, once learned there appears to be virtually no limits regarding the capacity to remember new sound-based items.

1.1.3 *Orders of Magnitude and Levels of Analysis*

As in other brain systems, auditory processing contains processing units that comprise many orders of magnitude from individual hair cells at the periphery, to single neurons in the auditory cortex, to large-scale neural networks in the auditory language system. The auditory system has been studied at each of these levels of analysis in both human and in

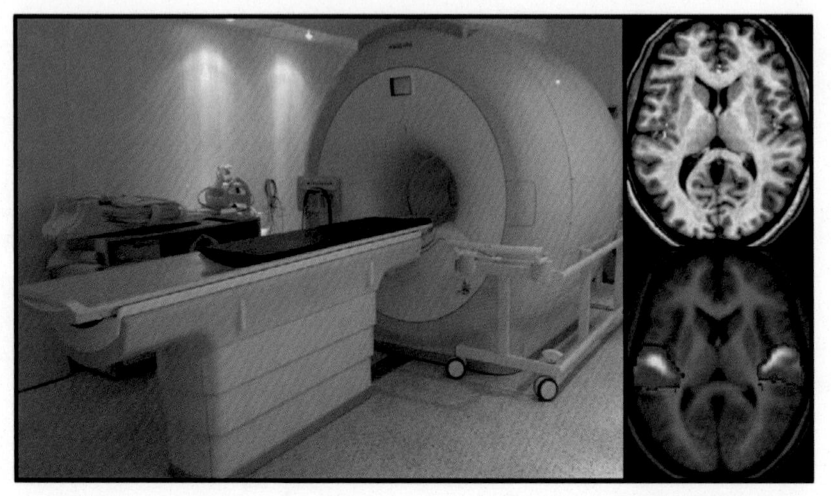

FIGURE 5.2 Left side of the figure: a magnetic resonance imaging machine. Functional magnetic resonance imaging studies have created a virtual explosion of new data about cortical processing of sounds. Right side of the figure: an axial slice showing left and right auditory cortical hemodynamic response (in *red*) to classes of sounds. Note that the front of the head is at the top of the image, the back of the head is at the bottom of the image. Auditory cortical regions in left and right temporal lobes are shown in *red*. *Source: Hall & Susi, in Celesia & Hickok, 2015.*

animal. In this chapter, we include information that we have learned at each of these levels of analysis. However, a large focus of the evidence presented in this chapter is on what we have learned about auditory processing at the system level from neuroimaging—magnetic resonance imaging (MRI), functional MRI (fMRI), magnetoencephalography (MEG), and electroencephalography (EEG)—studies. The advent of noninvasive measures to investigate cortical processing has revolutionized the field of cognitive neuroscience and psychology in general (Fig. 5.2). Previously, we relied on data from animal studies, made inferences from behavioral and psychophysical studies with healthy individuals, or investigated sound and language processing in humans who had suffered brain damage due to injury, disease, or stroke. The capability of investigating brain processes in healthy individuals has provided us with a wealth of new information about sound and language processing. It has also provided us with the ability to investigate brain-wide processes in large-scale systems that span multiple brain regions, such as the language system.

1.1.4 Time

Time is a critical aspect of auditory processing: the auditory system differs from the visual system in that all sound processing occurs over time. Nothing "stands still" in sound processing. Speech, the most complex signal that the auditory system must decode, has differences in speech sounds (phonemes) such as /b/ and /p/ that occur on a scale of 20–30 thousandths of a second (milliseconds), and yet our speech perceptual processes decode these transient differences with ease, even in relatively noisy environments (Gage, Poeppel, Roberts, & Hickok, 1998; Gage, Roberts, & Hickok, 2002).

Thus the speech decoding system has a high temporal resolution of fine-grained and transient changes at the level of the phoneme. However, the speech system also needs to decode information that changes over a longer time span than those contained within

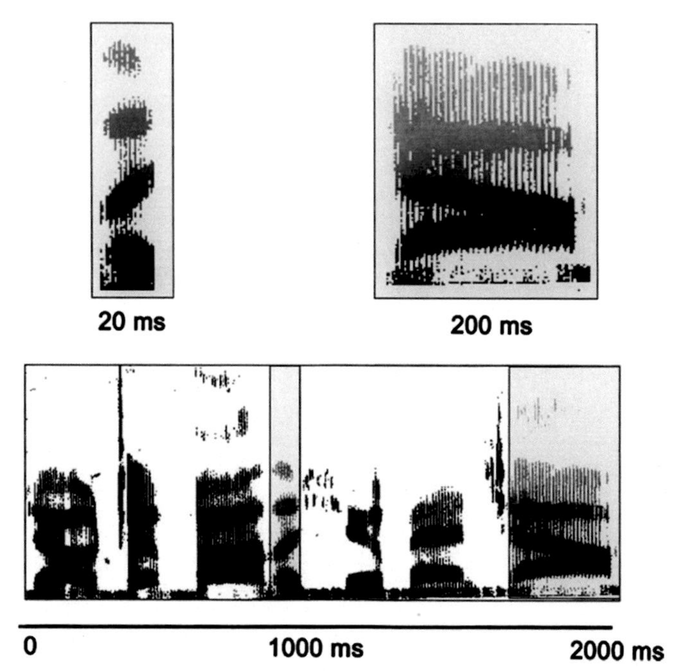

FIGURE 5.3 A spectrogram is a picture of the sound-based features in speech. Time is represented on the x-axis and frequency is represented on the y-axis. The *darker shading* represents higher intensity. Speech contains harmonic content (formants) at specific regions in the spectral (frequency-based) aspect of the spectrogram. Here we show a spectrogram showing three time scales critical for decoding speech. Upper left: detail of the transients at the onset of a consonant, with transitions that occur on a timescale of 20 ms. Upper right: detail of the formants in a syllable that occurs on a timescale of 200 ms. Bottom: a sentence that occurs on a timescale of 2000 ms. *Source: Gage with permission.*

phonemes: syllabic stress (such as the different pronunciation of "melody" and "melodic") is an important speech cue and occurs in a time window of approximately 200 ms. Other key information occurs over 1–2 s (1000–2000 ms) at the level of a sentence, such as the rising intonation that is associated with asking a question. Thus each of these time windows—20, 200, 2000 ms—is critical to the accurate decoding of speech, and information extracted from each of these decoding processes must be available for integration in the complex processes underlying the mapping of sound onto meaning (Fig. 5.3).

Before we begin our discussion of how the brain processes complicated listening environments, with human voices, complex environmental sounds, and music, we need to discuss some basic principles of sound and hearing. We will begin with the physical features of sounds and how these features correspond to psychological aspects of sounds. Next, we will step through the processes and stages of peripheral hearing and subcortical feature extraction.

1.2 Sound and Hearing Basics

Sounds have both physical—acoustic, objectively measured—and psychological—perceived, subjectively measured aspects. When studying sounds, researchers typically rely on one set or the other set of the aspects of sounds. For example, they may vary the pitch of a sound by changing the frequency of a tone and investigate the differences in responses

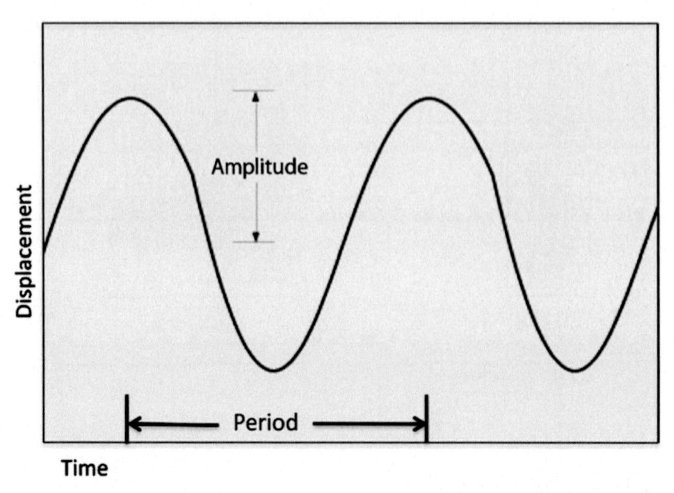

FIGURE 5.4 A sinusoid tone. Time is represented on the *x*-axis, displacement is represented on the *y*-axis. The frequency of the sinusoid is based upon the number of repeating periods (single cycle) per second, thus a 1000-Hz tone has 1000 cycles per second. *Source: Gage with permission.*

for this physical feature of a sound. Or, they may ask subjects in an experiment to make a subjective decision about the timbre of a sound. Both are reliable ways to study sound processing, but it is important to be clear just what aspect the researcher is using in the study: a physical feature or a psychological feature.

1.2.1 Physical Features of Sounds

How does the human auditory system transform sounds into comprehensible speech or recognizable melodies? Let us begin with how we encode simple sounds at the level of the ear. A physical definition of *sound* is the vibration that occurs when an object moves in space, producing an audible sound. What we hear is not the vibration itself but the effects of vibration in sound waves that move, or propagate, through space and make contact with our ears. The sinusoid is a basic building block of sound that has three main physical aspects: frequency, intensity, and time. The frequency of a sound is the rate of sound wave vibration and is measured as cycles completed per second, or *hertz* (Hz). A sinusoid with 1000 cycles per second has the frequency of 1000 Hz. The human auditory system can detect sounds across a wide range of frequencies, estimated at 20 to 20,000 Hz.

The intensity of a sinusoid reflects the amplitude (or displacement) of the wave within its cycle and over time. In Fig. 5.4, we show a 1000-Hz sinusoidal tone in the time domain, with time on the *x*-axis and displacement or amplitude on the *y*-axis. A single period of the sinusoid is shown extending from peak to peak. The spectral energy of a sinusoidal tone is limited to a single narrow band, so a 1000-Hz tone has energy centered only at 1000 Hz. This is why a sinusoidal tone is frequently referred to as a "pure" tone.

Of course, most sounds that we hear are more complex than a pure tone. A piano chord, a car horn honking, a person's voice—all of these have complicated structures. How do we describe these complex sounds in terms of the three physical parameters of frequency, intensity, and time? Joseph Fourier (1768–1830), a Frenchman who lived in the Napoleonic era, developed a series of theorems that describe how even complex signals can be separated into a series of simpler constituent parts through what is now called a *Fourier analysis* (Fourier, 1822). The work of Fourier was advanced by Georg Ohm (1789–1854), who proposed that the separation of complex sounds into simpler sinusoids occurred at the ear in hearing.

While we have mentioned the frequency, intensity, and time of a sound as comprising the basic physical features, sounds have other qualitative aspects. For example, if you heard someone play middle C (261 Hz) on a piano while an oboist played middle C at the same time, could you tell these sounds apart in spite of the fact that they are of identical frequency? Of course you could easily do so, suggesting that there must be many more dimensions in sound quality than just frequency. In this example, the *timbre* or quality of the note helps us distinguish between musical instruments, even when the notes they produce are identical in frequency. Timbre also allows us to distinguish human voices.

1.2.2 A Scale for Sound Intensity

The dynamic range of the human hearing system is extremely broad. We can hear barely perceptible sounds of very low intensity and very loud sounds that actually cause pain. This range has been calculated as ranging from 1 unit of intensity to 1,000,000,000,000,000 (10^{15}) units. This range is so large that it is difficult to deal with using normal numbering schemes. We typically use a logarithmic scale to deal more easily with the huge range in units of intensity, the *decibel* (dB) system. The dB scale is a relative (not absolute) scale and is based upon the ratio of two quantities: the relative intensity of a sound based on either the sound pressure level (SPL) in the air where hearing is occurring or based upon the hearing threshold or sensation level (SL) of an individual. (Note: There are many other ratios used in describing hearing. We use SPL and SL here because they are common ratios used to describe sound intensity). Human hearing ranges from 1 (threshold) to 150 dB SPL (Fig. 5.5). This figure

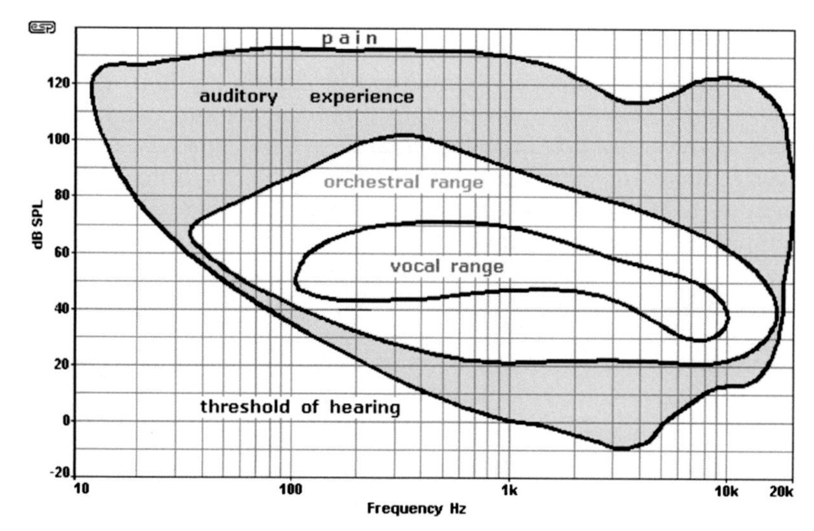

FIGURE 5.5 Hearing threshold and range of hearing for human listeners. Shown also are the ranges of frequency and sound pressure levels of common environmental sounds, including human speech. The most intense sounds are capable of damaging the inner ear receptor organ. The hearing sensitivity of the cat, a laboratory animal commonly used in studies of the peripheral and central auditory system, is illustrated as well. *Source: Public domain.*

also shows the hearing range in frequency from extremely low-pitched sounds (100 Hz) to very high-pitched sounds (10,000 Hz). The range of typical human speech is centered between 1000–6000 Hz and between 40 and 70 dB/SPL.

Sounds from about 120 dB/SPL are extremely loud, painful, and cause damage to the hair cells in the ear. You may experience temporary damage from attending a concert that has music that is amplified: you may feel a ringing in your ears or feel like you cannot hear well. While this feeling will go away as your ears recover, if you present very loud sounds to your ears on a regular basis, you are likely to suffer permanent hearing loss.

1.2.3 Psychological Aspects of Sounds

While sounds have physical parameters (frequency, intensity, time) that can be measured with a fine degree of accuracy, how do we know how they are perceived? The physical parameter of frequency, or cycles per second, corresponds to the psychological or perceptual quality of *pitch*. Pitch is a subjective perception, usually described as the "highness" or "lowness" of a sound—for example, the pitch of a person's voice. We use the physical and psychological terms differently when discussing sound perception. Here is why: while we may know the frequency of a sound because we have measured the cycles per second, we do not know the precise pitch that an individual experiences. A highly trained opera singer, for example, may have a very different sense of the differences in pitch between closely matched sounds than an untrained individual, even though both have normal hearing. This is also the case for the physical parameter of intensity, which corresponds to the subjective perception of *loudness*. Individual listeners have a wide variety in how they perceive the loudness of sounds, depending on many factors ranging from hearing loss to personal preference. Therefore it is important when describing sounds to be aware if you are describing the *measured* physical parameters or the *subjective* psychological features.

1.2.4 Auditory Processing Begins—and Ends—at the Ear

Let us begin with how sounds proceed from vibrations at the eardrum, through the fluid of the inner ear, to innervate fibers at the auditory brainstem on their way to the auditory cortex. Before we begin, however, we need to highlight a central feature of the human auditory system: there are approximately as many *feedback projections* from the auditory cortex to the ear as there are *feedforward projections* from the ear to the auditory cortex. The auditory system is highly bidirectional, with acoustic information being decoded and transmitted through a hierarchical system of *ascending pathways* that lead to auditory cortex. At the same time, there are multiple signals being sent from the auditory cortex down through the *descending pathways* and, quite literally, to the ear itself. Keep this dynamic aspect of auditory processing in mind as we discuss sound processing, speech perception and music processing in the brain.

Returning to our discussion of the ear, vibrating objects cause sound waves to move through air. When these sound waves reach the tympanic membrane, or eardrum, they propagate through the middle ear through the mechanical action of the three bones of the middle ear: the hammer, anvil, and stirrup, to the cochlea, the organ of hearing in the inner ear (Fig. 5.6).

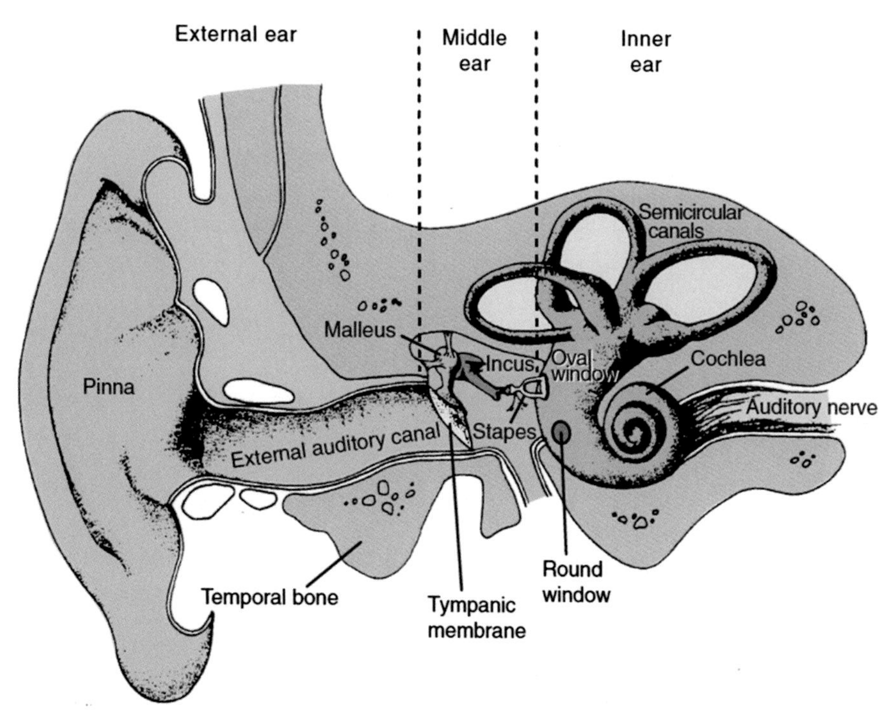

FIGURE 5.6 Drawing of the auditory periphery within the human head. The external ear (pinna and external auditory canal) and the middle ear (tympanic membrane or eardrum, and the three middle ear ossicles: malleus, incus, and stapes) are indicated. Also shown is the inner ear, which includes the cochlea of the auditory system and the semicircular canals of the vestibular system. There are two cochlear windows: oval and round. The oval window is the window through which the stapes conveys sound vibrations to the inner ear fluids. *Source: Brown & Santos-Sacchi, in Squire et al., 2013.*

At the stage of the *cochlea*, in the inner ear, the physical aspects of the sounds are encoded. The traveling wave of sound moves across the basilar membrane from the base to the apex. The basilar membrane is topographically organized in a frequency-specific manner, called *tonotopy*, with higher frequencies encoded at the base and lower frequencies encoded at the apex (Fig. 5.7).

How is the traveling wave converted to a neural code and transmitted to the brain? Within the cochlea, there are approximately 16,000 sensory receptors called the *hair cells*. The motion of the traveling wave along the basilar membrane sets the tiny hair cells into motion. The peak amplitude of the traveling wave causes maximal bending of the hair cells located in specific places of the basilar membrane, thus encoding the frequency of sounds. This is called the *place principle* of hearing and is based on the theory that the brain decodes the frequencies heard based upon which hair cells along the basilar membrane are activated.

At this stage of processing, the movement of the hair cells produced by the traveling wave of sound is transformed or transduced into electrical responses in fibers of the *auditory nerve* (Kelly, Johnson, Delgutte, & Cariani, 1996). Specific hair cells map onto to specific fibers in the auditory nerve, and these fibers have a *characteristic frequency* to which they are most sensitive.

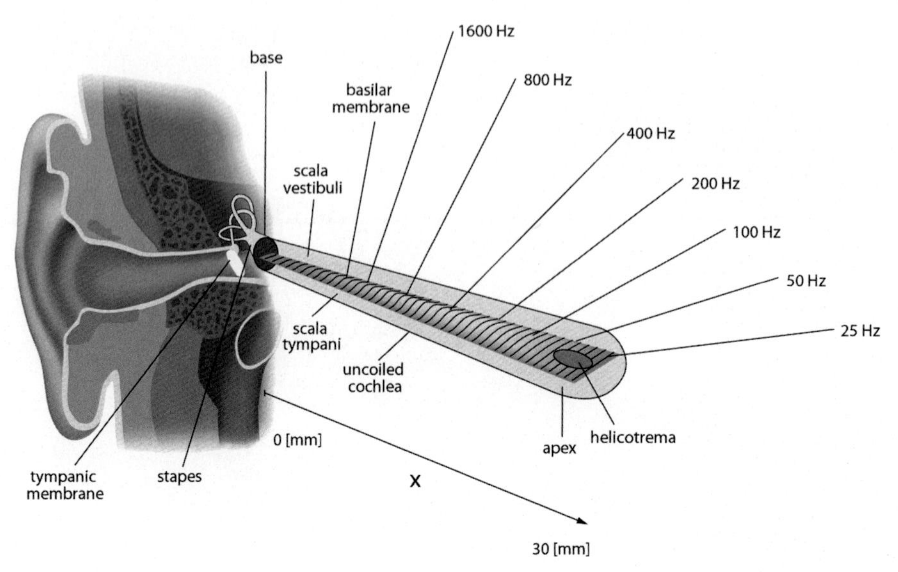

FIGURE 5.7 This figure depicts the basilar membrane uncoiled to show the frequency mapping lower frequencies near the base to higher frequencies near the apex. *Source: Kern, Heid, Steeb, Stoop, & Stoop. https://commons.wikimedia.org/w/index.php?curid=5929157.*

2. PATHWAYS TO AUDITORY CORTEX

The information in sound undergoes many transformations as it ascends to the auditory cortex. The auditory system is comprised of many stages and pathways that range from the ear, to the brainstem, to subcortical nuclei, and to the cortex. The three main divisions of the auditory system are the peripheral system, which we have already discussed, the pathways (ascending to the cortex, descending from the cortex, and parallel pathways across cortical sites), and the central (cortical) system. While each stage and pathway has functional significance in the decoding of sounds, it is important to consider the auditory system as a whole because of the complex interactions across and within its constituent parts. Throughout this system, there are feedforward projections on their way to the cortex, feedback projections from the cortex, and multiple parallel pathways at every stage and locus along the way.

2.1 Auditory Pathways

All sound processing occurs over time. The hallmark of the auditory system is its exquisite temporal resolution for decoding intricate information in sounds (Gage & Roberts, 2000; Gage, Roberts, & Hickok, 2006). One important aspect of the high temporal resolution of the auditory system is the fast and accurate transmission of sound information along its many pathways. Not only do transient features in complex sounds—such as the harmonic structure of consonants in speech or musical phrases—need to be conveyed rapidly from the eardrum to the cortex, but the information from the two ears needs to be combined and integrated in a meaningful way en route.

2.1.1 Ascending Pathways

The *ascending* (afferent) pathways transmit information about sounds from the periphery to the cortex. This pathway is not a simple delivery system but entails significant encoding and recoding of information in the sounds. The neural signal travels from the auditory nerve to the lower (ventral) *cochlear nucleus*. The cochlear nucleus is tonotopically organized. From the cochlear nucleus, the signal continues along the ascending pathway through the lateral lemniscus, inferior colliculus, thalamus, to the auditory cortex (Fig. 5.8). This is not a single pathway, but it is complex and includes many computational stages as well as the combination of sound inputs from the two ears.

A key function of the ascending pathway is to evaluate the information from the two ears to localize sounds in space. How does the brain locate sounds in space? Sounds are always changing in time, and the mapping of auditory space is a complex one. Here is how it works:

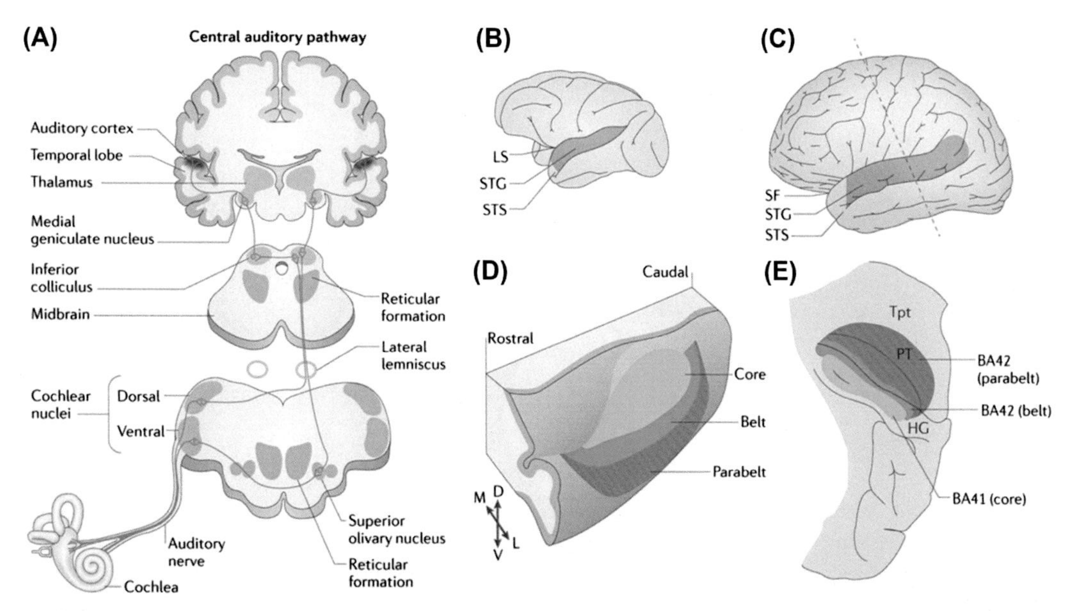

FIGURE 5.8 The human auditory system, showing pathways and subcortical nuclei in the ascending and descending pathways. (A) The ascending auditory pathways begin at the cochlea at the ear (lower left), continue to ascend through the auditory nerve (shown in blue at the bottom left of the figure), and continue to ascend, branching up through the dorsal and ventral cochlear nuclei at the left of the figure and across the superior olivary nucleus and reticular formation, shown at the bottom right of the figure. The pathways continue their ascent through the lateral lemniscus and onto the midbrain and upward to the inferior colliculus (center of the figure). Top of the figure shows the pathways extending through the medical geniculate nucleus of the thalamus and onto the auditory cortex in the temporal lobes in the left and right hemispheres. (B) The location of the auditory cortex in monkey is shown in blue in the superior temporal gyrus (STG), bordered at the top by the lateral sulcus (LS) and at the bottom by the superior temporal sulcus (STS). (C) The location of the auditory cortex in human is shown in blue in the STG, bordered at the top by the Sylvian fissure (SF) and at the bottom by the STS. (D). Close-up view of the auditory cortex within the STG in macaque monkey. The core region (light blue) corresponds to primary auditory cortex, and the belt and parabelt regions (darker blue) correspond to auditory association areas. (E) Close-up view of the auditory cortex within the STG in human. The Brodmann area 41 (BA41, light blue), located within Hesch's gyrus (HG), corresponds to primary auditory cortex in human and the core region in monkey, with BA42 (darker blue) in the planum temporale (PT) corresponding to the belt and parabelt regions in monkey and to auditory association areas. Located above the PT is the heteromodal temporoparietal region (Tpt). *Source: Javitt and Sweet (2015).*

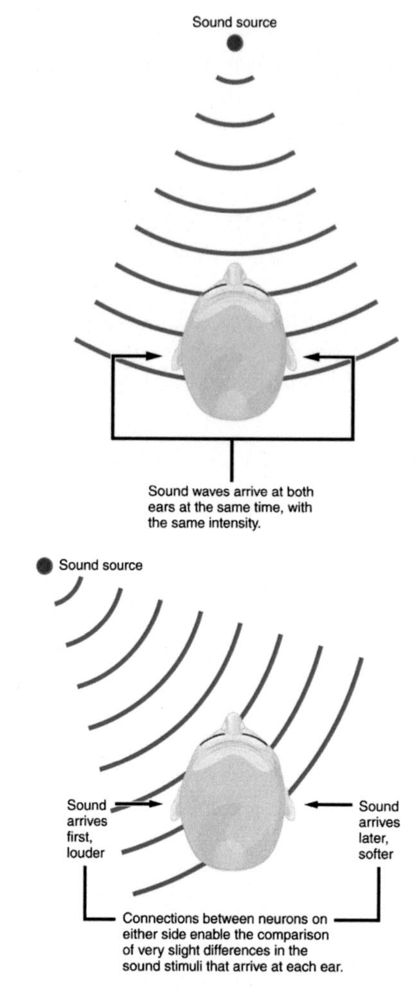

FIGURE 5.9 Sound localization. Top: when a sound source is located directly in front of the head, the sounds reach both ears at the same time and with the same intensity. Bottom: when a sound source is located to the side of the head, the sound arrives slightly sooner to the near ear (interaural time difference) and is slightly louder (interaural level difference) at that ear. These are key cues to sound localization. *Source: Illustration from Anatomy & Physiology, Connexions Web site.*

when a sound occurs, it will likely be off to one side or the other of you. It could also be behind you. To determine where the sound is in relation to you, your auditory system must make a very quick determination of the sound's arrival at the two ears. Two basic types of cues are used when our system localizes sound. The first is the *interaural (between ear) time difference*: the difference in time between a sound reaching your left ear versus your right (Fig. 5.9). A second important cue for localizing sounds is the *interaural level difference*. This is the small difference in loudness that occurs when a sound travels toward the head from an angle. The head produces a "sound shadow," so sounds reaching the far ear are somewhat quieter than the near ear, and the absolute level differences depend on the frequency of the sound.

The actual computations that produce sound localization functions involve complex algorithms called *head-related transfer functions* to calculate the location of sounds in auditory space.

2.1.2 Descending Pathways

The *descending* (efferent) pathways from regions in the cortical and subcortical auditory system to the periphery are under direct or indirect cortical control. Recent research indicates that this control extends all the way to the hair cells in the cochlea! One important function of the descending pathway is to provide "top-down" information that aids in selective attention processes and in perceiving sounds in a noisy environment.

The auditory pathways are not just ascending to or descending from the cortex, there are many important connections between the auditory cortices in the left and right hemispheres via the corpus callosum. There are also cortico-cortical pathways that provide integration of auditory processes with other sensory systems, as well as with working and long-term memory processes, and stored memories and knowledge.

2.2 Auditory Cortex

The auditory cortex is the region within the cortex specialized for sound processing. It is located in each hemisphere within the Sylvian fissure on the surface of the supratemporal plane and the upper banks of the superior temporal gyrus (Fig. 5.10). Information in sounds is transmitted from the ear to the auditory cortex via the ascending auditory pathways. Along the way, the signal is transformed and recomputed in many ways. The auditory cortex is not the end stage of this pathway but serves as a hub or nexus for sound processing, interacting dynamically with other systems within the cortex, across the hemispheres, and back down the descending pathways to the cochlea.

The auditory cortex is not a unitary brain area but is comprised of several structural (anatomical) areas that differ in their role in decoding sound. Early descriptions of these areas within the auditory cortex were made based on the structure, such as a gyrus within this cortical region, and by underlying neurophysiological features, such as the cytoarchitectonic classification. We discuss our current knowledge about the human auditory cortex in the following, with a description of the structure or anatomy, followed by details regarding the cellular organization and response properties, or neurophysiology.

2.2.1 Auditory Cortical Anatomy

Much of what we know about auditory cortical anatomy comes from work in nonhuman primates (Galaburda & Pandya, 1983). In macaque monkeys, the major regions of the auditory cortex are the core, belt, and parabelt regions. These distinct regions are distinguished by their cytoarchitectural, physiological, and connective properties.

In humans, the primary auditory cortex is located within *Heschl's gyrus* (Fig. 5.11) and is roughly analogous to core regions described in nonhuman primates. Typically, the primary auditory cortex comprises only a portion (one- to two-thirds) of the medial aspect of Heschl's gyrus. There is significant variability in the anatomy of Heschl's gyrus both in the two hemispheres and across individuals: Heschl's gyrus is typically located somewhat anterior in the right hemisphere than in the left, and some individuals have more than one Heschl's gyrus.

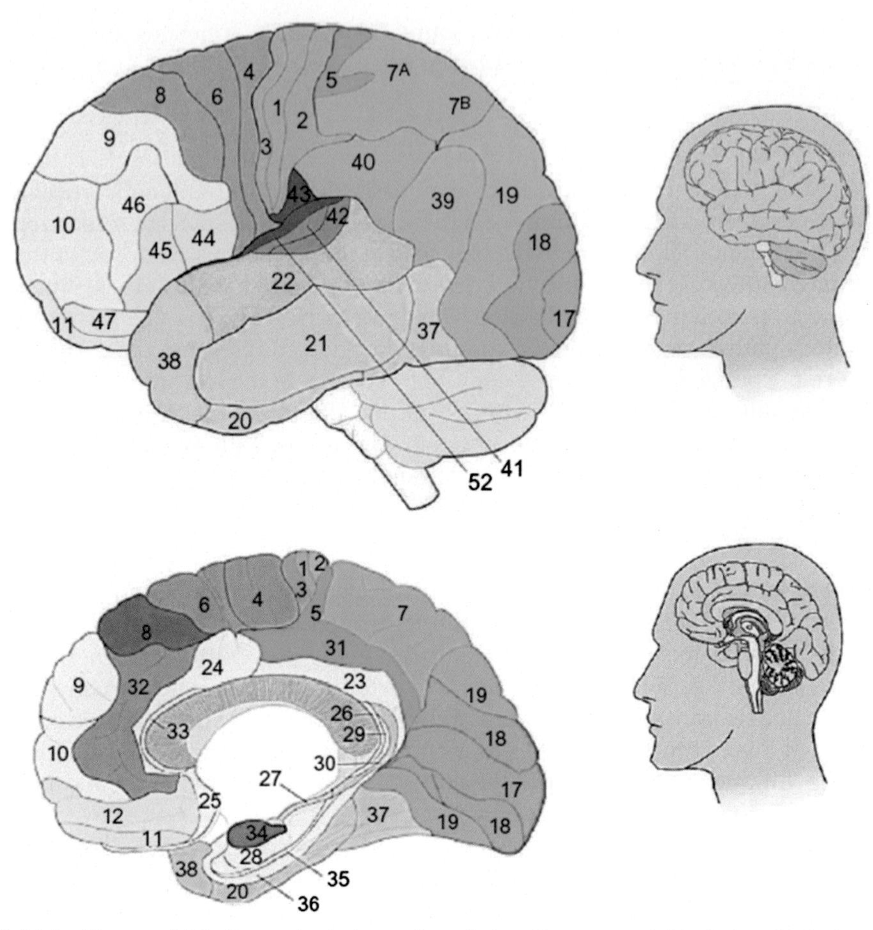

FIGURE 5.10 Top panel: The human brain from a lateral view. Bottom panel: The human brain from a medial view. Colored brain regions are adapted from Brodmann (1909). Auditory and receptive language cortical regions include Brodmann 22, 41, 42, and 52. Note, on the right side of each brain figure we show the brain's orientation within the head. *Source: Baars and Fu, with permission.*

The auditory cortex extends from Heschl's gyrus in the anterior-inferior direction and the posterior-superior direction along the supratemporal plane and the upper bank of the superior temporal gyrus.

A second important anatomical region in the human auditory cortex is the *planum temporale*, located just posterior to Heschl's gyrus. There are both hemispheric and individual differences in the planum temporale. However, unlike Heschl's gyrus, the differences fall into a general pattern: the *planum temporale* is typically much larger in the left hemisphere than in the right. In fact, the left planum temporale can be up to 10 times larger in the left hemisphere in right-handed individuals (Fig. 5.12). In Fig. 5.12, you can see a larger planum temporale in the left hemisphere versus the right (shown in red).

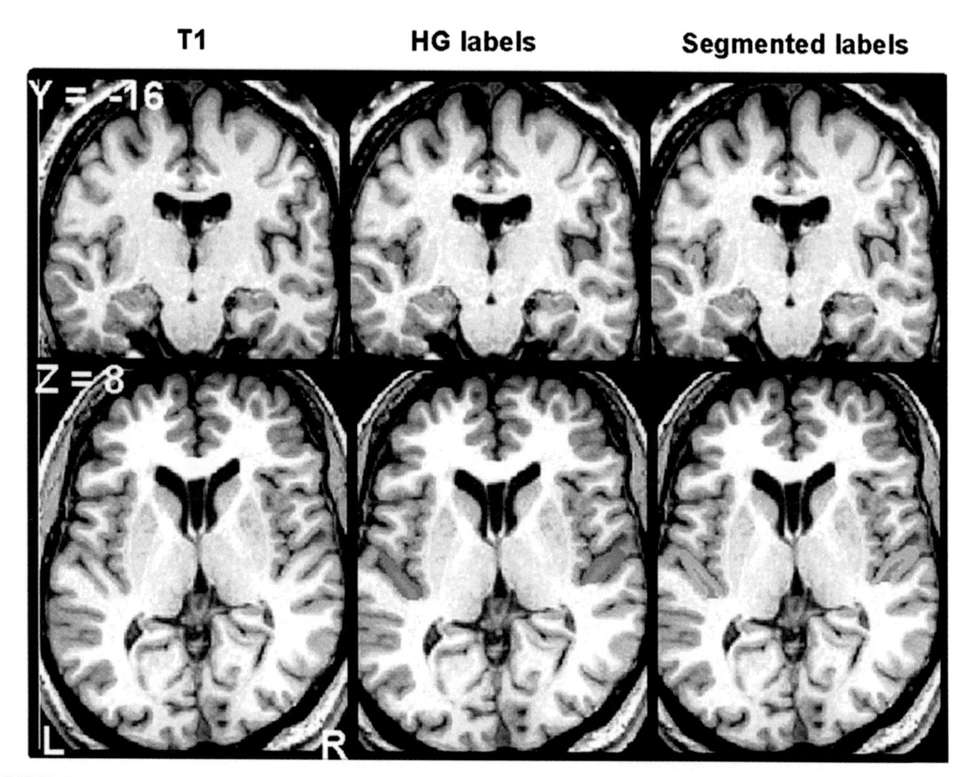

FIGURE 5.11 Heschl's gyrus. Coronal slices of one subject showing (left) unmarked magnetic resonance imaging image, (center) Heschl's gyrus marked in *red*, and (right) gray and white matter of Heschl's gyrus segmented with gray matter shown in *green* and white matter shown in *red*. The left side of the brain is the left hemisphere of the brain. *Source: Warrier et al. (2009).*

Planum temporale asymmetries were reported in a series of anatomical studies by Geschwind and Galaburda (1985a, 1985b, 1985c). These scientists noted that language function tends to be lateralized to the left hemisphere. They suggested that the larger left hemisphere planum temporale reflects its role in decoding auditory language. More recent neuroimaging studies, however, have provided a method to specifically test this hypothesis and provide differing views of the role of the planum temporale in sound processing.

Just anterior to Heschl's gyrus is the planum polare. This region has not been the focus of much study in humans, and little is known about its role in auditory perception. Posterior to the planum temporale and unimodal auditory areas is Brodmann area 22. This is the area that Carl Wernicke hypothesized played an important role in speech comprehension (Wernicke, 1874/1977). According to Wernicke, this region was not an auditory region per se but formed the language area for speech comprehension processes that were closely related (physically and functionally) to auditory processes. This region is typically referred to as Wernicke's area.

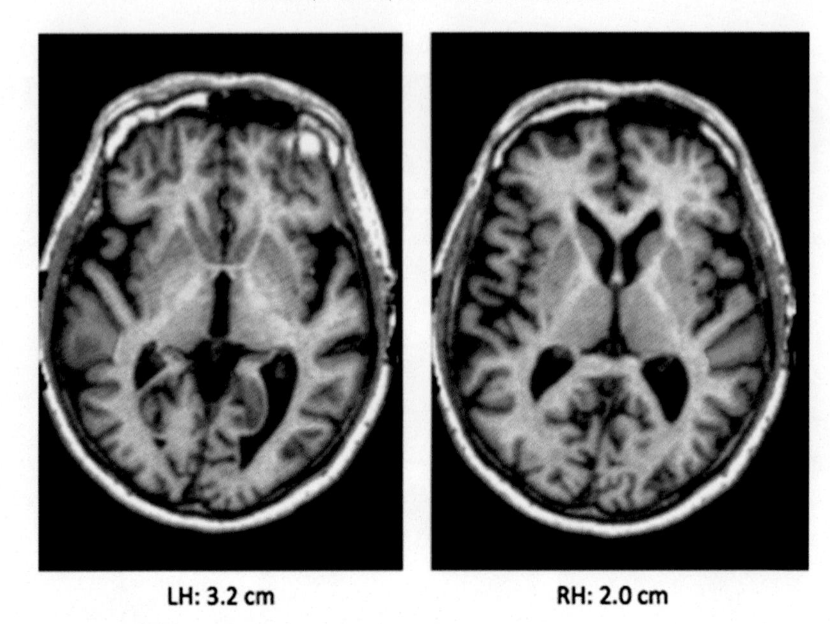

FIGURE 5.12 An axial slice of the brain with the left hemisphere shown on the left side of the image and the right hemisphere shown on the right side of the image. The planum temporale is highlighted in light red. This brain image shows the typical human planum temporale asymmetry with a much larger planum temporale in the left hemisphere. *Source: Marcus et al. (2007).*

2.2.2 *Neurophysiology*

Several guiding principles of auditory cortical organization have been established in studies of cats and nonhuman primates. The basic units of organization in the auditory cortex, as in other cortical sensory areas, are neurons, cortical columns, and neural networks (see Chapter 2, The Brain). There are several differing types of neurons in the auditory system. These neurons have different response properties for coding frequency, intensity, and timing information in sounds, as well as for encoding spatial information in processes for localizing sounds in space. Most cortical neurons respond to binaural inputs (inputs from both ears), demonstrating the importance of cortical processes for decoding binaural information for sound localization and other complex hearing processes.

2.2.2.1 TONOTOPY AS A GUIDING PRINCIPLE OF NONHUMAN AUDITORY CORTEX

Mapping receptive field properties of neurons in the auditory cortex has been the focus of many animal studies. A central guiding principle for the nonhuman auditory cortex is the *tonotopic* organization: according to this mapping principle, the *frequency of tones* or other sounds are represented in predictable maps in the auditory cortex (Schreiner & Winer, 2007) (Fig. 5.13). Within the core in the cat auditory cortex, for example, receptive fields of neurons reflect a tonotopic organization in primary (A1) regions that has a mirror image in adjacent (anterior and posterior) core regions.

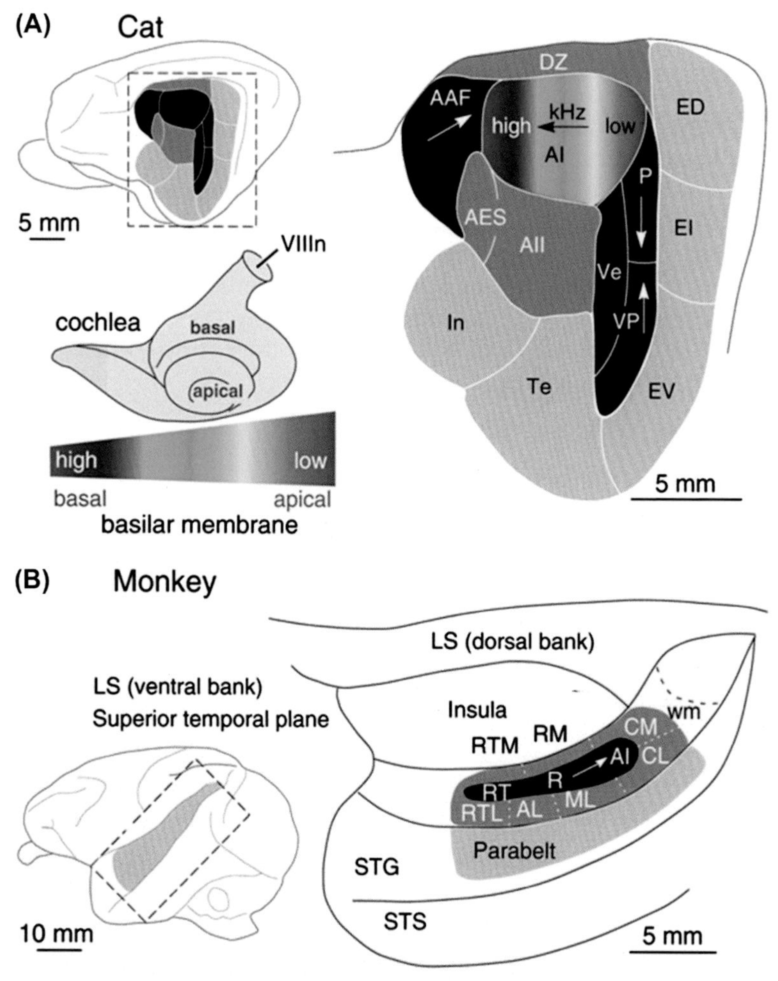

FIGURE 5.13 Tonotopy in cat and monkey auditory cortex. (A) Upper left. Cat auditory cortex has ~13 areas. Five of these areas (shown in *black*) have neurons that are tuned to frequencies in tones—they are tonotopic. Upper right: the color gradient shown in the center of the enlarged figure of cat auditory cortex shows a tonotopic organization, with responses for higher frequency shown in *purple*, midrange in *blue-green-yellow*, and lower frequency in *orange-red*. (B) Left panel: auditory cortex of the rhesus monkey with the superior temporal gyrus (STG) shown in *gray*. Auditory cortex in this monkey is located below the lateral sulcus (LS) in the STG. Regions within auditory cortex include the core, belt, and parabelt. Right panel: enlarged view of the tonotopic areas in the core (center), belt (peripheral), and parabelt (more peripheral) regions. *Source: Schreiner and Winer (2007).*

2.2.2.2 COMPLEX FEATURE MAPS IN HUMAN AUDITORY CORTEX

While the basic aspects of neurophysiology are likely similar for humans, the uniqueness of human speech, language, and music perception, as well as the substantially larger regions of the cortex devoted to the auditory cortex in humans, probably mean that there are neurons and networks that are specialized for these complex processes and specific to the human auditory cortex. For example, although many auditory cortical regions in nonhuman primates reflect a tonotopic organization, evidence for tonotopy has been less robust in human studies and may be limited to the primary auditory cortex and not represent the basic organizational principle in nonprimary auditory areas (Wessinger et al., 2001). Recent studies have contributed to this notion, showing that the mapping of sound-based features in human auditory cortex is a complex one that is not yet fully understood (see Fig. 5.14 and Moerel, de Martino, & Formisano, 2014 for a review).

The complex sounds that make up human speech and music contain feature dimensions such as melody, harmony, rhythm, and timbre. It is likely that the lack of tonotopy that has been found for simpler sounds (tones) in human auditory cortex reflects a more complex mapping system across multiple stimulus dimensions.

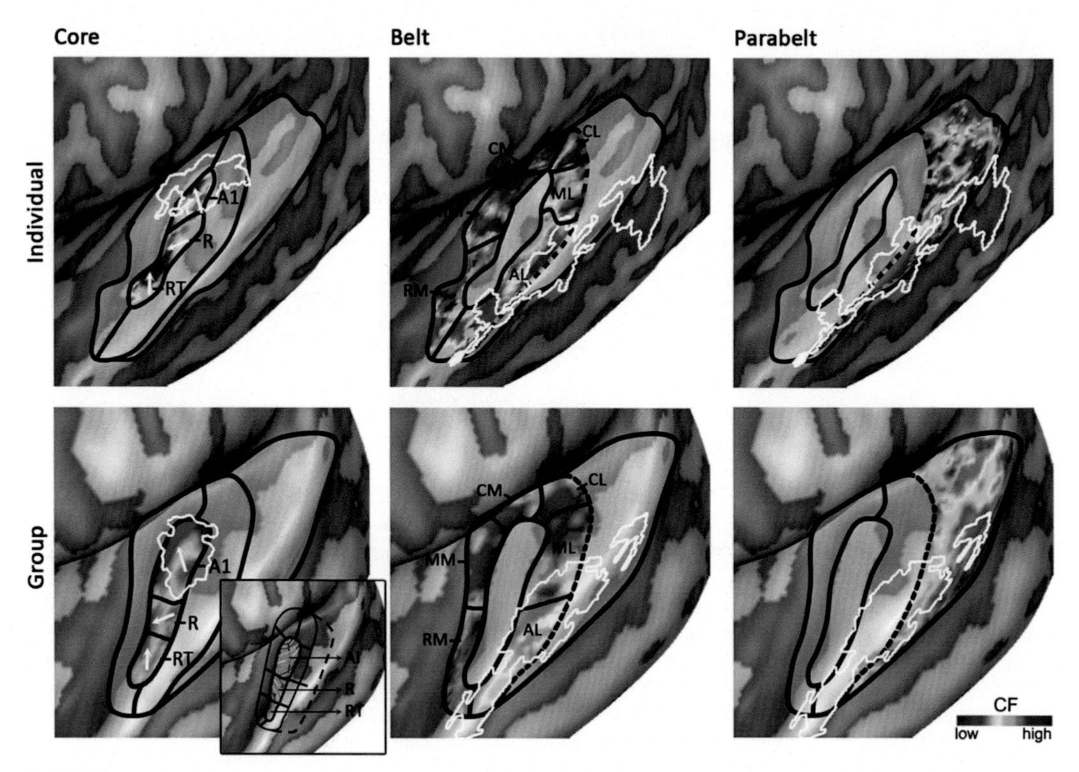

FIGURE 5.14 Human auditory cortical organization. For humans, tonotopic maps are less well established, likely due to the complexities of the sound signals (speech, music) that the human auditory system is tuned to. Top panel: results from a parcellation of auditory cortex for one subject for auditory core (left), belt (middle), and parabelt (right) areas. *Black lines* show the boundaries between these three types of auditory fields. Colors ranging from cool blues to warm reds show neural responses that are tuned to higher to lower frequency in sounds. Bottom panel: the same results shown for a group of subjects. *Source: Moerel et al. (2014).*

3. CORTICAL AUDITORY FUNCTIONS AND PATHWAYS: WHAT AND WHERE

Consider again the large college classroom where doors are opening and closing, students are talking, backpacks are being unzipped, and books are being dropped on desktops. All of these sound events are happening at the same time and are overlapping in frequency and intensity. How does the auditory system decode the individual auditory "objects" such as a friend's voice, a door shutting, and a cell phone ringing? To accomplish this, the auditory system must keep track of the many aspects of the complex auditory scene: *where* sounds are occurring in space and *when* sounds occur (are they simultaneous, or does one sound precede another?) to determine *what* the sound represents in terms of known auditory objects, such as speech or music or new auditory objects to be learned. Of course, these perceptual tasks are not limited to the auditory system but make contact with other sensory systems as your brain integrates what you hear with what you see, feel, and smell. These tasks also interact with your memories and learned information already stored regarding auditory objects that have taken a lifetime to develop.

3.1 Auditory Object Learning and Recognition

The auditory system must decode sounds "online" as they occur to form a percept of a sound event or auditory object. These objects are learned over time as we grow from infant to child to adult, and they change with experience throughout our lifetime. Auditory objects can take many shapes, similar to visual objects, and vary widely in complexity from a simple computer alert chime, to the slamming of a car door, to a friend's voice, to a symphony. It seems that the brain has a nearly limitless capacity for storing and retrieving auditory objects. Auditory objects are organized into categories, such as human voices, musical instruments, and animal sounds, that aid us in decoding learned objects as well as learning new ones. Over time, associations are formed between learned auditory objects and coinciding inputs from other sensory systems, and these different sensory memories become integrated in the conceptual representation system. Early work on describing how these sensory inputs are experienced and combined to form conceptual knowledge was provided by Carl Wernicke (1874/1977), who proposed that with experience, when you hear the sound of a bell, you will recognize it as such, and the sound of the bell will also bring to mind (activate) the visual features of a bell, the feel of a bell, and so on.

Where in the brain are auditory objects located? The lack of a widely agreed upon definition of *just what constitutes an auditory object* has led to some lack of consensus about where these auditory objects are stored and retrieved in cortex. Is an auditory object *that stream of sound* that identifies an item, such as a piano? Is it a *single note*? Or a *concerto*? Nevertheless, there is growing agreement in the field of auditory neuroscience that auditory objects are accessed via a hierarchical pathway that extends cortically from Heschl's gyrus, through the planum temporale, and then to the superior temporal sulcus (Kumar, Stephan, Warren, Friston, & Griffiths, 2007). We will revisit this question when we turn our attention to auditory streams later in the chapter.

3.2 Auditory Scene Analysis

We have described how a sound is decoded by the auditory system to be recognized or learned as an auditory object. This process seems relatively straightforward when you consider a situation where a single sound event occurs in a quiet environment. But how is this perceptual task accomplished in noisy environments, with complex sounds that occur simultaneously in time, with overlapping frequencies, and possibly coming from the same spatial location? How does the auditory system distinguish them as separate sound events? This perceptual task—called the "cocktail party problem" (Cherry, 1953)—has been the subject of many investigations of auditory perception from a theoretical perspective to understand how the auditory system extracts information from complex signals, as well as a practical perspective in designing speech recognition systems. Bregman (1990) provided a model to describe how the auditory system segregates the many different signals in a noisy environment. The four elements in this model are as follows:

1. The source
2. The stream
3. Grouping
4. Stream segregation

The *source* is the sound signal itself. The *stream* is the percept related to the sound. This distinction between the physical signal and the related perception is analogous to the relationship we described earlier in this chapter between the frequency (in Hz) of a sound and the pitch perceived by the listener. *Grouping* refers to how the signals are perceptually combined to identify and maintain attention to some aspects of the auditory scene (such as listening to one friend's voice in a crowd of people) (Fig. 5.15). Perceptual grouping processes create the stream. There are two basic types of grouping. One is *simultaneous grouping*, where if two or more sounds have common onsets and offsets, they may be grouped together. Think of a choir or an orchestra: you will not typically hear each individual voice or instrument but will group them into a single stream due to the beginning and ending of their music together, as well as their shared spatial location. The other is *sequential grouping*, which refers to the process in which features or properties are shared across sounds that occur over time. An example of this grouping process is if you are listening to a professor lecture and someone in front of you coughs. The stream coming from the professor is interrupted by the cough, but you will likely not notice an effect in hearing what is being said. *Stream segregation* uses the grouping processes to segregate separate auditory objects or events into streams.

How does the brain perform auditory scene analysis? Investigations of the neural substrates of perceptual organization have led to the formation of several theories of how and where perceptual streaming is decoded. One view holds that auditory stream segregation involves the primary auditory cortex and that the underlying mechanisms for this segregation involve neural suppression of information not contained within an auditory stream (Fishman et al., 2001). A second view holds that auditory stream segmentation exploits cortical change detector mechanisms in detecting aspects of the auditory scene that are not part of a single stream (Sussman, 2005). According to this view, an individual auditory stream is detected based on the acoustic aspects of the auditory sound, such as its frequency and location in space. Once these characteristics are formed into a neural

(A) Independent auditory stimuli are created by each of the three sources: the singer, banjo player, and bassist

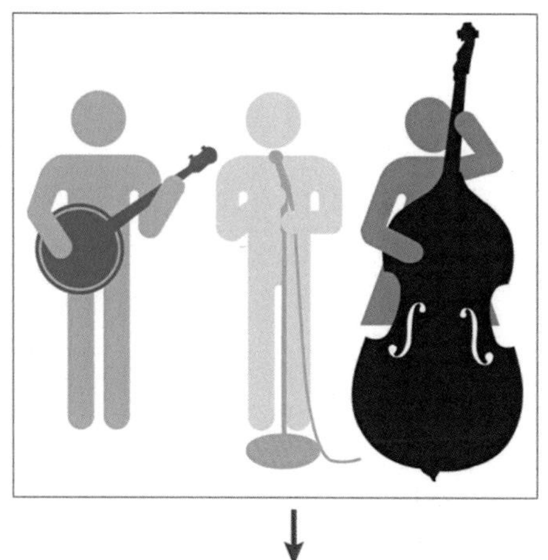

(B) The auditory stimulus that reaches a listener's ear is a complex mixture of these three sources

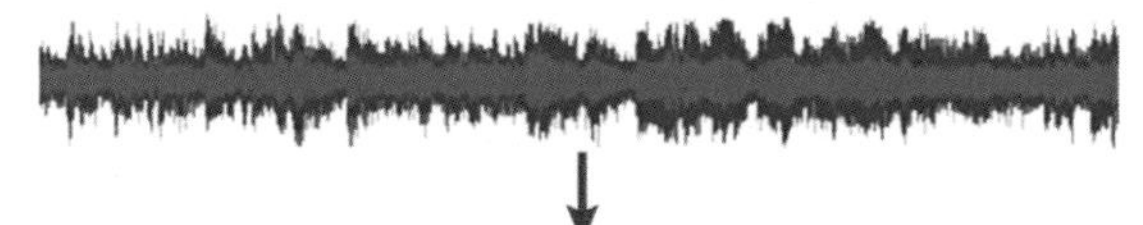

(C) A listener hears each source as a distinct auditory object

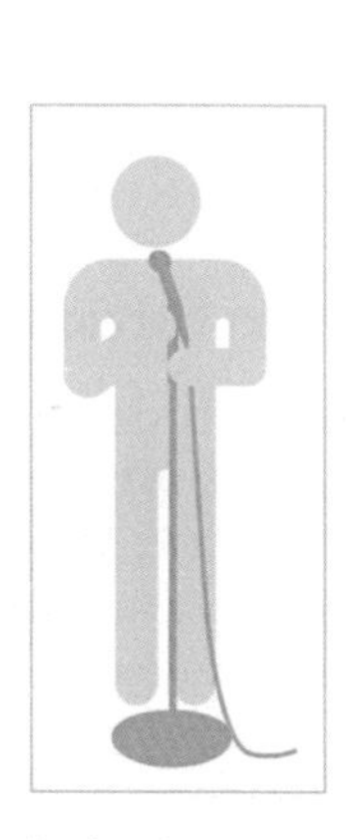
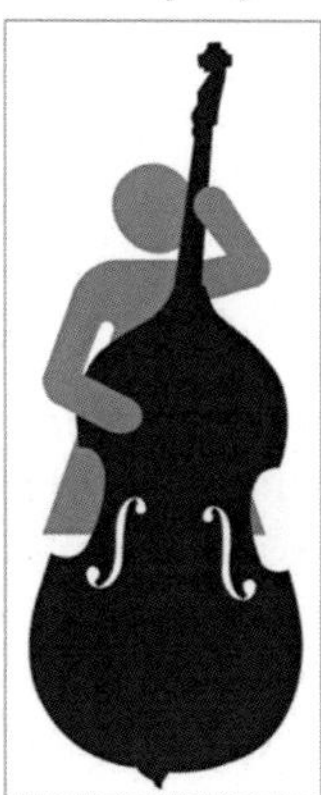

FIGURE 5.15 A cartoon showing the "problem" of auditory scene analyses mechanisms. The sounds emanating from the three sources shown at the top of the figure (a banjo player, a singer, and a bassist) produce complex and overlapping sound signals that must nevertheless be decoded and deciphered for accurate source detection of music coming from three discrete sources. *Source: Bizley and Cohen (2013).*

representation of the stream, inputs that do not match this stream are detected using auditory cortical change detection mechanisms. A third view is that the perceptual organization processes take place in an area of the cortex that is thought to underlie binding processes for visual and somatosensory input, the intraparietal sulcus (Cusack, 2005). In this view,

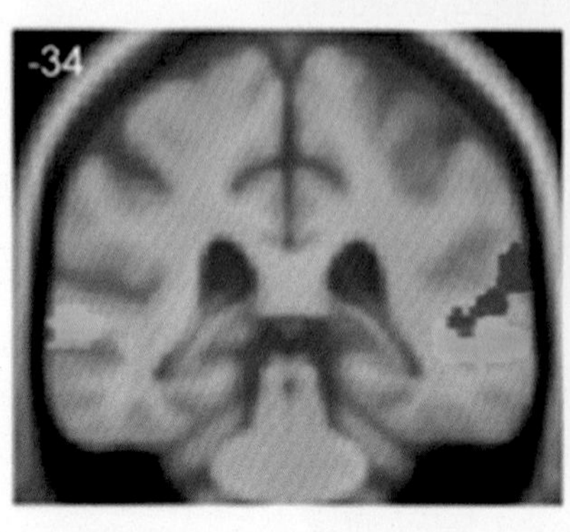

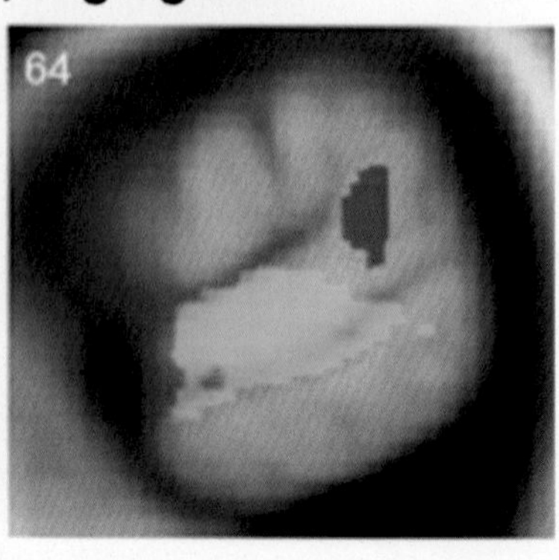

FIGURE 5.16 Auditory stream analysis in the brain. Left panel: coronal image showing auditory responses in the temporal lobe with the left hemisphere shown on the left side of the image. Right panel: sagittal image showing the right hemisphere responses. Cortical areas for auditory stream analysis: regions shown in cyan are in response to hearing one's own name (name identification). Regions shown in magenta are in response to auditory object segregation processing. Response to hearing one's own name activated regions bilaterally in the superior temporal lobe. The interaction of name and auditory object segregation (extracting one's name out of a stream of sound, the "cocktail party effect") showed right hemisphere activation. *Source: Modified from Golden et al. (2015).*

the perceptual organization of multiple auditory streams occurs external to the auditory cortex in neural territory that is implicated in the multimodal cortex.

An fMRI study of auditory scene analysis using a contrast of the sound of a person's own name versus a complex auditory stream of sounds showed activation bilaterally in the temporal lobe auditory and language cortex (Golden et al., 2015) (Fig. 5.16).

3.3 "What" and "Where" Processing Streams

When the auditory system "parses" or decodes a complex auditory scene, multiple aspects of that scene must be analyzed to determine which auditory objects are present, where they are in space, if they are overlapping in time, etc. To effectively analyze the auditory scene, the auditory system must know where objects are in space, if they are moving or are stationary, and what exactly those objects are: object recognition. Many years of investigations in nonhuman primates and in humans have provided evidence that the processing of *where* objects are and if and how they are in motion and *what* an object is, its identity, are subserved by different regions in the cortex. In other words, while the early processing stream is in the auditory cortex, later processing of that auditory scene diverges to one of two major pathways. This is the *dual stream model* of auditory processing.

3.3.1 *Dual Streams for Processing What and Where Information*

There is a large and growing body of evidence that is in support of distinct (but highly interactive) cortical networks for decoding spatial (where) and nonspatial (what) information in sounds. This dual pathway of auditory information processing has largely been explored in studies of human speech processing: more on that later in the chapter. However, a "where"

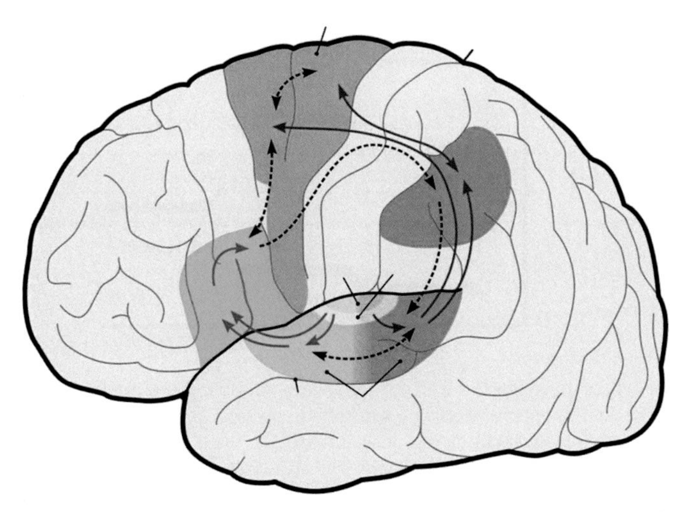

FIGURE 5.17 A model of dual auditory processing streams of the human brain. A ventral stream (shown in green) extends from the superior temporal regions in the auditory cortex anteriorly and then on to the inferior frontal cortex. The dorsal stream extends posteriorly through superior temporal regions to areas within the parietal lobe and then anteriorly to the premotor cortex. Note that the *arrows* show bidirectional, reflecting the feedforward, feedback, and parallel processing that is a hallmark of the auditory speech and language system. *Source: Rauschecker and Scott (2009).*

and "what" dual stream organization for nonspeech sounds has strong support across a wide range of methods and approaches. In experiments where sounds are in motion or have other spatial cues, a processing stream extends posteriorly from auditory regions and dorsally to the parietal and frontal lobes. This is the dorsal stream, or "where" stream. In experiments where auditory object recognition is the central goal, a processing stream extends from auditory regions in the superior temporal lobe ventrally to middle and inferior temporal regions. This is the ventral stream, or the "what" stream. In a recent review article, Rauschecker and Scott (2009) provide a summary of findings to date that provide hypothesized brain regions for "what" and "where" processing streams in the human brain (Fig. 5.17).

4. SPEECH PERCEPTION

The basic task of the speech perceptual system is to map sounds onto meaning. This seems to be a relatively straightforward process: when a speech sound, such as "d," is heard, the physical sound is mapped onto an abstract representation of that sound: the *phoneme*. The two main types of phonemes are consonants (such as "d") and vowels (such as "i"). Individual phonemes are stored in echoic memory while an entire word is being spoken—for example, "dim." To decode the spoken word "dim," you might imagine that the neural representations for "d," "i," and "m" are decoded individually and sequentially, and combined to map onto a sound representation of the word "dim." The result is that the word "dim" is activated in the semantic/conceptual knowledge system. Unfortunately, this description makes perfect sense,

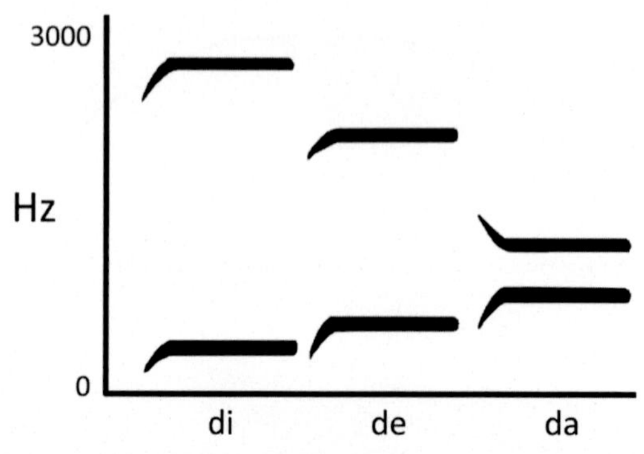

FIGURE 5.18 As we articulate consonant and vowel speech sounds, there are energy and frequency changes as we articulate each individual phoneme. Here is a simplified spectrogram for three syllables beginning with a d: the formant transitions (frequency sweeps) at the onset of each of these three syllables (/di/, /de/, and /da/) corresponds to the /d/ at the beginning of each syllable. The steady-state part of the formant reflects the following vowel. Note that each set of formants for the steady-state vowels are different: not surprising since the vowels are very different in how and where they are articulated in the mouth. Note also, however, that the initial consonant,/d/, represents a single phoneme in our language; however, the changing sweep of information at the onset of each of these syllables reflect that the acoustics of the onsets are quite different from one another despite the fact that they all three map onto the single phoneme /d/. This acoustic difference between tokens of a single consonant phoneme that differs sharply depending on following a vowel is referred to as the "lack of invariance" problem as there is no single unvarying acoustic representation of /d/. This is a fundamental problem to be solved for understanding how speech sounds are decoded. *Source: Hickok (2009).*

but it is not how the speech system actually works. Or at least that is the current view: take a look at Fig. 5.18: you will see that the simplified speech structure shown there for syllables /di/, /de/, and /da/ are quite different even though both syllables begin with the phoneme /d/. This difference in the acoustic features at the onset of these syllables is called the "lack of invariance problem" because the acoustic features underlying the /d/ in /di/ differ from the /d/ in /de/ and /da/. It has spawned many theories about how the auditory system decodes human speech.

Partly due to the lack of invariance problem, at present, there is little agreement in the field of speech perception regarding the basic "building blocks" of speech. Is an individual phoneme the smallest unit of analysis for speech systems, or is the syllable the appropriate unit?

The speech system must not only decode the individual phonemes in speech to map the sound information to meaning but it must also decode "who" information to know who is speaking and "when" to understand the temporal order of speech phonemes, syllables, words, and sentences. Think of speech processing as a special case of auditory scene analysis, where instead of *deciphering differing sources* in a complex listening environment, the speech system is *decoding differing aspects or features* in a complex speech signal such as phonemic contrasts, fundamental frequency which provides cues as to who is the speaker, and intonation contours which provide information about syllabic stress. As mentioned earlier in the chapter, the speech signal must be evaluated across multiple time scales (20, 200, 2000 ms) (see Fig. 5.3). This information must be decoded accurately regardless of the differences in human speech: whether we hear a high-pitched voice of a child or a low-pitched voice of a man, whether we are speaking very loudly or whispering, or whether we are speaking quickly or slowly. Obviously, the speech system is doing a lot more than a simple mapping of sound onto meaning, and it cannot rely solely on the physical aspects of speech, since they vary so widely both within and across speakers.

Recall that we mentioned earlier in the chapter that the auditory system is a complex one that includes ascending pathways, descending pathways, as well as auditory cortical regions. The auditory complex has massive interconnectivity, with feedforward, feedback, and parallel projections that combined to form a system with exquisite (fractions of a second) temporal resolution for decoding the speech signal. And this multidirectional system includes inputs from the cortical language system as well as other regions in sensory, association, and motor cortex. Thus despite the lack of consensus in the field as to what constitutes the basic building block of speech (phoneme? syllable?), there is growing agreement regarding the highly interactive nature of the cortical speech processing system (Box 5.1).

The advent of neuroimaging techniques such as fMRI and MEG have created a virtual explosion of studies investigating the brain bases for speech perceptual processing. In particular, fMRI (Fig. 5.2) has provided researchers the opportunity to investigate separable aspects of auditory and speech processing to investigate cortical regions that subserve specific aspects and stages of speech perception. See Fig. 5.20 for an example of fMRI responses in several cortical auditory regions of interest. Within speech processing, fMRI has allowed researchers to compare responses for specific aspects of speech decoding such as phonological processing: that is, the detecting and decoding of phonemes in human speech (Fig. 5.21).

Speech processing has dual roles in human language: when we hear someone speak, our speech decoding system must transform the acoustic signal of their spoken speech into the meaning of what they are communicating. Conversely, when we choose to speak to someone, our cortical language system must be able to take the idea, the meaning of what we wish to convey to someone, and transform those ideas into spoken speech.

4.1 Dual Processing Streams: Dorsal and Ventral

These dual roles for accessing the sound patterns of speech when decoding or when producing speech have motivated studies investigating dual streams in language cortex that subserve the two roles. A growing consensus implicates a dorsal stream that includes structures in the posterior planum temporale at the parietal-temporal junction (Fig. 5.22, Fig. 5.23) along with the posterior frontal lobe. This stream contains the neural mechanisms for translating the sound patterns of speech into the motor, articulatory-based representations that are used in speech production (Hickok, 2009) (Fig. 5.24). A ventral stream includes structures in the middle and superior temporal lobe. This stream contains the neural mechanisms for mapping the sound pattern of speech onto meaning. The dorsal stream is theorized to be left-dominant for most humans, while the ventral stream is organized bilaterally in both left and right temporal regions. Note how this model of speech and language processing builds on the dual processing streams proposed for general auditory processing by Rauschecker and Scott (Fig. 5.17). The white-matter pathways that subserve these dual routes are presented in Fig. 5.25.

BOX 5.1

BIONIC HEARING USING COCHLEAR IMPLANTS

Bionic hearing? Cochlear implants stimulate the cochlear in the inner ear to provide stimulation that restores partial hearing for some individuals with hearing loss. This is how they work—although there are many variants being designed each year. A sound processor is worn behind the ear. This processor picks up sounds in the environment and turns them into a digital code. This code is then transmitted to the cochlear implant (Fig. 5.19) that has been surgically inserted into the cochlear. The implant then converts the digitally coded sound into electrical impulses and sends them along the implanted array that has been placed in the cochlear. The electrical impulses are then transmitted along the hearing pathways to the brain.

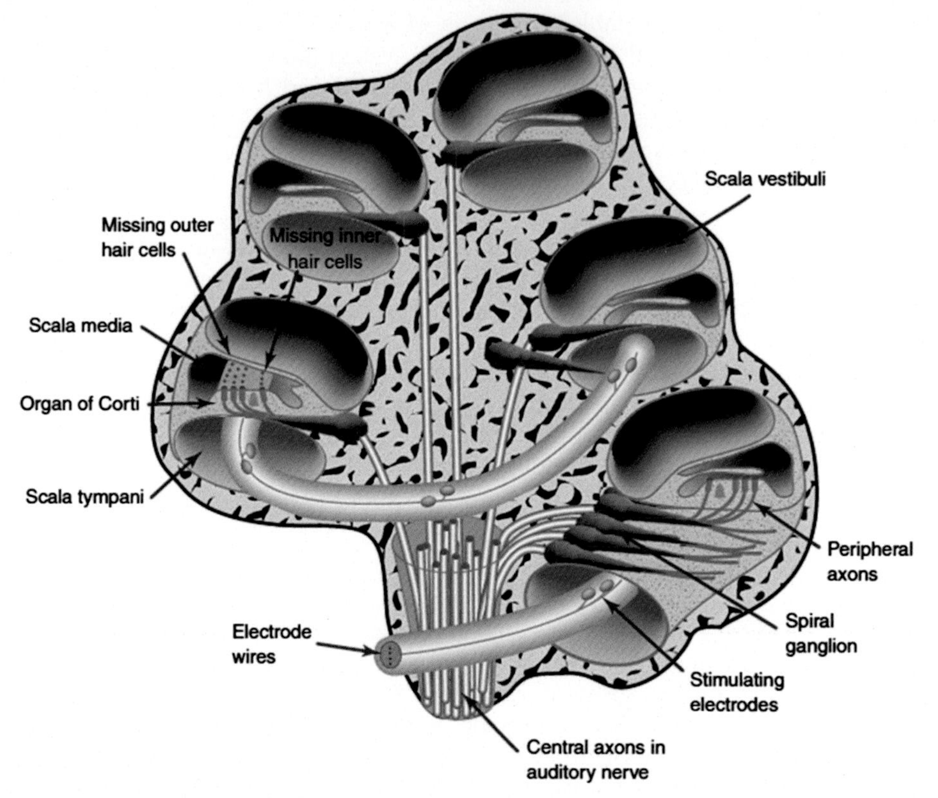

FIGURE 5.19 Cochlear Implants. The cochlear implant can provide partial restoration of hearing for individuals with hearing loss by stimulating the cochlear in the inner ear (see Fig. 5.6). *Source: Brown & Santos-Sacchi in Squire et al., 2013.*

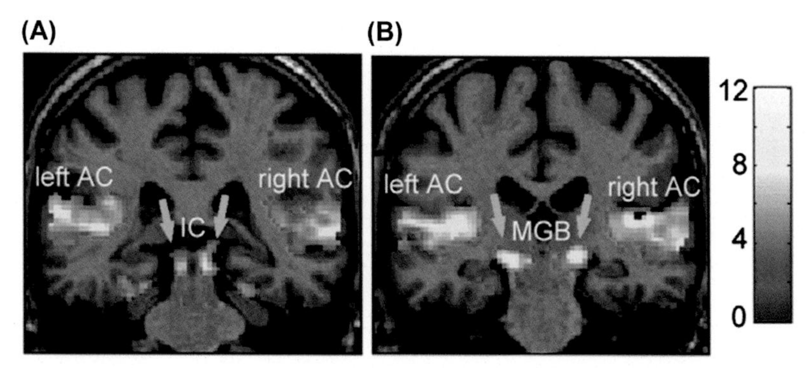

FIGURE 5.20 Functional magnetic resonance imaging responses for classes of sounds shown on a coronal slice, with left and right auditory hemodynamic response shown in red-yellow shading on the left and right side of each brain figure. (A) On this "slice," through the grouped brain responses, the subcortical inferior colliculus (IC) is shown being activated (*green arrows*). (B) On this "slice," through the same grouped brain responses, the medial geniculate body (MGB) can be seen to be activated as well (*green arrows*). *Source: Brown & Santos-Sacchi in Squire et al., 2013.*

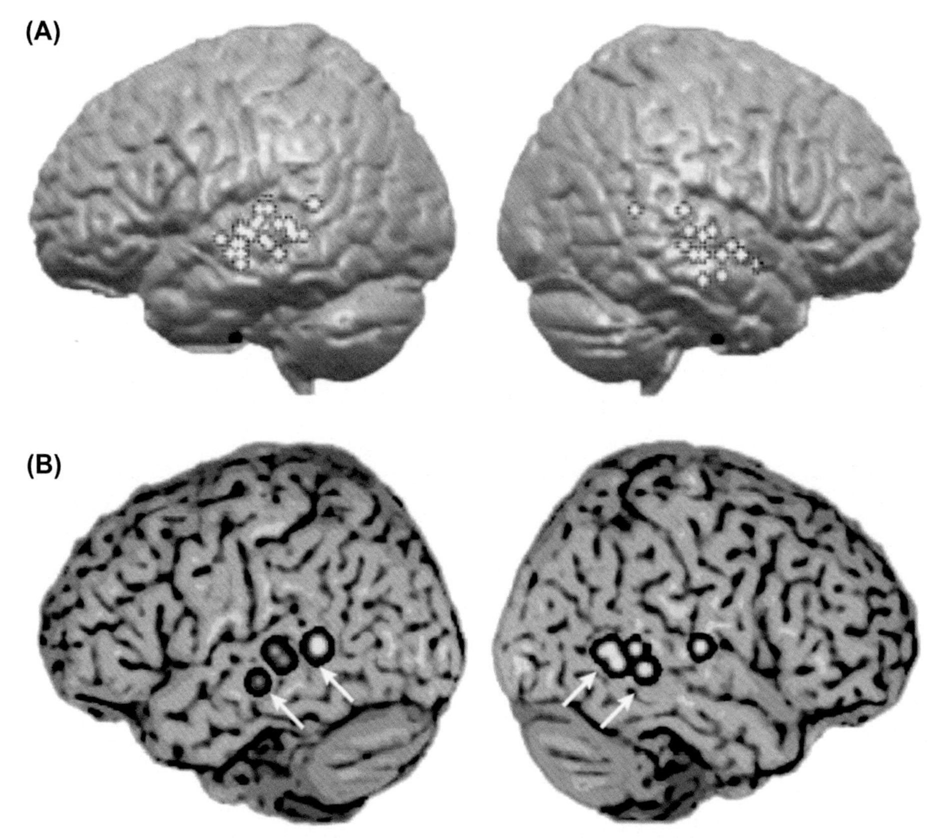

FIGURE 5.21 (A) Functional magnetic resonance imaging (fMRI) activations plotted for several recent studies using sublexical speech sounds (consonants, vowels, consonant-vowel syllables). Note the activation pattern for both left and right hemispheres in the temporal lobe. (B) fMRI activation for lexical items (words) showing bilateral but somewhat more inferior temporal lobe activation. These types of analyses provide evidence for just what aspect or stage of speech processing is occurring in left and right brain areas. *Source: Hickok and Poeppel (2007).*

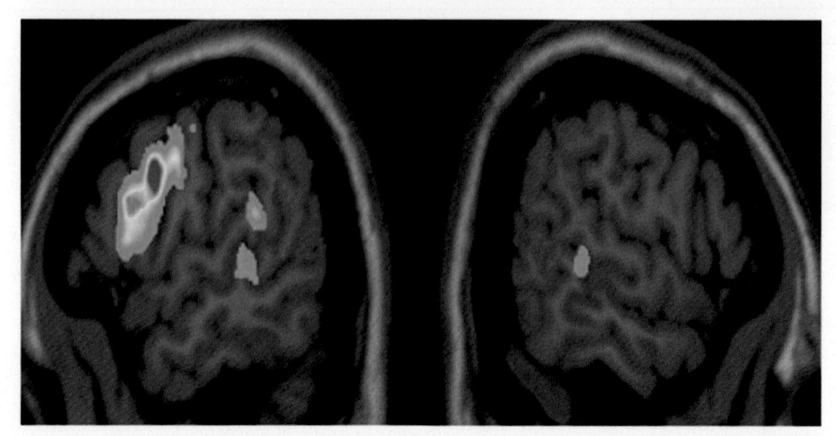

FIGURE 5.22 Functional magnetic resonance imaging results for both the perception of speech and subvocal reproduction of speech. The left hemisphere is shown on the left side of the figure, the right hemisphere is shown on the right side of the figure. Note that in the left hemisphere, there is activation in the frontal lobe as well as in area Sylvian-parietal-temporal (Spt). Note also that the superior temporal sulcus is activated in both hemispheres by these tasks. *Source: Hickok and Poeppel (2007).*

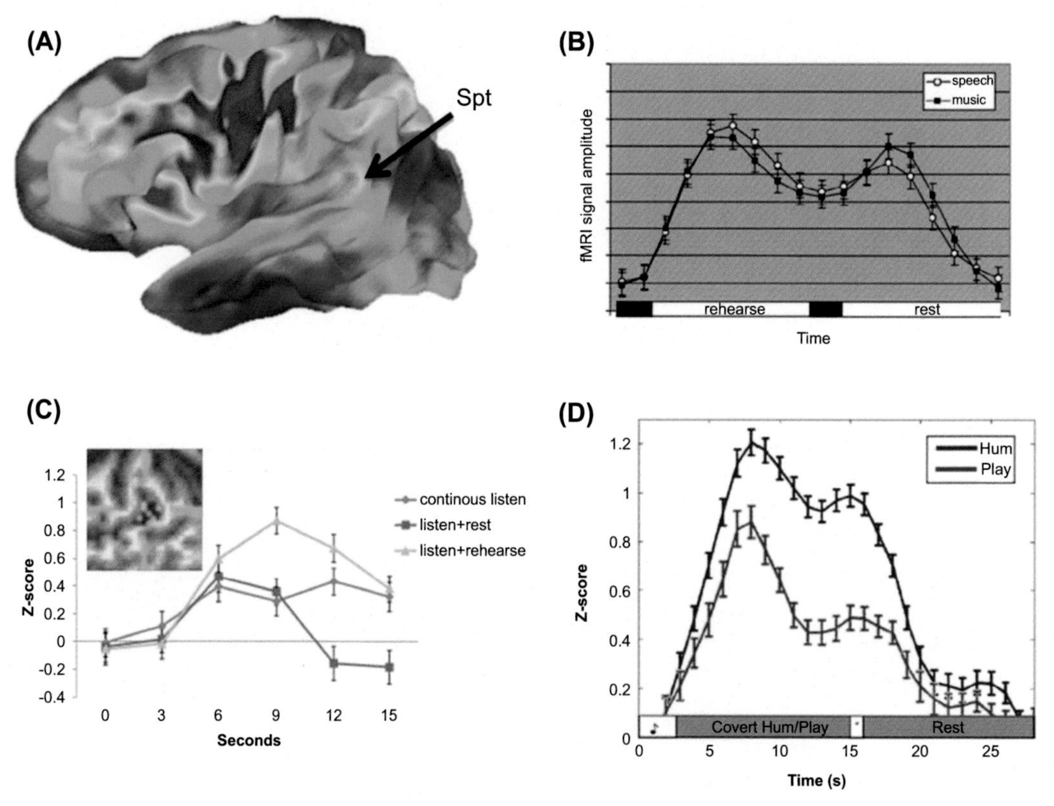

FIGURE 5.23 (A) Area Sylvian-parietal-temporal (Spt) is shown in a brain activation map for 102 subjects performing a task that involves both listening to and covertly rehearsing speech—thus activating both speech perceptual and speech production neural systems. (B) Time course of activation of Spt for both speech (open symbols) and music (closed symbols). (C) Activation time course in area Spt for three conditions: blue: listening to continuous speech, red: listening + rest, and green: listen + rehearse. Note that continuous listening and listen + rehearse have different activation patterns, with higher response amplitudes during the 6–12 s of listen + rehearse compared with continuous listening. (D) Activation time course in Spt in skilled pianists listening to novel melodies and humming along with them versus listening to novel melodies and imagining playing them on the piano. Results indicate that area Spt is more active for vocal tract activities than hand and finger activities. *Source: Adapted from Hickok (2009).*

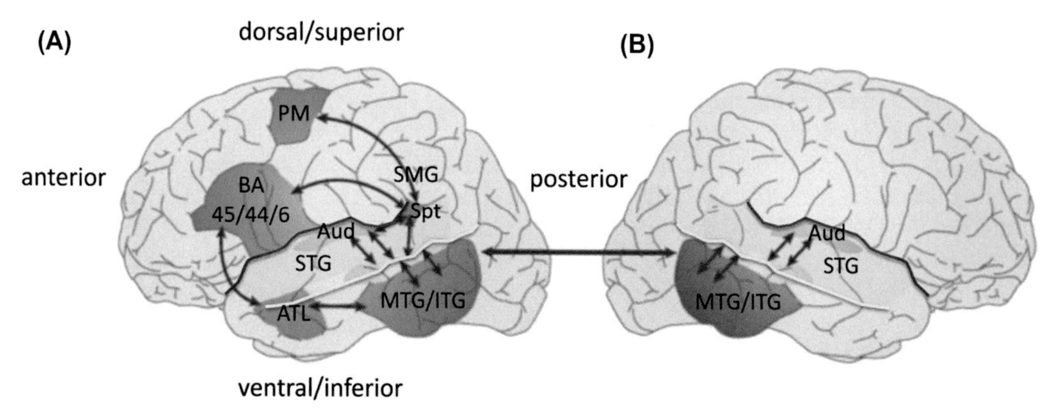

FIGURE 5.24 (A) A schematic for the model of auditory language processing proposed by Hickok (2009). (B) Brain regions proposed to reflect stages of the model. Note that early speech perceptual systems (shown in green and yellow) for mapping the acoustic-phonetic information in sounds onto meaning are proposed to be mediated bilaterally in left and right hemispheres, while later processes are proposed to be mediated by left hemisphere regions. *ATL*, anterior temporal lobe; *Aud*, auditory cortex; *BA45, BA44, BA46*, Brodmann areas 45, 44, and 46; *ITG*, inferior temporal gyrus; *MTG*, middle temporal gyrus; *PM*, premotor; *SMG*, superior marginal gyrus; *STG*, superior temporal gyrus. *Source: Modified from Hickok (2009).*

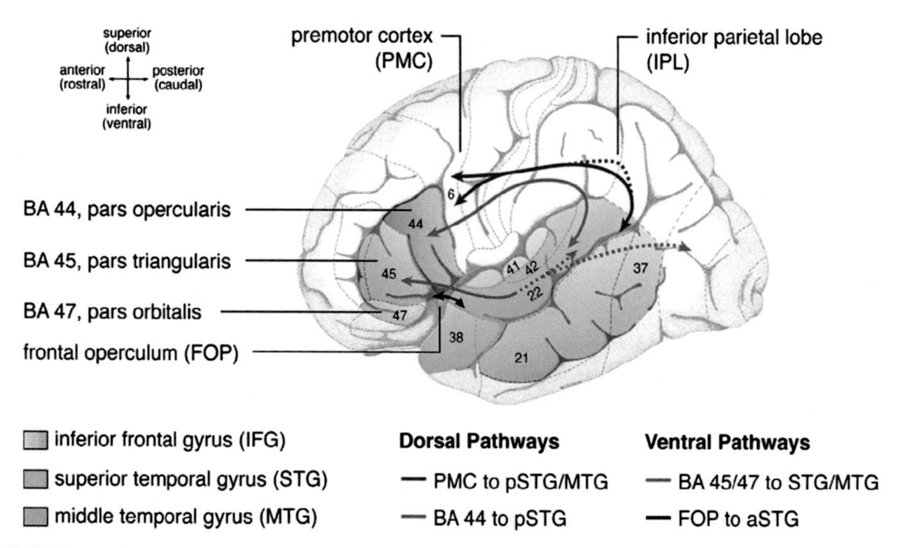

FIGURE 5.25 White-matter pathways of the speech and language system. Pathways that aid in the dynamic processing of speech and language. The numbers on the brain figure represent the cytoarchitectonically defined Brodmann areas (BAs). The major pathways include a bidirectional dorsal pathway between the superior temporal gyrus to premotor cortex (shown in purple) and a second pathway (shown in blue) between BA 44 in the inferior frontal lobe and the posterior region of the superior temporal gyrus. The ventral pathways include a pathway from BA 45/47 in the inferior frontal lobe and the superior temporal gyrus and the middle temporal gyrus (shown in pink). A second ventral pathway is the short pathway between frontal operculum and the anterior superior temporal gyrus (shown in green). *Source: Friederici, in Celesia & Hickok, 2015.*

5. MUSIC PERCEPTION

Like speech perception, music perception is uniquely human. There are many similarities in speech and music perception: music has complex phrase structures, and its perception involves the mapping of sound onto meaning (and emotion). Music perception allows for the recognition of melodies despite differences in instruments, keys, and tempos; thus it cannot be a system built on absolutes but must have relative representations. Thus music perception systems must have the ability to maintain a perceptual constancy in music representation. As in speech decoding, think of music perception as a special case of auditory scene analysis. In this case, where instead of *deciphering differing sources* in a complex listening environment, the music perception system is *decoding differing aspects or features* in a complex musical signal such as *meter* which provides cues to the temporal structure and rhythm, *intervals and chords* which aid in decoding melody, and *timbre* which aids in deciphering the musical instruments or human voices.

A central difference between speech and music perception is that all typically developing humans master speech perception. We are not only good at speech perception but we are masters! This is not the case in music perception: there is tremendously more variability in music perception abilities and significantly more explicit learning that goes along with musical acuity. The variability in music perception abilities, combined with the many levels of musical training and skill, has made the study of music perception difficult because of these inherent individual differences. These difficulties, however, provide a unique opportunity in that they provide an opportunity to understand the effects of learning and plasticity in the brain areas that decode music.

5.1 Stages of Music Processing

The perception of features in music involves many stages of processing within the auditory system as well as across brain regions (Fig. 5.26) (Zatorre, Chen, & Penhune, 2007). As we have stated about auditory and speech processing, music processing must include both feedback and feedforward systems, as well as making contact with both stored memories and experiences, and emotional systems.

While the music signal is complex, like all sound, music has basic physical elements: frequency, intensity, and time. The psychological aspects of frequency and time in music correspond to pitch (melody) and temporal structure (rhythm). Traditionally, melodic and temporal aspects of music have been investigated as separate features of music perception. However, they likely are not completely independent. Just as some speech scientists propose that speech may be processed in brain areas specialized just for speech, music scientists have theorized that there may be neural systems specialized for music. Evidence in support of music-specific systems in the brain has been provided in neuropsychological studies with patients who have suffered brain damage. Peretz et al. provided a series of

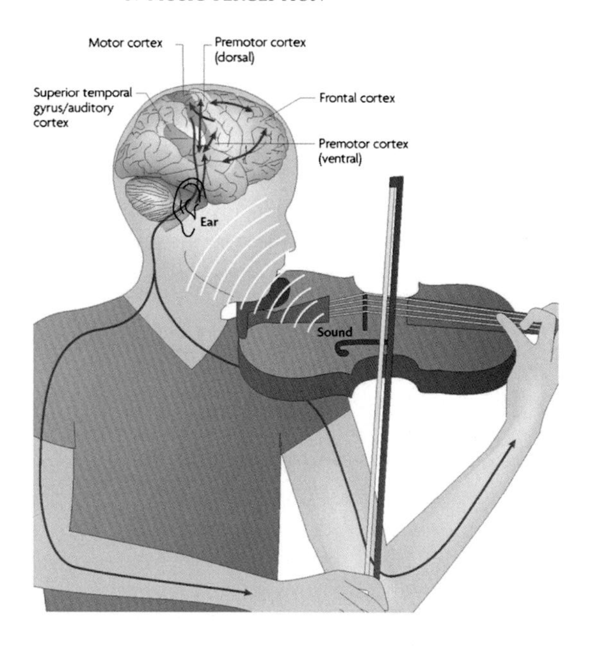

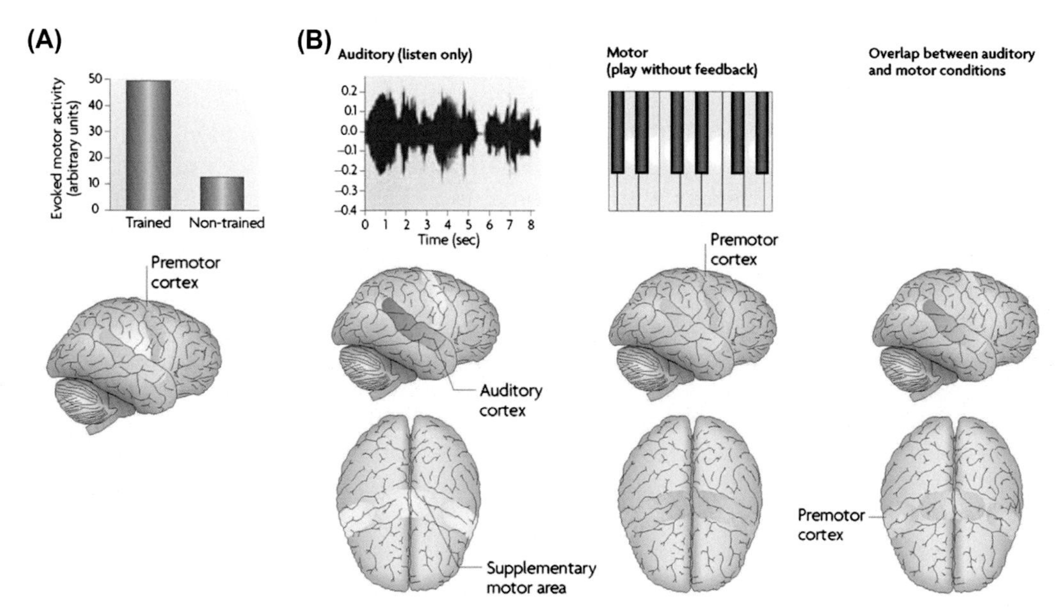

FIGURE 5.26 Left panel: have you ever wondered what areas of your brain "light up" when you play or hear music? Of course, the auditory regions do activate when you play an instrument, for example, but so do many motor regions involved in the production of the sound from the instrument. In fact, these motor and sensory (auditory) systems are tightly coupled and together form a neural circuitry that provides feedback and feedforward information for musical instrument playing. Right panel: results from several neuroimaging studies provide evidence for a tight coupling between activity in the auditory and premotor cortex (PMC). (A) People without musical training were taught to play a simple melody on a keyboard, and their brain activity for listening to that melody was compared in pretraining versus posttraining scans. Results showed activity in the auditory cortex, as expected, when listening to the melody, but there was also activity in the PMC but only in the posttraining condition. (B) In other studies, researchers compared the brain activity of musicians while they listened to a piece they knew how to play (left column) with the brain activity while they played the same piece but without auditory feedback (middle column). There was a significant overlap in the PMC and in the auditory cortex, suggesting that the auditory and motor systems interact closely during both perception and production. *Source: Adapted from Zatorre et al. (2007).*

investigations with brain-damaged individuals showing that, in some individuals, pitch or melody perception may be selectively damaged, leaving temporal structure perception intact, while in other individuals temporal perception may be damaged while pitch perception is intact. These findings have led to the development of a model for the brain organization for music perception (Peretz & Zatorre, 2005), where melodic features in music are processed preferentially in the right hemisphere and can be selectively impaired with right hemisphere brain damage, whereas temporal structure in music is decoded in a larger network of brain areas in both hemispheres.

5.2 A Separate System for Music Perception?

Is music perception a separable aspect of auditory processing? While the work of Peretz et al. provides compelling evidence that this is the case, a review (Koelsch, 2005) of neuroimaging studies of music perception describes a growing body of evidence in support of the view that some aspects of music perception, notably the musical structure or syntax and the musical meaning or semantics, share neural territory with brain areas for language processing. The studies reviewed by Koelsch provide compelling evidence for at least some shared systems in music and language processing. Language and music are both uniquely human and highly structured signals, with multiple dimensions along spectral and temporal axes for understanding their basic and complex structures. Perhaps there is no unitary brain region for either language or music: it may be the case that language and music systems have some neural territory that is specific for their processing and some neural territory that is shared.

A key aspect of music perception that differs from speech perception is the strong emotions that music raises in listeners. Orchestral music with no lyrics can evoke strong and varying emotions in listeners with no words. In fact, the emotions felt can frequently not be translated into words (Box 5.2). Music also seems to *enter into our brain* across many pathways so that we feel the emotion of the piece, our foot taps or our hands clap to the beat, and we may smile or frown in response to elements in the music. How does this seemingly brain-wide response occur? Koelsch (2014) has extensively studied music perception in the brain and has presented a model for music pathways that produce this widely active and varied response (Fig. 5.27). These pathways include an auditory–limbic pathway, a somatic motor system, a visceromotor (autonomic) system, and an acoustically activated vestibular pathway. Together, these complex pathways and systems combine to explain how music may make us soar with emotion, tap our foot lively, and *feel* the music throughout our brain and body.

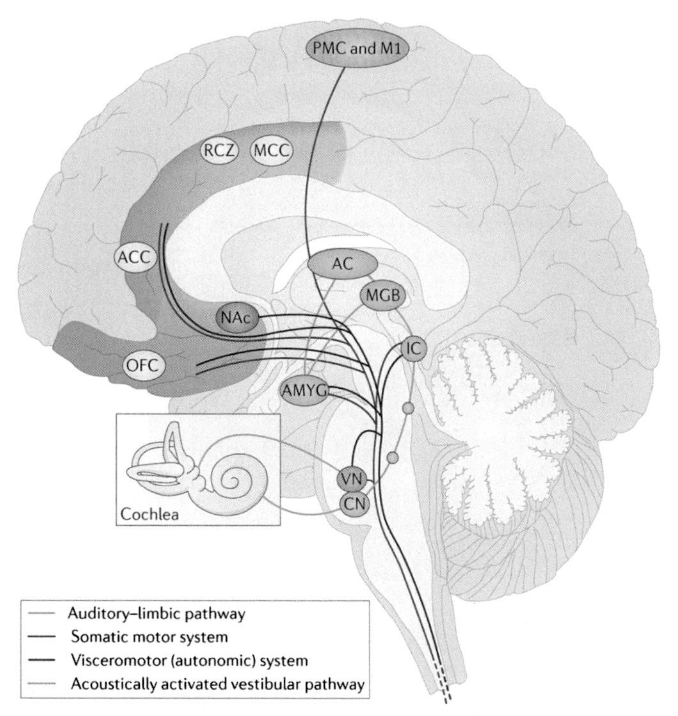

FIGURE 5.27 Koelsch (2014) has extensively studied music perception in the brain and has presented a model for music pathways that produce this widely active and varied response. These pathways include an auditory–limbic pathway (shown in turquoise), a somatic motor system (dark blue), a visceromotor (autonomic) system (red), and an acoustically activated vestibular pathway (orange). Together, these complex pathways and systems combine to explain how music may make us soar with emotion, tap our foot lively, and *feel* the music throughout our brain and body. *Source: Adapted from Koelsch (2014).*

BOX 5.2

THE NEURAL LANGUAGE OF MUSIC

Music can exhilarate us…scare us…calm us down or make us happy (Fig. 5.28). How does this happen and what parts of the brain are involved? Is it an essential part of being human? Or does it develop with experience? These are core questions being asked by neuroscientists studying music and emotion in the brain.

We have all had the experience…watching a suspenseful movie and just as the action is building, the music rises…What is it about the particular piece of music that increases our feeling of suspense? Is it learned through repeated association of watching a suspenseful movie and experiencing the musical background? Or is there something in the music itself that increases our feeling of suspense?

What Is Music Anyway?

Music comes in all shapes and sizes…from classical sonatas to jazz to rap, from country to good ol' rock and roll. Music arrives with lyrics and without. It differs in rhythm, tone, beat, tempo…across more parameters than you can image. So it is no wonder that it has been difficult to come up with simple ways to discuss music and human cognition. Some

Continued

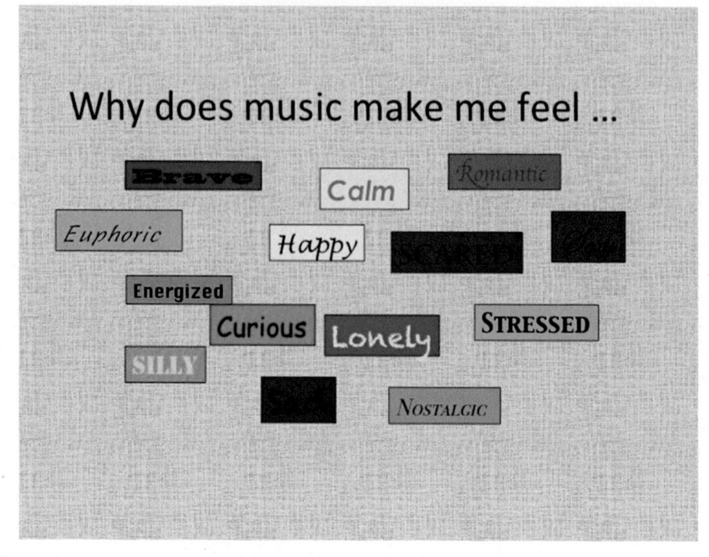

FIGURE 5.28 The many moods inspired by music. Music can exhilarate us…scare us…or calm us down. How does this happen and what parts of the brain are involved? *Source: Gage with permission.*

might argue that you cannot even discuss music using words. Louis Armstrong was famously quoted as saying "If you have to ask what jazz is, you'll never know" (1).

There is some central agreement about one aspect of music, however. It is emotional. We feel it. Music can evoke fear, sadness, joy, and rekindle long-past memories and events. How does this happen in the human brain?

Music and Emotion in the Brain

There is a lot of evidence that music is processed across widely varying brain areas and, in particular, in the limbic system of the brain where emotional processes occur. In a review article, Koelsch (2010) presented a consolidated view of activation of limbic and paralimbic brain regions by music (Fig. 5.29). See Chapter 11—Feelings—to learn more about the limbic and paralimbic systems.

The Neuroscience of Music and Emotion

While music and emotion seem tightly coupled from an intuitive sense, to date, there is not a unified theory of the neuroscience of music. And while most neuroscientists studying the brain basis of music perception and experience agree that emotion plays a key role, just what that role is and how it happens in the brain is hotly disputed. But it seems to us that Louie Armstrong had the sense of it…mere words do not really do justice to our musical experience—whether we are making music or listening to it.

Does music have a special language of its own as it is processed in our brains? That is an idea put forth by Large and his colleagues. They get right to the point in their review article "Dynamic musical communication of core affect," stating that "music speaks to the brain in its own language." (3). How does this communication take place?

Brain rhythms are a key aspect of how the brain works: oscillations between neural networks form a central form of communication across the thalamocortical core of the brain (see Chapter 1). Spatially distributed dynamic processes across the brain come together to form a dynamic core (4). These distributed neural processes provide a mechanism for conscious processing (5).

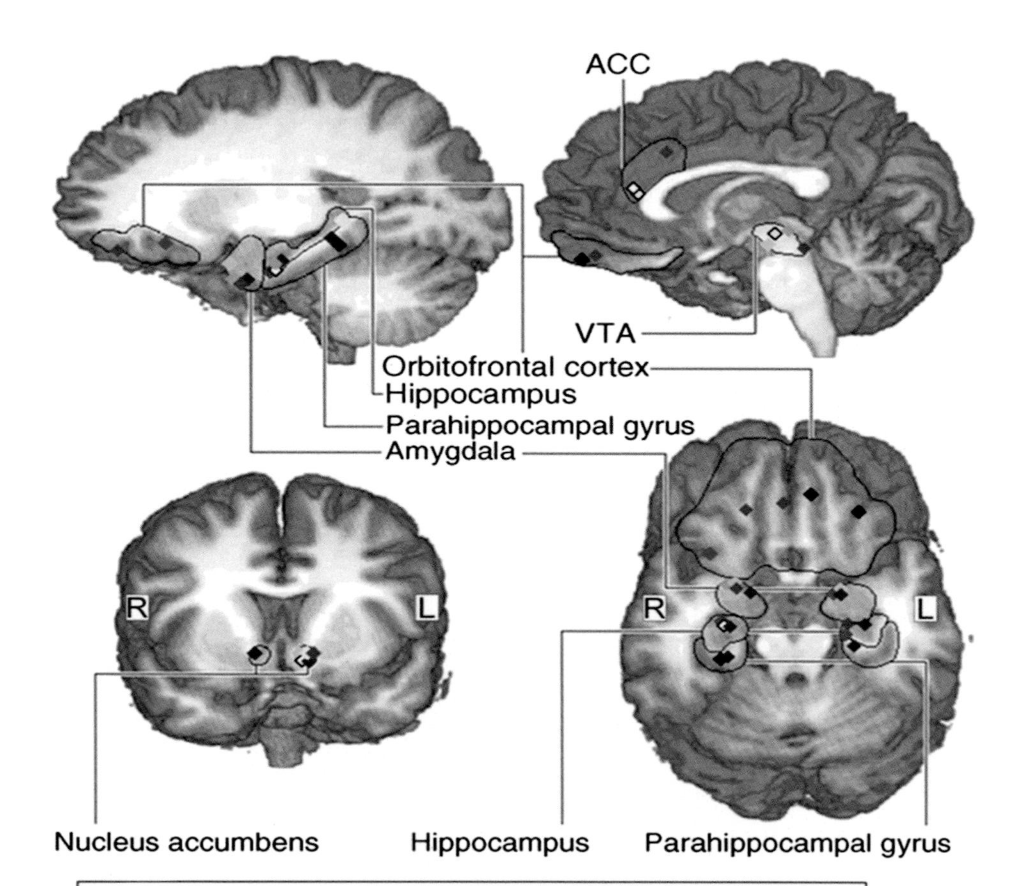

FIGURE 5.29 Brain structures in the limbic and paralimbic system that respond to music: a review of brain areas activated by music in 12 studies: note that areas reflect many regions within the human brain for emotional processing. The top right brain figures show the right hemisphere from a sagittal and a medial view. The bottom figures show a coronal view (left) and an axial view (right). *Source: Koelsch (2010).*

According to Flaig and Large (3), music "connects" with these core neurodynamic processes: the synchronous brain responses to music couple with the core dynamic processes to enable "music to directly modulate core affect" (Fig. 5.30).

Does it sound far-fetched to think of music connecting with brain rhythms? We have known for a long time that aspects of sound such as rhythm, pitch, and tone produce *neural entrainment* in the brainstem and cortical auditory regions. That is, the *rhythm of the*

Continued

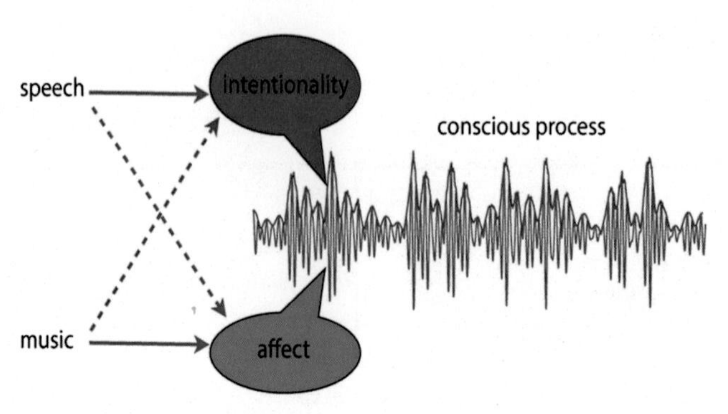

FIGURE 5.30 Speech and music enter into the contents of consciousness; however, it is theorized that they do so in differing ways. Speech in general reflects intentionality whereas music in general reflects affect (moods and emotions). *Source: Flaig and Large (2014).*

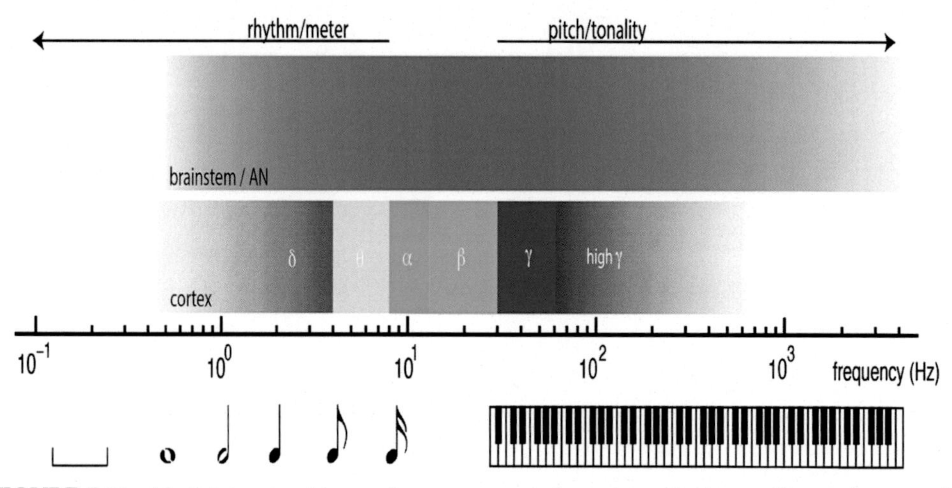

FIGURE 5.31 Music interacts with neural processes across large timescales from millisecond to seconds. *Source: Flaig and Large (2014).*

music affects the *rhythm of the neural response.* In Fig. 5.31, a schematic of the timescales of neural processing and music are shown.

Flaig and Large (3) argue that the distinct natures of music and neural dynamics afford a way for the emotional aspects of music to engage the brain in a direct way, without being translated by our cognitive selves into language or thought. This, they propose, may be a way to explain *musical qualia*—that aspect of our experience with music that is experientially based and not tied to the actual acoustical parameters of the sound.

> We suggest that, to support affective communication, music need not mimic some other type of social interaction; it need only engage the nervous system at the appropriate timescales. Indeed, music may be a unique type of stimulus that engages the brain in ways that no other stimulus can.

BOX 5.2 *(cont'd)*

The theory presented by Flaig and Large (4) provides an intriguing notion about how our brain decodes music...and may open the door for new insights about the neuroscience of music.

References

(1) Louis Armstrong Society Jazz Band. http://www.larmstrongsoc.org/quotes/.

(2) Koelsch, S. (2010). Toward a neural basis of music-evoked emotions. *Trends in Cognitive Sciences, 14*(3), 131–137.

(3) Flaig, N.K., & Large, E.W. (March 17, 2014). Dynamic musical communication of core affect. *Frontiers in Psychology, 5*, 72. https://doi.org/10.3389/fpsyg.2014.00072.

(4) Edelman, G.M. (2003). Naturalizing consciousness: a theoretical framework. *Proceedings of the National Academy of Sciences of the United States of America, 100*, 5520–5524. https://doi.org/10.1073/pnas.0931349100.

(5) Edelman, G.M., Gally, J.A., & Baars, B.J. (2011). Biology of consciousness. *Frontiers in Psychology*, 2(4), 1–7.

6. AUDITORY AWARENESS AND IMAGERY

We can close our eyes and shut out visual images, but we cannot close our ears to shut out auditory events. What is the effect on the auditory system? The auditory system is the last sensory system to fall asleep (or become unconscious with sedation) and the first to awaken. In this section, we highlight some recent studies of auditory awareness during less-than-conscious states, such as sleep or sedation. We also highlight studies of auditory activation for imagined-not-heard sounds.

6.1 Auditory Awareness During Sleep and Sedation

Think about the best way to wake up a sleepy friend: call his name! A neuroimaging study investigated brain responses in sleep and in wakefulness in two conditions: neutral, where the sound was a simple beep, and significant, where the sound was the subject's own name (Portas et al., 2000). Two main results of that study were that beeps and names activated the auditory cortex both when the subject was awake and when the subject was sleeping, indicating that the auditory cortex processes sounds even during sleep. A second key finding was that the auditory cortex response for the neutral tone versus the subject's name did not differ during sleep, indicating that auditory processing during sleep encodes the presence of sounds but did not differentiate between these very different sounds. Brain activation patterns for names versus tones did differ, however, in the middle temporal gyrus and frontal lobe regions. There were also areas in the amygdala that were more active during the presentation of the subject's own name during sleep

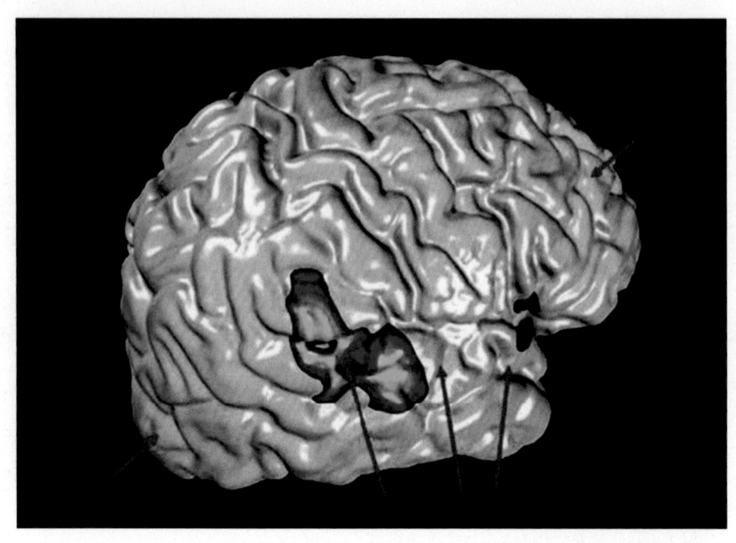

FIGURE 5.32 Brain areas active for imagined sounds. Although the brain scanning took place in silence while the subject was imagining sound, the posterior aspect of the superior temporal gyrus was activated (shown in blue and green colors). *Source: Modified from Zatorre and Halpern (2005).*

than when awake. Do these brain areas represent a circuit in the brain that alerts us to wake up when we hear our own name? More investigations are needed to support this theory, but these findings provide intriguing evidence for how the auditory system "wakes itself up."

6.2 Auditory Imagery

"Sounds not heard" are playing in our heads all day. Some sounds are uncalled for; they just seem to happen: a melody that spins around in your head, your inner voice talking to yourself. Other sounds that are not heard aloud are planned: practicing lines for a school play or rehearsing a phone number before dialing. Where are these sounds processed in the brain? We are aware that we actually seem to "hear" these inner sounds. Does that mean that the auditory cortex is activated when they are playing despite the fact that there is no actual sound? Zatorre and Halpern (2005) investigated this question using neuroimaging techniques to measure brain activation for imagined sounds versus heard sounds. Results (Fig. 5.32) show that the nonprimary auditory cortex is indeed active during imagined—and not heard—sounds.

A related finding was reported by Jancke et al. (Bunzeck, Wuestenberg, Lutz, Heinze, & Jancke, 2005). These investigators wanted to study auditory imagery for environmental sounds. Using fMRI, they recorded neural responses to subjects perceiving sounds and imagining those sounds. Results are presented in Fig. 5.33: the primary and secondary auditory cortexes in both hemispheres are active when perceiving sounds (left panel), while the secondary (and not primary) auditory cortex is active when imagining those same sounds (right panel). These findings provide compelling evidence that

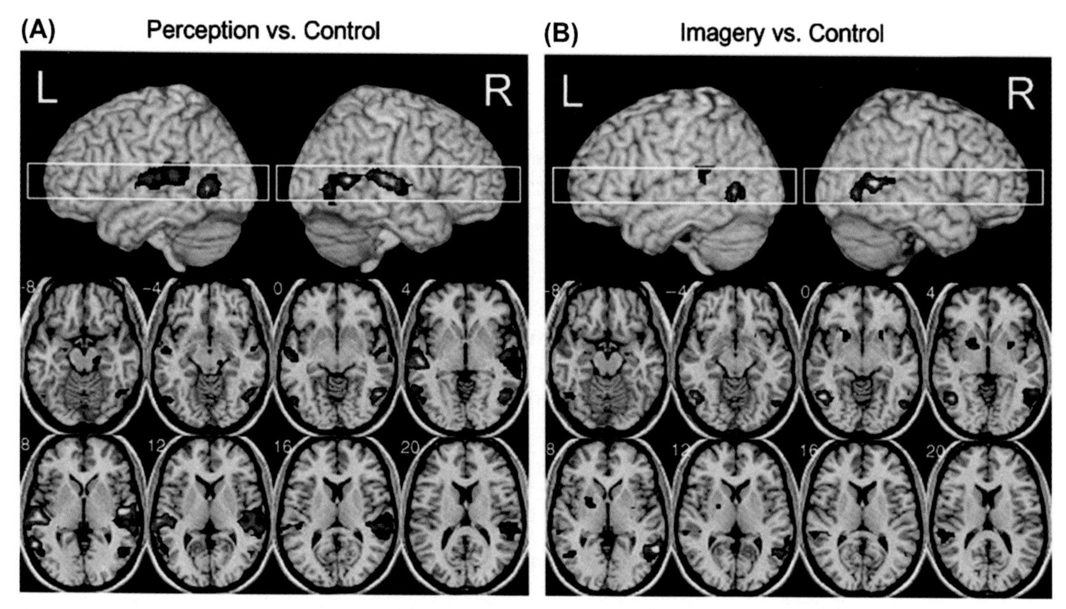

FIGURE 5.33 Functional magnetic resonance imaging study of perceived sounds versus imagined sounds. The sounds used in this study were neither language nor music, to determine the localization of imagined nonlinguistic or musical sounds. The primary auditory cortex was active during the perception phase (A) of the experiment but not during the imagery phase (B). *Source: Modified from Bunzeck et al. (2005).*

imagined sounds activate similar neural regions in the auditory cortex that are activated when sounds are heard. The findings presented here, while representing only a small proportion of the ongoing investigation of auditory imagery, indicate that similar processes occur in humans as well, with imagining and perceiving sounds sharing neural territory.

7. SUMMARY

In this chapter, we presented an overview of the complex auditory system, from hearing basics to music perception to auditory imagery. The advent of neuroimaging techniques has provided a wealth of new data for understanding the cortical auditory system and how it interfaces with other cortical regions. While we have made major inroads on understanding the puzzle of auditory perception, there is still much work to be done. For example, teasing apart neural systems that underlie music and speech perception is still in the early phases. There are many other key questions that are being addressed in the field of auditory brain science. For example, what are the differing roles of the left and right hemispheres in speech and music perception?

There is fruitful work in the investigations of processing streams in the auditory system and in the brain. And while the work in nonhuman primates has informed us greatly about the existence of "where" and "what" processing streams, these streams may be established differently for humans due to the unique and important roles of speech and music perception in the evolution and development of the human brain. The next time an uncalled melody plays inside your head, consider the areas that might be activated in your brain as you "hear" your silent song!

8. STUDY QUESTIONS AND DRAWING EXERCISES

8.1 Study Questions

1. What are the basic physical features and psychological aspects of sound?
2. What are the main parts of the auditory system and what are their roles in perception?
3. Briefly describe some differences between the "what" and "where" processing streams.
4. What are the basic units of analysis for speech perception?
5. What have new brain imaging techniques provided us in terms of investigating auditory function?
6. Music fills us with emotion: what are the neural pathways that enable this?

8.2 Drawing Exercises

1. Fill in the blank labels on this framework for auditory perception. Refer to Fig. 5.1 for help.

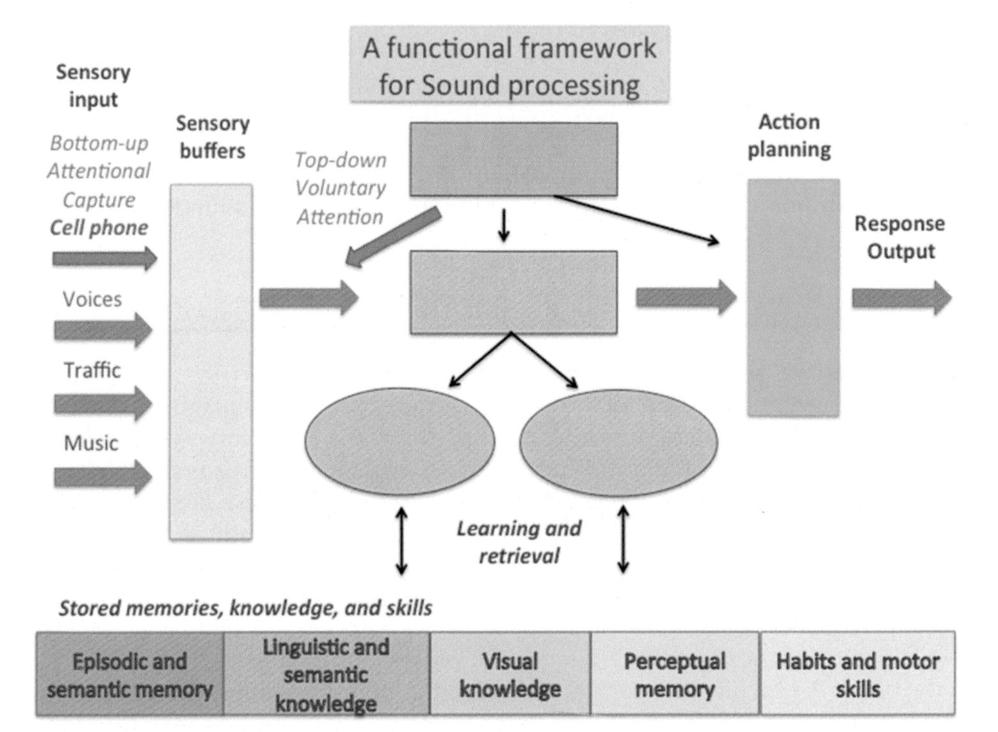

FIGURE 5.34 A functional framework for auditory processing, adapted from the general functional framework presented in Chapter 1. Fill in the blanks with the cognitive processes that are occurring. *Source: Gage with permission.*

2. Fill in the blank labels on this figure for sound localization. What are the two basic cues for localizing sound?

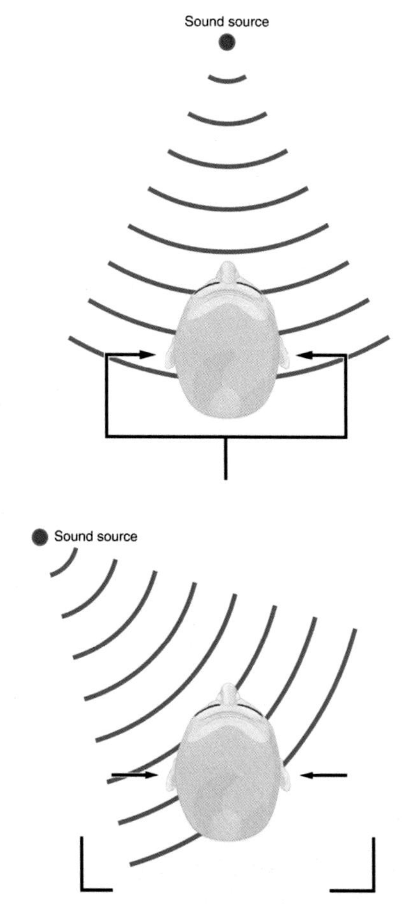

FIGURE 5.35 Sound localization. Complete the figure with labels describing the two basic cues for sound localization. *Source: Illustration from Anatomy & Physiology, Connexions Web site.*

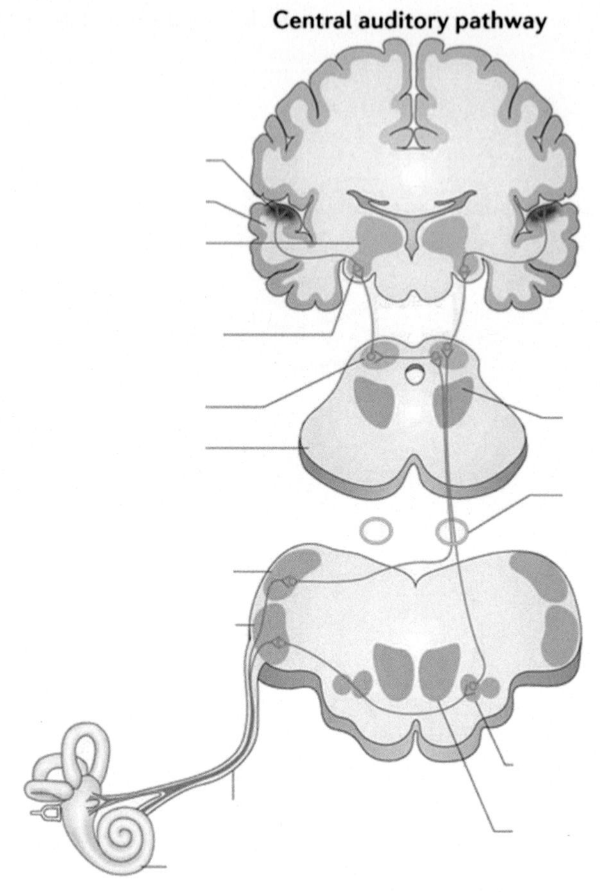

FIGURE 5.36 A schematic diagram of the auditory pathways from the ear to the brain. Complete as many of the missing labels as you can. Refer to Fig. 5.8 for help. *Source: Javitt and Sweet (2015).*

3. Fill in as many of the missing labels for the auditory system as you can. Refer to Fig. 5.9 for help.

CHAPTER

6

Language and Thought

MAN THE COMMUNICATOR

Language is a process of free creation: its laws and principles are fixed, but the manner in which the principles of generation are used is free and infinitely varied. Even the interpretation and use of words involves a process of free creation. ***Noam Chomsky, 1968, American Linguist.***

Lascaux cave drawings: man as a communicator. *Source: Prof. Saxx with permission.*

1. INTRODUCTION

Is it possible to separate the mind/brain processes of language and thought? Are words and ideas, sentences and concepts, language and thought the same or highly overlapping processes...or are they separate? This is an issue that has long been argued about by language scientists.

An early proponent of language and thought being highly overlapping processes in the brain was John Hughlings Jackson (1835–1911). Hughlings Jackson was an English 19th

Fundamentals of Cognitive Neuroscience
https://doi.org/10.1016/B978-0-12-803813-0.00006-4

FIGURE 6.1 John Hughlings Jackson in 1895. *Source: Wikipedia Public Domain.*

century neurologist and scientist interested in diseases of the nervous system and how they shed light on how a healthy nervous system was organized. His scientific method advanced the study of brain science dramatically and his writings remain influential in the field of neuroscience. He was clear in his writing on his findings from studying patients with aphasia—language dysfunction typically brought about by brain injury, stroke, or disease—that *mentation*, thought processes, and *speech*, the production of those mentations in spoken language, were tightly coupled in brain processes. He wrote that "*words serve us during reasoning; they are necessarily required in all abstract thought*" (Jackson, 1884, p. 704, in Franz & Gillett, 2011) (Fig. 6.1).

Hughlings Jackson's views were not shared by all language scientists in the 19th century, or in the present day. Opposing views were put forth by French scientist Pierre Paul Broca (1824–1880) and especially by German scientist Carl Wernicke (1848–1905), contemporaries of Hughlings Jackson and just as passionate about understanding brain function and indeed just as influential as, if not more than, Hughlings Jackson. We will explore more on this topic later in the chapter (Fig. 6.2).

Are language and thought, words and ideas highly interrelated? The answer seems to be a resounding "yes." However, when the strong notion that *language and thought, words and ideas are inseparable* is raised, the answer seems to an equally resounding "no." While there are many brain studies of individuals with aphasia and healthy adults that support this view, perhaps the strongest phenomenal evidence comes from our own intuitions: there seem to be many thoughts and ideas that we just cannot *put into words*. Thus it seems likely that at some stage in brain function, human language and human thought are indeed separable.

In this chapter, we will take the view that language and thought are separable but highly interconnected brain processes. We will begin with a discussion of the cortical organization of language and then continue with a discussion of how thought processes are organized in the brain.

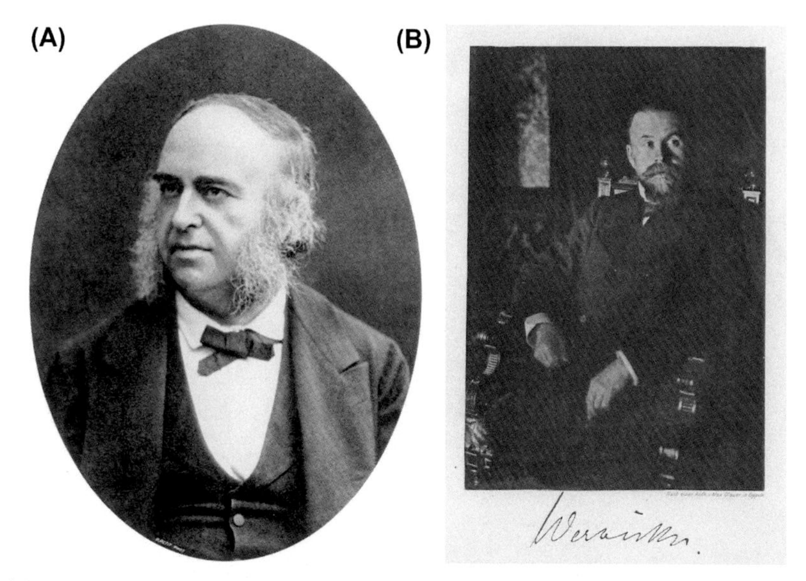

FIGURE 6.2 (A) Pierre Paul Broca c. 1860. (B) Carl Wernicke before 1905. *Sources: Wikipedia Public Domain.*

2. LANGUAGE...SPOKEN AND HEARD

Language is the foremost tool of human thought and culture. It is also one of the major landmarks of child development. Before our fourth birthday, we understand the specific sounds—*phonemes*—of our first language, our basic words—vocabulary, and grammar, the rules of combining words and phrases.

Language is not unitary, however, but consists of many stages and aspects of analysis.

Within a few seconds, the speech sounds we hear go through some of the following steps:

1. Sound analysis—decoding sounds into phonemes and syllables
2. Word identification—recognizing the words we know
3. Semantics—accessing a network of meanings
4. Syntax and grammar—identifying nouns, verbs, and phrases
5. Intonation and stress—decoding the contours of speech
6. Discourse—how do the meanings relate to previous ones in the conversation?
7. Purposes—what is the speaker's goal, and what does it mean for me?

In Fig. 6.3, we provide a rough schematic of the processes involved in decoding the speech signal (working from the top of the figure to the bottom) and in producing the speech signal (working from the bottom of the figure to the top).

2.1 The Sounds of Language—Phonology and Word Identification

Phonemes are the building blocks of individual languages: they consist of the sounds that are specific to an individual language. The study of the phonemes of languages is called *phonology*. While there are many similar sounds that the human vocal tract can produce, the individual sounds do not necessarily become cataloged into an individual phoneme. Here is an example: in Japanese, the sounds "r" and "l" are not differentiated as two

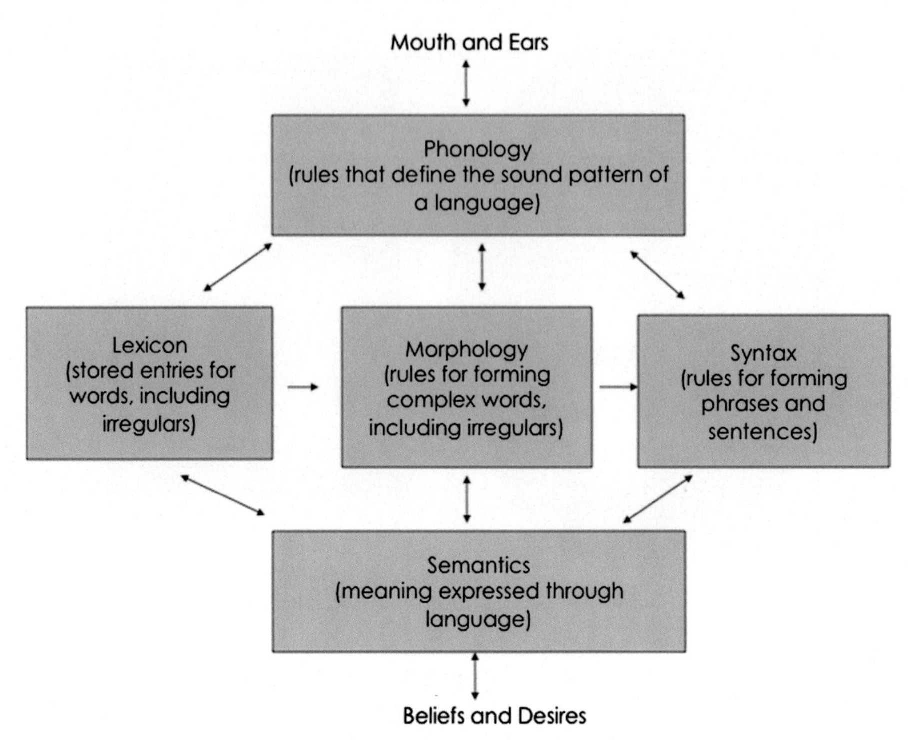

FIGURE 6.3 A schematic of the speech and language processes that are engaged when producing speech (beginning at the top of the figure) or understanding speech (beginning at the bottom of the figure). Many linguistic processes are thought to be largely parallel processes, indicated by the access to the lexicon, morphology, and syntax appearing in boxes that are side by side. *Source: Gage, with permission.*

phonemes. There is a single phoneme that sounds similar to both the "r" and the "l" as pronounced in English. In English, however, these two sounds are distinguished as individual phonemes, marked as /r/ and /l/ to show that they are phonemes and not just sounds. Thus the words "ray" and "lay" are clearly distinguishable to a native English speaker. Similar differences between sounds are classified into separate phonemes—or combined into a single phoneme—depending on the language. Some languages separate phonemes according to whether they are *aspirated* (produced with a breath of air) or not, whereas others such as English combined these sounds into one phoneme: they are *allophonic.*

Word identification/recognition is the learned process of identifying a cluster of sounds (phonemes in spoken speech) or letters (graphemes in reading and writing) that together form an individual word. These processes are essential for acquiring language in the spoken domain and in becoming literate in the reading and writing domains. Note that we separate the recognition of individual words due to the ordered cluster of their phonemes or graphemes from word meaning and conceptual knowledge: more on that shortly.

In this chapter we focus on the neural bases of word identification and recognition from a *phonological* or sound-based perspective: while any typically developing child acquires full spoken and heard language processes through mostly implicit (unconscious) means, reading and writing are acquired through explicit (conscious) formal educational processes. Put another way, we will explore the brain bases for spoken and heard language as they are similar processes and abilities across individuals and thus likely share their neural bases across individuals who otherwise vary greatly in education and intellectual levels.

2.2 Words and Meaning—Semantics

Words—or lexicons—and their meanings are studied in the field of *lexical semantics.* You may think of a word as a *signifier or symbol* that corresponds to what that word *means*: its denotation and conceptual significance. In this field of study, words or lexical units are categorized into classes and they are investigated frequently cross-linguistically (across various languages). How lexical knowledge is organized in the brain has long been a focus of study. The "tip-of-the-tongue (TOT)" phenomenon has shed light on the neural organization of lexical items: in TOT, there is typically a feeling that the word being searched for is known, is literally "on-the-tip-of-the-tongue" and may be partially recalled. Individuals experiencing TOT frequently can accurately describe what letter or sound the lexical item begins with and how many syllables it contains. TOT research has provided evidence that lexical retrieval is a staged process, with some general lexical knowledge, such as the onset of sound and length of a word, being accessed prior to the actual lexical item's retrieval.

2.3 Rules and More Rules—Syntax and Grammar

Human speech is a *combinatorial system,* in which words, phrases, and sentences may be combined in a seemingly infinite number of ways to communicate our thoughts and desires. While the number of unique sentences one person can produce is truly infinite, the way those words and phrases combine to form the sentences is actually quite limited. This aspect of human language has been termed "*discrete infinity*" (Chomsky, 1995). The rules underlying the combination of words and phrases to form sentences (and paragraphs) represent our language's grammar, and can vary across individual languages. The study of those rules or grammars that are common to all languages is the study of a *universal grammar.*

In English, words can be combined in certain ways, with some nouns that form the *Subject* of a sentence (as in "umpires" in Fig. 6.4) while other nouns form the *Object* of a sentence (as in "players" in Fig. 6.4). *Verbs*, or action words, reveal the action that takes place between the subject and the object, put simply. Spoken sentences can be infinitely long: one of the longest *printed* sentences in literature was written by Jonathan Coe in his *The Rotter's Club* and weighs in at 13,955 words. Of course, you can easily top that in a long *spoken sentence* by stating "Jonathan Coe wrote:..."!

While word learning is a relatively, intuitively obvious process and includes conscious or explicit effort, learning the rules of one's native language is far from obvious and contains many unconscious or implicit efforts. We can usually easily recognize if a phrase or sentence follows our language's syntactical rules by using our "grammatical ear": if it sounds ok to us, it is allowed! But it is far more difficult for us to explain *why* that usage is or is not grammatical: we have trouble stating just what "rule" it follows.

Fig. 6.5 shows a schematic of how language systems interface with other brain systems. The core processes of language, the syntax and lexical items (blue shape at the top of the figure), interface with the sensory-motor systems as the sounds of language are heard or produced (red shape at the left of the figure) and with internal concepts and thought processes (orange shape at the right of the figure).

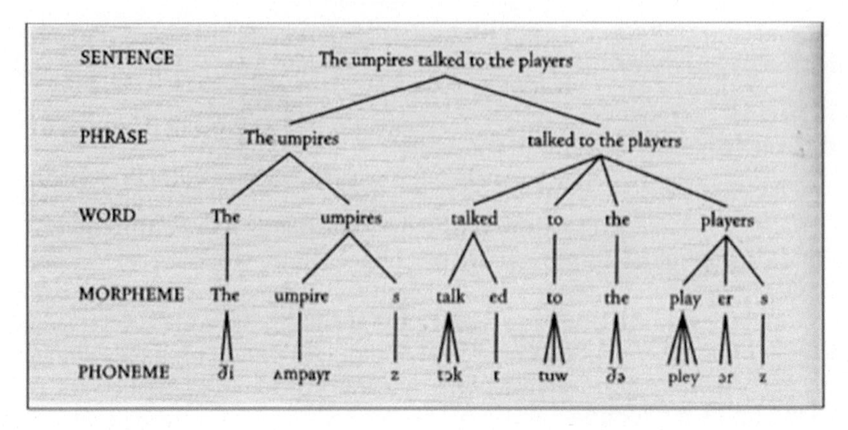

- Sentences
- Phrases
- Words
- Morphemes
- Phonemes

- Semantics
- Syntax
- Lexicon
- Morphology
- Phonology

FIGURE 6.4 Language is a hierarchical structure. The smallest unit in language, the phoneme, shown at the bottom of the figure, combines with morphemes to form words, which in turn form phrases and sentences. The combination of phonemes, morphemes, words, phrases, and sentences is bound by rules of a given language. The number of unique sentences that can be formed, however, is boundless. *Source: Gage, with permission.*

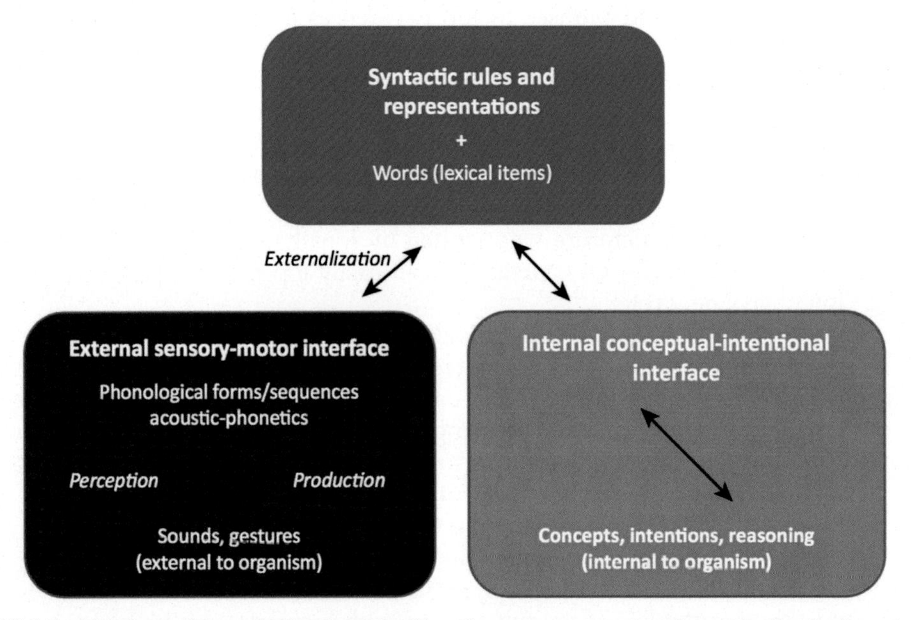

FIGURE 6.5 This figure shows a basic schematic of how language systems interface with other brain systems. The core processes of language, the syntax and lexical items (*blue shape* at top of figure), interface with the sensory-motor systems as the sounds of language are heard or produced (*red shape* at left of figure) and with internal concepts and thought processes (*orange shape* at right of figure). *Source: Berwick et al. (2013).*

2.4 The Melody of Language—Intonation and Stress

Speaking and singing are similar. Physically, singing is just a lengthening and tuning of vowels. If you stretch out the vowels of the words "cognitive neuroscience," you are already singing in a monotone. Then you can change the pitch of each syllable, and you have a little song. Like language, music and dance are species-specific capacities in humans. Other living primates do not have those specific skills, though they certainly share a great deal of semantic knowledge with us.

To successfully decode spoken language, we need to not only identify the phonemes that form the words, recognize sublexical units such as morphemes, and parse the grammar of the utterance: we also need to decode the *intonational contours* of the utterance. Let us begin with a sentence-level intonational contour. If you state: *You are going to the beach today.* versus *You are going to the beach today?* you are using the same words, word order, and sentence construction. You are only varying the intonation of the sentence, with a *rising tone* indicating that it is a question and not a statement of fact. In this case, the intonation provides information to the listener that you are asking a question. Intonation contours can vary to give a sarcastic edge to a sentence as well.

On a subsentence level, intonation carries information about the syllabic stress or lexical stress. Some languages have very regular lexical stress patterns. For example, Finnish and Hungarian words are almost always stressed on the first syllable. This regular stress pattern aids in word recognition. In other languages, such as English, the lexical stress patterns vary widely. Lexical stress patterns frequently change when a root word is modified: in English, the noun "syllable" (***syl**lable)* has the stress on the first syllable while the adjective "syllabic" (*syl**lab**ic)* has the stress on the second syllable.

2.5 The Goals of Language—Propositions and Discourse

We do not speak in single words but in *propositions*—that is, in semantically meaningful sentences that convey information. These propositions combine in conversations to become *discourse* with larger "chunks" of information obtained not just through series of sentences but through the *combined sentences among the speakers* involved in the discourse or discussion.

Thus while spoken and heard language is made up of strings of phonemes, which combine to become lexical items, and sentences are formed of strings of lexical items that convey a thought or idea, *discourse* contains information across multiple sentences and speakers. Understanding language at the level of discourse requires another level of analysis and processing beyond sentence-level decoding. An example provided by the Linguistic Society of America on their website (https://www.linguisticsociety.org/resource/discourse-analysis-what-speakers-do-conversation) describes the different meaning derived from analysis of these two signs posted at a swimming pool taken as a single discourse as compared to each sentence standing alone: (1) "Please use the toilet, not the pool." (2) "Pool is for members only."

Understanding the meanings and even subtle connotations gleaned from human discourse requires linguistic sophistication as well as social skills and emotional processing abilities.

2.6 The Learning of Language—Language Acquisition

We have discussed that language is not a unitary process, but has many stages and aspects from phoneme decoding to word identification to semantic meaning to grammatical structure to discourse processing (Fig. 6.3). It may not surprise you, then, that child language acquisition is not a unitary process but develops in a staged manner. In fact, child language acquisition follows some of the basic steps shown in Fig. 6.3, beginning at the top of the figure. While both receptive language—speech perception—and productive language — spoken—learning occurs in an overlapping manner in infants and young children, it is the *receptive* aspect of language that leads the way early in development (see Chapter 14, Growing Up, for more on this).

Babies begin to learn the specific phonemes of their language in the first months of life. By 7–8 months, infants begin to recognize words that they have become familiarized with ("baby," "bottle," for example). At this age, they are also beginning to recognize stress patterns in words. In English, many common, early-acquired two-syllable words have stress on the first syllable (e.g., "**Mom**my," "**Dad**dy," "**bot**tle," "**blan**ket," "**broth**er," "**sis**ter") and words with stress on the second syllable are less common in early word learning (e.g., "gui**tar**"). This helps guide the infant in learning word boundaries in spoken speech. Decoding intonational boundaries develops during the last months of the first year of life. Lexical item learning and their corresponding semantics begins to flourish by the end of the first year and develops rapidly during the first years of life. Syntax and sentence-level processing are slower to develop and typically begin to take shape during the early third year of life.

On the productive side, speech production lags behind speech perception. Babies understand a lot more language than they can produce. But by the age of 2 years, toddlers are typically forming 2–3 word utterances and then speech production literally explodes from the age of 3 onward. The grammatical correctness of these utterances improves with age as well, as syntactic knowledge grows in the 2–7 years age range.

While there are many differing views about how language is acquired by typically developing infants and children, there is substantial evidence that there is a *critical period* for learning language. If a child reaches a certain age (early adolescence) without access to spoken language—due, perhaps, to profound deafness or estrangement from people—that child will never acquire full language abilities. This is especially true for the syntax and grammar of one's first language: processes underlying rule learning are largely implicit or unconscious. Rule learning of the grammar of one's native language is typically matured during middle childhood and adolescence. And while these processes are largely implicit and early maturing, word learning continues throughout one's life and is far more explicit—conscious—in nature. If you are learning a second language as an adult, you may find that learning the grammar of that language is difficult while learning the vocabulary is far easier. This reflects the implicit versus explicit nature of language learning.

2.7 The Cortical Orgnization of Language

While the cortical organization for language is still being uncovered, many scientists have produced excellent models showing neural pathways and regions that form the language processing system in the brain (for a review, see Price, 2012). And although language scientists disagree about the details of just where and how the many subfunctions of language (i.e., speech perception, word recognition, syntactical analysis, semantic processing) are processed in the brain, they tend to agree on the basic arrangement of the brain pathways for processing the incoming speech signal and producing the outgoing spoken language.

2.7.1 Classical Theories: Broca's and Wernicke's Area

Two basic pathways for language were proposed in the mid-19th century. Two key findings form the basis for the "classical" model of language organization. First, in 1861, Pierre Paul Broca presented a paper detailing the case of a patient who, following brain damage, could only produce a single syllable "tan" (Broca, 1861). Upon autopsy, Broca described the patient's brain damage as focused in the left hemisphere in the inferior frontal lobe in a region now called Broca's area and proposed that this region was critical for speech production. Shortly thereafter, Wernicke published a monograph detailing a region in the posterior superior temporal gyrus (STG) in a region now called Wernicke's area that he proposed was critical for speech perception (Wernicke, 1874). The general notion was that cortical language processes were distributed within the left hemisphere, with a speech processing region in the superior temporal lobe and a speech production region in the inferior frontal gyrus (IFG) (Fig. 6.6).

2.7.2 Current Models

While modern neuroscience has expanded and clarified some of these early hypotheses, many aspects of the classical model are still valid today (Rauschecker & Scott, 2009). Current views hold that there are bidirectional *dorsal pathways* for language, which extend from the posterior STG anterior to the frontal lobe (shown in blue and purple on Fig. 6.7). This is similar to the classical view. Current models also hold that there is a *ventral pathway*, which extends between the STG anterior to the inferior frontal lobe (shown in pink on Fig. 6.7, Berwick, Friederici, Chomsky, & Bolhuis, 2013). This view is well supported by diffusion

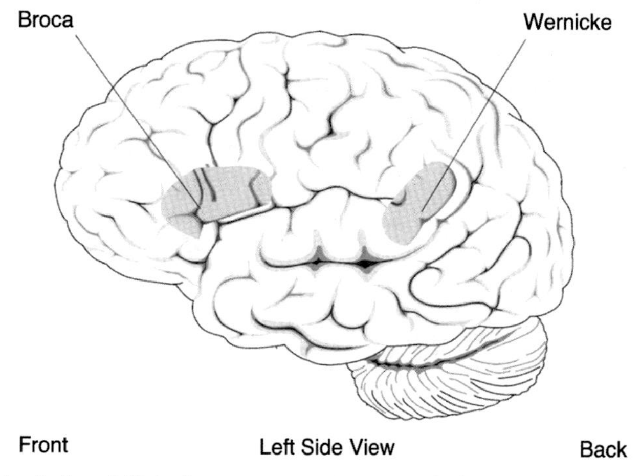

FIGURE 6.6 The "classical model" for language organization in the brain. The general notion was that cortical language processes were distributed within the left hemisphere, with a speech processing region in the superior temporal lobe referred to as "Wernicke's area" after Carl Wernicke and a speech production region in the inferior frontal gyrus (IFG) referred to as "Broca's area" after Pierre Paul Broca. *Source: Public Domain.*

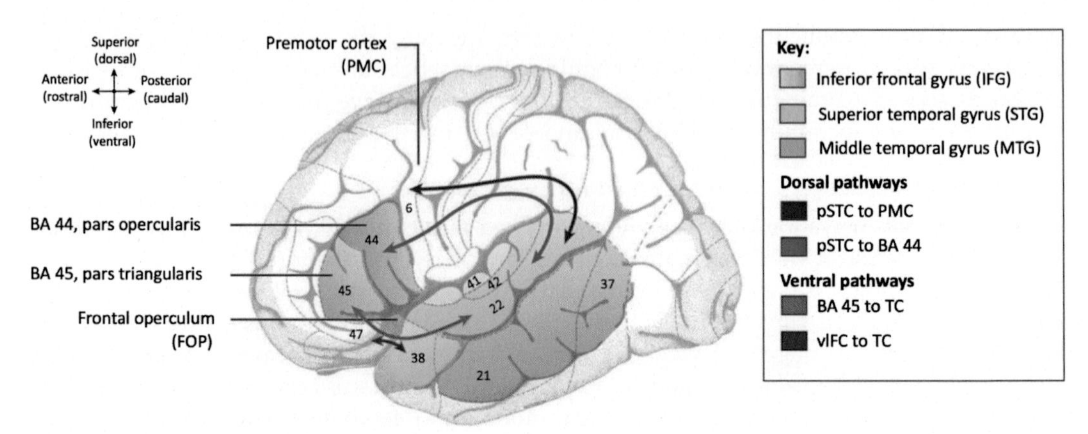

FIGURE 6.7 Current views hold that there are bidirectional *dorsal pathways* for language, which extend from the posterior superior temporal gyrus (STG) anterior to the frontal lobe (shown in *blue and purple*). This is similar to the classical view. Current models also hold that there is a *ventral pathway*, which extends between the STG anterior to the inferior frontal lobe (shown in *pink*). *Source: Berwick et al. (2013).*

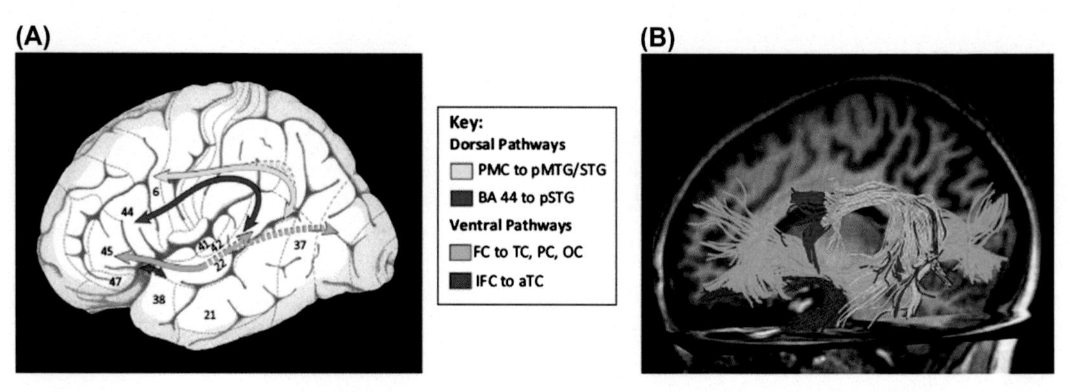

FIGURE 6.8 A current view of language organization in the brain, shown in Fig. 6.7, is well supported by diffusion tensor imaging of the white matter fiber tracts that extend from the temporal lobe. These fiber tracts include the arcuate fasciculus and the superior longitudinal fasciculus (shown in *blue and yellow*) in the dorsal stream and the inferior fronto-occipital fasciculus (shown in *green*) in the ventral stream. (A) A brain schematic view of the white matter fiber tracts (similar to those shown in Fig. 6.7). The dorsal or upper pathways include the bidirectional path from the premotor cortex (PMC) to the posterior middle temporal gyrus (pMTG) and superior temporal gyrus (STG) (shown in *yellow*). A second dorsal bidirectional pathway extends between the Brodmann Area BA44 in the frontal lobe and the posterior STG (pSTG) (shown in *dark blue*). Ventral pathways include the long bidirectional pathway between the frontal cortex (FC) and the temporal cortex (TC), parietal cortex (PC), and occipital cortex (OC) (shown in *green*). A short ventral bidirectional pathway extends between the inferior frontal cortex (IFC) and the anterior temporal cortex (aTC) (shown in *red*). (B) The anatomical view of the pathways shown in schematic view in (A). The same color coding is used for the four pathways. *Source: Friederici & Singer (2015).*

tensor imaging of the white matter fiber tracts that extend from the temporal lobe. These fiber tracts include the arcuate fasciculus and the superior longitudinal fasciculus (shown in blue and yellow in Fig. 6.8) in the dorsal stream and the inferior fronto-occipital fasciculus (shown in green in Fig. 6.8) in the ventral stream.

Although early language models focused on the left hemisphere as the home of cortical language regions based on observations that language production was impaired with the left but not the right hemisphere brain damage, more recent studies have shed light on the role of the right hemisphere in speech processing (for a review, see Hickok & Poeppel, 2007).

According to the view put forth by Hickok and Poeppel, early spectrotemporal sound signal analyses and phonological processing systems are organized *bilaterally* in the left and right dorsal STG and mid-post superior temporal sulcus (STS), respectively (Fig. 6.9, center of the figure shown in green and yellow shaded boxes on the upper section and in green and yellow shading in the left and right hemispheres in the lower section). These

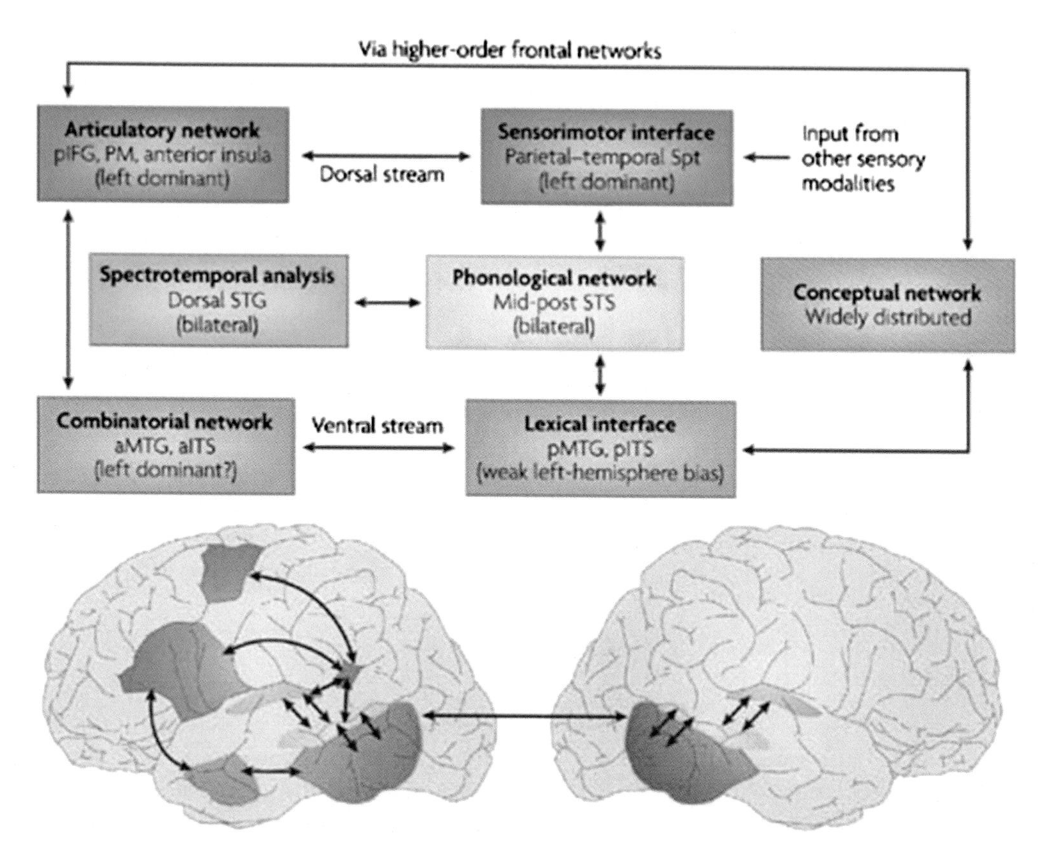

FIGURE 6.9 According to the view put forth by Hickok and Poeppel, early spectrotemporal sound signal analyses and phonological processing systems are organized *bilaterally* in the left and right dorsal STG and mid-post superior temporal sulcus (STS), respectively (center of the figure shown in *green and yellow shaded boxes* on the upper section and in *green and yellow shading* in left and right hemispheres in the lower section). These bilateral pathways then split into two left-dominant processing streams. The ventral stream includes a combinatorial network in anterior middle temporal gyrus (aMTG) and a lexical interface for decoding syntactical and semantic information in the posterior MTG and posterior inferior temporal sulcus (pITS). This left-biased ventral stream, shown in *pink shaded* regions, then, is home to speech recognition processes that map sound onto meaning. *Source: Hickok and Poeppel (2007).*

bilateral pathways then split into two left-dominant processing streams. The ventral stream includes a combinatorial network in the anterior middle temporal gyrus (aMTG) and a lexical interface for decoding syntactical and semantic information in the posterior MTG and posterior inferior temporal sulcus (pITS). This left-biased ventral stream, shown in pink shaded regions on Fig. 6.9, then, is home to speech recognition processes that map sound onto meaning.

A dorsal stream includes a left-dominant sensorimotor interface in area Sylvian-parieto-temporal (Spt) at the parietal-temporal boundary and a left-dominant articulatory network that includes regions in the anterior insula, premotor cortex, and the posterior inferior frontal gyrus (pIFG). This left-biased dorsal stream, shown in blue shaded regions on Fig. 6.9, then, is home to speech recognition processes that map sound to action.

Finally, Hickok and Poeppel propose that both the ventral speech network and the dorsal motor-articulatory systems must have interfaces with the widely distributed conceptual network in the cortex (shown in gray shading in Fig. 6.9). Conceptual knowledge, then, must be available both during the decoding of a spoken speech stream and during the planning for a spoken articulation.

2.8 Thoughts About Language and Thought

Cortical language systems are tightly coupled with cortical conceptual networks. Thus cortical systems for processing language and thought, words and ideas are closely interrelated but, we hold, are not the same. They are separable. Now let us turn to the cortical organization for thinking and problem-solving—perhaps the central "job" that the brain must accomplish for us to achieve our goals, both large and small.

3. THOUGHTS

If I had an hour to solve a problem I'd spend 55 minutes thinking about the problem and 5 minutes thinking about solutions. ***Albert Einstein.***

3.1 Thinking and Problem-Solving

If you enter a Google search for "thinking and problem-solving" you will find more than 65,000,000 hits. Links range from scientific journal articles about brain processes underlying thinking and problem-solving, to practical guides for thinking and problem-solving in the workplace, to mental activities and games that foster more effective critical thinking and problem-solving in aging. The point is clear: thinking and problem-solving are central aspects of human cognition and they are the key for making progress toward our goals.

Cognition involves problem-solving of all shapes and sizes. Some problems are massively complex, others are fairly routine. Large or small, problem-solving typically involves our attention, focus, and planning abilities. Most of our solution processes are conscious: we are aware of the mental steps we are taking to form a solution. As we will see later in this chapter, through experience and over time, some problems have been solved so often that we operate on a more unconscious basis.

Problem-solving implies an initial state (A) and a goal state (B), with some set of goal-driven procedures to get from A to B. Box 6.1 presents a "Tower" puzzle that is used in problem-solving experiments. Formally, problem-solving can be described in terms of a *problem space*, a kind of a maze, showing end goals, subgoals, and step-by-step progress through the maze. Mathematics, computer science, and psychology have added to our understanding of problem-solving, but its brain basis is only beginning to be understood.

Fig. 6.10 shows a *problem space* for this version of the Tower puzzle. In this case, all possible moves can be shown, as well as all the step-by-step pathways through the maze. Tower puzzles are useful, but we have to be careful about generalizing to real-world problems, which are often quite different. Real-life problems tend to be complex, vague, ambiguous, or poorly understood. Whole categories of scientific problems are not formally *tractable*—that is, they cannot be solved by conventional mathematical methods.

Problem spaces are *formal descriptions.* They tell us about the structure of a problem—not about the human brain. It is important not to confuse the formal structure of a problem space, like chess, with what human beings are actually doing when they play chess.

In many cases, what looks like a serial problem to newcomers may allow for fast shortcuts for experts, who can use their prior knowledge to leap to the solution without following a serial set of steps. Similarly, when we gain expertise in speech, reading, writing, and many other skills, we can process difficult tasks quite quickly. Furthermore, human beings pick

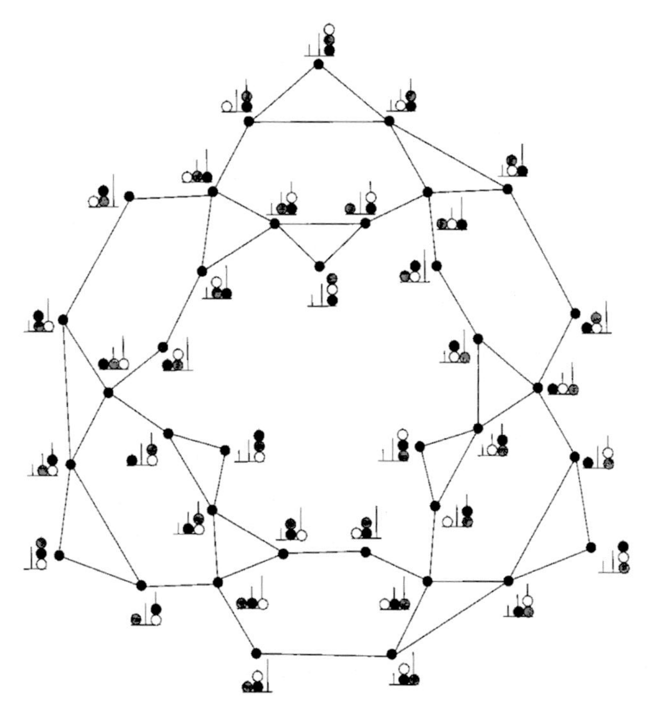

FIGURE 6.10 A problem space for the Tower of London test. The problem is defined by the joint selection of an *initial state* and a *goal*. In this case, all possible moves can be shown, as well as all the step-by-step pathways through the maze. *Source: Dehaene and Changeux (1997).*

BOX 6.1

A PROBLEM SPACE FOR PUZZLES

Fig. 6.11 shows a "Towers" puzzle that is used to test neurological patients. It is helpful to try a few moves for yourself. Towers puzzles can be adjusted in difficulty level. They are designed to be solved only by step-by-stop planning, a frontal lobe function.

In serial problems the time from start to finish is a linear function of the number of steps. Over multiple trials, solvers speed up as they develop skills. "Time to a correct solution" is therefore a reasonable measure of efficiency.

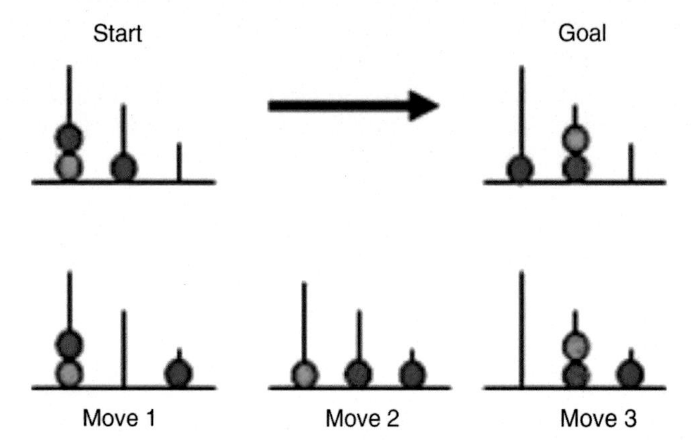

FIGURE 6.11 The Tower of London test. Solving the Tower of London test requires the solver to think ahead and to plan subgoals on the way toward a solution. The solver can only move one bead or ring (depending on the version of the test) at a time and the beads or rings cannot extend beyond the post. One post can hold three beads or rings, the center post can hold two beads or rings, and the third post can only hold one bead or ring. Patients with frontal lobe damage (and thus impaired executive functioning) have tremendous difficulty solving a Tower test. *Source: Miller & Wallis, in Fundamental Neuroscience, 4th edition.*

up all kinds of *heuristics*—tricks and tips that help us get to a "good enough" solution faster than Fig. 6.10 suggests. Human experts know many heuristics to solve practical problems. For humans to survive as hunter-gatherers during a severe drought, a formal "best solution" to the danger of starvation does not even exist. In Herbert A. Simon's terms, humans do not "optimize" in those circumstances—that is, look for the very best solution—but "satisfice." In other words, all we want is a "*good enough*" solution, one that will get us through the dry season with minimal losses (Simon, 1948). The formal problem space of Fig. 6.10 therefore might not tell us how people reach their goals in the real world.

The real world often gives us *advantages* we do not find in artificial puzzle problems, such as the benefits of *prior knowledge.* If we know all the houses on a street, we do not have to check each one on the way to a friend's home. A newcomer to the neighborhood may have to think serially by checking each street address, one by one, but with the right knowledge we can go straight to the goal. Navigational problems can often be solved by prior knowledge. If you are familiar with your destination, you may not even bother to pay attention to most of the route going there. You may be allocating most of your attention to other things.

3.1.1 *Conscious and Unconscious Processes*

Human thinking comes in two general varieties, *explicit* and *implicit*, roughly the same as "conscious" and "unconscious." In experimental practice they correspond to "reportable brain events" versus "unreportable" ones. Everyday problem-solving is a mixture of the two.

Studying explicit thinking seems fairly straightforward, because people can tell us their step-by-step thoughts in a game of chess, for example. By definition, people cannot tell us about their *implicit* brain activities, but the evidence for unconscious cognition has now become very strong, and better methods are available to study it.

3.1.2 *Explicit Thinking and Problem-Solving*

Mental arithmetic and chess playing use *explicit* thinking. We can walk through such problems consciously and deliberately, step by step. Explicit thinking often has clear, conscious subgoals that must be finished before the main goal can be reached.

Compared to implicit thinking, explicit cognition requires the following:

1. Greater executive control and decision-making
2. More conscious involvement
3. More intense judgments of mental effort
4. Much wider neural activity in the brain, as measured by functional MRI (fMRI), electroencephalography (EEG)/magnetoencephalography (MEG), or direct brain recording

However, after years of practice, explicit problem solvers, like chess experts and mental calculators, tend to become more intuitive and less conscious of the details of their thinking. Thus *explicit cognition* may become more *implicit* with practice.

Explicit thinking often uses symbols, like words or visual images. In neuroscience terms, symbols can be defined as conscious events that refer to accessible knowledge and/or a category of events in the world. A phrase like "this final exam" is an example of a symbolic string that evokes a meaning representation in the brain and that in turn refers to an event in the world. One rule of thumb of explicit problem-solving is to simplify a problem to a few symbols that can be held in working memory. Most fundamental equations in the sciences and mathematics can be simplified, like the famous Einstein mass–energy equation $e = mc^2$. Each symbol in the equation refers to a large network of other symbols and equations, which can often (but not always) be measured in nature.

3.1.3 *Executive Function, Attention, Working Memory, and Problem-Solving*

In a broad sense of the term, working memory is the domain of problem-solving, language, and thought. It is "the set of mental processes holding limited information in a temporarily accessible state in service of cognition" (Cowan, Izawa, & Ohta, 2005). We need working memory to perform mental arithmetic, to carry on a conversation, and to solve a path-finding problem. Executive functions such as planning and goal setting are also key

BOX 6.2

DEFINITIONS OF EXECUTIVE CONTROL FUNCTIONS

There are many names for executive control functions. These names include:

- Flexibility: the capacity to switch attention between different tasks.
- Goal setting: the capacity to set a goal that leads to effective action.
- Planning, including initiation and sequencing: the capacity to determine a series of steps necessary to reach a goal.
- Inhibiting interference: the capacity to suppress distracting or irrelevant information, thoughts, and actions.
- Monitoring: the capacity to check whether actions result in their intended outcome.
- Adjustment: the capacity to adjust a course of action in light of previous results.
- Maintenance: short-term memory for information needed to carry out executive functions. For example, to act on a goal we have to be able to keep it in short-term memory as long as it is needed to guide our actions. This may seem easy, but a large range of factors interferes directly with the ability to keep goals in mind.

to successful explicit problem-solving (see Box 6.2). Of course, we need to pay attention—and hold that attention—if we are going to come up with a solution for a problem. These aspects of cognition are not limited to the problem-solving domain, of course, but they are key to the success of migrating through a day full of problems to solve, big and small. This is *not* the case with less conscious, more implicit problem-solving, as we shall see.

3.1.4 *Using Inner Speech to Solve Problems*

If you try to mentally add $3 + 5 + 11$, chances are that you can hear your own inner speech. That means you have recoded the figures in the previous sentence into auditory/articulatory brain activities that are partly conscious. Most people seem to have a spontaneous inner speech. If we want to speak to ourselves, we do not even need a listener; we are our own audience.

We use inner speech for planning, self-reflection, and much more. It is often combined with visual imagery (as in remembering where your bicycle is parked) and probably with other "inner senses" and thus play an important role in explicit problem-solving.

3.1.5 *Expert Thinking and Cognitive Load*

It takes thousands of hours of practice, talent, and dedication to become an expert in chess, music, or athletics. One important part of chess expertise, for example, is having a large and accurate set of chess patterns in memory, along with many learned tactics and strategies that may become automatic after practice. These chess patterns are *chunks*—a way to store a large amount of learned knowledge into a single unit or chunk—in the same way that the word *avatar* is a memory chunk for a virtual life player.

Much of what is known about expertise goes back to studies of chess players by de Groot et al. (1966) and Chase and Simon (1973). de Groot demonstrated clear differences between levels of chess experts in a memory task, using brief pictures of chessboard positions taken from a tournament game. Typically, players at Masters levels recalled a flashed chess picture almost perfectly, but weaker players could not remember a flashed chess board.

Does this mean that expert chess players have better memories than others? Chase and Simon (1973) found no difference in the recall of *random* positions between three subjects: a chess master, a class A player, and a novice (Fig. 6.12) It therefore seems that chess experts do *not* necessarily have better memory skills than others. Rather, chess experts recognize possible board positions, an array of pieces on the board that makes sense in a game. This result has now been repeated a number of times. This is the nature of expertise and it generalizes to many areas of human life: if we are experienced drivers, we are expert at navigating a vehicle through complex situations with relatively little attention and mental effort. Similarly, if we are experienced musicians, we are expert at playing our instrument. And if we are experienced mathematicians, then we are expert at resolving complex mathematical equations with comparative ease.

Problem-solving usually entails adding a *cognitive load* to typical cognition (Sweller, 1988). That is, the attention, thought processes, and working memory aspects of problem-solving

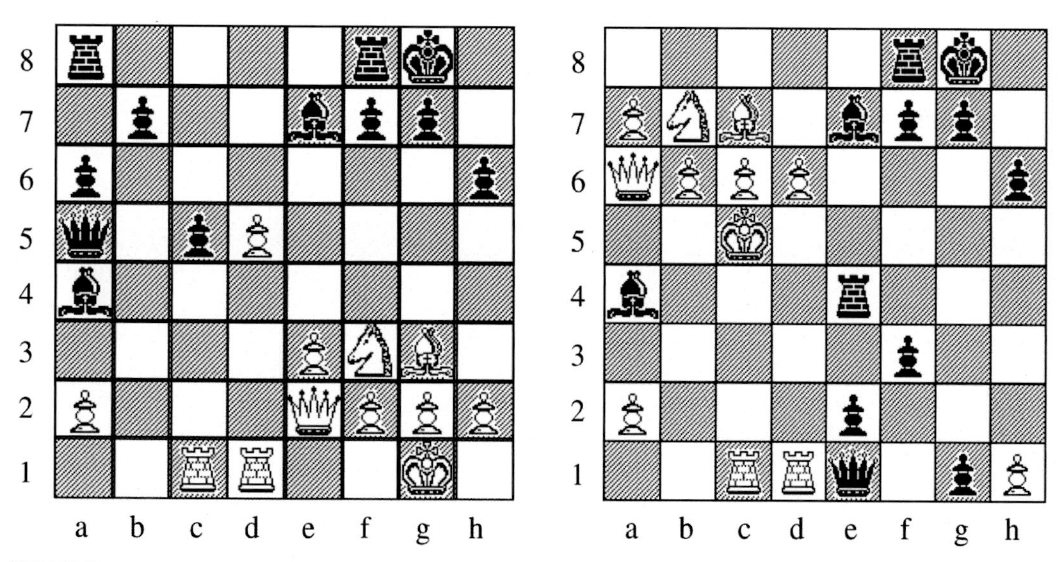

FIGURE 6.12 Expertise aids in memory load tasks. Simon and colleagues have investigated the role of expertise, in this case a chess expert, and memory skills. The chess pieces on these chess boards, from a 1997 experiment, have been systematically varied by swapping one quadrant at a time across the chess board. Chess experts were quite adept at recognizing known patterns for chess pieces (left panel) but fared no better than novice chess players on varied patterns for chess pieces that were not known (Gobet & Simon, 1997). *Source: Gobet and Simon (1997).*

can be quite heavy if the problem being addressed is weighty enough—or novel enough—to require high effort across these three aspects of cognition.

Expertise, then, *"lightens the load"* of problem-solving in that discipline or field.

3.1.6 Fixedness and Mental Flexibility

While our prior knowledge and expertise may dramatically improve our problem-solving abilities, they may actually impede our progress toward a solution. It is almost as if our expertise limits our approach to a given problem—due to our substantial experience, we may have a razor-sharp approach that skips important steps in analyzing the problem space of a problem to be solved. One way to describe this issue is to call this razor-sharp and usually effective problem-solving strategy as being a *fixed* one.

The word *fixedness* describes the experience of being "stuck" in a way of looking at a problem that does not always produce useful results. In the sciences we see some degree of fixedness whenever a major new theory or type of evidence becomes available. Pioneering scientists like Galileo or Einstein often provoke a period of debate between old and new perspectives, but they are not immune to having their own fixed beliefs. Fixedness seems a nearly universal problem. The science fiction writer Arthur C. Clarke pointed out that many important discoveries were declared to be impossible by famous scientists shortly before they were proven to be true.

Fixedness occurs in many kinds of problem-solving. It happens in chess games when one player tricks the other into making what looks like a safe move and then checkmates the opponent. Under experimental conditions, it is easy to demonstrate that people can indeed become "fixated" in a habitual approach that does not lead to solutions (Duncker, 1945). In competitive games, one side can take advantage of a routine habit on the other side. Some habitual fixedness seems to be practically universal.

Consider the quote by Albert Einstein at the beginning of the section on Thought: "If I had an hour to solve a problem I'd spend 55 min thinking about the problem and 5 min thinking about solutions." The point expressed by Dr. Einstein is that *correctly describing and identifying the actual problem to be solved* is just as important—or more so—than coming up with solutions. Too little time taken for understanding just what the problem to be solved actually *is* can lead to many trips down dead-end roads of reasoning. When that combines with a fixed approach that is poorly matched to the actual problem, time is wasted, frustration grows, and little or no progress is made.

Mental flexibility, the ability to switch between different viewpoints on a problem, is associated with healthy functioning. If an approach is sufficiently flexible, then time spent on developing solutions can be deftly switched toward new problem approaches and their solutions as they come to mind. This mental agility works well to help us through the daily problem-solving that confront us in life and is key to successful cognition. Patients with brain damage often have difficulty solving problems because they seem to become "stuck" in ideas that do not work.

Fig. 6.13 shows the Wisconsin Card Sorting Task, which is designed to induce a misleading mental "set" in a simple card sorting task. Subjects are asked to guess which card is "correct," using color, number, or shape. They are not explicitly told the "rule," but must infer it from the feedback they receive from the experimenter, who will tell them if their guess is correct or not. They are given feedback for each guess. Initially they are rewarded for one pattern—for

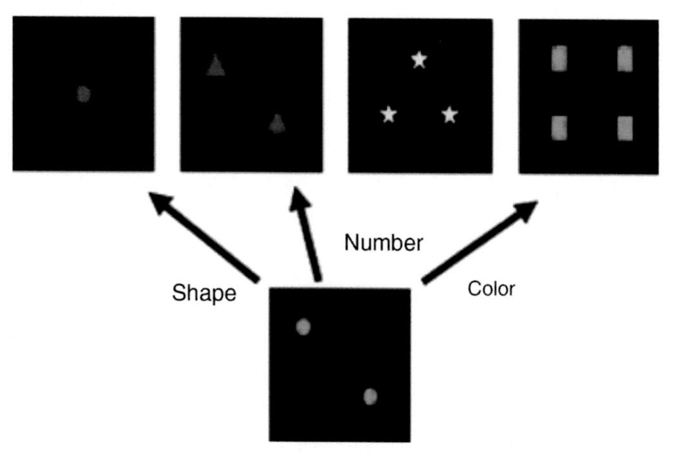

FIGURE 6.13 The Wisconsin Cart Sorting Task is designed to induce a misleading mental "set" in a simple card sorting task. Subjects are asked to guess which card is "correct," using color, number, or shape. They are not explicitly told the "rule," but must infer it from the feedback they receive from the experimenter: their guess is correct or not. They are given feedback for each guess. Initially they are rewarded for one pattern—for example, the rule that the color yellow is always correct. At some point the rule is changed without telling the subject. The time and number of missed trials needed for subjects to shift set are taken as an index of their ability to test different hypotheses, to demonstrate their mental flexibility. Typically, patients with frontal lobe damage are impaired on this task and perseverate on a fixed strategy despite being given feedback that their choices are incorrect. *Source: Miller & Wallis, in Fundamental Neuroscience, 4th edition.*

example, the rule that the *color yellow* is always correct. At some point, the rule is changed without telling the subjects: perhaps to *3 items*. The time and number of missed trials needed for subjects to *shift set* from the color yellow to the 3 item cards are taken as an index of their ability to test different hypotheses, to demonstrate their mental flexibility.

Changing rules is difficult when we are mentally fatigued or drowsy or otherwise impaired. Even switching from one task to another seems to require additional mental resources beyond those involved in routine and automatic actions. Thus drivers who are sleepy might find it harder to make fast decisions in unpredictable traffic situations.

3.1.7 *Mental Effort*

Explicit thinking often involves mental effort—a strong sense that we are really trying hard. Mental effort has emerged in the last decade as a major variable in psychology and brain science. There seems to be a limit to the amount of mental effort we can exert each day (Baumeister, Sparks, Stillman, & Vohs, 2008). This differs sharply from the experiences during implicit problem-solving—these processes seem to be occurring *"in the background"* of our mental set and do not seem to present a conscious mental effort. At least, we cannot *report consciously* on just what efforts went into solving problems when they occur implicitly.

3.2 Implicit Thinking

People cannot tell us about their unconscious thoughts. To study implicit cognition we can watch what people do, but their reports may be inconsistent with their actions. What people tell us may be what they *think* they are doing. Reliable studies of the neural bases of implicit problem-solving are therefore fairly recent, frequently conducted using neuroimaging techniques to assess brain regions active during implicit thinking-based tasks rather than relying on subject report.

Existing knowledge works together with conscious events during explicit problem-solving. It is possible that implicit thinking uses many of the same functions that have become highly practiced, automatic, and largely unconscious.

3.2.1 *Implicit Problem-Solving*

While we are far less aware of them, implicit problem-solving may happen more often than the explicit kind. Soon after birth we start learning new skills. Babies babble spontaneously. Childhood play helps to practice eye—hand coordination, social and emotional habits, and learning our own limitations. By age 10 we have many overlearned skills that are largely unconscious.

In general, implicit thinking involves the following:

1. Less executive control and decision-making
2. Less conscious involvement
3. Less mental effort
4. Less neural activity in the brain, as measured by fMRI, EEG/MEG, or direct brain recording

Although it takes years and years to become a fluent speaker and listener of one's native language, once acquired, speaking, listening, and even writing and reading are mostly implicit. We use these abilities largely without even knowing what we are doing. In reading the last sentence, you were probably unconscious about your own eye movements. Eye movements in reading are not random but very selective and purposeful. Moving your eyes in reading involves fast and highly skilled problem-solving.

Expert readers may be unaware of entire levels of language processing, like word identification, phrase comprehension, syntax, and semantics—processes that we may use every second. There are many ways to disrupt practiced skills, thereby turning implicit routines into explicit thinking. We can make words harder to perceive by crossing out letters, scrambling their order, or making their meanings harder to understand. ~~Notice that you have to pay more effortful attention when these letters are crossed out~~.

Nevertheless, in natural situations, implicit problem-solving takes less executive control than explicit thinking: it involves less conscious access, less subjective effort, and less cortical activation in fMRI studies. Implicit thinking depends more on long-term memory and highly practiced routines. Consistent practice and rehearsal can turn explicit problem-solving into the implicit kind.

3.2.2 *Implicit Problem-Solving: The* Aha! *Moment*

Have you ever tried to solve a problem and been stymied, frustrated … finding a solution beyond your reach when suddenly, the solution springs to mind? This is called the *"Aha!"* moment. It is characterized by no sense that you are getting closer to the solution to a problem. Instead, the solutions suddenly, unbidden, comes to mind. This type of problem solution experience is typically called *"insight."* We think of it as implicit because the steps and stages of solving the problem are not conscious to the solver, they are unconscious and hence implicit.

Insight in problem-solving has been studied for decades, with groundbreaking work accomplished by gestalt psychologists (Duncker, 1945; Köhler, 1921) and more recent cognitive scientists such as Jung-Beeman, Bowden, Davidson, Metcalfe, and Ollinger (Kounios & Beeman, 2014; Metcalfe, 1986; Öllinger & Knoblich, 2009; Sternberg, 1996) (Figs. 6.14 and 6.15).

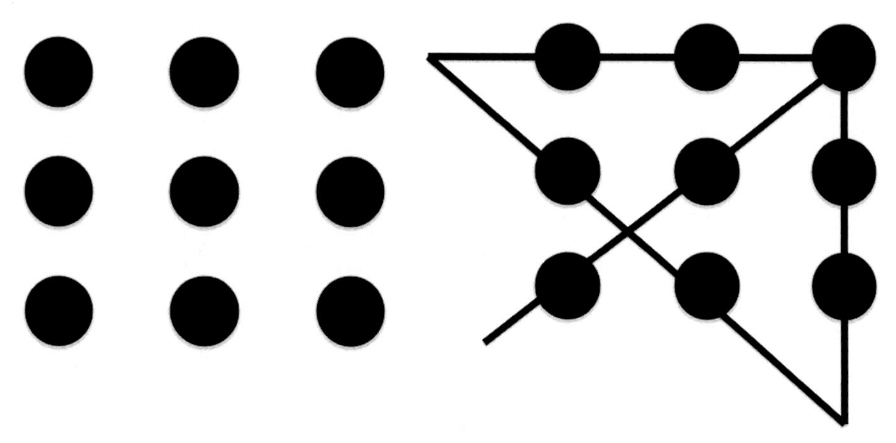

FIGURE 6.14 Left panel: the 9 dots puzzle used in many insight experiments. The task is to connect all 9 dots with 4 straight lines without lifting the pencil. Right panel: the solution to the 9 dots puzzle: solvers must "think outside the dots" to solve the puzzle. This solution frequently comes to solvers' minds as an Aha! experience. *Source: Gage, with permission.*

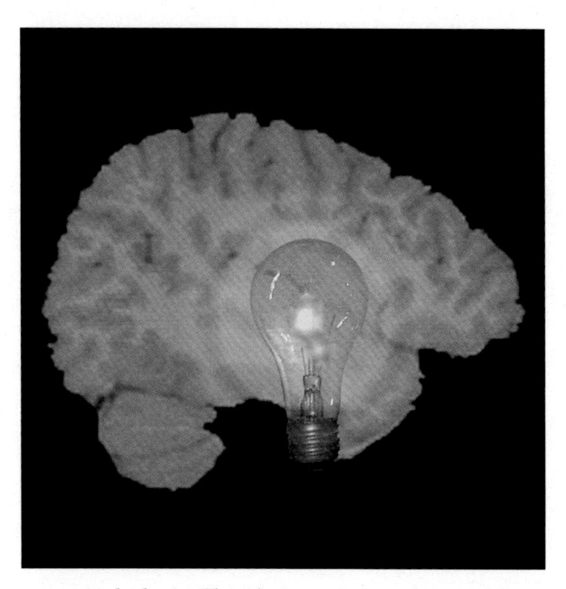

FIGURE 6.15 The light goes on in the brain. The Aha! experience is frequently reported as a *"light coming on"* in the mind/brain as the problem's solution comes to mind. *Source: Insight lights up the brain https://doi.org/10.1371/journal.pbio.0020111.g001, Public Domain.*

A defining feature of the Aha! moment in experimental studies is that it is *the subject*—not *the experimenter*—who dictates whether there was an Aha! moment...or not. Ollinger et al. (Danek, Fraps, von Muller, Grothe, & Ollinger, 2014) defined five dimensions of the Aha! experience to develop a multidimensional model to describe these dimensions and their relative importance in implicit problem-solving or insight solutions (Box 6.3). Subjects watched videos of a magician performing magic tricks and pressed a button if they *"solved the trick"*—that is, if they knew how the trick was done. If they experienced an Aha! moment, they were asked to rate that experience along five dimensions: *Suddenness, Surprise, Happiness, Impasse, and Certainty*. Two weeks later, they were asked to rerate their Aha! experience. As predicted, the subjects were very consistent in describing their Aha! experience across the two rating

BOX 6.3

FIVE DIMENSIONS OF THE AHA! MOMENT EXPERIENCE

Ollinger et al. have developed a multidimensional model to determine the interplay of five Aha! moment experiences and their relative importance to insight problem-solving (Danek et al., 2014).

1. **Suddenness**: That insightful solutions are experienced as very sudden was demonstrated by Metcalfe (Metcalfe, 1986; Metcalfe & Wiebe, 1987) who showed that although problem solvers are able to accurately judge their progress toward the solution (recorded as feeling-of-warmth ratings) for noninsight problems, they are unable to do so for insight problems.
2. **Surprise**: Based on introspection and informal observation, Gick and Lockhart (1995) suggested a division of the Aha! experience in two components: surprise and suddenness. In their account, the surprise aspect can vary by strength and it can be accompanied by either positive (delight) or negative (chagrin) emotions. To disentangle surprise from these accompanying emotions, we decided to assess the emotional components separately, adding "happiness" as a new dimension.
3. **Happiness**: Because Gick and Lockhart (1995) proposed the emotional response to vary between the positive and negative pole, we used a scale with "unpleasant" and "pleasant" as two extremes.
4. **Impasse**: Ohlsson postulated that prior impasse is a necessary precondition for Aha! experiences to occur (1992). An impasse is defined as a state of mind where problem-solving behavior ceases.
5 **Certainty**: Obviousness of solution, i.e., the certainty that an insightful solution is correct, was stressed as an additional aspect by Bowden and Jung-Beeman (2007). This "intuitive sense of success "related to insightful solutions is also often described in the context of scientific discoveries (Gick and Lockhart, 1995, p. 215).

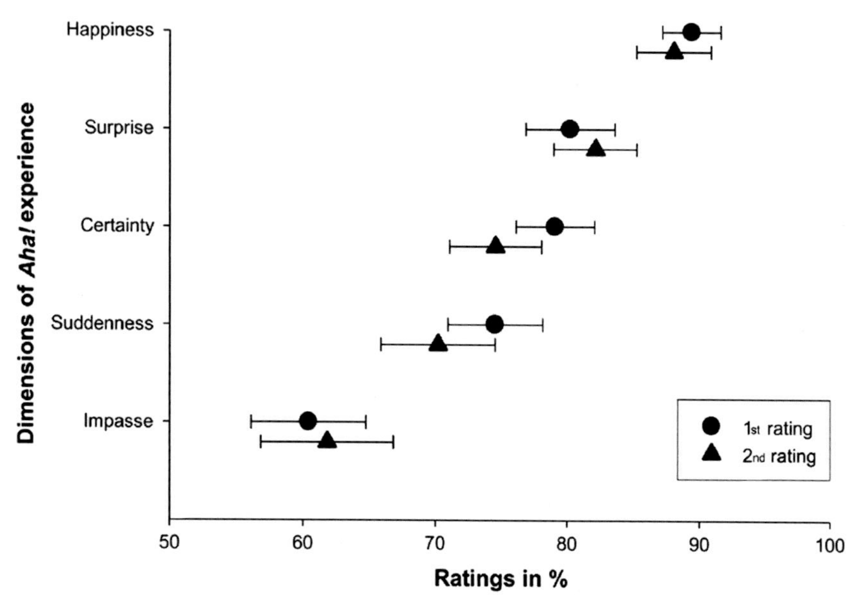

FIGURE 6.16 Ollinger et al. (Danek et al., 2014) defined five dimensions of the Aha! experience to develop a multidimensional model to describe these dimensions and their relative importance in implicit problem-solving or insight solutions (Box 6.2). Subjects watched videos of a magician performing magic tricks and pressed a button if they *"solved the trick"*—that is, if they knew how the trick was done. If they experienced an Aha! moment, they were asked to rate that experience along five dimensions: *Suddenness, Surprise, Happiness, Impasse, and Certainty*. Two weeks later, they were asked to rerate their Aha! experience. The initial rating is shown in black-filled circles, the second rating 2 weeks later is shown in black-filled triangles. As predicted, the subjects were very consistent in describing their Aha! experience across the two rating events. Not predicted, however, was the fact that raters had very different ratings of their feeling of the importance of each of the five dimensions in the Aha! experience of problem-solving: "Happiness" and "Surprise" were rated as very important by the majority of the subjects (~80%–90%) while the feeling of "Impasse" was rated far less important (~60%). *Source: Danek et al. (2014).*

events (Fig. 6.16). Not predicted, however, was the fact that raters had very different ratings of their feeling of the importance of each of the five dimensions in the Aha! experience of problem-solving: "Happiness" and "Surprise" were rated as very important by the majority of the subjects (~80%–90%) while the feeling of "Impasse" was rated far less important (~60%).

The findings of Danek et al. (2014), and many other similar studies investigating the Aha! experience, shed light on the implicit or unconscious aspect of this type of problem-solving, with *Surprise* coupled with *Happiness* as key dimensions of the emotional reaction to finding the solution. While a solver may feel happiness after solving a problem using explicit thinking, they are not typically surprised by the event. Not so for the Aha! moment: it is, in fact, defined by the sudden "light coming on" of the solution as it springs to mind.

3.3 Cortical Organization for Thinking and Problem-Solving

Explicit and effortful versus *implicit and sudden*: understanding where and how these very different problem-solving processes occur in the brain is a key goal of many cognitive neuroscientists. Far more is known about explicit problem-solving: the steps and stages of these processes are reportable by the solver and the major aspects of cognitive involved in their solution—that is, *attentional, working memory, and executive processes*—have long been studied.

Let us begin with a look at the neural basis of explicit problem-solving. Driving the brain activity for explicit problem-solving processes are the three key aspects of explicit thinking described above: attentional, working memory, and executive processes. Your attentional networks are activated as you focus on solving the problem and on resisting distraction by other things happening in your environment. Your working memory networks are engaged as you keep aspects of the problem and its solution in mind. Your executive control regions in the prefrontal cortex are active as you plan your strategy and then put it into action. Thus active brain areas during explicit problem-solving span these three large networks/systems in the brain that are highly overlapping (Fig. 6.17). We can almost think of attentional, working memory, and executive processes as the central core of conscious explicit thought.

Along with the core processes described above, the type of problem being solved will influence which brain regions are activated in the process. If you are searching for a friend in a crowded restaurant, your face perception systems will be activated whereas if you are trying to remember where you parked your car at the airport, your spatial relationships and perhaps inner speech regions will be activated.

Conceptual knowledge systems—which are widely distributed in the cortex—are also likely activated, depending on the problem to be solved. Subcortically, processing regions such as the limbic area will be engaged if there are emotional aspects to the problem and its solution. Thus brain areas active for explicit thinking and explicit problem-solving are widely distributed and located throughout the cortex and the subcortex.

Understanding how *implicit thinking* and *implicit problem-solving processes* are organized in the cortex has long been a difficult problem itself to solve since those processes are out of our conscious control and are not reportable. One approach to investigate the neural bases of implicit problem-solving is to study the process of insight, mentioned above. Experiments where subjects can report having—or not having—an Aha! moment of sudden insight provide researchers with the capability to begin to isolate which brain regions are correlated to the Aha! moment and thus with the insight process.

Jung-Beeman and colleagues used a combination of fMRI and EEG to investigate the neural substrates of insight accompanied with an Aha! experience (Jung-Beeman et al., 2004). fMRI results showed increased activity in the anterior superior temporal gyrus (aSTG) in the *right hemisphere*—but not in the left—for the insight versus noninsight conditions (Fig. 6.18). EEG results showed a burst of high-frequency (gamma band) neural activation in the same region of the brain, and again just in the right hemisphere. Jung-Beeman and colleagues interpreted their findings as indicating that right hemisphere aSTG—which has been shown to be involved in with integrating seemingly widely distributed semantic knowledge during language comprehension—may play an important role in the processes leading to the sudden solution of a problem using insight and accompanied with an Aha! experience.

Following up on this early finding, Jung-Beeman and colleagues investigated the neural processing that occurs *just before the Aha! moment* (Kounios et al., 2006). They found that the mental processes that successfully led to insight (the Aha! moment) were more

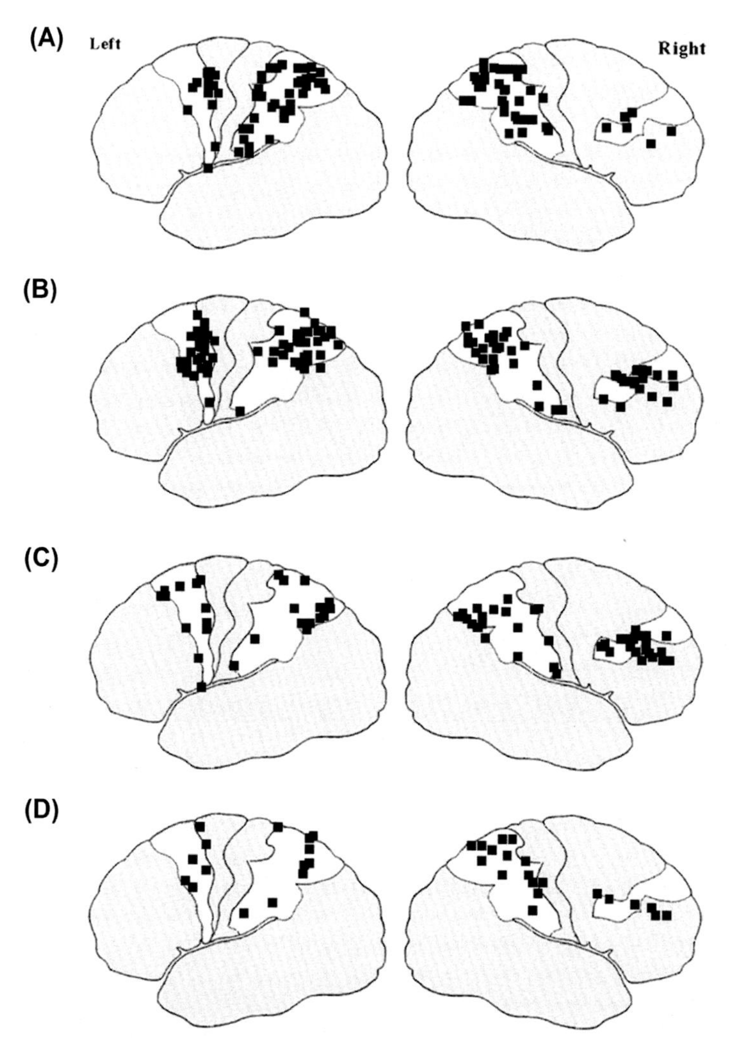

FIGURE 6.17 Overlapping brain regions for problem-solving processes. Driving the brain activity for explicit problem-solving processes are key aspects of explicit thinking such as attention (A), working memory (B), memory retrieval (C), and conscious perception (D). These processes tap common fronto-parietal regions in both hemispheres. *Source: Naghavi and Nyberg (2005).*

active in medial frontal lobe regions and in temporal lobe areas (Fig. 6.19). The authors hypothesized that the increased activation in frontal regions related to cognitive control (executive) processes and the increased activation in temporal regions related to semantic processing.

Another approach to investigating explicit versus implicit problem-solving processes is to consider the effect of *experience, overlearning, and expertise* on those processes. As our problem-solving abilities improve with experience, the conscious effort used to solve the problem reduces and the entire process becomes more and more implicit. This effect is

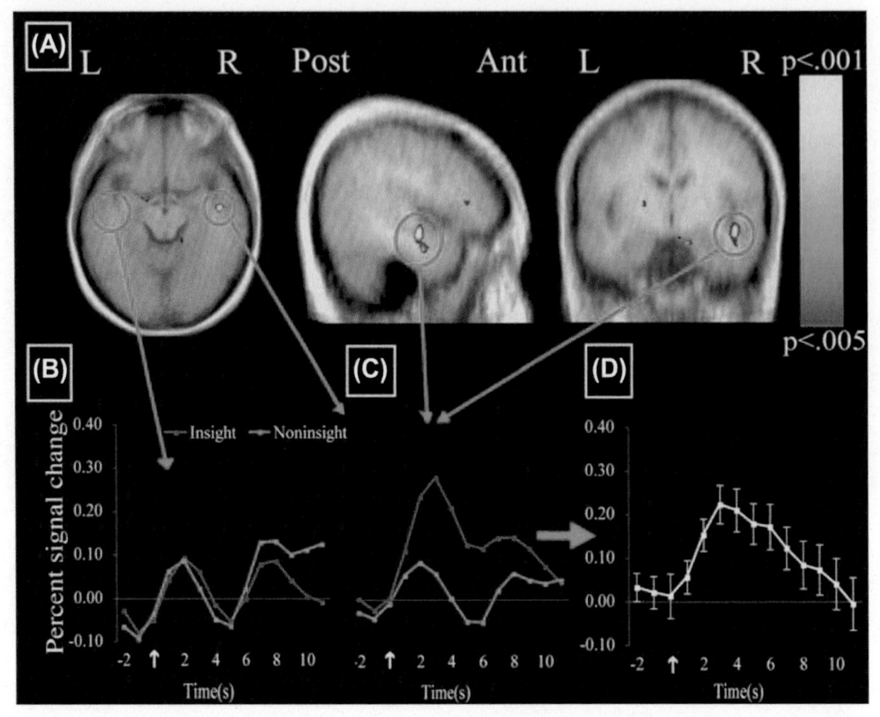

FIGURE 6.18 Brain regions active for Aha! experiences. Jung-Beeman and colleagues used a combination of functional MRI (fMRI) and electroencephalography (EEG) to investigate the neural substrates of insight accompanied with an Aha! experience (Jung-Beeman et al., 2004). fMRI results showed increased activity in the anterior superior temporal gyrus (aSTG) in the *right hemisphere*—but not in the left—for the insight versus noninsight conditions. (A) Top panel shows axial, sagittal, and coronal slices (from left to right) with aSTG activity shown in *red/yellow* patches circled in green. (B and C). Left bottom panel shows in the increased signal change recorded for the Insight condition in the right (C) but not the left (B) hemisphere. (D) Signal change for Insight minus Noninsight conditions, showing the net increased activation for the Insight condition. *Source: Jung-Beeman et al. (2004).*

present in video games as expertise develops, in driving a car as experience grows, and in most any task where a problem to be solved and its solutions have become well known. Thus another way to assess which brain processes are involved in explicit/effortful problem-solving versus implicit/unconscious or automatic problem-solving is to investigate the brain regions that differ between trials when a person is naive to a task or problem (Explicit condition) and trials when that person has become well trained in that task or problem (Implicit condition).

One of the most studied aspects of explicit learning and problem-solving becoming implicit with time and experience has been *motor skill learning*. As we develop from infants to children to adults, and all through our adult life, motor skill learning takes place. Initially effortful, motor skills become less and less effortful and indeed take little or no conscious effort once well learned. These motor skills develop in scope and complexity throughout life as we learn to walk, climb, run, ride a bike, snowboard, drive, play a musical instrument, and so on.

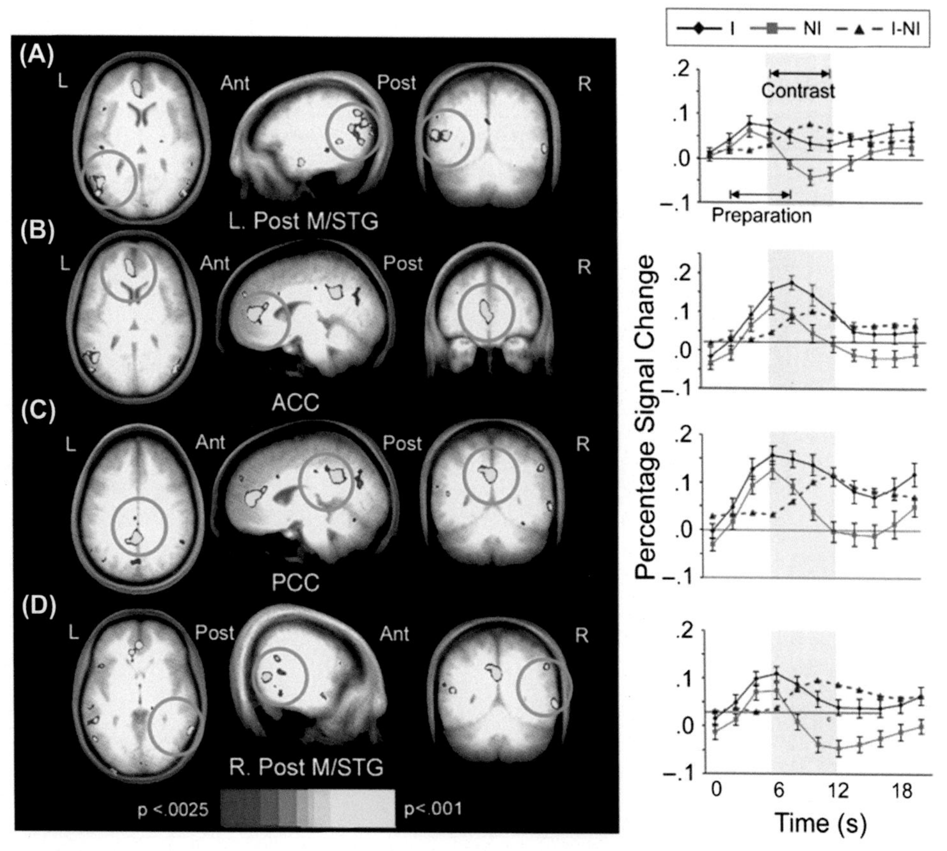

FIGURE 6.19 What brain regions are active just prior to the Aha! moment? Following up on their early finding, Jung-Beeman et al. investigated the neural processing that occurs just before the Aha! moment (Kounios et al., 2006). Group results for 25 subjects are shown in axial (left), sagittal (center), and coronal (right) slices, with the left hemisphere presented on the left and the right hemisphere presented on the right. Results (signal change prior to the Aha! experience) were located in (A) the left posterior middle temporal gyrus, (B) anterior cingulate cortex, (C) posterior cingulate cortex, and (D) right posterior middle/superior temporal gyri. The authors hypothesized that the increased activation in frontal regions related to cognitive control (executive) processes and the increased activation in temporal regions related to semantic processing. *Source: Kounios et al. (2006).*

Taubert, Draganski, and Anwander (2010) investigated brain changes that take place while learning a new motor skill (Fig. 6.20). Taubert and colleagues reported brain structural changes that occurred rapidly during learning a motor skill task and that corresponded to task performance (Fig. 6.21). In this case, results showed gray matter volume increases in the frontal and parietal lobes after only two practice sessions had been completed, indicating that the cortical processes related to skill learning adapt rapidly and reorganize during the learning process.

Both explicit and implicit thinking and problem-solving involve widespread cortical and subcortical regions. Although less is known about the implicit processes, it is likely that both explicit and implicit thinking processes are widely—but differentially—distributed in the brain.

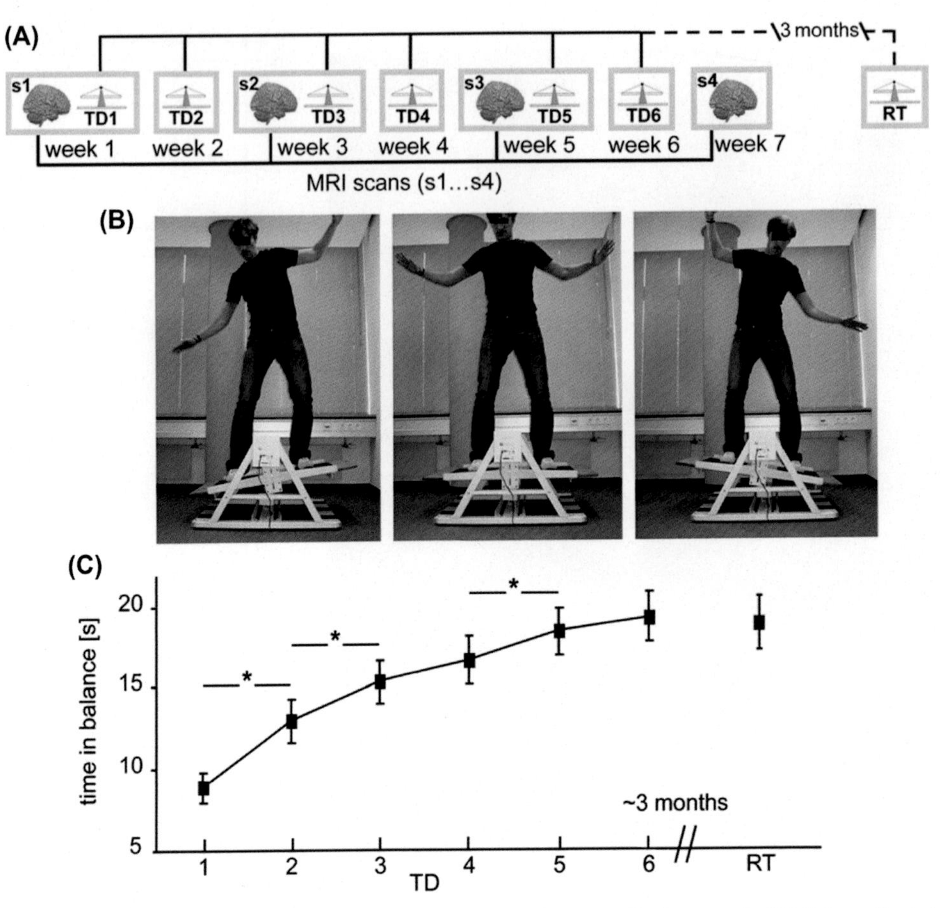

FIGURE 6.20 Brain changes occur rapidly with motor skill learning. Taubert et al. (2010) investigated brain changes while learning a whole-body dynamic balancing task (DBT) shown in (B) The experiment consisted of 2 groups: 14 subjects trained on the DBT and 14 served as controls. The DBT group had 6 training days (TDs), once a week, shown in (A), followed by a retention test (RT) 3 months later. Behavior results—improved balancing—are shown in (C), showing continued improvement during the 6 TD sessions that was maintained 3 months later during the RT. *Source: Taubert et al. (2010).*

4. SUMMARY

Language and thought, words and ideas: while we find substantial evidence that these are separable aspects of the mind/brain, there is no doubt that they are inextricably linked in our mental processes. While language and thought share many regions of activation in the brain, they are nonetheless nonidentical processes that can be disambiguated with careful experimentation. The fact that both language and thought processes seem to entail a significant amount of implicit, nonconscious processes lends a particularly intriguing aspect of their

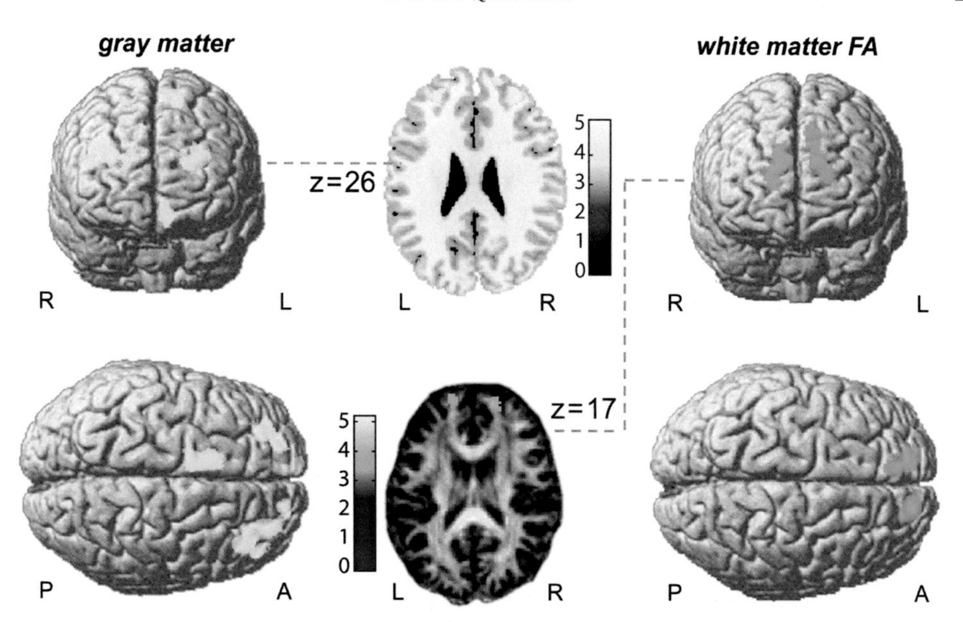

FIGURE 6.21 Gray matter changes across the learning period. Taubert et al. (2010) found that gray matter volume increased beginning even with the second training day (TD) for the dynamic balancing task (DBT) group—with no corresponding gray matter change found for the control group. Results are shown in *yellow* on axial slices on the left side of the figure. Corresponding white matter changes are shown in cyan on the same axial slices on the right side of the figure. *Source: Taubert et al. (2010).*

study. Perhaps we will never fully understand the implicit nature of language and thought, however, it is clear that we use both explicitly known knowledge and concepts along with their implicit counterparts in everyday life.

5. STUDY QUESTIONS

5.1 As the speech signal reaches our ears and is decoded in the brain, many aspects of spoken language are processes in nearly parallel fashion. List 3–4 aspects of spoken language that are decoded as we understand the ideas being put forth to our ear and mind.

5.2 Noam Chomsky described language as "discrete infinity." What did he mean by this? What aspect of language is discrete? What is infinite?

5.3 What are the two brain areas that are typically included in a "classical theory" of language organization in the brain? Where are they located?

5.4 According to Hickok and Poeppel (2007), some aspects of speech processing are mediated in both left and right hemispheres and some are left dominant. Name one or two processes that are organized bilaterally in the brain according to their model.

5.5 What is a "problem space" and what does it tell you about problem-solving?

5.6 What is a key difference between explicit thinking and problem-solving and implicit thinking and problem-solving?

5.7 What is the Aha! experience and how does it differ from explicit problem-solving experiences?

5.8 According to Jung-Beeman et al., what regions of the brain are active *just prior to* the Aha! experience? What is their hypothesized role in problem-solving?

CHAPTER

7

Learning and Remembering

PAINTED FROM MEMORY

Further, much recent evidence makes it clear that "memory systems" in both humans and animals are as engaged by thinking about the future as when retrieving the past. ***Lynn Nadel, American Memory Researcher.***

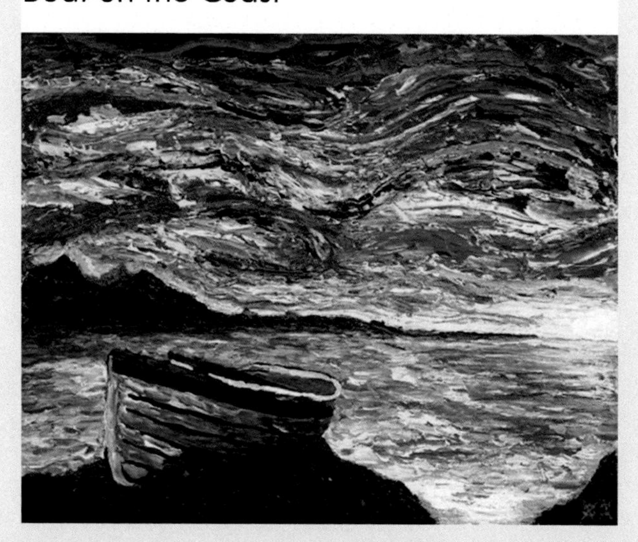

"Boat on the Coast". Blind artist paints from memory. John Bramblitt is a young man who has gradually lost his eyesight due to unknown reasons. As his vision was fading, John decided that he would not give up his art due to his infirmity. Consequently, he set about developing ways to continue to paint despite his vision loss. His approach includes using his hands to "visualize" a face. His collection of original art is large and absolutely awesome. This work is entitled "Boat on the Coast" and is painted from his memory. *Source: John Bramblitt, https://bramblitt.com/, with permission.*

Fundamentals of Cognitive Neuroscience
https://doi.org/10.1016/B978-0-12-803813-0.00007-6

1. INTRODUCTION

A *memory* can be defined as a lasting brain representation that is reflected in thoughts, experiences, or behaviors. *Learning* is the process of acquiring such representations. Most students would like to have a "better memory," which usually means being able to recall a large number of facts, like the ones we might need for a chemistry exam. But the human brain has not evolved for academic study and recall. Rather, it has evolved for the tasks of survival, which generally involve a lot of interaction with the things we need to know.

Our brains are adapted to deal with complex, urgent, vague, surprising, and novel life challenges, the kinds of problems that people have to solve in the real world. Perhaps for those reasons, our ability to *recognize* past events is excellent, but our ability to *recall* memories, based on partial cues, is not nearly as good. School exams are designed to test recall through partial cues, but formal schooling is a recent cultural practice. We can learn to memorize facts for recall on exams, but it takes hard work.

In contrast, learning how to throw a ball or how to recognize an old path through a forest is surprisingly easy. We can learn those things simply by paying attention (making ourselves conscious of something) and by interacting with whatever is to be learned. There is a good reason why teachers all over the world start their classes by asking students to "please pay attention." Attention and interaction are two keys to learning.

Humans are exceptionally flexible in the face of new conditions. Our neolithic brains work well in a world of computers, brain science, and academic challenges, even though we are not biologically adapted to computers. Learning provides that remarkable flexibility.

The most important brain structures in this chapter are the *cortex* (*or neocortex*, the terms are used interchangeably)—the visible outer brain—and the *medial temporal lobes* (MTLs), which contain the two hippocampi (Fig. 7.1). Together, these two brain areas form the major neural bases for learning and memory formation and retrieval.

1.1 Memory Is Not a Single Thing: A Useful Classification

Memory is not a single thing. Fig. 7.2A shows a standard view of known memory systems (Squire, 2004). There is an ongoing research about these questions, but the terms used in Fig. 7.2 are important. We will focus mostly on episodic and semantic learning which form our declarative memory, which have seen a great deal of brain research. Fig. 7.2 starts from the top by dividing memories into two types. *Declarative memory* is about events and concepts, the "what, where, and when" of stored information. We are conscious of many of these "what, where, and when" properties of the world. The word *conscious* can, therefore, be put in parentheses under the word *Declarative*. Keep in mind, however, that human beings always mix conscious and unconscious processes. One way to think about declarative memories is that we can talk about them, describe them, and point to them. That is a good behavioral index of the things we are conscious of.

Nondeclarative memory is about the "how" and "why"—knowing how to ride a bicycle, and the "whys"—our implicit goals, tendencies, and dispositions (Banaji & Greenwald, 1995). Notice that in the diagram the word *nondeclarative* can also be called *nonconscious*. That is certainly true for many memory types in the figure.

The brain regions theorized to be associated with declarative and nondeclarative memory are shown in Fig. 7.2B. We will discuss these brain regions in more detail throughout this

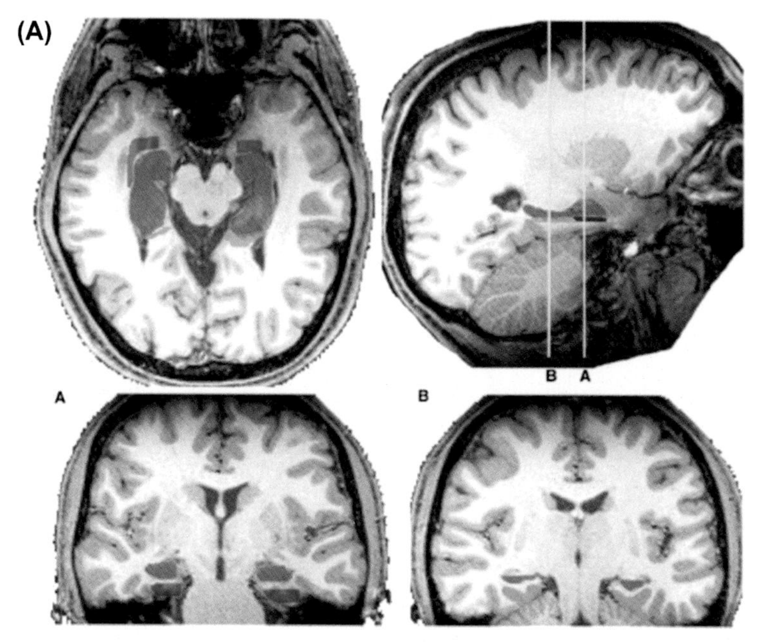

FIGURE 7.1 The hippocampus in the medial temporal lobe (MTL). These regions are spatially complex and difficult to visualize. (A) The two hippocampi in the left and right hemispheres are shown shaded in red on an axial brain image (upper left), a sagittal brain image (upper right), and on two coronal images (lower left and right). Other key structures in the MTL include the perirhinal cortex (shaded in blue) and the parahippocampal cortex (shaded in green). *Source: Diana, Yonelinas, and Ranganath (2007).*

chapter, and we will also present a diverging view of these brain regions and their role in memory and retrieval.

Would you agree with the labels presented in Fig. 7.2A based on your own experiences? Remember that categories like this are helpful approximations. They are not perfect. Most of our mental activities combine several types of memory. We will focus mainly on episodic memories (memories of conscious events) and semantic memories (memories about facts and concepts).

1.2 Episodic and Semantic Memory

Declarative memory can be divided into *episodic* and *semantic memory* (Tulving, 1972). *Episodic memory* is autobiographical. It refers to memories of conscious events in our lives...episodes...that have a specific source in time, space, and circumstances. We can try to travel mentally back in time to relive the original episodic experience, or something close to it. An example is your memory of what you had for breakfast today, or what happened on the day you graduated from high school. The words *episodic* and *autobiographical* are similar in meaning.

By contrast, *semantic memory* involves knowledge about the world, about ourselves, and about other people. It is often hard to remember when we learned some particular semantic

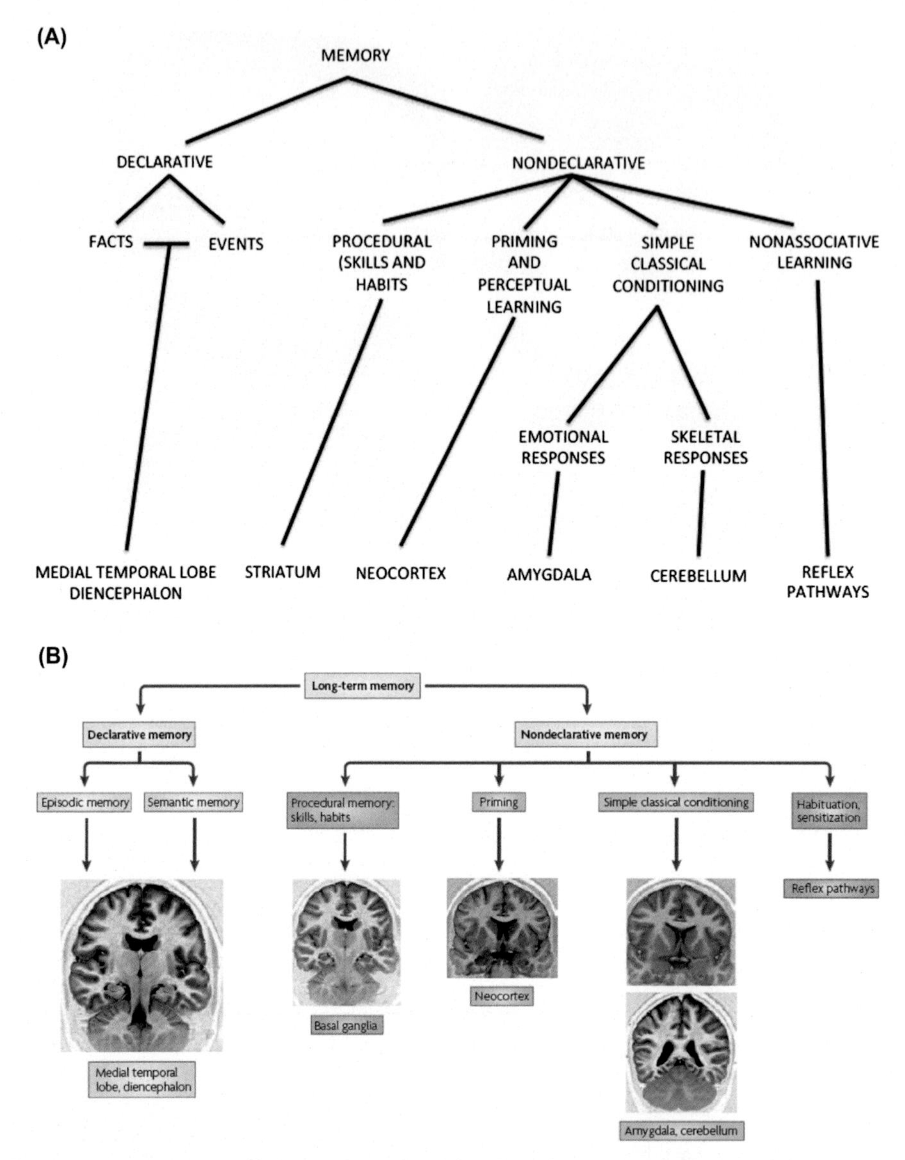

FIGURE 7.2 (A) Schacter and Tulving initially proposed this classification of memory types. Declarative memories have been studied in great detail and are believed to be explicit (conscious). Nondeclarative memory types are said to be unconscious or implicit, but this claim is still debated. We will focus mainly on semantic versus episodic memory, where a great deal of research has been done. (B) Brain regions hypothesized to subserve the memory classifications presented in (A). *(A) Source: Squire (2004) and (B) Source: Henke (2010).*

memory. You may not remember when you learned that Paris is the capital of France or when you understood a new idea about the human brain. In semantic memory, you do not need to remember the time and place when you learned it. All you need is a meaningful piece of information.

2. EPISODIC LEARNING AND MEMORY

Episodic memories may be learned simply by paying attention to some conscious event. Episodic learning can either happen *intentionally*—when we try to learn—or *incidentally*—without deliberately trying to memorize anything. We may remember a movie scene just by paying attention to it. Paying attention may be enough to establish an episodic memory for a romantic scene at the start of a movie or the exciting chase scene at the end. If you see the same video clip years later, you may be still able to recognize it.

Human beings are extraordinarily accurate when we test for *recognition* of episodic memories, even when subjects have no deliberate intention to memorize it. That basic fact is easy to verify by checking a high school yearbook to see if you can recognize the photos of classmates you have not seen for a long time.

The high accuracy of episodic memory *as tested by recognition measures* suggests that our brains have a very large storage capacity for the things we pay attention to. Recognition measures are yes/no measures ("Yes, it is familiar" vs. "No, it is not"). It is harder to *recall* episodic memories from partial hints (what happened on your friend's birthday 5 years ago?). Recall accuracy is generally lower than recognition accuracy.

Because school exams tend to use cued recall rather than recognition, it is difficult to score well on academic exams. But the effortless way in which we may score very highly on *recognition* measures suggests that a great many things we have paid attention to are "in storage" and potentially retrievable. In one classical experiment, college students were shown 10,000 pictures for 6 s each and then tested several days later, using recognition measures. They were given hundreds of "foils," pictures they had never seen before, and an equal number of pictures they had seen before. Subjects scored above 80% in accuracy (Standing, 1973). No instructions were given to memorize the pictures, and 6-s exposures are too brief to allow for much deliberate memorizing. Spontaneous episodic learning is therefore remarkably accurate, provided that we test for recognition rather than recall. Such evidence suggests there is much more information in memory than we usually bring to mind.

Episodic learning requires the hippocampus and its neighboring structures. A half-century of brain studies have shown that the hippocampal neighborhood (MTL) is needed for episodic learning—converting perceptual experiences into long-lasting episodic memories. The hippocampus is shaped like a thin tube inside each temporal lobe, but many small regions in this area play special roles. For that reason, the more inclusive term *MTL* is often used.

Fig. 7.3 shows our standard diagram of cognition with special emphasis on the long-term memories: the green boxes at the bottom. Episodic events are stored in autobiographical memory. The basic question in episodic learning is simply this: How do sensory and other experiences turn into long-term autobiographical memories that may be needed years later?

According to our functional diagram, when we pay attention to sensory input, it flows into *working memory* (WM), which, in turn, allows information to be actively maintained and manipulated (see Fig. 7.3). The act of paying attention tends to bring specific conscious contents to mind.

WM allows us to keep small amounts of information in an accessible form for 10–30 s. Many everyday tasks call upon this capacity, such as keeping a landmark in mind to find

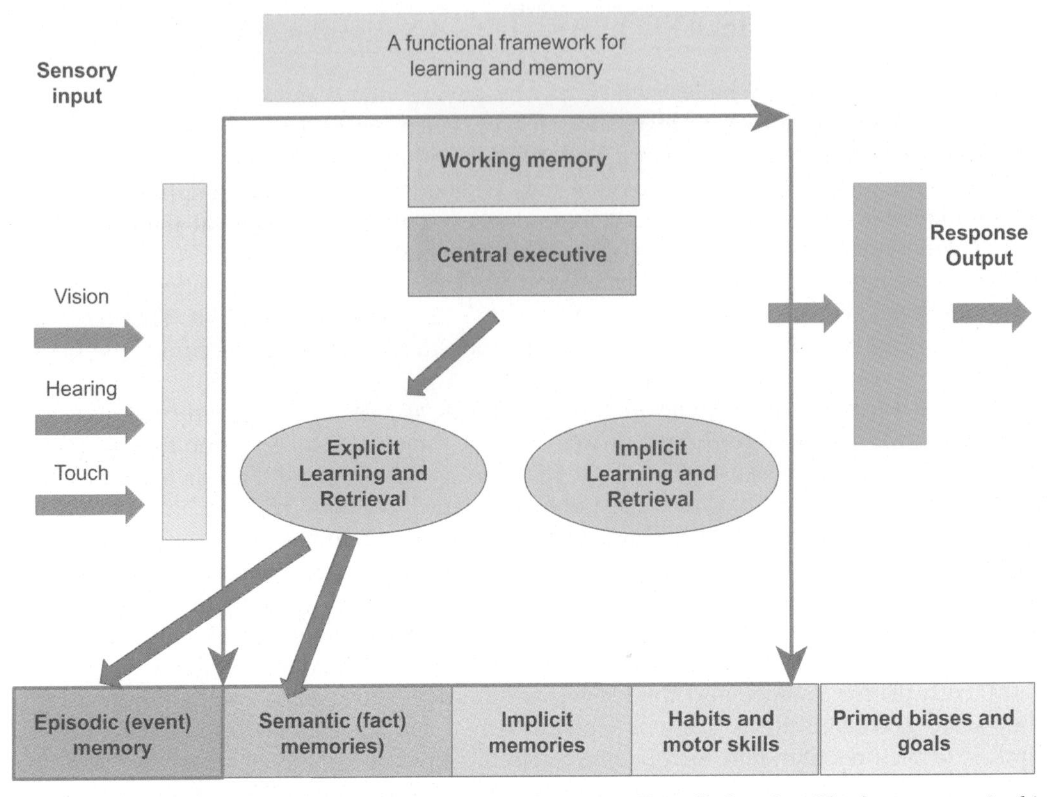

FIGURE 7.3 A functional framework for learning and memory. Episodic learning. The *brown arrows* in this diagram trace the route of episodic information from conscious sensory input to autobiographical long-term memory. *Source: Gage with permission.*

our way home in a strange city. WM gives us a sense of continuity by embedding our immediate conscious experiences into a psychological past, present, and future.

2.1 An Example of Episodic Learning

Consider what happens when we first see a specific coffee cup on a table (Fig. 7.4). From our chapter on vision we know that the visual brain begins by decomposing the retinal input into features in various maps in the visual hierarchy: roughly, "light points," line orientation, size (spatial frequency), color, local motion, and ultimately shapes and textures. After the basic features are decomposed, they are "bound" together so that linked feature maps define the object: a coffee cup on a table.

While the MTL is needed to bind and transfer episodic experiences into memory, the actual memory traces are believed to be stored in the cortex. Memory traces must be useful in the real world, and memory *retrieval* is therefore as important as learning. When we encounter

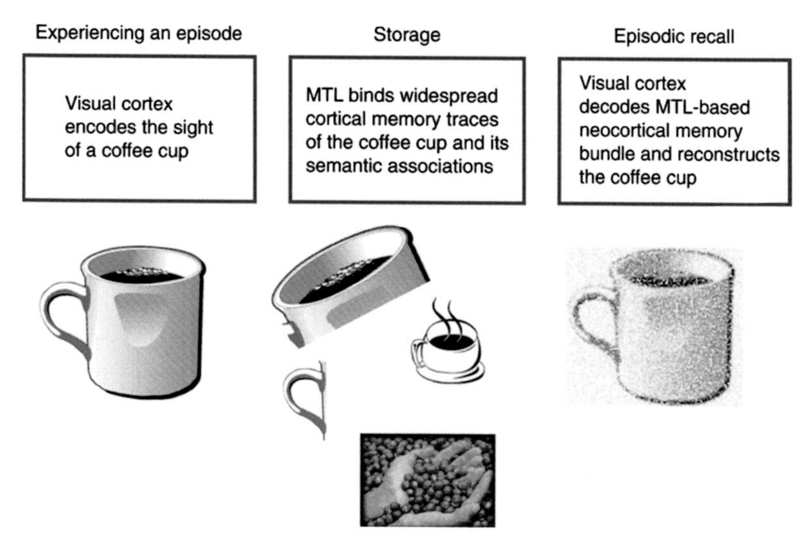

FIGURE 7.4 Episodic learning of conscious events. The sight of a specific coffee cup is encoded in visual maps for different features, like location, color, line orientation, motion, and so on. In the medial temporal lobe (MTL) the separate feature maps may be "bound" together into a single "gestalt." Notice that the coffee cup is associated with a handful of raw coffee beans and with the smell of hot coffee. During slow wave sleep, the active memory of the coffee cup is recoded into lasting synaptic changes in the cortex. Recall from episodic memory reverses these steps. *Source: Baars, with permission.*

a reminder of a specific experience of the coffee cup, the bound memory traces "light up" the corresponding traces in the cortex again. We thereby *reconstruct* some part of the original memory, again using the MTL to integrate memory traces into a coherent experience. That experience—of imagining the coffee cup—makes use of the visual cortex again. Because this is the central theme of this chapter, we begin with a cartoon version in Fig. 7.5.

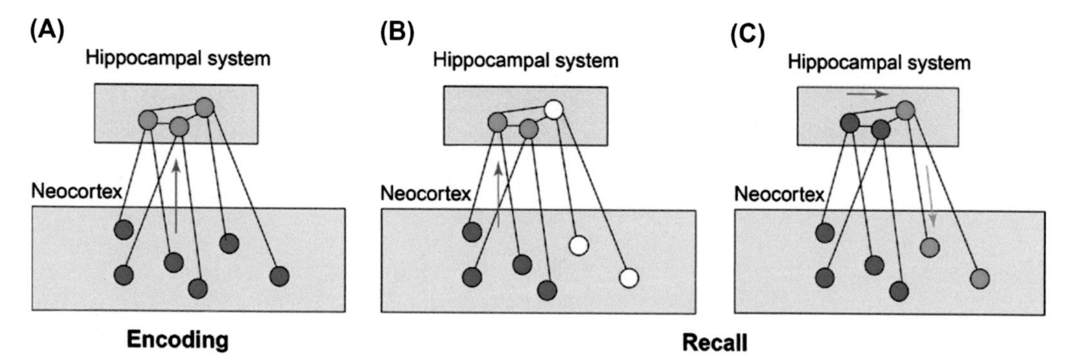

FIGURE 7.5 Hippocampus and cortex. The hippocampal area supports temporary episodic memories that activate and are stored in the cortex. In this cartoon, we show stages of encoding and recall and the roles of the hippocampal system and the cortex. (A) An episodic memory is formed (Encoded) in the cortex and gives rise to a pattern of activity with the hippocampal system. (B) During Recall, part of the pattern is active in the cortex, reactivating part or all of the hippocampal pattern. (C) The reactivation of the hippocampal pattern will, in turn, reactivate the remaining cortical pattern. *Source: Gluck et al. (2003).*

Visual features of the cup, like the handle, are also part of the associative complex that becomes activated. When the *episodic memory*—the sight of the coffee cup—is cued the following day—maybe by someone asking, "Did you like the way I made the coffee yesterday?"—the MTL is once again involved in retrieving and organizing widespread cortical memory traces. Visual cortex is therefore needed to reconstruct the sight of the coffee cup, which is never identical to the original cup but rather a plausible recreation of a pattern of visual activation that overlaps with the first one. Notice that visual cortex is involved in perception, learning, and episodic recall.

3. SEMANTIC MEMORY IS DIFFERENT

There is good evidence that semantic memories may be formed from repeated, similar episodes. Attending a high school is a long series of episodes. We may be able to recall dozens of those episodes, but much of the time they seem "smeared" together in memory in the semantic belief that "I attended such-and-such high school."

3.1 Episodic and Semantic Networks

Fig. 7.6 shows how episodic and semantic memories may be related in the brain. Specific episodic memories are shown in the pictures of a man presenting flowers to a young lady. This is a separate autobiographical memory, remembered as conscious events. Above these pictures, a small semantic network combines all these and other very specific and richly detailed episodes into a single figure: a semantic network of a man who cooks, loves, paints, and plays golf. The semantic network is more abstract and general than the episodes about

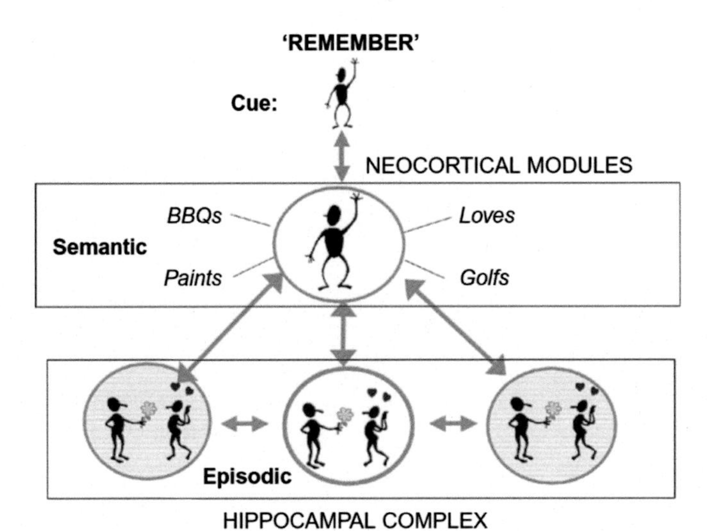

FIGURE 7.6 From episodic to semantic memories. Semantic memories are assessed by feelings of knowing, which can be very accurate. However, they do not require active reconstruction of the original episode and can apparently be cortex without the aid of the hippocampal complex. *Source: Moscovitch, with permission.*

particular events in the life of the person. Moscovitch et al. (2005) claims that the bottom row of episodes depends on the MTL, and the top figure depends on cortical modules.

Episodic memories typically:

1. have reference to oneself
2. are about a specific time
3. are *remembered* consciously, as if we reexperience them
4. can be forgotten
5. are *context-sensitive*, with remembered aspects of time, space, people, and circumstances

In contrast, semantic memories (Fig. 7.6) generally:

1. have reference to shared knowledge with others
2. are not organized around a specific time period
3. give a "feeling of knowing" rather than a fully conscious recollection of the original event
4. are less susceptible to forgetting than specific episodes
5. are relatively independent of context

3.2 The Hippocampus in Remember/Know Experiments

Tulving (1985) introduced the remember/know procedure, asking participants to report their conscious experience when they recognize studied items. If they believe an item was studied before, they must decide whether they *remember* the item (i.e., they can reexperience episodic details about the event) or whether they *know* the item (it feels familiar). Local hippocampus seems to affect only "remember" judgments. Memory based on feelings of knowing is spared after hippocampal damage (Moscovitch & McAndrews, 2002; Yonelinas, 2002). Similarly, in functional neuroimaging studies, hippocampal activation is associated more with remembering than familiarity (Eldridge, Sarfatti, & Knowlton, 2002; Yonelinas et al., 2005).

When you are asked if the face of a particular person is familiar to you, like the movie star Brad Pitt, on what basis are you making your judgment? Are you relying on your semantic memory, episodic memory, or both? Westmacott, Black, Freedman, and Moscovitch (2004) have shown that *both* systems may contribute because performance on semantic tests is better if the participant also has some episodic memory associated with the famous name.

3.3 Schemas and Memory

Humans are highly adept at extracting information from the world around them to learn and grow and achieve our goals, large or small. Throughout life, many situations are presented that are similar to previous situations. As learning "sponges" as young children, we develop *schemas:* large-scale knowledge systems that allow us to decode the scene around us and use preexisting knowledge to determine how and what we will do in a given situation.

Thus in addition to thinking about memory in the classifications presented in Fig. 7.2A, we need to enlarge the scope of those memories into the body of our conceptual knowledge. Schemas form a large part of that knowledge. They are more abstract than episodic memories and larger in scope that semantic memory—think of schemas as a superordinate level of memory, a higher level if you will. Schemas are dynamic, like our memory systems in general, they are constantly changing as we experience new situations and learn to adapt to our environment.

4. MEMORY TRACE FORMATION AND CONSOLIDATION

How—and *where*—are our memories formed and stored in the brain? Our understanding of *how* learning happens and memories are formed is expanding constantly. We will present some current theories about this process. As to *where* memories are stored in the brain, the current hypotheses support the notion that memories are widely distributed throughout the brain and likely activate, when recalled, the same or similar areas that were active during encoding. This may sound like a very new hypothesis; however, it dates back to at least the 19th century as Carl Wernicke described this notion of conceptual knowledge storage in his first monograph (Wernicke, 1874).

4.1 "Excitatory" and "Inhibitory" Memory Traces

Most synapses in the cortex are excitatory, using the neurotransmitter *glutamate*. A very large minority use inhibitory neurotransmitters like *GABA* (gamma-aminobutyric acid). To encode long-term memory traces in changed synaptic efficiency, these excitatory and inhibitory connections must somehow be made more permanent. These two processes are believed to occur in what is called long-term potentiation (LTP) for excitatory synapses and long-term depression (LTD) for inhibitory ones. These events, which have been observed in specific regions, are simply an increase and a decrease in the firing probability of a postsynaptic potential given a presynaptic spike.

LTP has been observed in the hippocampus, using single-cell recording in one of the neuronal layers of the hippocampus. Single-cell recording has been extensively done in animals, but there are cases of such recordings in human epileptic patients as well (Kreiman, Koch, & Fried, 2000). While we can observe LTP and LTD in specific locations like the hippocampus, the standard hypothesis about long-term memory involves billions of synapses in cortex and its satellites, amounting to literally trillions of synapses. We have no way of taking a census of all of the synapses in this system, or even a substantial fraction of them, at this time. Rather, we have a number of studies showing increased LTP and LTD, supplemented by studies of brain damage and of population activity among billions of neurons as measured by electroencephalography (EEG), event-related potential (ERP), functional magnetic resonance imaging (fMRI), and so on. In addition, we have evidence from stimulation studies, like temporal lobe stimulation of awake patients during neurosurgery and transcranial magnetic stimulation in normal subjects. What we know about memory is therefore an inferential picture, in which many hundreds of studies have been performed.

But we cannot yet come close to observing large numbers of changed synaptic connectivities directly at the submicroscopic level.

Our current picture suggests these points:

1. Episodic input is initially analyzed by the cortex.
2. It is integrated for memory purposes in the MTL, containing the hippocampi and related structures.
3. Consolidation: MTL and related regions then bind and integrate a number of cortical regions, a process that transforms temporary synaptic connectivities into longer-lasting memory traces in both the MTL and cortex. The main mechanisms for such changes are believed to be LTP and LTD.

4.2 Rapid Consolidation: Synaptic Mechanisms, Gene Transcription, and Protein Synthesis

Rapid or synaptic consolidation is accomplished within the first minutes to hours after learning occurs. Weiler, Hawrylak, and Greenough (1995) showed that it correlates with morphological changes in the synapse itself. Stimulus presentation initiates a cascade of neurochemical events at the synaptic membrane and within the cell, which increase the synaptic strength or efficiency with which neurons that form the memory trace can communicate with one another. The first of these processes involves local, transient molecular modifications that lead to an increase in neurotransmitter release at the affected synapse. If the stimulus is intense enough and/or repeated, additional processes are activated. These involve gene transcription and protein formation that lead to long-lasting cellular changes, including the creation of new synapses that support the formation and maintenance of long-term memory (Fig. 7.7). These processes may last from hours to days (Dudai, 2004; Lees, Jones, & Kandel, 2000; McGaugh, 2000).

Although we are well on our way to understanding the basic cellular and molecular mechanisms of synaptic consolidation, we are far from understanding prolonged or system consolidation, which is being debated heatedly in the literature.

4.3 System Consolidation: Interaction Between the Medial Temporal Lobes and the Cortex

System consolidation can take much longer to complete and may range from days to years or decades. Patients with MTL lesions show a retrograde memory loss that is temporally graded, so recent memory loss (before the amnesia) is greater than earlier memory loss. This temporal gradient is restricted to explicit memory, leaving implicit memory intact and stable over time (Scoville & Milner, 1957).

These observations suggest that the MTL forms a temporary memory trace needed for explicit memories until they are consolidated elsewhere in the brain, presumably in the cortex (Squire, 1992; Squire & Alvarez, 1995). This standard model of consolidation makes no distinction between various types of explicit memory. For instance, it predicts a similar pattern for episodic and semantic memory.

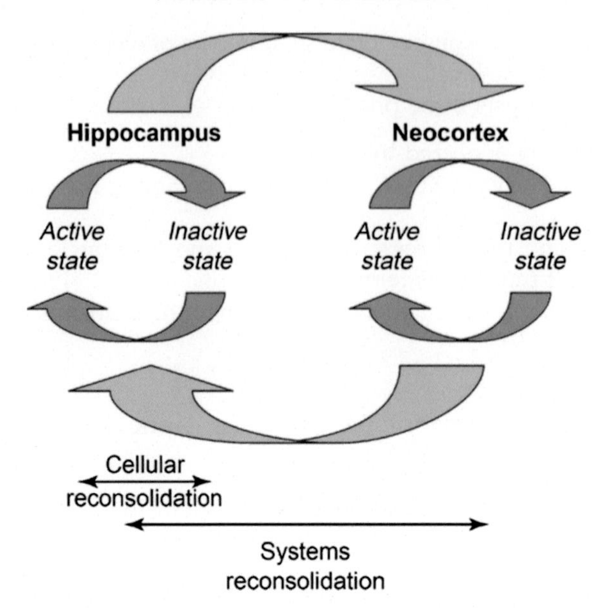

FIGURE 7.7 Consolidation turns active neuronal connections into lasting ones. Two kinds of consolidation are believed to exist: cellular and systems consolidation. Both are evoked by activation of the medial temporal lobe (MTL) and cortex. This diagram emphasizes the degree to which the MTL (also called the hippocampal complex) and the cortex establish active cell assemblies corresponding to the learned input, in which neurons resonate with each other until more permanent connections are established. *Source: Nader (2003).*

Nadel and Moscovitch (1997, 1998) concluded, contrary to the standard consolidation model, that the MTL is needed to represent even old episodic memories for as long as the memory exists (Moscovitch & Nadel, 1998; Nadel, Samsonovich, Ryan, & Moscovitch, 2000). The cortex, on the other hand, is sufficient to represent *repeated* experiences with words, objects, people, and environments. The MTL may aid in the initial formation of these cortical traces, but once formed they can exist on their own. Thus unique autobiographical memories are different from repeated memories in that they continue to require the MTL. Repeated experiences are proposed to create multiple traces, adding more traces each time the event is brought to mind.

Neuroimaging studies provide evidence for this interpretation. These studies found that the hippocampus is activated equally during retrieval of recent and remote autobiographical memories (for review, see Maguire et al., 2000; Moscovitch et al., 2005). These questions continue to be debated at this time.

5. MEMORY TRACES AND MODELS

5.1 Episodic Memory Traces and Consciousness

The episodic memory trace consists of an ensemble of the MTL and cortical neurons, while the MTL acts as a pointer to the neural elements in cortex for the event. Retrieval occurs when a conscious cue triggers the MTL, which in turn activates the entire cortical ensemble associated with it. When we recover episodic memories, we bring to mind something close to the original conscious experience (Moscovitch, 1995). The recovery of these experiences always depends on the hippocampus. As Moscovitch (1992) has argued, the hippocampal complex acts as a module whose domain is consciously apprehended information.

5.2 A Multiple Trace Theory Is Proposed

Nadel and Moscovitch (1997) proposed a *multiple trace theory,* suggesting that the hippocampal complex rapidly encodes all information that becomes conscious. The MTL binds the cortical neurons that represent the conscious experience into a memory trace. The MTL neurons act as a pointer, or *index*, to the cortical ensemble of neurons that represent the experience (Teyler & DiScenna, 1986). A memory trace of an episode, therefore, consists of a bound ensemble of cortical *and* MTL neurons. Formation of these traces is relatively rapid, lasting in the order of seconds or at most days (Moscovitch, 1995).

In this model, there is no prolonged consolidation process that slowly strengthens the cortical memory trace. Instead, each time an old memory is retrieved, a new hippocampally mediated trace is created, so that old memories are represented by more traces than new ones and therefore are less susceptible to disruption. Because the memory trace is distributed in the MTL, the extent and severity of retrograde amnesia is related to the amount and location of damage to the MTL (Fig. 7.8).

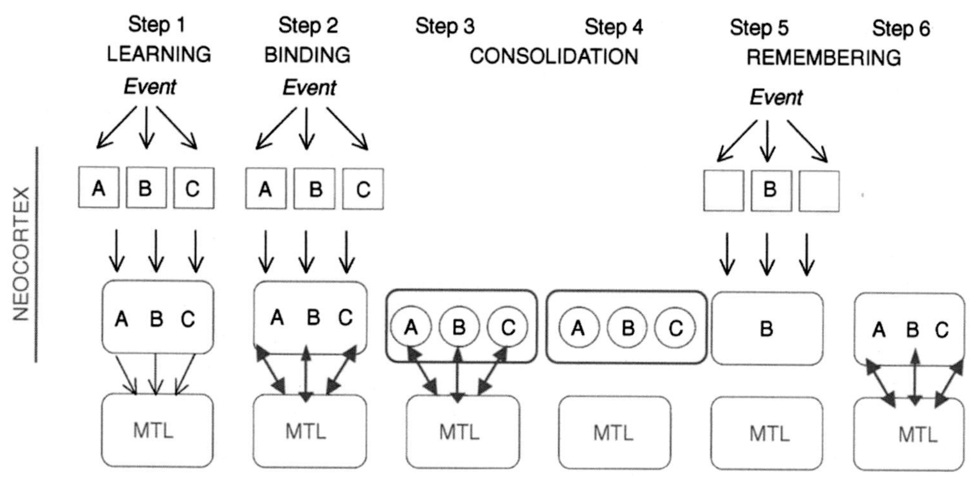

FIGURE 7.8 The steps of learning, binding, consolidation, and remembering. In this summary, Step 1 is the learning of an event, consisting of three elements, A, B, and C. It is initially encoded by the cortex (such as the visual cortex) and sent to the medial temporal lobe (MTL). In Step 2, the MTL and the cortex resonates with each other to begin establishing the memory trace. In Step 3, the stimulus event is no longer available, and the MTL-cortical resonance is now independent of external support. Step 4 shows how consolidation leads to permanent, separate memory traces (synaptic changes) in both the MTL and the cortex, which now exist separately from each other, while other input is being processed. In Step 5, element B of the original event (A-B-C) is presented as a reminder or recall cue. In Step 6, the memory traces of A-B-C are activated by resonating activity between the MTL and the cortex. At this point, the episodic memory has been retrieved in the absence of the original stimulus. *Source: Moscovitch, modified with permission.*

5.3 Challenges to the Multiple Trace Theory

Some recent studies, however, have questioned the multiple trace theory. Using fMRI, Henke et al. (2003; see also Degonda et al., 2005) showed that the hippocampus can be activated by subliminal presentation of faces and their associated professions. Moreover, these activations are correlated with performance on subsequent explicit tests of memory for face-profession pairs. Likewise, Daselaar, Fleck, Prince, and Cabeza (2006) found that the posterior MTL was activated more by old, studied items at retrieval, even when the person was not aware that the item was old. Finally, Schendan, Searl, Melrose, and Stern (2003) showed that the hippocampus was activated on a similar memory task if the repeated sequences were of a higher order of association.

There also have been similar reports from studies with amnesic patients. Ostergaard (1987) was the first to suggest that performance on some priming tests was related to the extent of medial temporal damage. More recently, Chun and Phelps (1999) showed that nonconscious context effects in visual search were not found in amnesic patients, suggesting that the MTL was needed for retaining contextual information of which the person was not aware. Likewise, Ryan, Althoff, Whitlow, and Cohen (2000) showed that amnesic people did not show the normal pattern of eye movements around the location where a change occurred in a studied picture, even though neither they nor the normal controls were consciously aware of the change. Thus the role of the hippocampus does not seem to be limited to consciously apprehended information, as proposed by Moscovitch (1992).

5.4 Some Current Views About Memory Processes

5.4.1 A New View About the Classic Long-Term/Short-Term Memory Distinction

We have discussed types of long-term memory (LTM) such as episodic and semantic memory along with short-term memory (STM) such as WM. In a recent review, memory researcher Nadel proposed a new way to think about memory. Rather than categorize memory into LTM and STM, Nadel proposes a processing model for what he refers to as a *representational view* of memory that is based on the *contents*—what memories are about, and *their processing function*—what they do (Nadel & Hardt, 2011). That is to say, a human being needs to have a knowledge store—memories—that provides information beyond the scope of a long-term versus short-term storage. Instead, the information needed in this knowledge store needs to include *what* the individual experiences, *where and when* things happened, *who* was involved in an event or experience, *the value or relevance* of the event of experience, and the extraction of *how to act or behave in future* similar events or experiences. These, according to Nadel, are the key aspects of memory and not their temporal status. Brain regions theorized by Nadel and colleagues to be involved in episodic memory and cortical and subcortical connectional pathways are shown in Fig. 7.9.

5.4.2 A Debate: Is Consciousness Needed for Episodic Learning?

Another new avenue of research in memory involves the question of whether episodic learning is mediated by conscious processes. As we stated earlier in the chapter, recent evidence has surfaced challenging this view. In a recent review of the roles played by conscious and nonconscious encoding and retrieval in the declarative and nondeclarative

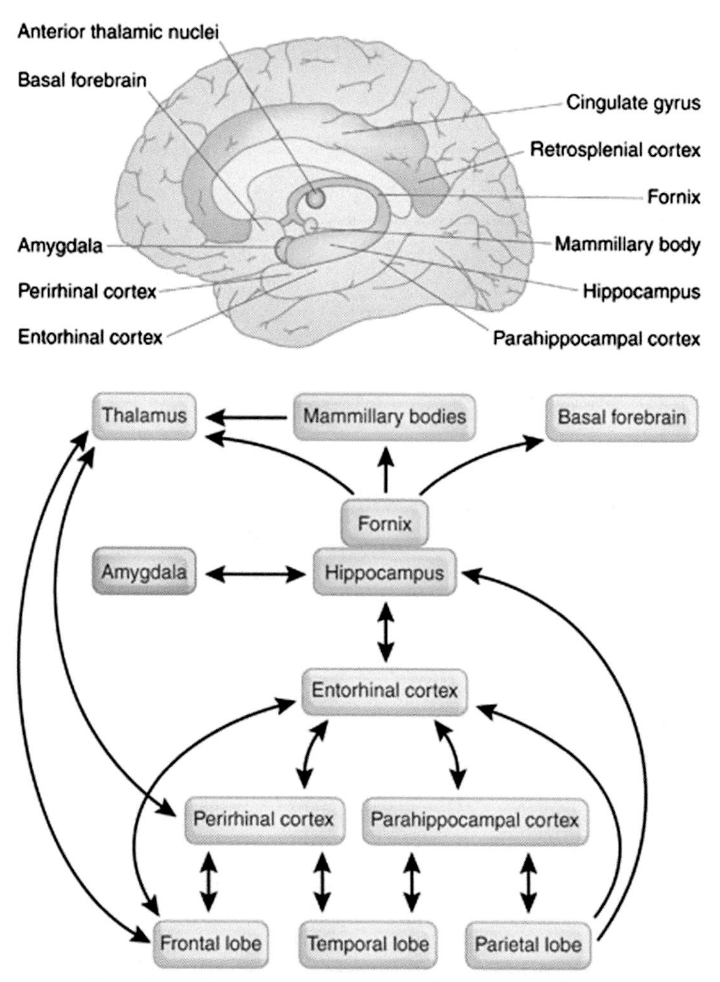

FIGURE 7.9 Top: Brain areas involved in episodic memory, including the hippocampus and fornix, shaded in green, and the amygdala, shaded in purple. Bottom: Unidirectional (one-way) and bidirectional (two-way) pathways between major regions of the medial temporal lobe are shown with *black arrows. Source: Nadel and Hardt (2011).*

memory systems, Henke (2010) proposed a revised classification system for memory processing that is divided into three processing models, each with their brain pathways (Fig. 7.10). This approach to thinking of memory classes in terms of *processing modes* rather than their *conscious or nonconscious aspects* (see Fig. 7.2B) better reflects the empirical findings of the past decade.

5.4.3 *The Role of Schemas in Memory Encoding and Retrieval*

While classically long-term memory stores are divided into episodic and semantic memory, it is clear from decades of research that memories are not sharply divided across

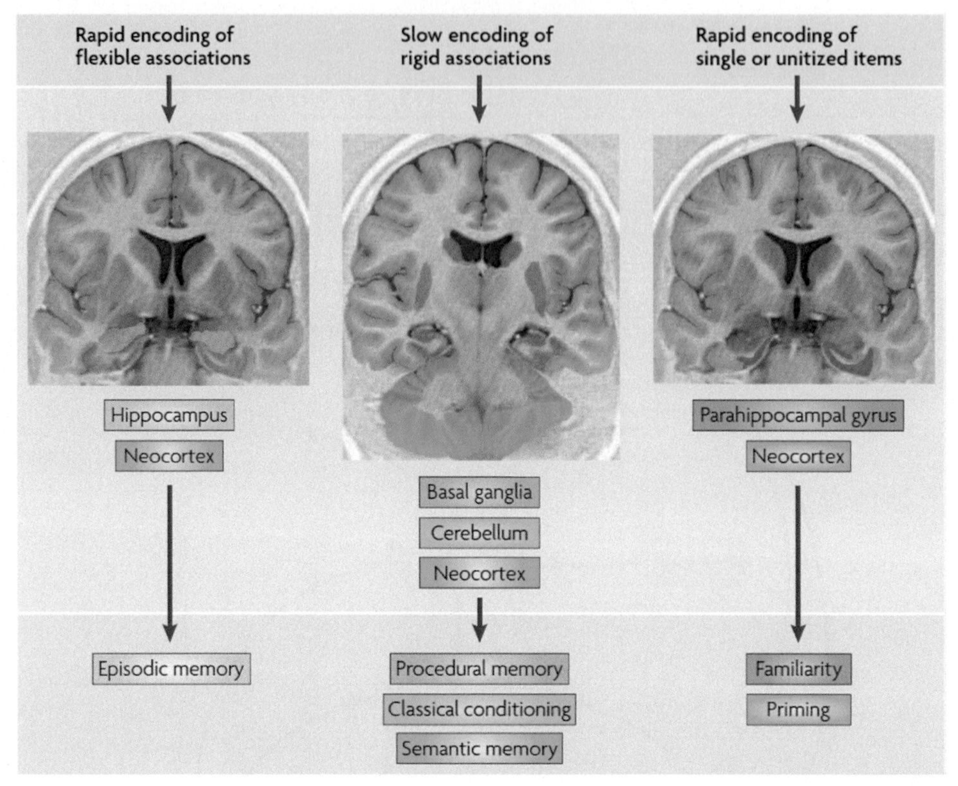

FIGURE 7.10 Henke (2010) proposed a revised classification system for memory processing that is divided into three processing models, each with their brain pathways. According to this view, some memory systems involve rapid encoding of flexible associations (left) and are subserved by the hippocampus and cortex; some systems involve slow encoding of rigid associations (center) and are subserved by the basal ganglia, cerebellum, and cortex; and some systems involve the rapid encoding of single or unitized items (right) and are subserved by the parahippocampal gyrus and the cortex. This approach to thinking of memory classes in terms of processing modes rather than their conscious or nonconscious aspects (shown in Fig. 7.2B) better reflects the empirical findings of the past decade. *Source: Henke (2010).*

this classification. As we experience the world around us, we are encoding episodes or events that occur daily and we are also updating our knowledge store about these events. Another approach to thinking about memory storage and knowledge representation is to consider the vital role schema formation and retrieval plays in our conceptual knowledge about the world.

In a recent review, Gilboa and Marlatte (2017) proposed a neurobiological model for where and how schemas interact in the brain during their instantiation and their retrieval. In this model, they propose that the ventromedial prefrontal cortex, the hippocampus, the angular gyrus, and association cortex together form the neural network that supports schema instantiation and storage (Fig. 7.11).

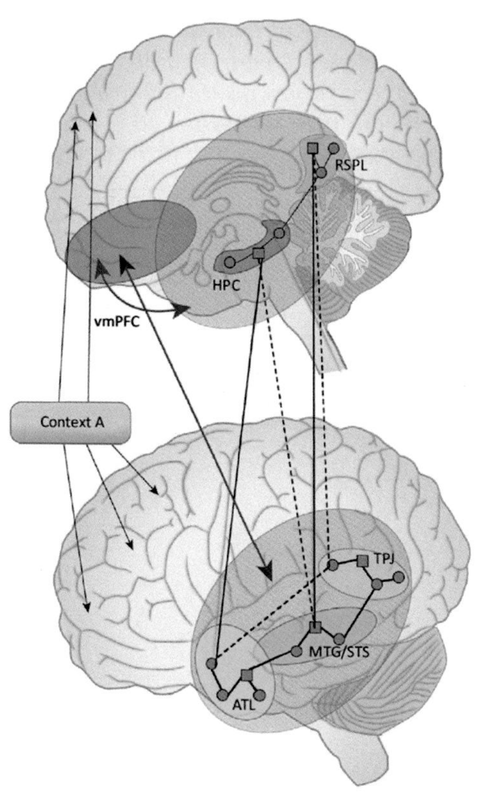

FIGURE 7.11 The role of schemas in memory encoding and retrieval. Memory schemas are formed of coactivated long-term representations in the posterior cortex, including the retrosplenial cortex (RSPL, shown in purple), middle temporal gyrus (MTG, shown in blue), superior temporal sulcus (STS, shown in blue), anterior temporal lobe (ATL, shown in green), and the temporoparietal junction (TPJ, shown in tan). These regions are temporally bound together by the ventromedial PFC (vmPFC, shown in red) to form the neural substrate of the schema knowledge. *Source: Gilboa and Marlatte (2017).*

5.4.4 Strengthening Memories: The Relative Effectiveness of Restudy Versus Retrieval

Another line of research investigates the best method for strengthening memory items to better retain them over long periods. Many studies have provided evidence that the process of *retrieving a memory* is more effective in strengthening that memory than *restudying it.* As an example, consider that you are studying for a geography test. You study the countries of Europe repeatedly. You have the intuition that you have learned them well. Your roommate, on the other hand, has been filling in blank maps of Europe, recalling the individual countries, and writing each one in the blank country outline.

Then, during the test, you are "blank" on where Hungary is located. How did that happen? By contrast, your roommate earns a perfect score. What went wrong?

A theory proposed by Antony, Ferreira, Norman, and Wimber (2017) provides a neural "*answer*" to this studying question. During *restudy* of an event, a set of facts, or whatever is being studied, cortical representations of *only those elements being restudied* are activated (Fig. 7.9). However, *retrieval* of whatever is being studied activates *not only the core representations* that were studied but also *coactivates related information. Strongly active memories* and their connections, in this process, are then strengthened. *Moderately coactivated information* is weakened. Together, these processes serve to aid in differentiating memories—increasing their specificity (Fig. 7.12).

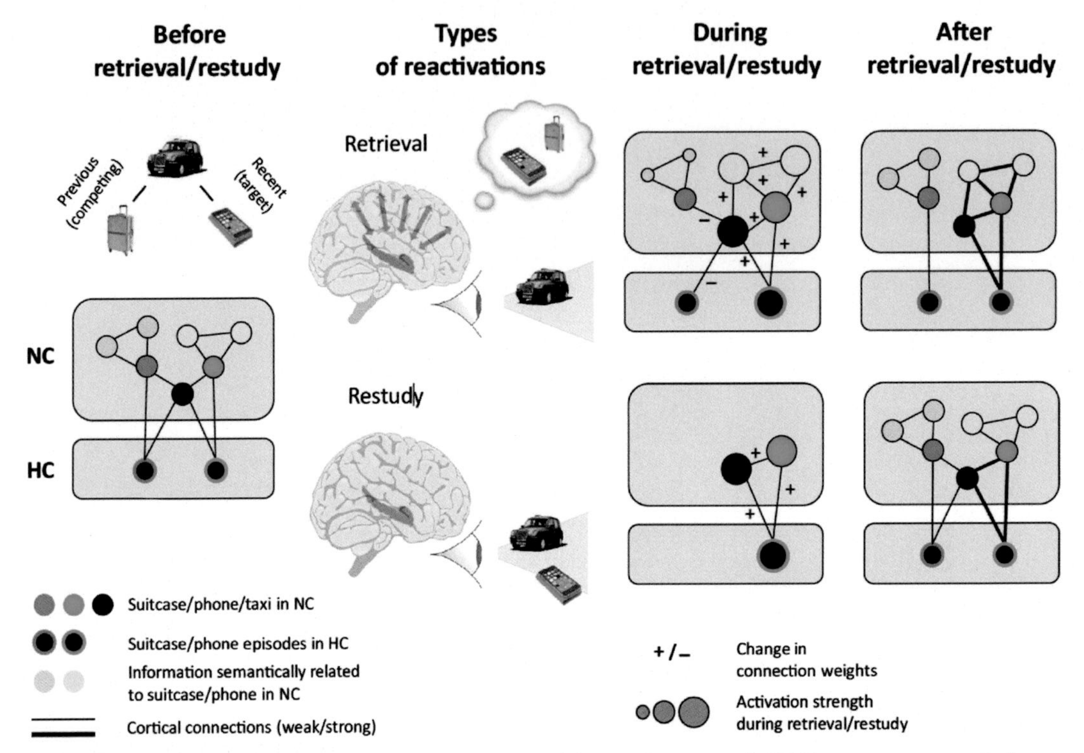

FIGURE 7.12 Retrieval as a fast route to memory consolidation. Antony et al. (2017) propose a network change model for retrieval versus restudy. Let's use an example of a new memory where you left your phone in a taxi. Two weeks before this event, you took a taxi because you had a very heavy suitcase to transport. According to Antony et al., these two episodes have separate, distinct, and nonoverlapping representations in the hippocampus (shown as HC in the figure) however, they have a shared element—"taxi"—that is stored in the cortex (shown as NC in the figure. Before retrieval/restudy: the elements of suitcase, phone, and taxi are all stored in the NC. Retrieval of the taxi + phone event will activate their core representations in the HC and NC, and will also cause coactivation of related information in the NC, such as suitcase + taxi, and any related semantic knowledge about phones, for example. Strongly active memories and their connections (blue elements) will be strengthened (indicated by +), supporting memory integration. Moderately coactivated information will be weakened (indicated by _), supporting differentiation, while no change in connection strength occurs for weakly active memories. By comparison, restudy reinstantiates the cortical representations of only those elements that are explicitly presented; connections between these elements will be strengthened but the memory will not be integrated with or differentiated from other stored knowledge. This allows memory specificity to develop and strengthen. *Source: Antony et al. (2017).*

5.5 A Continual Process

Memory formation and strengthening is a continual process: while each autobiographical memory trace is unique, the existence of many related traces facilitates retrieval. Episodic memories are integrated to form semantic memories. Thus facts about people and events that are learned in specific episodes become separated from their sources. This process may give the appearance of classical consolidation, but the brain mechanism is different from the classical view.

6. CONTROL OF MEMORY

How we interpret and deal with material in our memory is determined by our current goals and concerns, as well as by our existing knowledge. Sensory and internal information may be brought to consciousness using attention. Once it becomes conscious, a number of theorists maintain that information is rapidly *encoded* into long-term memory (e.g., Moscovitch, 1992). There is also evidence for some unconscious learning, but so far only unconscious fear conditioning has been shown to result in long-term memory (LeDoux, 1996). In general, conscious exposure correlates well with learning.

6.1 Working With Memory: The Frontal Lobe Works Purposefully With Memory

We have a measure of control over what we encode and what we retrieve from memory. How can we reconcile this with other facts we know about how memory works? One solution is that the frontal lobes control the information delivered to the medial temporal system, initiate and guide retrieval, and monitor and help interpret information that is retrieved. The frontal lobes act as *working-with-memory* structures that control the more automatic medial temporal system and give a measure of intelligence and direction to it. Such a complementary system is needed if memory is to serve functions other than mere retention and retrieval of past experiences (Moscovitch, 1992).

6.2 Prefrontal Cortex in Explicit (Conscious) and Implicit (Unconscious) Learning and Memory

WM may help us to learn both explicit (conscious) and implicit (unconscious) information. One of the functions often attributed to consciousness is the integration of information across domains. In a very illuminating study, McIntosh, Lobaught, Cabeza, Bookstein, and Houle (1998) had subjects perform in a trace conditioning task, which requires the person to make an association between a color and a tone separated by a blank delay of about a second. Using PET, McIntosh showed that learning, and the conscious awareness that accompanied it, was associated both with frontal activation and with coherence of activation across many areas of the cortex. McIntosh et al. speculated that consciousness is associated with activation in the prefrontal cortex (PFC), which, in turn, leads to a correlated pattern of activity across disparate regions of the cortex.

It remains to be seen, however, whether frontal activation preceded or followed conscious awareness of the association. If the PFC plays a pivotal role in consciousness, as many people have speculated, deficits on all memory tests dependent on consciousness should be observed in patients with frontal lesions. However, so far, the evidence indicates that the effects of frontal lesions are much more selective and not nearly as debilitating as lesions to the MTL and related nuclei of the thalamus.

The PFC contributes to implicit learning and memory if it requires search, sequencing, organization, and deliberate monitoring. Implicit learning of language is a good example. Even though we rarely try to make the rules of grammar conscious and explicit, we nevertheless need to direct our attention to the order of words in a sentence to learn a language implicitly. It is likely that unconscious inferences help us to discover rules or regularities, provided that we pay "conscious attention" to a series of words, for example, from which we can discover the implicit.

6.3 Prefrontal Cortex and Working Memory—Storage or Executive Control?

The PFC is an important site for WM function. According to one interpretation, this brain region participates directly in the storage of information. However, the PFC is also associated with *control of WM.*

PFC plays a role in WM. The PFC is situated in front of the motor cortex in both humans and other primates (Fig. 7.13). The macaque monkey has been the primary experimental animal in many studies of WM. Obviously, humans have other abilities, like language, that are not directly paralleled in other species. But in the case of WM studies, the macaque has constantly been an important source of evidence.

6.3.1 The Prefrontal Cortex and Animal Studies of Working Memory

Knowledge of a link between the PFC and short-term memory dates back to the 1930s, when it was first discovered that large bilateral lesions of the PFC in animals impaired performance on a delayed response task. In this task, a sample stimulus is presented (e.g., a color or location), and its identity must be maintained over a short delay period so it can guide a later response (Fig. 7.14). Using variants of this basic task with more recent neuroscientific techniques, modern research has firmly established the role of the PFC in active maintenance of WM information (Fig. 7.15).

6.3.2 The Prefrontal Cortex and Delayed-Response Tasks

Much of the animal research has focused on a specific frontal region called the dorsolateral prefrontal cortex (DLPFC; see Fig. 7.13). One of the key early findings came from the laboratory of Joaquin Fuster (Fuster & Alexander, 1971). Fuster et al. trained monkeys to perform a delayed-response task in which the monkeys had to remember a color over a brief delay and then point to the correct color when later presented with two alternatives. Since no information about the correct color was offered after the initial presentation, its identity had to be retained in WM. Using implanted electrodes to record neural activity during performance of the task (see Chapter 4, The Art of Seeing), it was found that individual neurons in the monkey DL-PFC exhibited sustained and persistent activity across the delay period. That is, after the color had been removed from the visual display, neurons in the DLPFC continued to fire at an increased rate, and this activity then subsided once the match/nonmatch response was made (see Fig. 7.14).

This pattern of sustained delay-period activity in the DLPFC has been replicated many times since, and in a wide variety of tasks. For example, to confirm that PFC contributions are truly memory-related and not simply a reflection of subtle preparatory motor gestures, Patricia Goldman-Rakic et al. developed a version of the task in which monkeys see a target presented briefly at one of several possible locations on a display and then, after a delay, must shift their gaze to that location to receive a reward (see Levy & Goldman-Rakic, 2000; for a review). Importantly, the monkey is required to look straight ahead until the end of the delay period, so neural activity during the delay cannot be simply a by-product of moving the eye but must instead reflect memory processes. Again, this paradigm produces sustained neuronal activity in the DLPFC and, what is more, the *amount* of delay-period activity predicts whether or not items will be remembered; when DLPFC delay-period activity is weak, there is a greater likelihood of forgetting (Funahashi et al., 1993).

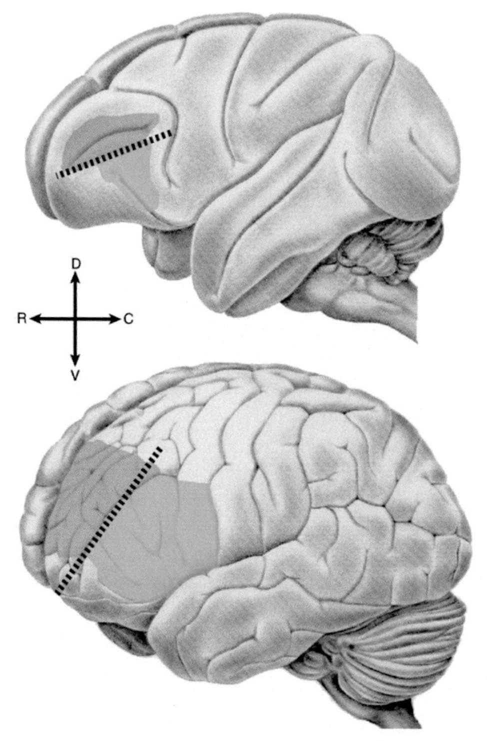

FIGURE 7.13 The prefrontal cortex in monkeys (top) and humans (bottom). The most common division is between the upper and lower halves of the prefrontal cortex (PFC), called the dorsolateral prefrontal cortex (DLPFC) for the light purple region and the ventrolateral prefrontal cortex (VLPFC) for the light green area. Also notice the orientation cross, pointing to dorsal (upper), ventral (lower), rostral (toward the nose in humans), and caudal (toward the back of the head in humans). *Source: Ranganath (2006).*

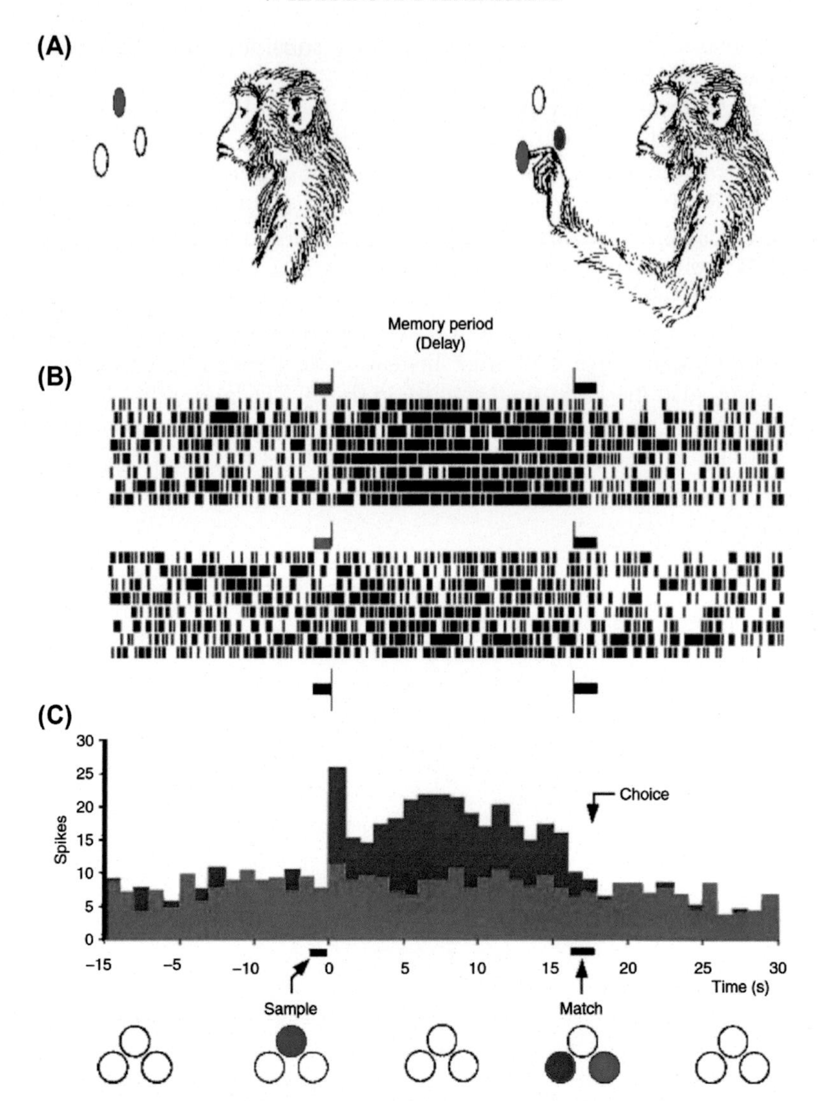

FIGURE 7.14 "Delayed match to sample (DMTS)" in the macaque. (A) In a classic experiment, a macaque monkey is trained to delay responding to a stimulus—in this case the location of a red, white, or blue light. The monkey shows recognition of the stimulus after delay by matching it in the display in a task called "DMTS." In effect, the monkey is communicating "This is what I saw." DMTS methods are widely used in animals, nonverbal babies, and other subjects.(B) During the delay while the monkey is holding in mind the blue light that he is to recall following the delay, some neurons in the dorsolateral prefrontal cortex (DL-PFC) fire actively only during the delay period (top row of black and white neural spiking pattern, while other neurons fire consistently during the presentation of the stimulus (blue light, top left), the delay (center), and the matching of the stimulus (blue light, top right). (C). A plot showing the enhancing activity for some neurons in the DL-PFC only during the delay period (shown in red) while others show continued response across the trial (shown in blue). Results like these provided evidence that some neurons in DL-PFC were active while an item was kept in memory and did not reflect the presence or absence of a given stimulus at that time—and thus these neurons may reflect working memory processes in the DL-PFC. *Source: Fuster (1997).*

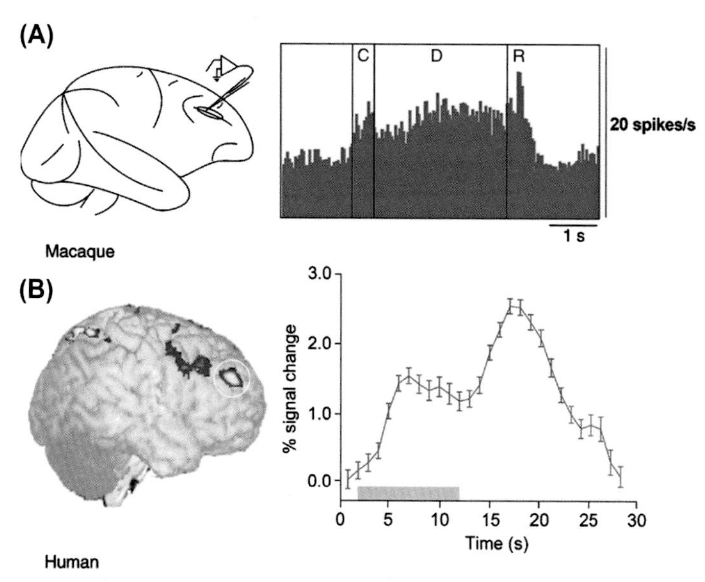

FIGURE 7.15 A delayed-response task to study working memory (WM) in monkeys and humans. It has been proposed that the prefrontal cortex (PFC) serves a specific role in the active storage of information in WM. That is, sustained activity in prefrontal neurons reflects this region's role in maintaining specific representations of the items that must be kept in mind over the delay. This interpretation is supported by the finding that individual neurons in the PFC are selective for particular target stimuli. For example, a given cell may fire strongly over the delay period when the target is in the upper left portion of the display but weakly when the target is elsewhere in the display. This pattern suggests direct involvement in the internal representation of target features. In this figure, neurons in the PFC respond during the delay period in a delayed-match-to-sample task. Results in macaques (A) and humans (B) are similar. *Source: Curtis and D'Esposito (2003).*

Monkey brain lesion studies have further implicated the PFC, and the DLPFC, in particular, in WM function. With very precise techniques for localizing experimentally induced lesions, it has been shown that damage isolated specifically to the DLPFC is sufficient to impair performance on WM tasks (Fuster, 1997). Such findings show a causal role for the PFC in WM. Not only are cells in this region active during a delay but their lesioning impairs WM. This impairment gets worse as the length of the delay increases, suggesting that there is more rapid forgetting when the PFC is prevented from sustaining them.

6.3.3 The Prefrontal Cortex and Human Studies

Studies in humans using neuroimaging have corroborated many of the findings from animal literature. Hundreds of imaging studies have shown PFC activity when participants are trying to maintain task-relevant information. Consistent with the animal work, fMRI studies in humans show that PFC activity persists during the delay period of a WM task.

Human neuroimaging studies have also varied the *WM load*—the number of items that must be held in immediate memory (Cohen et al, 1997; Rypma, Berger, & D'Esposito, 2002). In one study, memory load was varied between one and eight items, and subjects had to hold these items for a short delay. PFC activation was found to be positively correlated with the number of items in memory. Such "load dependence" in the PFC supports the notion that this part of the brain is involved in WM storage.

While PFC contributions to WM have been clearly demonstrated, its specific contribution to WM storage has been recently questioned. Several other cortical and subcortical areas exhibit similarly persistent stimulus-specific activity over short delays. It appears that PFC may be part of a more distributed brain network supporting WM. Other data suggest that the PFC may not be involved in storage per se but in providing top-down, or *executive*, support to other regions where information is actually stored.

6.3.4 *The Prefrontal Cortex in Patients With Brain Damage*

Patients with left temporoparietal damage may have a storage deficit in WM and cannot perform even simple maintenance tasks with auditory-verbal information. The findings from these patients can be contrasted with those from patients with damage to the PFC (D'Esposito & Postle, 1999). Some PFC patients show little impairment in passive maintenance of information over a delay. However, these patients were impaired in mentally *manipulating* or *acting upon* briefly stored information. Perhaps PFC supports the mental "work" performed on stored information, rather than as a site for storage itself (D'Esposito & Chen, 2006).

6.3.5 *The Prefrontal Cortex and Working With Memory*

One possibility is that different parts of the PFC do different things. This is the so-called "maintenance" versus "manipulation" distinction. It has been argued that *all* of the PFC has an executive function in WM but that different subdivisions do this at different levels of analysis (Ranganath, 2006). The PFC may enhance relevant information in other parts of the cortex. When the information is specific, more ventral PFC regions are engaged. When the information involves integration of multiple items in memory, the dorsal PFC regions are engaged. More frontal regions of the PFC may coordinate and monitor different PFC regions. If this is true, the main role of the PFC is not WM but *working with memory* (Moscovitch, 1992).

6.4 Real Memories Are Dynamic

Traditionally, a memory is considered to be a stable record of an event, which can be recalled in the same form it was learned. In this commonsense view, memories can be retrieved, examined, and played back like a high-fidelity recording. Memories can also be forgotten without affecting other cognitive systems.

There are reasons to question this idea. One is that real memories are rarely accurate. The *process view* considers memory to be a product of a dynamic process, *a reconstruction* of the past influenced by past and current conditions, anticipations of future outcomes, and other cognitive processes. In the process view, memory is based on stored information but is not equivalent to it. It is dynamic and mutable and interacts with other processes. Thus two people experiencing the same event may have different memories of it. It is not simply that one person is right and the other wrong, but each person's outlook, knowledge, motivation, and retentive abilities may alter what is retrieved.

Everyone's memory changes with time. We forget most of what has happened within minutes or hours, and what remains is commonly reorganized and distorted by other knowledge or biases. We would not want computer files or books to be that way. We do not want files to decay over time or to leak into neighboring files. Computers and libraries are designed to keep everything as distinct and stable as possible. Yet normal memories do fade and are often confused with others.

Try to reconstruct in as much detail as possible all the things you did two weekends ago, in the exact order in which you did them. To do that, most of us have to search for cues to determine exactly what we did. Having found a cue, there is a process of reconstruction, especially in trying to figure out the sequence of events. Did I meet my friend before I spoke to my parents or afterward? Did I go shopping, and what was the order of the stores I visited? In each store, in what order did I look at the merchandise and buy it? You can try this with a recent movie and then see how accurate your memory is by checking it against a copy of the movie.

To answer these questions, you must draw on a body of knowledge and inference that is unlike anything that is needed when you enter a file name to access a computer file or use a call number to find a library book. You may confuse what you did two weekends ago with what happened another time. As we will see, some patients with brain damage have a disorder called *confabulation*, in which they make up false memories without any intention of lying and without any awareness that their memories are incorrect.

7. WHEN MEMORIES ARE LOST

7.1 Hippocampal Versus Cortical Damage

The case of Clive Wearing has shed new light on the role of brain areas for memory. Wearing has lived with a dense amnesia since 1985, when a viral infection destroyed some brain areas for memory. Over a few days, Wearing was transformed from a rising young musician to a man for whom each waking moment feels like the first, with almost no recollection of the past and no ability to learn for the future.

Little has changed for Wearing since 1985. While he cannot recall a single specific event, some aspects of his memory are spared. He can carry on a normal, intelligent conversation. Some short-term memory is spared, allowing him to stay with a topic over several seconds. He has retained general world knowledge, an extensive vocabulary, and a tacit understanding of social conventions. Wearing also remains a skilled musician, able to play complex piano pieces from sheet music. Although he cannot remember specific events, he does recall a limited number of general facts about his life. Among the few memories that have survived is the identity of his wife, Deborah. He greets her joyfully every time she comes into his room, as though they have not met for years.

However, just moments after she leaves, Clive Wearing cannot remember that she was there at all. In a recent book, Deborah Wearing (2005) tells of coming home after visiting Clive at his care facility and finding these messages on her answering machine:

> Hello, love, 'tis me, Clive. It's five minutes past four, and I don't know what's going on here. I'm awake for the first time and I haven't spoken to anyone… .

> Darling? Hello, it's me, Clive. It's a quarter past four, and I'm awake now for the first time. It all just happened a minute ago, and I want to see you… .

> Darling? It's me, Clive, and it's 18 minutes past four, and I'm awake. My eyes have just come on about a minute ago. I haven't spoken to anyone yet, and I just want to speak to you.

Wearing's history suggests that amnesia is selective—certain kinds of memory may survive, while others are lost. Learned skills, like the ability to speak or play the piano, are different from episodic memories. Thus memory is not unitary but consists of different types.

Amnesic memory loss varies in degree. Clive Wearing's amnesia resembles that of other patients, but he is unusual in his nagging sense that he has just awoken from some unconscious state. He also "perseverates," repeating the same thoughts and actions over and over again, as in the repetitive telephone messages he leaves for his wife. These symptoms may result from additional damage to prefrontal structures that allow us to monitor our own actions.

Another patient, known as HM, is by far the best-studied victim of amnesia. In the case of HM, we know exactly where the lesion occurred (Figs. 7.16 and 7.17). This makes him very rare. Most brain injuries are very "messy," can spread wider than the visible lesions, and may change over time. Clive Wearing's viral infection apparently destroyed hippocampal regions on both sides of the brain but also some frontal lobe areas. Wearing may also have suffered brain damage that we simply do not know about. In the case of HM, however, because his lesion was carefully performed by a surgeon, we know that both sides of the MTL were removed as accurately as was possible at the time. The extent of HM's brain damage and functional deficits has been verified with great care in more than 100

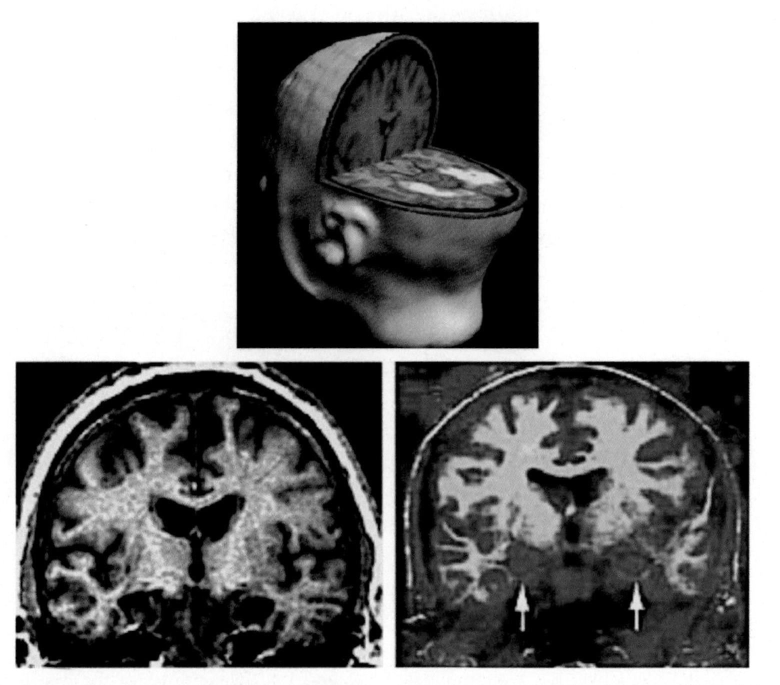

FIGURE 7.16 Lost brain tissue in the case of HM. The two *arrows* in the blue cross-section of the brain point to lost tissue in the hippocampus on both sides (hemispheres). The gray photo shows actual black regions where the tissue was lost. *Sources: Left: Aminoff and Daroff (2003); right: Corkin, Amaral, Gonzalez, Johnson, and Hyman (1997).*

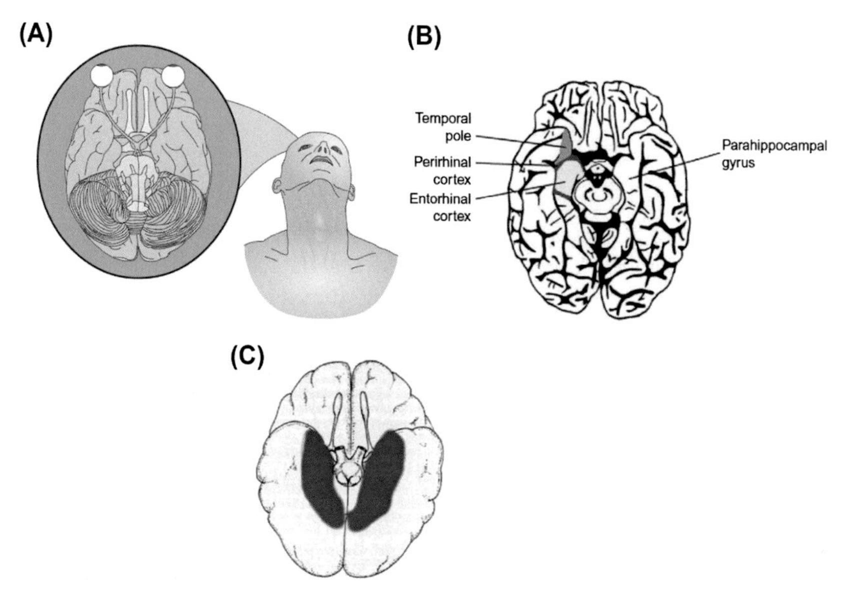

FIGURE 7.17 HM: Surgically removed medial temporal lobes (MTLs). The MTLs and HM's lesions, seen from below. (A) The orientation of the head and brain; (B) the bottom of the MTL, with major subregions for memory labeled. Notice that the rhinal (smell) cortices indicate the ancient origin of this region. In all figures you can see the two olfactory bulbs pointing upward, an important landmark for orientation. (C) The surgical lesion in HM's brain. The surgeon at the time was unaware of the importance of this region for memory. *Source: (A) Baars and Fu, with permission, (B) Buckley and Gaffan (2006); (C) Moscovitch, personal communication.*

published articles. This has made HM one of the most important patients in the history of brain science (Box 7.1).

7.2 Defining Amnesia

Amnesia is any major loss of memory while other mental functions are still working. The cause can be infection, stroke, tumor, drugs, oxygen loss, epilepsy, and Alzheimer's disease. Amnesia can also be *psychogenic*, resulting from trauma or hypnotic suggestion (Nilsson & Markowitsch, 1999).

As we have seen, amnesia can be caused by bilateral damage to the MTLs, which includes the hippocampal formation. While explicit memories are impaired, there are surviving functions:

1. Perception, cognition, intelligence, and action
2. Sometimes WM is spared
3. Remote memories
4. Implicit memory

As mentioned, implicit memory is commonly assessed by priming tasks. Perceptual priming is mediated by the sensory cortex, and conceptual priming is believed to involve both the temporal and prefrontal regions. Amnesics do not lose their capacity to perform well on

BOX 7.1

THE CASE OF HM

The science of memory received a major boost from Herbert Scoville and Brenda Milner's (1957) report of a memory disorder in HM after bilateral removal of his medial temporal lobes to control severe epileptic seizures. Throughout his lifetime, HM's privacy was protected by noone using his name: he was only case HM. However, upon his death in 2008 at the age of 82, HM became widely known as Henry Molaison—the case that made a lasting and influential mark on memory research (Fig. 7.18).

FIGURE 7.18 HM. A photo of Henry Molaison taken before his fateful surgery. *Source: https://en.wikipedia.org/wiki/File:Henry_Gustav_1.jpg.*

As a result of a head injury from a bicycle collision when he was a young boy, HM was beset with epileptic fits that increased in frequency and severity into his late 20s. As a treatment of last resort, Scoville performed an operation in which he removed tissue in and around the hippocampus on both sides of HM's brain (see Fig. 7.16). While the surgery reduced HM's seizures, it had a profound and unexpected impact on his memory. This result was unknown at the time, and Scoville would undoubtedly have changed the procedure had he known about this harmful result. The report of HM's case by Scoville and Brenda Milner was the first to demonstrate directly the importance of the hippocampus and surrounding structures for memory.

As a result of the operation, HM could not remember any of the events of his life thereafter—the people he met, the things he did, events taking place in the world around him. As an adult, he could not keep track of his age and could no longer recognize himself in the mirror because he was unfamiliar with his changed appearance. In addition, to this *anterograde* (postdamage) memory deficit, HM also could not remember events or experiences from the years immediately before the surgery, a *retrograde* amnesia. While his episodic (autobiographical) memory loss was acute, other cognitive functions seemed to be intact. He could reason, solve problems, and carry on a normal conversation. His intelligence was normal, and he retained his language abilities.

HM had intact short-term memory. He performed like healthy controls on tests of working memory, such as the digit span task. HM had been under the care of doctors from a young age, and his intellectual abilities before surgery were well documented. The specific locus of damaged tissue was both limited and well characterized. In most amnesias, the damage is more widespread and difficult to identify. HM had been tested and imaged a number of times since his surgery, giving a very complete picture of his condition.

As you can tell from Figs. 7.15 and 7.16, HM had an apparently intact cortex—the outer structures in the brain scan. Like Clive Wearing, HM could carry on a normal

BOX 7.1 *(cont'd)*

conversation. He could discuss the immediate present, using his general knowledge of the world. He was conscious, had normal voluntary control over his actions, and appeared to be emotionally well adjusted. It was only when his episodic memory was tested that he revealed that he simply could not remember the past or learn new memories for the future.

priming tasks, such as word-fragment completion. For example, subjects may study a list of words (such as *metal*) and are tested with fragments of the words they studied to see if they can complete them (*met__*). The study phase increases the speed of completion. On such tasks, amnesic patients can perform as well as normals.

Functional neuroimaging studies confirm the perceptual locus of perceptual priming. Unlike tests of explicit memory, which are associated with *increased activation* during retrieval in regions that support memory performance, such as the MTL and the PFC, perceptual priming is associated with *decreased activation* on repeated presentations in regions believed to mediate perceptual representations. Thus repeated presentation of faces and words leads to decreases in activation in inferior temporal and extrastriate cortex, which mediate face and word perception (Schacter, Dobbins, & Schnyer, 2004; Wigg & Martin, 1998).

In conceptual priming, the relationship between study and test items is meaning-based. Conceptual tasks include word association ("Tell me the first word that comes to mind for *elephant*"), category exemplar generation ("Provide examples of animals"), and general knowledge ("What animal is used to carry heavy loads in India?") (Roediger & McDermott, 1993). Conceptual priming occurs if studied words (e.g., *elephant*) are retrieved more frequently than unstudied ones. Because conceptual priming depends on meaning, a change in the physical form of stimuli has little influence on conceptual priming.

Conceptual priming is impaired in people with damage to regions of the cortex mediating semantics. Thus patients with semantic dementia, whose degeneration affects lateral and anterior temporal lobes, show an inability to recognize repeated objects that share a common meaning—for example, two different-looking telephones. But they have no difficulty recognizing the object if it is repeated in identical form (Graham, Patterson, & Hodges, 2000).

Likewise, patients with Alzheimer's disease show preserved perceptual priming but impaired conceptual priming. Functional neuroimaging studies of conceptual priming implicate semantic processing regions, such as prefrontal and lateral temporal areas. As in tests of perceptual priming, tests of conceptual priming lead to decreases in activation in those regions (Buckner et al, 1998; Schacter et al., 2004). Asking people repeatedly to generate examples of "animals" results in *reduced* activation in the same regions.

Numerous studies have shown that priming on a variety of tests is normal in amnesic patients. This applies to most tests of perceptual and conceptual priming, indicating that the MTL does not contribute to them.

7.3 Amnesia Can Impair Working Memory

We have already encountered examples of patients, like Clive Wearing and HM, who seem to have spared short-term memory but impaired long-term memory. Tim Shallice and Elizabeth Warrington (1970) reported one of the earliest recognized cases of a patient, KF, with the opposite pattern of impairment: a severely impaired short-term memory but apparently intact long-term memory. For example, when asked to recall short lists of spoken digits, the *digit-span task*, KF could recall only one or two items reliably (as compared to a typical digit-span of around seven items). Still, KF had comparatively normal speech-production abilities and could learn and transfer new information into long-term memory. The finding that a patient with severely impaired short-term memory could still transfer information into long-term memory presented a challenge to the standard hypothesis posited that a unitary short-term memory serves as the gateway into long-term memory. Baddeley's WM model suggested that if verbal rehearsal is impaired, the visuospatial sketchpad might be used to compensate (see Fig. 9.3).

Indeed, the short-term memory impairment in patient KF, and a number of similar patients reported since, seems to be tied to particular types of information. For example, while these patients struggle to remember verbal items when presented auditorily, their performance is considerably improved when the items are presented visually. What might account for this pattern of findings? Baddeley's answer is that visually presented items can be coded directly into the visuospatial sketchpad, thus avoiding the damaged verbal rehearsal loop.

Neuroimaging has helped to clarify different kinds of memory. These include the distinction between verbal and visuospatial maintenance subsystems (e.g., Smith, Jonides, & Koeppe, 1996), the dissociability of storage and rehearsal in verbal maintenance (Awh et al., 1996; Paulesu, Frith, & Frackowiak, 1993), and the assumption of a central executive processor that mediates the behavior of the subsidiary maintenance subsystems (e.g., Curtis & D'Esposito, 2003). In general, neuroimaging studies have tended to support the basic model (Hartley & Speer, 2000; Henson, 2001; Smith & Jonides, 1998).

7.4 Habits and Implicit Memory Tend to Survive Amnesia

One of the earliest demonstrations of preserved memory in amnesia was on tests of learning perceptual motor skills called *procedural memory*. Corkin (1965) and Milner (Milner, Corkin, & Teuber, 1968) showed that HM was able to learn and retain a pursuit-rotor task, keeping a pointer on a moving target. HM showed improvement on these tasks even months later, though he could not recall doing it even minutes afterward if he was distracted. These findings have been repeated in other cases.

Procedural memory depends on regions like the basal ganglia, which interact with the cortex. Patients with impaired basal ganglia due to Parkinson's or Huntington's disease show little or no improvement after practicing sensorimotor tasks (Gabrieli et al., 1994; Kaszniak, 1990).

Functional neuroimaging studies also show that learning on the implicit serial reaction-time (SRT) task[1] is associated with activity in the basal ganglia but not with MTL activity.

Suppose you are asked to read a set of words and then your memory for them is tested a week later. We could test your memory directly by asking you to recall as many of the studied words as you can remember or to recognize the words by picking them out from a list of old and new words. If the interval is long enough, you are likely to recall or recognize only a small subset of the items and mistakenly classify old words that you studied as new ones that did not appear on the list.

However, if your memory is tested indirectly by asking you to read the words as quickly as you can, you will be able to say old words faster than new words. The old words are *primed*. On such an indirect test no mention is made of memory, and the subject is typically not even aware that memory is being tested. Yet by looking at how quickly subjects read a word, we can tell whether the previous experience left a residue in memory. The same result can be seen in amnesic patients who cannot recall studying the words at all.

In the case of conceptual or semantic priming, words such as *food* may increase the processing efficiency of words like water, even though they share little perceptual content. Priming can be viewed as a way of tapping into the general tendency of the brain to engage in predictive processing at every moment.

Perceptual priming is based on alterations of perceptual representation in posterior cortex associated with perceptual processing. Conceptual priming is associated with alterations of conceptual systems in PFC.

In summary, amnesia due to bilateral damage to the MTL seems to be primarily a disorder of *episodic memory*, resulting from impaired transfer of information from WM into long-term memory. Because memories acquired long before the onset of amnesia are relatively spared, it is believed that the hippocampus and related structures in the MTL are needed only temporarily to hold information in memory until they are consolidated elsewhere in the brain, presumably in cortex.

8. SUMMARY

Classical models of memory categories have long served as the vital underpinning of memory research, providing a basis for understanding aspects of learning and memory that are vastly different. Early categorization divided memory into two major divisions: declarative or conscious memory and nondeclarative or nonconscious memory.

More recent research is shedding new light on these categories, with new data showing a far more blurred line between the classic declarative and nondeclarative memory typology. Memory and our knowledge store are thought to be organized more along lines of their relevance to daily life than their temporal bases (long vs. short term). Further, the conscious/nonconscious line proposed for episodic versus habit learning, for example, also seems to not hold up under careful experimentation. Thus the approach to thinking

[1]The serial reaction-time task has been widely used to study cognition. It uses a series of presentations of a stimulus along with a response—for example, 1-2-3-4. This task is used to assess aspects of learning and memory such as the role of context in learning.

of memory classes in terms of *processing modes* rather than their *conscious or nonconscious aspects* better reflects the empirical findings of the past decade (Henke, 2010). Continuing the theme of knowledge stores reflecting their relevance and processing modes rather than their temporal and conscious aspects, theories of the neural organization of higher order schemas add new ways of thinking about the cortical organization and maintenance of our knowledge (Gilboa & Marlatte, 2017). Finally, new neural models providing evidence that it is the retrievals not necessarily the restudying of material that strengthens their memory provides useful guidance not only for memory scientists but also for students of life everywhere (Antony et al., 2017).

9. STUDY QUESTIONS

1. Describe two or three types of memory that have been discussed in the chapter. Discuss whether they are thought to be mediated by conscious or nonconscious processes. How have these process categorizations been challenged in recent years?
2. What is the difference between episodic and semantic memory? How do they differ? How are they similar?
3. What is a schema? What brain regions are thought to support their instantiation and retrieval?
4. What is a memory trace? Describe two types of connections that underlie memory traces in the brain.
5. What is memory consolidation?
6. How has the role of the PFC in working memory been investigated? Describe two lines of research.
7. What are some defining aspects of amnesia?

CHAPTER

8

Attention and Consciousness

THE VAST SEA OF CONSCIOUSNESS

Consciousness is the water in which we swim. Like fish in the ocean, we can't jump out to see how it looks from the outside. ***Bernard Baars, American Scientist.***

A large beautiful wave symbolizing the "sea of consciousness" we live within during our waking state. *Source: public domain image.*

1. INTRODUCTION

Consciousness is the water in which we swim. Like fish in the ocean, we cannot jump out to see how it looks from the outside. As soon as we lose consciousness, we can no longer see anything. Ancient people knew about waking, sleep, and dreaming because they experienced those states in themselves, and they could see other people when they were sleeping and waking. Sometimes clan members would wake up from a dream and tell others about their inner journeys. All human cultures know about the three basic states.

https://doi.org/10.1016/B978-0-12-803813-0.00008-8

We begin our discussion about attention and consciousness by examining these three global brain states: *waking, sleeping, and dreaming*. Brain activity in each of these states shifts dramatically. We will focus on the *waking state* in this chapter on attention and consciousness and address *sleeping and dreaming states* in Chapter 12, Sleep and Levels of Consciousness.

Next, we will turn to the very complex and highly interactive phenomena, human attention and consciousness, and examine the brain networks that are hypothesized to subserve them. We will end the chapter with a look at exceptional states of mind, such as psychoses and out-of-body experiences (OBEs).

1.1 Three Global Brain States

Sleep, waking, and dreaming are easy to identify based on our own experiences—and also by the electrical activity all over the brain. Steriade (1997), a leading scientist in this field, wrote, "The cerebral cortex and thalamus constitute a unified oscillatory machine displaying different spontaneous rhythms that are dependent on the behavioral state of vigilance." We will take Steriade's one-sentence summary as the theme of this chapter. As we will see, the cortex and thalamus work very much like an oscillatory medium, like the ocean. In addition, we know from earlier chapters that the cortex consists of flat arrays of neurons. Indeed, we can think of the cortex and its related brain regions as a huge, oscillatory array of many hundreds of arrays.

For many years the underlying mechanisms for these electrical patterns were unknown. There was debate about whether electroencephalography (EEG) was even a useful measure. However, basic research has now shown that scalp EEG reflects the fundamental "engine" of the thalamus and cortex. Although the brain's electrical field is only a side effect of the normal working of the core brain, we can use EEG to understand how the thalamus and cortex do their work.

Fig. 8.1 shows the three daily states recorded from one electrode on the scalp. One electrode is enough because states of consciousness involve *global* activities, like day and night traffic in a large city. If daytime traffic is very heavy and nighttime traffic is very light, only one observation point can show the difference.

The daily (circadian) cycle is controlled by precise biological mechanisms, stimulated by daylight and darkness, and by eating, activity patterns, sleep habits, and other factors. A small group of receptors in the retina detect daylight and darkness, signaling the suprachiasmatic nucleus (SCN), the pineal gland, hypothalamus, and deep brain nuclei to release state-specific chemicals called neuromodulators. Melatonin is an important sleep hormone triggered by the onset of darkness. Chemical neuromodulators spread very widely in the brain and help to trigger global states (Figs. 8.2 and 8.3, see Chapter 12 for more on this topic).

Keep in mind that the thalamus is the "gateway to the cortex." Most thalamic nuclei are constantly signaling back and forth to corresponding parts of the cortex. The thalamus and the cortex are often considered to be one large functional "engine." When we shift between major circadian states, the engine changes gear.

Fig. 8.4 shows a neural control circuit of the thalamus and cortex, consisting of three connected cells in the cortex, thalamus, and the reticular nucleus of the thalamus. The upper neuron in Fig. 8.4 is a pyramidal cell in the cortex. Together, this circuit works as a "pacemaker" for the major states of the circadian cycle. Signaled by neurochemicals from the

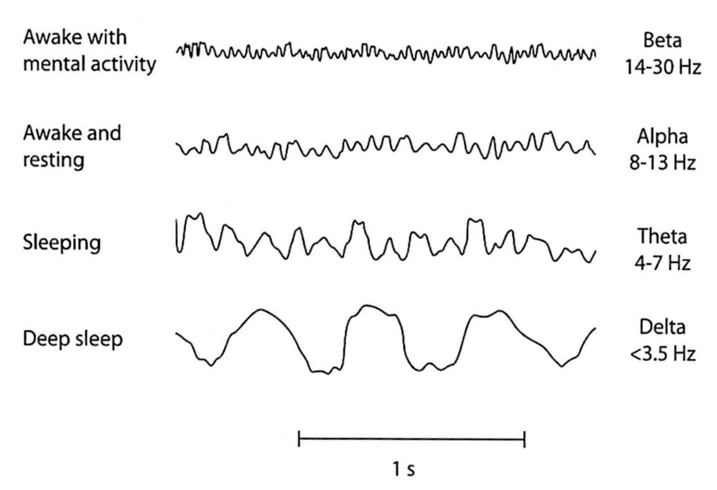

FIGURE 8.1 Electroencephalography (EEG) of waking, dreaming, and deep sleep. To identify brain states, we only need one electrode on the scalp (and a reference electrode that is often attached to the ear). Global states can be identified just by looking at EEG. Waking consciousness and REM dreaming look remarkably similar, consistent with the fact that we experience a rich flow of conscious events during both waking and dreams. Deep sleep looks very different: it is high in voltage, slow, and regular. *Source: Stock photo.*

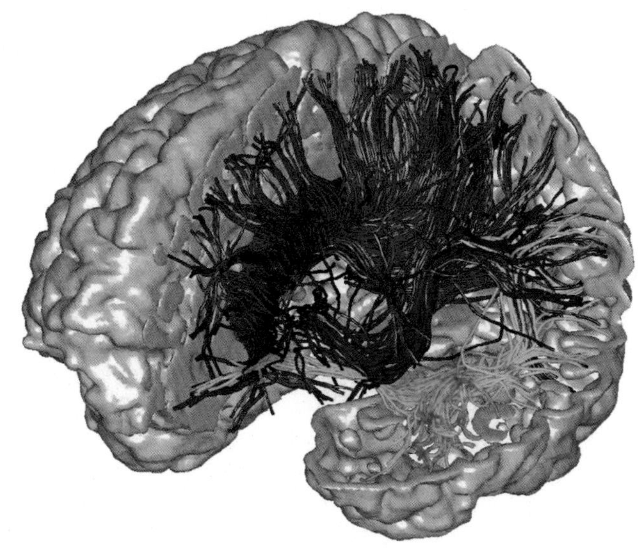

FIGURE 8.2 The cortex and thalamus make up one unified system. This figure shows the massive size of axonal fiber tracts between the thalamus and the outer layers of cortex. Since axons coming from each cell body are wrapped in glial cells that are filled with lipid molecules, they appear white to the naked eye and therefore were called the "white matter" by traditional anatomists. These tracts are artificially colored in this computer-generated graphic to show the different major pathways. All the fiber tracts seem to emerge from the left thalamus from the point of view that is shown here, but at least equal numbers of axons travel from cell bodies in cortex to the corresponding nucleus of the right thalamus. Traffic is always two-way or "reentrant" in the thalamocortical system, a fundamental fact that requires explanation. Notice that there is no cross-hemisphere traffic in this image. *Source: Izhikevich and Edelman (2008).*

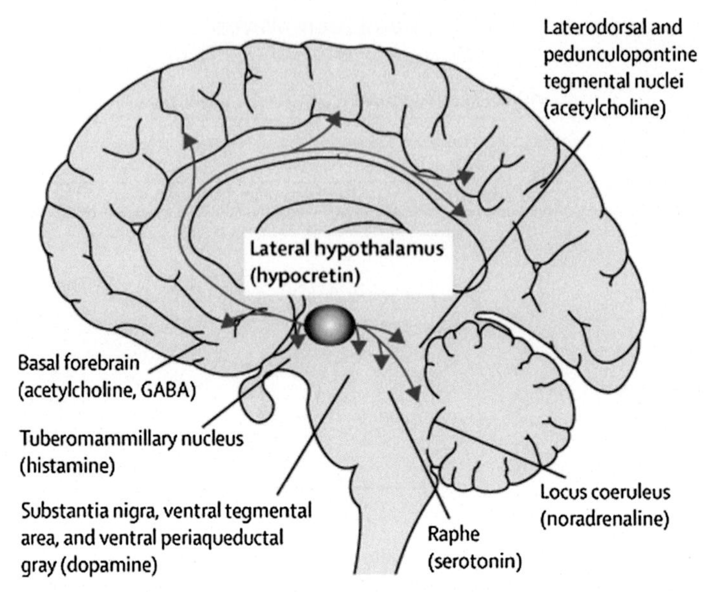

FIGURE 8.3 Chemical switching controls the state circuit. The daily states of consciousness are turned on and off by surprisingly small numbers of neurons, located at the bottom of the brain in nuclei like the substantia nigra (SN). Their widespread axons spray special neurochemicals called "neuromodulators" to modify local neurotransmitters. The bundle of axons that project from the small SN in the bottom of the brain are called the nigrostriatal neurons because they start at the SN and terminate in the striatum ("striped region") of the basal ganglia. *Source: Maski and Owens (2016).*

brainstem, the three-neuron circuit "pumps" the giant corticothalamic core. There are millions of such circuits, but the simplicity of this loop is striking. Notice that the circuit is itself triggered by chemical neuromodulation from the brainstem.

2. WAKING: PURPOSEFUL THOUGHTS AND ACTIONS

Waking is our time to open up to the world around us, perceive and explore, think about ourselves and one another, learn and prepare for the future, cope with challenges, express our emotions, and advance our personal and social goals. From a biological point of view, all of our *purposeful* survival and reproductive activities take place during the conscious state.

We have used a "functional diagram" of human cognition in this book (Fig. 8.5). What we have not said is that the waking state is *the necessary condition* for all those mental functions. That fact has many surprising consequences for our understanding of the brain. For example, the global features of the conscious state (Figs. 8.1–8.4) may not be obvious, but we know they are necessary for us to have conscious sensations, working memory, and voluntary control of our muscles.

It will help to take another look at our basic functional diagram and think about the boxes that work only in the waking state. They are generally the colorful ones. The green boxes (long-term memories) continue to store information 24 h/day. As we will see, conscious experiences we learn during waking periods are often consolidated in a slow wave sleep (SWS).

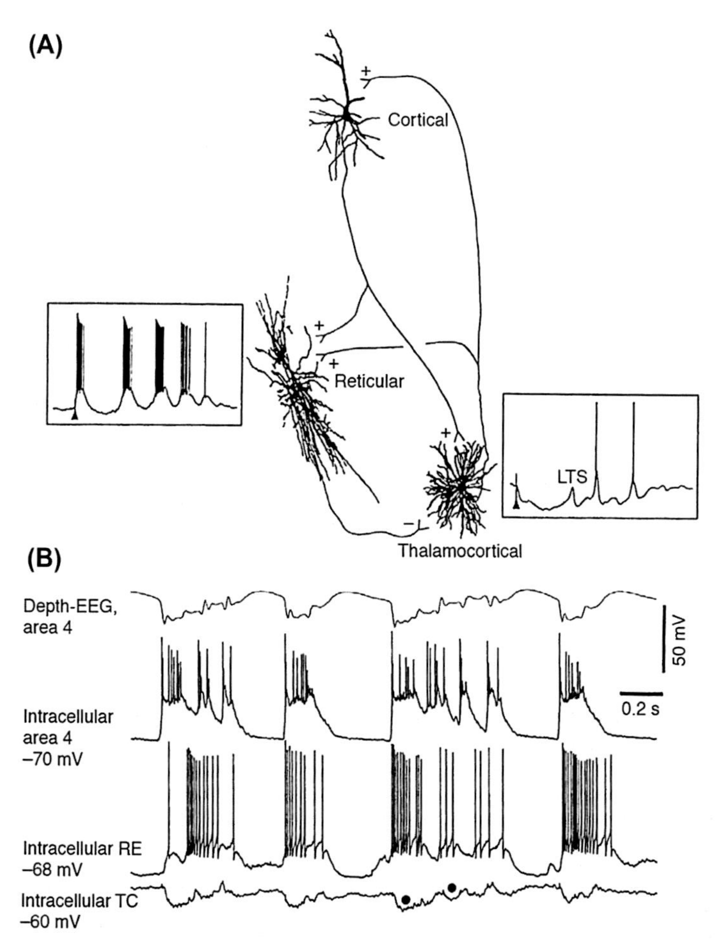

FIGURE 8.4 (A) The state control circuit: three neurons in a thalamocortical loop. The basic rhythmic "pump" of the brain. Waking, sleep, and dreaming are driven by thalamocortical oscillations. Thalamic nuclei interact closely with corresponding regions of cortex. (B) Similarity of responses for the three types of neurons shown in (A) across recording sites in the cortex, reticular (RE), and thalamocortical (TC) regions, with cortical area four responses shown in the first two tracings (Depth-EEG area 4 and Intracellular area 4), RE responses shown in the third tracing (intracellular RE), and TC responses shown in the fourth tracing (intracellular TC). This core brain is shared by other mammals and birds. *Source: Adapted from Steriade (2005).*

Sensory consciousness obviously also depends on the waking state, as do the "inner senses" of verbal rehearsal (inner speech), the visuospatial sketch pad (imagery), and the like. Normal, voluntary control is mainly limited to the waking state.

Voluntary selective attention also occurs primarily during waking. Voluntary attention is shown in Fig. 8.5 as an arrow running from the central executive (roughly, the frontal lobes) to brain activities that are enhanced by attention. Spontaneous attention is not under voluntary control, but if someone yells out loud, a large dog barks unexpectedly from a few feet away, or a truck looms into your field of vision, attention will be "stimulus-driven."

Many biologically significant events trigger a "bottom-up"attention, such as the smell of food when you are hungry. We spontaneously pay attention to personally significant stimuli

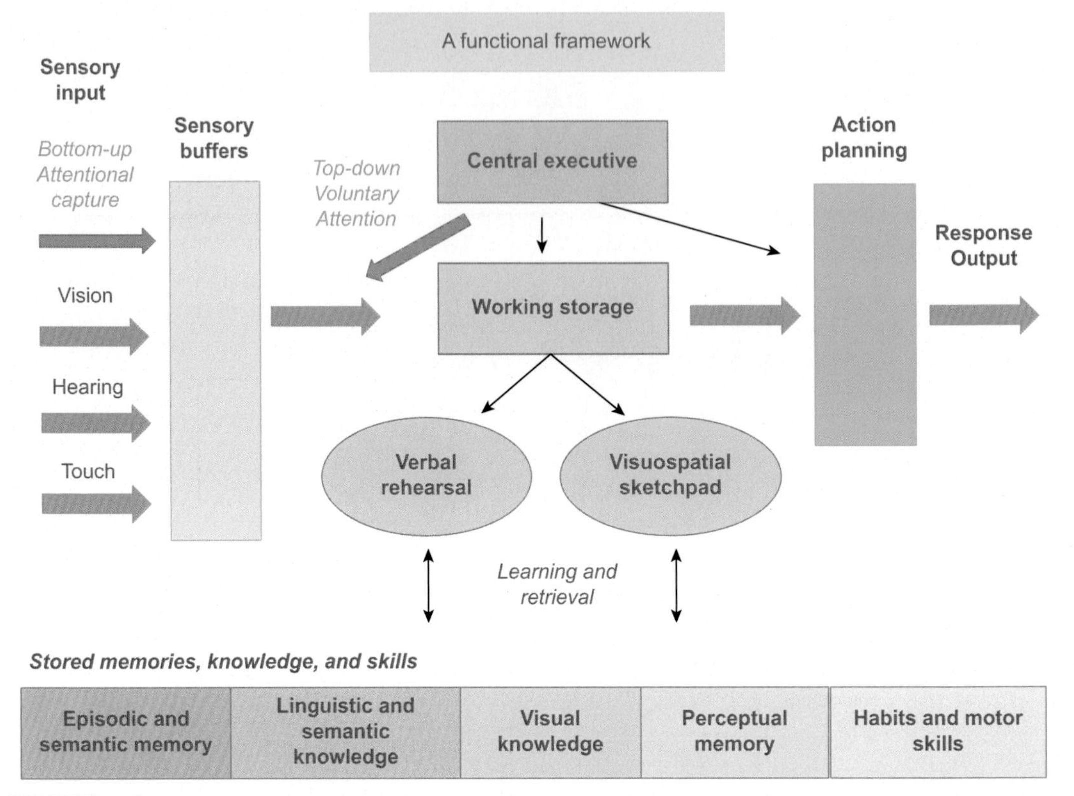

FIGURE 8.5 We are *not* conscious of all cognitive functions at the same time, but we can get rapid conscious access to all the *colored boxes* in the diagram. The *green boxes* at the bottom are never directly conscious. However, we can *retrieve* conscious information from long-term memories. Note that the diagram would look very different in dreams and sleep. *Source: Gage, with permission.*

as well, like the sounds of our own names. In the figure we symbolize stimulus-driven attention with an arrow coming from the sensory boxes of the diagram to suggest that some stimuli are inherently attention-capturing.

Our goal-directed actions happen during the waking state, including thinking and problem-solving, food gathering, social behavior, and mate seeking. It makes sense therefore that task-related signaling in the brain is mostly found in waking consciousness.

2.1 The Stadium Analogy: Chattering, Cheering, and Chanting

Waking EEG has puzzled scientists since 1929, since it looks very irregular, even random, as if it is a kind of "white noise," such as the random noise we hear from waterfalls and ocean waves. If we record the sound of a waterfall and then average separate stretches of sound, the average will look like a flat line with zero voltage. That is because random activity is so unpredictable that it adds up to zero. The chance of a voltage at any moment being above and below zero is about equal. As we have seen with the averaged evoked potential (AEP), we can use that fact to obtain beautiful AEP curves that are time-locked to a stimulus. The random EEG just drops out of the averaging process.

2.1.1 *Chattering in the Waking Brain*

If we think of the brain as a huge football stadium with thousands of people just chattering with one another, the averaged sound is so irregular that it resembles white noise. However, every conversation in the stadium is very meaningful to the people doing the talking. We see *local synchrony* between two individuals in a conversation and *global randomness* because none of the conversations are linked to one another. It is convenient to call this the "chattering" state of the football stadium.

Fig. 8.6 shows that during the waking state, neurons in different parts of the cortex and hippocampus are "phase-locked" to one another. (Phase-locked simply means "synchronized with a small lag time.")

The evidence comes from Nair, Klaassen, Poirot, Vyssotski, and Rasch (2016), who recorded from hundreds of neurons directly in the brain of a cat. Notice that the "WAKE" figure shows activity in the gamma power range (red) as opposed to the SWS AND REM sleep figures (Fig. 8.6). That is analogous to spectators in the stadium talking "in sync" with one another. But like the chattering people in the stadium, their conversations are independent of one another. If we record the overall sound in the stadium, it seems random, but if we record the talk of two people in a conversation, we can see the synchronized activity.

Although the stadium analogy is not supposed to be taken literally, two people talking to each other do "dance" in synchrony with each other, so a slowed-down video of a conversation will show them "microdancing" with each other. We can think of Fig. 8.6 as showing relatively localized synchrony between regions of the brain that are working together. During deep sleep and dreaming, that kind of synchrony breaks down, as you can see in the second and third column of the figure.

The conscious state therefore seems to support local synchrony (or phase-locking), while deep sleep and dreaming do not. As we will see next, there is very direct evidence that synchrony serves as an important coordinating rhythm for neurons that may be widely dispersed but are supporting the same cognitive task, whether it is sensory perception, motor control, memory storage, or the other active tasks in our functional diagram (see Fig. 8.5).

2.1.2 *Cheering in the Waking Brain*

Football spectators are not always chattering to one another locally. They do at least two massively coordinated actions. One we will call *cheering*, such as when one team scores a thrilling goal. Ten thousand people suddenly applaud (or boo) out loud. Since this is an event-related cheer, it is analogous to the event-related potential. In the brain, we commonly average evoked potentials over a number of triggering events, like the thrilling football play that triggers a cheer. That is the AEP. Sending a big flash or a loud noise through the brain is very much like a simultaneous cheer going through a football stadium. Suddenly all the noisy background chat turns into a giant "hooray!"

2.1.3 *Chanting in the Unconscious Brain*

What about unconscious periods, like deep sleep (SWS)? You can see in Fig. 8.1 that it consists of very large, very slow (by brain standards), and very coordinated brain activity. We know that billions of neurons are highly coordinated during SWS because all their activity adds up during the UP state (the peak of the slow wave), and nothing seems to be happening during the DOWN swing of the slow wave.

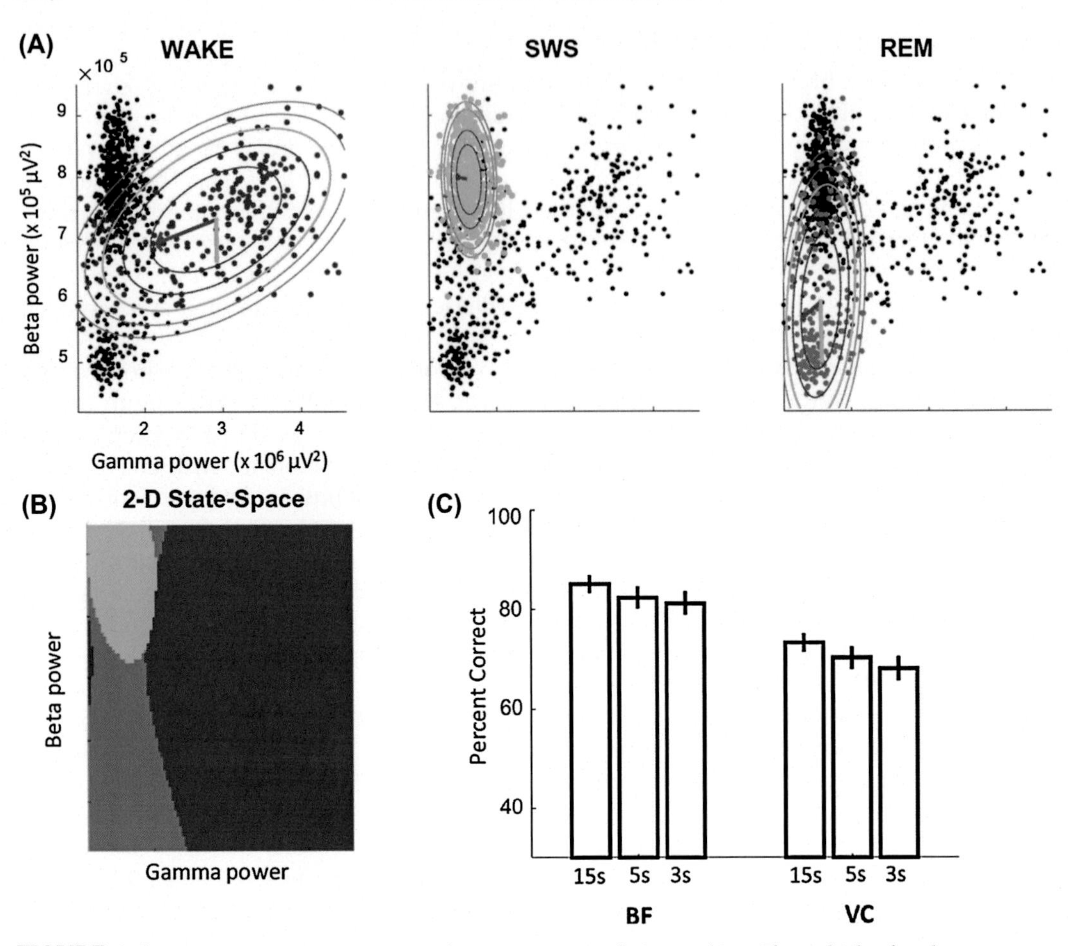

FIGURE 8.6 Synchronized "conversations" between neurons during waking. If we think of each neuron as a spectator in a large football arena, during the conscious state thousands of spectators are chattering with one another at the same time. Each dialogue synchronizes just two people, but hundreds of different conversations are happening at the same time. There is *local* synchrony but not *global* synchrony. Here we show neural synchronization in the gamma band between the frontal cortex and visual cortex across Wake, Sleep, and Dream states. (A) Synchronized neural responses in basal forebrain (BF) and vision cortex (VC) in rat in the beta and gamma frequency range (shown in colored dots) for the Waking, Slow Wave Sleep (SWS), and Rapid Eye Movement (REM) states. Notice that the distribution of the responses are more focal and smaller in extent for the sleep stages (SWS, REM) versus Wake stage, where there is a wider distribution, reflecting the increased neural activity during Wake versus sleep. (B) A 2-dimensional "decision-space" shows the relative effects of beta and gamma power on BF and VC synchronization. (C) Estimates of behavioral state classification performance are shown for BF (left) and VC for 15, 5, and 3 s duration recordings. Note that the classification predictions were most accurate for BF with longer recording durations. *Source: Nair et al. (2016).*

In fact, direct brain recordings show that during the DOWN swing of the global wave most neurons in the cortex are pausing, while during the UP phase billions of neurons are firing. This is called "buzz-pause" activity, and it appears to be controlled by the neurons shown in Fig. 8.4: the state control circuit. Notice that some of these neurons are located in the thalamus and others in the cortex.

The stadium analogy holds nicely for SWS, where billions of nerve cells are going "buzz-pause" over and over again ("*chanting*"), from 0.5 to 3.5 Hz (Fig. 8.7). That is presumably why waking cognitive functions are largely lost during SWS.

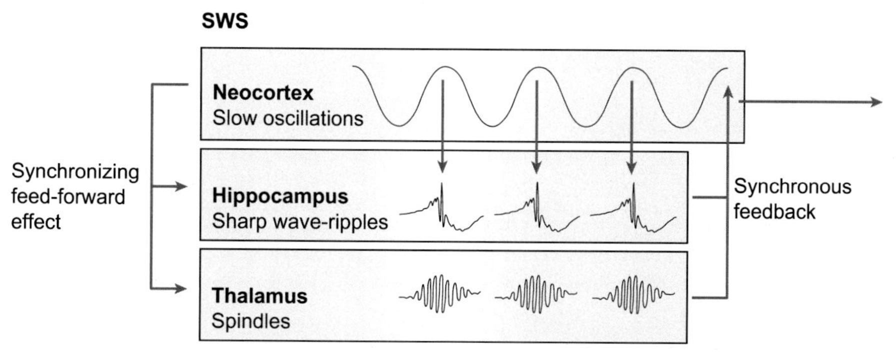

FIGURE 8.7 Chattering and chanting neurons. The waking electroencephalography (EEG) looks small, irregular, and faster than slow wave sleep (SWS). It is believed that waking (and REM dreaming) therefore involves more differentiated information processing in much the same way that a stadium full of talking people serves to process more information than the same people all chanting in unison. The unison chanting is largely redundant (you can predict the crowd chants from just one person) so the information content is lower. During SWS, slow cortical oscillations move in large UP and DOWN swings to form a "chanting" process. Because the chanting brain stops working from 0.5 to 3.5 times per second, we cannot do any cognitive work during the DOWN swings of the slow waves. Neurons can fire and even synchronize during the UP phase, but any task longer than a fraction of a second will be disrupted. You can think of it as rapidly switching a computer on and off, several times a second. The computer might be working when you turn it on, but by switching back and forth, you disrupt any long-lasting computational processes. Here we are showing three characteristic neural activities that occur during SWS: (top) the slow oscillations of the cortex that give SWS its name, (middle) sharp wave-ripples coming from the hippocampus in the medial temporal lobe, and (bottom) spindles coming from the thalamus in the subcortex. We will discuss these characteristic patterns more deeply in the chapter on Sleep later in the book. *Source: Adapted from Diekelmann and Born (2010).*

3. CONSCIOUSNESS

The three global states describe well the general stages a human being experiences on a daily basis as we wake in the morning, work through the day, and sleep and dream at night. However, consciousness is not an on/off phenomenon consisting of two discrete states: consciousness and unconsciousness. Rather, there are trading threads of conscious cognition and unconscious processes that seem to interact throughout our daily waking state and during sleep and dreaming. Only when we are experiencing dreamless sleep or are under general anesthesia are we truly unconscious.

3.1 Waking Has Conscious and Unconscious Threads

Waking cognition is woven of *both* conscious and unconscious threads, constantly weaving back and forth. For example, the process of reading *these words* is only partly conscious. You are a highly practiced reader, and you have learned over many hours of practice to automatically convert *these tiny marks on paper* into your own inner speech and then into unconscious processes like word recognition, grammar, and meaning.

We do not become conscious of every mental step in reading a sentence. In fact, there is a great deal of scientific evidence today that all cognitive tasks have many conscious and unconscious components. Fig. 8.8 shows an example of unconscious detection of the picture of a snake. Human beings are attuned to detecting some things unconsciously, it seems,

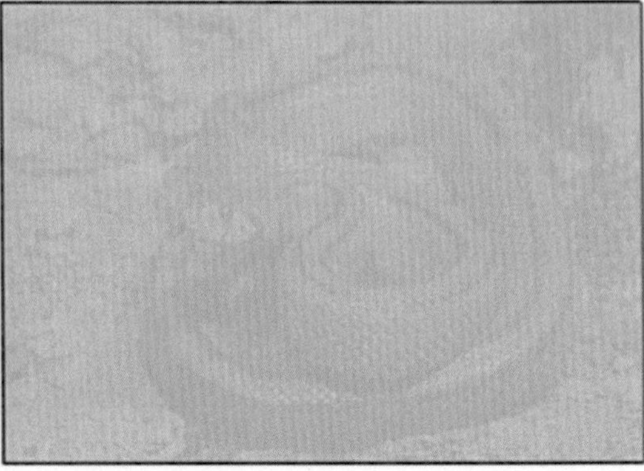

FIGURE 8.8 Comparing conscious versus unconscious events during the waking state. Some experiments comparing conscious and unconscious aspects of perception compare brain responses to stimuli that are clearly viewed and therefore enter consciousness and stimuli that have been masked or in some manner made not quite consciously visible. The subject can report the snake on the left side but not on the right side, even though the experimental stimulus is identical. This can be done using binocular rivalry, for example. In this experiment the emotional brain (the amygdala) can recognize the unconscious snake. Humans evolved in environments where deadly snakes were not unusual. *Source: Gage with permission.*

including snakes and even facial expressions, which are processed through the visual system, and trigger the fear-sensitive amygdala, a major subcortical center for processing emotions.

One can imagine how that ability may have evolved, since humans (and our ancestors) lived for millions of years in environments where snakes were an everyday, deadly threat. Human beings learn a vast amount of information from other people, such as the danger of crossing busy streets without looking. We are not innately attuned to cars rushing through a street, although we may be afraid of "looming" objects like a fast-approaching elephant.

Until recently, however, the evidence was still hotly debated on whether humans respond to genuinely unconscious stimuli. More recent studies seem to prove the point convincingly.

3.2 What We Expect From Conscious People

While there exist conscious and unconscious threads during our waking state, we are mostly conscious and aware of our surroundings during this state. From a medical perspective, there are clear aspects of what this conscious waking state should "look like" in a healthy person. People with head injuries are often given a *mental status examination*, which tests for abilities we expect normal, conscious people to have (McDougall, 1990). The test includes the following:

1. Orientation to time, place, and person: "What day of the week, date, and season is it?" "Where are you now?" "Who are you?" "Who am I?" (asked by the examiner)
2. Simple working memory: "Remember these three objects…"
3. Attention and calculation: "Count down by sevens starting from 100" (100, 93, 86, 79, 72 …).
4. Intermediate recall: "What were the three objects I asked you to remember a few minutes ago?"

TABLE 8.1 Major Features of the Conscious State

1. A very wide range of conscious contents
2. Widespread brain effects beyond the specific contents of consciousness
3. Informative conscious contents
4. Rapid adaptation (learning and coping)
5. The "fleeting present" of a few seconds
6. Internal consistency
7. Limited capacity and seriality
8. Sensory binding
9. Self-attribution
10. Accurate reportability
11. Subjectivity
12. Focus-fringe structure
13. Consciousness facilitates learning
14. Stability of conscious contents
15. Object and event-centered
16. Consciousness is useful for voluntary decision making
17. Involvement of the thalamocortical core

Source: Modified from Seth and Baars (2005).

5. Language: Asking the patient to identify a wristwatch and a pencil, to repeat a sentence, to follow three simple commands, to read and understand a simple sentence, and to write their own sentence.
6. A basic visuomotor skill: Asking the patient to copy a drawing of a line pentagon.

The mental status exam gives a broad overview of normal mind and brain functions. Some disorders involve disorientation, others affect short-term memory, and still others impair visual and motor coordination.

We also expect healthy, conscious adults to do *realistic thinking*. While dreams are unrealistic, waking consciousness has been believed to be necessary for logical, mature, and reality-oriented thought. Nevertheless, we routinely experience waking fantasies, unfocused states, daydreams, emotional thinking, and mind wandering.

Table 8.1 presents 17 properties of consciousness that are generally recognized in scientific and medical literature. (Of course, the humanities, arts, philosophy, and religion have a long history of exploring consciousness, too.)

3.3 The Scientific Rediscovery of Consciousness: A Global Workspace Theory

For many years in the modern science world of the 20th century, the study of human consciousness was thought to be too *philosophical*, too *anecdotal*, too *unscientific* to be addressed by neuroscientists in any meaningful way. The trend has turned, however, and the scientific study of human consciousness has emerged and become stronger and stronger in recent years.

Baars has reviewed seven predictions and their evidence about the brain bases of human consciousness and developed a theory of consciousness based on that evidence. This theory, the *global workspace theory*, forms a key basis for most current views of the neural correlates of consciousness (NCC). The central hypothesis of the global workspace theory developed by Baars is "the notion that consciousness facilitates widespread access between otherwise independent brain functions" (Baars, 2002, p. 47).

Consciousness, according to Baars' and many others' view, performs an *integrative function* in the brain. The brain bases of this integrative, widespread activity known as consciousness is theorized to be supported in part by the vast cortico-thalamo-cortical hub that forms the bases of neural communication throughout the brain.

According to Baars (2002), the seven predictions that were controversial in the 1990s that have since been well supported by current evidence are:

1. Conscious perception involves more than sensory analysis: it enables access to widespread brain sources, whereas unconscious input processing is limited to sensory regions.

2. Consciousness enables comprehension of novel information, such as new combinations of words.

3. Working memory depends on conscious elements, including conscious perception, inner speech and visual imagery, each mobilizing widespread function.

4. Conscious information enables many types of learning, using a variety of brain mechanisms.

5. Voluntary control is enabled by conscious goals and perception of results.

6. Selective attention enables access to conscious contents, and vice versa.

7. Consciousness enables itself to "self:" executive interpretation in the brain.

Thus the global workspace theory presents consciousness as performing an integrative function that allows for widespread activation in the brain, under conscious control, and typically purposeful and goal-oriented.

3.4 Components of Consciousness and Unconsciousness: Awareness and Wakefulness

It is obvious when a normal, healthy, awake person is conscious. We can easily see that they are alert, aware of their surroundings, and wakeful. However, when an individual has suffered head injury or other types of brain damage, clarifying their conscious state can be more difficult. For example, if an individual has suffered a head injury and appears to be unconscious, can we be certain if they are truly unconscious or just unable to move or open their eyes?

To address these types of conditions, and to help with end-of-life decisions, Laureys (2005) has developed a method to describe two separable aspects of consciousness: *awareness* and *wakefulness* (Fig. 8.9).

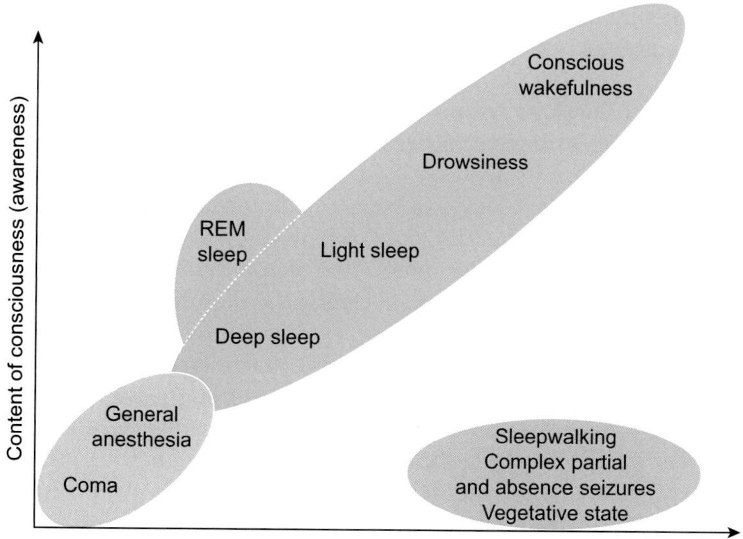

FIGURE 8.9 A simplified version showing levels of the two major components of consciousness: awareness and wakefulness. On the vertical y-axis, levels of awareness are shown beginning at the bottom of the axis with no level of awareness, as in a coma or when under anesthesia (shown in *pink*). Moving up the y-axis, level of awareness increases as one moves through heavy sleep, light sleep, and into drowsiness and finally to full conscious wakefulness (shown in *blue*). Note, however, that during REM sleep, there is a higher level of awareness than during deep or light NREM sleep stages. The shape for REM sleep is sitting above the other sleep stages with a higher level of wakefulness, although well below conscious wakefulness. On the horizontal x-axis, levels of wakefulness are shown beginning with coma and moving similarly, through stages of sleep to conscious wakefulness. According to this simplified version, levels of awareness and wakefulness are roughly similar as one moves through the stages from coma to deep sleep to wakefulness, represented by the *pink and blue* shapes forming a diagonal from the bottom left corner, where no awareness or wakefulness are observed in coma, to the upper right, where both levels of awareness and wakefulness are high during conscious wakefulness. Sharply contrasting to this balance of levels of awareness and wakefulness are abnormal states of consciousness such as sleepwalking, epileptic seizures, and vegetative states (shown in *purple* on the x-axis). In these states, the individual shows a high level of wakefulness with an extremely low level of awareness shown by the *purple* shape resting on the bottom of the y-axis but far to the right of the x-axis. For more on this topic, see Chapter 13, Disorders of Consciousness. *Source: Laureys (2005).*

To understand Fig. 8.9, it is best to begin with one axis. Let us start with the y-axis on the left side of the figure where we are showing the *contents* of consciousness: *levels of awareness*. The top of the axis represents a fully aware state during conscious wakefulness. As you look lower and lower down the y-axis until it meets the x-axis, there are decreasing levels of awareness across states of drowsiness, slight sleep, deep sleep, and ultimately to general anesthesia and coma where there is little or no awareness at all. Now look at the x-axis on the bottom of the figure where we are showing the *level of* consciousness: *wakefulness*. Again, wakefulness is highest—shown on the far right side of the axis, when we are in a state of conscious wakefulness. Wakefulness decreases as you move to the left of the x-axis toward where it meets the y-axis. Thus wakefulness decreases as we become drowsy, fall into light sleep, then deeper sleep. And wakefulness is ultimately lost under general anesthesia and in coma.

Now let us look at the combination of the y- and x-axes: generally, we have a similar level of awareness and wakefulness as we move from our most wakeful state of full conscious wakefulness. Next, we tend to have a rather even balance between levels of awareness and wakefulness as we drop down to drowsy, light sleep, REM sleep, and deep sleep. In other words, Fig. 8.9 is a schematic that shows a balance between awareness and wakefulness in healthy individuals. However, there are some exceptions. Take another look at REM sleep and you will see a lower level of wakefulness—the shape denoting REM sleep is moved to the left of the figure to represent this lower level of wakefulness versus awareness. This should feel intuitively correct: you have a higher level of awareness during REM dream sleep than during deep sleep, yet your wakefulness levels are similar. There is a second exception for healthy individuals: this occurs during *sleepwalking*. Although sleepwalking is unusual, it is not considered a sleep disorder. Take a look at the shape for sleepwalking in Fig. 8.9: you will see that it is very high on the x-axis wakefulness scale and very low on the y-axis awareness scale. If you have ever seen someone sleepwalk—or if you have sleepwalked yourself—this should sound right to you because a sleepwalker has no awareness of what they are doing; however, they look like they are wide awake. We will discuss the other exceptions to this general balance between levels of awareness and wakefulness and how it is abnormal in disorders of consciousness in Chapter 13.

4. ATTENTION

Common sense makes a distinction between attention and consciousness. The word *attention* seems to imply an ability to bring something to mind. If you cannot remember a word in this book, for example, you might try to "pay more attention." What trying to pay more attention comes down to is allowing the forgotten word to be in consciousness for a longer time. So we rehearse the forgotten word (consciously), or we make a note of it (making it conscious again), or we write a definition about it (same thing). The traditional "law of effect" about learning states that the more we make something conscious, the more we will learn it. When we call someone's attention to a speeding car, we expect him or her to become conscious of it.

In everyday language, "consciousness" refers to an experience—of reading this sentence, for example, or conscious sensory perception, conscious thoughts, feelings, and images. Those are the experiences we can talk about to one another. Selective attention implies a *selection among possible conscious events*. When we make an attentional selection, we expect to become conscious of what we have chosen to experience.

With careful studies we can separate the phenomena of attention and consciousness. To focus on conscious events "as such," we typically study them experimentally in contrast with closely matched unconscious events. By contrast, experiments on attention typically ask subjects to select one of two alternative stimuli. "Attention" is therefore concerned with the process of selection and consciousness, with reportable experiences themselves.

4.1 Attention Selects Conscious Events

Attention has two main aspects: *the source* of attentional control, which decides what to pay attention to, and *the target* of attention, which is selected for additional processing. Consider the case of a college student sitting in a lecture room, with many sensory inputs at the same time and many simultaneous tasks to perform. The student must stay alert, orient to the visual and auditory input, keep track of the lecture, take notes, and more. In fact, students are always multitasking, and as we know, that is inherently difficult. That is why it is important to review lectures. Live lectures can easily overwhelm our attentional capacities.

4.2 Voluntary and Automatic Attention: "*Top-Down*" Control and "*Bottom-Up*" Capture

The term *attention* is used most intuitively when there is a clear voluntary or executive aspect. We ask people to pay attention to things, which implies they can choose to do so or not, depending on some decision-making processes. Voluntary attention is the kind that is studied most often, and as you might guess from the other chapters, it is likely to engage the prefrontal cortex (PFC) processes in humans.

Corbetta, Kincade, and Shulman (2002) wrote that voluntary attention "is involved in preparing and applying goal-directed (top-down) selection for stimuli and responses." Automatic attention, on the other hand, "is not involved in top-down selection. Instead, this system is specialized for the detection of behaviorally relevant stimuli, particularly when they are salient or unexpected." When we hear a sudden loud noise, our attention is "captured," even without executive guidance. As you might expect, visual attention can be captured by human faces, emotional expressions, and bodies, when compared with neutral stimuli. Intense or sudden stimuli or unexpected events generate larger brain responses than control stimuli. Thus we can talk about "bottom-up" capture of selective attention, driven by stimuli.

In the real world, voluntary and automatic types of attention are generally mixed. We can train ourselves to pay attention to the sound of the telephone ringing. When it rings and we suddenly pay attention to it, is that voluntary or automatic? Well, it began being voluntary and became more automatic. The dimension of voluntary versus automatic attention is therefore a continuum. Perhaps the strongest case of voluntary attention is the one where we must exert intense mental effort over a period of time. A clear example of the opposite pole of the continuum might be a case of a loud sound or a biologically important event like a crying baby, which is hard *not* to pay attention to.

Therefore attention is defined here as the ability to select information for cognitive purposes. Selection may be shaped by emotion, motivation, and salience, and it is at least partly under executive control. Thus selective attention works closely with all the other components of our framework diagram (see Fig. 8.5). Without flexible, voluntary access control, human beings could not deal with unexpected emergencies or opportunities. We would be unable to resist automatic tendencies when they became outdated or change attentional habits to take advantage of new opportunities.

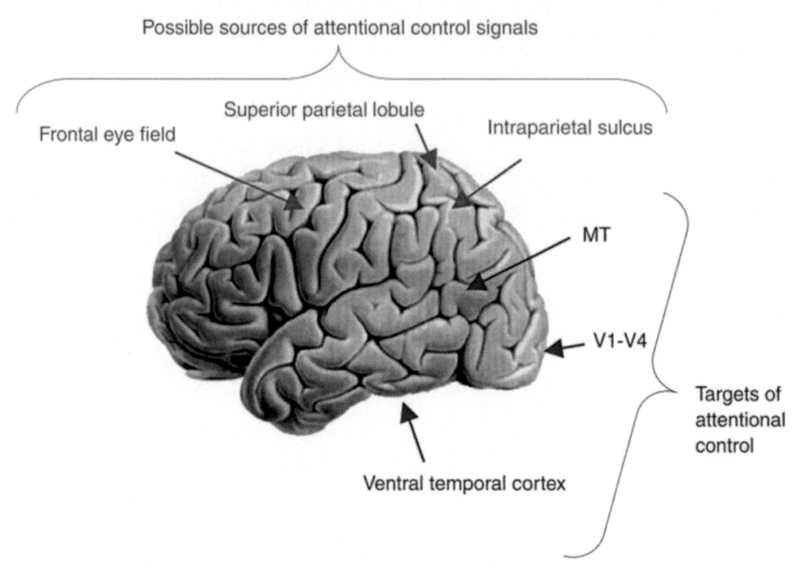

FIGURE 8.10 Voluntary attention. From frontoparietal to sensory cortex. Voluntary attention in perception is directed *to* sensory cortex *by* frontal and parietal regions. Parietal regions are believed to be involved in spatially directed attention. Visual regions (in *red*) are enhanced by attentional mechanisms, such as gamma synchrony among multiple visuotopic maps for the selected spatial location and visual features. *Source: Yantis (2008).*

As Fig. 8.10 shows, attention is a selective capacity, either under voluntary control or driven by a stimulus. The result of selective attention is to enhance the quality of the selected information or at least try to do so.

4.3 The "Spotlight of Voluntary Attention": Global Workspace Theory Revisited

Several aspects of the contents of consciousness are important to note: they are limited in capacity, controlled by voluntary attention, and directed by executive functions. Baars developed a "theater of consciousness" analogy to help understand the many "players" on the stage of consciousness. According to this analogy, the entire theater—stage, audience, players, and backstage areas—forms the basis of conscious and unconscious brain processes. The theater stage represents working memory. A spotlight on the stage represents voluntary attention. Only the stage contents *illuminated by the attentional spotlight* are conscious. The rest of the theater represents the vast unconscious store of knowledge and memories that can enter the contents of consciousness once they are on the stage *and* under the spotlight. A key point here is that the spotlight of attention on the stage is very *limited in capacity*: it represents just a small portion of the stage (working memory), which in turn represents a small portion of the vast theater (unconscious knowledge and processes) (Fig. 8.11).

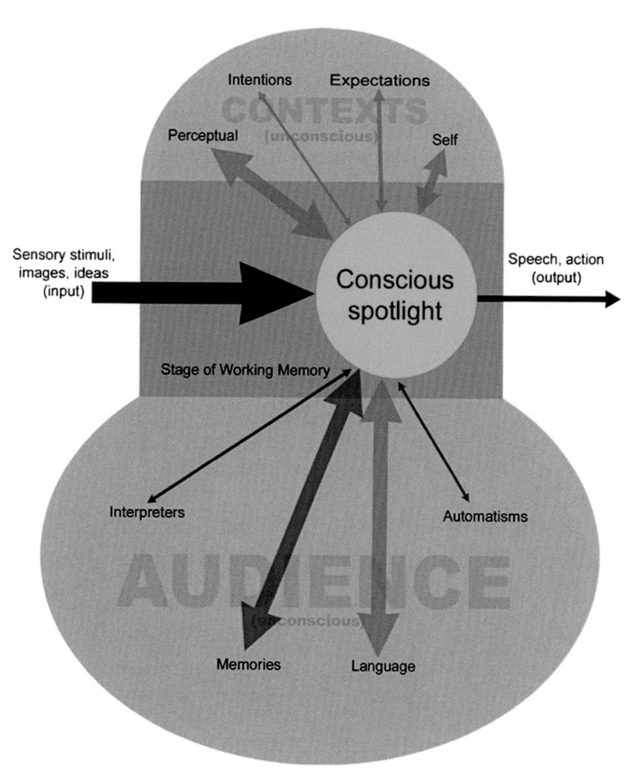

FIGURE 8.11 The theater analogy for the global workspace theory. According to this analogy, the entire theater—stage, audience, players, and backstage areas—form the basis of conscious and unconscious brain processes. The theater stage (shown in teal) represents working memory. A spotlight (shown in turquoise) on the stage represents voluntary attention. Only the stage contents illuminated by the attentional spotlight are conscious. The rest of the theater (shown in *purple*) represents the vast unconscious store of knowledge and memories that can enter the contents of consciousness once they are on the stage *and* under the spotlight. A key point here is that the spotlight of attention on the stage is very limited in capacity: it represents just a small portion of the stage (working memory), which in turn represents a small portion of the vast theater (unconscious knowledge and processes). *Source: Baars with permission.*

4.4 Neural Bases of Voluntary Attention: Alerting, Orienting, and Executive Control

We will focus on *voluntary attentional processes* in our description of brain areas that subserve attention. These processes are closely aligned with our discussion of consciousness and will factor in how we describe the neural bases of consciousness.

An early investigation used PET to show localization patterns for sustained voluntary attention and provided important new information about prefrontal regions that subserved voluntary attentional processes (Pardo et al., 1991). At about the same time, seminal work on the voluntary attentional networks involved in perceptual tasks was provided by Michael Posner and Steve Petersen (Petersen & Posner, 2012; Posner & Petersen, 1990), where they detailed an anterior attentional system, which they described as having three major separable networks that perform the *alerting, orienting, and executive* (conflict resolution) voluntary

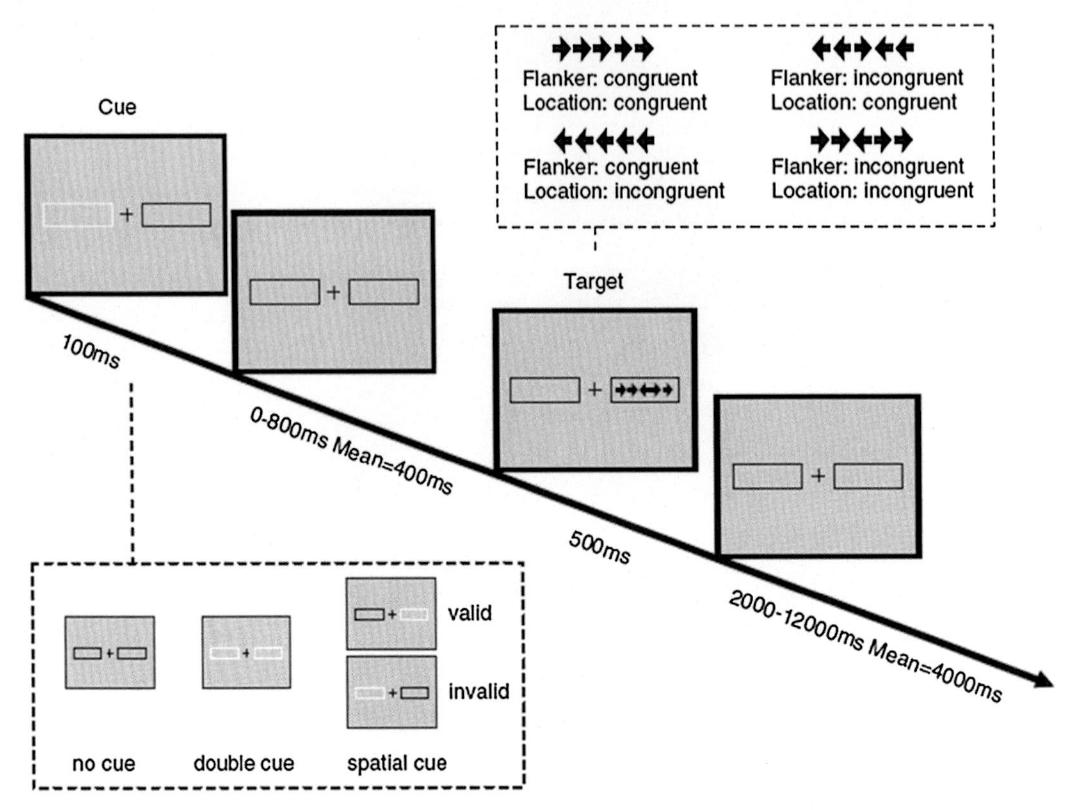

FIGURE 8.12 ANT-R is a revised test to assess the interaction among attentional functions: alerting, orienting, and executive control. A schematic shows the experimental design. The subject fixates on a central cross and in each trial, one of three cues is presented: no cue, double cue, and spatial cues (valid, invalid). After a duration of 0, 400 or 800 milliseconds (ms), a Target appears that is either congruent (matching the cue) or incongruent. Thus the design allows for the subject to *alert* to the Target, *orient* to the Target once it appears, and use *executive control processes* to adapt to the actual Target once it is perceived to perform the task demands. *Source: Xuan et al. (2016).*

attentional functions that are important to many tasks. Over the years, Posner et al. developed an attention network test (ANT) that provided them with ways to separately measure and record brain regions that subserve these voluntary attentional networks. The anterior attentional system they describe has three key networks: alerting, for maintaining an alert state, such as being ready to look at visual images in the previous scanner example; orienting, such as preparing to see a new visual image of a face that appears in the scanner; and executive, such as deciding whether the face is male or female.

Brain areas that subserve alerting, orienting, and executive control of attention have been investigated by Fan et al. (Fan, McCandliss, Fossella, Flombaum, & Posner, 2005; Xuan et al, 2016). Xuan et al. (2016) used a revised ANT (shown in Fig. 8.12) and functional MRI (fMRI) to investigate the neural bases of these aspects of voluntary attention as *interactive attentional networks*. Results showed activation in an extended frontoparietal network (FPN), including the thalamus, dorsolateral prefrontal cortex (DLPFC), the anterior cingulate cortex (ACC), and the anterior insular cortex (AIC) (Fig. 8.13).

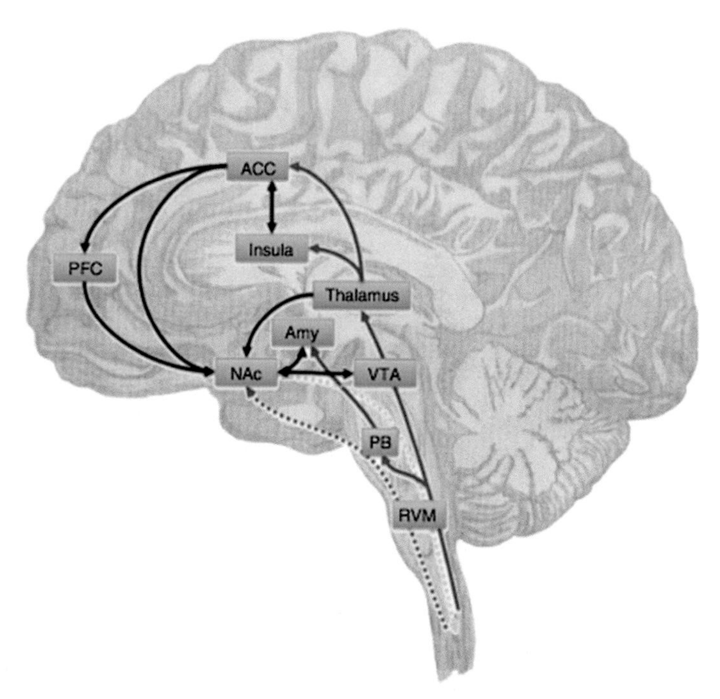

FIGURE 8.13 The frontoparietal network (FPN) for attention includes the prefrontal cortex (PFC), and anterior cingulate cortex (ACC), the insula, and the thalamus. Other brain regions shown include the parabrachial area (PB), the amygdala (Amy), the nucleus accumbens (NAc), the ventral tegmental area (VTA), and the rostral ventromedial medulla (RVM) (all shown in *blue boxes*). *Source: Navratilova and Porreca (2014).*

Specific results showed the *alerting* function activating the FPN and the locus coeruleus (LC) (Fig. 8.14). The *orienting* function activated frontal eye fields (FEFs) and the superior colliculus (SC) (Fig. 8.15). The *executive control* function activated the FPN and the cerebellum (Fig. 8.16). Simply put, the FPN and the thalamus and other subcortical regions and bodies form a cortico-subcortical interactive network that provides a dynamic and highly flexible human attentional network.

4.5 Neural Oscillations Underlying Sustained Attention

Figs. 8.14–8.16 show a multitude of brain regions that support voluntary attention, forming a distributed network across wide regions of the cortex. How do to these disparate regions in the attentional network synchronize during sustained attention? Clayton et al. (Clayton, Yeung, & Kadosh, 2015) investigated the ways in which cortical oscillations sustain attention across brain regions. *Neural oscillations* reflect the rhythmic activity of populations of neurons in the brain. These oscillations can be recorded from the surface of the brain using EEG (Fig. 8.17A). The recorded signal can then be analyzed in the power domain for each frequency (Fig. 8.17B). The signal can also be analyzed in the temporal (time) domain for each frequency (Fig. 8.17C). Characteristic neural oscillation frequency bands are delta (1–4 cycles per second, Hertz, Hz), theta (4–8 Hz), alpha (8–14 Hz), beta (14–30 Hz), and gamma (>30 Hz).

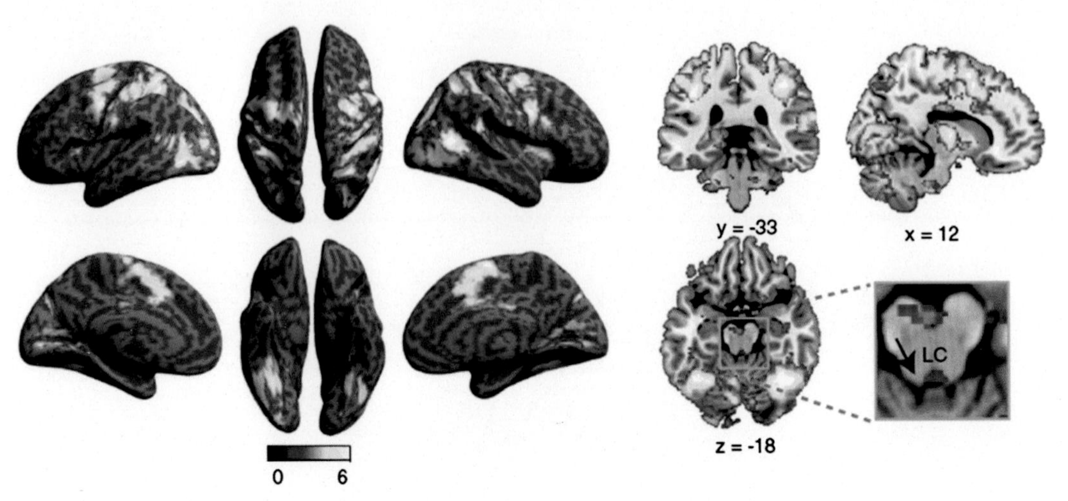

FIGURE 8.14 Alerting in the attention network test-revised (ANT-R). Specific results showed the *alerting* function activating the frontoparietal network (FPN) in the left hemisphere (top left sagittal images), the midline (center axial images) and the right hemisphere (top right sagittal images) and the locus coeruleus (LC) (far right coronal images). *Source: Xuan et al. (2016).*

According to the model proposed by Clayton et al., *monitoring of attention* is supported by theta oscillations in posterior medial frontal cortex (pMFC, shown in purple in Fig. 8.18). The neural communication between the pMFC and lateral prefrontal cortex (lPFC) is aided by phase synchronization in the theta frequency band (shown in the *purple arrow* in Fig. 8.18). Next, neural communication between the frontal lobe lPFC and the posterior sensorimotor regions is supported by phase synchronization in the lower frequency bands (theta, alpha, shown in *gray arrows* in Fig. 8.18).

Sustaining attention during a cognitive task requires the excitation of task-relevant areas in the brain with the inhibition of task-irrelevant areas (Fig. 8.19). Clayton et al., (2015) proposes that this is accomplished through neural oscillations from the frontal lobe to the posterior sensory cortex. Thus the influence of the frontal cortex over posterior regions provided by the low frequency bands of oscillations described above allows frontal processes to promote high-frequency gamma band oscillations in a task-relevant area of the brain (in this case, visual cortex shown in green in Fig. 8.18) and lower frequency alpha band oscillations in a task-irrelevant area of the brain (in this case, auditory cortex shown in orange in Fig. 8.18). The influence of the frontal cortex over visual cortex shown in Fig. 8.18 allows the continuous activation of a task-relevant activity (in this example, a visual task with visual cortex activation) and suppression of task-irrelevant activity (in this example, auditory cortex with suppression of auditory processing).

The model proposed by Clayton et al., fits well with the existing literature on the role of neural oscillations across frequency bands and their role in synchronization across large networks within the cortex. New approaches to understanding the neural bases of attention are developing continuously and will no doubt shed new light on just how voluntary attention is sustained in the brain.

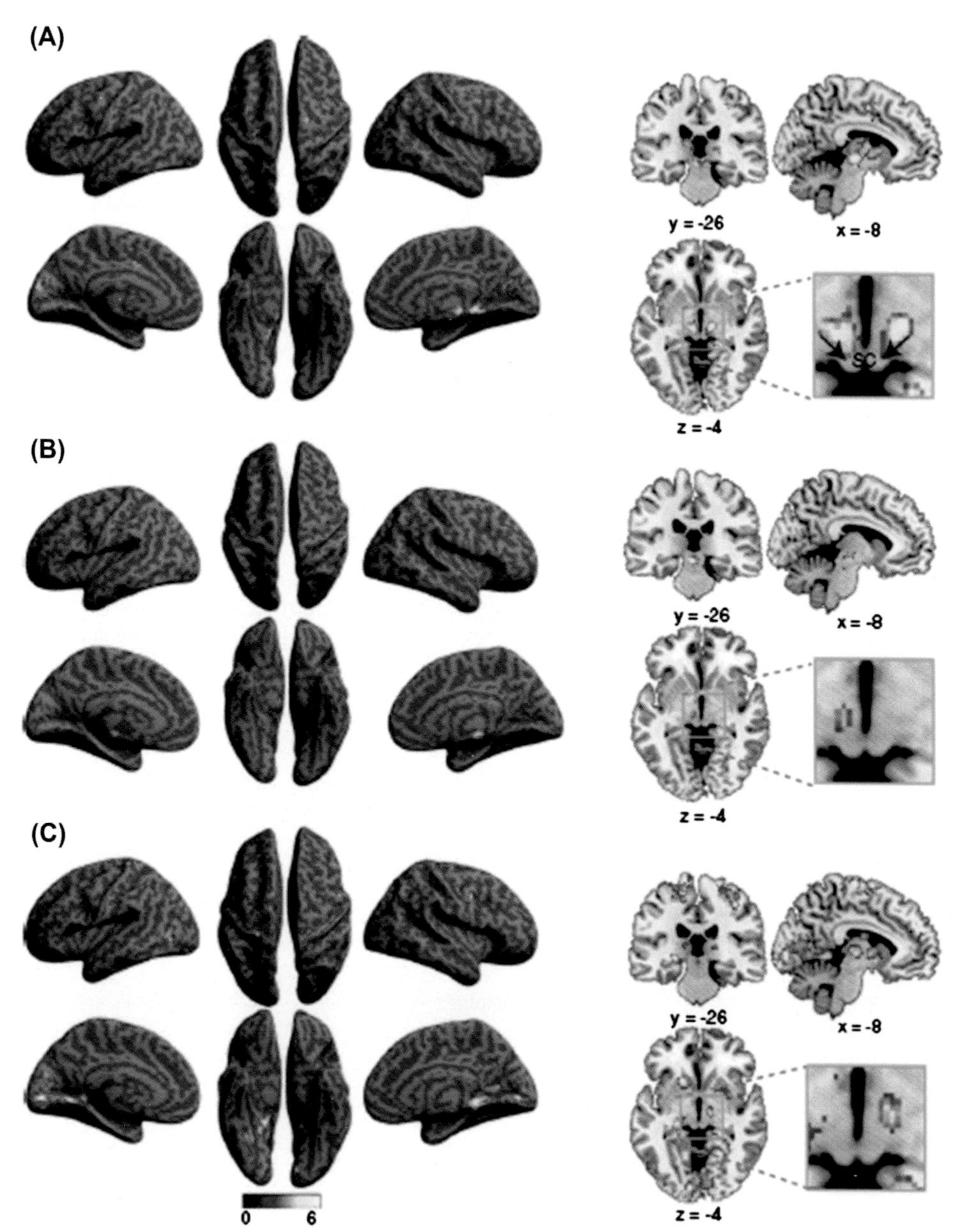

FIGURE 8.15 Orienting in the attention network test-revised (ANT-R). The *orienting* function activated frontal eye fields (FEFs) and the superior colliculus (SC). The activity shown reflects (A) disengaging, (B) engaging and movement to the target (orienting), and (C) the validity effect (was the cue accurate?). *Source: Xuan et al. (2016).*

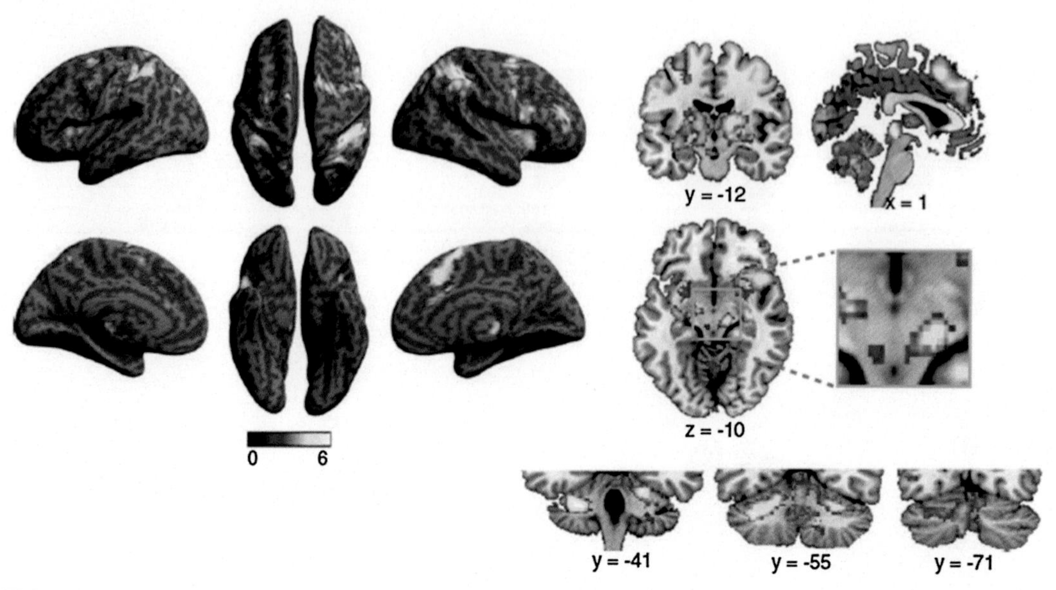

FIGURE 8.16 Executive control in the attention network test-revised (ANT-R). The *executive control* function activated the frontoparietal network (FPN) (Shown for the left hemisphere) (left sagittal images), midline activity (center axial images) and right hemisphere (right sagittal images) and the cerebellum (lower right coronal images). *Source: Xuan et al. (2016).*

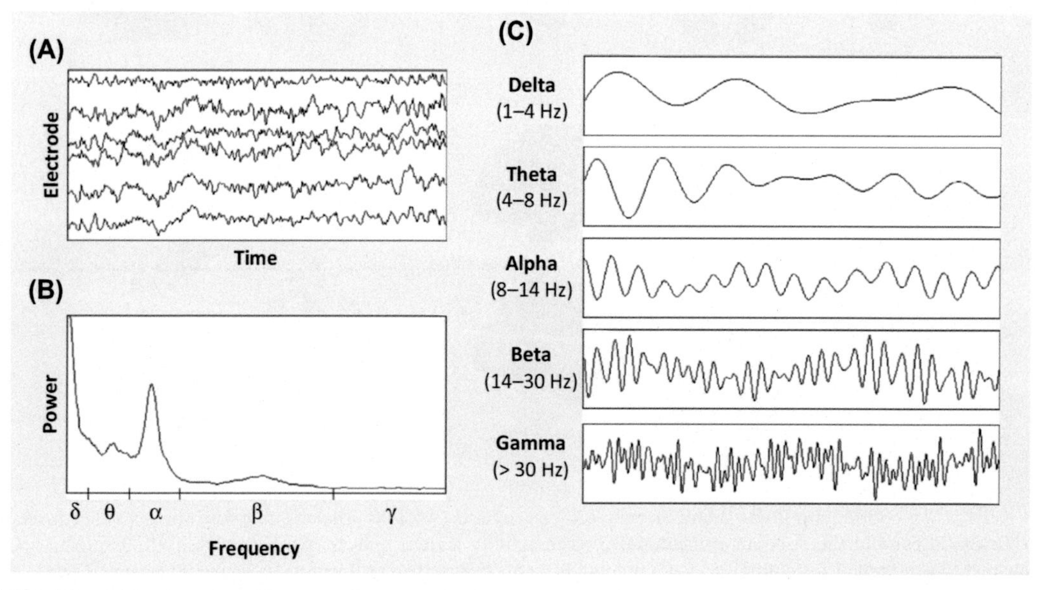

FIGURE 8.17 *Neural oscillations* reflect the rhythmic activity of populations of neurons in the brain. These oscillations can be recorded from the surface of the brain using electroencephalography and recording electrodes (EEG, A). The recorded signal can then be analyzed in the power domain for each frequency (B). The signal can also be analyzed in the temporal (time) domain for each frequency (C). Characteristic neural oscillation frequency bands are delta (1–4 cycles per second, Hertz, Hz), theta (4–8 Hz), alpha (8–14 Hz), beta (14–30 Hz), and gamma (>30 Hz). *Source: Clayton, Yeung and Kadosh (2015).*

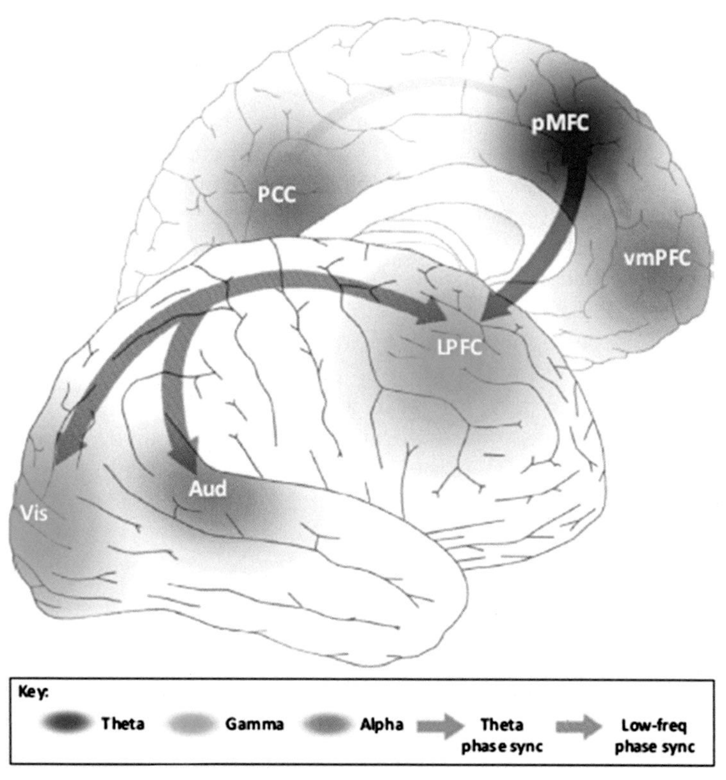

FIGURE 8.18 According to the model proposed by Clayton et al., *monitoring of attention* is supported by theta oscillations in posterior medial frontal cortex (pMFC, shown in *purple*). The neural communication between the pMFC and lateral prefrontal cortex (lPFC) is aided by phase synchronization in the theta frequency band (shown in the *purple arrow*). Next, neural communication between the frontal lobe lPFC and the posterior sensorimotor regions is supported by phase synchronization in the lower frequency bands (theta, alpha, shown in *gray arrows*). Sustaining attention during a cognitive task requires the excitation of task-relevant areas in the brain with the inhibition of task-irrelevant areas (Fig. 8.19). Clayton et al. (2015) proposes that this is accomplished through neural oscillations from the frontal lobe to posterior sensory cortex. Thus the influence of the frontal cortex over posterior regions provided by the low frequency bands of oscillations described above allows frontal processes to promote high frequency gamma band oscillations in a task-relevant area of the brain (in this case, visual cortex shown in *green*) and lower frequency alpha band oscillations in a task-irrelevant area of the brain (in this case, auditory cortex shown in *orange*). The influence of the frontal cortex over visual cortex allows the continuous activation of a task-relevant activity (in this example, a visual task with visual cortex activation) and suppression of task-irrelevant activity (in this example, auditory cortex with suppression of auditory processing). *Source: Clayton, Yeung and Kadosh (2015).*

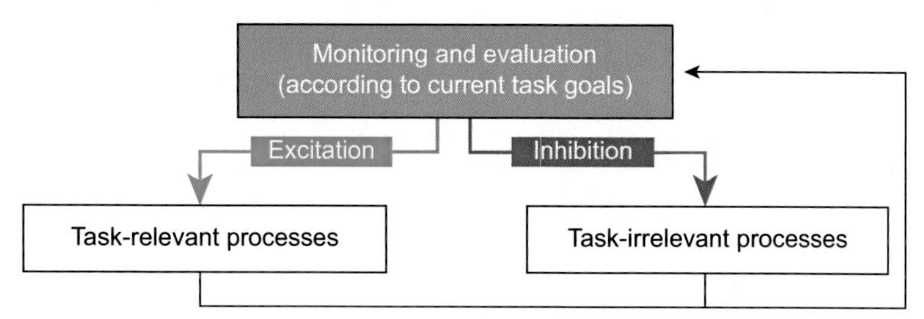

FIGURE 8.19 According to cognitive models of sustained attention, the ongoing cognitive processes are *monitored and evaluated* (*purple box*) according to the task goals at that moment. If required, attention is selectively adjusted with *excitation (shown in green) of task-relevant processes* and *inhibition (shown in red of task-irrelevant processes*. The output of these adjusted processes (shown as a *black line*) is fed back to the monitoring and evaluation systems so that the continuous assessment and control of attention can occur. *Source: Clayton, Yeung, and Kadosh (2015).*

5. ATTENTION AND CONSCIOUSNESS: SEPARABLE BUT CLOSELY INTERTWINED PROCESSES

What we pay attention to *enters our consciousness*. When we turn our attention to other matters, that item fades from consciousness. Thus it is difficult to clearly separate voluntary attention and consciousness: they are clearly closely intertwined. Functional neuroimaging studies of conscious processes have provided evidence for a combination of a FPN (Fig. 8.13) along with posterior cortical regions that are theorized to subserve consciousness perception—the *contents of our consciousness* at a given moment (frequently tested using visual stimuli). However, it is difficult if not impossible to factor out aspects of the neural activity measured for conscious perception that are actually due to experimental task demands such as selective attention, self-monitoring, and other executive control processes.

Is there a way, experimentally, to tease apart these closely interrelated brain processes and networks to assess the neural substrates of consciousness itself? Koch et al. have investigated the NCC using a wide variety of techniques and paradigms (reviewed in Koch, Massimini, Boly, & Tononi, 2016). To focus on the NCC of consciously narrowly defined, and not including the effects of voluntary attention and executive control mechanisms, Koch et al. defined the NCC of the contents of consciousness as "*the minimal neuronal mechanisms jointly sufficient for any one specific conscious percept*" (Koch et al., 2016, p. 308). A central goal of their studies has been to identify the NCC while paring away brain activity related to the attentional and executive functions described above.

5.1 Disentangling Attention and Conscious Perception: The *No-Report Paradigm*

Koch et al. argue that traditional methods for assessing the neural bases of conscious perception frequently employ a *report paradigm* in which the subject reports that they *do see* a stimulus presented or they *do not see* a stimulus. According to Koch, the report paradigm taps into attentional and executive control processes that are not, as strictly defined, part of the NCC. Support for this notion comes from results of these neuroimaging studies that typically report a broad FPN activated for the conscious perception: a network very similar to the network reported for tasks that tap attentional processes. Koch et al. developed a *no-report paradigm* in which the subject makes no overt report of being conscious of seeing a stimulus or not. According to Koch et al. (2016), the no-report paradigm "*allows the NCC to be distinguished from events or processes that are associated with, precede or follow conscious experience*" (Koch et al., 2016, p. 308). Results of neuroimaging studies using the no-report paradigm provides evidence for a more constrained neural basis for content-specific NCC, which are typically posterior cortical regions *but not* the PFC (Fig. 8.20).

Note that these findings are for *content-specific* NCC (seeing or being conscious of a stimulus, for example) and *not* for *consciousness as a whole*, which is theorized to include regions in the parietal, temporal, and occipital lobes, as well as the brainstem and other subcortical brain areas.

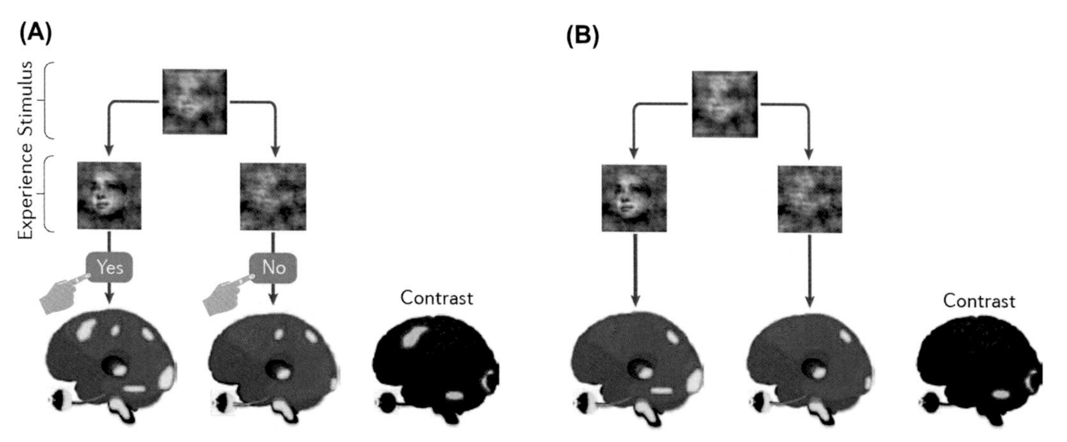

FIGURE 8.20 Results of neuroimaging studies using the no-report paradigm provides evidence a more constrained neural basis for content-specific neural correlates of consciousness (NCC), which are typically posterior cortical regions *but not* the prefrontal cortex (PFC). For example in this figure using functional MRI (fMRI) to record hemodynamic activity in a report paradigm (A) and a no-report paradigm (B) shows a clear difference in brain activation that depends on whether or not the report is involved. (A) In the report paradigm, brain activation for the "*YES* I did see it" trials includes the frontal and parietal networks and sensory (visual) regions (shown in *orange and yellow* on the *blue-shaded* brain image). In the report paradigm, brain activation for the "*No* I did not see it" trials includes the less frontal and parietal networks and sensory (visual) regions (shown in more *orange* and less *yellow* on the *blue-shaded* brain image). The contrast measure (the "*YES* I did see it" trials minus the "*NO I did not see it*" trials) shows activation in the frontal network and the posterior sensory (visual) regions (shown in *yellow and orange* highlighted on the *black-shaded* brain). (A) In the no-report paradigm, brain activation for the seen activation trials includes less frontal network and sensory (visual) regions (shown in *orange and yellow* on the *blue-shaded* brain image on the left). In the no-report paradigm, brain activation for the not seen trials includes less frontal networs and sensory (visual) regions (shown in more *orange* and less *yellow* on the *blue-shaded* brain image). The contrast measure (the seen minus not seen trials) shows activation in only the posterior sensory (visual) regions (shown in *yellow and orange* highlighted on the *black-shaded* brain). *Source: Koch et al. (2016).*

5.2 Neural "Switches" for Consciousness Versus Unconsciousness

Which brain areas support consciousness *as a whole*? Where are the neural "switches" that turn regions of the brain *off* during unconsciousness during sleep, anesthesia, or brain injury, and *back on* during the regaining of consciousness during waking, rearousal following anesthesia, or recovery from brain injury. One way to investigate the brain areas that underlie consciousness is to turn to individuals whose job it is to render people unconscious: anesthesiologists. Results from studies investigating the brain areas that are *turned off* (deactivated) with anesthetics to render a patient unconscious and *turned back on again* (reactivated) to wake the patient back up to full consciousness point to the corticothalamic circuitry in underlying the integration of wide-ranging cortical networks with the mighty thalamus. Specifically, researchers refer to the loss of consciousness—due to sleep, anesthesia, or brain injury—as a loss of *cortical integration* (Fig. 8.21) in key brain areas involved in consciousness, including the frontal lobe, the precuneous, and the thalamus among others (Fig. 8.22, Alkire, Hudetz, & Tononi, 2008).

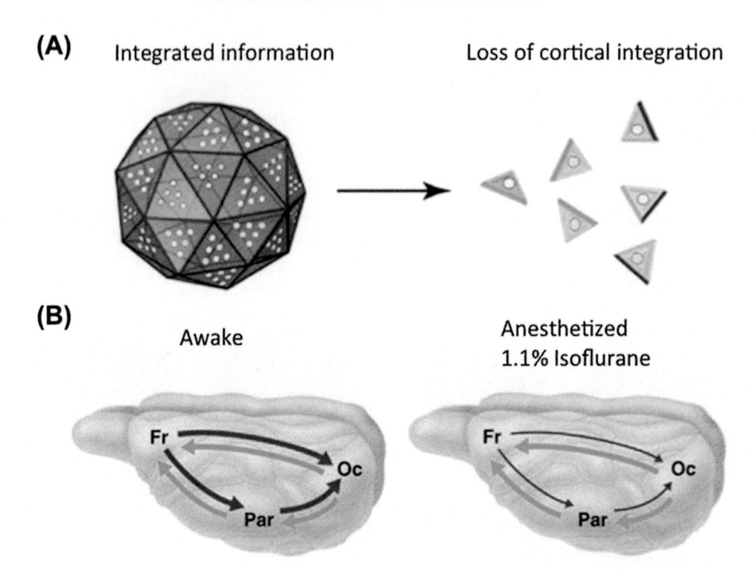

FIGURE 8.21 According to Alkire et al. (2008), unconsciousness is associated with a loss of cortical integration. (A) the *green* "die of consciousness" shown on the left has many faces which each correspond to brain firing patterns during conscious brain activity. (B) the "die" shatters apart during unconsciousness resulting in a loss of integrated information across the die faces and their proposed firing patterns. (B) In a rat, anesthesia reduces cortical integration by reducing feedback processes (shown with *red arrows*) between the frontal (Fr), occipital (Oc) and parietal (Par) lobes. *Source: Alkire et al. (2008).*

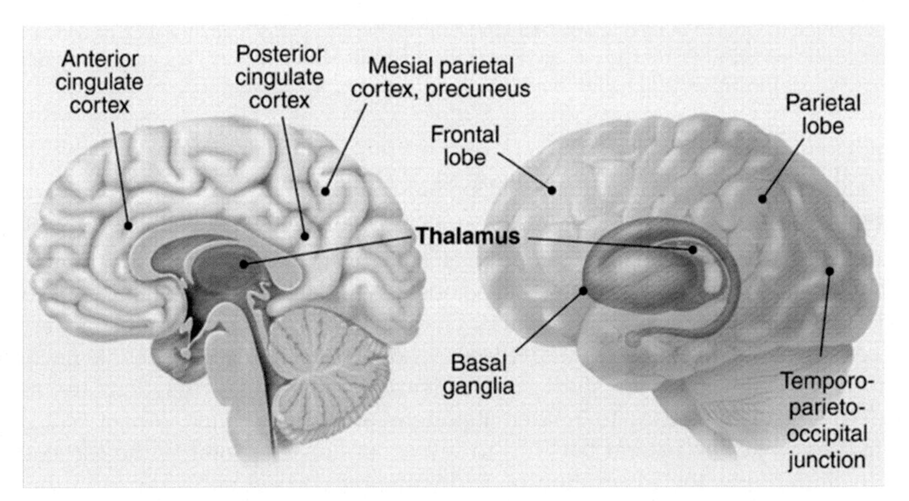

FIGURE 8.22 Researchers refer to the loss of consciousness—due to sleep, anesthesia, or brain injury—as a loss of *cortical integration* in key brain areas involved in consciousness, including the frontal lobe, the precuneous, and the thalamus among others. *Source: Alkire et al. (2008).*

Another approach to determining the NCC for consciousness as a whole has been to investigate it by comparing healthy adults with patients in a vegetative state. Progress on this front is presented in Chapter 13.

6. EXCEPTIONAL STATES OF MIND

We have all experienced our day-to-day conscious selves as we wake, work, and sleep and dream. Our conscious states are similar as we go through everyday life. But there is significant evidence for a *much-altered conscious self,* either through drugs, mental illness, or even meditation. Whether purposeful or not, whether we are aware of it or not, our conscious experience can vary sharply from the actual world around us.

To achieve altered states, humans have used a huge variety of methods: taking psychoactive plants, fasting, drinking alcohol, special exercises, visualization, disorientation, hypoxia, self-mutilation, sexual practices, dramatic rites, lucid dreaming, dancing, whirling, chanting and music, suggestion, trauma, sleep deprivation, and social isolation. Traditions like Buddhism and Vedanta Hinduism believe that humans should change their states of consciousness and self. That belief can be found in many times and places.

6.1 Epilepsy, Drugs, and Psychoses

Some mystical experiences are associated with epilepsy. Epileptic patients sometimes describe altered states as their EEG begins to show slow synchrony, even without visible seizures. Since epileptic synchrony alters the workings of the thalamocortical core, there could be a link between altered subjective states and brain rhythms.

Psychedelic drugs have been compared to dreams, often showing vivid visual hallucinations, delusions, dreamlike actions, emotional encounters, time loss, and discontinuities. The LSD molecule resembles serotonin, which is involved in REM dreaming. "Recreational drugs" by definition are taken with the goal of having unusual experiences.

In the psychoses, delusions and hallucinations are often extremely upsetting, frightening, and unwanted. They occur at unexpected times, interfere with normal life, and can develop into painful persecutory voices and distressing beliefs. The degree to which these experiences are unwanted is one major difference between psychedelic drugs and psychotic experiences. Mental disorders are not a matter of choice, and people who suffer from them cannot stop at will.

6.2 Out-of-Body Experiences

Direct stimulation of the right posterior parietal cortex can evoke dramatic changes in experienced body space (Fig. 8.23), including OBEs. Some of the most reliable results come from studies of OBEs, the experience of looking at one's own body from the outside. Some epileptics experience alterations in body space, perhaps because of hypersynchronous activity affecting the parietal cortex.

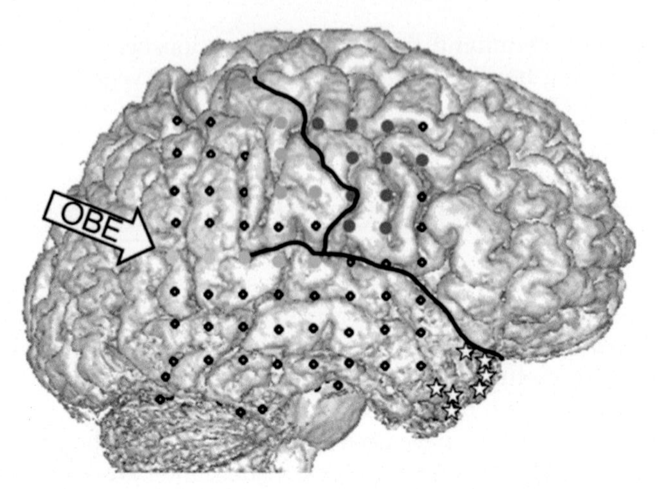

FIGURE 8.23 Electrical stimulation of the colored locations on the right parietal cortex and evoke out-of-body experiences. *Source: Tong (2003).*

6.3 Neurofeedback

Neurofeedback training is defined as learning to control brain events by giving sensory (conscious) feedback, contingent on the event. Simply by holding a thermometer one can learn to increase warmth, for example, which seems to help people relax and lower blood pressure. Neurofeedback studies of animals have shown positive results over several decades. In a brain with billions of neurons, simple sensory feedback can allow training of voluntary control over otherwise involuntary neuronal firing.

EEG neurofeedback show significant results, but long-term studies are often lacking. Because neurofeedback may work in cases where other medical treatments have failed, long-term trials would seem to be vitally important.

6.4 Sensorimotor Rhythm Feedback

The sensorimotor response (SMR) is an alpha-like EEG pattern found over the motor cortex when people voluntarily suppress some planned action. SMR feedback training has been shown to be an effective treatment in cases of human epilepsy, ADHD, and impulse control disorders (Sterman and Egner, 2006). On average, 80% of patients trained to enhance SMR had significant improvements.

Drug treatments for epilepsy do not always work, and having another treatment is helpful for many patients. Scientifically, because epilepsy shows slow, synchronous, and high waves in the EEG, the ability to modify it by training reveals an interesting fact about the voluntary control of brain rhythms.

6.5 Rhythmic Entrainment

If brain rhythms constitute a basic neural code, it would be interesting to know if we can drive rhythms externally. If we listen to a 10-Hz auditory stream, will alpha waves increase?

Brain wave entrainment has been described with transcranial magnetic stimulation and -transcranial alternating current. However, although there are many popular claims about the effects of auditory entrainment, they have not been demonstrated to work by demanding medical and scientific standards so far.

6.6 Hypnosis and Conversion

About one-fourth of the normal population is highly hypnotizable, as assessed by standard hypnotic procedures. Hypnotic suggestions can change brain events, as in reducing the AEP to a flashing light simply by "hallucinating" a cardboard box covering the stimulus (Spiegel, 2003). We do not know how hypnotic induction works. About 5% of the population are said to be "hypnotic prodigies," who can hallucinate perceptual events, such as the buzzing of a fly; they can also enter into dissociated identity states (Hilgard, 1977). fMRI studies show that hypnotically suggested pain activates some of the same brain regions as real pain. Hypnotic procedures can alleviate chronic pain (Spiegel, 2003).

Hypnosis may involve a dissociation between voluntary control (dorsolateral prefrontal cortexDL-PFC) and the ability to monitor errors (the ACC). Egner, Jamieson, and Gruzelier (2005) reported an fMRI-EEG study of hypnosis in the Stroop task showing that "hypnosis decouples cognitive control from conflict monitoring processes of the frontal lobe."

Mild conversion symptoms are quite common in medicine. Medical students who are studying serious diseases may become worried when they seem to notice "real" symptoms in themselves. "Medical student syndrome" is common and generally fades over time. Conversion disorders might be a result of the general human tendency toward autosuggestion. The placebo effect is a positive version of autosuggestion.

6.7 Meditation and Yoga

The term *meditation* covers many mental techniques. One is silent repetition of a word called a "mantra." Asian and other traditions describe mantra meditation as a method for changing mind states, as taught in Vedanta, Buddhism, and Chinese medicine. It has also been widely practiced in Europe and the Middle East.

Meditation methods have been reported to increase coherence (synchrony), especially in theta, alpha, and beta-gamma EEG. Frontal-lobe coherence is also reported, as well as improved attentional functioning (Lazar et al., 2000; Lutz, Greischar, Rawlings, Ricard, & Davidson, 2004; Tang et al., 2007).

One of the surprises with mantra repetition is a significant drop in metabolic activity, reflected in "breath suspensions"—spontaneous stopping of the breath without compensatory breathing afterward. This is different from holding our breath voluntarily. In swimming, for example, we may take a deep breath before diving and then feel the need to take some extra breaths after coming up for air. The lack of compensatory breathing suggests that energy demand has indeed dropped more than an ordinary resting state, as has been demonstrated by measuring O_2/CO_2 exchange (Benson, Alexander, & Feldman, 1975).

Herbert Benson et al. (1975) proposed that these results represent a "relaxation response," like other physiological reflexes. Spontaneous breath suspensions are reported to be associated with reports of periods of "pure consciousness," defined as relaxed alertness without any specific mental contents. Converging evidence has come from measures of metabolism, sympathetic nervous system tone and, recently, large-scale changes in gene expression. Because the exact functions of "relaxation-related" genes are not yet clear, this promising direction requires additional studies (Dusek et al., 2006; Jacobs, Benson, & Friedman, 1996).

A different procedure is called "mindfulness meditation." Cahn and Polich (2006, 2009) describe it as "allowing any thoughts, feelings, or sensations to arise while maintaining a specific attentional stance: awareness of the phenomenal field as an attentive and nonattached observer without judgment or analysis." Mindfulness meditation has been shown to improve depression and even suicidal thinking.

7. SUMMARY

Consciousness has intrigued human beings since the beginning of history. With improved scientific tools we have seen considerable progress in recent years. Attention and consciousness can be considered complementary processes, although with careful experimental work they can be disentangled. Conscious contents often are thought to involve the widespread distribution of focal contents, like the sight of a rabbit. As soon as we see a rabbit (consciously), we can also judge whether it's real or if it is somebody's pet. We can laugh or scream for fear of being bitten and try to remember if rabbits carry rabies. The point is that a great variety of brain events can be evoked by a conscious object, including a great variety of unconscious processes.

Conversely, we can *select* a source of information to focus on. Selective attention allows us to choose what we will be conscious of. In order to learn about the brain, we direct attention to it over and over again. Learning a difficult subject involves allocating attentional resources over time.

While the NCC are still being uncovered, real progress has been made in determining the neural underpinnings of the closely defined aspect of our awake selves that reflects the contents of our consciousness at any given moment. A central challenge to clinicians is to understand the level (and contents) of consciousness when brain damage has occurred and the patient cannot respond or report their experiences, as in the case of vegetative state patients. Are they conscious? Are they not? These are key end-of-life questions, addressed in more detail in Chapter 13.

8. STUDY QUESTIONS

1. What brain region triggers the circadian rhythm of waking, sleep and dreaming?
2. What are the main effects of selective attention? What part of the brain appears to control voluntary, selective attention?
3. What are neural oscillations and what role do they play in sustained attention?
4. What are the features of the conscious waking state? What kind of cortical waveforms seem to occur?
5. Are voluntary attention and consciousness the same processes or separable?
6. According to the theater analogy for the global workspace theory, what does the stage represent? the spotlight? the area on the stage illuminated by the spotlight?

7. Describe neurofeedback training. What is it useful for?
8. Is there convincing evidence that hypnosis reflects a brain state?
9. Meditation methods have been described for thousands of years. What are some common features? What evidence do we have pro or con?

CHAPTER

9

Decisions, Goals, and Actions

THE ROUTE TO SUCCESS

Our goals can only be reached through a vehicle of a plan, in which we must fervently believe, and upon which we must vigorously act. There is no other route to success. ***Pablo Picasso, Spanish artist.***

Young ladies on the banks of the Seine, Pablo Picasso (1950). *Source: Public Domain.*

1. INTRODUCTION

What do these behaviors have in common? Remembering your friend's new cell phone number while looking for a piece of paper. Deciding to study first and play basketball second. Planning a new route when a road has been closed for repair. Paying attention while you are reading this sentence. Changing your mind about raising the stakes in a poker game. All of these behaviors—and many more—are guided by the frontal lobes.

Fundamentals of Cognitive Neuroscience
https://doi.org/10.1016/B978-0-12-803813-0.00009-X

1.1 The Many and Complex Frontal Lobe Functions

The frontal lobes have been described as a control center for functions such as paying attention selectively to one item rather than another, making plans and revising them when needed, monitoring the world around us—complex functions that are part of our everyday life. The frontal lobes also are described as the "action lobes" where physical actions are planned and motor system activity is initiated. The frontal lobes also are described as the "home" of our personality and social morality. How does one brain region provide these diverse and complex functions?

The answer to this question is still being investigated, but we have learned about frontal lobe function through many investigative pathways. In animal research, we have learned about the role of the frontal lobe using single unit recordings. In humans, we have learned about the many roles of the frontal lobe through brain damage and disease. A key figure in our knowledge about frontal lobe function is Phineas Gage, the railroad worker who had major frontal lobe damage due to a railroad accident. A tamping iron blasted through Phineas's cheek and out his skull, producing massive damage to his frontal lobe (Box 9.1). Phineas's sudden personality change following the accident shed light on the role of frontal regions in personality formation and social cognition. Studies of people who have had frontal lobe damage through disease, brain injury, or disorder have aided us in developing categories of frontal lobe function based on where the brain damage or disease has occurred and the resultant change in behavior. More recently, neuroimaging studies of healthy individuals have provided new information about the localization of patterns of frontal lobe activity occurring during tasks that tap frontal lobe functions, such as voluntary or executive attention and decision-making.

A recent increase in symptoms showing deficits in the functions theorized to be mediated by the frontal lobes has been observed in former National Football League (NFL) football players. Like Phinease Gage, although not as suddenly occurring, they were no longer "*themselves*," exhibiting personality changes, dementia, depression, and aggressive behavior.

BOX 9.1

A RECONSTRUCTION OF THE INJURY TO PHINEAS GAGE

He is fitful, irreverent, indulging at times in the grossest profanity (which was not previously his custom), manifesting but little deference for his fellows, impatient of restraint or advice when it conflicts with his desires.... A child in his intellectual capacity and manifestations, he has the animal passions of a strong man.... His mind was radically changed, so decidedly that his friends and acquaintances said he was "no longer Gage." *John Martyn Harlow, American Physician who attended to Phineas Gage.*

A reconstruction of Phineas Gage's railroad accident in 1848, when he was 25 years old. Notice likely damage to the orbitofrontal and medial frontal regions as well. Injuries like this create damage from swellings, bleeding, heat, infection, inflammation, and physical twisting of tissues that extend far beyond the immediate region of impact. Thus we do not really know the extent of brain damage in this classic neurological patient (Fig. 9.1).

BOX 9.1 *(cont'd)*

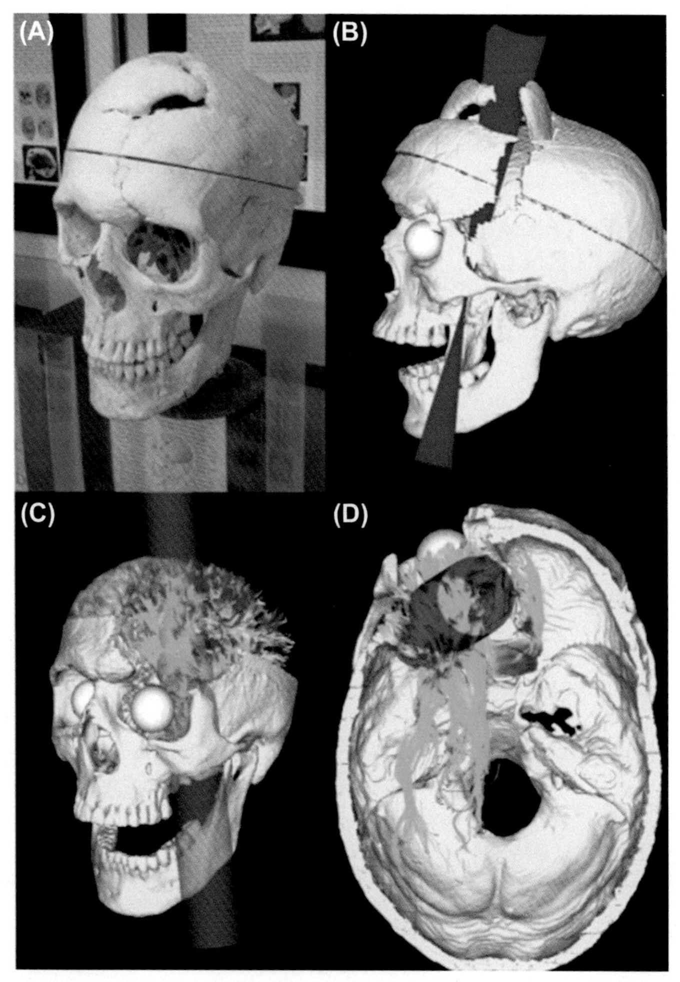

FIGURE 9.1 The reconstruction of the damage to Phineas Gage's head from the tamping iron. (A) The actual skull of Phineas Gage is on display at the Warren Anatomical Museum at Harvard Medical School. Note the damage done by the tamping iron at the top of the skull where the iron exited. (B) Computerized tomography (CT) scans were made of the skull and a reconstruction was made showing the hypothesized location of the tamping iron (shown in *red*) as it passed through the skull and brain. (C) A rendering of the tamping iron and the brain fibers that were likely damaged in the accident (shown in *green*). (D) A top view of the fibers that were likely damaged in the accident—in this view, the left hemisphere is shown on the left of the image, and the forehead and face areas are shown at the top of the image. *Source: van Horn et al. (2012).*

In these cases, however, a *tamping iron* was not to blame. It was the *gridiron*. Take a look at Box 9.2 to learn more.

The frontal lobes are a vast mosaic of cell types and cortical regions, diverse in their cell structures, anatomical features, and connectivity patterns. Unlike regions in the sensory cortex, the frontal lobes do not have a single job to do; they are not specialized for decoding

BOX 9.2

THE NATIONAL FOOTBALL LEAGUE VERSUS NEUROSCIENCE: THE CHRONIC TRAUMATIC ENCEPHALOPATHY CONTROVERSY

America loves football. *The most watched TV* event of all time in early 2016 was the 2015 Super Bowl, with more than 114 Million viewers (1). That is about one out of every three people in the United States (2)! Clearly many of those viewers tuned in to check out the new Super Bowl ads that have become as famous as the game itself…or to watch the Super Bowl Halftime show. But it is still a huge number of viewers. And…the *next* most watched TV event EVER was…Super Bowl 2014. And the third most watched TV event EVER was, well yes, another Super Bowl. America loves football. And the National Football League (NFL, Fig. 9.2) controls American football.

On Monday, March 14, 2016, something groundbreaking happened at the NFL. Nope—no announcement of a new expansion team or the retirement of a key quarterback…No new coaches hired or fired. Not even an announcement of the location of the next Super Bowl.

What happened on March 14 was groundbreaking because it was the first time that the NFL officially acknowledged the connection between football-related head injuries and neurodegenerative disease, specifically chronic traumatic encephalopathy (CTE). It happened at a roundtable discussion on concussions convened by the US House of Representatives Committee on

FIGURE 9.2 The National Football League (NFL) logo. *Source: Public domain.*

BOX 9.2 *(cont'd)*

Energy and Commerce. Jeff Miller, the NFL Senior Vice President for Health and Safety, was asked the key question: was there any relationship between football and CTE. Miller responded: "Well, certainly Dr. McKee's research shows that a number of retired NFL players were diagnosed with CTE, so the answer to that question is certainly yes". What was groundbreaking? And who is Dr. McKee?

It makes intuitive sense that a physical contact sport such as football would cause players to have head injuries that could be damaging to the brain. Especially head injuries such as a concussion. Why would they be denied by the very league that was supposed to sponsor the safe playing of our nation's favorite sport?

It turns out that the backstory is a bit complicated. Here is a brief version of it.

What Is Chronic Traumatic Encephalopathy?

CTE is a neurodegenerative disease that at present can only be diagnosed through autopsy. It was first noticed in boxers in the 1920s and was referred to as "punch drunk" syndrome (3). It became more formally described as CTE in the 1950s and 1960s (4). Clinical symptoms of CTE vary widely in breadth and severity, but typically include memory loss and dementia; mood and personality disorders, including aggression and depression and even suicide; and problems with executive functions such as problem-solving and decision-making. It is a progressive disease, and as the brain damage continues, the clinical symptoms worsen and worsen. And there is no cure.

What causes the wide array of brain damage? A key factor of CTE is the damage to tau proteins in the cortex due to repeated—even if mild—brain trauma. Healthy tau protein in the brain contributes to connectivity throughout the brain. In CTE, the tau protein causes it to become misshapen and clump together. These "clumps" of tau proteins kill healthy neurons. Why is it only diagnosed after death? Although the clinical symptoms of neurodegenerative disorder and, in particular, CTE are fairly well agreed upon, their neural bases are still being worked out. At the time this book is going to press, no current biomarkers or tests can diagnose CTE in a living individual despite the observance of the clinical symptoms. Indeed, even after death in patients with severe symptoms, the brain can look normal to the naked eye. But it is far from normal.

National Football League and Chronic Traumatic Encephalopathy

The key research on CTE in former football players was pioneered by Dr. Bennet Omalu, a neuropathologist in Pittsburgh. Dr. Omalu is a native of Nigeria and had little knowledge of the American sport of football. His life changed—and the future of the NFL as well—on a day in 2002 when Mike Webster's body was brought in to Dr. Omalu's lab. Webster had been a legendary Pittsburgh Steeler Center for 15 years. Nicknamed "Iron Mike," he was a nine-time Pro Bowler and a Hall of Famer. And now he was dead at the age of 50. Mike had been homeless and deeply disturbed emotionally before his death…a long fall from his glory days at the Steelers. What happened to him? Dr. Omalu was curious. Especially when his first

Continued

BOX 9.2 *(cont'd)*

observances of Mike's brain showed a normal-looking brain. Although it was clearly far from normal based on Mike's severe clinical symptoms.

What did he find? Dr. Omalu's careful study of the entire brain showed a diffuse distribution of neurofibrillary tangles and amyloid plaques. Tangles and plaques are found in patients with Alzheimer's disease (AD); however, this former football player had no symptoms of AD and the distribution of the tangles and plaques was distinctly different from those seen in AD. Omalu reported his findings as a case study published in 2005 in the journal Neurosurgery (6). He was direct and to the point: in the opening of the article, he states "This case draws attention to the need for further studies in the cohort of retired National Football League players to elucidate the neuropathological sequelae of repeated mild traumatic brain injury in professional football."

In Fig. 9.3, the brain of a deceased former NFL player with CTE is shown on the left. It looks normal; however, the tau-stained section shown on the right reveals the tangles. These tangles were found throughout the brain.

Omalu's case study did not go unnoticed—especially at the NFL—however, the response was far from what he anticipated. Instead of embracing the findings and the need for a concerted effort to determine the root causes of CTE, the NFL pushed back with a strong effort to have the article retracted. Additionally, the chairman of the Mild Traumatic Brain Injury Committee of the NFL published two papers rejecting the hypothesis put forward by Omalu that head injury causes CTE (7, 8).

For more about the story of Omalu versus the NFL, see the article published in Gentleman's Quarterly (GQ) in 2009 (9). And you might want to check out the 2015 film "Concussion" starring Will Smith.

Through the pioneering and ongoing research of Dr. Omalu, coupled later with the groundbreaking research conducted by

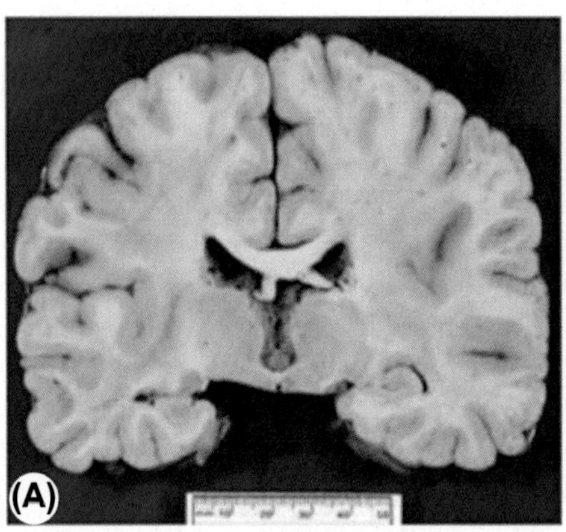

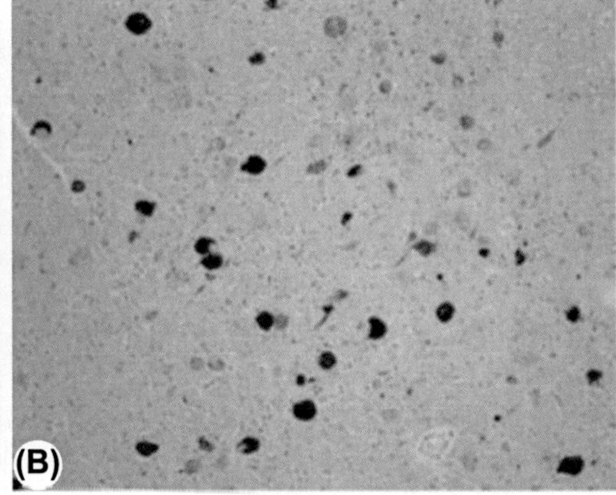

FIGURE 9.3 (A) A coronal section of the brain of a former National Football League player who was diagnosed with chronic traumatic encephalopathy. The brain looks normal. (B) A photomicrograph of a brainstem nucleus that shows the tangles distributed in that brain region. Tangles were found throughout the brain of this former player. *Source: Omalu (2014).*

BOX 9.2 *(cont'd)*

Dr. Ann McKee et al. at the Center for the Study of Traumatic Encephalopathy in Boston, we now know a lot more about how the brain is affected by CTE. The central cause of CTE is not one or even a few major concussions—although those events clearly can cause brain damage. Rather, the central cause is consistent mild brain trauma through small- to medium-level hits to the head. Think about a lineman—in today's football, a lineman may have his head/helmet hit in almost every play of the game. Consistent, repetitive mild brain injury causes the abnormal tau proteins, which begin to become distorted...and then kill healthy neurons and eventually cause severe brain damage. And there is no way for the players to know if this is happening to their brain during their lifetime.

Despite the publications of case studies of former NFL players by Omalu in 2005 and 2006, the NFL did not take on the issue of CTE. Nor did they acknowledge it as a problem for current and future players. Despite the significant list of hundreds of former NFL players, boxers, wrestlers, rugby players, hockey players, soldiers, and others diagnosed in postmortem by Dr. McKee et al., the NFL did not formally acknowledge the problems raised by CTE.

Until March 14, 2016.

What Happened Next?

Since that acknowledgment, much has changed in the NFL. And much has stayed the same. More and more young NFL players are opting to retire early. The link between CTE and head injury sustained during repeated play of football is just too clear to ignore. And many players are playing on.

A new study from Dr. McKee was published in July of 2017 (10). In that study, Dr. McKee et al. examined the brains of 111 former NFL players whose families had donated their deceased family member's brain to them. Of the 111 former players, 110 were found to have CTE. Virtually every player.

Which football positions were most at risk? Of the 110 players, 44 were linemen. Linemen typically have hits to the head on most every football play they line up for throughout their career. Of the 110 players, 20 were running backs. Again, they were likely tackled on almost every play. Seventeen were defensive backs. Thirteen were linebackers. You see the trend here: Of the 110 brains showing CTE, 94 were from linemen and backs. Only one placekicker and one punter were included in the sample and both had CTE. These players are rarely tackled at the professional level (but we do not know what positions they played as children, high school, or college athletes). And yet, both showed the brain damage. It makes one wonder, are any players *not at risk*?

It is important to note that Dr. McKee's sample was a *convenience sample* of NFL players. They and/or their families volunteered to donate their brains to the project, presumably because the players had exhibited symptoms of behavior similar to those that had been observed in previous cases where CTE was later discovered in pathological examination. It was *not a random sample*. To establish a clear link between playing in the NFL—or any sport—and brain

Continued

BOX 9.2 *(cont'd)*

damage due to CTE, the science has to be far more rigorous. To determine if the head trauma experienced during football play *caused* CTE, the studies have to go far beyond the *correlational* finding in this new study. That is to say, there was an *extremely high co-occurrence* of *NFL play* and *CTE,* but there may be other factors beyond football play that caused the development of CTE. To determine this, random samples of football players, other athletes, and nonathletes need to be done to answer questions such as: can you acquire CTE as a nonathlete? Is there some other factor that coassociates with playing football that is actually the root cause of the development of CTE? Future studies will shed light on these questions and their answers. For now, it is enough to know that there is a very high correlation between football and other sport play, repeated head trauma, and the development of CTE.

How Can Chronic Traumatic Encephalopathy Be Prevented?

What is the secret to preventing the agony of CTE? Is it a better helmet…or does it mean the end of football as we know it?

New helmet technology is changing the game, literally. Many are addressing the key issue in TBI. TBI—even if mild—caused by frequent head hits causes the brain, floating in its cerebral fluid, to sway and torque, twisting, and hitting the rock-hard skull. This is the cause of the injury and until we can somehow prevent the force of a head hit to be dampened so that the brain does not hit against the skull, then we will continue to see neurodegenerative disorder due to brain damage. Helmets designed to absorb the force of the hit—so that skull and brain do not—are hopefully providing the pathway to making our favorite sport safe for our players.

However, the risk of brain damage may change football—and other contact sports—forever. Evidence of this is already being provided by the growing ranks of young NFL players opting to retire early to avoid future cognitive decline.

What would you do?

References

(1) Casson, I. R., Pellman, E. J., Viano, D. C. (2006). Chronic traumatic encephalopathy in a National Football League player. *Neurosurgery, 58*, Comment.

(2) asson, I. R., Pellman, E. J., Viano, D. C. (2006). Chronic traumatic encephalopathy in a National Football League player. *Neurosurgery, 59*, E1152. Comment.

(3) Corsellis, J. A., Bruton, C. J., Freeman-Browne, D. (August 1973). The aftermath of boxing. *Psychol Med, 3*(3), 270–303. PMID: 4729191.

(4) Critchley, M. (1949). Punch-drunk syndromes: the chronic traumatic encephalopathy of boxers. In *Hommage a Clovis Vincent*. Paris: Maloine.

(5) Laskas, J. M. (2009). Bennet Omalu, Concussions, and the NFL: How one doctor changed football forever. GQ: http://www.gq.com/story/nfl-players-brain-dementia-study-memory-concussions.

(6) Mez, J., Daneshvar, D. H., Kiernan, P. T., Abdolmohammadi, B., Alvarez, V. E., Huber, B. R., et al. (2017). Clinicopathological evaluation of chronic traumatic encephalography in players of American football. *Journal of the American Medical Association, 318*(4), 360–370.

BOX 9.2 *(cont'd)*

(7) Omalu, B. (2014). Chronic traumatic encephalopathy. In Niranjan, A., Lunsford, L. D. (Eds.), *Concussion, Prog Neurol Surg* (Vol. 28) (pp. 38–49). Basel: Karger.

(8) Omalu, B. I., DeKosky, S. T., Minster, R. L., Kamboh, M. I., Hamilton, R. L., Wecht, C. H. (July 2005). Chronic traumatic encephalopathy in a National Football League player. *Neurosurgery, 57*(1), 128–134 (discussion 128–134).

(9) Super Bowl TV Ratings Source. Retrieved from: http://tvbythenumbers.zap2it.com/2015/02/02/super-bowl-xlix-is-most-watched-show-in-u-s-television-history/358523/.

(10) US Census Bureau. Retrieved from: http://www.census.gov/topics/population.html.

speech sounds or recognizing faces. Rather, the frontal lobes are engaged in almost all aspects of human cognitive function.

In this chapter, we present the results of neuroimaging studies that have provided new data on where, what, and how specific regions in the frontal lobes are activated during cognitive tasks. Next, we look at frontal lobe syndromes that are observed in patients with damage to the frontal lobe and connected regions. Finally, we examine how the results of neuroimaging and patient studies combine to inform us about the role of the frontal lobes in human behavior.

1.2 From the Silent Lobes to the Organ of Civilization

It took scientists many years to begin to appreciate the importance of the frontal lobes for cognition. But when this finally happened, a picture of particular complexity and elegance emerged. The frontal lobes used to be known as "the silent lobes" because they are not easily linked to any single, easily defined function. Now it is known that the frontal lobes play a key role in almost every aspect of human behavior, including decision-making, social and personality behaviors, and planning strategies. The frontal lobes are not silent after all; they are the organ of civilization!

If the frontal lobes are the "organ of civilization," then what exactly is their function? What is their "civilizing" effect? The functions of the frontal lobes defy a simple definition. They are not invested with any single, ready-to-label function. A patient with frontal lobe damage will typically retain the ability to move around, use language, recognize objects, and even memorize information.

For the most part, this chapter focuses on the prefrontal cortex (PFC), the most anterior part of the frontal lobes, in front of the motor areas. There are many subregions in the PFC, but four regions most typically are identified when assessing their functional role in cognition: *dorsolateral PFC* (DLPFC), *ventrolateral PFC* (VLPFC), *anterior PFC* (APFC), and *medial PFC* (MPFC) (Fig. 9.4). PFC is located in front of the primary motor cortex, sometimes called the motor strip.

PFC plays the central role in forming goals and objectives and then in devising plans of action required to attain those goals. It selects the cognitive skills needed to implement the plans, coordinates these skills, and applies them in a correct order. Finally, the PFC is responsible for evaluating our actions as success or failure relative to our intentions. See Table 9.1 for other common prefrontal functions.

1.3 "Memory of the Future"

David Ingvar (1985) coined the phrase "memory of the future." Ingvar was referring to one of the most important functions of advanced organisms: making plans and then following the plans to guide behavior. Unlike primitive organisms, humans are active, rather than reactive, beings. The transition from mostly reactive to mostly proactive behavior is among the central themes of the evolution of the nervous system. We are able to form goals, our visions of the future. Then we act according to our goals. But to guide our behavior in a sustained fashion, these mental images of the future must become the content of our memory, thus the "memories of the future" are formed.

Human cognition is forward-looking, *proactive* rather than *reactive*. It is driven by goals, plans, hopes, ambitions, and dreams, all of which pertain to the future and not to the past. These cognitive powers depend on the frontal lobes and evolve with them. The frontal lobes endow the organism with the ability to create neural models as a prerequisite for making

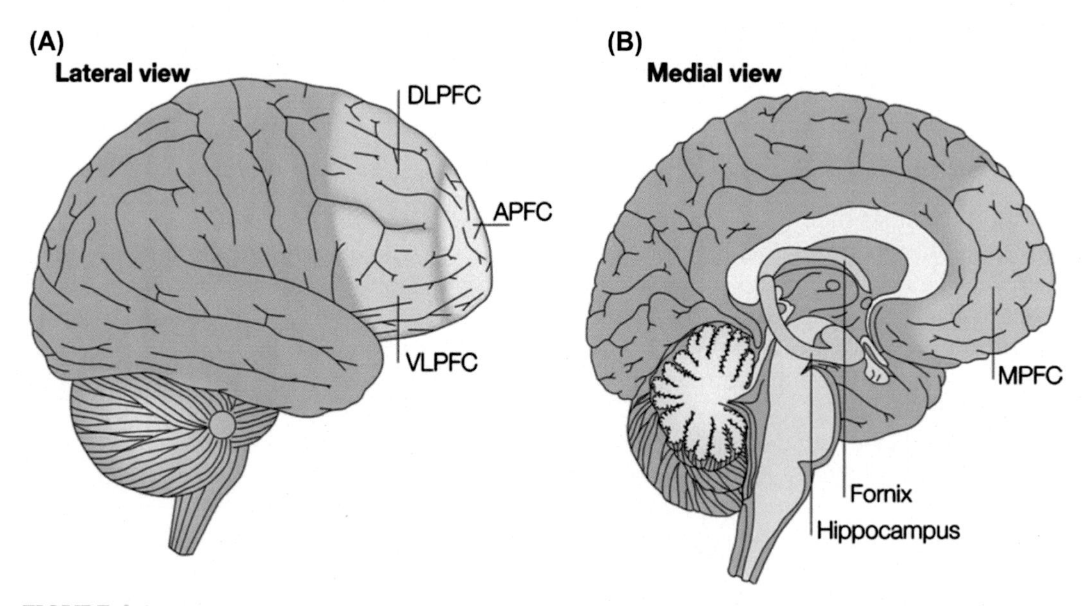

FIGURE 9.4 The major divisions of the prefrontal cortex (PFC) shown from a lateral sagittal view (A) and a medial sagittal view (B) of the brain. The PFC can be divided into lateral (side), medial (midline), ventral (bottom), and frontal regions. The lateral division divides into upper (dorsal) and lower (ventral) halves, separated by a major horizontal fold, the inferior lateral sulcus. *Source: Simons and Spiers (2003).*

TABLE 9.1 Some Common Prefrontal Functions

1. Planning, setting goals, and initiating action
2. Monitoring outcomes and adapting to errors
3. Mental effort in pursuing difficult goals
4. Interacting with other regions in pursuit of goals (basal ganglia, thalamic nuclei, cerebellum, motor cortex)
5. Having motivation, being willing to engage in action
6. Initiating speech and visual imagery
7. Recognizing other people's goals, engaging in social cooperation and competition
8. Regulating emotional impulses
9. Feeling emotions
10. Storing and updating working memory
11. Active thinking
12. Enabling conscious experiences
13. Sustained attention in the face of distraction
14. Decision-making, switching attention, and changing strategies
15. Planning and sequencing actions
16. Unifying the sound, syntax, and meaning of language
17. Resolving competition between plans

things happen, models of something that, as of yet, does not exist but that you *want* to bring into existence.

To makes plans for the future, the brain must have an ability to take certain elements of prior experiences and reconfigure them in a way that does not copy any actual past experience or present reality exactly. To accomplish that, the organism must go beyond the mere ability to *form* internal representations, the models of the world outside. It must acquire the ability to *manipulate and transform* these models. We can argue that tool making, one of the fundamental distinguishing features of primate cognition, depends on this ability, since a tool does not exist in a ready-made form in the natural environment and has to be *imagined* to be made. The neural machinery for creating and holding "images of the future" was a *necessary prerequisite* for tool making and thus for launching human civilization.

We can also argue that the generative power of language to create new ideas depends on this ability as well. The ability to manipulate and recombine internal representations depends critically on the PFC, which probably made it critical for the development of language.

1.4 Self-awareness and Executive Function

Goal formation is about "I need" and not about "it is." Therefore the ability to formulate goals must have been inexorably linked to the emergence of the mental representation of self.

It should come as no surprise that *self-awareness* is also intricately linked to the frontal lobes. All these functions can be thought of as metacognitive rather than cognitive, since they do not refer to any particular mental skill but provide an overarching organization for all of them. For this reason, some authors refer to the functions of the frontal lobes as *executive functions*, by analogy with a governmental or corporate executive.

1.5 Frontal Lobe Development

If it seems to you that the role the frontal lobes play in cognition seems uniquely human, you are right! The vast expansion of the frontal lobes during evolution and their maturational path during the lifetime in humans are unique among living creatures.

In evolution, the frontal lobes accelerated in size only with the great apes. These regions of the cortex underwent an explosive expansion at the late stage of evolution. According to Brodmann (1909), the PFC or its analogs account for 29% of the total cortex in humans, 17% in the chimpanzee, 11.5% in the gibbon and the macaque, 8.5% in the lemur, 7% in the dog, and 3.5% in the cat (Fig. 9.5). While whales and dolphins have large brains, it is the *parietal* rather than the frontal cortex that has expanded in these aquatic mammals.

As the seat of goals, foresight, and planning, the frontal lobes are perhaps the most uniquely human of all the components of the human brain. In 1928, the neurologist Tilney suggested that all human evolution should be considered the "age of the frontal lobe," but scientific interest in the PFC was late in coming. Only gradually did it begin to reveal its secrets to the great scientists and clinicians like Hughlings Jackson (1884/2011) and Alexander Luria (1966), and in the last few decades to researchers like Antonio Damasio (1995), Joaquin Fuster (1997), Patricia Goldman-Rakic (1987), Donald Stuss and Frank Benson (1986), and others.

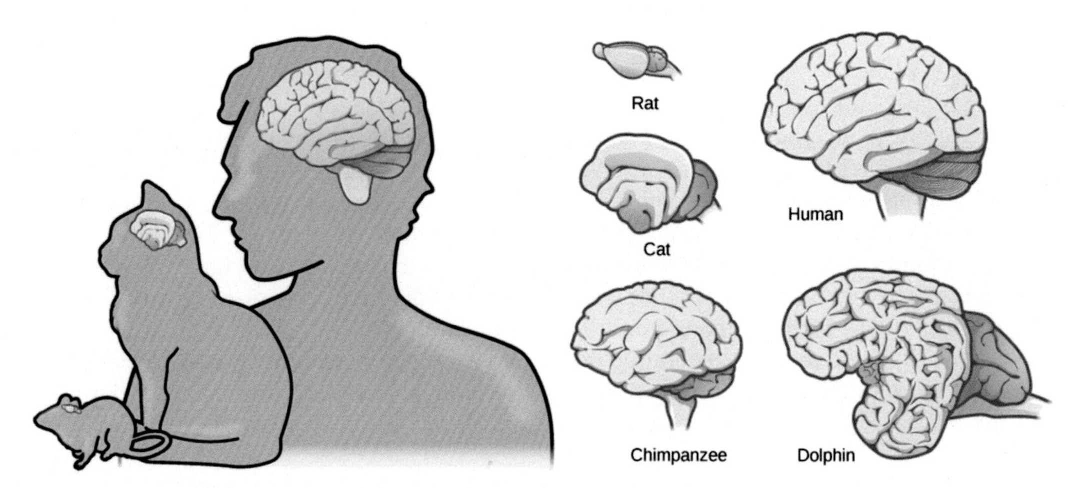

FIGURE 9.5 The prefrontal cortex (PFC) expands over mammalian and primate evolution. A greatly enlarged PFC is a distinctively human and primate feature. Other large-brained mammals like whales and dolphins have expanded parietal rather than prefrontal regions. Bottom right, a human brain, with a chimp brain on the bottom left. *Source: https://commons.wikimedia.org/.*

2. STRUCTURE OF THE FRONTAL LOBES

As we have mentioned before, the neuroanatomy and neurophysiology of the frontal lobes reflect their diverse and complex functions, with many distinct regions that differ sharply in their cellular and anatomical structure. Adding another layer of complexity is their high degree of connectivity across many brain regions. In this section, we highlight the main subdivisions of the frontal lobes and show the amazing neural highways that connect them to the rest of the brain.

2.1 Neuroanatomy and Neurophysiology of the Frontal Lobes

The boundaries of the frontal lobes typically are described using gross anatomical structures and Brodmann areas to define boundaries of frontal lobe regions. A simple way to think of the location of the frontal lobes is that they lie *in front of* the central sulcus that separates them from the parietal lobes and *above* the Sylvian fissure that separates them from the temporal lobes.

2.2 How Prefrontal Cortex Is Defined

A more precise definition of PFC can be accomplished by using *Brodmann area* maps (Brodmann, 1909). Brodmann areas are based on the types of neurons and connections that typically are found there within. According to this definition, the PFC consists of Brodmann areas 8, 9, 10, 11, 12, 13, 44, 45, 46, and 47 (Fig. 9.6) (Fuster, 1997). These areas are characterized by the predominance of the so-called granular neural cells found mostly in layer IV (Campbell, 1905, in Fuster, 1997).

Another method of outlining the PFC is through its subcortical projections. Fuster (2008) has hypothesized a perception–action cycle including PFC and posterior regions of the brain and subcortical nuclei including the thalamus (Fig. 9.7). Note that PFC has *bidirectional* connections between many subcortical regions.

Finally, the PFC sometimes is delineated through its biochemical pathways. According to that definition, the PFC is defined as the area receiving projections from the mesocortical dopamine system. These various methods of delineating the PFC outline roughly similar territories.

2.3 The Vast Connective Highways of the Frontal Lobes

The PFC is connected directly with every distinct functional unit of the brain (Nauta, 1972). It is connected to the highest levels of perceptual integration, and also with the *premotor cortex*, *basal ganglia*, and the *cerebellum*, all involved in aspects of motor control and movements. PFC also is connected with the *dorsomedial thalamic nucleus*, often considered to be the highest level of integration within the thalamus; with the *hippocampi* and *medial temporal structures*, known to be critical for memory; and with the *cingulate cortex*, believed to be critical for emotion and dealing with uncertainty. In addition, PFC connects with the *amygdala*, which regulates most emotions and social cognition, and with the *hypothalamus*, which is in charge

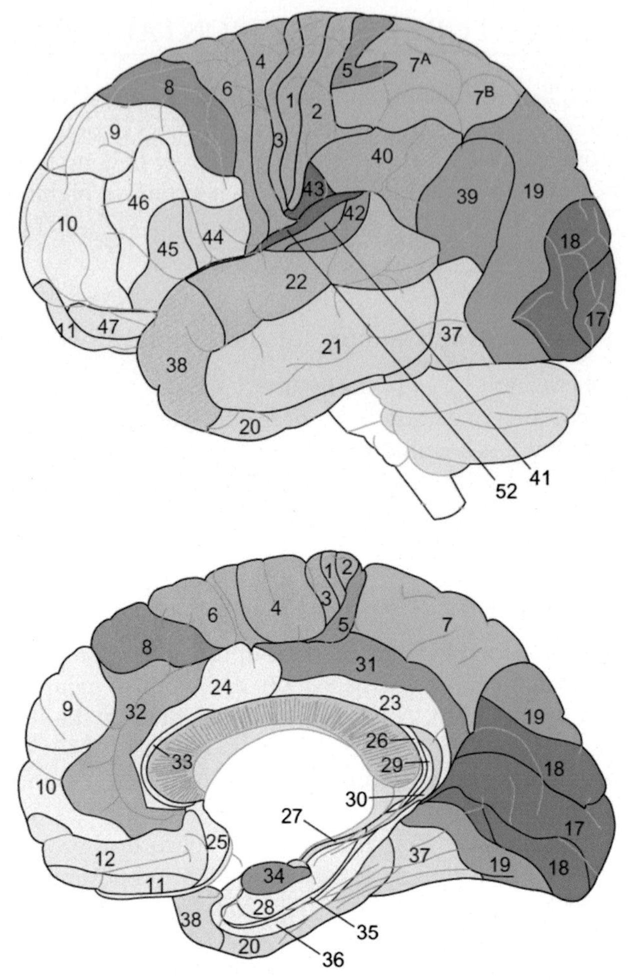

FIGURE 9.6 Brodmann areas in the frontal lobes. Areas forward of the motor cortex are considered to be prefrontal (Brodmann areas 4 and 6 are motor and premotor regions.) However, the boundary is not rigid. It is often useful to think of a gradual transition between more "cognitive" areas and primary motor cortex (BA 4), which directly controls voluntary muscles. *Source: Adapted by Bernard J. Baars from M. Dubin, with permission; drawn by Shawn Fu.*

of control over the vital homeostatic functions of the body. Finally, PFC is also well connected with the *brainstem* nuclei involved in wakefulness, arousal, and overall alertness, regulation of sleep and rapid eye movement dreams.

This unique connectivity makes the frontal lobes singularly suited for coordinating and integrating the work of other brain structures (Fig. 9.8). This extreme connectivity also puts the frontal lobes at a particular risk for disease. Some scientists believe that the PFC contains a map of the whole cortex, an assertion first made by Hughlings Jackson (1884/2011) at the end of the nineteenth century. This hypothesis asserts that prefrontal regions are needed for normal consciousness. Since any aspect of our mental world

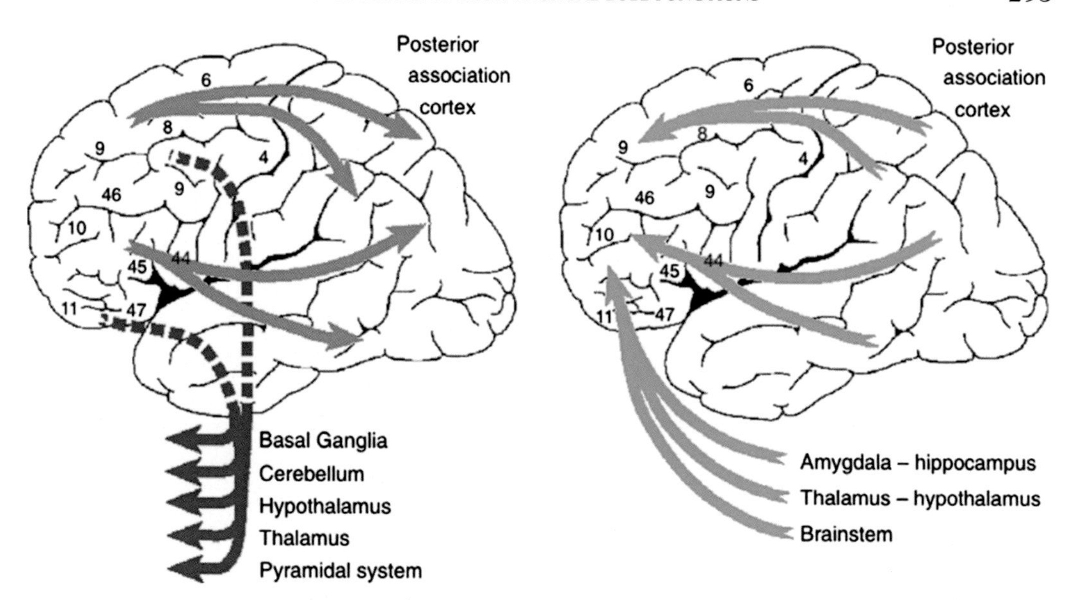

FIGURE 9.7 A schematic diagram of connections *from* (shown in *red*) the prefrontal cortex (PFC) to posterior brain regions and to subcortical nuclei. Connections to the PFC are shown in *blue*. *Source: Fuster (2008).*

may, in principle, be the focus of our consciousness, it stands to reason that an area of convergence of all its neural substrates must exist. This leads to the provocative proposition that the evolution of consciousness, the highest expression of the developed brain, parallels the evolution of the PFC.

3. A CLOSER LOOK AT FRONTAL LOBE FUNCTIONS

3.1 Executive Functions

The concept of executive functions is inextricably linked to the function of the frontal lobes. The groundwork for defining the executive functions system was laid by Alexander Luria (1966) as early as 1966. At the time, Luria proposed the existence of a system in charge of intentionality, the formulation of goals, the plans of action subordinate to the goals, the identification of goal-appropriate cognitive routines, the sequential access to these routines, the temporally ordered transition from one routine to another, and the editorial evaluation of the outcome of our actions.

Subsequently, two broad types of cognitive operations linked to the executive system figured most prominently in the literature:

1. One's ability to guide one's behavior by the formulation of plans, and then guiding behavior according to these plans
2. One's ability not only to guide one's behavior but also having the capacity to "switch gears" when something unexpected happens

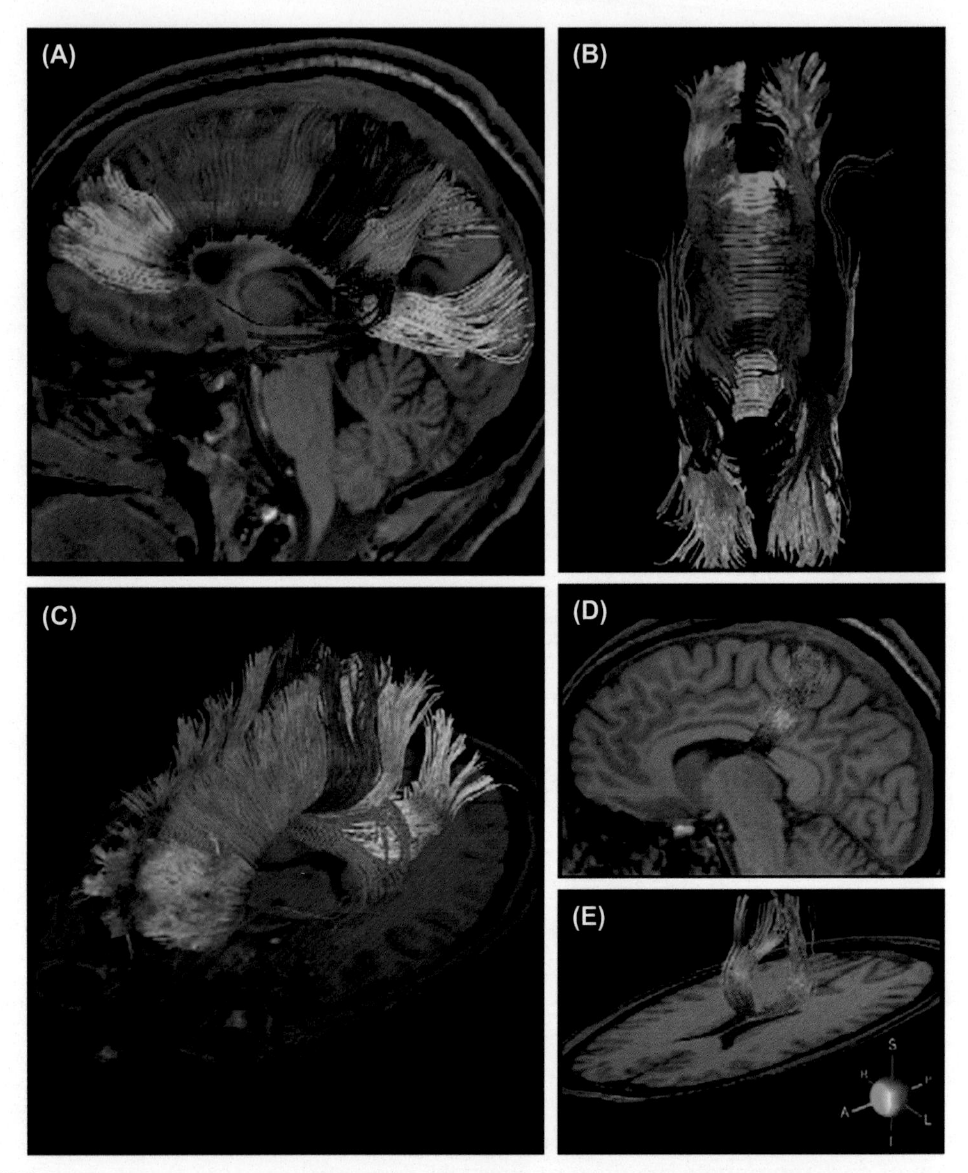

FIGURE 9.8 Fiber bundles in the brain are shown in midsagittal, horizontal, and coronal sections. (A) Centerline of the brain (midsagittal) view of the fiber bundles extending into the prefrontal cortex, shown in *green*, and into the premotor and motor cortex, shown in *blue*. (B) A horizontal slice (axial view) of the same fiber bundles. (C) A coronal slice showing the same *green and blue* fibers at the front of the brain. (D) and (E) show sagittal (D) and oblique (E) views of callosal fiber tracts that project into the primary motor cortex. The color coding reflects the diffusion direction as indicated in the lower right corner, with *A*, anterior; *I*, inferior; *L*, left; *P*, posterior; *R*, right; *S*, superior. *Source: Hofer and Frahm (2006).*

To deal effectively with such transitions, a particular ability is needed—*mental flexibility*—which is the capacity to respond rapidly to unanticipated environmental contingencies. Sometimes this is referred to as an ability to shift *cognitive set*. Additionally, the executive system is critical for planning and the generative processes (Goldberg, 2001).

Fuster (1997) enlarged on the premise originally developed by Luria by suggesting that the so-called executive systems can be considered functionally homogeneous in the sense that they are in charge of actions, both external and internal (such as logical reasoning). In general, the executive functions are not unique to humans. However, the uniqueness of the human executive functions is in the *extent* to which they are capable of integrating such factors as time, informational novelty, and complexity, and possibly ambiguity.

3.2 Social Maturity and Moral Development

While many neuroimaging and behavioral studies have investigated attention, working memory, and executive control processes in the PFC, the frontal lobes also play a critical role in the development of social cognition—a key link to the role of the frontal lobes as the "organ of civilization." The capacity for volitional control over one's actions is not innate, but it emerges gradually through development and is an important, perhaps central, ingredient of *social maturity*.

3.2.1 *Early Life Experience and Orbitofrontal Cortex Development*

Parent–infant interactions during the first months of life are thought to be key to the normal development of the orbitofrontal cortex (Figs. 9.9 and 9.10) (Schore, 1999). By contrast, early-life stressful experiences permanently damage the orbitofrontal cortex, predisposing the individual to later-life psychiatric diseases. This implies that early social interactions help shape the brain. One critical element in this learning is eye gaze. When the infant and parent have mutual eye gaze, they establish the neural underpinnings of understanding another's intentionality and empathy that are key to social development (see Chapter 13, Disorders of Consciousness, and Chapter 14, Growing Up).

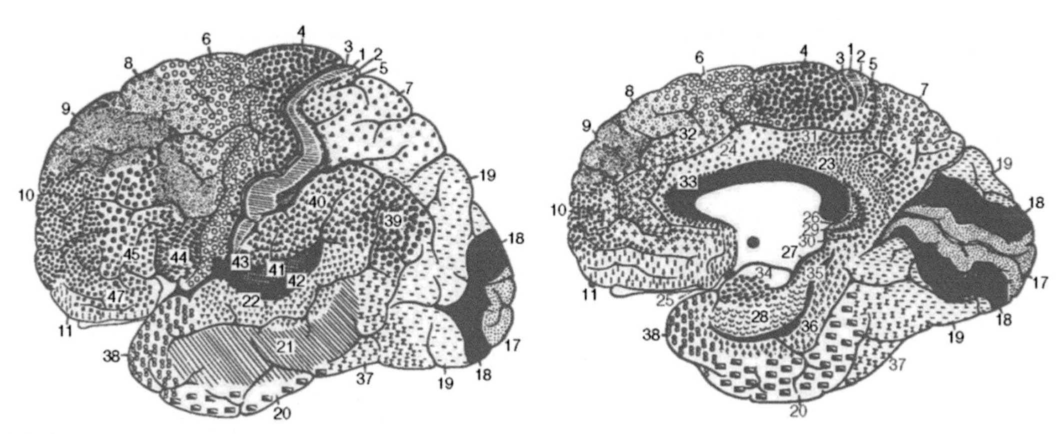

FIGURE 9.9 Cytoarchitechtonic maps of the brain—areas 10, 11, and 47 comprise the orbitofrontal cortex. *Source: Kringelbach (2005).*

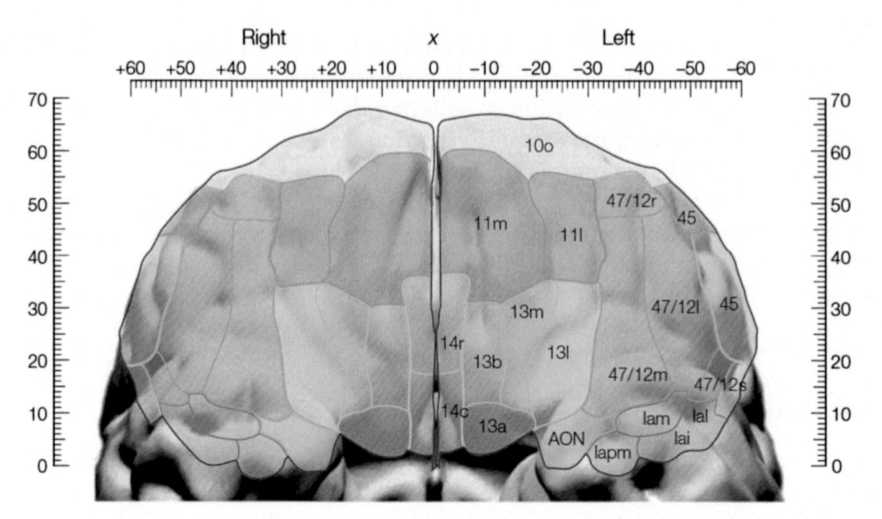

FIGURE 9.10 Another view of the orbitofrontal cortex looking from the front of the brain. *Source: Kringelbach (2005).*

The orbitofrontal cortex is perhaps the least well-understood region in the PFC. Recent neuroimaging advances are providing new data for understanding this region's role in human behavior and cognition, but its role is still being elucidated. A current view of the role of the orbitofrontal cortex in adults is that it plays a key role in emotional and reward-related behaviors such as making predictions, risk assessment, and decision-making (Kringelbach, 2005). These emotional and reward-related behaviors are frequently associated with aspects of human behavior that involve risk, such as gambling, illegal drug use, and thrill-seeking activities. According to our current knowledge, the orbitofrontal cortex is a very slow-to-mature region of the brain, with maturation continuing throughout adolescence (Fig. 9.11). The late-to-mature aspect of the orbitofrontal cortex may be associated with risk-taking behaviors observed during the teenage years (Galvan et al., 2006).

3.2.2 Moral Development and the Frontal Cortex

Furthermore, following this logic, is it possible that moral development involves the frontal cortex, just as visual development involves the occipital cortex and language development involves the temporal cortex? The PFC is the association cortex of the frontal lobes, the "action lobes." The posterior association cortex encodes generic information about the outside world. It contains the taxonomy of the various things known to exist and helps recognize a particular exemplar as a member of a known category. By analogy, the PFC may contain the taxonomy of all the *sanctioned moral actions and behaviors*. And could it be that, just as damage or maldevelopment of the posterior association cortex produces *object agnosias*, so does damage or maldevelopment of the PFC produce, in some sense, *moral agnosia*?

A report by Damasio et al. lends some support to this idea (Anderson, Bechara, Damasio, Tranel, & Damasio, 1999). Damasio studied two young adults, a man and a woman, who suffered damage to the frontal lobes very early in life. Both engaged in antisocial behaviors: lying, petty thievery, truancy. Damasio claims that not only did these patients fail to act according to the proper, socially sanctioned moral precepts but they even failed to recognize them as morally wrong.

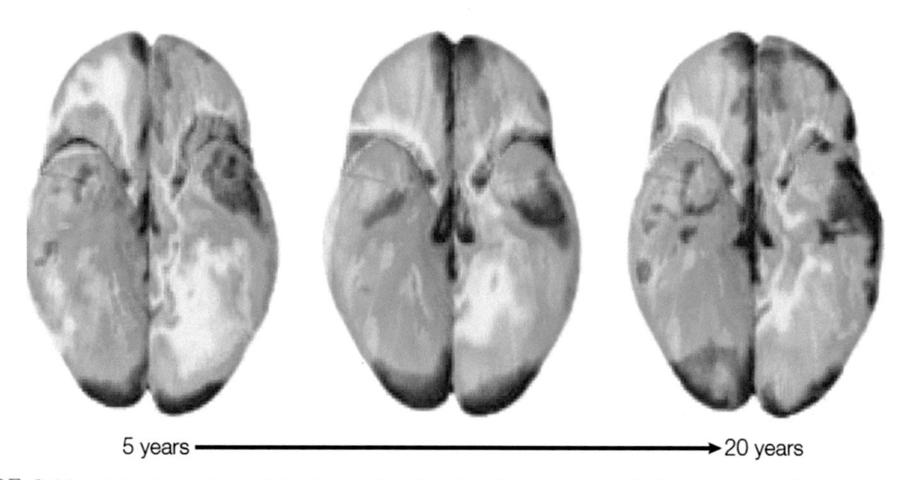

FIGURE 9.11 A bottom view of the brain showing the changes in cortical gray matter from young children (5 years, left brain figure) as compared to young adults (20 years, right brain figure). The color shows gray matter volume from low (*blue*) to high (*pink*). The orbitofrontal cortex is later to mature than neighboring regions. *Source: Kringelbach (2005).*

The orbitofrontal cortex is not the only part of the frontal lobes linked to socially mature behavior. The *anterior cingulate cortex* (ACC) occupies a midfrontal position and is closely linked to the PFC. The ACC traditionally has been linked to emotion. According to Michael Posner, it also plays a role in social development by regulating distress (Posner & Rothbart, 1998).

3.2.3 *Age of Maturity and Frontal Lobe Development*

The implicit definition of social maturity changes throughout the history of society, and so does the time of "coming of age." In modern Western societies, the age of 18 (or thereabouts) has been codified in the law as the age of social maturity. This is the age when a person can vote and is held responsible for his or her actions as an adult. The age of 18 is also the age when the maturation of the frontal lobes is relatively complete. Various estimates can be used to measure the course of maturation of various brain structures. Among the most commonly used such measures is pathway myelinization (Yakovlev & Lecours, 1967). The frontal lobes cannot fully assume this role until the pathways connecting the frontal lobes with the far-flung structures of the brain are fully myelinated.

The agreement between the age of relatively complete maturation of the frontal lobes and the age of social maturity is probably more than coincidental. Without the explicit benefit of neuroscience, but through cumulative everyday common sense, society recognizes that an individual assumes adequate control over his or her impulses, drives, and desires only by a certain age. Until that age, an individual cannot be held fully responsible for his or her actions in either a legal or moral sense. This ability appears to depend critically on the maturity and functional integrity of the frontal lobes.

4. NEUROIMAGING THE EXECUTIVE BRAIN

The two broad types of cognitive operations linked to the executive systems in the frontal lobe have been extensively investigated using functional neuroimaging techniques such as

PET and functional magnetic resonance imaging (fMRI). We summarize some general findings here, separating the results into three sections: *attention and perception*, *working memory*, and *executive function and motor control*. But first a word of caution about functional neuroimaging studies of complex processes: As we have described, the PFC is intricately involved in many cognitive and executive processes such as paying attention, holding something in mind for a few moments, switching attention when needed, and making decisions. Thus *any task* used in a neuroimaging study will necessarily involve these complex and overlapping processes in the frontal lobe and elsewhere in the brain. In fact, just participating in the study engages the subject—and their frontal lobes—in complicated ways (for a good discussion of these issues, see Fuster, 2008).

To disentangle the many processes engaged in any task, investigators studying frontal lobe function need to carefully design their studies so they can identify which processes are specifically related to, for example, attention versus working memory, which are aspects of almost any task. In the following section, we present a brief summary of many neuroimaging studies that have looked at these prefrontal processes.

4.1 Attention and Perception

Imagine that you are a subject in an fMRI study. Your task is to look at a screen, and when you see a picture of a male face, you are to press one button, and if you see a picture of a female face, you are to press another button. Easy, right? What parts of the brain will be activated by this task? You might suggest that the visual cortex will be busy, based on your knowledge of sensory activity in the brain. You might also suggest that the fusiform face area (see Chapter 4, The Art of Seeing) will "light up" for this task. And you would be correct. But there are other aspects of this task that will activate the brain: paying attention to the screen, making a decision about whether the face is male or female, preparing to press a button, and then actually pressing the button—all key aspects of frontal lobe function. In fact, this is a central finding in neuroimaging studies of visual or auditory perception; many areas in the brain outside of the sensory cortex are activated.

But first, a word about attention. In this chapter about frontal lobe functions, we will be discussing *voluntary attention*. This is the aspect of attention in which we are in control of what we decide to pay attention to. When reading this sentence, for example, you are deciding to focus on the words and their meaning and not on the font type or size, or the sound of the clock ticking at your elbow. As we described in Chapter 8, Attention and Consciousness, voluntary attention often is called *executive attention* or *top-down attention* to indicate that it is the class of attentional processes that are under our control. Another type of attention is involuntary; these are automatic, frequently stimulus-induced processes whereby our attention is "grabbed" by something in our environment. For example, the ringing of your cell phone, the sudden bright light when the sun comes from behind a cloud, the aroma of coffee brewing all may temporarily attract your attention without any conscious effort on your part. Automatic attentional processes can serve to disrupt our voluntary attention, but both are critical for being able to plan and initiate behavior—whether that behavior is finishing an essay for class in the morning or jumping out of the way when a car horn sounds behind you. The push and pull of voluntary and involuntary attentional processes are key functions of the frontal lobes.

An early investigation used PET to show localization patterns for sustained voluntary attention and provided important new information about prefrontal regions that subserved voluntary attentional processes (Pardo, Pardo, Janer, & Raichle, 1990). At about the same time, seminal work on the voluntary attentional networks involved in perceptual tasks was provided by Michael Posner and Steve Petersen (1990), where they detailed an anterior attentional system, which they described as having three major separable networks that perform the alerting, orienting, and executive (conflict resolution) voluntary attentional functions that are important to many tasks. Over the years, Posner et al. developed an attention network test (ANT) that provided them with ways to separately measure and record brain regions that subserve these voluntary attentional networks. The anterior attentional system they describe has three key networks: *alerting*, for maintaining an alert state, such as being ready to look at visual images in the previous scanner example; *orienting*, such as preparing to see a new visual image of a face that appears in the scanner; and *executive*, such as deciding whether the face is male or female.

Brain areas that subserve alerting, orienting, and executive control of attention have been investigated by Fan et al. (Fan, McCandliss, Fossella, Flombaum, & Posner, 2005; Xuan et al., 2016). Xuan et al. (2016) used a revised ANT (shown in Fig. 9.12) and fMRI to investigate the neural bases of these aspects of voluntary attention as *interactive attentional networks*. Results showed activation of an extended frontoparietal network (FPN), including the thalamus, dorsolateral prefrontal cortex (DLPFC), the ACC, and the anterior insular cortex (AIC) (Fig. 9.13).

Specific results showed the *alerting* function activating the FPN and the locus coeruleus (Fig. 9.14). The *orienting* function activated frontal eye fields and the superior colliculus (Fig. 9.15). The *executive control* function activated the FPN and the cerebellum (Fig. 9.16). Simply put, the FPN and the thalamus and other subcortical regions and bodies form a cortico-subcortical interactive network that provides a dynamic and highly flexible human attentional network.

Another central finding from these and other studies of anterior voluntary attention networks is that the *level of activity* in the PFC corresponds to the *level of attention* required by the task, with more activity for tasks with higher attentional demands (Pardo et al., 1990; Posner, Petersen, Fox, & Raichle, 1988). A part of the prefrontal region that typically is activated by tasks requiring focused attention is the anterior cingulate gyrus in the ACC (Posner et al., 1988; Raichle et al., 1994).

4.2 Working Memory

The ability to keep something in mind for a limited amount of time is a central function in cognition. This ability—working memory—is closely associated with voluntary attentional systems. In fact, one way of describing working memory is that it serves as an *inward directed voluntary attention* system, directing attention to internal representations (Fuster, 2008). A central issue in investigating brain areas that subserve working memory is, how do you study working memory and separate findings from other executive processes such as attention and decision-making? To accomplish this, D'Esposito et al. (1995) developed a dual task paradigm using two tasks that, individually, did not have working memory demands. Together, however, they did produce working memory demands, and in this way, D'Esposito et al. were able to isolate processes that were specific to working memory function and not to general task performance. They identified regions in the PFC that were specifically involved in working memory processes, providing important new data on separable aspects of central executive functions in the PFC.

In his book, *The Prefrontal Cortex*, Fuster (2008) presents a metaanalysis of several neuroimaging studies of working memory to provide a schematic summary of brain areas involved

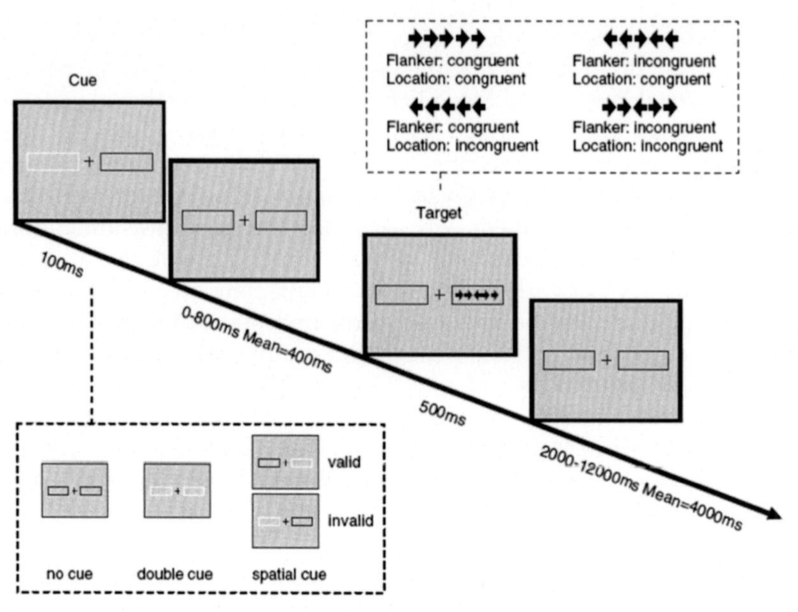

FIGURE 9.12 Attention network test-revised is a revised test to assess the interaction among attentional functions: alerting, orienting, and executive control. A schematic shows the experimental design. The subject fixates on a central cross, and in each trial, one of three cues is presented: no cue, double cue, and spatial cues (valid, invalid). After a duration of 0, 400 or 800 milliseconds (ms), a target appears that is either congruent (matching the cue) or incongruent. Thus the design allows for the subject to *alert* to the target, *orient* to the target once it appears, and use *executive control processes* to adapt to the actual target once it is perceived to perform the task demands. *Source: Xuan et al., 2016.*

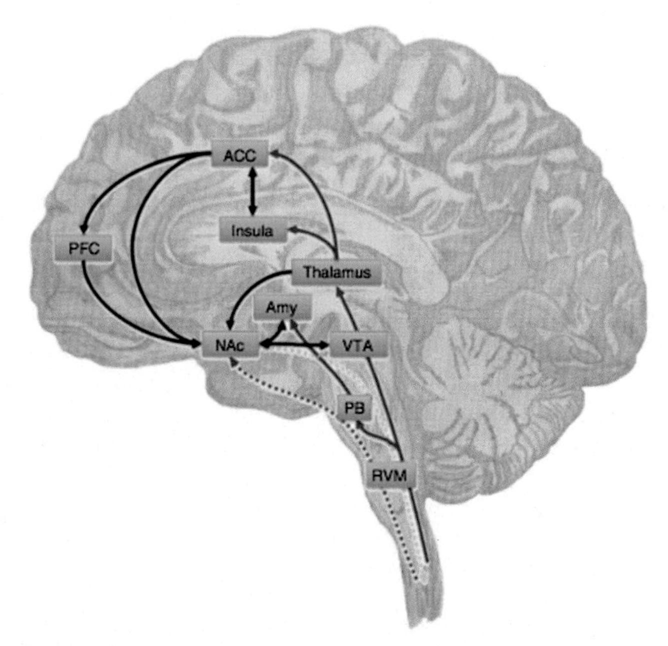

FIGURE 9.13 The frontoparietal network (FPN) for attention includes the prefrontal cortex (PFC), and anterior cingulate cortex (ACC), the insula, and the thalamus. Other brain regions shown include the parabrachial area (PB), the amygdala (Amy), the nucleus accumbens (NAc), the ventral tegmental area (VTA), and the rostral ventromedial medulla (RVM) (all shown in *blue boxes*). *Source: Navratilova and Porreca (2014).*

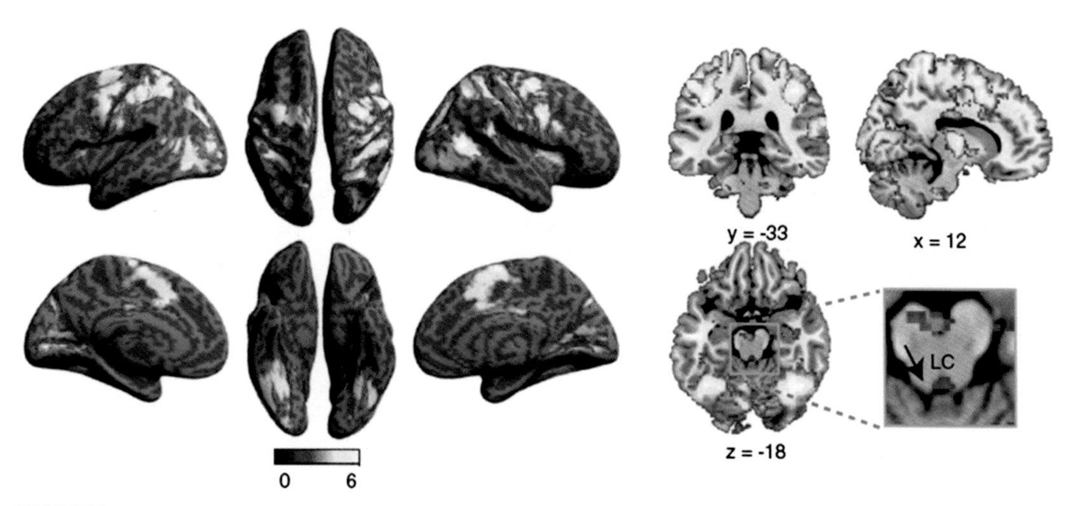

FIGURE 9.14 Alerting in the attention network test-revised. Specific results showed the *alerting* function activating the frontoparietal network (FPN) in the left hemisphere (top left sagittal images), the midline (center axial images), and the right hemisphere (top right sagittal images), and the locus coeruleus (LC) (far right coronal images). *Source: Xuan et al. (2016).*

in studies tapping visual versus verbal working memory. Results are presented in Figs. 9.17, 9.18, and 9.19.

Fig. 9.17 shows a schematic of brain areas—both visual from the lateral view of the brain and those tucked inside in the mesial cortex—that are active in experiments that tap working memory processes. Fig. 9.18 shows activation patterns for a visual memory task, with activity in the occipital lobe as expected, for an experiment using visual stimuli, and frontal lobe regions that are active due to the nature of the task. Fig. 9.19 shows a similar pattern of activation, but this time it is in response to a verbal memory task, so in this case, temporal lobe areas are active due to the auditory stimuli, along with frontal lobe regions.

One hypothesis that has been put forth about the role of working memory systems/networks in the PFC is that their function may be to *select* the material or information required for the task at hand, whereas areas in the posterior, sensory areas of the brain are involved in the actual *active maintenance* of that material or information while it is being used in a given task (see Curtis & D'Esposito, 2003; Wager & Smith, 2003, for reviews).

4.3 Executive Function and Motor Control

We have described a central function of the PFC as the ability to plan our actions—whether mental or physical—and then to follow out that plan. The mental planning of motor action—from initial abstract representations to the actual motor codes—takes place in the frontal lobes. A current view of the neural organization for these processes is that the more abstract representations/planning activities occur in the anterior portions of the PFC, moving more posterior (and thus closer to the motor regions) as the activities become less abstract and move toward motor codes for movement (for a review, see Fuster, 2008). The level of brain activity for planning and executing complex behaviors corresponds to the level of difficulty of the action. A task frequently used to test frontal lobe executive

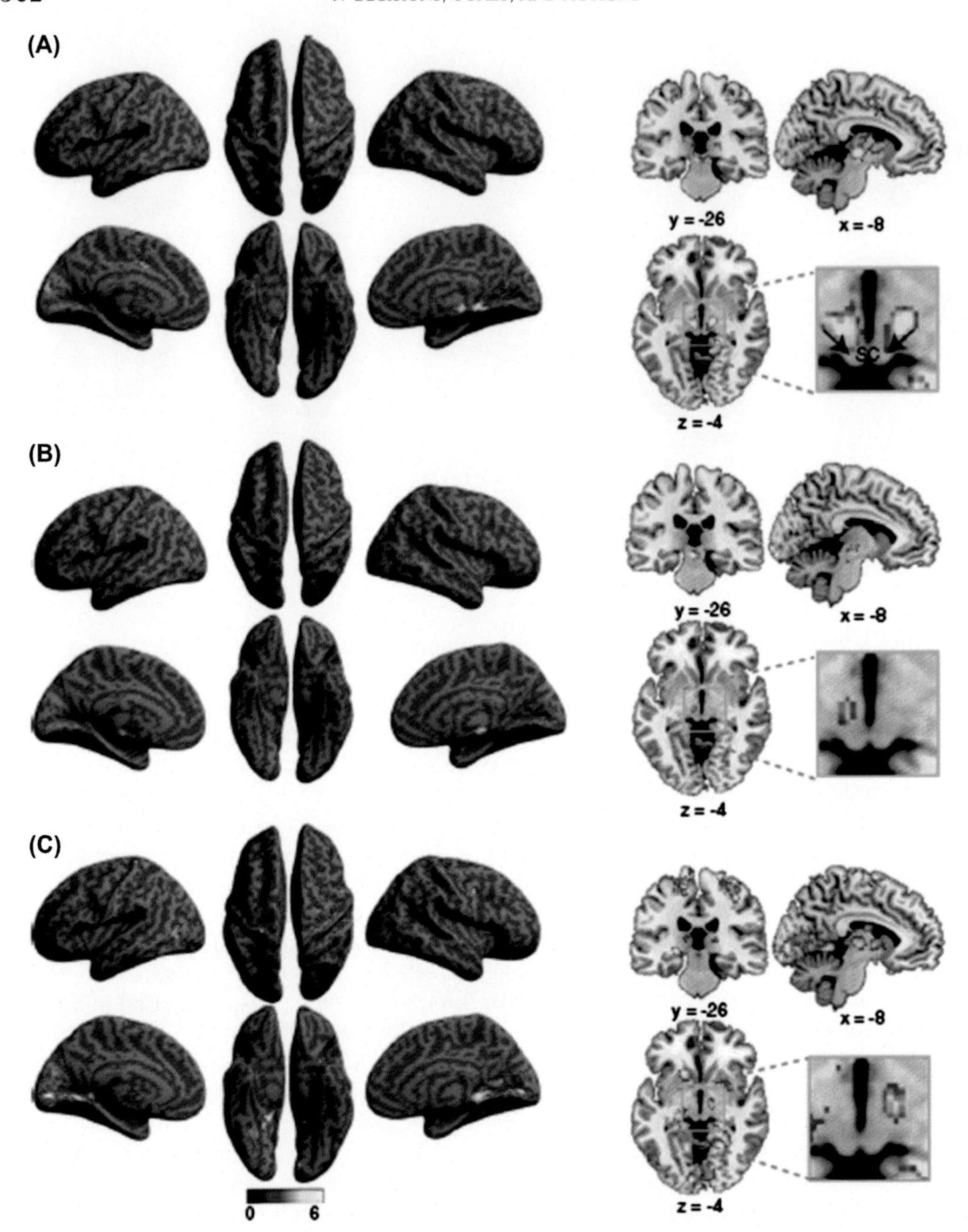

FIGURE 9.15 Orienting in the attention network test-revised. The *orienting* function activated frontal eye fields and the superior colliculus (SC). The activity shown reflects (A) disengaging, (B) engaging and movement to the target (orienting), and (C) the validity effect (was the cue accurate?). *Source: Xuan et al. (2016).*

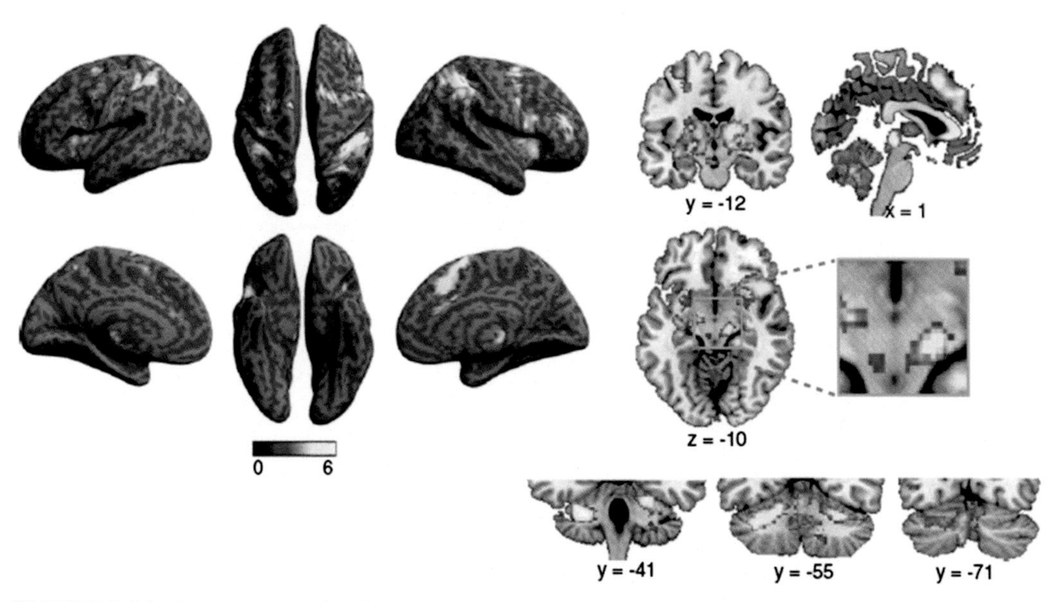

FIGURE 9.16 Executive control in the attention network test-revised. The *executive control* function activated the frontoparietal network (Shown for the left hemisphere (left sagittal images), midline activity (center axial images), and right hemisphere (right sagittal images), and the cerebellum (lower right coronal images). *Source: Xuan et al. (2016).*

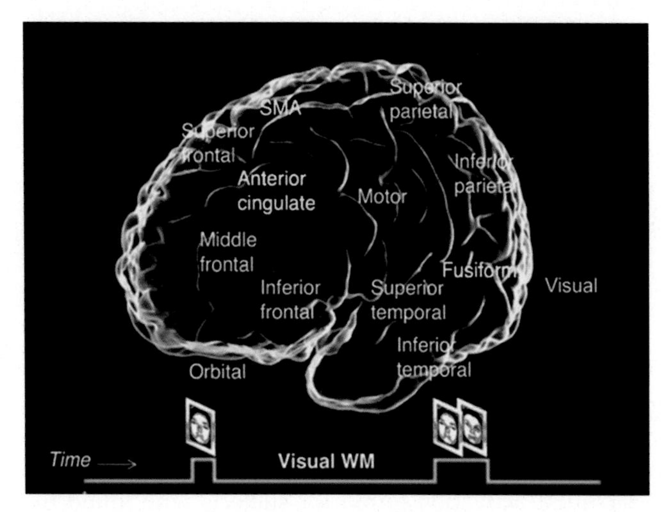

FIGURE 9.17 Outline of the left cortex used in subsequent figures to mark areas activated in working memory. Areas in convexity cortex are designated with white labels, and those in mesial cortex with yellow labels (supplementary motor area [SMA]). Below, temporal display of a trial in a typical visual working-memory (WM) task, delayed matching-to-sample, with faces. First, upward inflexion of timeline marks the time of presentation of the sample face, and second, inflexion of the choice faces. Delay-memory-period, between sample and choice, lasts 20 s. *Source: Fuster (2008).*

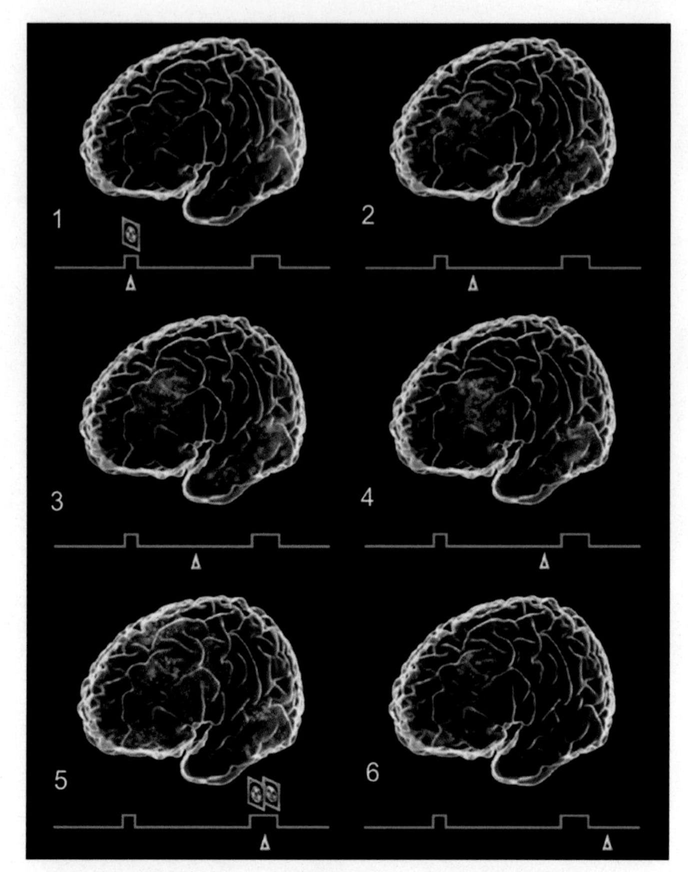

FIGURE 9.18 Relative (above-baseline) cortical activation at six moments in time (marked by *yellow triangle*) in the course of the visual memory task in Fig. 9.11. Activations of convexity cortex in *red*, of medial cortex in *pink*. (1) At the sample, the activation is restricted to visual and posterior inferotemporal cortex; (2) in early delay, it extends to lateral prefrontal cortex, anterior cingulate, anterior inferotemporal cortex, and fusiform cortex; (3) in mid-delay, it persists in prefrontal, inferotemporal, and fusiform cortex; (4) in late delay, it migrates to premotor areas, persisting in inferotemporal and fusiform cortex; (5) at the response (choice of sample-matching face), it covers visual, inferotemporal, and fusiform cortex in the back, and extends to motor areas (including frontal eye fields), supplementary motor area, and orbitofrontal cortex in the front; (6) after the trial, activation lingers in anterior cingulate and orbitofrontal cortex. *Source: Fuster (2008).*

function is the Tower of London task (Fig. 9.20) (Shallice, 1982). This task is like a puzzle with many steps for successful completion. To solve it, the subject needs to develop a plan. Researchers have found activation in DLPFC in the left hemisphere when solving this task, with higher levels of DLPFC activation found for subjects who found the task challenging (Morris, Ahmed, Syed, & Toone, 1993).

However, learning and practice change this effect, with well-learned—although complex—behaviors producing lower levels of brain activity (Fig. 9.21) (Poldrack et al., 2005; Posner & Raichle, 1997). Highly automatic behaviors like tying your shoe or locking a door produce little PFC activation. Thus the *feeling* you may have that these behaviors

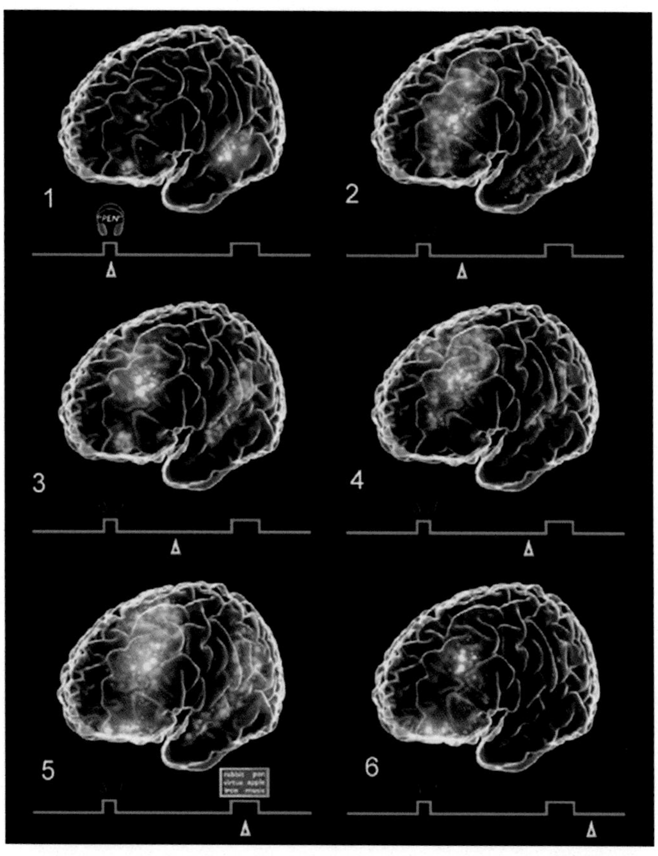

FIGURE 9.19 Cortical activation at six moments in time (*yellow triangle*) in the course of a verbal memory task: The memorandum, in 1, is a word through earphones. Activations of convexity cortex are in orange, and of medial or sulcal cortex in *yellow*. (1) At the cue-memorandum, the activation is restricted to auditory cortex, superior temporal gyrus, and inferior frontal cortex; (2) in early delay, it extends to lateral prefrontal, anterior cingulate, and superior-temporal and parietal association cortex; (3) in mid-delay, it persists in prefrontal and temporoparietal cortex; (4) in late delay, it persists in prefrontal and migrates to premotor areas, while persisting in temporoparietal cortex; (5) at the response (signaling whether cue word is on the screen), it covers visual and temporoparietal cortex in the back, and extends to frontal eye fields, supplementary motor area, inferior frontal and orbitofrontal cortex in the front; (6) after the trial, activation lingers in anterior cingulate, orbitofrontal cortex, and language areas. *Source: Fuster (2008).*

are easy, requiring little effort on your part, corresponds to brain studies showing little *PFC activation* for them.

We have briefly discussed the ACC (Brodmann area 24 in Fig. 9.6). What is the role of the ACC in executive function? Although the role of the ACC in executive functions is still being elucidated, one hypothesis is that it has an inhibitory effect on frontal lobe processes. If this is the case, then the ACC may represent a functional part of the orbito-medial PFC that helps in reducing the effects of distracting influences on the executive planning function (Fig. 9.22) (Bush, Luu, & Posner, 2000). Fuster (2008) hypothesizes that this inhibitory control that resists distracting influences may serve as the flip side of executive attentional processes. Thus in

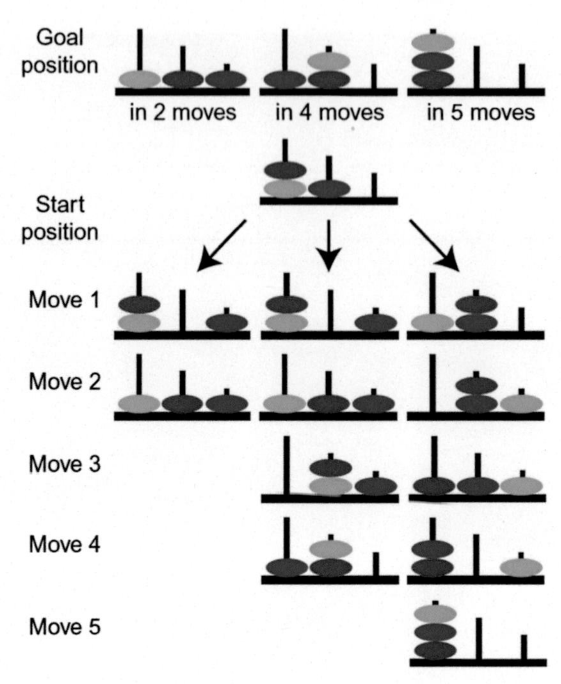

FIGURE 9.20 The Tower of London test. This task is like a puzzle with many steps for successful completion. To solve it, the subject needs to develop a plan. The goal position is shown at the top of the figure. Follow the moves from the Start position down through Move 5. Could you reach this Goal position with fewer moves? *Source: Fuster (2008).*

this way your success at focusing on a task at hand is aided both by your ability to pay attention to it and by your ability to ignore distractions.

4.4 Decision-Making

Where do the processes underlying decision-making occur in the frontal lobes? Converging evidence implicates the orbito-medial PFC, especially when there are emotional factors in the decision-making process. Think about playing a hand of poker. You must constantly make decisions about what to do next with incomplete information. Should you fold, raise,...what will your competitors do? To do well at this game, you need to assess the risk factors and the rewards. Bechara, Damasio, Damasio, and Anderson (1994) investigated these processes using the Iowa gambling task. They and many other researchers who have investigated the neural bases for assessing risk and reward have shown that the orbital or medial PFC is activated during these tasks.

Have you ever been in a state of internal conflict over a decision that you are making? Did you feel like "part of you" wanted you to listen to your *heart* and another "part of you" wanted you to listen to your *head*? This type of internal conflict between emotional feelings and rational thoughts is proposed to reflect the trading relationship between the orbitomedial PFC, with its connections to subcortical emotional regions, and lateral PFC, with its connections to executive control regions (Box 9.3).

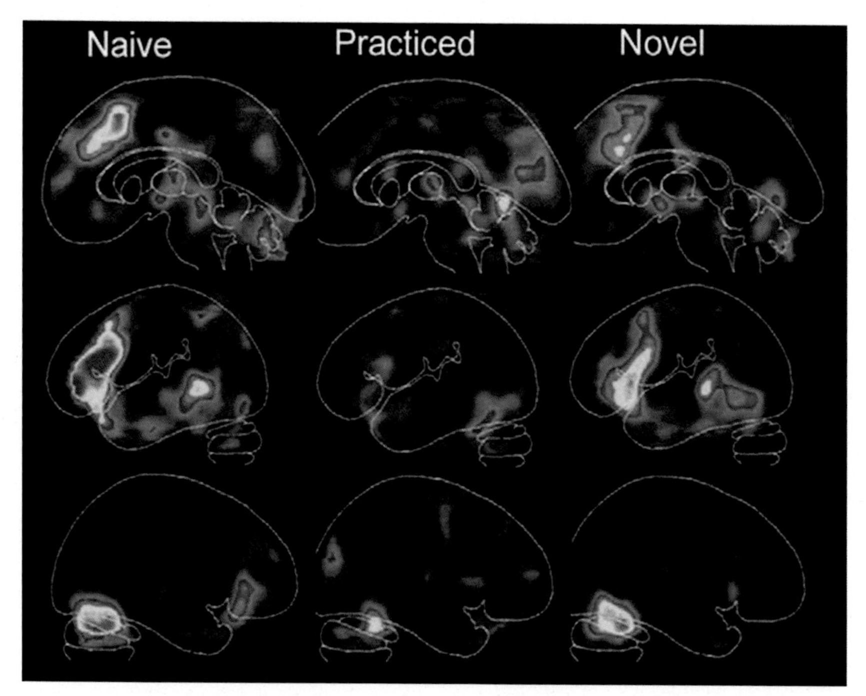

FIGURE 9.21 PET activation during performance of a verb generation task. The subject is given a series of nouns by the investigator and required to produce a verb appropriate to each. The control task was simply the reading aloud of the names as they appeared on a TV monitor. Subtractive scans of three brain slices in three different conditions (*left to right*): performance by a naive subject, subject practiced with the same list of nouns, and subject presented with a new list of nouns. *Source: Fuster (2008), with permission.*

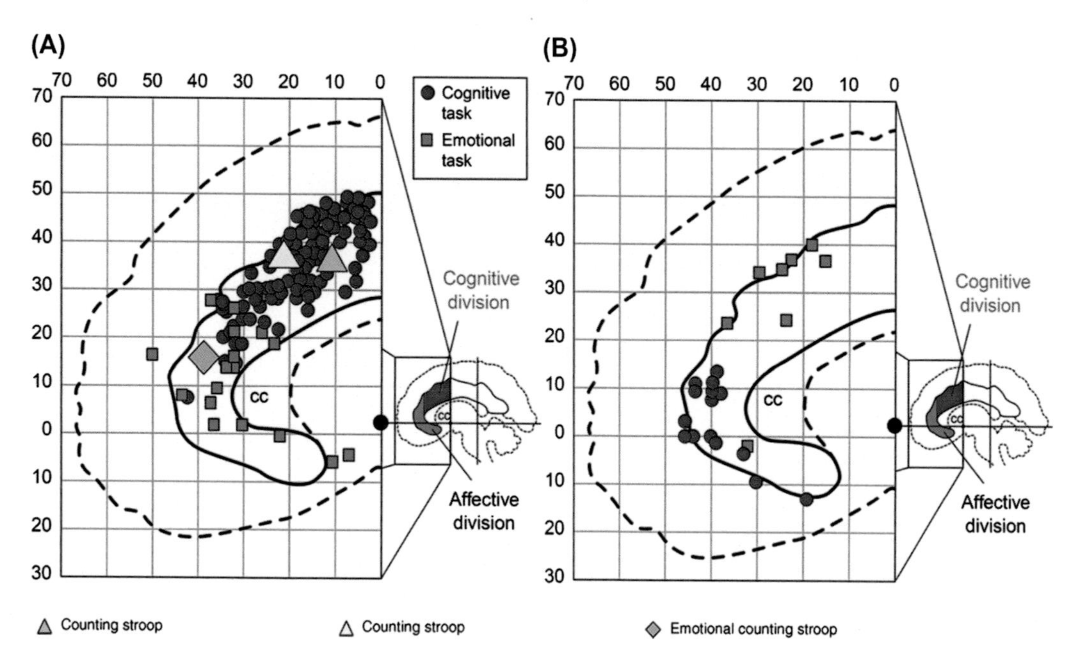

FIGURE 9.22 Confirming separate cognitive and emotional conflict regions. A summary of many studies of the anterior cingulate cortex (ACC), showing different cognitive and emotional regions. (A) Brain areas in the cognitive division of the ACC that were shown to be *activated* in previous studies. (B) Brain areas in the cognitive that were *deactivated* by tasks in previous studies that tapped emotional processes. *Source: Bush et al. (2000).*

BOX 9.3

PROBLEMS FOR THE MIRROR NEURON THEORY OF ACTION UNDERSTANDING

Mirror neurons. This fascinating discovery by Rizzolatti et al. blazed into neuroscience glory in the 1990s and has fueled revised motor theories of mammalian cognition for monkey and man ever since.

What is a mirror neuron and why is it so captivating? Here is the description of the *mirror mechanism* in the scientists' own words:

> Mirror neurons are a distinct class of neurons that discharge both when individuals perform a given motor act and when individuals observe another person performing a motor act with a similar goal. Mirror neurons were first discovered in the ventral premotor cortex (PMv) of the macaque monkey (area F5) (1, 2, 3). Neurons with mirror properties have subsequently been found in many brain cortical areas of monkeys and other species, including humans.
>
> The discovery that a large number of cortical areas that are involved in the production of certain motor behaviours selectively respond to those behaviours irrespective of whether they are being performed or observed indicates that the mirror mechanism, far from being a specific characteristic of the premotor cortex, is a basic principle of brain functioning. This statement becomes less surprising once it is acknowledged that the brain acts, first and foremost, as a planning and control system for organisms whose main job is exploring their surrounding world and facing its challenges and that are able to catch positive opportunities and escape threats. *(4), p. 757.*

According to Rizzolatti et al., during *action observation,* the mirror neurons encode not just the actions themselves—finger and hand movement, for example—but the *action goals.* That is, these neurons encode *the outcome of the action.* In humans, brain areas involved in the grasping-observation network of the mirror neuron system include inferior frontal gyrus, dorsal premotor cortex, and parietal and temporal lobe areas (Fig. 9.23).

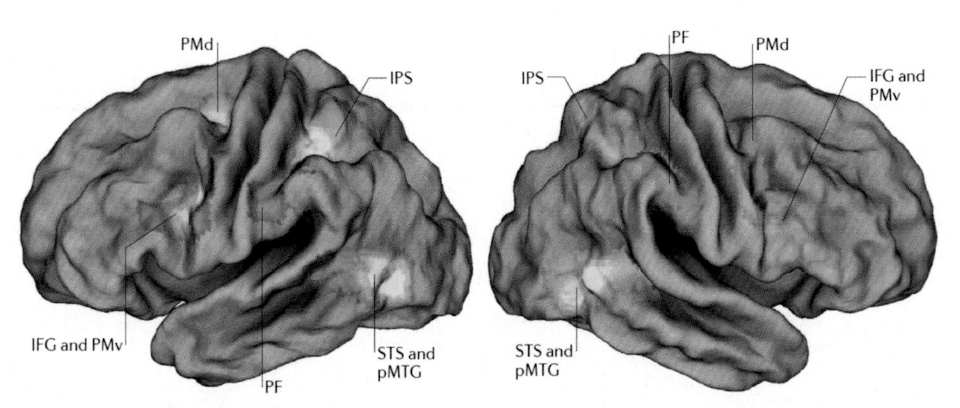

FIGURE 9.23 The human grasping-observation network. A metaanalysis of many investigations of the grasping-action observation function shows left (on the left side) and right (on the right side) hemisphere activation in the inferior frontal gyrus (IFG), ventral premotor cortex (PMv), the dorsal premotor cortex (PMd), the parietal area F (PF), intraparietal sulcus (IPS), superior temporal sulcus (STS), and posterior middle temporal gyrus (pMTG). *Source: Rizzolatti and Sinigaglia (2016).*

BOX 9.3 (cont'd)

While, as we stated above, the Mirror Neuron Theory of Action Understanding put forth by Rizzolatti et al. has sparked an enormous outflow of interest, replication studies, and new lines of research, it has also met with some pushback by scientists looking carefully at the observed data and their inferences.

One of these scientists, Greg Hickok, wrote a critical review article entitled *"Eight problems for the Mirror Neuron Theory of Action Understanding in Monkeys and Humans"* (2009). As the title states, Hickok reviews the literature and comes away with a strong case against the theory. In his words:

> The discovery of mirror neurons in macaque frontal cortex has sparked a resurgence of interest in motor/embodied theories of cognition. This critical review examines the evidence in support of one of these theories, namely that the mirror neurons provide the basis of action understanding. It is argued that there is no evidence from monkey data that directly tests this theory, and evidence from humans makes a strong case against the position. ***(5), p. 1229.***

Hickok steps through those eight problems with the theory, carefully citing specific studies and assessing the findings with respect to the theory. What are those eight problems? Here they are briefly; we encourage you to read the entire review. It is an education.

Eight problems for the Mirror Neuron Theory of Action Understanding in Monkeys and Humans:

1. There is no evidence in monkeys that mirror neurons support action understanding
2. Action understanding can be achieved via nonmirror neuron mechanisms
3. M1 [primary motor cortex] contains mirror neurons
4. The relation between macaque mirror neurons and the "mirror system" in humans is either nonparallel or underdetermined.
5. Action understanding in human dissociates from neurophysiological indices of the human "mirror system."
6. Action understanding and action production dissociate
7. Damage to the inferior frontal gyrus is not correlated with action understanding deficits.
8. Generalization of the mirror system to speech recognition fails on empirical grounds.

Hickok's review is not only a good read but also an excellent example of a truly careful approach to science. The explosion of enthusiasm and support for the appealing mirror neurons and the action observation theory has distracted the field from understanding their actual function in cognition. The lack of sufficient scientific rigor in data analysis and interpretation has confounded the problem.

Again in Hickok's own words:

> Unfortunately, more than 10 years after their [mirror neurons] discovery, little progress has been made in understanding the function of mirror neurons. I submit that this is a direct result of an overemphasis on the action understanding theory, which has distracted the field away from investigating other possible (and potentially equally important) functions. ***(5), p. 1238.***

References

(1) Gallese, V., Fadiga, L., Fogassi, L., & Rizzolatti, G. (1996) Action recognition in the premotor cortex. *Brain, 119*, 593–609.

Continued

BOX 9.3 *(cont'd)*

(2) Hickok, G. (2009). Eight problems for the mirror neuron theory of action understanding in monkeys and humans. *J Cogn Neuroscience, 21*(7), 1229–1243.

(3) di Pellegrino, G., Fadiga, L., Fogassi, L., Gallese, V., & Rizzolatti, G. (1992). Understanding motor events: a neurophysiological study. *Exp. Brain Res, 91*, 176–180.

(4) Rizzolatti, G., Fadiga, L., Gallese, V., & Fogassi, L. (1996) Premotor cortex and the recognition of motor actions. *Cogn. Brain Res, 3*, 131–141.

(5) Rizzolatti, G., & Sinigaglia, C. (2016). The mirror mechanism: A basic principle of brain function. *Nature Reviews Neuroscience, 17*, 757–765.

4.5 Rule Adoption

To navigate our way through our complex daily lives, it is critical to develop ways to shortcut all the things that we need to plan for and carry out. Humans are wonderful rule adopters; we develop and learn strategies for streamlining our busy lives. Like a strategic plan or a schema, rules help us increase our efficiency. The Wisconsin Card Sorting test (shown in Fig. 9.24) is a good example of the mental flexibility humans have in acquiring rules and, importantly, in changing them when needed.

Neuroimaging studies of rule learning in PFC have shown that, in a manner similar to attentional and working memory demands, neural activity in the frontal regions increases with the complexity of the rule set to be learned or carried out (Fig. 9.25) (Bunge, 2004).

Neuroimaging studies have shed new light on the many and diverse operations carried out—or directed—by the PFC, from paying attention to a stimulus in your environment, to monitoring how it is changing, to keeping something in mind, to complex decision-making. Many of these processes are highly overlapping in time and neural regions, so we are still elucidating which frontal lobe areas contribute to these processes. Although we are still in the early stages of understanding just how and where executive processes are being done in the PFC, converging evidence from neuroimaging studies are beginning to present a clearer picture of PFC function.

5. DAMAGE TO THE EXECUTIVE BRAIN

We have discussed many functions of the frontal lobe, including voluntary attention, working memory, decision-making, and even your personality. What happens when this

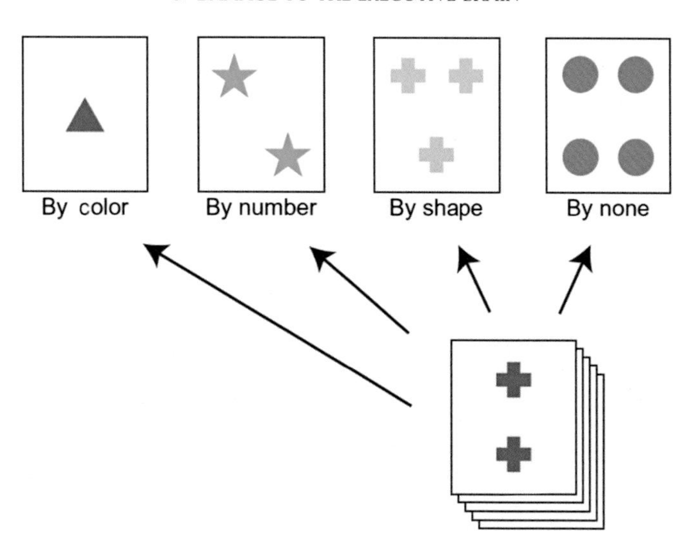

FIGURE 9.24 The Wisconsin Card Sorting Test (WCST). The cards can be sorted by matching the color of the item on the card, or the number of items, their shape, or even if they do not match on any of these features. At the beginning of the game, the experimenter determines which matching "rule" to be used (such as "match the new card by color"), but he does not tell the player the rule; the player must learn it by trial and error. During the game, the experimenter will change the rule, and again, the player must learn the new rule through trial and error. This is a test of mental flexibility: the ability to learn a new rule and to adapt to a new rule when needed. *Source: Fuster (2008).*

critical area of the brain is damaged? Or when it fails to develop in a typical manner? The answers to these questions are as complex as the frontal lobes themselves.

5.1 The Fragile Frontal Lobes

Frontal lobe dysfunction often reflects more than direct damage to the frontal lobes (Goldberg, 1992). The frontal lobes seem to be the bottleneck—the point of convergence of the effects of damage virtually anywhere in the brain. There is a reciprocal relationship between the frontal and other brain injuries. Damage to the frontal lobes produces wide ripple effects through the whole brain. At the same time, damage anywhere in the brain sets off ripple effects that interfere with frontal lobe function. This unique feature of the frontal lobes reflects its role as the "traffic hub" of the nervous system, with a singularly rich set of connections to and from other brain structures. This makes frontal lobe dysfunction the most common and least specific finding among neurological, psychiatric, and neurodevelopmental conditions (Goldberg, 1992).

The frontal lobes' exceptionally low "functional breakdown threshold" is consistent with Hughlings Jackson's concept of "evolution and dissolution" (1884/2011). According to Jackson's proposal, the evolutionary "youngest" brain structures are the first to succumb to brain

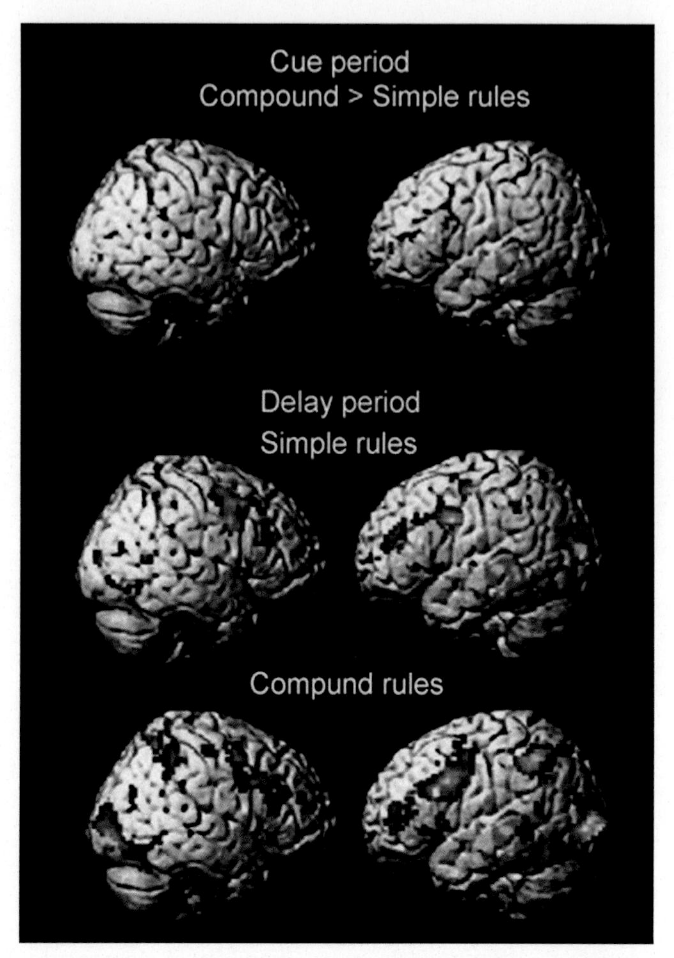

FIGURE 9.25 In this study, the experimenters wanted to see which brain areas were active while subjects implemented either simple or compound abstract "rules." A rule cue was provided followed by a long (7–15 s) delay. Brain areas more active for compound versus simple cues are shown at the top of this figure. Areas active during the delay (while the subject is keeping the rule in mind) are shown in the middle (simple rule) and bottom (compound rule) of the figure. *Source: Bunge (2004), as adapted in Fuster (2008).*

disease. The frontal lobes' unique vulnerability is probably the flip side of the exceptional richness of their connections. A frontal lobe dysfunction does not always signify a direct frontal lobe *lesion*. Instead, it may be a remote effect of a diffuse, distributed, or distant lesion.

5.2 Frontal Lobe Syndromes

The importance of executive functions can be best appreciated through the analysis of their disintegration following brain damage. A patient with damaged frontal lobes retains, at least to a certain degree, the ability to exercise most cognitive skills in isolation (Luria, 1966). Basic abilities such as reading, writing, simple computations, verbal expression, and movements may remain largely unimpaired. Deceptively, the patient will perform well on the psychological tests measuring these functions in isolation. However, any synthetic activity requiring the coordination of many cognitive skills into a coherent, goal-directed process will become severely impaired.

Damage to different parts of the frontal lobes produces distinct, clinically different syndromes. The most common among them are the *dorsolateral* and *orbitofrontal prefrontal syndromes* (Fig. 9.26) (Goldberg & Costa, 1985).

5.2.1 Dorsolateral Prefrontal Syndromes

Most common symptoms of dorsolateral prefrontal syndromes are *personality changes, high levels of distractibility,* and much reduced *mental flexibility*. Clinically, a patient with dorsolateral prefrontal syndrome is neither sad nor happy; in a sense, he or she has no mood. This state of indifference persists no matter what happens to the patient, good things or bad. However, the most conspicuous feature of a patient with dorsolateral prefrontal syndrome is a drastically impaired ability to *initiate* behaviors. Once started in a behavior, however, the patient is equally unable to terminate or change it on his or her own. Such combined "inertia of initiation and termination" is seen in various disorders affecting the frontal lobes, including chronic schizophrenia.

Another common symptom of dorsolateral prefrontal syndromes is that the patient is at the mercy of incidental distractions and thus is unable to follow internally generated plans. In extreme cases it may take the form of field-dependent behavior. A frontal lobe patient will drink from an empty cup, put on a jacket belonging to someone else, or scribble with a pencil

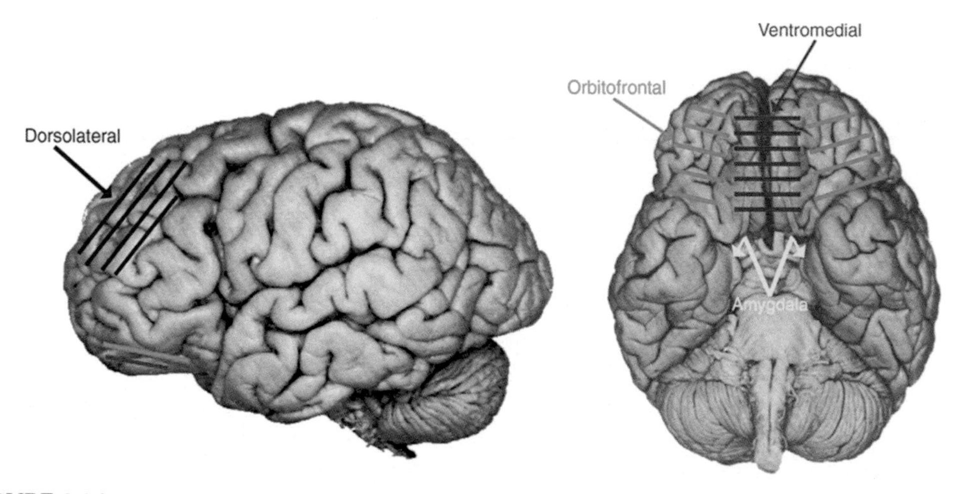

FIGURE 9.26 The orbitofrontal region is just above the orbits of the eyes. The orbitofrontal cortex (*green stripes*) can be distinguished in the ventral prefrontal lobe. Orbitofrontal cortex is involved in understanding future rewards, changes in reward contingencies, and goal selection. Damage to the orbitofrontal region may lead to a loss of behavioral inhibition. *Source: Davidson and Irwin (1999).*

on the table surface, merely because the cup, the jacket, and the pencil are there, even though these actions make no sense. Easy distractibility is a feature of many neurological and psychiatric disorders, and it is usually associated with frontal lobe dysfunction. For example, *attention deficit (hyperactivity) disorder (AD(H)D)*, with its extreme distractibility, is usually linked to frontal lobe dysfunction (Barkley, 1997).

Our ability to maintain mental stability has to be balanced by *mental flexibility*. No matter how focused we are on an activity or a thought, there comes a time when the situation calls for doing something else. Being able to change one's mindset is as important as staying mentally on track. The capacity to switch with ease from one activity or idea to another is so natural and automatic that we take it for granted. In fact, it requires complex neural machinery, which also depends on the frontal lobes. Frontal lobe damage often produces extreme mental rigidity, which may severely undermine the patient's cognition. Quite often, a closer look at a frontal lobe patient's performance on a number of tasks shows that complete transition from one task to another is impossible, and fragments of a previous task attach themselves to the new one. This phenomenon is called *perseveration*.

5.2.2 Orbitofrontal Prefrontal Syndromes and Self-control

The *orbitofrontal prefrontal syndrome* is, in many ways, the opposite of the dorsolateral prefrontal syndrome. The patients are behaviorally and emotionally *disinhibited*. Their affect is rarely neutral, constantly oscillating between euphoria and rage, with impulse control ranging from poor to nonexistent. Their ability to inhibit the urge for instant gratification is severely impaired. They do what they feel like doing, when they feel like doing it, without any concern for social taboos and legal prohibitions. They have no foresight of the consequences of their actions. A patient afflicted with orbitofrontal syndrome may engage in shoplifting, sexually aggressive behavior, reckless driving, or other actions commonly perceived as antisocial. These patients are known to be selfish, boastful, puerile, profane, and sexually explicit. If the dorsolateral patients are in a sense devoid of personality, then orbitofrontal patients are conspicuous for their "immature" personality.

5.2.3 Reticulofrontal Disconnection Syndrome

In cases when the frontal lobes themselves are structurally intact but the patient presents with frontal lobe symptoms, the problem may lie with the pathways connecting frontal lobes to some other structures. Damage to these pathways may result in a condition known as the *reticulofrontal disconnection syndrome* (Goldberg, Bilder, Hughes, Antin, & Mattis, 1989).

The brainstem contains the nuclei thought to be responsible for the arousal and activation of the rest of the brain. A complex relationship exists between the frontal lobes and the brainstem reticular nuclei, which are in charge of activation and arousal. The relationship is best described as a loop. On the one hand, the arousal of the frontal lobes depends on the ascending pathways. On the other hand, there are pathways projecting from the frontal lobes to the reticular nuclei of the ventral brainstem. Through these pathways the frontal lobes exert their control over the diverse brain structures by modulating their arousal level. If the frontal lobes are the decision-making device, then the brainstem structures in question are an amplifier helping communicate these decisions to the rest of the brain in a loud and clear voice. The descending pathways are the cables through which the instructions flow from the frontal lobes to the critical ventral brainstem nuclei. We can easily see how damage to the pathways between the brainstem and the frontal lobes may disable executive functions without actually damaging the frontal lobes per se.

5.3 Frontal Lobe Damage and Asocial Behavior

The relationship between frontal lobe damage and asocial behavior is particularly intriguing and complex. It has been suggested, based on several published studies, that the prevalence of head injury is much higher among criminals than in the general population, and higher in violent criminals than in nonviolent criminals (Raine, Buchsbaum, & LaCasse, 1997; Volavka et al., 1995). For reasons of brain and skull anatomy, closed head injury is particularly likely to affect the frontal lobes directly, especially the orbitofrontal cortex. Furthermore, damage to the upper brainstem is extremely common in closed head injury, even in seemingly mild cases, and it is likely to produce frontal lobe dysfunction even in the absence of direct damage to the frontal lobes by producing the "reticulofrontal disconnection syndrome" (Goldberg et al., 1989).

5.4 Other Clinical Conditions Associated With Frontal Lobe Damage

PFC is afflicted in a wide range of conditions (Goldberg, 1992; Goldberg & Bougakov, 2000), and it is not necessary to have a focal frontal lesion to have prefrontal dysfunction. The frontal lobes are particularly vulnerable in numerous nonfocal conditions. Such disorders as schizophrenia (Ingvar & Franzen, 1974), traumatic brain injury (TBI) (Deutsch & Eisenberg, 1987), Tourette's syndrome (Sacks, 1992; Shapiro & Shapiro, 1974), and AD(H)D (Barkley, 1997) are known to involve frontal lobe dysfunction. Executive functions are also compromised in dementia and in depression.

5.4.1 *Attention Deficit (Hyperactivity) Disorder*

The PFC and its connections to the ventral brainstem play a particularly important role in the mechanisms of attention. When we talk about AD(H)D, we usually implicate these systems. Any damage to the PFC or its pathways may result in attentional impairment. The exact causes of damage to these systems vary. They may be inherited or acquired early in life. They may be biochemical or structural. The way the diagnosis of AD(H)D is commonly made, it refers to any condition characterized by mild dysfunction of the frontal lobes and related pathways in the absence of any other, comparably severe dysfunction. Given the high rate of frontal lobe dysfunction due to a variety of causes, the prevalence of genuine AD(H)D should be expected to be very high.

6. A CURRENT VIEW OF ORGANIZING PRINCIPLES OF THE FRONTAL LOBES

After decades of research into frontal lobe function in nonhuman primates and in humans, using many experimental techniques and methods, some organizing principles have emerged. A leader in the field of frontal lobe research is Joaquin Fuster (2008), who proposed a model for frontal lobe function. He states the following:

1. The entirety of the cortex of the frontal lobe is devoted to the representation and production of action at all levels of biological complexity.
2. The neuronal substrate for the production of any action is identical to the substrate for its representation.
3. That substrate is organized hierarchically, with the most elementary actions at low levels of the hierarchy, in orbitofrontal and motor cortex, and the most complex and abstract actions in lateral PFC.
4. Frontal lobe functions are also organized hierarchically, with simpler functions nested within, and serving, more global functions.

The notion of a hierarchical organization for simple versus complex, abstract versus concrete frontal lobe function has been developed in neuroimaging studies of human frontal lobe processes. A review article by Badre (2008) provided a summary of findings to date, including models developed by Koechlin et al., showing data supporting the view of a hierarchical organization of the frontal lobe, with an anterior-to-posterior organization that is based on the degree of abstraction (Fig. 9.27).

Does the anatomy and the neurophysiology of the frontal lobe support this view? That question was addressed in a follow-up review by Badre and D'Esposito (2009). They reported evidence for a rostral–caudal (anterior–posterior) gradient in the degree of abstraction that neurons respond to, supporting their hierarchy view from a cell physiology perspective for both monkey and human (Fig. 9.28). They also reported neuroimaging evidence supporting a rostral–caudal gradient in function in human (Fig. 9.29). Together, this evidence lends strong support for the notion of a hierarchical organization for the frontal lobe. Like all theories, however, it is important to continually test the underlying hypotheses and verify the data to fully assess how the massive and highly differentiated frontal lobes are organized.

7. TOWARD A UNIFIED THEORY OF FRONTAL LOBE FUNCTION

While many more details clearly need to be discovered before we have a unitary explanation of frontal lobe function, recent advances in neuroimaging techniques and methods have helped to increase our knowledge of the complex role the frontal lobes, and in particular the PFC, play in human cognition. However, even though neuroimaging studies may serve to guide us as to specific regions where certain executive processes—such as voluntary attention or working memory—may be located, they show a dramatically different picture of human executive function than is seen when observing a patient with frontal lobe damage. How do

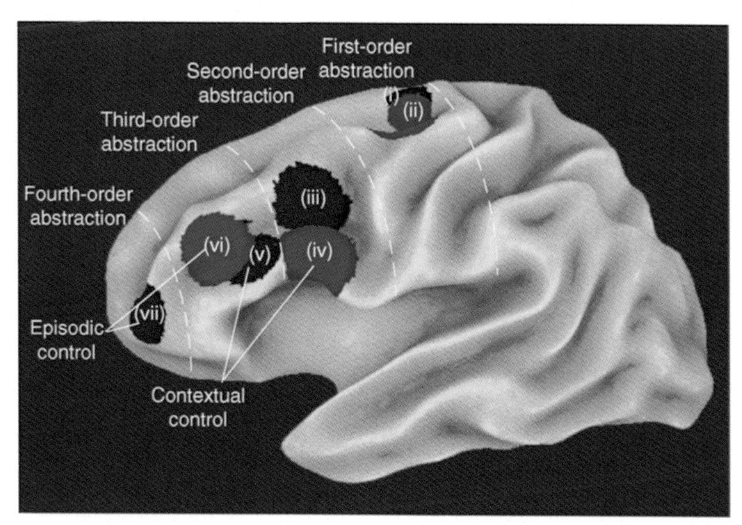

FIGURE 9.27 In a review article by Badre (2008), a proposed hierarchical organization of the frontal lobes is presented. According to this view, there is an anterior-to-posterior organization in the frontal lobes that is based on the degree of abstraction involved in the task. Moving from the front of the frontal lobes toward the primary motor areas, activity in the frontal lobes becomes less abstract and more concrete, until they relate to the planning and execution of motor-related movements for performing a given task. Results from the response, feature, dimension, and context experiments and comparison to the model of Koechlin et al. (2003). Abstract relational hierarchy seems to provide a parsimonious account of rostrocaudal gradient across the models of Koechlin et al. (2003) and Badre and D'Esposito (2007). Spheres with diameters of 8 mm (within the smoothing kernel of each experiment) were centered on maxima from response (i), feature (iii), dimension (v), and context (vi) manipulations of Badre and D'Esposito (2007) (shown in *red*), and on the sensory (ii), context (iv), and episodic (vi) manipulations of Koechlin et al. (2003) (shown in *blue*). These spheres were rendered on an inflated Talairach surface. Note that the spheres are for precise illustration of proximity but do not represent actual spread of activation in each experiment. Broken lines separate manipulations at equivalent levels of abstraction in a representational hierarchy. Equivalent episodic and contextual control manipulations across the two experiments are also labeled. *Source: Badre (2008), adapted from Badre and D'Esposito.*

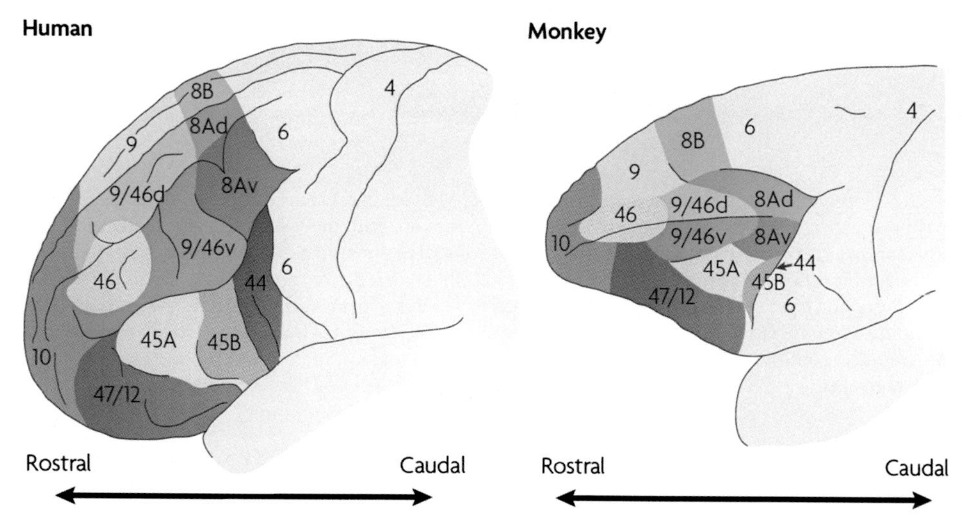

FIGURE 9.28 Neurophysiology of the frontal lobe in human (left) and monkey (right). Brodmann areas in the frontal lobe have been worked out through the years, with this version provided by Petrides and Pandya (1994) providing a well-accepted way to compare comparable Brodmann areas across human and monkey brains. Using these Brodmann area definitions, Badre and D'Esposito (2009) reported evidence for a rostral–caudal (anterior–posterior) gradient in the degree of abstraction that neurons respond to, supporting their hierarchy view from a cell physiology perspective for both monkey and human. *Source: Badre and D'Esposito (2009).*

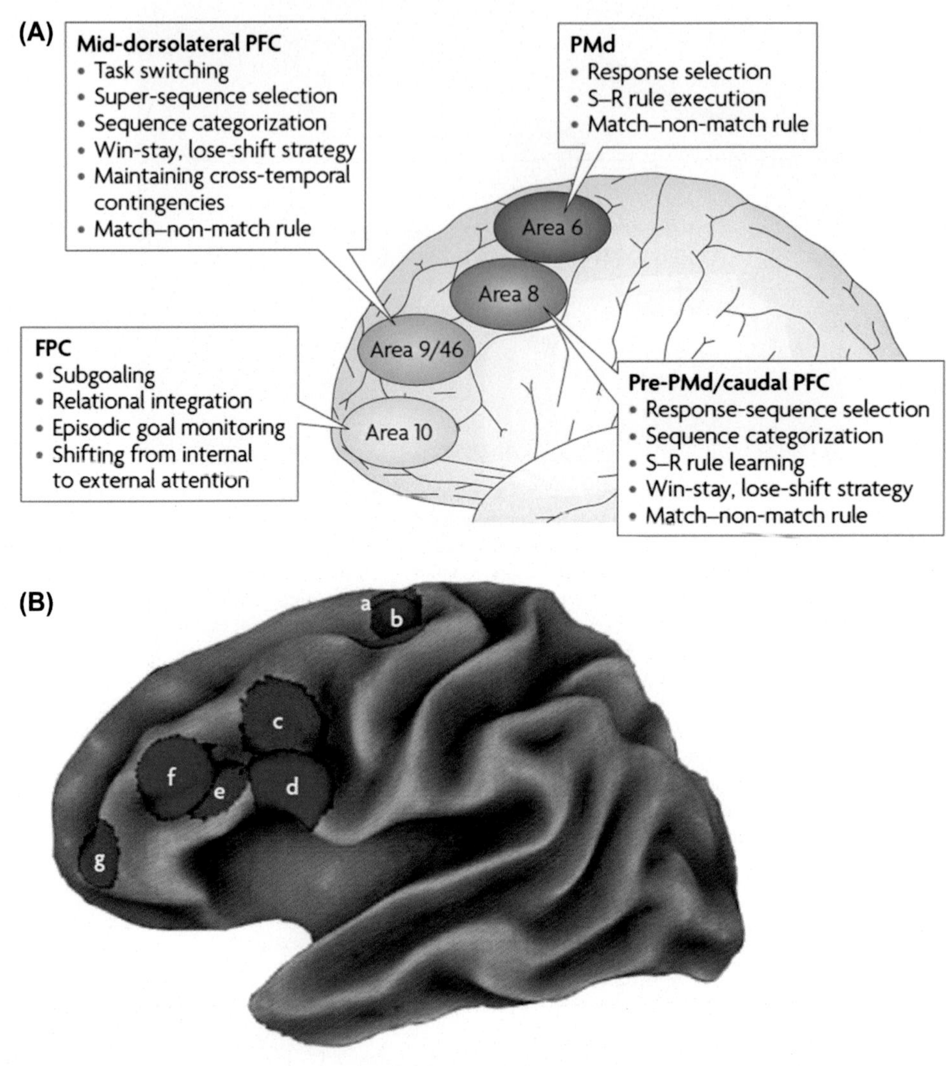

FIGURE 9.29 A hierarchy of function in the frontal lobes. (A) Badre and D'Esposito (2009) reported neuroimaging evidence supporting a rostral–caudal gradient in function in human, with more rostral regions (areas 9/46 and 10), supporting more abstract action rules and subgoals and more caudal regions (areas 6 and 8) supporting less abstract and more motor-based functions. (B) Studies by Koechlin et al. (2003) (shown in *blue*) and Badre et al. (2007) (shown in *red*) investigated the notion of a functional gradient in the frontal cortex by using tasks that varied in the amount of cognitive control at more and more abstract levels. Less cognitive control with lower abstract level–activated regions caudally (a, b), more cognitive control with moderate abstract level–activated regions midlobe (c, d, e), and even more cognitive control with high abstract–activated regions rostrally (f, g). *Source: Badre and D'Esposito (2009).*

results from neuroimaging studies correspond to studies of patients with frontal lobe syndromes? These two sets of findings are not easy to bring together into a cohesive whole at present. Neuroimaging studies have provided a wealth of data about specific regions and locations that activate for various aspects of frontal lobe processes, but the results do not correspond directly to what we know about human behavior when frontal lobes are damaged. This is the challenge for cognitive neuroscientists: to continue to provide converging evidence across many techniques and subject groups to elucidate the underlying organizational principles for the complex and uniquely human PFC, the "organ of civilization."

8. STUDY QUESTIONS

1. What are some of the functions attributed to the prefrontal lobe?
2. What are mirror neurons and the action observation theory?
3. How is mental flexibility important in everyday life?
4. What type of frontal lobe damage causes mental rigidity?
5. What is one frequent way for testing mental flexibility/rigidity?
6. How do findings from neuroimaging research differ from studies of patients with frontal lobe damage?

CHAPTER

10

Humans Are Social Beings

OUR SOCIAL NATURE

We are an intensely social species — it has been argued that our social nature defines what makes us human, what makes us conscious or what gave us our large brains. As a new field, the social brain sciences are probing the neural underpinnings of social behavior and have produced a banquet of data that are both tantalizing and deeply puzzling. We are finding new links between emotion and reason, between action and perception, and between representations of other people and ourselves. ***Ralph Adolphs, 2003, American Scientist.***

The Banquet at the Eglinton tournament, Ayrshire, Scotland, by James Henry Nixon. *Source: Public domain.*

https://doi.org/10.1016/B978-0-12-803813-0.00010-6

1. INTRODUCTION

Is there a social brain? Are there networks and regions in the brain that are specialized for understanding social signals—such as direct eye gaze or averting one's eyes? Are faces that display different emotions—angry versus happy, for example—processed in the brain differently? These are questions that social cognitive neuroscientists are addressing in current research in this very new field. The advent of neuroimaging methods has helped researchers answer these questions in new and intriguing ways.

In this chapter we discuss *social cognition*—the behavioral and brain bases of human social knowledge. In Chapter 11, Feelings, we will discuss the *emotional aspects* of social cognition along with other aspects of cognition, language, and consciousness. Thus, in this chapter we will discuss *human face perception*, and in the following chapter, we will discuss perception of *emotional facial expression*.

What is a social brain? These are brain regions that process important social information such as facial expressions, eye gaze, and intentionality. Body posture and position are also key to understanding social cues and interpreting—and predicting—actions. The study of *where* these social processes are decoded in the brain is a new area of focus in this science.

The study of social cognitions seems quite different from the study of language, attention, or learning and memory. Social cognition incorporates these aspects of human cognition, but somehow social cognition goes *beyond* each of these domains of cognition. For example, you may listen to a friend tell you a story, and your brain easily decodes the sounds, words, phrases, and sentences of your native language. But you will also be recognizing the influences of intonation and facial expression to decode the "real message" being expressed. You may be paying attention to your friend as the story unfolds, but you will also be aware—sometimes unconsciously—of their body posture, eye gaze, and emotional state. And remembering your wedding day or the day you graduated from college seems much more salient—emotional—than remembering other days in your life. Social cognition spans these and other aspects of human cognition. The scientific study of brain areas that underlie these complex processes is still a new field; however, much progress is being made.

1.1 Terms That Are Used to Refer to Social Cognition

In the research literature, terms that refer to aspects of social cognition are often used interchangeably and in different ways by different researchers. *Empathy* carries the sense of feeling the feelings of others. In Latin, the word means "feeling inside" or "feeling with." On the other hand, theory of mind (TOM) is often used to highlight the idea that we normally have complex metacognitive understandings of our own minds, as well as the minds of others—including cognitive and affective aspects. Similarly, Frith and Frith (1999) introduced the term *mentalizing* to capture the idea that when we have a well-developed TOM, we understand ourselves and others not just as sensory objects but also as subjective beings with mental states. We understand others as having mental states that we can anticipate and use to guide our own behaviors. Mind reading, like mentalizing, identifies our ability to attune our own behaviors to the minds and anticipated actions of others.

One of the most difficult aspects of the concept of TOM is understanding the difference between seeing others as sensory objects versus seeing others as subjective beings with minds and mental states. Having a complete TOM gives us the ability to go beyond the sensory into the mental. We can do things that those with deficient TOMs cannot do. Once we have a

TOM, we can pretend, lie, deceive, guess, play hide-and-seek, and predict and understand the full range of human emotion. People who have deficits in TOM have limited abilities to do these things, as we will see later in the chapter.

Philosophers use the term *intentionality* when they want to speak about how minds and mental states are always "*about* something else" in a way that other physical objects, such as body parts, are not. Our thoughts always have an object. For example, we think "about" the chair, the book, or the idea in a way that our stomach, arm, or tooth is not about anything other than itself. Minds have mental states; minds *represent* objects and events outside themselves. It is not clear whether other species comprehend the intentional nature of minds in their conspecifics. Humans seem to have an implicit understanding of the contents of others' minds.

A separate concept is the psychological term *intention*, which is our ability to form an image of a goal state and to organize action in pursuit of that goal state. You must be careful not to confuse these two very similar terms! TOM abilities allow us to read the intentions of others and to share attention with others about a common focus.

Finally, the term *intersubjectivity* emphasizes our ability to coordinate mutual interactions in light of our perception of the subjectivity and intentionality of others. When this ability is absent, we readily recognize the deficiency in the social exchanges of others. Examples are found in autistic spectrum disorders, in the sometimes deficient emotion recognition of schizophrenia, and in the empathic failures of psychopathic and borderline personalities (Box 10.1).

BOX 10.1

#YOURBRAINONTWITTER—YOUR BRAIN ON SOCIAL MEDIA

Three Out of Four American Adults Are on Social Media Sites

Have you been on the Internet today? Sent any texts? Shared any photos? Looked at your Facebook page or sent an email or two? Most Americans answer those questions with a resounding YES (Figs. 10.1 and 10.2). In fact, recent analyses show that about three out of every four American adults use social media sites (1).

And while younger adults are the most likely to use social media sites—90% of them according to a recent study (1)—older adults are beginning to become a large segment of the users. In fact, a tiny fraction (2%) of Americans older than the age of 65 used social media sites in 2005, whereas today that number has risen to 35% and is still growing (1).

When Does It Begin?

"Nearly Constant" Online Access by Teenagers as Young as 13

The early teen years are an incredibly social time—new friendships and social group memberships are literally exploding as teens move to middle school and on to high school. With smartphone usage among this age group at an all-time high, it is no wonder that these teenagers are accessing the Internet and social media sites in record numbers. In fact, while its not surprising that more than 90% of teenagers between the ages of 13 and 17 years go online daily, it might surprise you to find that nearly 25% of them are online "almost constantly" (2). Or maybe that is not surprising—when is the last time you saw a teenager without a smart phone in his/her hand?

Continued

BOX 10.1 *(cont'd)*

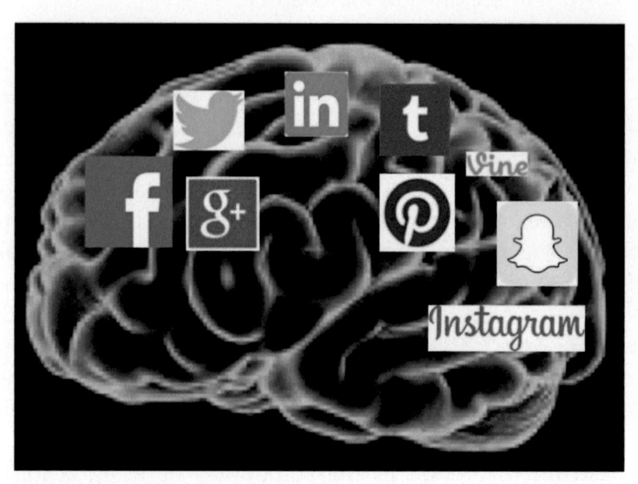

FIGURE 10.1 While new social media sites are springing up almost daily, Facebook, Twitter, LinkedIn, and other early-established sites are maintaining their hold on the American public. *Source: Nicole Gage with permission.*

Social Media Sites' Effects on the Brain—Healthy or Dangerous?

Ok. Accessing social media sites is at an all-time high—for children, teens, adults, and seniors—and there is no sign of that trend changing any time soon.

What is the problem?

It depends on who you ask—there are many positive aspects of social media. Just going online and searching the Internet has been found to stimulate the brain in important and healthy ways for older adults (3). However, there are some less positive effects of heavy social media site usage. Do you have a friend who is proud of his or her multitasking abilities—with multiple smart phone conversations happening at the same time such as person-to-person conversations, searching online, etc.? A recent study comparing the performance of self-described multitaskers versus nonmultitaskers showed that the multitaskers actually had more difficulty shifting between cognitive tasks and were more distractible (4).

Social media use feeds into two central needs for human beings: self-presentation[1] and belonging. Accessing social media sites can be quite fulfilling to our social selves; however, there are as many potentially negative effects as positive. While social media sites are rewarding to visit—how many Likes did I get today?—they can also be so rewarding that we become almost addicted to them, spending hours each day. And while we enjoy checking out what our friends and family are posting, it can become an almost voyeuristic behavior if unchecked. As oxymoronic as it sounds, heavy social media site usage can actually lead to a sense of isolation (Fig. 10.3). For a thoughtful review of the positive and negative aspects of social networking sites, see Ref. (5).

A Balancing Act

For most of us, accessing social media sites is an enjoyable activity, allowing us to connect with friends and family who live far away, letting us have "face time" with loved ones, and

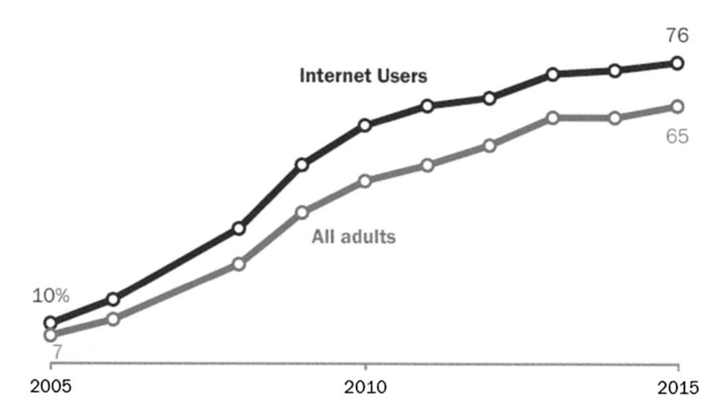

FIGURE 10.2 Social networking has risen dramatically from only 7% of American adults using it in 2005 to more than 75% of adults using Internet in 2015. *Source: Andrew Perrin. "Social Networking Usage: 2005–15." with permission.*

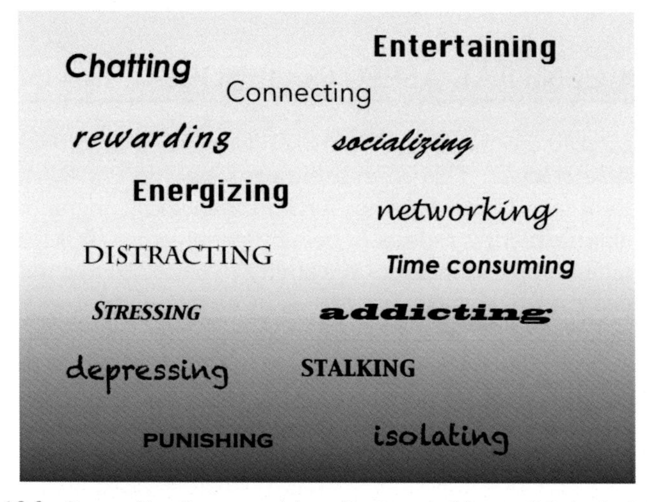

FIGURE 10.3 Is spending time on social media sites good for you? Or bad—it depends on who you ask. *Source: Nicole Gage with permission.*

Continued

BOX 10.1 *(cont'd)*

maintaining friendships new and old. However, the studies being released to date provide evidence that accessing social media sites can be damaging—with the risks ranging from mere distraction to time wasting to actual addiction.

The solution? Moderation, my friend, moderation. See you on Facebook!

References

(1) Perrin, A. (October 2015). *Social Networking Usage: 2005–2015.* Pew Research Center. Available at http://www.pewinternet.org/2015/10/08/2015/Social-Networking-Usage-2005-2015/.

(2) Lenhart, A. (April 2015). *Teen, Social Media and Technology Overview 2015.* Pew Research Center. Available at http://www.pewinternet.org/2015/04/09/teens-social-media-technology-2015.

(3) Small, G. W, Moody, T. D, Siddarth, P., & Bookheimer, S. Y. (2009). Your brain on Google: Patterns of cerebral activation during Internet searching. *American Journal of Geriatric Psychiatry, 17*:2, 116–126.

(4) Ophir, E., Nass, C., & Wagner, A. D. (2009). Cognitive control in media multitaskers. *Proceedings of the National Academy of Science, 106*(37), 15,583–15,587. https://doi.org/10.1073/pnas.0903620106.

(5) Mantymaki, M., & Najmul Islam, A. K. M. (2016). The Janus face of Facebook: Positive and negative sides of social networking site use. *Computers in Human Behavior, 61*, 14–26. https://doi.org/10.1016/j.chb.2016.02.078.

[1]Self-presentational behavior is the behavior that we use to control how we are presenting ourselves to other people. It can be very consciously done, such as preparing for an important job interview—checking out how we are dressed, speaking, etc. It can also be an almost unconscious behavior, such as smoothing our hair or straightening our tie.

2. FOUR CENTRAL ASPECTS OF SOCIAL COGNITION

This chapter is organized around four central aspects of social cognition: *TOM, empathy, social perception*, and *social behavior*. This is an arbitrary separation of the many and complex cognitive and behavioral factors that together form human social cognition—however, it is a useful way to think about differing aspects of being a social being. TOM, as described briefly earlier and in more detail to follow, includes the ability to recognize the mental and emotional states of others. A closely related aspect, *empathy*, is more than just *recognizing* someone else's pain; it is *feeling that pain as well.* Think of *social perception* as the nonverbal aspects of communicating with others: it includes decoding social and emotional information in others' faces, voice intonations, bodies, and other nonverbal cues. Correctly decoding these cues can be key to survival, especially recognizing threat or aggression in others' face and body expressions. Face perception is a critical aspect of social perception: the inability to recognize faces and the emotional and social cues in faces is a central deficit in some disorders of social cognition, which will be discussed later in the chapter. This inability or deficit makes interacting with others a very difficult and uncomfortable situation—and clearly, if an individual is unable to decode emotional cues in another's facial expressions, this will severely hinder the ability to feel empathy for that individual. Lastly, we will discuss *social behaviors* and how they relate to social cognition. The knowledge of how to act socially is a central part of being a social human: extracting cues for appropriate social behavior from others is a process that occurs early in life for typically developing humans. Inappropriate social behaviors are also a key aspect of disorders of social cognition.

2.1 Brain Areas Subserving the Four Central Aspects of Social Cognition

The advent of neuroimaging techniques such as functional magnetic resonance imaging (fMRI) has caused an explosion of new studies investigating where and how these aspects of social cognition are processed in the brain. As we will see, these are complex questions to address experimentally. We will discuss brain areas involved in each of the four aspects we focus on here, but before we do, we will present a brief summary of the major brain areas that—cumulatively, across many studies—are activated during certain types of tasks that tap into the four aspects (see Henry, von Hippel, Molenberghs, Lee, & Sachdev, 2015; for a review).

TOM includes many stages and types of social and mental processes; thus, finding a single TOM place in the brain is not a reasonable or attainable goal. We will discuss this in more detail later in the chapter. However, two key brain regions systematically are activated for TOM tasks: the temporoparietal junction (TPJ) at the border of the temporal and parietal lobe and the dorsomedial prefrontal cortex (dmPFC) (Fig. 10.4).

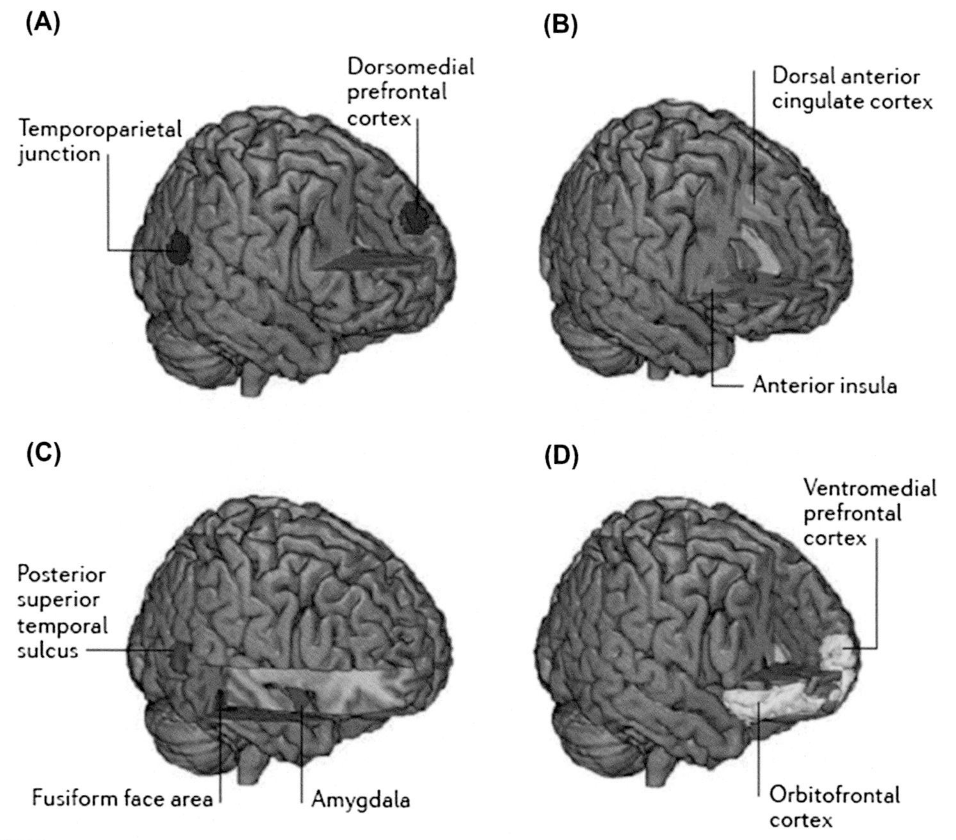

FIGURE 10.4 Brain regions that are involved in four key aspects of social cognition. (A) Theory of mind experiments consistently show activation in the dorsomedial prefrontal cortex and the temporoparietal junction. (B) Experiments investigating the brain bases of empathy show dorsal anterior cingulate cortex and anterior insula activation. (C) Studies of social perception, including face and biological motion perception, show activations in the fusiform face area and the posterior superior temporal sulcus, respectively. (D) Social behavior studies reliably activate the ventromedial prefrontal cortex and the orbitofrontal cortex. *Source: Henry et al. (2015).*

Empathy can be broken down into two rudimental parts: the *cognitive* aspects of evaluating and understanding someone else's feelings and *emotional* aspects of feeling the same feelings. Brain areas that are activated in studies investigating the neural bases of empathy are the anterior midcingulate cortex (aMCC) and the anterior insula in both left and right hemispheres (Fan et al., 2011; Kanske, Bockler, Trautwein, & Singer, 2015). The activations in the cingulate cortex are thought to correspond to cognitive aspects of the feeling of empathy, and the activations in the insula are thought to correspond to emotional aspects (Fig. 10.4).

Two aspects of social perception that have been the object of much study are *biological motion* and *face perception*. Biological motion refers to motions that a body can make as a set of articulated limbs, trunk, and head. Understanding the social implications of body movements is key to detecting any sign of danger or threat—as well as helping understand another person's intentions, emotions, and state of mind. *Face perception* is a core human ability that allows us not only to recognize someone we already know but also to learn new faces and to detect social cues in their features and facial movements. Biological motion perception taps regions within the superior temporal sulcus (STS) (Kanske, Bockler, Trautwein, & Singer, 2015; Grossman, Jardine, & Pyles, 2010). Face perception taps a small region in the fusiform gyrus (Fig. 10.4).

Social behavior is perhaps the most difficult aspect of social cognition to discuss in concrete terms. How does a child learn to behave in a socially acceptable way? Of course, the child's parents teach the child good manners, how to behave in certain social situations such as attending a birthday party or going to a restaurant. However, social behavior happens in both far more subtle ways and far more extreme ways. One way to think about social behavior is to think about social misbehavior: aggressive or violent behavior, odd or unusual behavior, and impulsive and addictive behaviors. All of these aspects of social behavior—and misbehavior—fall under the research that is conducted on social behavior. You can imagine that the brain bases for aspects of human behavior ranging from addiction to violence may be highly variable—and that would be correct. However, two areas in the brain are frequently cited in studies that investigate social behavior: the ventromedial prefrontal cortex (vmPFC) and the orbitofrontal cortex (OFC) (Fig. 10.4). Their role in social behavior will be discussed in more detail later in the chapter: briefly, the vmPFC is an area implicated in *social decision-making* and the OFC for processing the rewards associated with those decisions.

3. THEORY OF MIND MODEL: AN ORGANIZING FRAMEWORK FOR SOCIAL COGNITION

Once past our fourth birthday, we human beings give indications of understanding others' minds—"mentalizing," as Chris Frith has called it. We can recognize and respond to the invisible, internal subjective regularities that account for the behaviors of others. We will call the full-fledged ability to understand and predict our own and others' minds *TOM*. TOM has been explained by three kinds of theories: module theories, theory theories, and simulation theories.

According to *module theories*, such as that of Simon Baron-Cohen, human beings develop a theory of mind module (TOMM) that is separate from but builds on other mental abilities that may be shared with nonhuman primates and other mammals; only humans are presumed to have a complete TOMM. This kind of theory fits well with findings from the study of autism.

Theory theories suppose that TOM capabilities develop as a primitive, implicit theory over the course of development, much like Piaget's conservation theories. Such implicit theories predict abrupt changes in behavior as new knowledge is added, as is seen in the abrupt change in children's understanding of their own minds between ages three and four.

Simulation theories suppose that we understand other minds by internally simulating or "running off-line" the mental states of others in each situation. The dual responsiveness of mirror neurons to self- and other-generated action could be taken as support for simulation theory.

It seems very likely that all three kinds of theories are needed to account for human "mentalizing" abilities. As we will see, there are separable skills that develop in mammals and humans, which operate much like *modules*; we can lose one module but still have the other. The system that allows us to imitate others seems to operate through *internal simulation* of the actions of others. Finally, adult human beings have sophisticated social perception abilities that allow us to reason about other people's internal states; we act as if we have a complex set of rules about our own and others' mental states that could be called an *implicit theory*.

Simon Baron-Cohen (1995) hypothesized that a fully developed TOM is composed of four kinds of skills that develop independently. These skills are *detection of intentions of others, detection of eye direction, shared attention*, and a complex repertoire of implicit knowledge about others, which he called the *TOMM*. Some of these skills are observed in mammals and nonhuman primates as well as in humans. However, only the TOMM appears in normally developing human beings. We will first introduce Baron-Cohen's model (Fig. 10.5) and then use it as a way to organize the larger body of social cognition research.

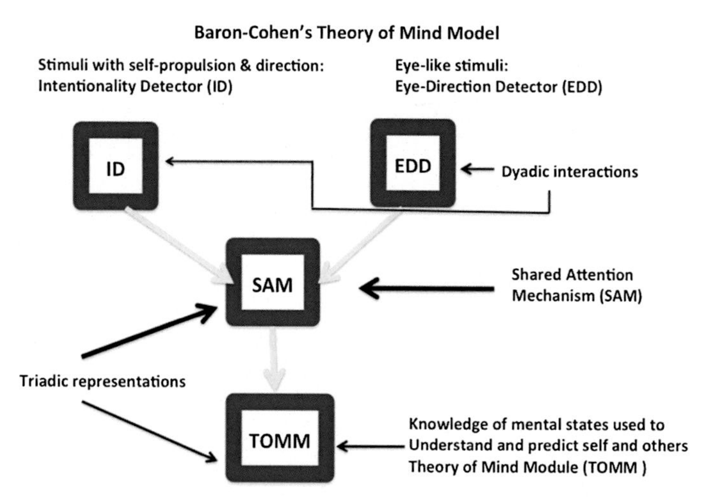

FIGURE 10.5 A schematic diagram of Baron-Cohen's theory of mind model with the EDD, shown on the upper right side, sending inputs to the ID, upper left, and to the SAM, shown in the center of the diagram. The SAM also receives inputs from the ID and interacts with the TOMM. *Source: Adapted from Baron-Cohen (1995).*

3.1 Intention

The first component of Baron-Cohen's TOM is called the *intentionality detector* (ID). This is the ability to perceive intention or purposeful action in many forms of biological and nonbiological movement. For example, when we watch leaves swirling in a parking lot, we have a tendency to see the leaves as "wanting to go together." We ascribe common purpose to the pile of leaves. Or when we watch pieces of modeling clay being moved around an artificial landscape in claymation films, we readily attribute intentions and other mental states to the pieces of clay. Likewise, when we watch people and animals engaged in behaviors, we seem to understand their goals and the desired outcomes of their actions. We interpret *action* as intention.

3.2 Eye Detection

The second component of the model is the *eye-direction detector* (EDD), which is the skill to detect eyes and eyelike stimuli and to determine the direction of gaze. Many mammals seem to have the ability to notice and use information about eye direction. Cats, for example, use eye direction as part of their social dominance behavior with other cats; the nondominant cat must avert its eyes in the face of the dominant cat. Humans, from the first hours of life, search for and focus on the eyes of their caregivers. We also have a strong tendency to see nonliving stimuli as eyelike; hence, we see the "man in the moon" and faces on automobiles, gnarled trees, and mountains. The "language of the eyes" seems to be a fundamental means of communicating mental states among humans.

Both the ID and the EDD involve *dyadic* (two-way) interactions. That is, there is *one perceiver* and *one object of perception*. As yet, no sharing of mental states is necessarily involved. Both EDD and ID are found in nonhumans as part of their social perception abilities. It is the third module of TOM that is unique to human social cognition.

3.3 Shared Attention Mechanism

The *shared attention mechanism* (SAM) is the ability we have by the end of our first year of life to understand that when someone else shifts his or her direction of gaze, he or she is "looking at" something. We seem to learn that looking leads to seeing—an advance over the simpler signal of eye direction. We realize that we can look, too, and see the same thing. Gaze shifting and social pointing of fingers are ways we learn to direct the attention of a companion (Fig. 10.6).

Infants before 1 year of age, most other primates, and other mammals do not have a shared attention ability. We can see this in our much-loved companion animals. While our family dog may chase a ball and bring it back, he will not follow our gaze if we look toward a ball lying in the grass. He will not follow our pointing finger when we try to direct his gaze toward the ball. The dog has considerable intelligence but does not have shared attention. Similarly, an infant at 6 months does not turn his/her head to follow the caregiver's gaze, but a 1-year-old does. Shared attention abilities mark the human species (Fig. 10.7).

FIGURE 10.6 A 14-month-old pointing. Pointing is a sign of triadic interaction. *Source: Public domain image.*

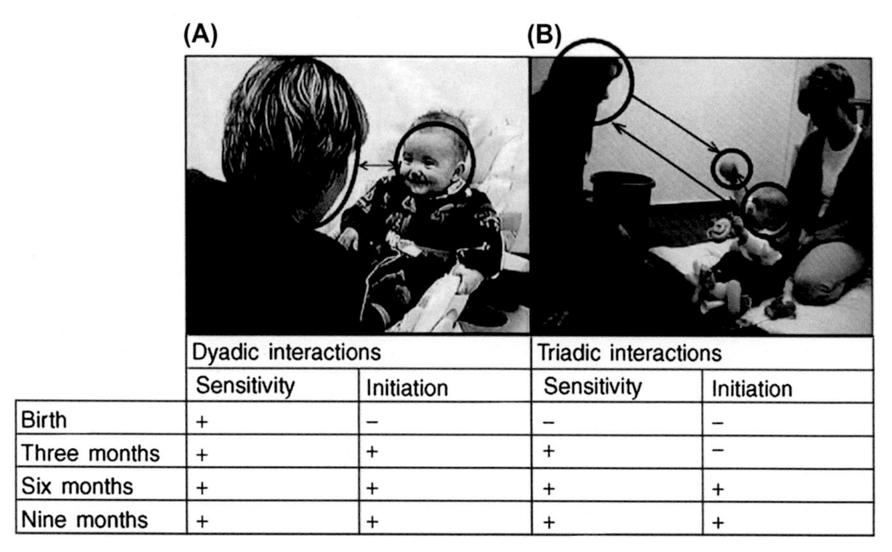

	Dyadic interactions		Triadic interactions	
	Sensitivity	Initiation	Sensitivity	Initiation
Birth	+	−	−	−
Three months	+	+	+	−
Six months	+	+	+	+
Nine months	+	+	+	+

FIGURE 10.7 There is a developmental progression in the sensitivity of infants to dyadic (two-way) and triadic (three-way) social interactions. (A) Dyadic (person-to-person) exchange. (B) Triadic (person–object–person) exchange. The plus (+) sign indicates evidence for that skill (dyadic or triadic exchange), and the minus (−) sign indicates no evidence for that skill. *Source: Striano and Reid (2006).*

3.4 Higher-Order Theory of Mind

The final component of full-fledged TOM is what Baron-Cohen has called the *TOMM*, a complex knowledge base containing rules of social cognition that develops by the time we reach our fourth birthday. TOMM tells us the following:

- Appearance and reality are not necessarily the same: A rock can look like an egg but not be an egg; I can pretend to be a dog but not be a dog.

BOX 10.2

TESTING THEORY OF MIND—THE SALLY–ANN TASK

Have you wondered how you investigate a young child's theory of mind abilities? The Sally–Ann task is often used (Fig. 10.8). In this scenario, Sally has a doll carriage and

FIGURE 10.8 In this scenario, Sally has a doll carriage and Ann has a toy box (top panel). Second and third panel: Sally puts a toy into the carriage and walks away. Fourth panel: Ann takes the toy out of the carriage and puts it in her toy box. Fifth panel: When Sally returns, where will she look for the toy? A normally developing child from the age of four or so will respond that Sally will look for the toy in the baby carriage because that is where she would think it was. This child has successfully attributed the mental state of Sally in this situation—despite the fact that the child knows where the toy actually is. This is a key step in developing a full theory of mind. *Source: Adolphs (2003).*

BOX 10.2 *(cont'd)*

Ann has a toy box. Sally puts a toy into the carriage and walks away. Ann takes the toy out of the carriage and puts it in her toy box. When Sally returns, where will she look for the toy? A normally developing child from the age of four or so will respond that Sally will look for the toy in the baby carriage because that is where she would think it was. In other words, the child being assessed for theory of mind abilities has correctly put aside their knowledge of the toy being placed in the toy box by Ann when Sally was out of the room because they have understood that Sally would not know that—and that Sally would look in the carriage because that is what Sally would understand. Thus the child has been able to put herself in Sally's place and is able to correctly predict what Sally would do: she or he has *attributed a mental state* to Sally. These abilities are important steps in developing a full theory of mind in a young child and are key for good socialization skills.

- A person who is sitting still in a chair may be "doing something"—that is, thinking, imagining, or remembering (young children do not appreciate this).
- Other people can have mental states as well as physical states.
- Other people can know things that I do not know: I can be fooled or deceived; I can detect deception.
- I can know things that other people do not know: I can fool or deceive others; I understand the point of games such as hide-and-seek.
- My mental state in the past was different from how it is now.
- Facial expressions are indicators of mental states as much as they are indicators of physical states: I can distinguish a surprised face from a yawning one.

TOM is not the same as intelligence or IQ. Developmentally delayed children and adults display complete TOM abilities despite low IQs, whereas people living with ASDs may have high IQs but markedly deficient TOM abilities. We can now use Baron-Cohen's four TOM skills as a way to organize and guide our study of social cognition.

3.5 Brain Areas for Theory of Mind Processes

The number of journal articles investigating TOM has skyrocketed in recent years (Fig. 10.9). Some of this increased attention to TOM investigation is due the development of neuroimaging techniques such as fMRI that allow this type of study to be feasible. A second reason for the increased attention is the investigation of TOM in developmental disorders such as autism.

Before we begin to discuss in more detail the brain bases for TOM and other aspects of social cognition, we need to mention a caveat about these types of neuroimaging studies

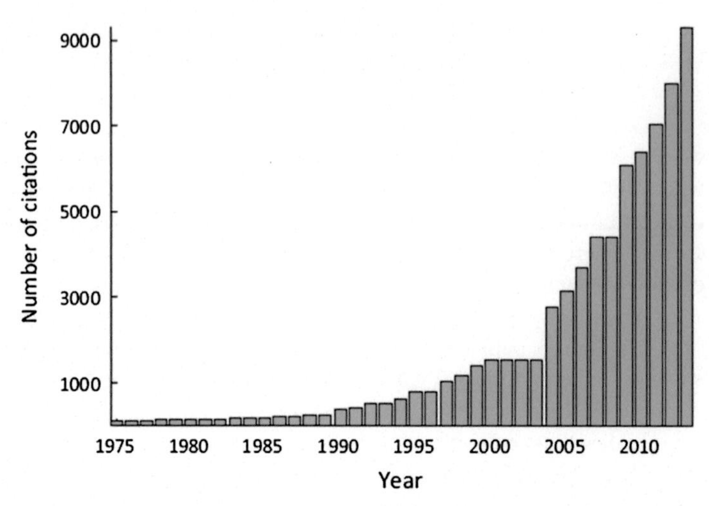

FIGURE 10.9 There has been a virtual explosion of articles referring theory of mind in recent years. *Source:- Schaafsma, Pfaff, Spunt, & Adolphs (2015).*

of higher-order human cognition: setting up experiments that provide precise information about the neural mechanisms at work in aspects of social cognition is a challenge. Think about one aspect of social cognition: TOM. This theory includes many differing abilities and skills that may tap similar or overlapping brain regions. Neuroimaging studies almost always incorporate tasks. The specific aspect of TOM and the precise tasks used will have a major effect on the brain findings. And, if neural regions activate for more than one task, and if a subtractive analysis is used to compare activation in one task versus another, then the activity that has activated for both tasks will be zeroed out and thus lost in terms of identifying it as a social cognition region. Despite these challenges, much progress has been made in studying the neural basis of social cognition.

3.5.1 Intention

You see your roommate come out of his room wearing sweatpants, a T-shirt, and running shoes. He is resetting his watch. Ah, you think, he is going for a run. But how do you know his intention based on the simple actions of him leaving a room and setting a watch? Understanding intentions is a key aspect of action perception, and humans are amazingly adept at it. Converging evidence from animal studies using single-cell recordings and human neuroimaging studies implicate the posterior part of the STS (pSTS) in interpreting actions.

3.5.2 Eye Detection and Gaze Perception

Perception of eyes in conspecifics (individuals of the same species: such as human–human, dog–dog) and guiding of social behavior in light of that perception occurs in many classes of animals from reptiles to humans. Although fish, reptiles, birds, and other mammals have some ability to process eyelike stimuli and perceive gaze direction, only great apes (gorillas, chimpanzees, bonobos, and orangutans) and humans have shared attention and can use the "language of the eyes" to understand the mental states of others.

Several types of eye and gaze processes are shown in Fig. 10.10. In the first example (A left), mutual gaze is shown where two individuals are making direct eye contact. In some animals, this type of direct eye contact can be threatening. In humans, however, direct

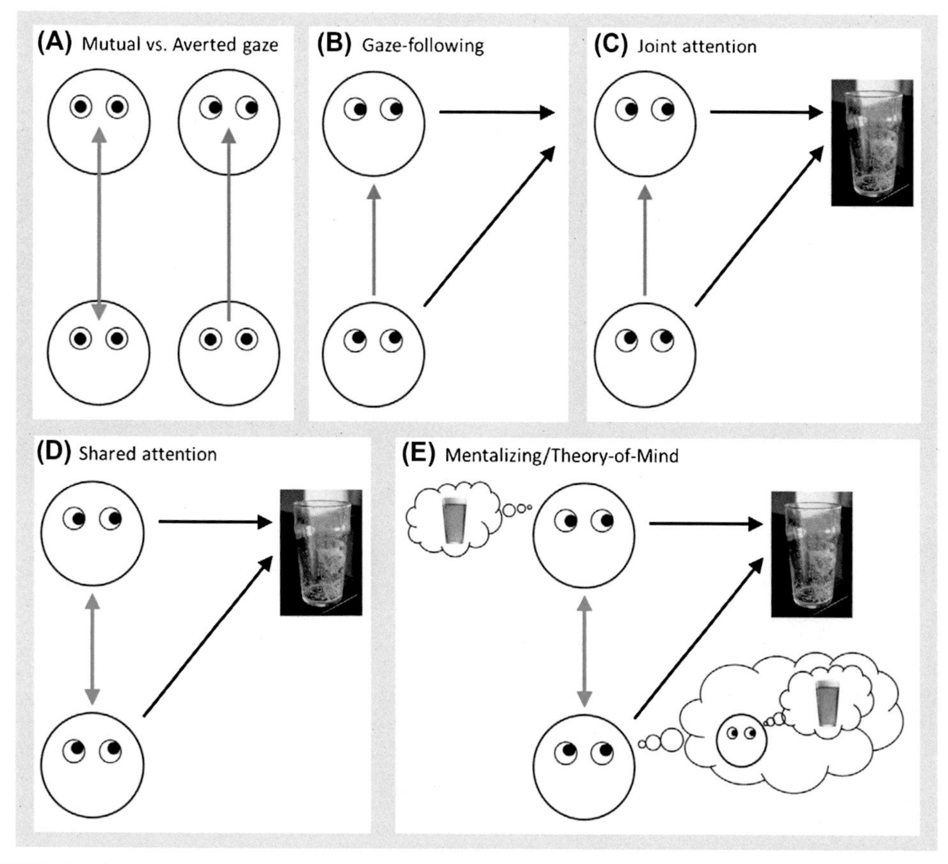

FIGURE 10.10 Eye and gaze processing. (A) Mutual gaze is where the attention of individuals A and B is directed to each other. Averted gaze is when individual A is looking at B, but the focus of B's attention is elsewhere. (B) Gaze following is where individual A detects that B's gaze is not directed toward him and follows the line of sight of B to a point in space. (C) Joint attention is the same as gaze following except that there is a focus of attention (an object) so that individuals A and B are looking at the same object. (D) Shared attention is a combination of mutual attention and joint attention, where the focus of attention of A and B is on an object of joint focus and on each other ("I know you're looking at X, and you know I'm looking at X"). (E) Theory of mind relies on 1–4 as well as higher-order social knowledge that allows individuals to know that the other is attending to an object because that person intends to do something with the object. *Source: Pfeiffer et al. (2013).*

eye contact and mutual gaze are important bases for social communication (see the review by Senju & Johnson, 2009).

Where do these processes take place in the brain? If you think about conducting a neuro-imaging study of gaze perception, you would need to bear in mind that while some eye detection and gaze perception situations occur when someone is looking straight at you, many occur when the head is turned to one side or the other. To determine what areas of the brain are activated for gaze perception regardless of the position of the head and eyes, Carlin, Calder, Kriegeskorte, Nilli, and Rowe (2011) used an experimental design that included many head rotation possibilities (Fig. 10.11). They found that across all possible head rotations, an area in the anterior STS (aSTS) was reliably activated.

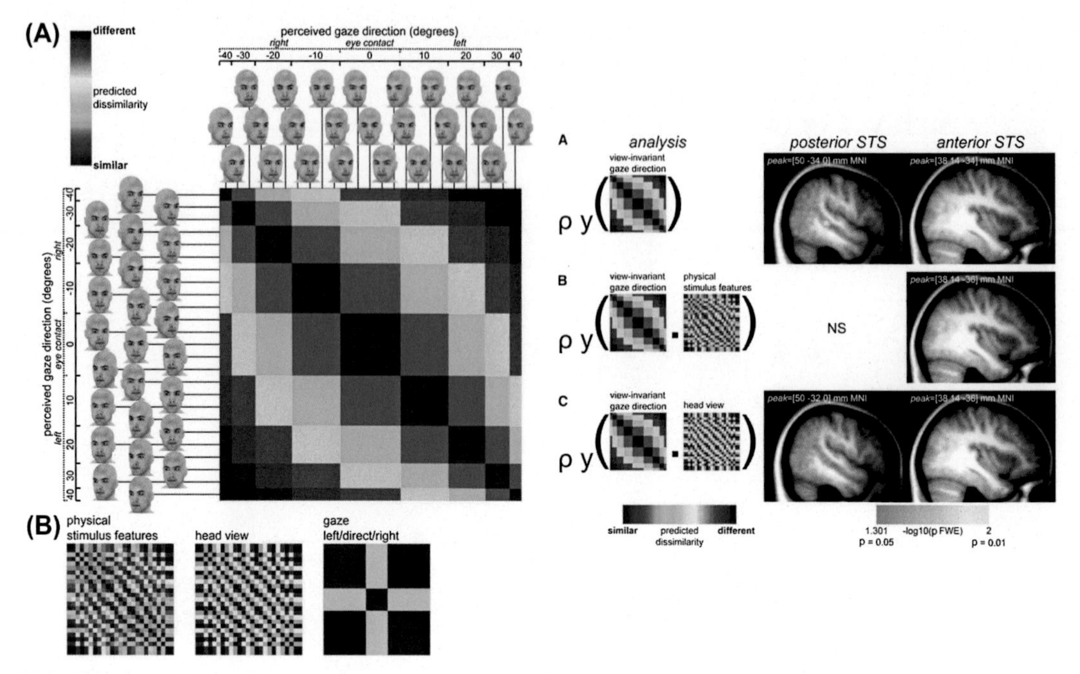

FIGURE 10.11 To determine what areas of the brain are activated for gaze perception regardless of the position of the head and eyes, Carlin and colleagues (2011) used an experimental design that included many head rotation possibilities. Left side of the figure: (A) The head stimuli used in the experiment, showing the many types of views and head rotations. The colored square in the center of (A) shows the predicted dissimilarity of the many gaze directions shown, with the color blue associated with predicted similar gaze directions and the color red with predicted dissimilar gaze directions. (B) Predicted dissimilarity for physical stimulus features (left), head views (center), and gaze (right). Right side of the figure: Brain regions showing reliable activation across all possible head rotations were found in an area in the anterior superior temporal sulcus (aSTS). (A) View invariant gaze direction showed reliable activation in both posterior and anterior STS (left and right brain images). The aSTS activation was also reliable when controlling for physical stimulus features (B) and head views (C). *Source: Carlin et al. (2011).*

3.5.3 Shared Attention

Shared attention seems to be a social skill that is unique to great apes and humans. Remember that shared attention is more than simply looking at the same thing that another person is looking at. Shared attention involves the additional qualification that the two observers not only observe the same object but also know that the other is looking at the object. It is a triadic (three-way) activity (Fig. 10.10). It has two stages: first, one observer *initiates the shared attention* by gazing at an object. Second, the second observer *responds to the shared attention* initiation by gazing at the same object. Shared attention allows us implicitly to recognize that "*I* know that *you* are looking at *that*." Apes and humans seem to know that when conspecifics are gazing at something, they are also internally representing it. Looking leads to seeing. If I want to see what you see, I can follow your gaze.

How do these shared processes get set up in humans? Brooks and Meltzoff (2003) have shown that human infants begin to follow the direction in which an adult turns his or her head by the age of 9 months; however, at this age, the infant follows head direction despite knowing whether the model's eyes are open or closed. By 12 months of age, the infant will follow gaze more often when the model's eyes are open than when they are shut. The infant follows gaze rather than head direction now. Shared attention has developed. The infant seems to know implicitly that open eyes allow looking, and looking leads to seeing.

What areas of the brain support shared attention? We know that STS supports eye detection. To move from simple detection to shared attention, areas in the prefrontal cortex become involved. Redcay, Kleiner, and Saxe (2012) investigated both the initiation of shared attention and the response to shared attention initiation in an fMRI study. They found a shared attention network for both initiating and responding to shared attention that included the dmPFC in both hemispheres and the pSTS in the right hemisphere (Fig. 10.12).

Take a careful look at Fig. 10.12, and you will see these areas shown in pink. Consider that an experiment investigating brain areas involved during shared attention will also involve the visual cortex and the executive attentional system. Redcay et al. were able to identify the brain regions activated for each of these separate aspects of their task and identified them with color coding in Fig. 10.12: the social network regions for shared attention are shown in pink, the attention and control network regions are shown in green, and the visual network regions are shown in blue.

3.5.4 Higher-Order Theory of Mind Abilities': Attribution of Mental States to Ourselves and Others

Higher-order TOM abilities include a broad brush of social cognitive processes, which are shown in Fig. 10.13 (Schaafsma, Pfaff, Spunt, & Adolphs, 2015).

How can we isolate distinct brain regions that contribute to these broad and overlapping mental processes? Before we begin, let us take a look at the neuroanatomy of regions implicated in these higher-order TOM processes. They include the medial wall of the prefrontal cortex and surrounding areas including the anterior cingulate cortex (ACC) and the TPJ (Fig. 10.4A). The medial wall of prefrontal cortex can be divided into three segments from top to bottom: dmPFC, medial prefrontal cortex (MPFC), and vmPFC. dmPFC includes the cortex at the top of the medial wall of the prefrontal cortex. MPFC is the middle section of the medial wall of prefrontal cortex (PFC). vmPFC is composed of the bottom of the prefrontal lobe and the lower inside wall of the prefrontal cortex (Fig. 10.14).

The cingulate and paracingulate gyri, which are sometimes spoken of collectively as the ACC, form a belt (*cingulum* in Latin) around the corpus callosum. In addition, depending on how the gyri of individual brains are folded, the paracingulate gyrus may be folded into a sulcus. Fig. 10.15 helps us see the anatomy of the cingulate gyrus.

Looking at the many TOM processes detailed in Fig. 10.13, you might wonder if—despite their diverse nature—they activate a core region or set of regions in the brain. Looking at a metaanalysis of more than 70 neuroimaging studies of TOM, you will see that across many differing tasks and experimental conditions, these core regions described earlier, namely the medial wall of the PFC and the TPJ, are activated reliably again and again (Fig. 10.16, top). However, if you were to conduct that metaanalysis again, but instead of plotting the *pooled brain regions* activated across all tasks, you plotted the *brain regions activated separately* by each of a set of six tasks frequently used in TOM experiments—how would the results differ? Schurz, Radua, Aichhorn, Richlan, and Perner (2014) did just that. Look again at Fig. 10.16, but this time, look at the bottom of the figure. The different colors plotted on the MPFC and the TPJ show the activation pattern for six TOM tasks. See Box 10.3 and Fig. 10.17 for more details.

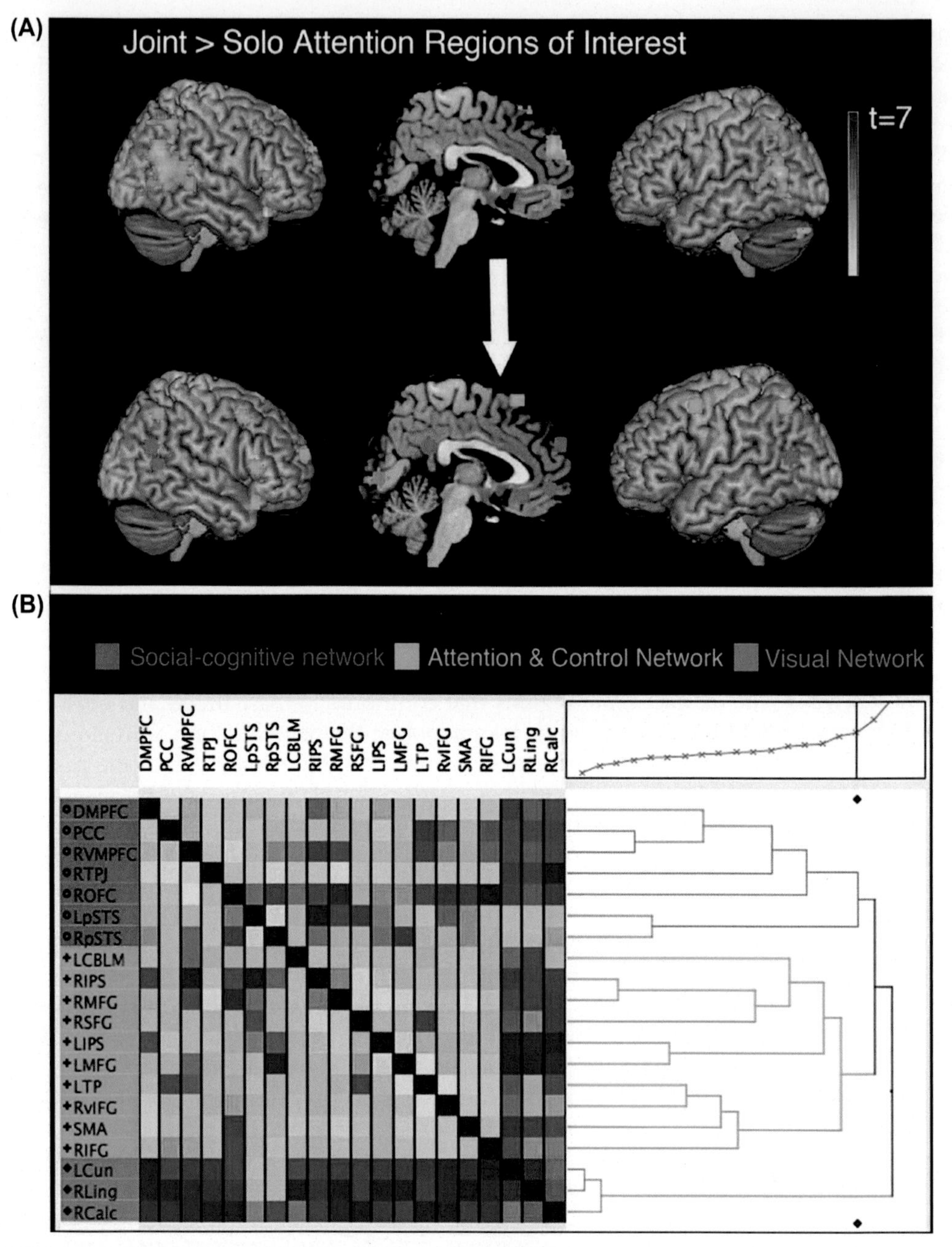

FIGURE 10.12 Redcay et al. (2012) investigated both the initiation of shared attention and the response to shared attention initiation in a functional magnetic resonance imaging study. (A) They found a shared attention network for both initiating and responding to shared attention that included the dorsomedial prefrontal cortex (DMPFC) in both hemispheres and the posterior superior temporal sulcus (pSTS) in the right hemisphere. Consider that an experiment investigating brain areas involved during shared attention will also involve the visual cortex and the executive attentional system. (B) Redcay et al. were able to identify the brain regions activated for each of these separate aspects of their task and identified them with color coding: the social network regions for shared attention are shown in pink, the attention and control network regions are shown in green, and the visual network regions are shown in blue. *Source: Redcay et al. (2012).*

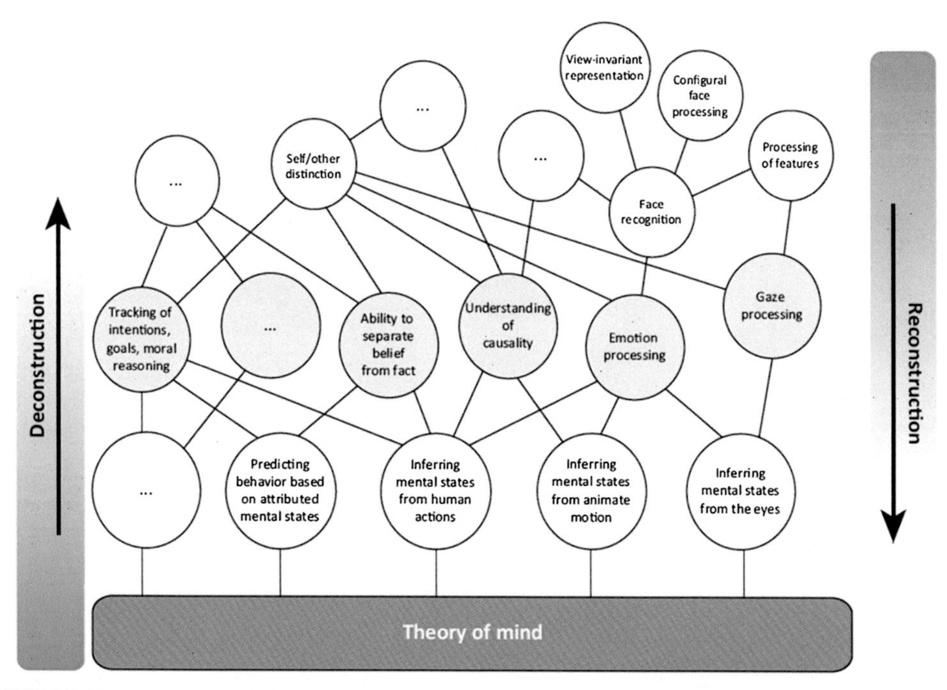

FIGURE 10.13 Higher-order theory of mind (TOM) abilities include a broad brush of social cognitive processes. The authors present an example of "rethinking" theory of mind functions to deconstruct them into basic components and then reconstruct them to form new ways of investigating TOM. *Source: Schaafsma et al. (2015).*

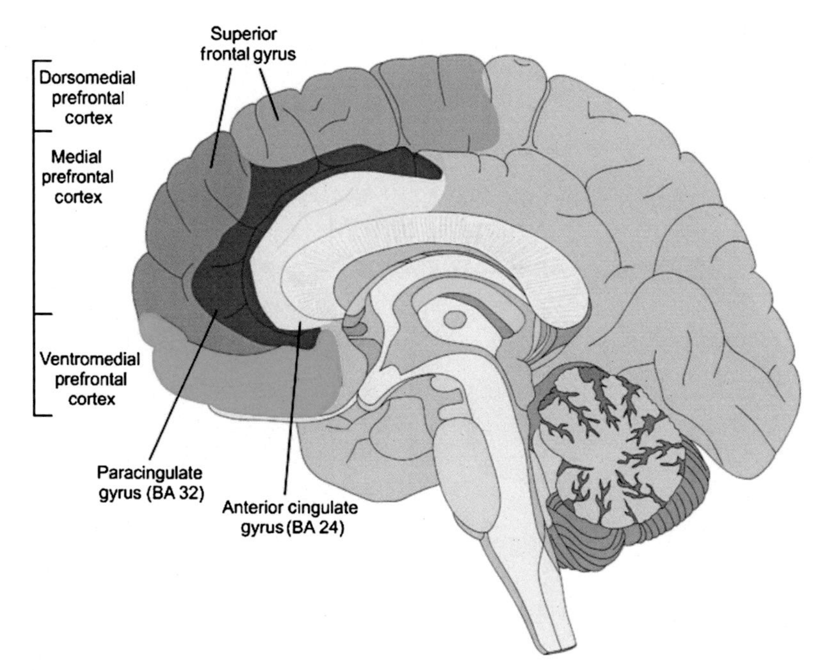

FIGURE 10.14 Divisions of prefrontal cortex: dorsomedial prefrontal cortex, medial prefrontal cortex , and ventromedial prefrontal cortex. Superior frontal, cingulate (BA 24), and paracingulate (BA 32) gyri are areas shown to be important in attribution of mental states. *Source: Baars and Fu with permission.*

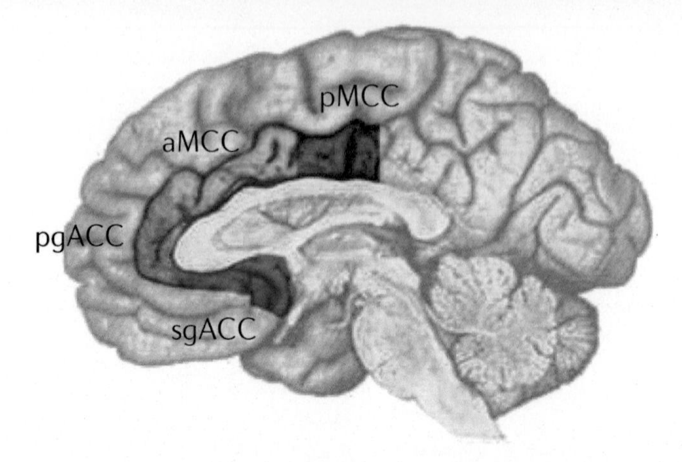

FIGURE 10.15 Left panel: The four major subdivisions of the cingulate are shown. The midcingulate cortex (MCC) is divided into two regions: the posterior MCC (pMCC) shown in red and the anterior MCC (aMCC) shown in green. The anterior cingulate cortex (ACC) is also divided into two regions: the pregenual ACC (pgACC) shown in orange and the subgenual ACC (sgACC) shown in blue. Right panel: The functions ascribed to the MCC and the ACC: the dorsal MCC including the aMCC and pMCC is thought to be involved in cognitive control, and the ACC including the pgACC and sgACC is thought to be involved in affective processes such as motivation. *Source: Shackman et al. (2011).*

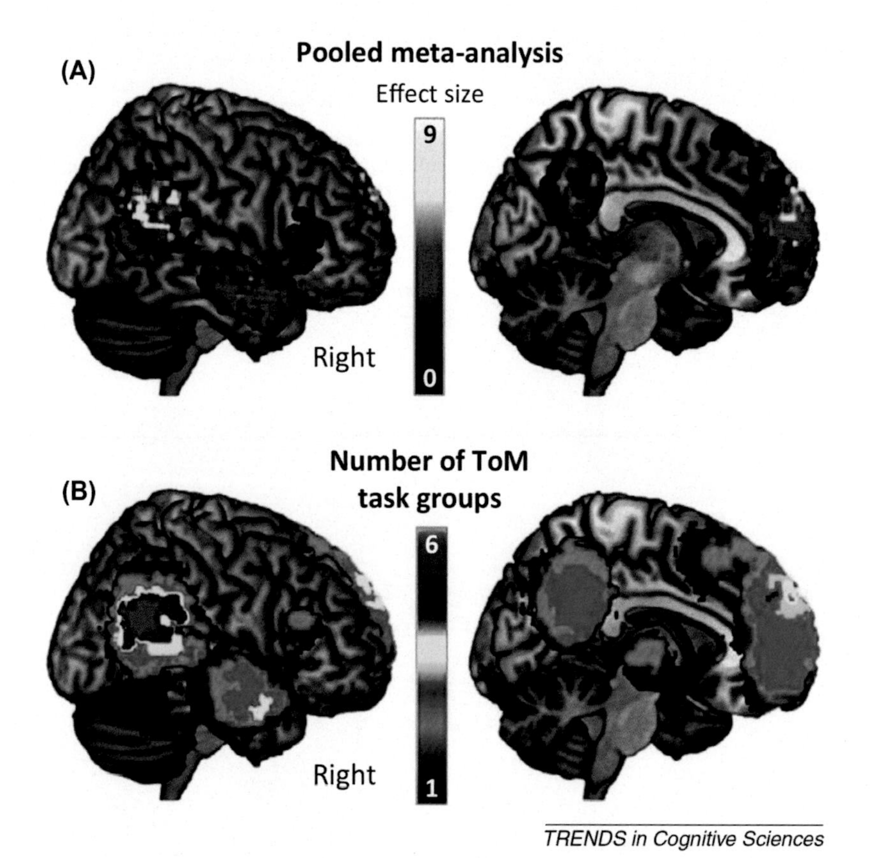

FIGURE 10.16 Top panel: (A) Data from a pooled metaanalysis of more than 70 neuroimaging studies investigating the theory of mind using one of six frequently used tasks is shown for the right hemisphere on a sagittal (left) and midsagittal (right) view. Bottom panel: (B) If the brain activations were separated by task (see tasks detailed in Box 10.3), the regions activated show distinctly differing underlying distributions. See Fig. 10.18 for more details. *Source: Schaafsma et al. (2015).*

BOX 10.3

FRACTIONATING BRAIN IMAGING OF THEORY OF MIND (TOM)—THE EFFECT OF TASKS

Higher-order TOM processes include a wide array of mental steps and are investigated using an equally wide array of experimental tasks. To determine the potentially separable brain activation effects of experimental tasks Schurz et al. (2014) conducted a metaanalysis of 73 neuroimaging studies of TOM. They selected six tasks that are in wide use in TOM research and analyzed the brain regions that were activated according to the task used.

The six tasks selected were false belief, trait judgment, strategic games, social animations, mind in the eyes, and rational actions. *False belief* tasks use an example of a person holding a false belief and then ask the subject to predict the behavior of that person or to answer a question about that person. An example of a false belief task is shown in Box 10.2 and is illustrated in Fig. 10.8. *Trait judgment* tasks tap into the brain processes underlying the formation of trait knowledge about fellow humans. Typically some set of personality traits are used such as honesty, courage, friendliness, etc. *Strategic games* tasks are tasks that use social abilities for success—such as predicting the behavior of your opponent while playing *rock, paper, scissors. Social animation* tasks use geometric shapes whose actions can be interpreted as implying intentions and goals. *Mind in the eyes* tasks present a set of eyes and ask the observer to make decisions about the mental state of the person behind those eyes. For example, a set of eyes and eyebrows are presented, and the observer is asked to decide if the person is (1) angry, (2) afraid, (3) jealous, or (4) unfriendly. Finally, *rational actions* tasks tap into our knowledge of gestural and situational social information to convey predicted actions. For example, a person is shown seated at a table in a restaurant with empty plates and glasses. A waiter walks by. What is the next logical thing that an observer may predict about the behavior of the person at the table?

You can tell by these very brief descriptions that these tasks tap aspects of social cognition—TOM—that are quite overlapping but also distinctly different. Schurz et al. (2014) presented the brain regions that were activated by each of these tasks in their metaanalysis (Fig. 10.17).

Take another look at Fig. 10.16—you will see that the pooled activation plot for all 73 studies shown at the top of the figure does not really tell the whole story of TOM brain networks. While some regions do overlap for several tasks—such as prefrontal cortex for false belief, trait judgments, and strategic games—they do not reliably activate for the remaining three tasks. On the other hand, the TPJ shows activation across most tasks.

Reference

(1) Schurz, M., Radua, J., Aichhorn, M., Richlan, F., and Perner, J. (2014). Fractionating theory of mind: A meta-analysis of functional brain imaging studies. *Neuroscience and Biobehavioral Reviews*, *42*, 9–34.

Continued

BOX 10.3 *(cont'd)*

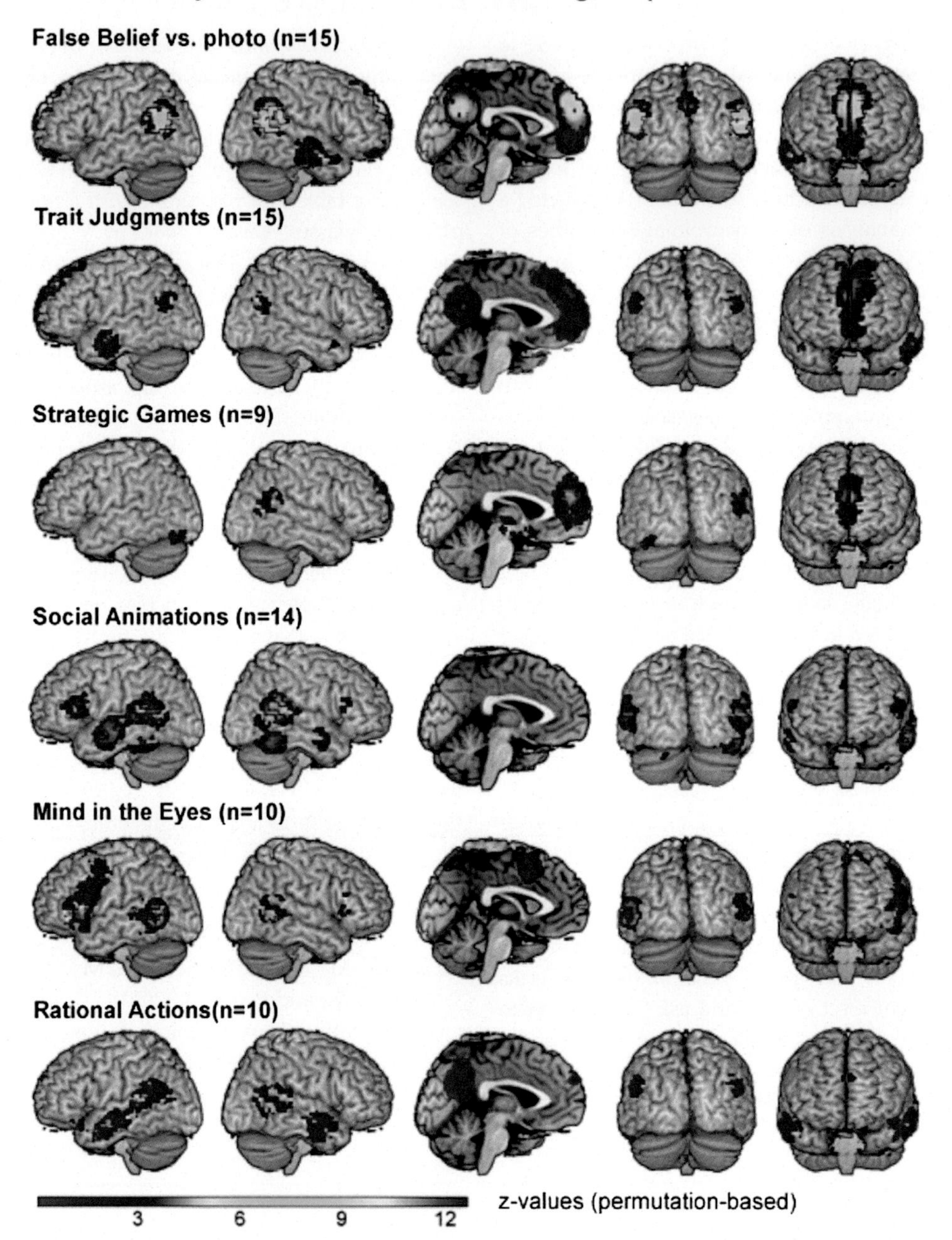

FIGURE 10.17 Instead of pooling brain activation patterns across all the six theory of mind tasks, Schurz et al. (2014) analyzed activation patterns individually for each of the six tasks. Notice that the activation patterns are starkly differing for the six tasks. *Source: Schurz et al. (2014).*

4. EMPATHY

Empathy is more than just recognizing someone's pain or pleasure—and feeling sympathetic or happy about it. Empathy goes beyond an *abstract sense of identifying someone else's emotional state*—it is the act of *experiencing that state oneself*. Think of it as *feeling someone else's pain*. Empathy is a core ability of the human social animal. Lack of empathy, as we will see later in the chapter in the section on disorders of social cognition, reflects a critical breakdown in social cognition.

Empathy can be broken down into two rudimental parts: the *cognitive* aspects of evaluating and understanding someone else's feelings and *emotional* (or affective) aspects of feeling the same feeling. Yan, Duncan, de Greck, and Northoff (2011) conducted a metaanalysis of neuroimaging studies investigating the brain bases of empathy. Studies were separated into two types to assess both the cognitive aspects of empathy and the emotional aspects. Studies tapping into the cognitive–evaluative aspects of empathy employed tasks where participants were asked to evaluate the emotional states of others (typically through the presentation of photos). Studies investigating the affective–perceptual aspects of empathy typically did not have an explicit task but rather passively viewed photos or films that depicted differing emotional states of others.

Results for the cognitive–evaluative and affective–perceptual forms of empathy are shown in Fig. 10.18. The cognitive–evaluative studies' brain responses are shown in green,

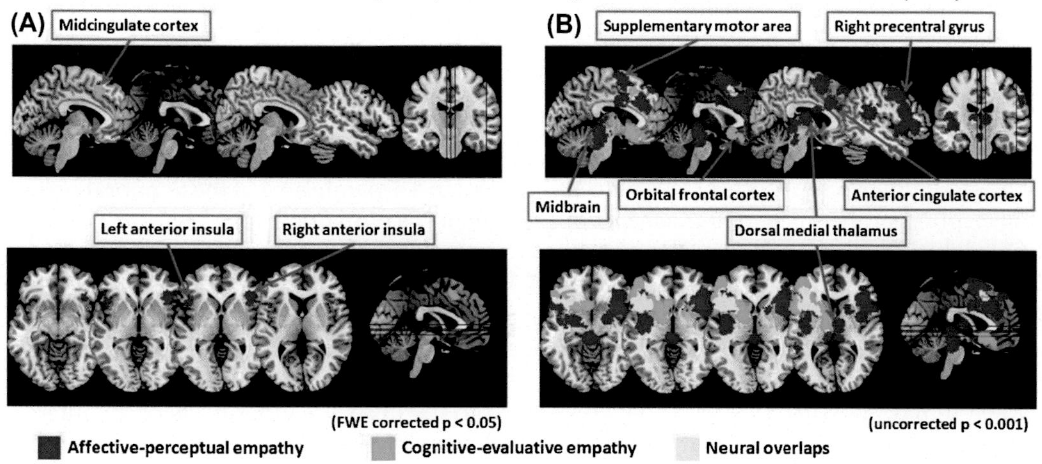

FIGURE 10.18 Yan et al. (2011) conducted a metaanalysis of neuroimaging studies investigating the brain bases of empathy. Studies were separated into two types to assess both the cognitive aspects of empathy and the emotional aspects. Studies tapping into the cognitive–evaluative aspects of empathy employed tasks where participants were asked to evaluate the emotional states of others (typically through the presentation of photos). Studies investigating the affective–perceptual aspects of empathy typically did not have an explicit task but rather passively viewed photos or films that depicted differing emotional states of others. The cognitive–evaluative studies' brain responses are shown in green, the affective–perceptual studies' brain responses are shown in red, and those brain areas that were activated for both sets of studies are shown in yellow. Key brain areas that were activated across all tasks and studies included the dorsal anterior cingulate cortex (dACC), the anterior midcingulate cortex, the Supplementary motor area, and the bilateral anterior insula (AI). The dACC was activated more frequently in the cognitive–evaluative studies, whereas the AI in the right hemisphere was activated more in the affective–perceptual studies. *Source: Yan et al. (2011).*

the affective–perceptual studies' brain responses are shown in red, and those brain areas that were activated for both sets of studies are shown in yellow. Key brain areas that were activated across all tasks and studies included the dorsal ACC (dACC), the aMCC, the supplementary motor area (SMA), and the bilateral anterior insula (AI). The dACC was activated more frequently in the cognitive–evaluative studies, whereas the AI in the right hemisphere was activated more in the affective–perceptive studies. One set of cognitive and emotional processes that have been proposed for empathy is that used in *simulation*—that is, those brain areas that would be active in ourselves if we were feeling the emotion we are observing in someone else. If that is the case, then it makes some intuitive sense that a broad range of brain areas spanning the SMA, insula, and cingulate cortex may together form a neural network for empathy processing.

5. SOCIAL PERCEPTION

Humans are social beings, and the ability to understand the mental and emotional states of other humans is a central aspect of social cognition. Social perception includes the ability to decode cues from another person's facial expressions, vocal intonation or prosody, and body movements. Together, these face, voice, and body cues provide a wealth of information about how that person is feeling and interacting in any given social situation.

Let us begin with face perception—a central part of social perception. For this chapter, we will focus on the perception of unchangeable aspects of the face. In Chapter 11, we will discuss emotional face perception.

5.1 Face Perception

Social information from eyes and gaze direction comes from the *changeable aspects of the human face*. We can also use visual information to detect the *invariant aspects of individual faces*, such as identity. These aspects of face perception occur in a separate area of the temporal cortex.

Perception of the unchanging aspects of the human face occurs in the fusiform face area (FFA), which is part of the inferior temporal lobe (Fig. 10.19). As an example of how the FFA looks in a brain image, we can look at the PET/MRI image from Caldara et al. (2006) (Fig. 10.20). These researchers compared cortical activation when participants observed objects versus human faces. In the image, we are looking at the bottom of the brain. The temporal lobes take up most of the outside areas of the image. The right and left FFAs are clearly marked in red, showing face receptive areas. Next to the FFAs are other parts of the inferior temporal lobe: the right and left parahippocampal gyri that respond to inanimate objects, such as houses or shoes.

Haxby et al. (2002) put the results of numerous studies together to create a model of face perception areas in the brain (Fig. 10.21). They proposed a hierarchical system of interconnected brain areas to account for both the changeable and invariant aspects of face perception that have been discussed in this chapter.

In this model, early visual analysis of facial features occurs in the visual cortex, inferior occipital gyrus (IOG). The IOG sends information to the STS, where changeable aspects of

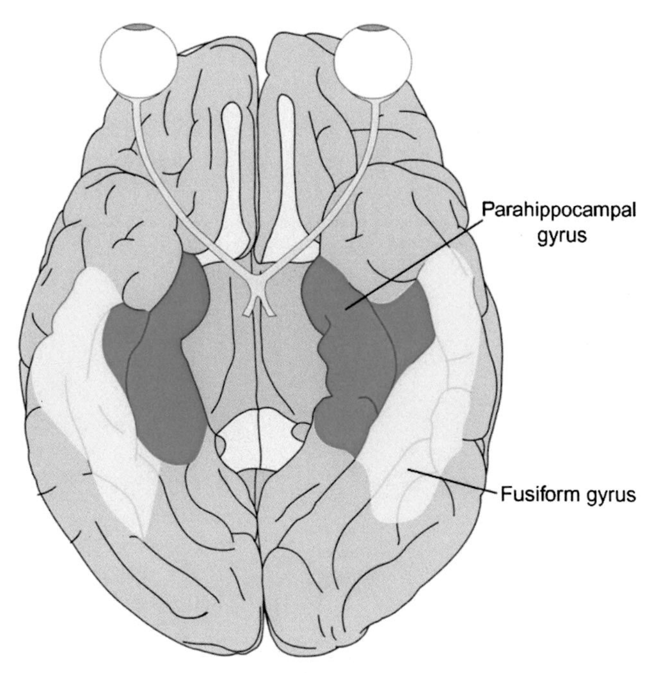

FIGURE 10.19 View from the underside of the theme brain, showing where the fusiform face area is located. *Source: Baars and Fu with permission.*

faces, such as eyes, are processed; from there, information about eyes is joined with spatial information in the intraparietal sulcus to generate gaze direction information. Information from STS can also be sent to the amygdala, where social and affective meanings are attached, and to the auditory cortex, where lip movements are registered. Invariant aspects of faces such as personal identity are processed in the lateral fusiform gyrus (the FFA) that is interconnected with the temporal lobe, where specific information about name and biographical data are retrieved.

We have often discussed in this book that neural processing includes both feed-forward information and reentrant and parallel processing. Together, these dynamic pathways and networks provide the incredibly rapid information processing that is the core of the human brain. Here is another model for face perception—this model includes the proposal of a dynamic interactive model for face perception. According to Freeman and Johnson (2016), both bottom-up (feed-forward) and top-down (reentrant and parallel) processes are at work in face perception (Fig. 10.22). This model makes intuitive sense: when we perceive a face—whether a new face or the face of a well-recognized friend—there are many cues that come from our stored knowledge of faces. According to the dynamic interactive model proposed by Freeman and Johnson, key brain regions for face perception include the fusiform gyrus, the OFC, and the anterior temporal lobe.

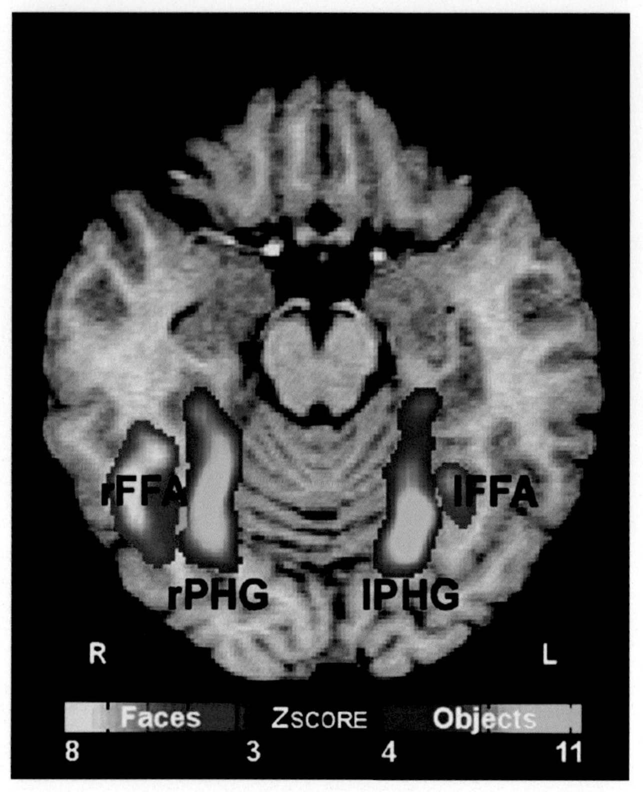

FIGURE 10.20 A view of the underside of the brain with fusiform face areas (FFAs) shown in red and parahippocampal gyri (PHG) shown in blue. *Source: Caldara et al. (2006).*

5.2 Biological Motion Perception

Think about a human being dressed all in black in a dark room: on each articulating joint—each shoulder, elbow, wrist, knee, and ankle—on both the left and right sides of the body, there is a single point of light. When this person moved around the dark room, your only cues to perceiving and understanding that motion would be the many points of light coming from the joints of the person's body. If you did this experiment, you might be amazed at how easy it is to understand this biological motion with only the barest bits of information coming from the points of light. As it turns out, experiments like this are a central way to investigate the neural bases of biological motion, although instead of having a person dressed in black with lights on their joints, modern experiments use computer generated point-light stimuli.

Where in the brain is the biological motion processed? There is a large and growing consensus that the STS is the region that responds reliably to biological motions—not only including the body as investigated using points of light experiments but also including eye, mouth, and hand movements (Deen, Koldewyn, Kanwisher, & Saxe, 2015). This region can also be activated by TOM tasks, voices, faces, and language (Fig. 10.23). The OFC and the amygdala are also part of this social perception network (Fig. 10.24).

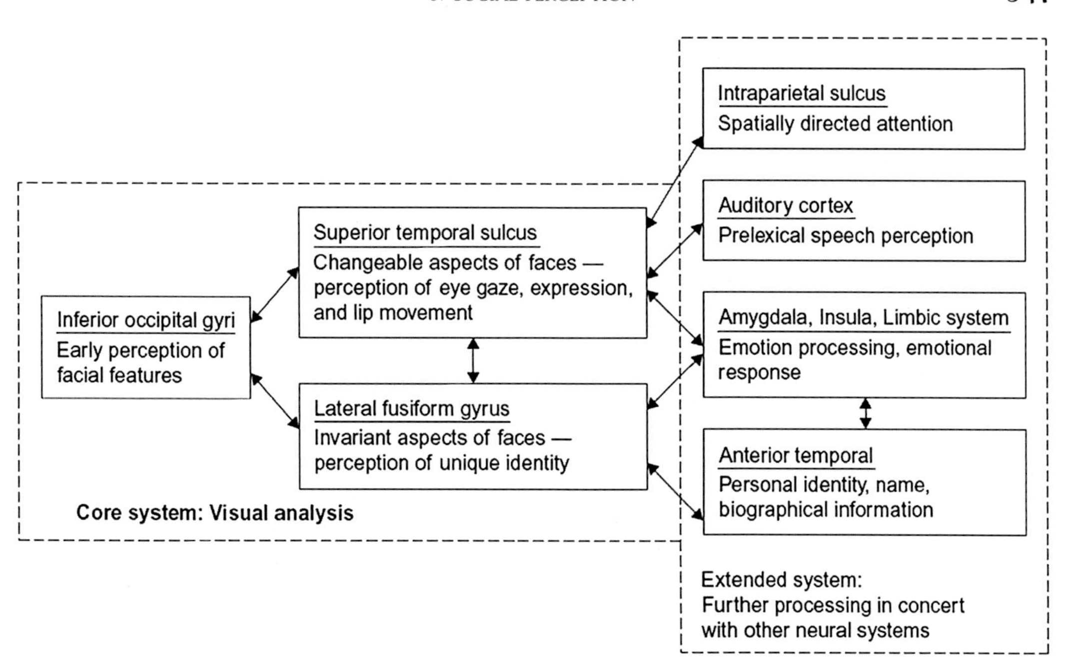

FIGURE 10.21 Model of facial perception developed by Haxby etal. The left side of the model shows early visual regions for face perception, the center shows brain regions for changing and nonchanging features in faces, and the right side shows further processing of facial features. *Source: Haxby et al. (2002).*

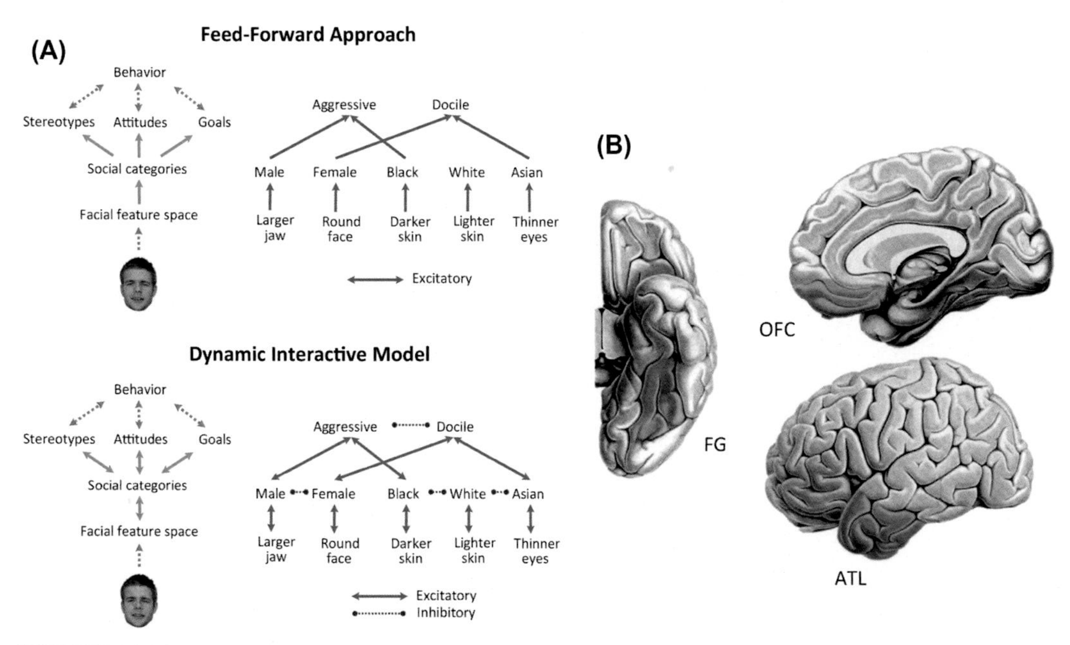

FIGURE 10.22 (A) According to Freeman and Johnson (2016), both bottom-up (feed-forward) and top-down (reentrant and parallel) processes are at work in face perception. This model makes intuitive sense: when we perceive a face—whether a new face or the face of a well-recognized friend—there are many cues that come from our stored knowledge of faces. (B) According to the dynamic interactive model proposed by Freeman and Johnson, key brain regions for face perception include the fusiform gyrus (FG), the orbitofrontal cortex (OFC), and the anterior temporal lobe (ATL). *Source: Freeman and Johnson (2016).*

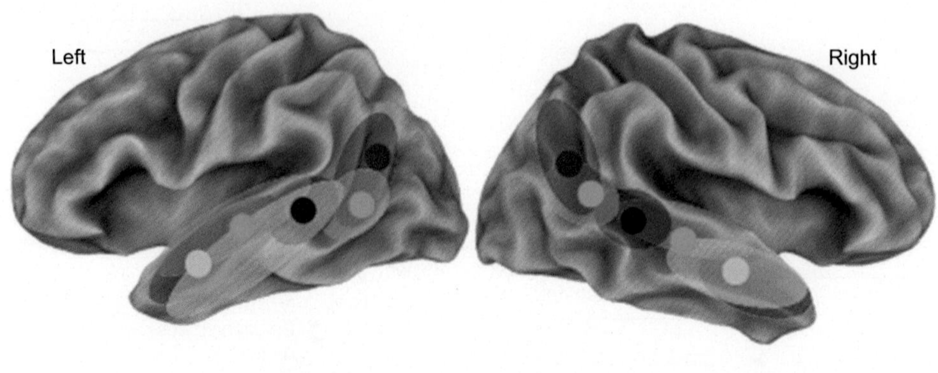

FIGURE 10.23 Social perception and cognition organization within the superior temporal sulcus. Results are shown on an inflated cortex to show the regions within the sulcus that respond to theory of mind false belief stories versus physical stories (red), biological motion (point-light humans vs. point-light objects) (teal), moving faces versus moving objects (purple), voices versus environmental sounds (green), and language (stories vs. nonsense speech) (yellow). *Source: Beauchamp (2015).*

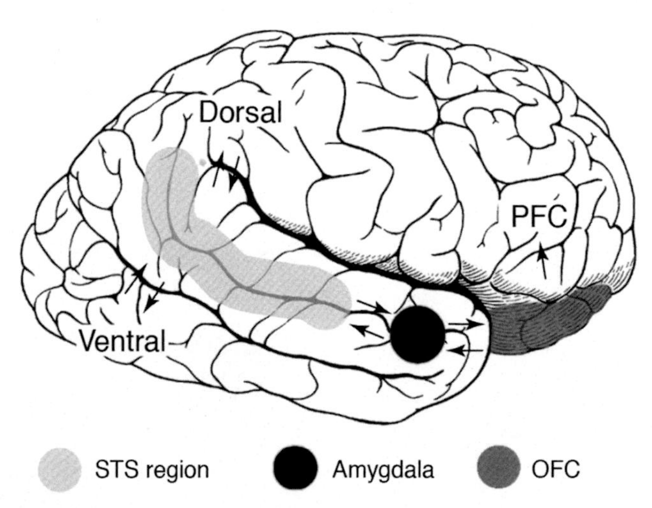

FIGURE 10.24 Where in the brain is biological motion processed? There is a large and growing consensus that the superior temporal sulcus (STS) is the region that responds reliably to biological motion—not only including the body as investigated using points of light experiments but also including eye, mouth, and hand movements (Allison et al., 2000). This region can also be activated by static images of the face and body: these findings imply that the STS (shaded in gray) is sensitive to stimuli that signal action, even if they do not contain the action themselves. The orbitofrontal cortex (OFC) (shaded in dark gray) and the amygdala (shaded in black) are also part of this social perception network. *Source: Allison et al. (2000).*

6. SOCIAL BEHAVIOR

Humans interact socially. It is a central part of being human: infants make eye contact, recognize voices, and begin to establish social relationships as a normal part of development. Where in the brain does social behavior "live"? Although social behavior is complex and has many aspects to, two key brain regions tend to appear over and over as those regions that mediate human social behavior: the vmPFC and the OFC (Fig. 10.4). The vmPFC is an area implicated in *social decision-making* and the OFC, for processing the rewards associated with those decisions.

6.1 The Ventromedial Prefrontal Cortex

The vmPFC has been implicated in diverse functions ranging from emotion and emotion regulation to episodic and semantic memory to economic valuation (Fig. 10.25). So what exactly

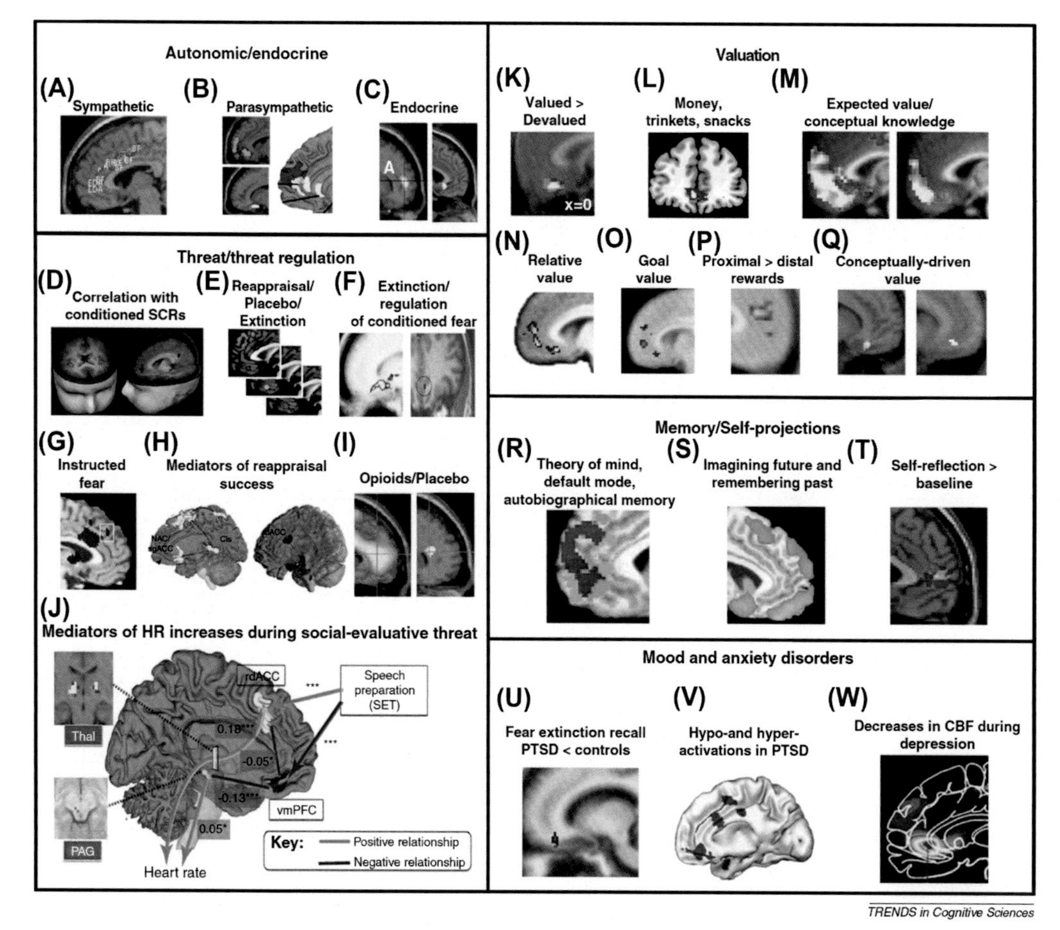

FIGURE 10.25 The ventromedial prefrontal cortex has been implicated in diverse functions ranging from emotion and emotion regulation to episodic and semantic memory to economic valuation. *Source: Roy et al. (2012).*

does the vmPFC do? Roy, Shohamy, and Wager (2012) proposed that the vmPFC is an integrative hub for emotional, sensory, social, memory, and self-related information processing. Situated in the medial portion of the prefrontal cortex, the highly interconnected vmPFC serves as a region for binding together the large-scale networks that subserve emotional processing, decision-making, memory, self-perception, and social cognition in general.

Thus it is not surprising that the vmPFC is activated across a wide range of experiments and experimental constructs ranging from memory to mentalizing to reward processing and even to pain (Fig. 10.26).

6.2 The Orbitofrontal Cortex

Early studies of the role of the OFC in behavior linked it to the rewards system in the brain. Many brain areas are active in the processing of rewards, and we will discuss that system in more detail in Chapter 11. However, as research into the functional role of the OFC continued, a new theme emerged: the OFC was integral to the determination of *value*—and the estimation of value not only had much in common with reward but also offered a much more complex aspect to it. The value one places on a reward is key to the decisions that person will make about obtaining the reward. Values add an abstract dimension to decision-making and rewards: values can be relative, and they satisfy a need. In the real

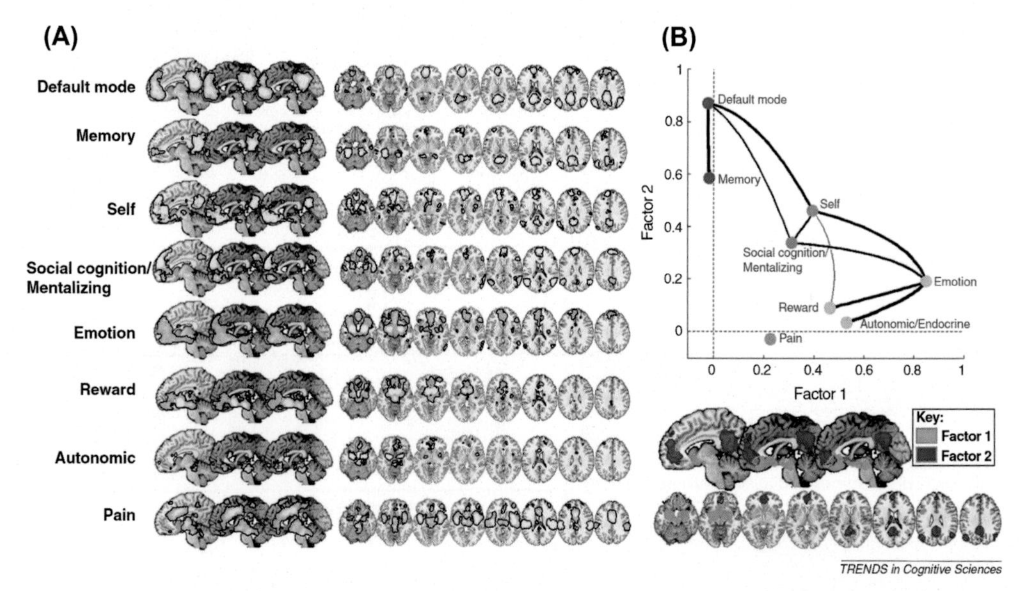

FIGURE 10.26 Roy et al. (2012) proposed that the ventromedial prefrontal cortex (vmPFC) is an integrative hub for emotional, sensory, social, memory, and self-related information processing. Situated in the medial portion of the prefrontal cortex, the highly interconnected vmPFC serves as a region for binding together the large-scale networks that subserve emotional processing, decision-making, memory, self-perception, and social cognition in general. (A) Results of a metaanalysis showing the probability of vmPFC activation for various aspects of social cognition, shown on left, including memory, self, and emotion. (B) Top panel: A two-factor analysis for brain activation showing the relationship between and across key terms such as default mode, memory, self, emotion, etc. Bottom panel: Color coding shows the brain areas associated with the factors presented above in (B), with Factor 1 comprising a large ventrocaudal portion of the vmPFC (shown in orange) and including the amygdala, and Factor 2 comprising a more rostral and dorsal portion of the vmPFC (shown in purple), including the posterior cingulate cortex. *Source: Roy et al. (2012).*

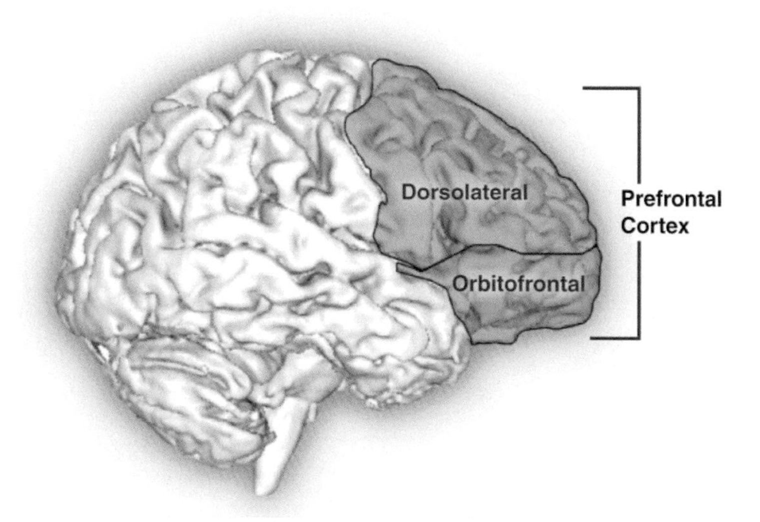

FIGURE 10.27 A lateral view of the right hemisphere showing the dorsolateral (shaded in pink) and orbitofrontal (shaded in blue) regions within the prefrontal cortex. *Source: Zahr and Sullivan (2008).*

world, we may not know if we have made the right decision immediately—we may have to wait and see. The OFC is thought to be that part of the prefrontal cortex that is active in integrating abstract and concrete aspects of decision-making to assess the relative value of available choices (Fig. 10.27).

The connectivity of the OFC is just what you might imagine for an integration area: the OFC has projections from all sensory systems (vision, audition, somatosensory, taste, and smell) and is highly interconnected with the limbic system, including the amygdala, cingulate gyrus, and hippocampus. The OFC does not have direct connections with the motor cortex: thus it appears that the OFC aids in calculating relative values during decision-making but is not active in the motor plan to carry out that decision.

7. AN INTEGRATED MODEL FOR SOCIAL COGNITION

As we have shown in Fig. 10.4 and throughout this chapter, there are many regions in the brain that constitute the brain bases of our social cognition. Key "players" in social cognition are the prefrontal cortices, TPJ, and anterior and posterior STS. Yang, Rosenblau, Keifer, and Pelphrey (2015) conducted a metaanalysis of more than 120 neuroimaging studies to determine which brain areas were common to various aspects of social cognition. Specific aspects of social cognition investigated were social perception, action observation (observance and predictions of others' actions), and TOM. Results are shown in Fig. 10.28: the one brain area that was common to all three aspects of social cognition was the pSTS. pSTS is highly interconnected with the other brain regions that are key to social cognition: a schematic is shown in Fig. 10.29.

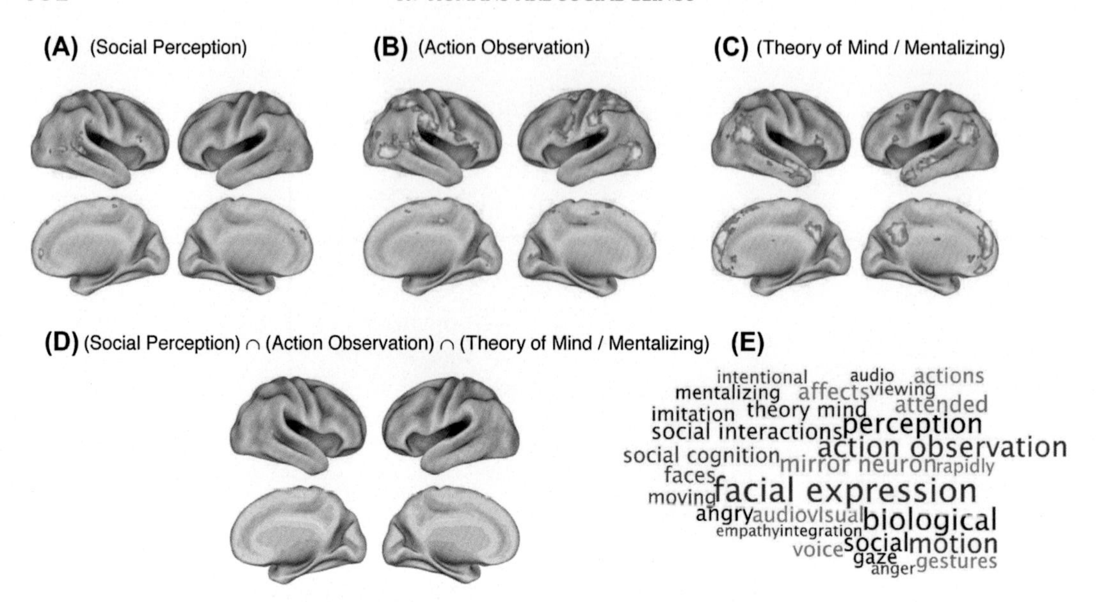

FIGURE 10.28 Yang et al. (2015) conducted a metaanalysis of more than 120 neuroimaging studies to determine which brain areas were common to various aspects of social cognition. Specific aspects of social cognition investigated were (A) social perception, (B) action observation (observance and predictions of others' actions), and (C) theory of mind(TOM). (D) The one brain area that was common to all the three aspects of social cognition was the posterior superior temporal sulcus. (E) A word cloud of words associated with TOM. *Source: Yang et al. (2015).*

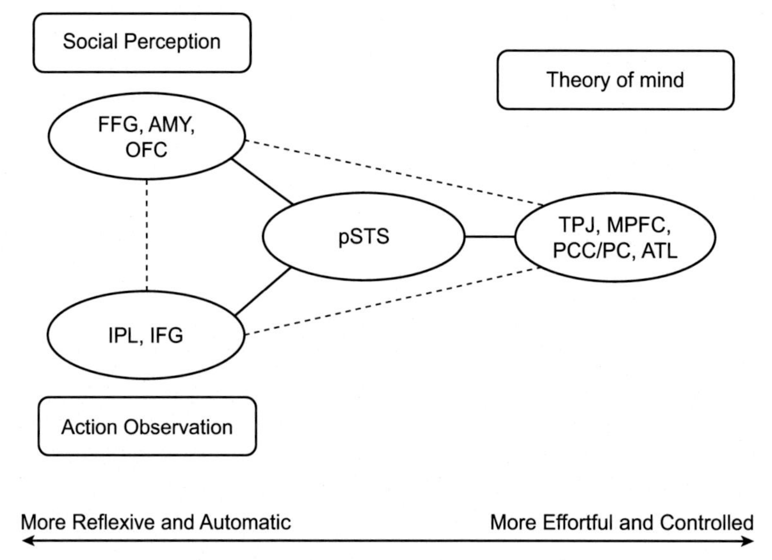

FIGURE 10.29 Posterior superior temporal sulcus (pSTS) is highly interconnected with the other brain regions that are key to social cognition: a schematic is shown here with pSTS. *Solid lines* indicate connections between brain areas. *Dashed lines* indicate possible connections. *AMY*, amygdala, *ATL*, anterior temporal lobe; *FFG*, fusiform gyrus; *IFG*, inferior frontal gyrus; *IPL*, inferior parietal lobule; *MPFC*, medial prefrontal cortex; *OFC*, orbitofrontal cortex; *PCC/PC*, posterior cingulate cortex/precuneus; *pSTS*, posterior superior temporal sulcus; *TPJ*, temporoparietal junction. *Source: Yang et al. (2015).*

8. DISORDERS OF SOCIAL COGNITION

Although human social cognition is an early-developing aspect of our cognitive and social selves, it can be impaired through brain disease, disorder, or damage. Indications of impairments in social cognition are shown in Table 10.1. Social cognition impairment can range from slight deviations from typical behavior—such as individuals who are "loners" or who make little eye contact—to violent antisocial behavior such as a serial killer.

The investigation of social cognition disorders provides new information about specific brain regions and their social and cognitive functions. Social cognition impairments occur in a wide array of psychiatric, developmental, and neurodegenerative disorders as well as in acute brain damage (Table 10.2). Studying patients who have had brain damage due to strokes, tumors, or head injury helps us understand both their specific impairments and what those injured regions provide in a healthy brain.

Since social cognitive impairments are so widely found across many types of brain disorders, disease, and damage, it is critical to have an assessment of social cognition included in general assessments to determine the specific type of impairment to aid in treatment and intervention plans (see Fig. 10.30).

TABLE 10.1 Indications of Social Cognitive Impairment

- Social withdrawal or avoidance of social contact
- Loss of social grace
- Limited eye contact
- Rude or offensive comments without regard for the feelings of others
- Loss of etiquette in relation to eating or other bodily functions
- Extended speech that generally lacks focus and coherence
- Neglect of personal appearance (in the absence of depression)
- Disregard of the distress or loss of others
- Inability to share in the joy or celebration of others when expected or invited
- Failure to reciprocate socially, even when obvious social cues are given
- Poor conversational turn-taking
- Overtly prejudiced or racist behavior
- Increased or inappropriate interpersonal boundary infringements
- Failing to understand jokes or puns that are clear to most people
- Failure to detect clear social cues, such s boredom or anger, in conversational partners
- Lack of adherence to social standards of dress or conversational topics
- Excessive focus on particular activities to the exclusion of important social or occupational demands

Source: Henry et al. (2015).

TABLE 10.2 Disorders with Social Cognitive Impairment

Disorders with Social Cognitive Impairment
PSYCHIATRIC DISORDERS
• Schizophrenia • Bipolar disorder • Antisocial personality disorder • Major depressive disorder • Posttraumatic stress disorder • Social phobia • Anorexia nervosa • Personality disorders (for example, borderline, antisocial, narcissistic, schizoid, avoidant personalities)
DEVELOPMENTAL DISORDERS
• Autism spectrum disorder • Fragile X syndrome • Williams syndrome • Angelman syndrome • Prader–Willi syndrome • Turner syndrome • Rett syndrome • Attention deficit hyperactivity disorder • Severe conduct disorder • Fetal alcohol syndrome
NEURODEGENERATIVE DISORDERS
• Frontotemporal dementia • Alzheimer disease • Amyotrophic lateral sclerosis • Parkinson disease • Huntington disease • Progressive supranuclear palsy • Corticobasal degeneration • Multiple sclerosis
ACUTE BRAIN DAMAGE
• Traumatic brain injury • Stroke

Source: Henry et al. (2015).

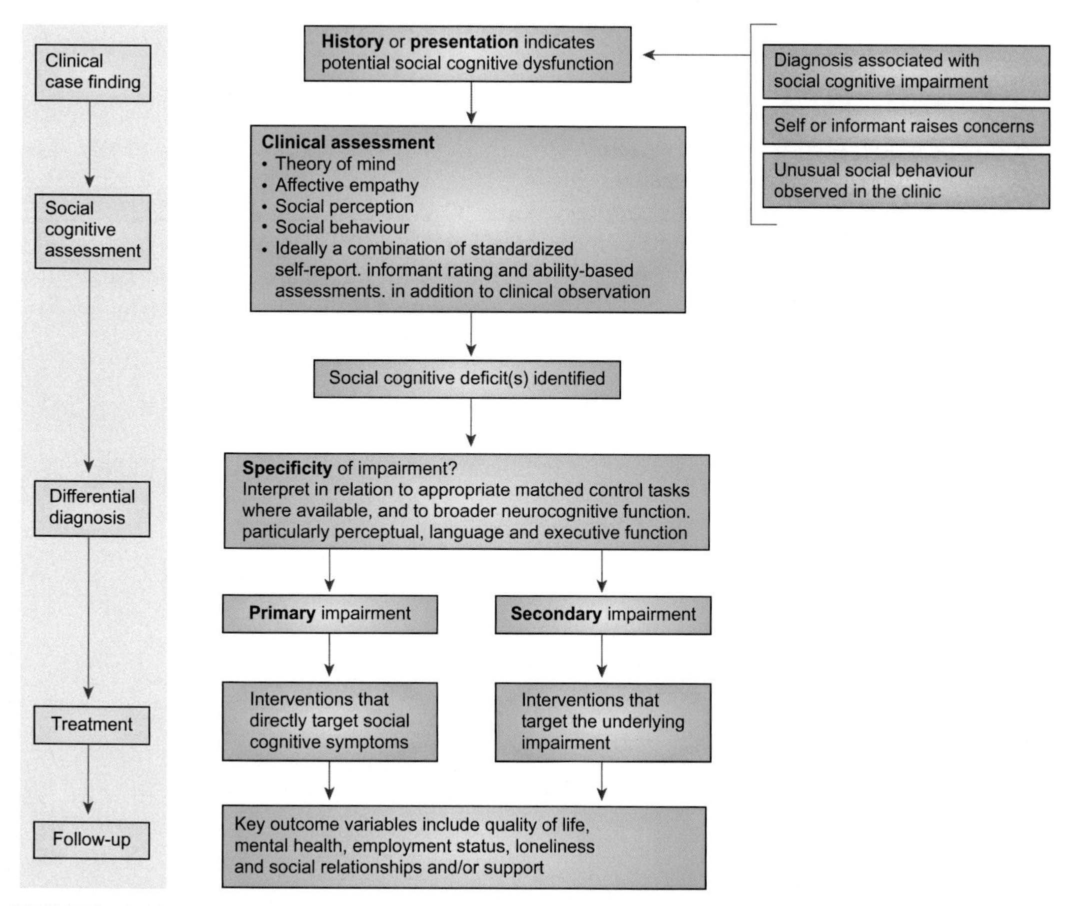

FIGURE 10.30 A schematic is proposed for assessment of social cognition in individuals with brain disorders, disease, or damage (*green shaded boxes*). If social cognition impairments are observed (*red shaded boxes*), it is recommended that each of the four domains (theory of mind, empathy, social perception, and social behavior) be evaluated (*blue shaded boxes*). *Source: Henry et al. (2015).*

9. SUMMARY

The social brain has amazing abilities, from a "fast path" for recognizing faces to special circuits for recognizing biological motion to detecting eye gaze and intent. Humans are excellent predictors in social situations, understanding other's intentions, sharing their attention, and understanding what others know or feel. These abilities develop early in life and are important stages of cognitive development. The advent of neuroimaging techniques has provided social cognition neuroscientists new ways to investigate the brain bases for these complex social abilities and processes. The brain areas for these social abilities are distributed throughout the cortex and into subcortical regions. Many open questions still remain to be discovered in the new field of social cognitive neuroscience, with new techniques being developed for uncovering the social brain's mysteries.

10. STUDY QUESTIONS AND DRAWING EXERCISES

10.1 Study Questions

1. Briefly describe what is meant by a *TOM*.
2. According to Frith, what is mentalizing?
3. Why are SAMs important for human development? When do they develop?
4. What aspect of social cognition do point-light experiments investigate?
5. What are the examples of social perception? Where in the brain are they processed?
6. Describe three impairments in social cognition and which brain disorder, disease, or damage they might relate to.

10.2 Drawing Exercises

1. Complete the schematic figure according to Simon Baron-Cohen's model for TOM (Fig. 10.31).

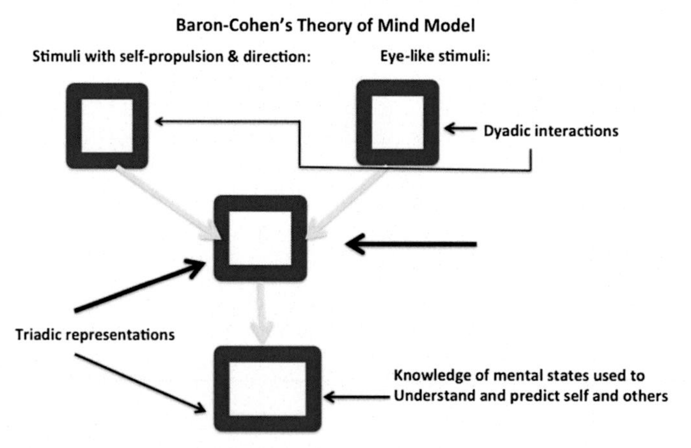

FIGURE 10.31 Complete the schematic diagram according to Simon Baron-Cohen (1995) model for a theory of mind module. *Source: Adapted from Baron-Cohen (1995).*

CHAPTER

11

Feelings

HUMAN EMOTION: FEELING HOW OTHERS FEEL

In highly social species, emotions are contagious and serve important communicative roles. Empathy is one example of such contagion; it regulates social behavior as well as providing us with a mechanism for figuring out how others feel. According to one currently influential line of thinking, we do not need to figure out how others feel by elaborate deductions from their observed behavior. All we need to do is let empathy do its automatic work and look inside ourselves: if we know how we feel, we will know how the other person feels. ***Ralph Adophs (2012), American Scientist.***

Frontpiece. Sisters showing empathy. *Source: Public domain.*

1. INTRODUCTION

Human emotions: a complex set of reactions and responses to events around us in our environment, to facial expressions of people we love, to our own moods, feelings, and thoughts. Reactions and responses to these emotional inputs range from a quiet acceptance to a flight-or-fight response…to a quick reaction followed by a rapid reappraisal and calming of the emotion invoked.

Fundamentals of Cognitive Neuroscience
https://doi.org/10.1016/B978-0-12-803813-0.00011-8

How do you go about understanding the neural machinery that supports these many processes? This is precisely what scientists who study emotional processing in the brain are tackling as a central challenge every day. It is a daunting task: how do you study—in the laboratory—in a neuroimaging scanner—or with a brain-damaged person—the complex brain processes that are called forth by an emotional stimulus? And importantly, how do these well-controlled experiments relate to the actual brain mechanisms that are invoked every day by real life?

1.1 Emotions—Categories and Dimensions

A reasonable beginning might be to describe the emotions that we are studying. Having a general agreement about just which emotions we are investigating seems to be an important first step. However, there is significant disagreement about how to characterize the many emotions we feel every day—from cheerfulness to envy to irritation. Even this first step is steeped in controversy and contains hugely varying approaches.

One way to begin to understand why investigations of human emotion are so difficult—and so varied in their approach—is to recognize the many differing disciplines that are at work in deciphering the brain bases of emotions: these disciplines range from neuroscientists who are studying an animal model of the neurobiology of emotion, to cognitive neuroscientists who frequently use neuroimaging techniques to investigate the neurobiology of emotion using a human-based model, to behavioral scientists and social and affective psychologists, to psychiatrists who use a combination of techniques to understand the brain bases of emotion in healthy individuals and in individuals with an emotional disease, disorder, or brain damage.

1.2 Emotions: Continuous or Discrete?

Depending on the discipline of the scientists investigating—and to be honest, even among scientists within a given discipline—the definitions and categorization of emotions range broadly from a short list of 5–7 basic emotions (Ekman, 1992) to a multidimensional and expandable set of features that represent abstract neural representations of emotion (Skerry & Saxe, 2015). Two differing theories have long been held in the investigation of emotion—there are many others as well, but these are the predominant ones: one is that emotions can be described along *continuous dimensions* such as positive or negative, pleasant or unpleasant; the second is that there are *core, discrete, basic emotions* that are universal to human beings. These basic emotions typically include *anger, fear, sadness, enjoyment, disgust, and surprise* (Ekman, 1992). The sense of the word *basic* is not to imply a strong sense of *simplicity* to these powerful emotions, rather that they are *fundamental to human life* and the challenges it presents.

Some investigators use a combination of the discrete basic emotions with the more continuous dimensions of *arousal* (low to high) and *valence* (pleasant to unpleasant) to describe emotional reactions and responses (Adolphs, 2002; Hamann, 2012) (Figs. 11.1 and 11.2). For example, you may experience more emotional arousal at seeing a smiling baby than a beautiful sunset due to the cuteness of the baby—although both might elicit feelings of happiness.

Take a look at Table 11.1 where we present some other ways to categorize emotions (Adolphs, 2002). The left side of the table shows the *Behavioral State* (Approach, Withdrawal) and *Motivational State* (Reward, Punishment, Thirst, Hunger, Pain, Crazing) categorizations of emotions. These emotional categorizations come to us primarily from animal studies

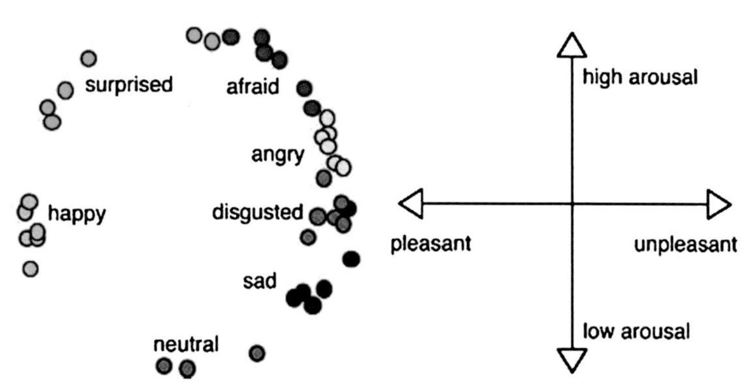

FIGURE 11.1 Emotions can be categorized as *discrete basic emotions*, as those shown on the left side of this figure, or as *continuous dimensions*, such as the two dimensions of Arousal (low to high) and Valence (pleasant to unpleasant) shown on the right side of this figure. Emotions can also be categorized using both the discrete and continuous scales: the basic emotions shown on the left side of this figure are plotted according to the Arousal and Valence strengths shown on the right side of the figure. Thus "surprised" is an emotion with a similar level of pleasantness as "happy" but with a higher level of arousal. Similarly, "surprised" has a similar level of arousal as "afraid" but is higher in pleasantness. *Source: Adolphs (2002).*

of emotion. These aspects of emotion are *under the control of the observer/investigator*: the *Behavioral State* emotions of the animal can be seen by looking at the animal's actions—their behavior; the *Motivational State* of the animal can be controlled through experimental manipulations such as offering (or not offering) a reward for specific behavior and by increasing the emotional level by inflicting pain.

Let us bypass the *Moods, Background Emotions* for the moment and move to the *Emotional System* (Seeking, Panic, Rage, and Fear) emotions in Table 11.1—we will return to it in a moment. These *Emotional Systems* have been much studied using an animal model for the

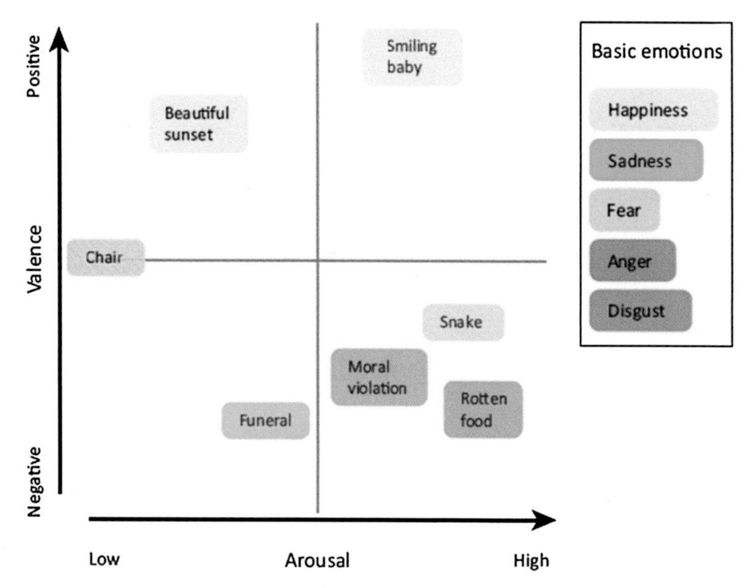

FIGURE 11.2 Similar to the emotions plotted in Fig. 11.1, in Fig. 11.2 we show examples of items that elicit differing levels of Arousal (low to high) and Valence (in this case, Positive and Negative). Thus a "smiling baby" and a "snake" both have high Arousal levels but differ sharply in their Valence. *Source: Hamann (2012).*

TABLE 11.1 Ways to Categorize Emotions

Behavioral State	Motivational State	Moods, Background Emotions	Emotion System	Basic Emotions	Social Emotions
Approach	Reward	Depression	Seeking	Happiness	Pride
Withdrawal	Punishment	Anxiety	Panic	Fear	Embarrassment
		Mania	Rage	Anger	Guilt
	Thirst		Fear	Disgust	Shame
	Hunger	Cheerfulness		Sadness	Maternal Love
	Pain	Contentment		Surprise	Sexual Love
	Craving	Worry		Contempt	Infatuation
					Admiration
					Jealousy

Source: Adolphs (2002).

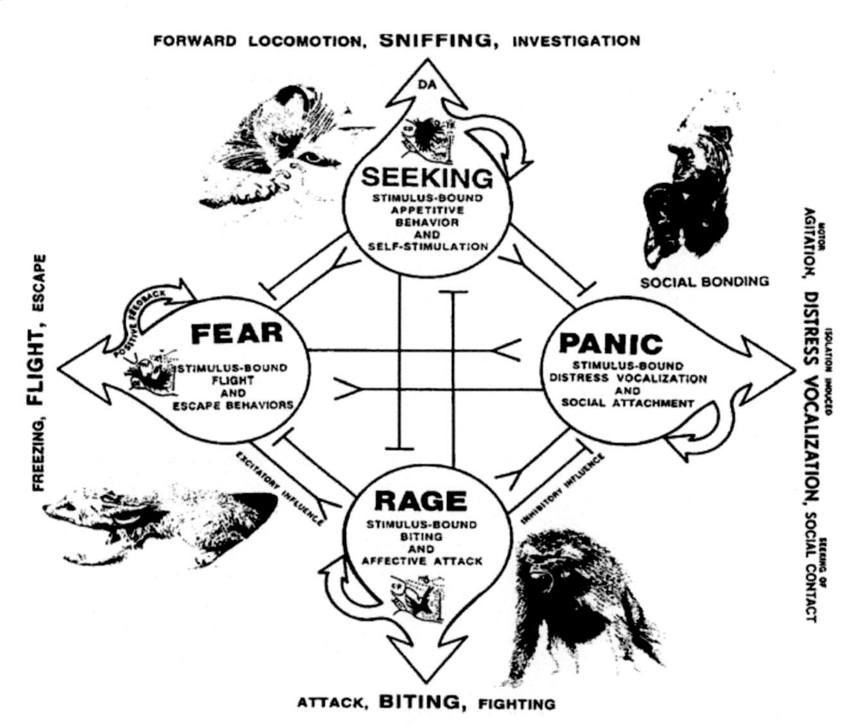

FIGURE 11.3 Panksepp's categorization of mammalian emotional operating systems include SEEKING, PANIC, RAGE, and FEAR. Each operating system includes both excitatory and inhibitory interactions; however, each can be isolated into a concrete set of brain regions. *Source: Panksepp (2011).*

brain bases of emotions. A long-standing model of *Emotional Systems* comes from the work of Panksepp (1998, 2005, 2006, 2011) who has provided a wealth of information about mammalian emotional behaviors (Fig. 11.3) and their corresponding emotional pathways in the brain (Fig. 11.4).

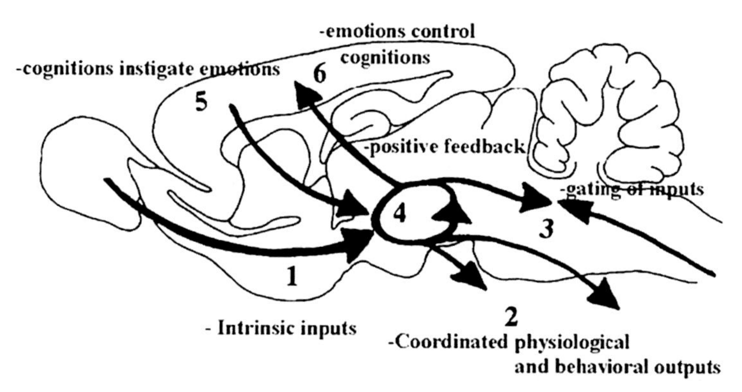

FIGURE 11.4 A schematic of the characteristics of basic instinctual emotional systems in mammals. These systems include: (1) Intrinsic inputs—some information from the environment; (2) Instinctive physiological and behavioral outputs—such as reacting with fear to a loud sound; (3) Inputs from higher brain regions such as the cortex—bringing in past learned behaviors or reactions; (4) Emotions—which can long outlast whatever stimulus produced them, and will interact with future behavior; (5) Higher cortical areas such as the frontal lobe, which will aid in activating or inhibiting emotion; (6) Emotional systems that can control or modify higher brain function; and (7) Reflects the full operation of this type of system with inputs from the environment that can signal an emotional response, produce higher brain responses and interact with emotional systems. *Source: Panksepp (2011).*

Turning back to Table 11.1, now let us take a look at *Moods, Background Emotions* (including Cheerfulness and Worry) and *Basic Emotions* (which contains the seven basic human emotions in long use in the field of affective science). Do these emotions seem more multidimensional to you? Finally, look at the final category, *Social Emotions* (including Embarrassment and Admiration). These last sets of emotions are also used in studies of human emotion and brain pathways; however, they carry more social cognition influences: it is easy to see how complex this study can be!

Adding to the complexity of studying human emotions and their brain bases are the differing types of stimuli used (very often faces with differing facial expressions) and a large variety of tasks employed in experiments investigating emotion pathways in the brain. Nevertheless, we have discovered many important features of how the brain processes emotional situations. Let us step through what we have found to date, beginning with the rather long list of brain regions that are involved in emotional processing.

2. THE EMOTIONAL BRAIN CIRCUITRY

The key emotional regions within the brain are frequently referred to as *the limbic system.* The word "limbic" comes from the Latin word *limbus* and means "border, edge, or hem"—it refers to the location inside the cerebral hemispheres around the edge of the lateral ventricles (fluid-filled spaces). The limbic system was first recognized in the late 1800s, but an understanding of its function in emotion did not develop until the work of neuroanatomist Papez was published in 1937. Another older name for the limbic system is the Papez circuit.

The list of component parts of the limbic system varies depending on the researcher that we consult—there seems to be no universal agreement about what the limbic system actually consists of! Some neuroscientists think that we should no longer speak of a limbic system at all. We will retain the term as a useful organizing concept for a set of related subcortical brain parts that support our emotional life.

Let us begin with the cortical limbic system—those regions associated with emotion processing that are in the cortex rather than the subcortex (Fig. 11.5). These regions surrounding the corpus callosum include the posterior part of the *orbitofrontal cortex* (pOFC), the *anterior, mid, and posterior cingulate cortex* (ACC, MCC, PCC), the *ventral anterior insula* (vaI), the *parahippocampal gyrus* (PHG), and the *temporal pole* (TP) (Chanes & Barrett, 2016).

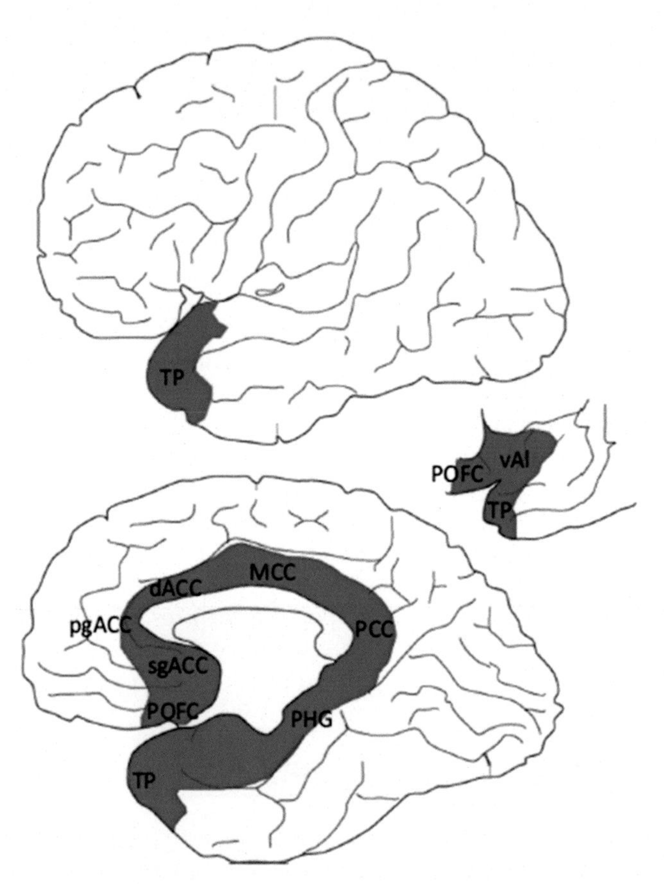

FIGURE 11.5 The cortical limbic system—those regions associated with emotion processing that are in the cortex rather than the subcortex (shown in blue). The top brain figure is a lateral view of the left hemisphere. The bottom brain figure is a midsagittal view of the right hemisphere. These regions surrounding the corpus callosum include the posterior part of the *orbitofrontal cortex* (pOFC), the *anterior, mid, and posterior cingulate cortex* (ACC, MCC, PCC), the *ventral anterior insula* (vaI), the *parahippocampal gyrus* (PHG), and the *temporal pole* (TP). *Source: Chanes and Barrett (2016).*

Subcortical regions associated with emotional processing include the "key player" in emotion, the *amygdala* (AMY in Fig. 11.6). The *basal ganglia* (BG) also play a role in emotional processing, as does the brainstem *periaqueductal grey* (PAG) region, as well as in pain processing (Bushnell, Ceko, & Low, 2013).

Just as you may be thinking that emotional processing includes a long list of brain regions, let us add more to that list! A simplified version of connections between and among the amygdala and the cortex is shown in Fig. 11.7. While we will describe these connections and their role in emotional processing in more detail later in this chapter, briefly, the amygdala has reciprocal connections with the anterior cingulate cortex (ACC) and the hippocampus, there are also reciprocal connections between the ACC and dorsolateral prefrontal cortex (dlPFC), the ACC and the hippocampus, and between the dlPFC and the hippocampus (Leisman, Machado, Melillo, & Mualem, 2012). This highly interconnected set of regions in the frontal lobe and the subcortical amygdala along with the hippocampus in the medial temporal lobe combines to form a network for emotional processing.

In Fig. 11.8 we show an expansion of these brain areas to include some of the more detailed aspects of emotional processing such as decoding facial expressions and emotional regulation (Green, Horan, & Lee, 2015). We have added the ventral regions of the prefrontal cortex (PFC) to our list of brain areas, as well as some regions you may be familiar with from the Social Cognition chapter, including the temporoparietal junction (TPJ) and the fusiform face area (FFA).

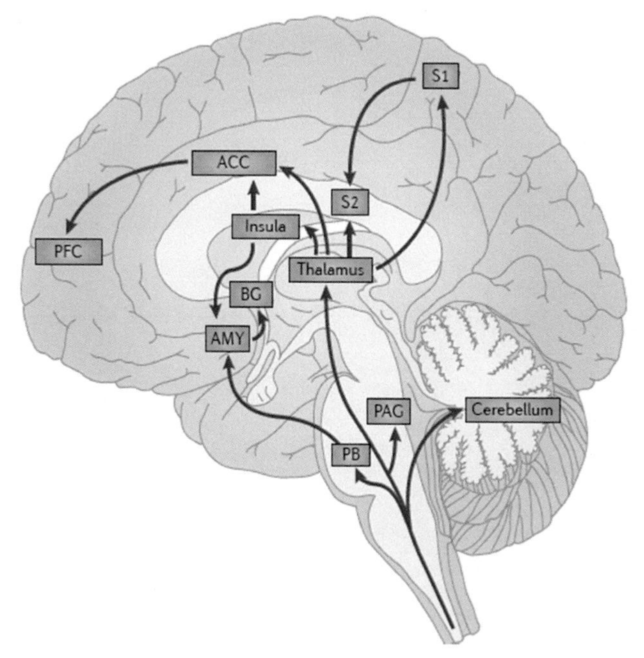

FIGURE 11.6 Subcortical regions associated with emotional processing include the "key player" in emotion, the *amygdala* (AMY), shown in this midsagittal view of the right hemisphere. The *basal ganglia* (BG) also play a role in emotional and in pain processing, as does the brainstem *periaqueductal grey* (PAG) region. Other key brain regions in the emotional brain are the insula, thalamus, anterior cingulate cortex (ACC), and the prefrontal cortex (PFC). S1 and S2 shown pain pathways. *Source: Bushnell et al. (2013).*

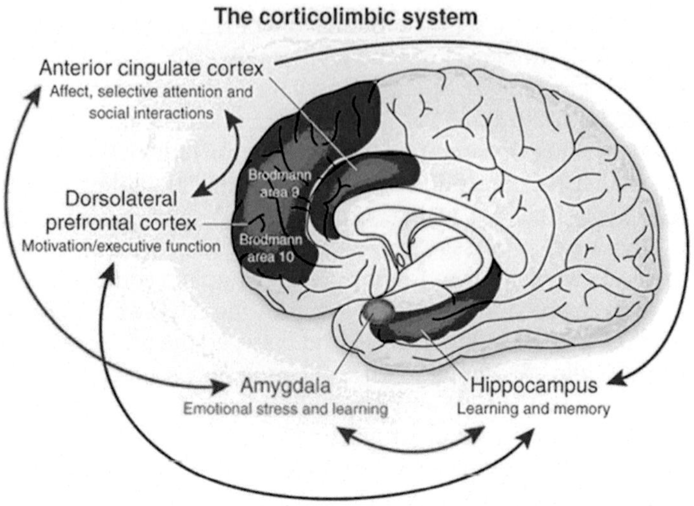

The corticolimbic system

Anterior cingulate cortex
Affect, selective attention and social interactions
Brodmann area 9
Dorsolateral prefrontal cortex
Motivation/executive function
Brodmann area 10
Amygdala
Emotional stress and learning
Hippocampus
Learning and memory

FIGURE 11.7 More "key players" in the emotional brain shown in this midsagittal view of the right hemisphere: the amygdala is at the heart of the emotional brain and has reciprocal connective pathways both to the anterior cingulate cortex (ACC) and to the hippocampus (shown in *red lines* with *bidirectional arrows*). The ACC plays a role in processing emotional experience and in controlling selective attention mechanism. The hippocampus interacts with the basolateral section of the amygdala for learning and memory processes. The dorsolateral prefrontal cortex (dlPFC) connects reciprocally with the ACC and the hippocampus (shown in *blue bidirectional lines*). *Source: Leisman et al. (2012).*

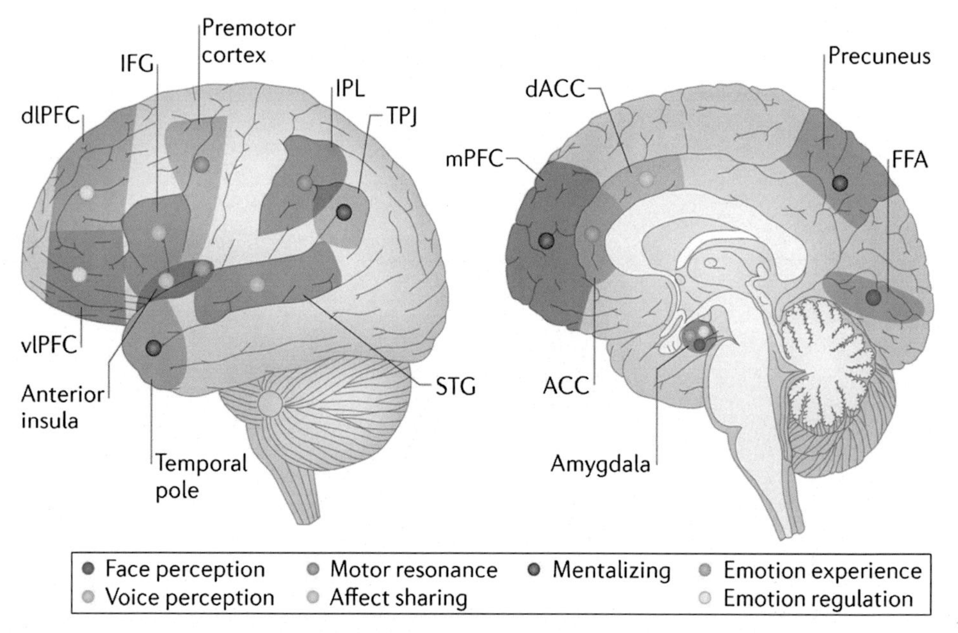

FIGURE 11.8 Here is an expansion of brain areas to include those that are involved in some of the more detailed aspects of emotional processing such as decoding facial expressions and emotional regulation. The left brain figure is a lateral view of the left hemisphere. The right brain figure is a midsagittal view of the right hemisphere. We have added the ventral regions of the prefrontal cortex (PFC) to our list of brain areas (vlPFC), as well as some regions you may be familiar with from the Social Cognition chapter, including the temporoparietal junction (TPJ) and the Fusiform Face Area (FFA). *Source: Green et al. (2015).*

3. EMOTIONAL REGULATION AND COGNITIVE REAPPRAISAL

3.1 Generation and Regulation of Emotions

Emotions frequently are *generated* outside of our direct control: they arise rapidly as a reaction to certain types of stimuli in the environment. For example, a loud noise may cause you to startle and you may feel fearful and experience more rapid heartbeat and breathing. These reactions can be based on previous experiences with events in the environment that were fear-inducing or for which you have a bad memory—these are *acquired emotional responses*. They are also very automatic reactions that we have as human beings—these are *innate emotional reactions*. They can occur even when we know we are not in any real danger—such as hearing an explosion while watching a movie.

Emotions flow through us and from us—positive emotions in response to watching a puppy race around the room, negative emotions when seeing a friend fall and get hurt—emotional responses are part of our everyday life. Emotions may arise from external events, such as the loud noise or cute puppy, but they can also arise from internal events, such as remembering a difficult breakup or the death of a relative. Emotions may flow consciously through our mind but they can also be on the fringe of consciousness. *Regulating* these almost constant emotional situations is a key aspect of cognitive control.

3.2 Valuation: Is This Bad for Me or Good for Me?

While there is no central, agreed-upon model for how the mind/brain regulates emotion, there are some basic core elements that are in wide use. A large part of our emotional reactions and responses have to do with a *valuation* of the situation at hand: Is it bad for me? Is it good for me? Based mostly on experience as we interact with our environment throughout our life—and partially through core reactions to things such as pungent smells or harsh sounds—we are taking stock of whether a given situation is good for us or bad for us. These valuations are key to survival—it is critical for us to be able to recognize threatening faces and gestures from others so that we may be prepared to fight or flee. It is also key to recognize fearful faces so that we may determine if there is a threat nearby. Understanding social emotions is a core aspect of social cognition—knowing when someone else is sad or mad or happy will help us to interact with them in an appropriate way.

3.3 Cognitive Reappraisal

A common strategy for human emotional regulation is *cognitive reappraisal*. Cognitive reappraisal involves changing one's interpretation of an event or a situation. Here is an example of cognitive reappraisal: Let us say you are driving to work and are blocked by a very slow freight train. Knowing this train is very l-o-n-g, you realize that you will be at least 10 min late to work. You start to seethe: it is so frustrating, you knew better than to drive this route, you are angry at yourself and at the annoying train. However, you catch yourself before you launch into a major anger event: you are always on time for work, this will be a rare occurrence, and there is nothing you can do about it now that you are stopped at the railway crossing. You decide that you need to look at better routes for going to work rather than going the same way every day. And then you decide to sit back and cue up that podcast you have been meaning to listen to.

In this way, an individual can calm himself/herself down when anger, fear, or any other emotion is inappropriate or unhelpful. Notice that cognitive reappraisal *does not* mean that you ignore your anger, or suppress it. Nor does it mean that you create a distraction for yourself (although turning to a podcast is a good idea in this situation). Cognitive reappraisal involves *actually changing one's cognitive appraisal* of a situation: you will still be late for work, you are still stuck at a railway crossing—however, in this example, you have changed your reaction to the events and have decided to change your work commute methods. By doing this, you have managed to tone-down—or downregulate—your emotional response and hopefully make a lasting improvement on your work commute.

Typically, but not always, cognitive reappraisal involves downregulating a *negative emotion* such as fear, anger, stress, or disgust. We tend to downregulate positive emotions less frequently; however, we would put this strategy into effect, for example, when at a serious or somber event. It would be inappropriate to walk around the event laughing and smiling, even if we had just received some very good news.

3.4 Brain Regions Are Involved in Emotion Regulation

As we have discussed, the limbic system is implicated in emotional processing. Key regions include the amygdala, insula, periaqueductal grey, and the dorsal ACC. Conscious or explicit regulation of emotions involves the dorsal lateral PFC, the ventrolateral PFC, parietal cortex, and supplementary motor areas (SMAs) (Etkin, Büchel, & Gross, 2015) (Fig. 11.9). Some emotional regulation occurs without conscious control: it is implicit. For example, some fear suppression occurs without our voluntary control. Implicit regulation is thought to include ventral ACC and ventromedial PFC. In this way, the core limbic regions of the amygdala, insula, and ACC interact with PFC as well as with parietal and SMA cortex to downregulate emotion. The PFC, with executive control functions, aids in controlling the arousal and reactivity to emotional situations or stimuli.

4. EMOTIONAL PERCEPTION

Are there specific brain pathways for basic emotions such as fear, disgust, sadness, happiness? One of the first and most studied emotions—in animals as well as in humans—has been fear. Why? It is a critical emotion for understanding threatening situations, threatening individuals, and for survival in general. Early studies in animals have shown a strong response to experiments using fearful stimuli to invoke a reaction in their subjects. In short, fear is a strong, easily evoked, and key emotion. Understanding the neural pathways that subserve fear processing can aid us in understanding this basic emotional cue in healthy individuals and aid us in determining how it is abnormal in individuals with emotional disorders such as phobias.

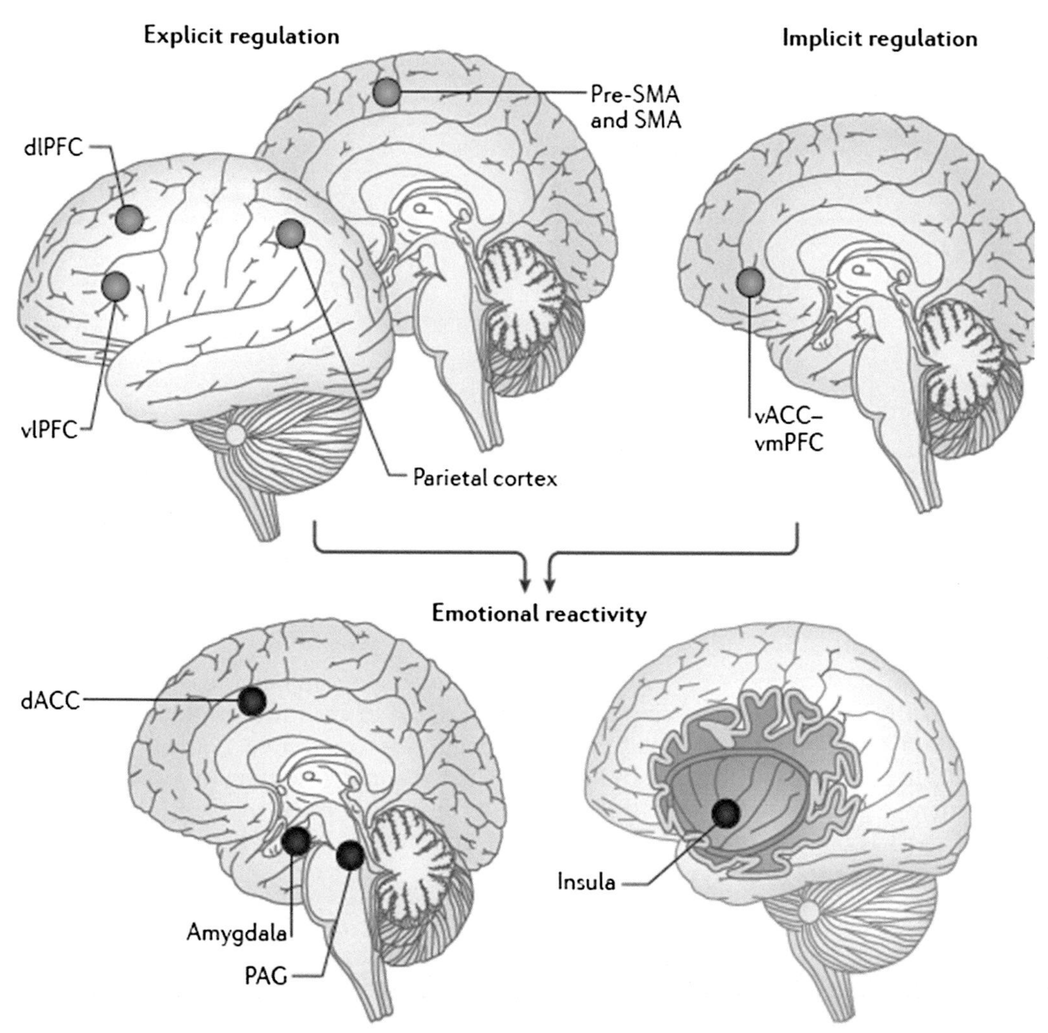

FIGURE 11.9 Conscious or explicit regulation of emotions involves the dorsal lateral prefrontal cortex (PFC), the ventrolateral PFC, parietal cortex, and supplementary motor areas (SMA), shown in a lateral view of the left hemisphere on the top leftmost figure, with a midsagittal view of the right hemisphere shown immediately next to it. The top right figure is also a midsagittal view of the right hemisphere. The bottom left brain figure is a midsagittal view of the right hemisphere. The bottom right brain figure is a lateral view of the left hemisphere with a window to expose the insula, which is buried deep within the cortex. Some emotional regulation occurs without conscious control: it is implicit. For example, some fear suppression occurs without our voluntary control. Implicit regulation is thought to include ventral ACC and ventromedial PFC. In this way, the core limbic regions of the amygdala, insula, and ACC interact with PFC as well as with parietal and SMA cortex to downregulate emotion. The PFC, with executive control functions, aids in controlling the arousal and reactivity to emotional situations or stimuli. *Source: Etkin et al. (2015).*

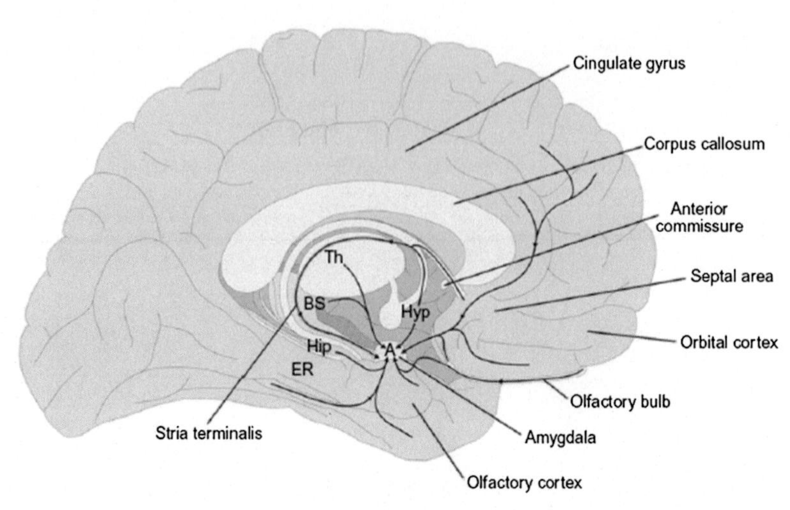

FIGURE 11.10 There are many projections to the amygdala and other regions in limbic cortex. Here is a midsagittal view of the left hemisphere showing the *afferent* projections to the amygdala. These *afferent* projections to the amygdala are a key form of neural communication about emotional information in our environment, forming the connective pathways between subcortical regions for taste, odor, and the brainstem, for example, and on to cortical regions. *Source: Baars with permission.*

There are many projections to the amygdala and other regions in limbic cortex (Fig. 11.10). These *afferent* projections *to the amygdala* are a key form of neural communication about emotional information in our environment, forming the connective pathways between subcortical regions for taste, odor, and the brainstem, for example, and on to cortical regions. Just as many projections lead *from the amygdala* to other regions in the subcortex and the cortex (Fig. 11.11). These *efferent* projections send information from the amygdala to other regions in both the subcortex and the cortex.

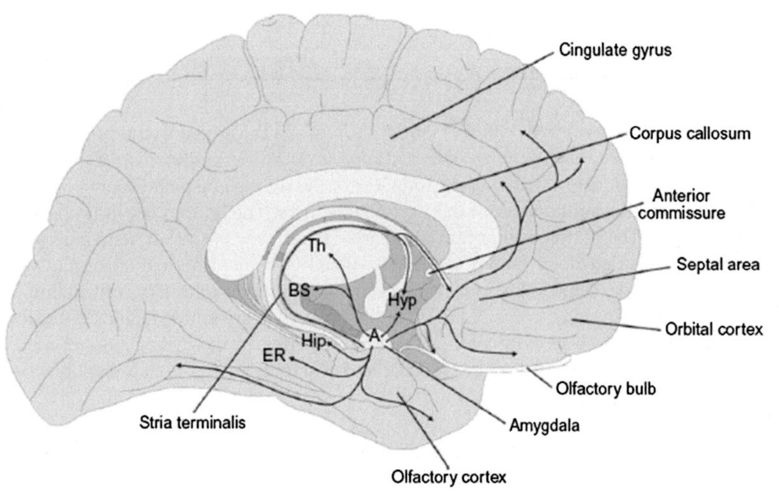

FIGURE 11.11 Here is a midsagittal view of the left hemisphere showing the *efferent* projections. Just as many projections lead from the amygdala to other regions in the subcortex and the cortex. These *efferent* projections send information from the amygdala to other regions in both the subcortex and the cortex. *Source: Baars with permission.*

4.1 Specific Pathways and Activation Patterns for Basic Emotions

While fear has been much studied in animal models of the neurobiology of emotion, studies in humans frequently incorporate neuroimaging techniques for investigating the brain bases across a range of basic emotional perceptions. Differing emotions—such as happiness, sadness, and anger—tend to activate differing sets of regions in the brain (Fig. 11.12). While there has been significant progress in understanding which brain areas are activated in response to the basic categories of emotion, there is at present not enough evidence to support a one-to-one mapping of a *basic emotion* to *a particular brain pathway or selection of activations* (Hamann, 2012). Instead, it is likely the case that differing emotions include a different set of regional neural networks within the cortical and subcortical emotion processing regions (Fig. 11.13). These emotional brain responses likely include overlapping regions, thus using a subtractive technique to compare brain response for fear versus disgust, for example, may not show the entire *fear* or *disgust* networks since both emotions activate the same or overlapping regions which would cause them to cancel out in the analysis. For these

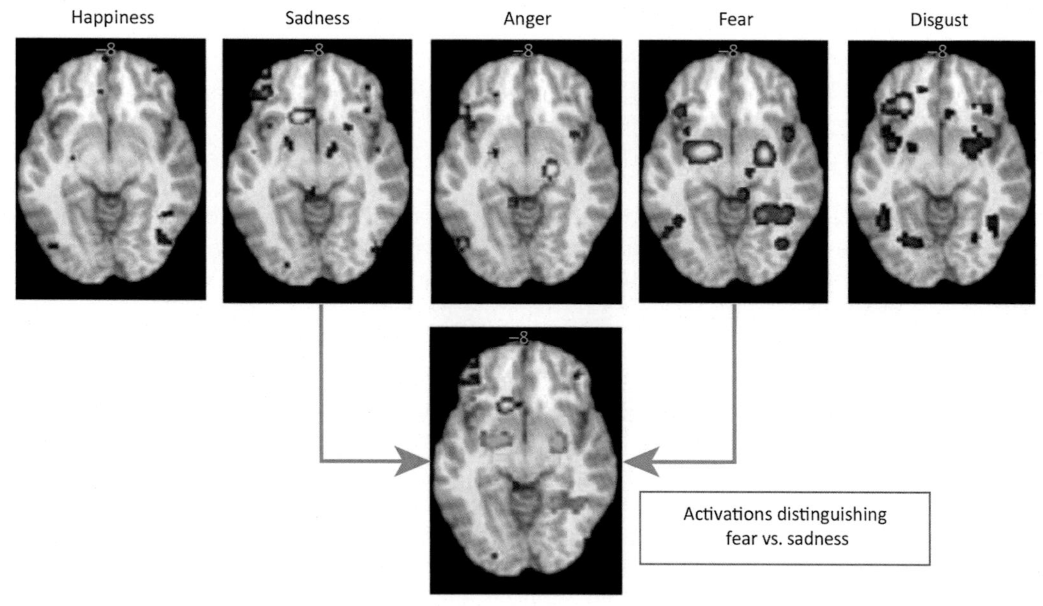

FIGURE 11.12 Neuroimaging basic emotions in the brain. Activated brain regions for the basic emotions—happiness, sadness, anger, fear, and disgust—shown on an axial slice of the brain. The frontal lobe is located at the top of each brain figure, the temporal lobe is on each side, and the occipital lobe is at the bottom of each brain figure. The left hemisphere is shown on the left side of each figure, the right hemisphere on the right side. Note that the arousal level previously discussed for the basic emotions is reflected in the amount of brain activity produced: thus "disgust" and "fear," which have higher arousal levels than "happiness," also shown more widely distributed brain activity. The bottom brain image is a contrast of the brain activation for fear versus sadness. While some common patterns of brain activity can be shown for the basic emotions, there is no support currently for a one-to-one mapping of a specific pattern of brain activation to a single emotion: individuals are highly variable in their reactions to emotions and this is reflected in a highly variable brain response. *Source: Hamann (2012).*

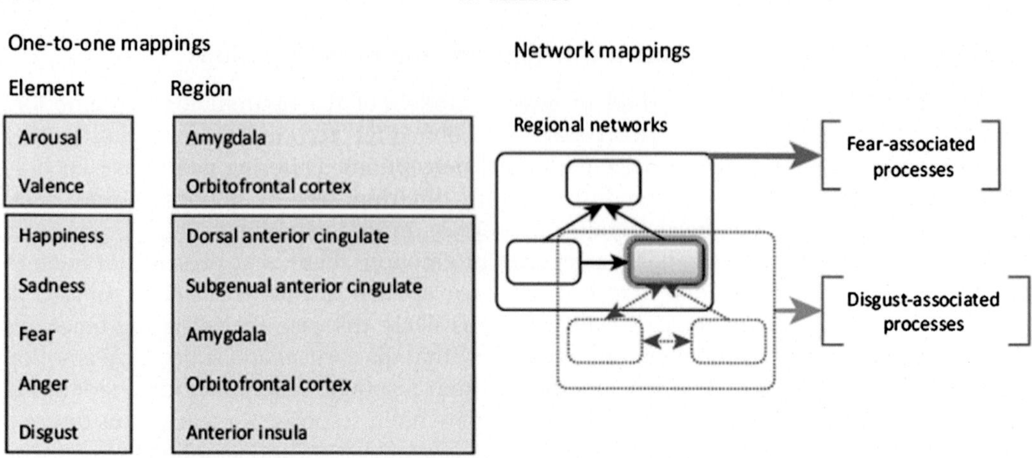

FIGURE 11.13 Basic emotions have been demonstrated to show activation in certain brain regions: for example, the insula is activated by the emotion "disgust" and the amygdala is activated by the emotion "fear." However, the emotional brain includes many regions and subregions in the brain that are activated in a network fashion in responding to emotional states. *Source: Hamann (2012).*

and many similar reasons owing to the complexity of mapping human emotions, there is not at this time a clear theory of brain areas that activate for *specific* emotions.

4.2 Emotional Facial Expression Processing

The most frequently studied aspect of emotional processing is the decoding of facial expressions. Understanding what a facial expression is saying about an individual's demeanor, feelings, and current situation are critical social cues that can be useful in social interactions and also key to survival in the case of a threatening or dangerous situation.

We discussed the brain regions for decoding human faces in Chapter 10, Humans Are Social Beings, they include the fusiform face area (FFA), and other regions within the visual system. The decoding of and recognition of emotional facial cues is a rapid process, with initial brain responses arising within fractions of a second. These rapid responses include a wide range of brain regions tuned to decoding emotional facial expressions, including the FFA, other visual areas in striate cortex, and the amygdala (Fig. 11.14).

It may not be surprising that fearful facial expressions are extremely salient ones and are processed somewhat differently than neutral facial expressions: since fearful faces carry important survival information, it is no wonder they seem to have an expedited route through the brain! Vuilleumier and Pourtois (2007) used fMRI to compare hemodynamic responses in the FFA to houses, neutral faces, and fearful faces. They found that both neutral and fearful faces activated the FFA, as expected, while the images of houses did not (Fig. 11.15). Critically, they found that the hemodynamic response for fearful faces was both more rapid and larger/stronger than the neutral faces. This study provides evidence for our intuition—that a fearful face gains access to our awareness and to our cognitive processes in a way that a neutral face does not.

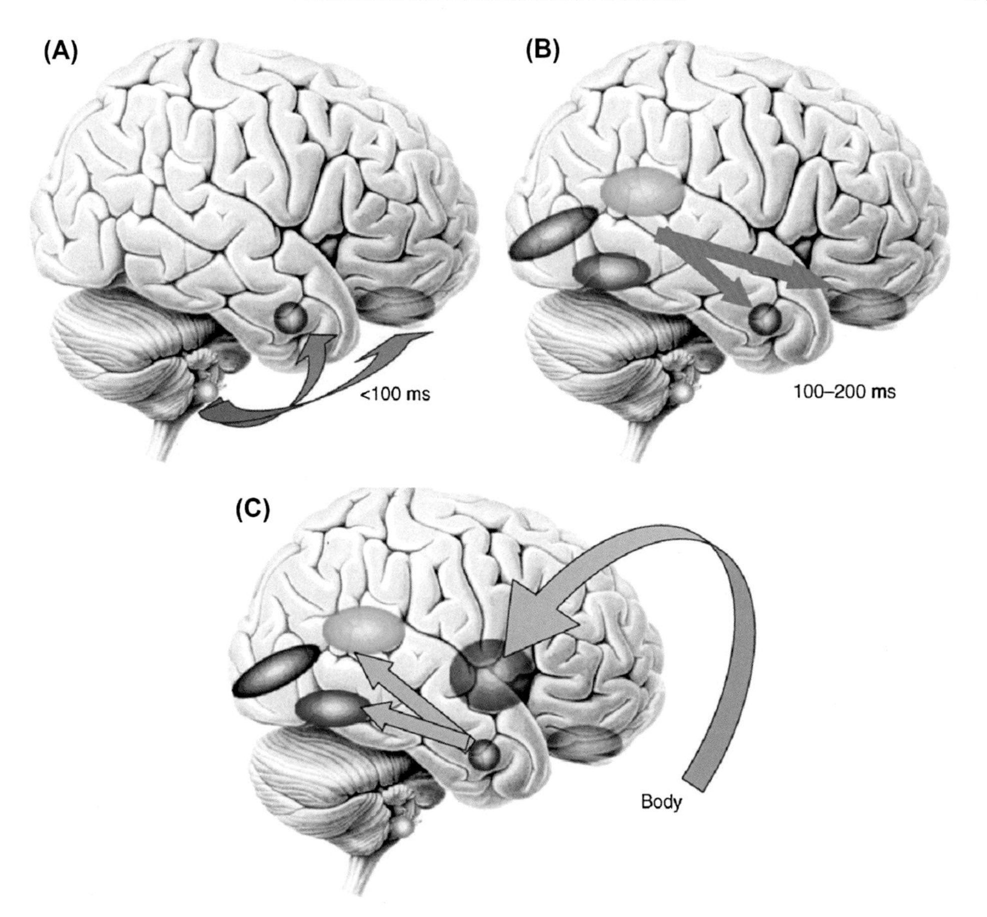

FIGURE 11.14 The time course of emotional face perception. The lateral view of the brain, showing the right hemisphere, shows many brain areas that are activated when perceiving and recognizing emotional facial expressions. (A) In under 100 ms, the prefrontal cortex and the amgydala (shown in red) respond to visual face stimuli, activated in part by subcortical regions. (B) within 100–200 ms, visual areas (shown in black) and other sensory regions are activated and provide more detail to the emotion-encoding regions such as the amygdala (red). The superior temporal cortex (shown in green) encodes dynamic face features and the fusiform gyrus (shown in blue) is important for encoding face identity. (C) Once the emotional meaning of the face stimulus has been deciphered, emotional responses are activated from the amygdala to the body via the brainstem. *Source: Tsuchiya and Adolphs (2007).*

5. EMOTIONAL CONTAGION AND EMPATHY

In Chapter 10, we discussed feelings of empathy and how they were important aspects of social cognition and behavior. Understanding how someone else is feeling, and feeling sympathetic, when that feeling is a negative or sad one is a key aspect of social behavior. However, empathy carries it a step further than just recognizing someone else's pain or sadness—in empathy, we *feel* that same pain…we can relate to the feelings and this relatedness invokes similar emotional processes in us.

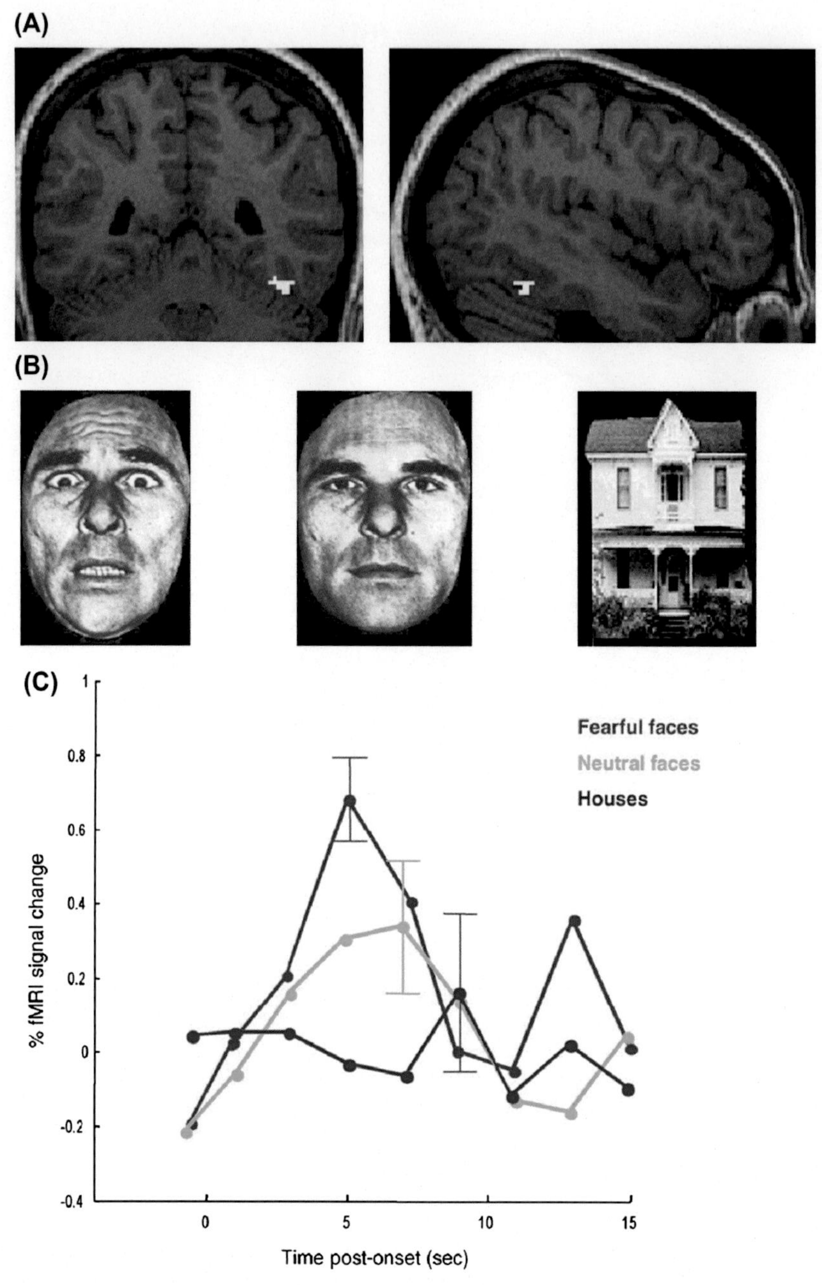

FIGURE 11.15 The fusiform face area (FFA) responds not only preferentially to face stimuli, but the amount of response corresponds to the emotion in that face. (A) Left panel: A coronal view of the brain showing the FFA in the right hemisphere activated for face stimuli. Right panel, The same activation in the FFA shown on a sagittal view of the right hemisphere. (B) The stimuli presented included fearful faces, neutral faces, and houses. (C) The brain signal in the FFA was higher for both fearful (red) and neutral (green) faces than for houses (blue). The FFA brain signal for the fearful faces was much higher than for the neutral faces, indicating the FFA response not just to faces but also to the emotional salience of those faces. *Source: Vuilleumier and Pourtois (2007).*

Emotional contagion: some very automatic responses to emotional expressions in another person have been described as *emotional contagion* (Hatfield, Cacioppo, & Rapson, 1993). One view is that emotional contagion is a primitive form of the more developed sense of empathy. One very basic emotional contagion response is when an adult hears an infant cry. There is typically a visceral reaction to the cry—experiments in emotional contagion have documented increases in cortisol level in those adults hearing an infant cry, along with heightened alertness (Fig. 11.16). Thus the adult is *feeling* some of the infant's stress and neediness, and responding in an affective-perceptive manner—as opposed to a cognitive-evaluative manner as we detailed in Chapter 10. Emotional contagion is common in social settings and can be as simple as responding automatically with a smile to someone who is smiling at you.

Emotional contagion is thought to arise from facial and body expression cues as well as verbal and nonverbal sounds or utterances. Think about being in a large gathering of people, say at a park. If a few people were to suddenly start screaming and running, with frightened faces, this emotion would definitely spread contagiously through the larger group of individuals. The combination of the frightened vocalizations, body language, facial expressions would produce a highly salient cue that would likely be extremely contagious.

However, a recent large study of Facebook users dares to differ: they investigated the effect of changing the positive or negative affect of the Facebook News Feed (Kramer, Guillory, & Hancock, 2014). Although there were no facial, verbal, or other nonverbal cues, the almost 700,000 individuals who were part of this study tended to send more positive posts when their News Feed had been filtered to increase the positive news and tended to send more negative posts when their News Feed had been filtered to increase the negative news. It seems that emotional contagion can be in effect even without the classic cues recognized for their transmission!

FIGURE 11.16 Emotional contagion: some very automatic responses to emotional expressions in another person have been described as *emotional contagion*. One view is that emotional contagion is a primitive form of the more developed sense of empathy. One very basic emotional contagion response is when an adult hears an infant cry. There is typically a visceral reaction to the cry—experiments in emotional contagion have documented increases in cortisol level in those adults hearing an infant cry, along with heightened alertness. Thus the adult is *feeling* some of the infants stress and neediness, and responding in an empathetic manner. *Source: Public domain image.*

Social scientists attribute emotional contagion as a basic step in forming the more complex form of *empathy*. Brain regions which are active for affective-perceptual empathy include left and right hemisphere insula and the ACC (see Fig. 10.18 in Chapter 10.

6. EMOTIONAL MEMORIES

Do you remember your high school graduation? Your senior prom? The day you found out that someone you loved had died? If you are like most people, these memories are vivid for you, even if they occurred years ago. Why are these memories so vivid while others, for example, where you had lunch 3 weeks ago, less clear? The limbic system, and especially the amygdala, plays an important role in both emotional memory formation and retrieval, along with other key brain areas such as the hippocampus. The basolateral area of the amygdala (BLA) projects to the hippocampus, the basal forebrain, the nucleus accumbens (NAc), the striatum, and regions in the cortex (McGaugh, 2002) (Fig. 11.17, see also Figs. 11.10 and 11.11).

How do some memories get to be so emotionally charged? It is theorized that the typical memory consolidation processes that involve the hippocampus and the cortex (see Chapter 7, Learning and Remembering) are also involved in the consolidation of emotional memories. However, in addition to these processes, evidence from animal studies and studies with

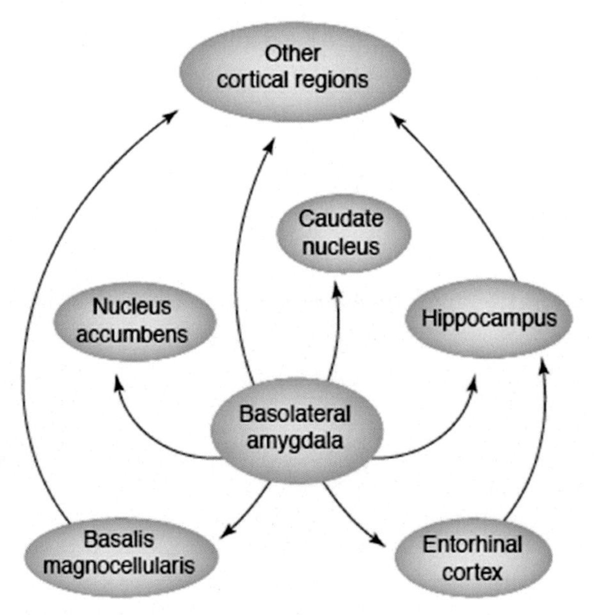

FIGURE 11.17 Modulation of memory consolidation by the amygdala: the amygdala plays a central role in the memory formation of emotionally significant experiences. The basolateral complex of the amygdala projects to other brains areas involved in memory consolidation, including the nucleus accumbens, the hippocampus, the caudate nucleus, the basalis magnocellularis, and the entorhinal cortex along with other cortical regions. *Source: McGaugh (2002).*

humans implicate the amygdala and the activation of β-adrenergic stress hormone systems during and after the emotional experience and memory formation. In a now-classic study, Cahill and colleagues (Cahill, Prins, Weber, & McGaugh, 1994) investigated the memory for an emotionally arousing story and an emotionally neutral story in human subjects who were assigned randomly to one of two experimental groups: placebo—with no medication given, and propranolol—with a dose of propranolol hydrochloride given, which is a β-adrenergic receptor antagonist or β-blocker. Subjects in both groups viewed a slide show of either an *emotionally arousing story* or an *emotionally neutral story*. All subjects' memories for the stories' events was tested 1 week later. Findings showed that all subjects—in both the placebo and the propranolol groups—performed similarly, on a memory test 1 week later for the emotionally neutral story. However, for the emotionally arousing story, the subjects in the propranolol showed much poorer memory than the subjects in the placebo group.

Cahill and colleagues interpreted these findings with human subjects, along with previous work with animals, as supporting the hypothesis that the emotional memory storage is modulated by β-adrenergic systems and further that these systems are not implicated in storage of emotionally neutral information or memories. These early findings have been well supported by more recent studies; however, a follow-up investigation by Cahill and colleagues (Cahill, Uncapher, Kilpatrick, Alkire, & Turner, 2004) has shed new light on the role of the amygdala in emotional memory processing: in an fMRI study of human subjects viewing pictures that varied in their emotional arousing ratings from tranquil to highly arousing. The amygdala region was activated by the arousing pictures, as hypothesized, but the central finding of the study was a strong *sex × hemisphere* interaction. For men, the *right hemisphere amygdala* was more activated for the emotional pictures; however, for the women, the *left hemisphere amygdala* was more activated (Fig. 11.18). This result has been replicated and reproduced by other experimenters—providing important new data about sex differences in the brain bases of emotional processing.

Reliving happy memories, including feelings of emotional arousal, can be a pleasant experience. However, when frightening or horrifying memories are reexperienced, along with the negative emotions felt when they were encoded, they can produce a life-altering state of stress and trauma. This is what happens in cases of posttraumatic stress disorder, or PTSD, discussed later in this chapter.

7. REWARDS AND MOTIVATION

When a behavior or a situation is highly rewarding to us, we tend to repeat that behavior or strive to have that situation occur again. If you are a chocolate lover, you do not need to be convinced that a bite of a brownie still warm from the oven sends you craving for more! This is a simplified explanation for how *reward* and *motivation* work together to influence behavior—we experience a *rewarding event* and this experience puts into place the *motivation to repeat the reward*...and if this reward is not able to be repeated, it may also begin a process of *craving for that reward*.

Brain pathways for this reward-inducing experience include the ventral tegmental area (VTA), the NAc, and the PFC (Laviolette & van der Kooy, 2004) (Fig. 11.19). Dopamine (DA) neurons from the VTA project to the NAc and PFC: this is a main *dopaminergic reward*

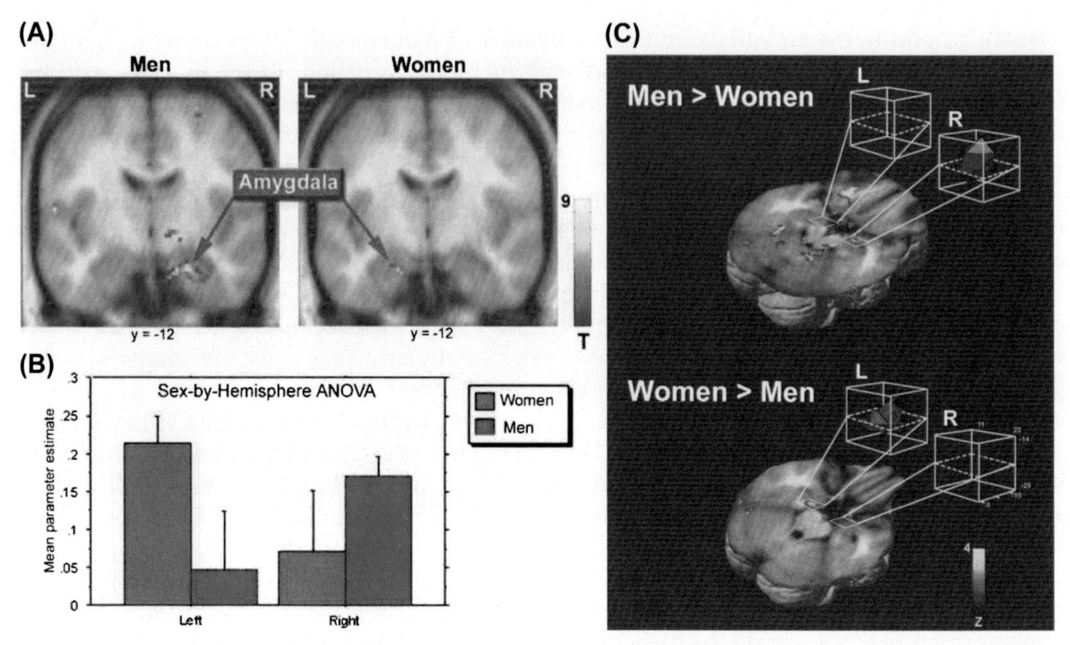

FIGURE 11.18 Cahill and colleagues investigated the activation of the amygdala in a functional magnetic resonance imaging (fMRI) study of human subjects viewing pictures that varied in their emotional arousing ratings from tranquil to highly arousing. The amygdala region was activated by the arousing pictures, as hypothesized, but the central finding of the study was a strong *sex × hemisphere* interaction. For men, the *right hemisphere amygdala* was more activated for the emotional pictures however, for the women, the *left hemisphere amygdala* was more activated. (A) A coronal view of the brain activity in the amygdala for the men in the study (left panel) and the women in the study (right panel). (B) Men and women showed a strong difference in amygdala activation. (C) Another view of the hemisphere asymmetry for men versus women in this study. This view shows coronal slices of the brain presented in a horizontal presentation. Shown in the small boxes is (top panel) the difference in activation for Men versus Women in the right amygdala and (bottom panel) the difference in activation for Women versus Men in the left amygdala. This result has been replicated and reproduced by other experimenters, providing important new data about sex differences in the brain bases of emotional processing. *Source: Cahill et al. (2004).*

signaling pathway. DA neurons from the VTA project to the PFC: this is the *mesocortical pathway*. They also project to the NAc: this is the *mesolimbic pathway*. Together, these two reward pathways are called the *mesocorticolimbic projection*. DA neurons' projections are shown in Fig. 11.20. DA neurons also innervate the amygdala and the hippocampus. GABA neurons can inhibit the DA projections. These DA and GABA pathways form a highly complex circuitry underlying our feelings of reward, motivation, and craving.

In addition to the naturally occurring rewarding experiences that produce DA, there are many other DA inducers: they include nicotine found in cigarettes, alcohol, and many drugs of abuse including cocaine. When these systems and pathways are functioning in a healthy way, they provide the brain bases for motivation and reward that influences our behavior in many positive ways. However, when these systems are altered or impaired, they can lead to addiction and mood disorders, discussed later in this chapter.

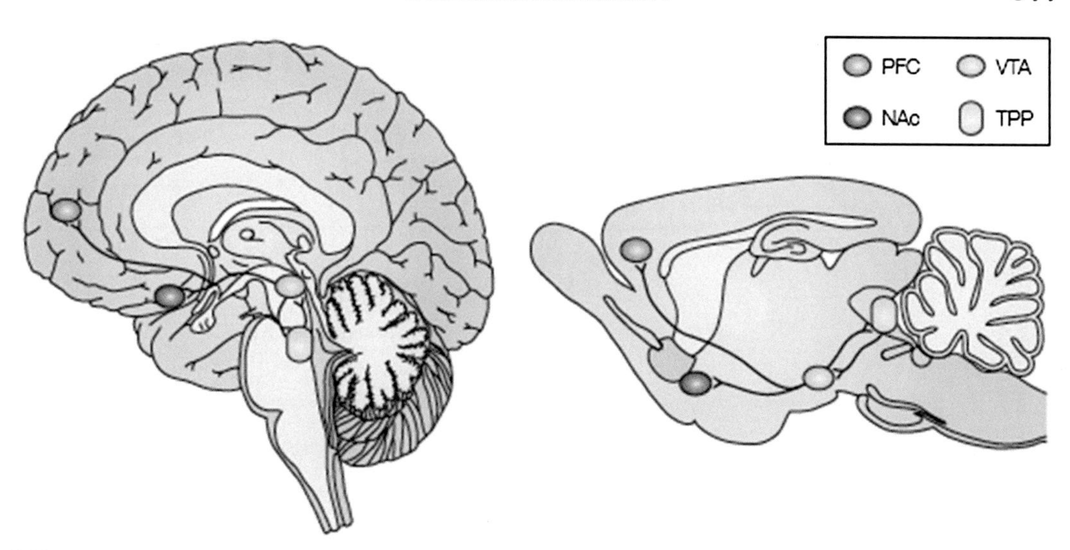

FIGURE 11.19 Human (left) and rat (right) brain pathways for this reward-inducing experiences include the ventral tegmental area (VTA), the nucleus accumbens (NAc), and the prefrontal cortex (PFC), shown on this midsagittal image of the right hemisphere of the brain. *Source: Laviolette and van der Kooy (2004).*

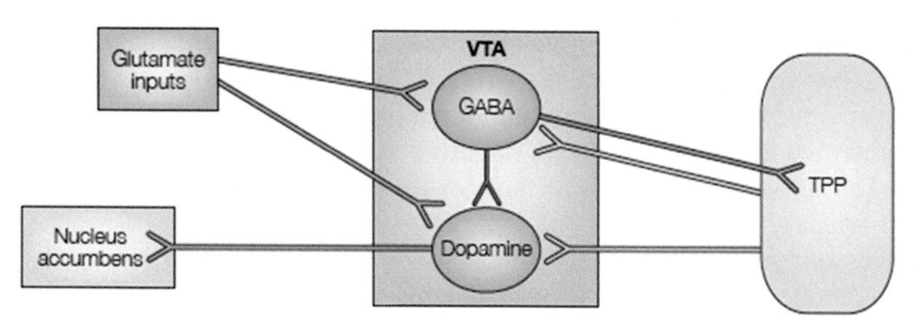

FIGURE 11.20 Dopamine (DA) neurons from the ventral tegmental area (VTA) project to the nucleus accumbens (NAc) and prefrontal cortex (PFC): this is a main *dopaminergic reward signaling pathway*. DA neurons from the VTA project to the PFC: this is the *mesocortical pathway*. They also project to the NAc: this is the *mesolimbic pathway*. Together, these two reward pathways are called the *mesocorticolimbic projection*. DA neurons also innervate the amygdala and the hippocampus. GABA neurons can inhibit the DA projections. These DA and GABA pathways form a highly complex circuitry underlying our feelings of reward, motivation, and craving. *Source: Laviolette and van der Kooy (2004).*

8. MOODS AND PERSONALITY

Emotions affect our daily lives in many ways, from brief feelings of emotions based on the context you are currently in, such as a brief feeling of irritation when just missing a traffic light on the way to work, to long-term emotional states that shape our moods and personalities in profound ways (Oatley & Johnson-Laird, 2014) (Fig. 11.21). While some emotions are fleeting and may be related to the context you are in, *moods* may vary due only to internal factors—but of course can also be affected by our environment. We usually describe moods

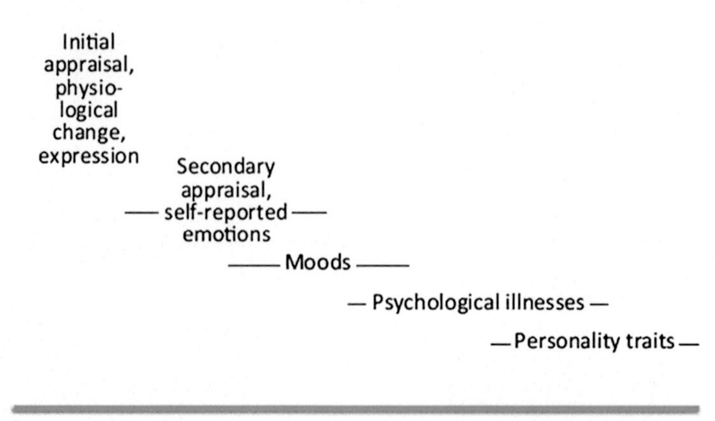

FIGURE 11.21 Emotions affect our daily lives in many ways, from brief feelings of emotions based on the context you are currently in, such as a brief feeling of irritation when just missing a traffic light on the way to work, to long-term emotional states that shape our moods and personalities in profound ways. While some emotions are fleeting and may be related to the context you are in, *moods* may vary due only to internal factors—but of course can also be affected by our environment. Over a lifetime, these emotional experiences help to form our personality. *Source: Oatley and Johnson-Laird (2014).*

simply as—we are in a good mood or we are in a bad mood. Our mood is important to how we interact with our environment: when we are in a good mood, positive memories are more available to us and we tend to look upon circumstances and events in a positive way. On the other hand, when we are in a bad mood, negative memories seem to be available to us and we look through dark-colored glasses at the world around us. This is typically referred to as *state-dependent memory*.

What controls our mood? In the most basic way, sleep and nutrition—or lack thereof—contribute to our mood. Lack of sleep can cause us to be irritable, as does low blood sugar when we are hungry. But there are many factors than come together to form our personality, among them the traits that we inherit from our parents. Throughout life, we make decisions, respond to the environment, react to social cues, and regulate our emotions and our behaviors. The combined force of these decisions and behaviors come together to form our personality.

Where does this personality formation happen in the brain? One answer comes to us from a bizarre accident in 1848. You may have heard of the strange case of Phineas Gage. He was a railway construction foreman and was struck with a tamping iron in a freak accident. The iron actually went through his head, from his chin to the top of his head (Fig. 11.22). On its way, the iron destroyed much of the medial PFC. Amazingly, he survived. However, after the accident, Phineas was no longer the Gage everyone knew. His behavior was erratic, he cursed and swore, he was impulsive, and he was unable to plan and organize as before.

From the brain damage suffered by Gage to present-day studies of individuals with PFC damage, we have pieced together what roles the medial regions of the PFC play in human cognitive function. We focus on the ventromedial PFC (vmPFC, see Fig. 11.23).

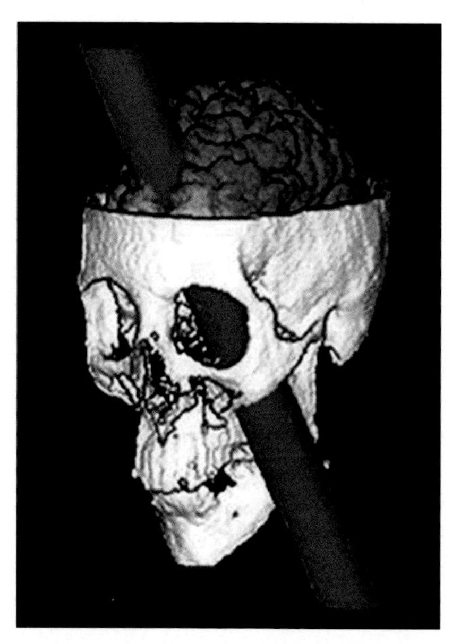

FIGURE 11.22 You may have heard of the strange case of Phineas Gage. He was a railway construction foreman and was struck with a tamping iron in a freak accident. The iron actually went through his head, from his chin to the top of his head. On its way, the iron destroyed much of the medial prefrontal cortex (PFC). Amazingly, he survived. However, after the accident, Phineas was no longer the Gage everyone knew. His behavior was erratic, he cursed and swore, he was impulsive, and he was unable to plan and organize as before. *Source: Miller and Wallis (2013).*

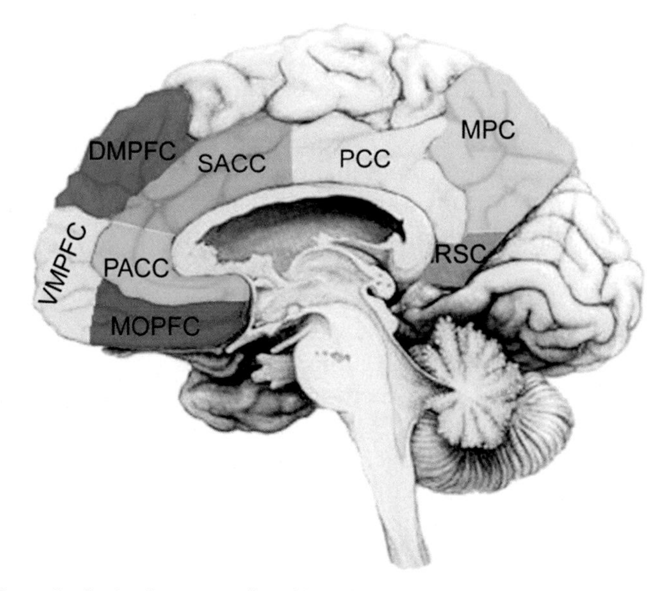

FIGURE 11.23 From the brain damage suffered by Phineas Gage to present-day studies of individuals with prefrontal cortex (PFC) damage, we have pieced together what role the medial regions of the PFC plays in human cognitive function. We focus on the ventromedial PFC (vmPFC, shown in light yellow just below the dorsomedial PFC (dmPFC, in blue): this region is involved in many cognitive, social, and emotional functions and processes ranging from decision-making to social/moral behavior. *Source: Northoff (2012).*

This region is involved in many cognitive, social, and emotional functions and processes ranging from decision-making to social/moral behavior (Koenigs, 2012). Much of what we know about the vmPFC comes from the work by Damasio and colleagues (Bechara et al., 1994; Damasio, 1995). These studies began by focusing on individuals with damage to the vmPFC. They developed the "somatic marker hypothesis," which was formed to explain the behavioral changes that patients with vmPFC damage experienced, including poor decision-making and uninhibited behavior. The idea behind this theory is that the vmPFC links the *somatic state* of an individual—their emotional state—with *potential outcomes of decisions*. Think of it as a connection between the emotional state of a person while making a decision and thinking about the various outcomes of that decision: I would like to eat that entire plate of brownies—it would feel good right now—but I know that the outcome of that decision would be a stomach ache and a few new pounds...According to Damasio and colleagues, the vmPFC plays a role in understanding—predicting the emotional effects of decision-making.

There are many more recent theories about the role of the vmPFC, especially with the advent of neuroimaging techniques so that the function of the vmPFC can be studied in healthy individuals. While there is no widely agreed-upon set of functions that the vmPFC performs, it is clearly a key player in social and moral decisions, self-reflection, and related thought processes that are key to effective living. More evidence about what role the vmPFC plays in human cognition comes from studies of psychopathy: a psychopath is a person who exhibits antisocial behavior—often violently—with no seeming sense of remorse.

9. CONFLICT AND CONFLICT RESOLUTION

A closely associated concept to decision-making is the all-too-familiar concepts of *conflict* and *conflict resolution*. Many times when we are making a decision—even a minor one—there are two or more possible decision/outcomes that come in conflict with one another. How do we decide which way to go in this process? The ACC is a key player in conflict resolution (Fig. 11.24). How is conflict measured? In neuroimaging studies, there are typical tasks set up that involve a conflict—or suppressing a behavior such as in a Stroop task where you must suppress reading a color word and name the color instead (Fig. 11.25). During these types of tasks, the dorsal part of the ACC (dACC) is activated when there is a high degree of conflict. The theory is that as the ACC is more activated due to conflict monitoring or resolution; the ACC activates regions in the dorsolateral PFC (dlPFC) where goal-directed behavior processes take place (Amodio, 2014). The rostral ACC (rACC) is thought to be involved in monitoring external cues that relate to the conflict resolution. The give-and-take between the dACC, rACC, and the dlPFC is thought to provide the neural machinery for determining if there is a conflict, monitoring cues that are relevant, and resolving the conflict, typically with some form of decision-making.

When conflict resolution mechanisms are not functioning in a typical manner, social, behavioral, and emotional processing is affected: we will discuss abnormalities and deficits in these processes later in the chapter in the section on Addiction.

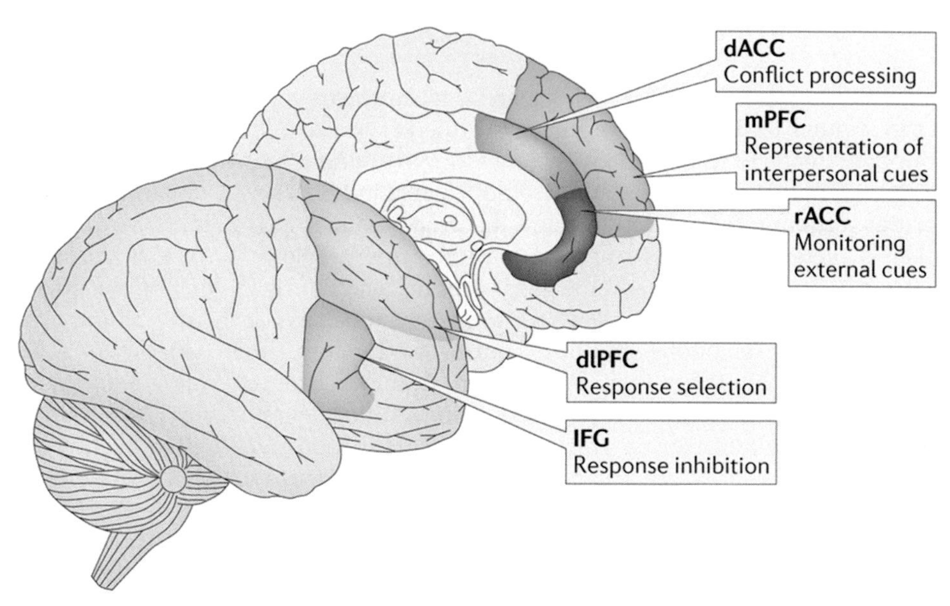

FIGURE 11.24 Many times when we are making a decision—even a minor one—there are two or more possible decision/outcomes that come in conflict with one another. How do we decide which way to go in this process? The anterior cingulate cortex (ACC) is a key player in conflict resolution. How is conflict measured? In neuroimaging studies, there are typically tasks set up that involve a conflict—or suppressing a behavior such as in a Stroop task where you must suppress reading a color word and name the color instead (Fig. 11.25). During these types of tasks, the dorsal part of the ACC (dACC, shown in light red) is activated when there is a high degree of conflict. The theory is that as the ACC is more activated due to conflict monitoring or resolution, the ACC activates regions in the dorsolateral PFC (dlPFC, shown in teal) where goal-directed behavior processes take place. The rostral ACC (rACC, shown in dark red) is thought to be involved in monitoring external cues that relate to the conflict resolution. The inferior frontal gyrus (IFG, shown in green) is thought to be involved in response inhibition. The give-and-take between the dACC, rACC, and the dlPFC is thought to provide the neural machinery for determining there is conflict, monitoring cues that are relevant, and resolving the conflict, typically with some form of decision-making. *Source: Amodio (2014).*

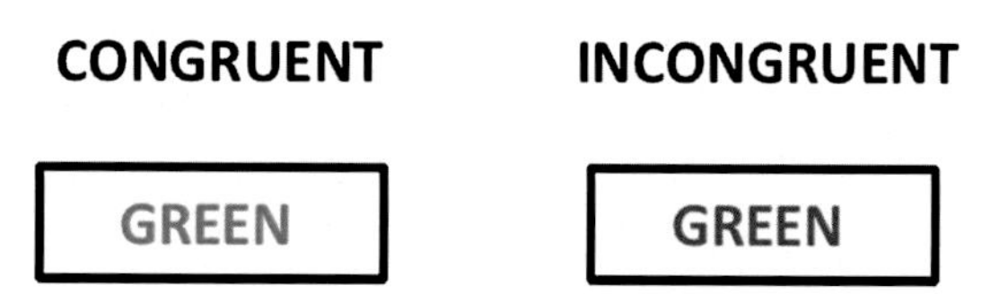

FIGURE 11.25 Typical stimuli used in a Stroop task: this type of task is used to investigate conflict and conflict resolution. When asked to name the color of the font presented in each of these words, the Congruent word—where the word is "GREEN" and the font is green, is typically provided faster and with more accuracy than the Incongruent word—where the word is "GREEN" and the font is red. The response "GREEN" must be suppressed to accurately name the color—red—for the Incongruent item. *Source: Gage.*

10. STRESS

Stress. Our lives are full of some form of stress: from the stress felt by a student when exam time comes around to the stress felt during a difficult workout. Whatever the source of the stress, our brain tends to react in foreseeable ways. When under high levels of stress, we often feel that we cannot think well, that we are not clear minded. There is a neurophysiological basis for this intuition—let us step through the processes that lead to it.

The key players in the stress response are three brain regions that are highly interconnected: the PFC, the amygdala, and the basal ganglia. The PFC is the hub of our intellectual brain: it contains the "control centers" for executive functions such as planning, making decisions, regulating emotions, impulse control, and many other aspects of a healthy functioning individual. When we are stressed from an external source—an exam, for example—or an internal source—worry about one's future—stress affects the brain. When we experience normal, unstressful emotions, the PFC functions well with a healthy balance of neuromodulators such as DA and norepinephrine (NE) (Fig. 11.26, left). Moderate levels of NE actually strengthen the PFC while weakening the role of the amygdala.

When we are stressed, the neuromodulators DA and NE are increased (Arnsten, Raskind, Taylor, & Connor, 2015) (Fig. 11.26, right). These high levels of neuromodulators actually weaken PFC function and strengthen the emotionally charged responses of the amygdala

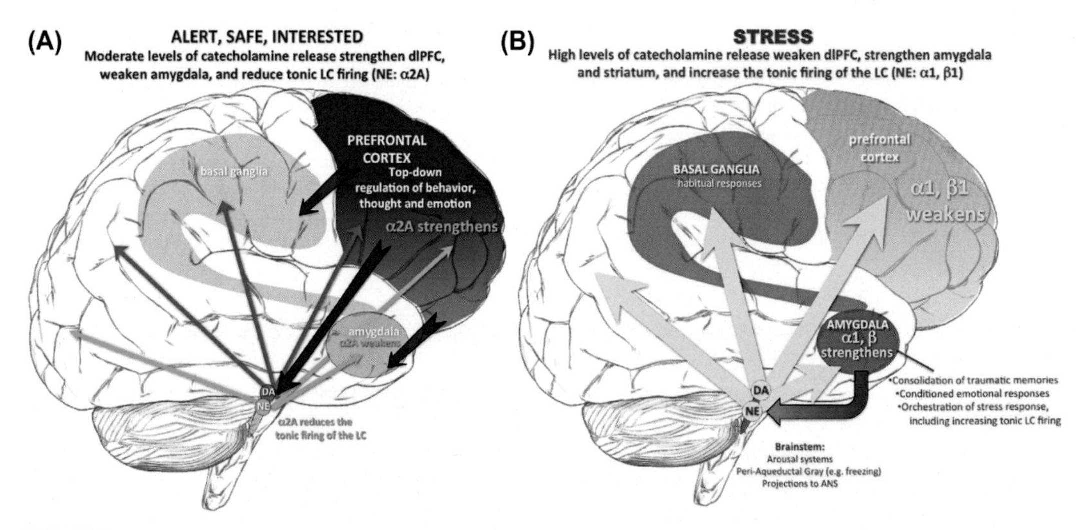

FIGURE 11.26 (A) A lateral view of the right hemisphere of the brain. When we experience normal, unstressful emotions, the prefrontal cortex (PFC) functions well with a healthy balance of neuromodulators such as dopamine (DA) and norepinephrine (NE). Moderate levels of NE actually strengthen the PFC while weakening the role of the amygdala. (B) The same lateral view of the right hemisphere. When we are stressed, the neuromodulators DA and NE are increased. These high levels of neuromodulators actually weaken PFC function and strengthen the emotionally charged responses of the amygdala and the basal ganglia. While this is a strong natural and biological response that allows us to perform physically well under periods of high stress—for example, running away from a danger source or dealing with a wild animal—prolonged states of stress can have a long-lasting and harmful effect. *Source: Arnsten et al. (2015).*

and the basal ganglia. While this is a strong natural and biological response that allows us to perform physically well under periods of high stress—for example, running away from a danger source or dealing with a wild animal—prolonged states of stress can have a long-lasting and harmful effect. Long-lasting periods of stress can have a negative effect on working memory and decision-making. Chronic stress—as in PTSD—can lead to loss of grey matter in the PFC and to strengthening of the amygdala and its connected regions in the limbic system. Thus in small doses, stress can be a good thing that aids in our escaping from danger. In large doses, or in cases of chronic stress, stress has a very harmful effect on the brain and especially in the executive cortex in the PFC.

11. EMOTIONAL IMPAIRMENTS AND DISORDERS

Emotions play a central role in our daily lives. Controlling our emotions, regulating our emotional responses, and understanding other's emotions are key to functioning as a healthy individual. An imbalance in emotional controls can lead to mental disorders that can be debilitating and even life threatening.

11.1 Posttraumatic Stress Disorder

Exposure to prolonged stress can lead to a disorder aptly named PTSD. Just as the name implies, there is a persistent traumatic stressful response following periods of prolonged or acute stress. PTSD can occur in children, teens, adults, and seniors. We are beginning to have a better understanding of its causes and treatments (see Box 11.1).

11.2 Addiction

Drug addiction—including addition to alcohol, cocaine, and other drugs—typically involves a continually occurring cycle of *Intoxication, Bingeing, Withdrawal,* and *Craving* (Goldstein & Volkow, 2011) (Fig. 11.29). During the Intoxication stage, the user has an impaired self-awareness. The danger or damage that the drug use can bring is overwhelmed by the rewards provided by the drug. During the Bingeing stage, there is a loss of control. The Withdrawal stage includes a lack of motivation and negative emotions—this is the "low" following the "high" of the drug use. And finally, during the Craving stage, there are strong feelings of expectation of the drug, with an attentional bias focused on the next usage.

As you read the preceding paragraph, which brain regions came to mind as likely involved in these stages? If you thought of the PFC, then your thinking is in line with current models for understanding the brain bases of addiction.

The regions in the brain typically discussed when addiction is studied are those regions that are involved in the reward and motivation pathways—the VTA, the NAc, and the PFC. More recent neuroimaging studies with humans have shifted the focus away from the VTA and NAc and onto the PFC. Always included in the reward circuitry, the PFC is taking a more central role in studies of addiction. This shift in focus is due to the many studies that have shown that drug addicts show differences in PFC function and activation levels as compared to nonaddicted adults. A key role played by the PFC in

BOX 11.1

WHEN THE WAR NEVER ENDS ... POSTTRAUMATIC STRESS DISORDER

In the past decade, more and more American troops have returned home from war only to face a new reality—the war has returned home with them (Fig. 11.27). These individuals suffer from recurring emotion-charged flashbacks of traumatic events that impair their ability to reconnect with civilian life and can lead to long-term depression and anxiety. This disorder, posttraumatic stress disorder or PTSD, has become so pervasive in the military that the United States Department of Veteran Affairs has created a National Center for PTSD to provide resources for individuals suffering from PTSD, their loved ones, caregivers, and clinicians (1).

What is Posttraumatic Stress Disorder?

While almost everyone who suffers a trauma—whether it is due to serving in a warzone, being in a natural disaster, surviving a sexual attack, or having a serious car accident—will have some aftereffects of that trauma, with frightening memories and difficulty sleeping, these aftereffects typically disappear over time following the traumatic event. This is not the case for individuals suffering from PTSD. For these individuals, the traumatic events are relived, reexperienced over and over again along with the very emotions felt at the time they initially lived through the trauma.

FIGURE 11.27 Soldiers serving in combat experience heightened emotions and stress, sometimes for long periods of time. *Source: Public domain image.*

BOX 11.1 *(cont'd)*

PTSD currently is estimated to afflict roughly 7%–8% of the US population (1). As mentioned above, it is not limited to former soldiers: victims of any type of trauma can suffer from PTSD, even young children. However, the large number of American troops serving abroad in war zones over the past 10 years or so who suffer from PTSD has heightened our awareness of its life-changing nature.

Symptoms of Posttraumatic Stress Disorder

PTSD affects individuals in several key ways: first, many PTSD sufferers relive the traumatic events—this is frequently referred to as having flashbacks. Importantly, these flashbacks evoke not only the memory of the events but also the raw emotions felt during the event. Second, not surprisingly, many PTSD sufferers tend to avoid situations that might remind them of the traumatic event. A war veteran may avoid Fourth of July festivities, for example, because of the sound of the fireworks. A car accident survivor with PTSD may stop driving a car altogether. Some PTSD individuals will refuse to discuss anything about the traumatic event. Third, many PTSD individuals will have a change in their beliefs and feelings—their personality may change due to the posttraumatic stress they are experiencing. They may no longer be interested in things that used to be important to them. Fourth, individuals with PTSD frequently suffer from hyperarousal—a feeling of being jittery, on the alert for any danger. This causes sleeping difficulties, difficulty concentrating, and a general feeling of stress.

Brain Bases of Posttraumatic Stress Disorder

A central challenge to understanding the brain circuitry underlying PTSD has been the high level of heterogeneity observed in individuals with PTSD: while a formal description of PTSD has been developed for clinical use, the specific forms of the disorder vary widely across individuals. This has provided a challenge for understanding the brain mechanisms that underlie PTSD; however, nonetheless, significant progress has been made over the past several decades.

A well-accepted view at present—although bear in mind that new research is changing this accepted view almost daily with new findings—puts forth a theory that PTSD stems from an overreactive response in brain regions that regulate fear, anger, stress, and other emotions—such as the amygdala in the limbic system, see Section 10—coupled with a reduced response in the prefrontal cortex (Mahan & Ressler, 2012) (see Fig. 11.28), thought to be critical for modulating or inhibiting the responses released in the amygdala and related regions. This is an extremely simplified description—what actually happens in the brain is far more nuanced, involved both local and long-range cortical and subcortical connections. In fact, there is evidence that the neural bases for PTSD are not so "localized" as including primarily the limbic regions and the prefrontal cortex: in this view, the neural bases for PTSD are widely distributed in dynamic neural circuitry throughout the brain (2).

Continued

BOX 11.1 (cont'd)

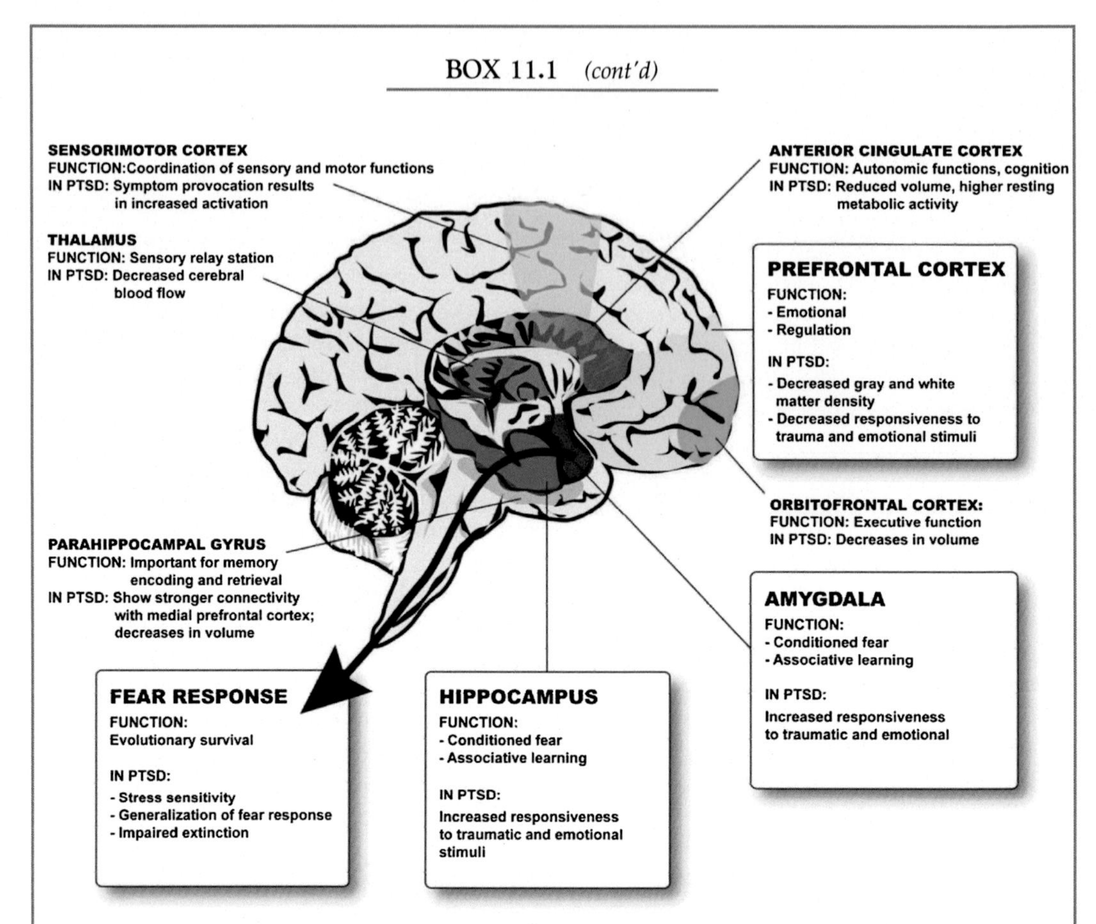
SENSORIMOTOR CORTEX
FUNCTION:Coordination of sensory and motor functions
IN PTSD: Symptom provocation results in increased activation
THALAMUS
FUNCTION: Sensory relay station
IN PTSD: Decreased cerebral blood flow
ANTERIOR CINGULATE CORTEX
FUNCTION: Autonomic functions, cognition
IN PTSD: Reduced volume, higher resting metabolic activity
PREFRONTAL CORTEX
FUNCTION:
- Emotional
- Regulation
IN PTSD:
- Decreased gray and white matter density
- Decreased responsiveness to trauma and emotional stimuli
ORBITOFRONTAL CORTEX:
FUNCTION: Executive function
IN PTSD: Decreases in volume
PARAHIPPOCAMPAL GYRUS
FUNCTION: Important for memory encoding and retrieval
IN PTSD: Show stronger connectivity with medial prefrontal cortex; decreases in volume
AMYGDALA
FUNCTION:
- Conditioned fear
- Associative learning
IN PTSD:
Increased responsiveness to traumatic and emotional
FEAR RESPONSE
FUNCTION:
Evolutionary survival
IN PTSD:
- Stress sensitivity
- Generalization of fear response
- Impaired extinction
HIPPOCAMPUS
FUNCTION:
- Conditioned fear
- Associative learning
IN PTSD:
Increased responsiveness to traumatic and emotional stimuli

FIGURE 11.28 The many brain regions we have discussed in this chapter that form the limbic system are shown in the lateral view of the right hemisphere. The high interconnectivity of the amygdala with the hippocampus and prefrontal cortex (PFC) forms the basis for some of the abnormal and prolonged stress response observed in individuals with posttraumatic stress disorder (PTSD). For example, in PTSD the amygdala shows increased responsiveness to traumatic and emotional stimuli/experiences, as does the hippocampus. Together they form a heightened system for encoding and retrieving emotionally charged memories. *Source: Mahan and Ressler (2012).*

The Balancing Act Between Emotional and Executive Brain Areas

Why does an overreactive amygdala and underreactive prefrontal cortex give rise to PTSD? Think about it this way: you are watching a scary movie. Suddenly a bad guy leaps out of nowhere and starts brandishing a knife at the heroine…who is screaming at the top of her lungs. Your heart races in response to the sudden noise caused by the leaping bad guy, the rise in the music that *always* seems to go along with bad guy leaps and the screaming of the victim—a fellow human being in clear distress. But your brain (think prefrontal cortex) kicks in to help you cognitively control your racing heart—it is only a movie after all, not real.

BOX 11.1 *(cont'd)*

This give and take between the emotional brain and the executive brain occurs continuously and serves to guide us as we make it through a busy day. It helps us distinguish between *truly terrifying situations* (say, the 405 Freeway in Southern California at rush hour) and not truly terrifying situations (Nightmare on Elm St. I, II, III, etc.).

References

(1) U.S. Department of Veterans Affair National Center for PTSD. http://www.ptsd.va.gov/index.asp.

(2) Suvak, M.K., & Barrett, L.F. (2011). Considering PTSD from the perspective of brain processes: A psychological construction approach. *Journal of Traumatic Stress, 24*(1): 3–24.

(3) Mahan, A. L., & Ressler, K. J. (2012). Fear conditioning, synaptic plasticity, and the amygdala: Implications for posttraumatic stress disorder. *Trends in Neuroscience, 35*(1), 24–35.

human function is to make decisions, assess potential negative and positive outcomes of those decisions, and to control impulses. More and more data are shedding light on how these aspects of PFC function are impaired or abnormal in addicts (Goldstein & Volkow, 2011) (Fig. 11.30).

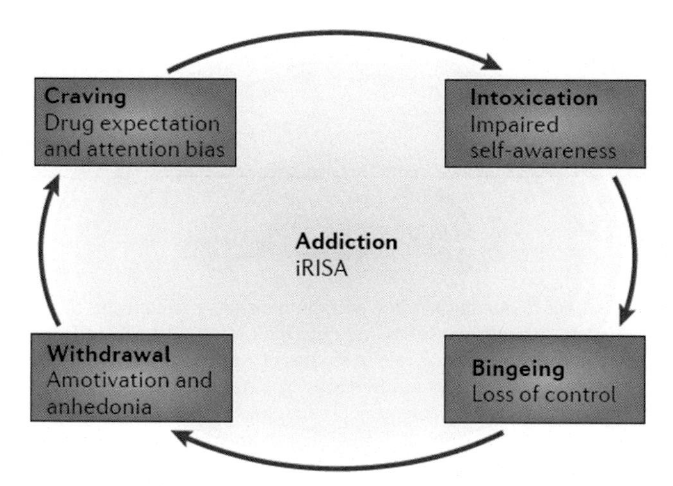

FIGURE 11.29 Drug addiction—including addition to alcohol, cocaine, and other drugs—typically involves a continually occurring cycle of *Intoxication, Bingeing, Withdrawal,* and *Craving*. During the Intoxication stage, the user has an impaired self-awareness. The danger or damage that the drug use can bring is overwhelmed but rewards provided by the drug. During the Bingeing stage, there is a loss of control. The Withdrawal stage includes a lack of motivation and negative emotions—this is the "low" following the "high" of the drug use. And finally, during the Craving stage, there are strong feelings of expectation of the drug, with an attentional bias focused on the next usage. *Source: Goldstein and Volkow (2011).*

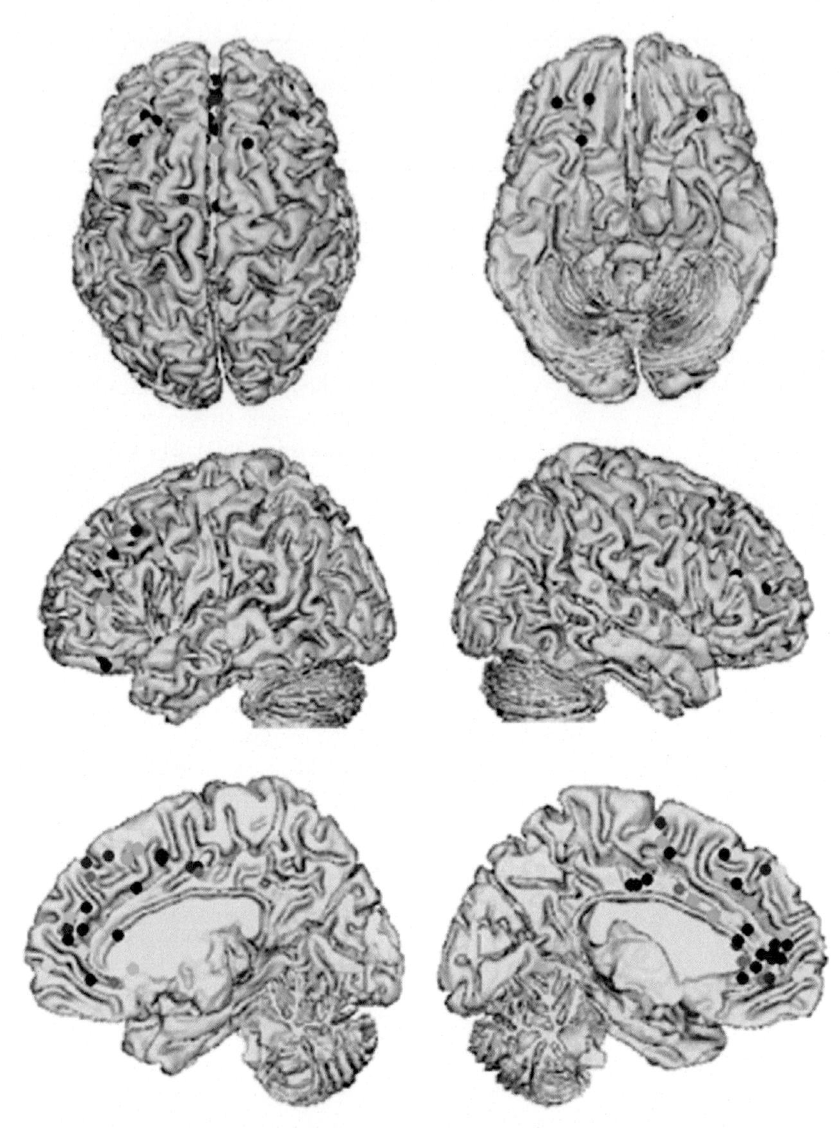

FIGURE 11.30 More recent neuroimaging studies with humans have shifted the focus away from the ventral tegmental area (VTA) and nucleus accumbens (NAc) and onto the prefrontal cortex (PFC). Always included in the reward circuitry, the PFC is taking a more central role in studies of addiction. This shift in focus is due to the many studies that have shown that drug addicts show differences in PFC function and activation levels as compared to nonaddicted adults. In this meta-analysis of many neuroimaging studies with individuals with addictions, the regions in the PFC that have been shown to differ for individuals with addiction versus healthy controls correspond to areas for attention and working memory (green), emotion and motivation (red), decision making (turquoise), and inhibitory control (yellow)—all key aspects of addictive behaviors. Results are shown (top panel) on axial images showing the brain as seen from above, with the frontal lobe at the top of the image; (middle panel) on lateral images showing the brain from the side with the left hemisphere on the left and the right hemisphere on the right; (bottom panel) on midsagittal images with the left hemisphere on the left and the right hemisphere on the right. *Source: Goldstein and Volkow (2011).*

Thus it is hypothesized that the control provided by the PFC—which allows nonaddicted individuals to manage their use of drugs, alcohol, and other potentially addictive substances—is weakened or impaired in addicted individuals. The PFC helps stabilize and reduce feelings of craving and supplies the brain basis of the wherewithal to "just say no." It is hoped that these findings can help provide more effective treatments for drug abuse: accurately determining the brain regions that correspond to addictive behavior may help pinpoint potential clinical interventions.

11.3 Mood Disorders

11.3.1 Depression

By far, the most common mood disorder is depression, which affects roughly one in five people in the United States at some time in their life (Kessler et al., 2005). Many individuals with depression do not get relief from the commonly used medications (Culpepper, 2010). In addition to suffering from depression, many of these individuals also suffer from anxiety. Another large percentage of these individuals have abnormal reward/motivation behaviors, such as addiction. The combination of depression, anxiety, and drug addiction motivates an assessment of the reward system in the brain. As discussed earlier in the chapter, the key brain areas involved in the reward pathways are the VTA, NAc, and PFC (see Fig. 11.19 and 11.20). The neuromodulators, specifically DA and NE, that this reward circuitry control, take center stage when assessing the brain bases of depression. While we have presented a simplified diagram of this circuitry for teaching purposes, the pathways and interconnections throughout the brain are highly complex. Results of neuroimaging studies of patients with depression have presented highly varied findings, which reflect the high level of heterogeneity found in this mood disorder. Two key findings, however, have shown that activity levels in the VTA and the NAc are reduced in individuals with depression as compared to healthy controls. Secondly, findings in the PFC have shown smaller volumes in key regions in the reward pathway, possibly indicating a loss of grey and white matter. While these studies are still in their initial phases, these results may reflect the neural bases for the depression.

11.3.2 Bipolar Disorder

Bipolar disorder, also referred to as manic-depressive disorder, is associated with large shifts in mood. Typically, an individual with bipolar disorder has a stage of mania where he or she is extremely "*up*," elated, and very excited about everything. The individual may report that they feel "wired" and may have trouble sleeping. These manic stages can last a week or so. They are followed by a depressive stage, with another large shift in mood. In this case, the mood changes to depressive, sad, "*down*." Again, the individual during the depressive stage may have trouble sleeping, which will increase the sadness or depressed feelings. Some individuals have suicidal thoughts during their depressive stage. The depressive stage may last for weeks. A period with neither manic nor depressive states can last for weeks as well.

What causes these abrupt and widely shifting mood changes? Bipolar disorder typically presents itself in teens and young adults, although it can be present in children. There are genetic factors involved in the risk of developing bipolar disorder. There are also environmental factors such as stress, child neglect or abuse, and trauma. A central hypothesis of the brain bases of bipolar disorder has to do with neural plasticity: changes that occur during development that can have long-lasting effects (Manji & Duman, 2001). In the case of bipolar disorder, these changes may have taken place in the emotional brain regions that we have been discussing in this chapter, notably, the amygdala and the PFC.

Bipolar disorder is typically treated with mood-stabilizing drugs. Lithium is one of the few options for treating bipolar disorder and many studies have been conducted to understand the neurobiology of both the diseases and the treatment. Glutamate levels, an excitatory neurotransmitter, are suspected to be altered in individuals with bipolar disorder, leading to the abrupt shifts in mood. Lithium has direct effects on glutamate level, and this may be why lithium is one of the successful treatments for bipolar disorder. Nevertheless, it is important to keep in mind that the exact neural circuitry that is abnormal in bipolar disorder has not been elucidated.

Brain areas that are implicated in bipolar disorder and other mood disorders are shown in Fig. 11.31. They include regions discussed in this chapter such as the PFC, the amygdala, and the anterior cingulate cortex (Schloesser et al., 2011).

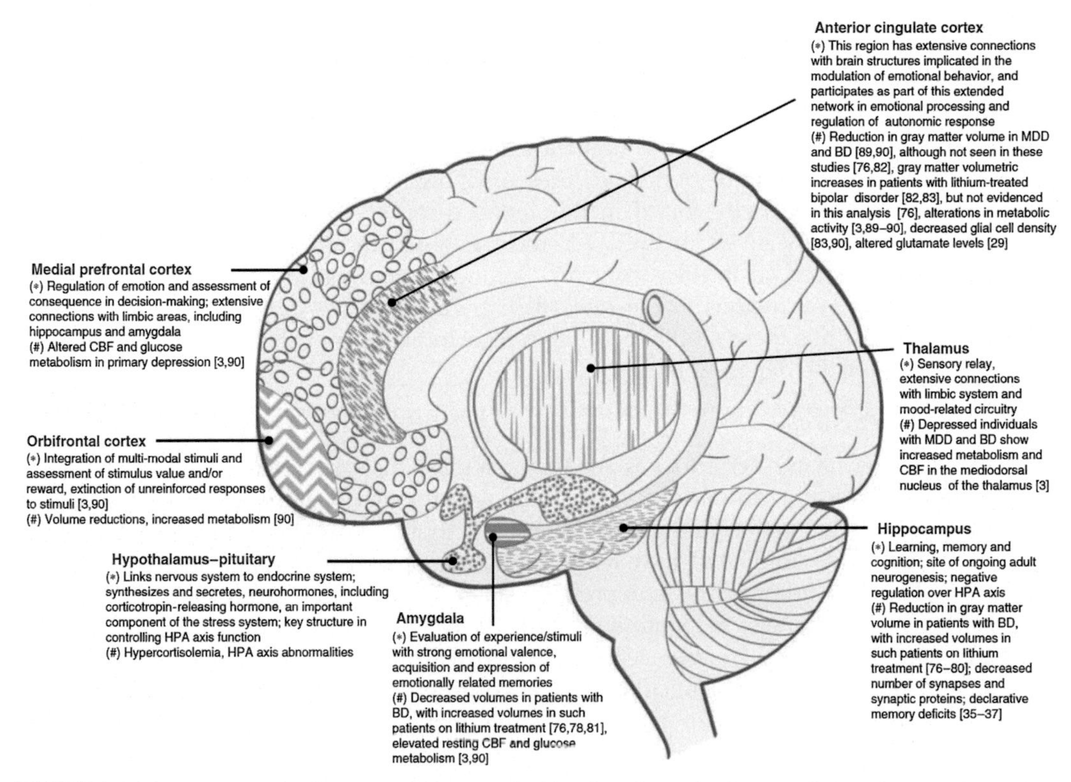

FIGURE 11.31 Brain areas that are implicated in bipolar disorder and other mood disorders are shown on a midsagittal view of the right hemisphere. They include regions discussed in this chapter such as the prefrontal cortex (PFC), the amygdala, orbitofrontal cortex, the hippocampus, and the anterior cingulate cortex. *Source: Schloesser et al. (2011).*

12. SUMMARY

In this chapter, we have discussed the complex role emotional processing plays in human cognition. While the prefrontal cortex is typically linked to executive functions such as decision-making, working memory tasks, and planning, we have seen over and over in this chapter that the PFC plays a central role in emotional functions such as emotional regulation and impulse control. The strongly interconnected roles of social cognition and emotion are also tightly interconnected with regions in the frontal lobe. The neural circuitry that underlies the healthy as well as the abnormal brain regulation of emotion is complex and involves many regions throughout the cortex and the subcortex.

Deficits or disorders in emotion can lead to devastating impacts on human life, such as addiction and mood disorders. While the precise neurobiology of these disorders is still being unraveled, progress to date has been positive and more and more individuals who suffer from them are getting relief through pharmaceutical treatments.

13. STUDY GUIDE

1. Describe how emotions can be characterized as discrete or continuous.
2. Where is the amygdala in the brain and what is its role in emotional processing?
3. How are facial emotions processed? What brain areas are involved in decoding emotional facial expressions?
4. What are key brain regions involved in reward and motivation? How do they differ for addicted individuals?
5. Describe the neural mechanisms that are believed to correspond to bipolar disorder.

CHAPTER

12

Sleep and Levels of Consciousness

SLEEP AND DREAMING

Death, so called, is a thing which makes men weep, And yet a third of life is passed in sleep.
Lord Byron, English poet.

Sleeping girl, c. 1660–1622. *Attributed to Domenico Fetti (1589–1623).*

https://doi.org/10.1016/B978-0-12-803813-0.00012-X

1. INTRODUCTION

1.1 Sleep Is Crucial to Healthy Cognition

Humans need sleep. It is crucial for our survival—try going for more than 24 h without sleep and you will see the immediate consequences of sleeplessness! When you think about the role of sleep in your cognitive selves, you can fairly easily describe what happens to your brain and body *without sleep*: you are irritable, unfocused, slow-moving, without energy. Can you as easily describe what happens to your brain and body *with sleep*? Scientists have been trying to unravel the mysteries of just what the brain does during sleep for years. We are making good progress but still have a distance to cover. We definitely know more about the effect of sleep deprivation on our emotional, cognitive, and somatic (body-related) selves (Fig. 12.1) than we understand how the brain actually functions during sleep. Even brief periods of sleep deprivation or shortened sleep can have rapid effects on our well-being. Chronic sleep deprivation causes even more stunning effects and can lead to early death.

We know, then, that sleep is critical for remaining healthy. Further evidence that it is vital for our survival comes from the daily evidence of how much sleep we need: a full 8 h—or one-third of each day—is spent in sleep (Table 12.1). During sleep, we are unaware of our environment, are in potential risk to any prey, and are literally paralyzed for large portions of each sleep session (more on that later when we discuss *rapid eye movement [REM] sleep*). And yet, despite the risk to ourselves, this amount of sleep has been a stable need for human beings for as long as we have investigated sleep. Thus sleep *must be critical for human survival* despite the long (8 h) time during each 24-h period when we are unconscious and defenseless. The evidence for the rapid cognitive, emotional, and body deficits we experience with the loss of even a partial night's sleep is clear evidence in support of the vital role sleep plays in human health and well-being.

1.2 Three Global Brain States: Awake, Asleep, and Dreaming

Most scientists agree to the existence of three global brain states: *waking, sleeping, and dreaming*. These states have largely been investigated using *electroencephalography (EEG)*, where brain rhythms are recorded and evaluated across levels of awareness in human subjects. We will discuss this more in Section 3.1, but briefly, EEG signatures of the three global brain states show a clear brain-wide difference in oscillatory brain activity across these states (Fig. 12.2).

Notice how different the EEG waveforms look across the three global brain states: during waking, there is a lot of uncorrelated activity across brain regions with different regions oscillating at differing rates. The sum effect when the EEG is measured at the skull is that you see a lot of wavy lines signifying that these brain regions have active brain rhythms that are widely varied in their speed of oscillation. In contrast, when you look at *nonrapid eye movement (NREM) sleep*, you begin to see the slow-moving oscillations of a brain in deep sleep. This is why some stages of NREM sleep are referred to as *slow wave sleep (SWS)*. Now look at the wave patterns for REM/dream sleep. Does it look more like the waking state than the sleep state of SWS? These findings from early EEG studies taken when subjects were

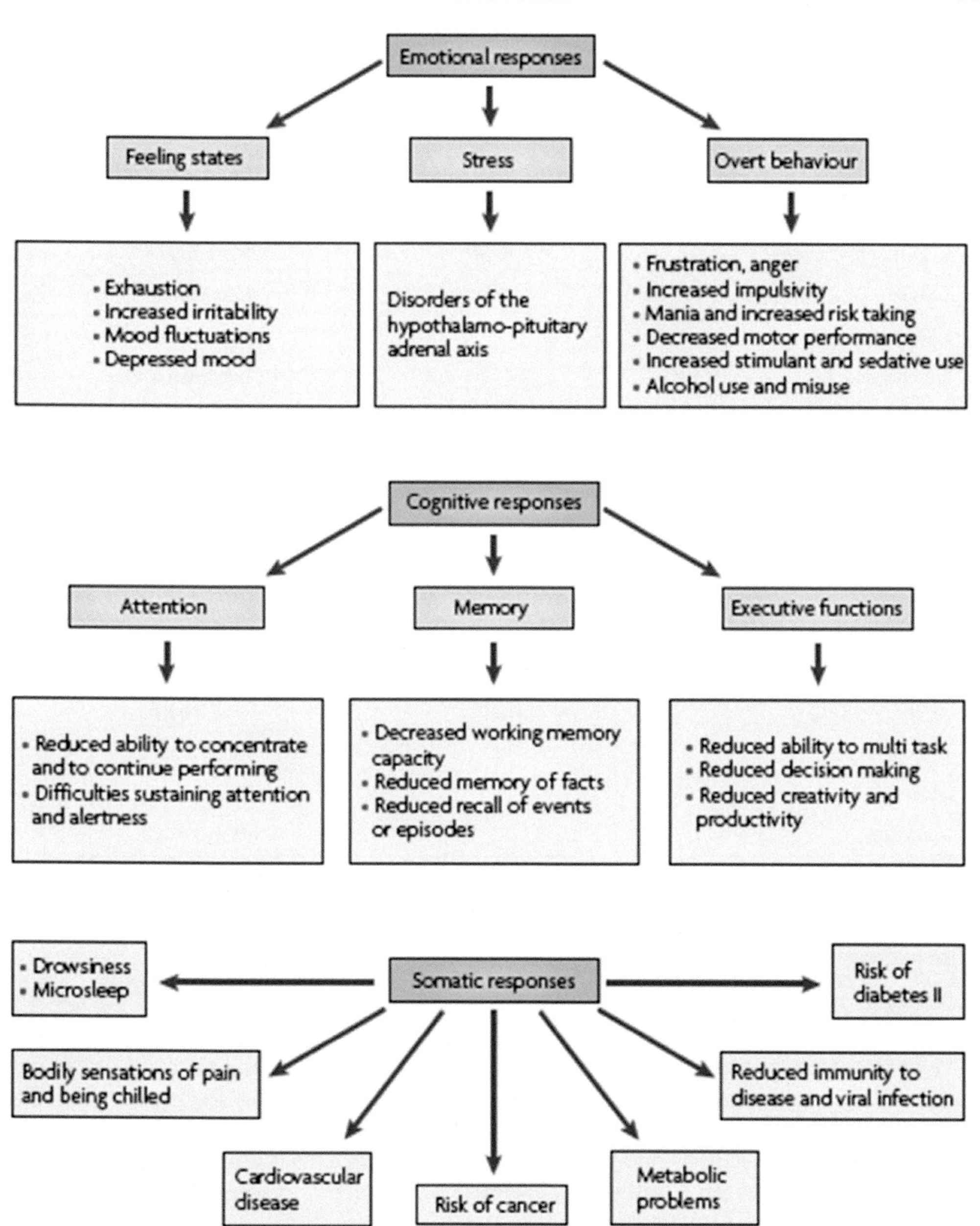

FIGURE 12.1 Lack of sleep affects many aspects of daily living, from our emotional well-being, to our cognitive abilities, to our body-related health. Emotional responses to sleep loss (top panel) include moodiness, stress, and feelings of frustration. Cognitive responses to sleep loss (center panel) include decreased working memory capacity and reduced decision-making abilities. Somatic or body-related responses to sleep loss (bottom panel) include drowsiness, reduced immunity, and can even lead to diseases such as cancer and diabetes. *Source: Wulff, Gatti, Wettstein, and Foster (2010).*

TABLE 12.1 Expert Panel—Recommended Sleep Durations

Age	Recommended (h)	May Be Appropriate (h)	Not Recommended (h)
Newborns	14–17	11–13	Less than 11
0–3 months		18–19	More than 19
Infants	12–15	10–11	Less than 10
4–11 months		16–18	More than 18
Toddlers	11–14	9–10	Less than 9
1–2 years		15–16	More than 16
Preschoolers	10–13	8–9	Less than 8
3–5 years		14	More than 14
School-aged children	9–11	7–8	Less than 7
6–13 years		12	More than 12
Teenagers	8–10	7	Less than 7
14–17 years		11	More than 11
Young adults	7–9	6	Less than 6
18–25 years		10–11	More than 11
Adults	7–9	6	Less than 6
26–64 years		10	More than 10
Older adults	7–8	5–6	Less than 5
>65 years		9	More than 9

Source: Hirshkowitz et al. (2015).

awake, sleeping, or dreaming have long informed us about the global changes in the brain's rhythms across the three states.

1.3 Levels of Consciousness: Awareness and Wakefulness

Awake or asleep, conscious or unconscious, aware or unaware: the typical human being crosses back and forth across these states during a typical day as he/she moves from sleep to waking and back to sleep. It is easy to think of the state of being awake as being conscious and the state of being asleep as being unconscious. However, it is not quite that simple: think about when you are dreaming, are you conscious? Most sleep and consciousness researchers agree that you are in a conscious dream state. Are you aware during dreaming? Again, most sleep and consciousness researchers would agree that you are aware—of your dream that is, although you will likely forget it later. You are not aware of your environment. So it seems that there is no easy way to subdivide our awake and asleep states into being conscious or unconscious states, or states of full awareness or full unawareness. To make things even more complex, while we are awake and conscious, there are many things in our immediate environment that we are unaware of.

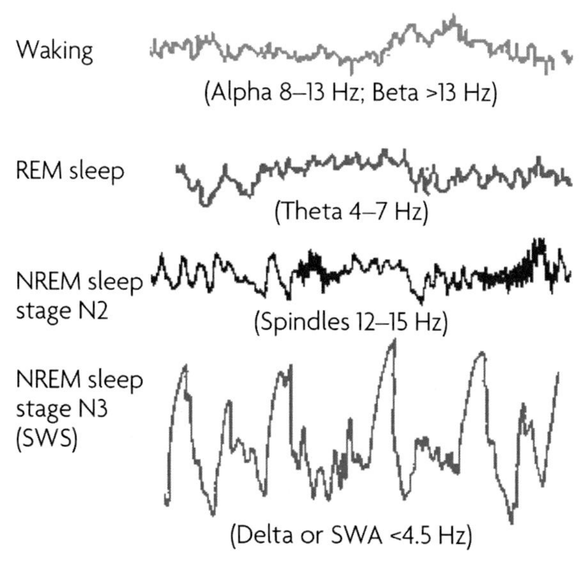

FIGURE 12.2 Electroencephalography (EEG) signatures of the three global states of waking, sleeping, and dreaming differ in important ways. During waking (*top green line*), there is widespread activity in the Alpha (8–12 Hz) and Beta (>13 Hz) bands across many brain areas. During slow wave sleep (SWS, bottom blue lines), brain rhythms slow dramatically with activity in the Delta (<4.5 Hz) band. During dreaming, most of which takes place during rapid eye movement (REM) sleep (*purple lines*) activity is somewhat slower than in waking in the Theta bank (4–7 Hz); however, the EEG waveforms look strikingly similar to waking. *Source: Cirelli (2009).*

Let us take a look at a comparison of two aspects of consciousness: *level of awareness* and *level of wakefulness* (Fig. 12.3). These aspects of consciousness help us understand how they vary during waking, sleeping, and dreaming. They are also key to understanding disorders of consciousness, such as coma or vegetative state, which we will discuss in detail in Chapter 13, Disorders of Consciousness.

To understand Fig. 12.3, it is best to begin with one axis. Let us start with the y-axis where we are showing *levels of awareness*. The top of the axis represents a fully aware state during conscious wakefulness. As you look lower and lower down the y-axis until it meets the x-axis, there are decreasing levels of awareness across states of drowsiness, slight sleep, deep sleep, and ultimately to general anesthesia and coma where there is little or no awareness at all. Now look at the x-axis where we are showing *levels of wakefulness*. Again, wakefulness is highest—shown on the right side of the axis, when we are in a state of conscious wakefulness. Wakefulness decreases as you move to the left of the x-axis toward where it meets the y-axis. Thus wakefulness decreases as we become drowsy, fall into light sleep, then deeper sleep. And, wakefulness is ultimately lost under general anesthesia and in coma.

Now let us look at the combination of the y- and x-axis: generally, we have a similar level of awareness and wakefulness as we move from our most wakeful state of full conscious wakefulness. Next, we tend to have a rather even balance between levels of awareness and wakefulness as we drop down to drowsy, light sleep, REM sleep, and deep sleep. In other words, Fig. 12.3 is a schematic that shows a balance between awareness and wakefulness

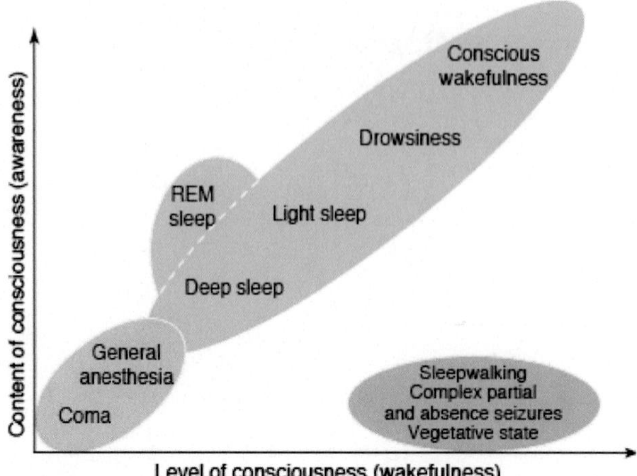

FIGURE 12.3 A simplified version showing levels of the two major components of consciousness: awareness and wakefulness. On the vertical y-axis, levels of awareness are shown beginning at the bottom of the axis with no level of awareness, as in a coma or when under anesthesia (shown in pink). Moving up the y-axis, the level of awareness increases as one moves through heavy sleep, light sleep, and into drowsiness and finally to full conscious wakefulness (shown in blue). Note, however, that during REM sleep, there is a higher level of awareness than during deep or light NREM sleep stages. The shape for REM sleep is sitting above the other sleep stages with a higher level of wakefulness, although well below conscious wakefulness. On the horizontal x-axis, levels of wakefulness are shown beginning with coma and moving similarly through stages of sleep to conscious wakefulness. According to this simplified version, levels of awareness and wakefulness are roughly similar as one moves through the stages from coma to deep sleep to wakefulness, represented by the *pink and blue shapes* forming a diagonal from the bottom left corner, where no awareness or wakefulness are observed in a coma, to the upper right, where both levels of awareness and wakefulness are high during conscious wakefulness. Sharply contrasting to this balance of levels of awareness and wakefulness are abnormal states of consciousness such as sleepwalking, epileptic seizures, and vegetative states (shown in purple on the x-axis). In these states, the individual shows a high level of wakefulness with an extremely low level of awareness shown by the purple shape resting on the bottom of the y-axis but far to the right of the x-axis. For more on this topic, see Chapter 13. *Source: Laureys (2005).*

in healthy individuals. However, there are some exceptions. Take another look at REM sleep and you will see a lower level of wakefulness—the shape denoting REM sleep is moved to the left of the figure to represent this lower level of wakefulness versus awareness. This should feel intuitively correct: you have a higher level of awareness during REM dream states than during deep sleep, yet your wakefulness levels are similar. There is a second exception for healthy individuals: this occurs during *sleepwalking*. Although sleepwalking is unusual, it is not considered a sleep disorder. We will discuss sleepwalking in more detail later in the chapter. Take a look at the shape for sleepwalking in Fig. 12.3: you will see that it is very high on the x-axis wakefulness scale and very low on the y-axis awareness scale. If you have ever seen someone sleepwalk— or if you have sleepwalked yourself—this should sound right to you because a sleepwalker has no awareness of what they are doing, however, they look like they are wide awake. We will discuss the other exceptions to this general balance between levels of awareness and wakefulness and how it is abnormal in disorders of consciousness in Chapter 13.

Fig. 12.3 is a useful figure for understanding the normal balance between awareness and wakefulness, and the exceptions to it that occur in normal healthy individuals and in brain disease or disorder.

2. FROM WAKEFUL TO SLEEPY...AND BACK AGAIN: DAILY RHYTHMS FOR SLEEP

How does the brain signal the changes that bring on sleep at night or wake us up in the morning? The Swiss scientist Alexander Borbély first proposed the *Two-Process Model of Sleep Regulation* in 1982 (Borbély, 2009) and since that time it has remained a central model for describing how brain processes lead us to sleep each night and work together to wake us up each morning. According to the Two-Process Model, Process S—this is the homeostasis process, which we will describe shortly—interacts with the circadian rhythms Process C, producing a daily push and pull of neural mechanisms leading us to sleep and then back to wakefulness (Fig. 12.4).

2.1 The Two-Process Model of Sleep—Wake Regulation: Circadian Rhythms and Homeostasis

Here is how it works: the brain has a sleep—wake homeostasis system, which seeks to maintain a balance or equilibrium between sleep and wakefulness. This system, called Process S by Borbély, reflects biochemical processes that represent sleep debt. Process S increases during wakefulness and declines during sleep. Thus the longer you are awake, the more sleep debt increases, the sleepier you get. Similarly, the longer you are asleep, the more sleep debt reduces or dissipates, and the likelihood of your arousing and awakening increases. Circadian rhythms operate quite differently. Named Process C by Borbély, they reflect the effects of the body's circadian clock.

The Two-Process Model for sleep—wake regulation proposes that during the wakefulness state—typically the daytime into evening—the *homeostatic sleep drive Process S* increases with the accompanying sense of sleepiness (Fig. 12.5). Process S interacts with *Process C of the circadian rhythms*, which begin to signal a shift toward sleep in a rhythmic way as the evening

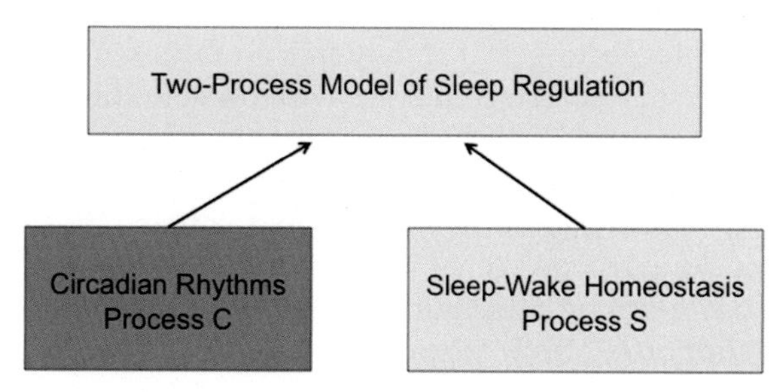

FIGURE 12.4 The Two-Process Model of Sleep Regulation proposed by Borbély (1982) in which sleep is regulated by a combination of circadian rhythms (Process C shown in the *orange box*) and sleep—wake homeostasis processes (Process S shown in the *yellow box*). *Source: Gage with permission.*

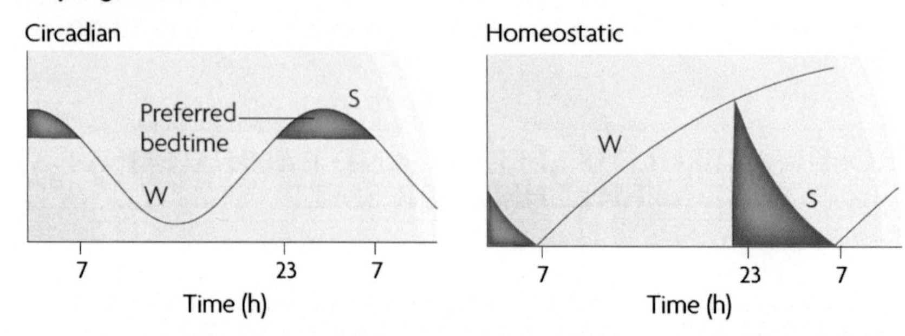

FIGURE 12.5 Sleep regulation as described initially by Borbély (1982) consisted of the interaction of two sets of processes. Left panel: Process C: Circadian rhythms begin to signal sleep onset as the evening approaches (shown in *orange shading*). Right panel: Process S: homeostatic processes increase sleep debt during the day, maximizing at the end of the day at the same time that circadian rhythms are signaling sleep. *Source: Cirelli (2009).*

progresses from light to darkness. Thus the circadian arousal drive of Process C is reduced in the evening and, interacting with the homeostatic sleep drive of Process S, sleep begins. This basic process is reversed in the morning as the sleep drive of Process S is much reduced after sleep and the circadian rhythms of Process C are signaling arousal or wakefulness.

2.2 Brain Bases of Process S and Process C

What are the brain bases for Process S and C? The homeostatic sleep drive described as Process S actually reflects a complex set of brain regions and neurons that are not well understood at present. What aspects of the brain share the processes described in Process S? One theory that addresses this question is the "sleep factor theory" (see Porkka-Heiskanen & Kalinchuk, 2011; for a review). The idea is that there must be substances—sleep factors—in the brain that must follow the processes detailed in Process S: they must be higher in concentration during waking time than during sleep, they must increase during wakefulness (increasing the sleep need, or debt) and decrease during sleep (recovering from sleep need). One brain substance that is a candidate to be a sleep factor is the neuromodulator adenosine (Landolt, 2008; for a review, see Porkka-Heiskanen & Kalinchuk, 2011). In studies with animals, many studies have provided converging evidence that adenosine levels rise during wakefulness and decline during sleep. Adenosine levels are also related to energy levels throughout the day, suggesting that it may indeed represent a sleep factor.

Interestingly, another line of research converges with these findings, pointing to adenosine as one of the sleep factors involved in the brain basis of sleep homeostasis: adenosine actions are blocked by a substance we are all far too familiar with—caffeine. A key effect of this stimulant is to *block the actions* of adenosine. For some people, caffeine causes sleep problems and the reason may be the blocking reaction it has on adenosine. That fact that a well-known stimulant such as caffeine has a strong blocking reaction to adenosine is an interesting bit of experimental evidence supporting both adenosine's role in the homeostasis of sleep and in caffeinated beverages' role in keeping us awake!

Process C—circadian rhythms are controlled by a relatively small (~20,000–50,000) set of neurons within a region in the anterior hypothalamus called the suprachiasmatic nuclei

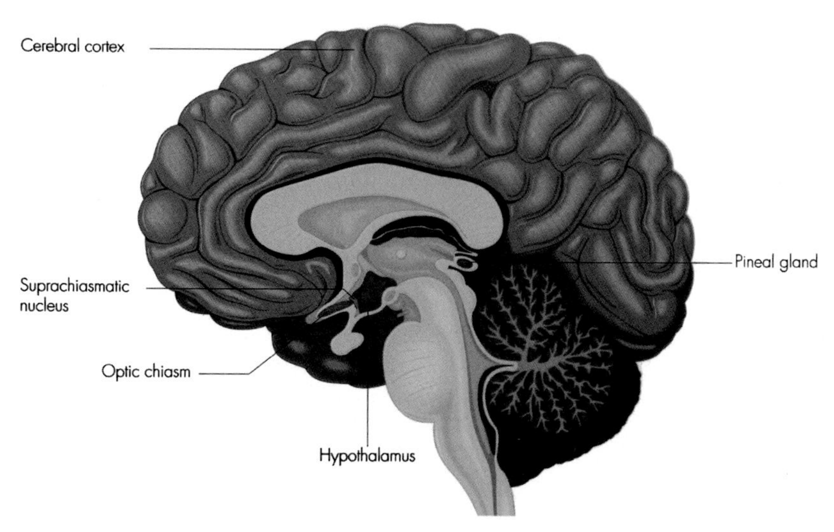

FIGURE 12.6 Circadian rhythms are controlled by neurons in the suprachiasmatic nuclei (shown in blue). Note that these nuclei are in close proximity to the optic chiasm, which brings visual signals of increasing or decreasing light. *Source: open access figure from https://commons.wikimedia.org/wiki/File:Suprachiasmatic_Nucleus.jpg.*

(SCN, Fig. 12.6). The SCN works as the body's central clock or pacemaker. The SCN and the circadian rhythms related to sleep–wake regulation are affected by light and entrain to daylight/darkness through their connections with the nearby optic chiasm, which in turn receives signals from some of the photoreceptors in the retina. Light and darkness signals the release of hormones and neurotransmitters to activate waking or sleeping processes. Human beings naturally wake up at first light and get sleepy when it gets dark in the evening. Naturally, that is, before the invention of the electric light! However, despite some individual variability, we humans operate naturally this way and have been since the beginning of recorded history.

Cortisol is one of those hormones and it is key to the circadian rhythms, wake, and sleep. Secreted by the adrenal glands, cortisol levels begin to increase at early morning. They increase by about 50% in the first half hour after awakening, triggering signals throughout the brain and body for arousal. This is termed the "cortisol awakening response." Cortisol levels decrease throughout the day into the evening, reaching their lowest levels between midnight and 4 a.m. during NREM and SWS stages. Cortisol levels increase again in the early morning during REM sleep and as dawn approaches, they ramp up for another day (Box 12.1).

3. THE ARCHITECTURE OF SLEEP

The stages of sleep are very similar across humans: while they vary a bit across individuals, as a whole we humans exhibit a very stable pattern of sleep that is highly structured. Human sleep is so structured, in fact, that it is typically described as the architecture of sleep.

BOX 12.1

GO AHEAD AND SLEEP IN ... OR WHY IT MIGHT BE A BAD IDEA TO SIGN UP FOR THAT 8 A.M. CLASS

We all know the feeling...whether it is sitting in that zero period class in high school, or in the 8 a.m. chemistry lab in college, or at the airport waiting to board the 6 a.m. flight. Sleepiness. Grogginess. That *I-do-not-want-to-be-here-ness.*

1. Most Adults Have Similar Daily Rhythm Patterns

A lot of people spend their entire lives trying to avoid early risings...And just as similar than different, and dictate an amazingly similar pattern for adults (Fig. 12.7).

1.1 But Not for Teenagers and Young Adults...

However, this circadian rhythm pattern changes sharply during adolescence—beginning at the onset of puberty and continuing into the early 20s, adolescents have a shift in this rhythm pattern that spans between 2 and 3 h. That is, *2–3 h later* than the adult rhythm. Look again at Fig. 12.7. See that

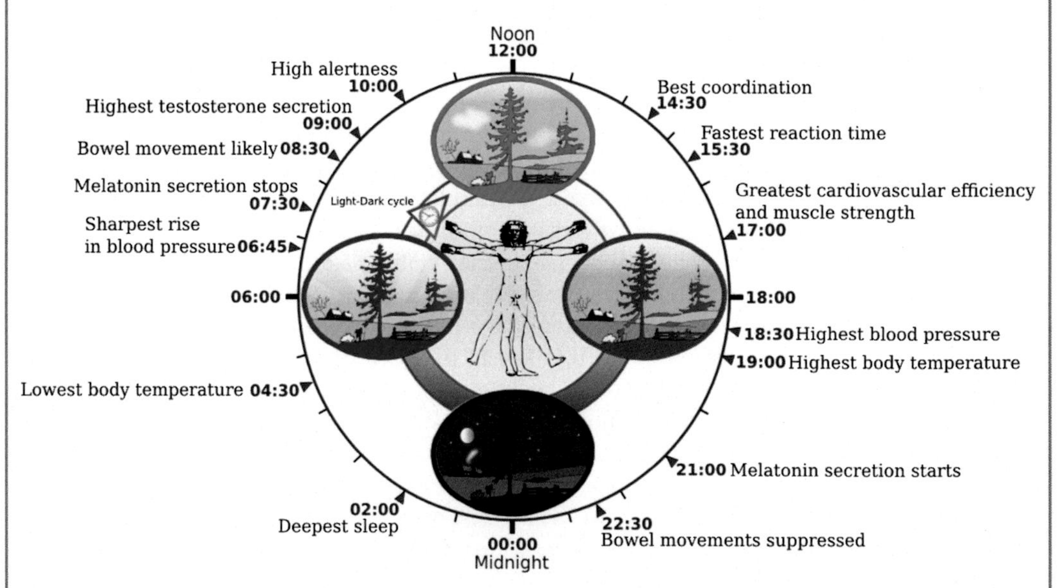

FIGURE 12.7 Most adults have a daily pattern of energy and arousal that are similar to those pictured here, with the sharpest rise in blood pressure around 6:45 a.m. and high alertness around 10 a.m. Secretion of melatonin begins around 9 p.m. and leads to sleepiness. *Source: Open source image. https://commons.wikimedia.org/wiki/File:Biological_clock_human.PNG.*

many people are "morning people" who love to be up with the sun. But did you know that although we all have our best time of day, our natural circadian rhythms are more peak in blood pressure at 6:45 a.m.? Move it ahead to 9:45 a.m. And that time of high alertness at 10 a.m....move that ahead to noon or 1 p.m. Similarly, the onset of

BOX 12.1 *(cont'd)*

melatonin secretion that aids in sleep onset is shifted as well—from 9 p.m. to closer to midnight. This shift dovetails well with findings that adolescents have difficulty falling asleep before 11 p.m. and waking up before 8 a.m. (Fig. 12.8) (1).

by the American Academy of Pediatrics highlighted the excessive sleepiness in adolescents and their inherent dangers and health concerns (2). Results of this report and the studies since are dramatic: almost all adolescents are sleep-deprived. Adolescents

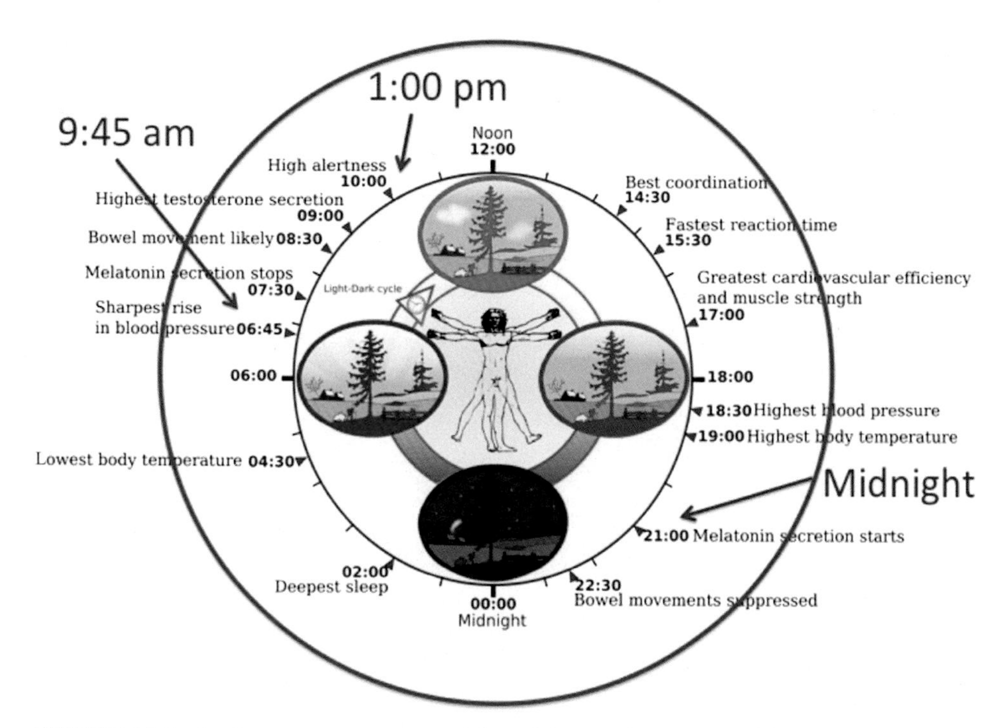

FIGURE 12.8 In sharp contrast to most adults, adolescents have a several hour shift in the daily pattern of energy and arousal, with the sharpest rise in blood pressure around 9:45 a.m. and high alertness around 1 p.m. Secretion of melatonin also begins much later in the evening, around 12 p.m. *Source: Modified Open source image. https://commons.wikimedia.org/wiki/File:Biological_clock_human.PNG.*

1.2 Sleep Deprivation in Adolescents and Young Adults

The shift in circadian rhythms in adolescents—and its effect on their education and health—has been almost ignored until recent years. In 2005, a technical report need at least 8 h of sleep a night and many need more, 9–10 h or so. And they are just not getting that sleep. Not just in the United States. Adolescent sleep-deprivation has been reported all around the globe (3)!

Continued

BOX 12.1 *(cont'd)*

1.3 Why the Sleep Deprivation? One Culprit—Early School Start Times

If we now know that teens have a circadian rhythm shift and a need for sleep that makes early morning wake-ups difficult, why then do our middle and high schools begin at 7:30 a.m. or even earlier? An obvious help to the widespread adolescent sleep-deprivation is to respect the very real circadian rhythm shift and begin classes later in the day.

In fact, the American Academy of Pediatrics issued a policy statement in 2014 (1) recommending later school start times (after 8:30 a.m.) to ensure that adolescents get sufficient sleep:

> The American Academy of Pediatrics strongly supports the efforts of school districts to optimize sleep in students and urges high schools and middle schools to aim for start times that allow students the opportunity to achieve optimal levels of sleep (8.5–9.5 h) and to improve physical (e.g., reduced obesity risk) and mental (e.g., lower rates of depression) health, safety (e.g., drowsy driving crashes), academic performance, and quality of life.

Is it happening? Not nearly often enough, however, those schools who have shifted to later start times have reported improved grades, better school attendance, better performance overall, and even reduced automobile accidents during school time commutes (5).

1.4 Why the Sleep Deprivation? Other Culprits…

Of course, many other factors besides what time classes begin in the morning are at work in adolescent sleep deprivation. Unlike our ancient ancestors who took to bed when it got dark, we modern-day humans have the glory of the electric light! YES, we can stay up all night any time we want. And to make things even better, we can keep ourselves awake with all matter of caffeinated beverages and spend endless hours on our computers, tablets, smart phones, watching television…these all contribute to the sleep-deprivation in teens and young adults. Social situations—whether on social media or in person—also form a huge distraction for young adults and tend to keep them up late. And, oh yes, there is also school work: adolescents and young adults are typically in some sort of classroom and have homework to be done each evening. And did we mention sports and other extracurricular activities?

1.5 Sleep Hygiene Helps a Lot

Here are some tips from the National Sleep Foundation for improving sleep health:

- Go to bed at the same time each night and rise at the same time each morning.
- Make sure your bedroom is a quiet, dark, and relaxing environment, neither too hot nor too cold.
- Make sure your bed is comfortable, and use it only for sleeping and not for other activities such as reading, watching TV, or listening to music. Remove all TVs, computers, and other "gadgets" from the bedroom.
- Avoid large meals a few hours before bedtime.

Avoiding caffeinated and alcoholic beverages helps a lot too, as does increasing exercise!

References

(1) American Academy of Pediatrics Adolescent Sleep Working Group. (2014). School start times for adolescents. *Pediatrics*. https://doi.org/10.1542/peds.2014-1697.

BOX 12.1 *(cont'd)*

(2) Millman, R. P. (2005) Working Group on Sleepiness in Adolescents/Young Adults; AAP Committee on Adolescence. Excessive sleepiness in adolescents and young adults: causes, consequences, and treatment strategies. *Pediatrics, 115*(6), 1774–1786.

(3) Hagenauer, M. H., Perryman, J. I., Lee, T. M., & Carskadon, M. A. (2009). Adolescent changes in the homeostatic and circadian regulation of sleep. *Developmental Neuroscience, 31*, 276–284. https://doi.org/10.1159/000216538.

(4) Owens, J., (September 2014). Adolescent Sleep Working Group, Committee on Adolescence. Insufficient sleep in adolescents and young adults: An update on causes and consequences. *Pediatrics, 134*(3), e921–e932. https://doi.org/10.1542/peds.2014-1696.

(5) Minges, K. E., & Redeker, N. S., (2016). Delayed school start times and adolescent sleep: A systematic review of the experimental evidence. *Sleep Medicine Reviews, 28*, 82–91. https://doi.org/10.1016/j.smrv.2015.06.002.

We will begin with a look at how sleep is studied and then move on to describe the stages of sleep and how they relate to human behavior.

3.1 How Sleep Is Studied

When Doctor Hans Berger (b.1873 d. 1941) recorded the first brain waves in 1924 from the head of a 17-year-old boy, he likely had no idea what he discovered or how it would change the medical and scientific world. Dr. Berger was actually looking into the brain basis for mental telepathy. He was a firm believer in mental telepathy: when he was a young man in the cavalry, he had a near-fatal accident with a military cannon. Although he was unhurt, at the moment of the accident he thought he was going to die. Far away, his sister had a sudden feeling of danger for her much-loved brother and insisted that their father telegraph to ensure that Hans was safe. This incident was central to Hans' later investigations as a neurologist into what he termed "psychic energy," which he felt transmitted the fear he felt in his mind to his sister's mind. In the process, Dr. Berger recorded his first recording of brain waves and termed it the *Elektrenkephalogramm*. We now refer to these brain recordings as the EEG.

The EEG reflects the brain's activity as recorded outside the brain on the skull. The EEG can be analyzed so that investigators, scientists, neurologists, and others can determine what bands of energy are active in selected brain regions, as discussed in Chapter 3, Observing the Brain. EEG is particularly helpful when investigating sleep because we have learned,

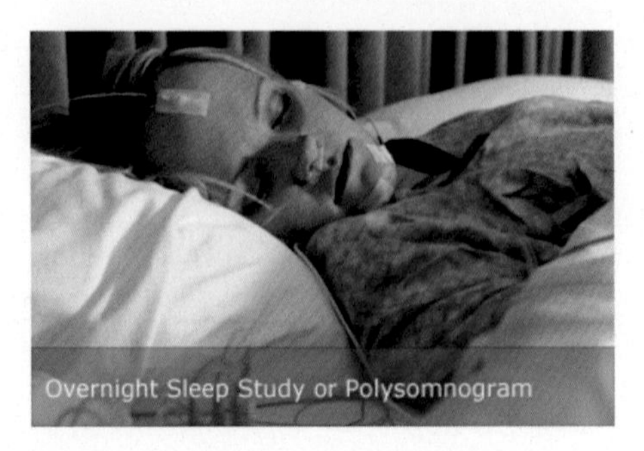

The Polysomnography Core In Action

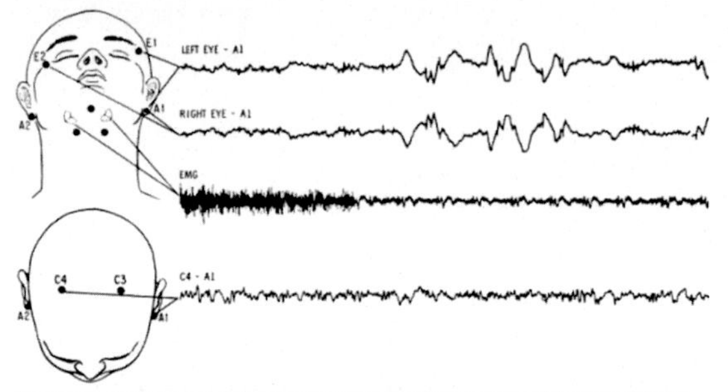

This image shows the standard electrode placement sites used in polysomnography, in addition to some typical polygraph readings to the right. The numbers indicate locations that are part of the more complex, standardized system that is used around the world. In non-sleep clinical tests that attempt to detect abnormal brain waves, as many as 20 or more electrodes are used on the scalp. In routine sleep studies, however, only one EEG tracing is necessary.

FIGURE 12.9 During a sleep study, a polysomnogram is recorded using a combination of four or more measures: electroencephalography (EEG) recordings measure brain waves, electrooculography (EOG) recordings measure eye movements that signify rapid eye movement sleep (REM) sleep, electromyography (EMG) recordings measure muscle activation and deactivation which also signifies REM sleep, and electrocardiography (ECG) recordings measure heart rhythms. *Sources: Upper http://www.ucmc150.uchicago.edu/sleep/studies.html. Lower figure: http://www.end-your-sleep-deprivation.com/polysomnography.html.*

thanks to Dr. Berger, that brain rhythms change in predictable ways as we move from one sleep stage to another and to waking. EEG is combined with an assortment of eye, muscle, and heart measures using a technique termed *polysomnography* (Fig. 12.9). The recordings produced using polysomnography—the polysomnogram—are used to diagnose sleep disorders such as sleep apnea, narcolepsy, and insomnia. The sleep study patient has electrodes placed on their head (EEG) to record brain waves, on their temples (electrooculography, EOG) to record eye movements, on their chin (electromyography, EMG) to record muscle activation, and on their chest (electrocardiography, ECG) to record heart rhythms.

As we will see as we continue our discussion of how the brain changes during sleep and dreaming states, the collection of brain wave, eye movement, and muscular activity measures are quite helpful in identifying the sleep stage progression in a healthy individual and in diagnosing sleep disorders in an individual who is struggling to obtain normal sleep patterns.

3.2 Sleep Stages: Nonrapid and Rapid Eye Movement Sleep

Two of the global brain states are invoked during normal sleep: sleeping and dreaming. Let us work through the stages of sleep, their levels of awareness and wakefulness, and how they correspond to dreaming. Using EEG and the polysomnogram to investigate human sleep patterns, sleep researchers have been able to characterize sleep into discrete stages. These sleep stages are typically shown on a *hypnogram* (Fig. 12.10). Sleep falls into two general patterns: NREM and REM sleep. Within NREM sleep, there are I–III or I–IV stages depending on whose theories you follow. Let us take the assumption that there are four NREM sleep stages. Sleep stages are sometimes labeled as stages 1–4 rather than I–IV. We will use the roman numerals in this chapter but either is acceptable and both refer to the same sleep stages: e.g., Stage III = Stage 3.

As we begin to feel drowsy and start to fall asleep, there are many brain responses and triggers that are occurring to bring us to the global brain state transition to sleep. Neuromodulators are at work, as are hormones, and complex patterns of activation and deactivation begin to trigger sleep. As we begin to fall asleep, the brain and body begin to slow down. Brain waves begin to slow and the characteristic slow waves begin to form. The heart rate and breathing slows, muscles relax, and energy use slows down.

As you fall asleep and enter Stage I sleep, your brain slows a bit from the alpha waves (8–12 Hz cycles per second, Hz) seen when you are awake to theta waves with a frequency of 4–7 Hz. Interestingly, you are actually somewhat conscious during Stage I sleep. In fact, someone who is in Stage I may claim they are not asleep at all because of this level of consciousness. If you are watching someone falling into Stage I sleep, however, you are not fooled. They will show the sure signs of early sleep with their muscle tone relaxing, their

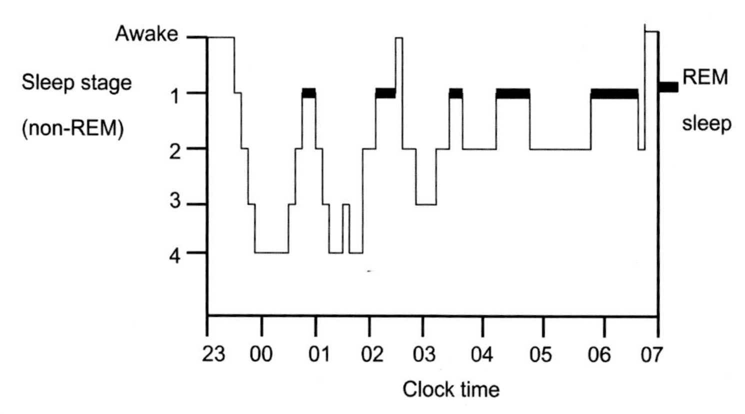

FIGURE 12.10 A hypnogram of sleep stages for a healthy young adult. The vertical y-axis on the left side of the figure shows the sleep stage from Awake to Stages I–IV (also described as Stages 1–4). The horizontal x-axis shows the time of sleep beginning at 11 p.m. or 23 h. Rapid eye movement (REM) sleep is shown on the vertical y-axis on the right side of the figure and is marked in solid black on the hypnogram. Note that the early hours of sleep are dominated by nonrapid eye movement (NREM) sleep while the later hours of sleep are dominated by REM sleep. *Source: Hall (2014).*

head may fall backwards or to the side, their breathing will slow, and of course their eyes will close. Stage I sleep is brief, lasting only about 10 min or so. The brain continues to slow and move toward deeper sleep during Stages II and III.

SWS is observed through EEG and polysomnography with characteristic slowly oscillating brain waves (Fig. 12.2). When you are in Stage IV SWS, your brain is oscillating very slowly, your breathing rate and heart rate are slow, and your muscles are relaxed. You may make some involuntary movements, especially if you are cold or hot. If you were awakened during Stage III or IV sleep, you would feel very groggy and it would be difficult to wake up. Sleepwalking typically occurs during Stage III or IV—we will discuss this later in the chapter under sleep disorders. After about 1 h of sleep, you ascend back up through Stages III, II, and I until you reach your first stage of REM Sleep, somewhere around 90–120 min after you fell asleep (Fig. 12.10). As you move up through these stages of NREM sleep, the brain oscillations begin to increase. Slow waves are less evident in the EEG and brain wave patterns begin to look more and more like an awake state, even though you are still asleep.

REM sleep involves a major shifting of brain processes, as can be seen in the changing EEG to a pattern that is similar to an awake state (Fig. 12.2). Another major change occurs in the body during REM sleep: the body enters a state of *atonia* or paralysis. Atonia is perfectly normal and it is thought to be a protection from a person acting out a dream during REM sleep. You may try to move—and in fact many dreams involve a feeling of slow motion or actual paralysis, reflecting the true state of the body even though in our REM dreams we move about freely.

3.3 Measuring Sleep Stages: Polysomnography

As we have discussed, polysomnography provides sleep researchers and clinicians with measurements of brain oscillations, eye movements, and muscular activity that aid in identifying sleep stages in a sleeping person (Fig. 12.9). Decreased brain oscillations found in the EEG, little eye movement found in the EOG, and slowed or relaxed muscular activity found in the EMG are signatures of NREM sleep. Increased brain oscillations, increased eye movements, and very limited muscular activity are signatures of REM sleep. Along with these polysomnogram signatures of NREM and REM sleep, sleep studies with subject interviews report any dreams that occur during NREM sleep tend to be logical, while dreams that occur during REM tend to be illogical and bizarre. See Fig. 12.11 for a description of one's behavioral state during Waking, NREM, and REM states, and the associated polysomnogram.

3.4 Neurophysiology of Sleep Stages: Slow Wave Sleep and Rapid Eye Movement Sleep

As we described earlier in this chapter, sleep is separated into stages characterized by NREM and SWS sleep and REM sleep. The early part of the night's sleep—the first 4 h or so—is dominated by SWS (shown in green Fig. 12.12). The later part of the night's sleep—the final 4 h or so—is dominated by REM sleep (shown in solid gray rectangles and encircled in pink in Fig. 12.12). Note that NREM sleep stages occur before the first REM stage of sleep.

Behavioural state		Wake	NREM	REM
Cognitive consequences		Acquisition of information	Iteration of informatiion	Integration of information
Conscious experience	Sensation and perception	Vivid, externally generated	Dull or absent	Vivid, internally generated
	Thought	Logical progressive	Logical perseverative	Illogical bizarre
	Movement	Continuous voluntary	Episodic involuntary	Commanded but inhibited
Surface recordings	EMG			
	EEG			
	EOG			

FIGURE 12.11 The three global states of waking (Wake), sleeping (NREM), and dreaming (REM) are shown with their characteristics in terms of cognition, consciousness, and measurements. Notice the different cognitive consequences of being awake—with the acquisition of new information, asleep—with the iteration of information leading to memory consolidation, and dreaming—with the integration of information. Sensation and perception processes are dull or absent during nonrapid eye movement (NREM) sleep while they are vivid during both Wake and rapid eye movement (REM) sleep. The difference between wake and REM sensation and perception is that during wake, the information is typically externally generated while in REM, the information is internally generated. Vivid, hallucinatory, and bizarre features characterize REM dreams while more logical features characterize NREM dreams. See Fig. 12.9 for a discussion of the electromyography (EMG), electroencephalography (EEG), and electrooculography (EOG) recordings. *Source: Hobson and Pace-Schott (2002).*

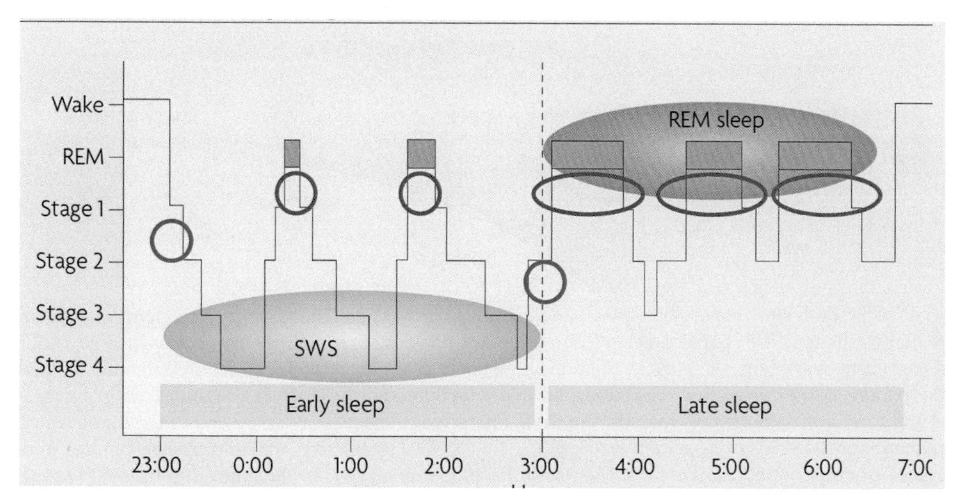

FIGURE 12.12 A hypnogram of a typical sleep stage sequence showing that the early part of sleep is dominated by slow wave sleep (SWS) (*green shading*) while later sleep is dominated by rapid eye movement (REM) sleep (*pink shading*). *Source: Diekelmann and Born (2010).*

3.4.1 *The Neural Symphony of Brain Oscillations During Slow Wave Sleep*

SWS and REM have characteristic field potential oscillations that are shown in Fig. 12.13. During SWS, slowed oscillations are observed over many areas in cortex. In humans, the typical peak frequency of these slow oscillations is ~0.8 Hz (cycles per second) (Diekelmann & Born, 2010). These slow oscillations are thought to reflect neuronal networks that are slowly oscillating in up and down states that are thought to be vital for memory redistribution. A second key SWS oscillation is the spindle. Spindles are typically observed during Stage II and SWS sleep with a regular frequency of ~10–15 Hz—much more rapid than the slow oscillations. Spindles originate in the thalamus, the subcortical hub that is highly interconnected with the cortex. The thalamocortical spindle activity during SWS is also thought to be part of the neural bases for long-term memory consolidation during sleep. The third type of field potential oscillation typically observed during SWS is the sharp wave–ripple. Generated in the hippocampus in the medial temporal lobe, sharp wave–ripples have a frequency of 100–300 Hz.

In summary, during SWS there are slow oscillations in the field potential across the cortex, which reflect the brain's up-and-down states during NREM and SWS sleep. In contrast to the very slow oscillations measured at the cortex, the higher-oscillating spindles are generated at the thalamus and even higher-oscillating sharp wave–ripples are generated in the hippocampus. Together, the very slow oscillating cortex combines with the relatively active thalamus and hippocampus to form the complex neural symphony that corresponds to a key aspect of human SWS: memory stabilization and consolidation. We will discuss more on that topic later in the chapter as we investigate the role of sleep in memory consolidation.

Our next question is: how do field potential oscillations observed during SWS compare to those observed during REM sleep?

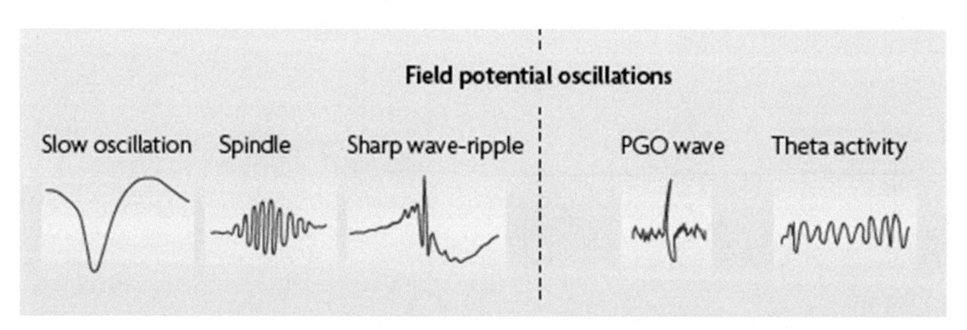

FIGURE 12.13 Both slow wave sleep (SWS) and rapid eye movement (REM) sleep have characteristic electrical field potential oscillations. Left panel: During SWS, three major types of oscillations are observed: slow oscillations, spindles, and sharp wave–ripples. Slow oscillations are slowly changing (0.8 Hz) brain rhythms that are observed over many regions in the cortex. Spindles have a far faster oscillation at ~10–15 Hz. Spindles are generated in the thalamus and form a thalamocortical loop during SWS. Sharp wave–ripples are generated in the hippocampus and have a frequency of 100–300 Hz. Together during SWS, slow cortical oscillations, thalamocortical spindles, and sharp wave–ripples from the hippocampus form the neural bases for memory consolidation. Right panel: During REM sleep, ponto-geniculo-occipital (PGO) waves are generated in the brainstem and travel via the thalamus to the occipital vision regions. Little is known about PGO activity in humans; however, animal studies have linked PGO waves to learning improvements following sleep. Theta (4–8 Hz) activity during REM sleep is theorized to be related to memory processing during sleep but the evidence to date is not strong enough to be convincing. *Source: Diekelmann and Born (2010).*

3.4.2 *The Lively Brain During Rapid Eye Movement Sleep*

During REM sleep, ponto-geniculo-occipital (PGO) waves reflect bursts of synchronized neural activity. Although not typically observed in humans, PGO waves measured in animals are generated in the *pontine* part of the brainstem (the P in the PGO wave), propagate to the lateral *geniculate nucleus* in the thalamus (the G in the PGO wave), and continue to the visual cortex in the *occipital lobe* (the O in the PGO wave) (Fig. 12.13). In studies with animals, PGO waves correspond to learning improvements measured in postsleep performance. A second field potential oscillation observed during REM is theta oscillations, so-called because they have a frequency of 4–8 Hz, which is in the theta band. Theta oscillations are observed mostly in the hippocampus. Theta oscillations during REM sleep are theorized to be related to memory processing during sleep, but the evidence to date is not strong enough to be convincing. An interesting observation is that theta and other oscillations during REM sleep are not as well coordinated across hippocampal and thalamocortical circuits as observed during SWS, with the hypothesis that memory systems that are active during SWS in these circuits are disengaged during REM sleep (Diekelmann & Born, 2010).

3.5 The Ups and Downs of Sleep Stages: Neuromodulators

Earlier in the chapter, we discussed the Two-Process Model of Sleep–Wake Regulation in which the brain shifts from wakefulness to sleep and back to wakefulness. However, we know that during sleep, there are differing sleep stages and brain processes that are shifting during the night's sleep. How does the brain "switch gears" from early sleep onset to deep sleep during SWS stages, to REM sleep and back to waking states? One key aspect of these changes that occur during sleep stages comes from neurons that act as brain modulators: *neuromodulators*. These shifts in wakefulness and sleepiness are affected by a myriad set of neuromodulators that function in a complex and highly interactive way to promote arousal or sleep in the brain.

Let us consider two key chemicals that are released by neuromodulators that affect sleep stages: acetylcholine (ACh) and noradrenaline/norepinephrine (NE) (Fig. 12.14). Most of the neurons that act as neuromodulators are located in the brainstem, but they extend their influence throughout much of the cortex and subcortex (Fig. 12.15).

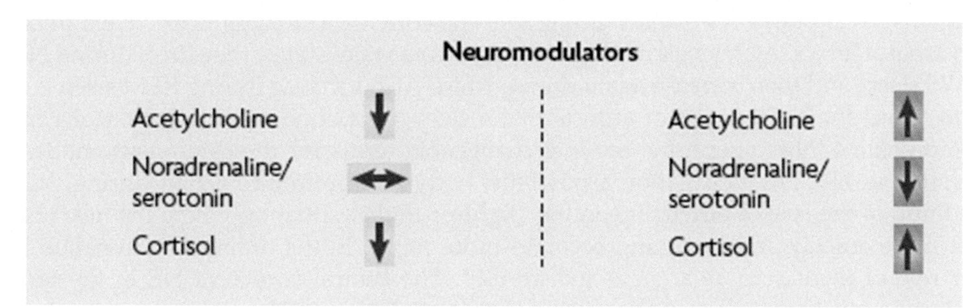

FIGURE 12.14 Key neuromodulator activity that changes during sleep stages includes acetylcholine (ACh), noradrenaline (NA), and serotonin. ACh levels are low during slow wave sleep (SWS, shown in *green arrows*) and increase during the transitions to rapid eye movement (REM, shown in *pink arrows*) sleep. NA and serotonin levels are decreased during REM sleep. Cortisol changes also coincide with sleep-stage transitions, decreasing in SWS and increasing during REM sleep. *Source: Diekelmann and Born (2010).*

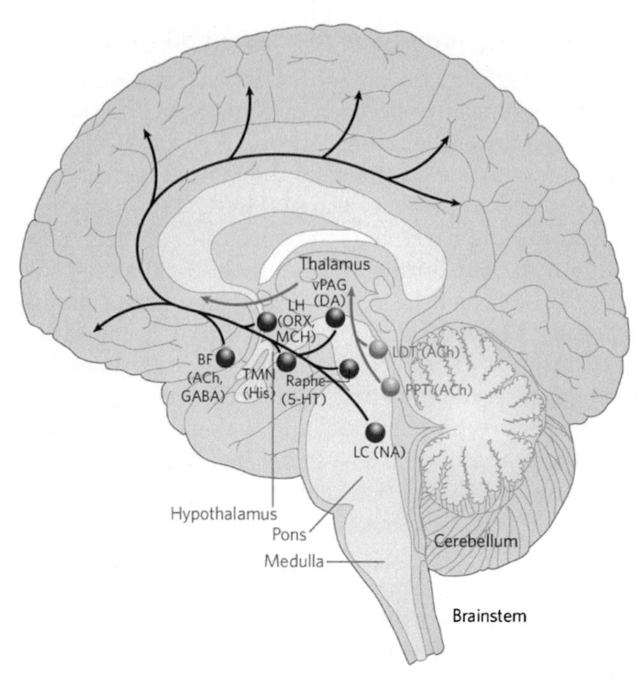

FIGURE 12.15 Most of the neuromodulators we are discussing in this section come from the brainstem region, but their influence is in the thalamus and throughout the cortex. Acetylcholine (ACh) pathways extend from the pedunculopontine (PPT) and laterodorsal tegmental (LDT) nuclei to the thalamus (yellow pathways), facilitating thalamocortical signaling. The red pathways activate the cortex with dopamine (DA), noradrenaline (NA), and serotonin (5-HT), and other modulating neurons (red pathways). Other abbreviations used: *BF*, basal forebrain; *GABA*, γ-aminobutyric acid; *His*, histamine; *LC*, locus coeruleus; *LHA*, lateral hypothalamus; *MCH*, melanin-concentrating hormone; *TMN*, tuberomammillary nuclei; *ORX*, orexin. *Source: Saper, Scammell, and Lu (2005).*

ACh functions as a modulator of arousal, memory, and attention during waking states. ACh is generated in the brainstem in the pedunculopontine nucleus and laterodorsal tegmental nucleus and in the frontal lobe in the nucleus basalis (Fig. 12.16). From these two regions, ACh pathways extend throughout the cortex and thalamus to have a major effect on arousal levels. ACh levels are high during active wake states, they drop during NREM and SWS sleep, and then increase again during REM. ACh increase during REM sleep is theorized to reflect the higher state of arousal that a sleeper experiences during REM dreaming.

Noradrenaline/norepinephrine are interchangeable terms for the same neuromodulator, abbreviated as NE. NE modulation arouses the body and brain into action during times of stress: think of the role of adrenaline in the "fight-or-flight" situation where the heart begins to beat more quickly and breathing becomes more rapid. In the brain, noradrenaline has a similar role of increasing awareness and arousal. The central source of NE in the brain is in the locus coeruleus (LC) (see Fig. 12.15 for a sagittal view of the brain showing the LC). The LC has a key role in promoting wakefulness, thus it is not surprising that NE modulation during NREM sleep is low and even lower during REM sleep (Fig. 12.14). Serotonin levels also decrease during REM sleep. The brain areas on the NE and serotonin pathways are shown in Fig. 12.17, left and right panel, respectively. Serotonin originates in the raphe nuclei

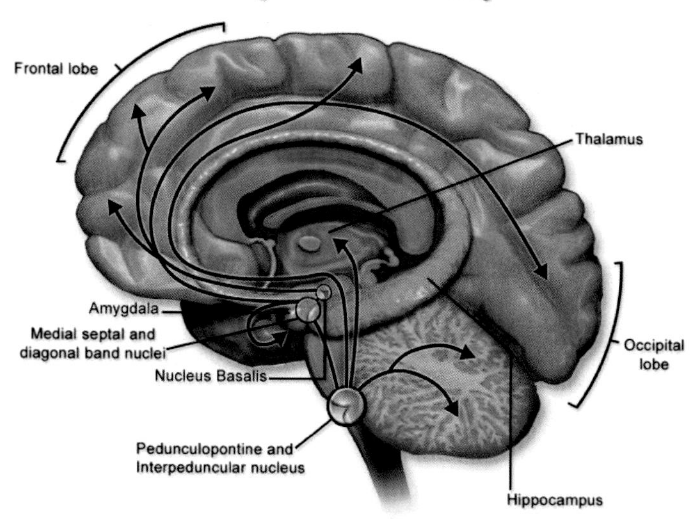

FIGURE 12.16 Acetylcholine (ACh) is generated in the brainstem in the Pedunculopontine nucleus and laterodorsal tegmental nucleus and in the frontal lobe in the nucleus basalis. From these two regions, ACh pathways extend throughout the cortex and thalamus to have a major effect on arousal levels. *Source: Open source https://commons.wikimedia.org/wiki/File:Acetylcholine_Pathway.png, Bruce Blaus.*

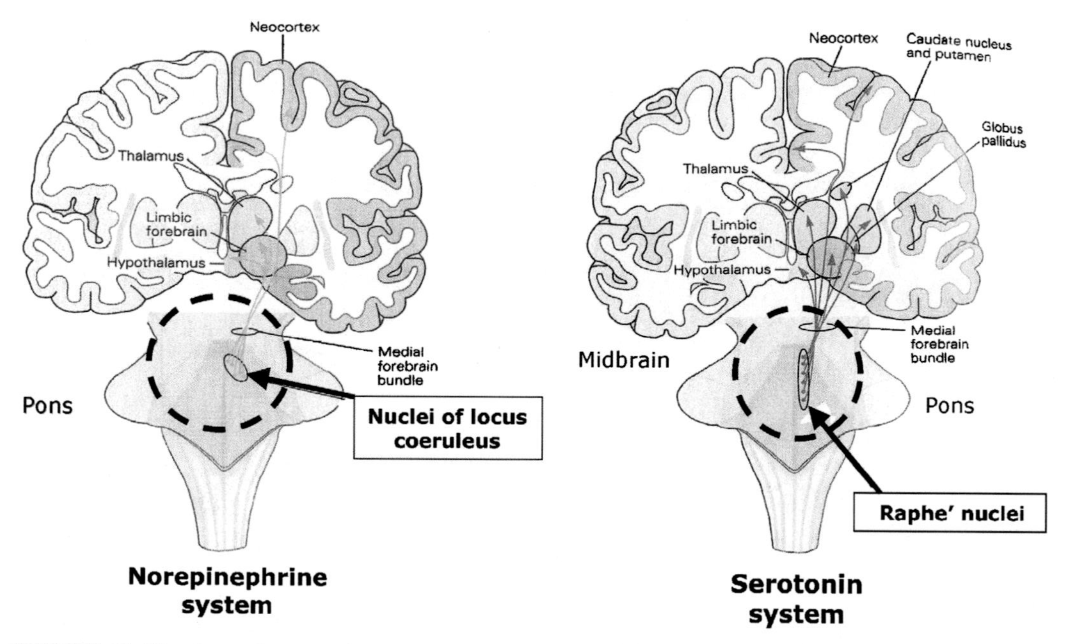

FIGURE 12.17 Coronal views of the brain showing the pathways for norepinephrine (NE) and serotonin. The locus coeruleus (LC) (left panel) plays a key role in promoting wakefulness through the neuromodulator NE. Targets of the NE pathway from the LC include the hypothalamus, thalamus, and the cortex. Serotonin originates in the Raphe nuclei in the brainstem (right panel). Note the complex pathways for serotonin, including the hypothalamus, limbic regions, the globus pallidus, the caudate nucleus, the putamen, the thalamus, and the cortex. *Source: Mirskey and Duncan (2005).*

in the brainstem. Note that the brain views in Fig. 12.17 are a coronal slice through the brain. Compare the location of the LC in Fig. 12.17 to the sagittal slice shown in Fig. 12.15.

3.6 Brain Activation and Deactivation During Rapid Eye Movement Sleep

As you can likely infer from the EEG tracings shown in Fig. 12.2, brain processes slow dramatically as you enter SWS. They speed up again during REM sleep later in the night. A central question for sleep researchers is: what areas of the brain are active during SWS, NREM sleep and REM sleep? Determining which areas of the brain are activated or deactivated during distinct sleep stages may aid in uncovering which brain areas are actively involved in brain processes that occur during sleep—such as memory consolidation and dreaming, as we will see later in the chapter.

We have mentioned that the EEG for wake versus REM sleep appears to be similar, reflecting a high degree of brain activity as compared to SWS. There are key differences in brain activity between wake and REM states, however, as you might expect since what you experience during a dream during REM can be very different from what you experience when you are awake. The main brain regions that are deactivated during REM sleep versus wake states are the dorsolateral prefrontal cortex (DLPFC), the posterior cingulate cortex (PCC), and primary visual cortex (V1, not shown on this figure) (Fig. 12.18).

Let us step through the roles these brain areas play during the wakeful state. The DLPFC plays a key role in executive function—controlling, planning, and coordinating activity and actions (see Chapter 9, Decisions, Goals, and Actions). The PCC's role in cognition is less well understood, but a central theory is that the highly interconnected PCC plays a role in internally directed cognition and thought (Binder et al., 1999; Buckner, Andrews-Hanna, & Schacter, 2008; Raichle et al., 2001). Research suggests that the connectivity between the PCC and DLPFC plays a key role in attention, arousal, and conscious thought. The deactivation of both the PCC and the DLPFC during REM sleep versus the wakeful state likely relates to the bizarre and hallucinatory nature of dreams during REM sleep. Primary visual cortex—

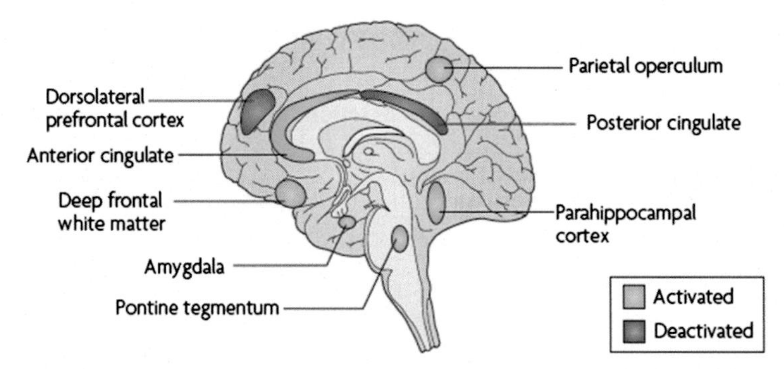

FIGURE 12.18 A sagittal view of the brain showing brain areas that are activated (shown in blue) and deactivated (shown in red) during rapid eye movement (REM) sleep versus waking. Areas more activated during REM sleep versus waking include the anterior cingulate, the amygdala, the pontine tegmentum, the parahippocampal cortex, and the parietal operculum. Areas deactivated during REM sleep versus waking include the dorsolateral prefrontal cortex, and the posterior cingulate. *Source: Hobson (2009).*

V1—is also deactivated during REM. While dreams during REM sleep typically include rich visual scenes, primary V1 is not processing external visual stimuli as the eyes are closed during sleep. Together, the deactivation of DLPF, PCC, and the activation of visual association (but not primary) cortex combine to provide a brain signature of the descriptions of REM sleep, with vivid and intense sensory imagery coupled with illogical and bizarre thought (refer back to Fig. 12.11).

In Fig. 12.19, we present a list of brain areas and their role in REM sleep and dreaming, that has been developed by Hobson and Pace-Schott (2002). Note that the DLPFC and V1 are shown in pink, reflecting their deactivation during REM sleep (PCC is not shown on this figure; however, it is shown on Fig. 12.18). However, while VI is deactivated during REM sleep, the visual association cortex—reflecting more complex visual processing—is relatively activated during the REM dream states (Item 11 on Fig. 12.19).

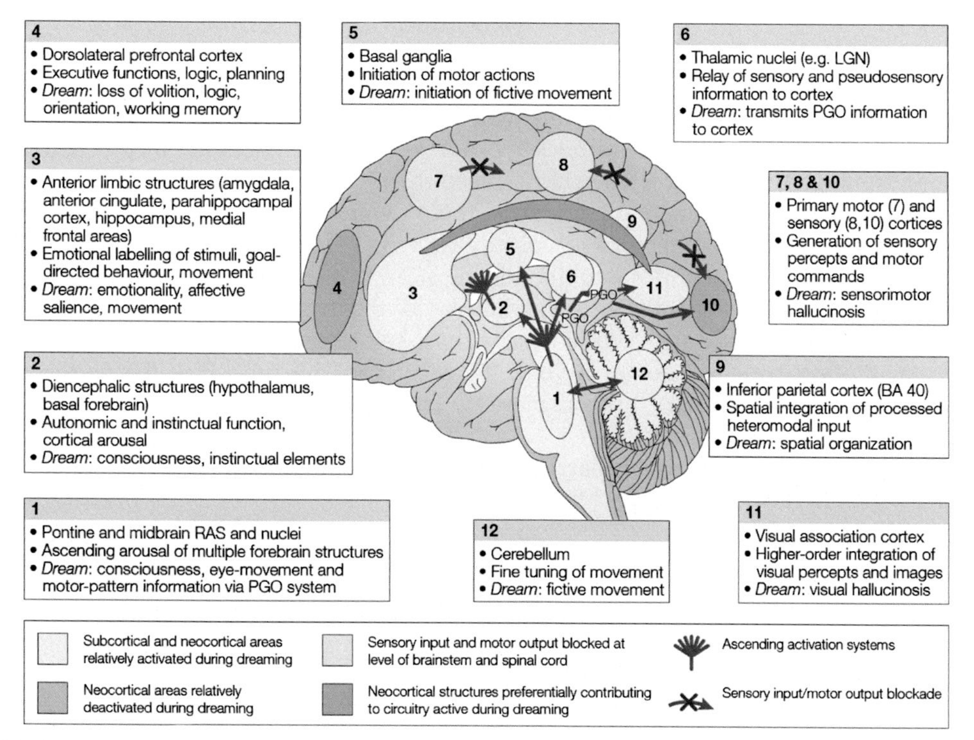

FIGURE 12.19 Brain areas and their role in rapid eye movement (REM) sleep and dreaming developed by Hobson and Pace-Schott (2002). Areas relatively activated during REM sleep dreaming are shown in light yellow. Areas relatively deactivated are shown in pink, including the dorsolateral prefrontal cortex (DLPFC) (item 4) and visual cortex (V1) (item 10). However, while VI is deactivated during REM sleep, visual association cortex—reflecting more complex visual processing—is relatively activated during the REM dream states (Item 11). The limbic region (item 3), including the amygdala, is also more activated during REM dream states than during normal waking, reflecting the highly emotional states than can occur during REM dreaming. Note that sensory inputs and motor outputs are blocked during REM sleep (shown in green). *Source: Hobson and Pace-Schott (2002).*

4. MEMORY AND SLEEP

We have long known that we function better when we are well rested. Sleep loss has an effect on most cognitive and health-related aspects of everyday life (see Fig. 12.1). What exactly does sleep do for us? We know that our bodies feel rested and restored after a good night's sleep. But we also know from EEG sleep studies that the brain is active during sleep—and that the brain is differentially activated during NREM and REM sleep. Sleep researchers have long studied the role of sleep in learning and memory processes—partly because of the many behavioral studies in humans and animals showing improved learning and memory following restful sleep and partly due to neuroscientific studies of brain modulation and activation across differing sleep stages. At present, sleep and memory researchers are pulling together what we know from behavioral and brain studies, across human and animal species, to learn what memory processes occur during sleep.

4.1 The Interaction of Sleep and Memory Processes

Memory consolidation—the stabilizing of memories over time—has long been a central concept in sleep research. The basic idea is that we form memories throughout the day as we experience our daily life. At night, while we are asleep, these memories are stabilized, connections to existing memories are formed, and our recent experiences are incorporated into our long-term memory and knowledge store. Sleep researchers who investigate the brain bases of sleep and memory have long focused on these two aspects of memory change during sleep: the stabilizing of new memories, typically referred to as *memory consolidation*, and the integration of these new memories into the existing store of memories, typically referred to as *plasticity*. Just how the brain deals with the interplay of the formation of new memories with the ongoing changes to existing stored memories is a key focus of research. And while there has been a lot of progress, we do not have all the answers at this time. Hampering this effort has been the complex nature of human memory and the many brain areas and functions involved in human sleep.

To quote Robert Stickgold, a psychiatry professor at Harvard and director of the Center for Sleep and Cognition:

> It would be nice if we could talk about sleep and memory as if there were only one type of memory and one type of sleep. But this is far from the case. Sleep and memory each comes in many forms, and furthermore, memories can go through multiple forms of postencoding processing that must be individually addressed **Stickgold (2013, p. 847).**

Let us step through these processes on a conceptual level and then discuss their brain bases.

4.2 Selective Memory Consolidation

Recent theories put forth the concept that while we are almost constantly encoding new memories throughout the waking part of our day, not every memory formed will last through the remaining wakefulness period and through the sleep stages when memory

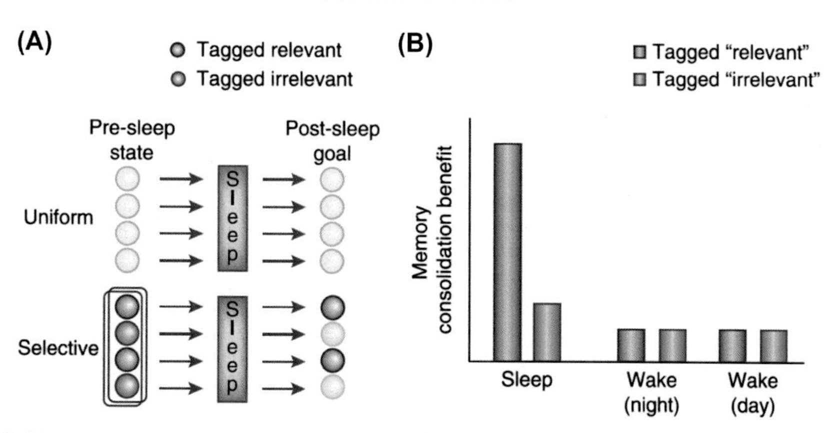

FIGURE 12.20 According to Stickgold et al., some new memories are more important, more relevant to us than others. This process is called *selective consolidation* and is in contrast to *uniform consolidation* theories. In selective consolidation, the more important memories get "tagged" (shown in red) as more relevant during wakeful learning or encoding, and then these tagged memories are selectively consolidated during sleep. Other memories that are tagged as irrelevant (shown in blue) are not consolidated, lost, or forgotten. This view is in contrast to theories of uniform consolidation, where all memories are consolidated (shown in yellow). *Source: Stickgold and Walker (2013).*

consolidation processes are occurring. Some new memories are more important, more relevant to us than others. These memories get "tagged" as more relevant during wakeful learning or encoding, and then these tagged memories are selectively consolidated during sleep (shown in red in Fig. 12.20) (Stickgold & Walker, 2013). This view is in contrast to theories of uniform consolidation, where all memories are consolidated (shown in yellow in Fig. 12.20).

The process of consolidating preferentially more salient or relevant new memories makes intuitive sense when you consider how we tend to recall gist information about events and occurrences that have taken place on any given day and we typically do not recall many highly detailed events—unless they are relevant to us!

Reflecting the complexity of human memories and how they are modified during sleep, new terms are being used currently that better describe what is actually thought to be occurring in the stabilizing and integrating processes. One such term is *memory evolution* rather than consolidation. This term better describes the concept that not only do new memories get integrated into the existing memory store but old memories can be changed or forgotten during these processes as well—thus memory can evolve over time.

The concept of *memory evolution through sleep processes* can take many forms. A recent theory put forth by Stickgold and Walker (2013) describes three forms or categories of memory evolution (Fig. 12.21). The first form is termed *item consolidation*. This refers to individual remembered items to be consolidated and stabilized if deemed relevant or forgotten if not (Fig. 12.21A). Next, in *item integration*, individual new remembered items can be integrated into existing memory stores, with new associations and pathways formed (Fig. 12.21B). Finally, in *multiitem generalization*, new remembered items encoded within a short time span can be integrated into long-term memory stores. These multiitem memories—which reflect a description of how multiple new things are learned with the association of various

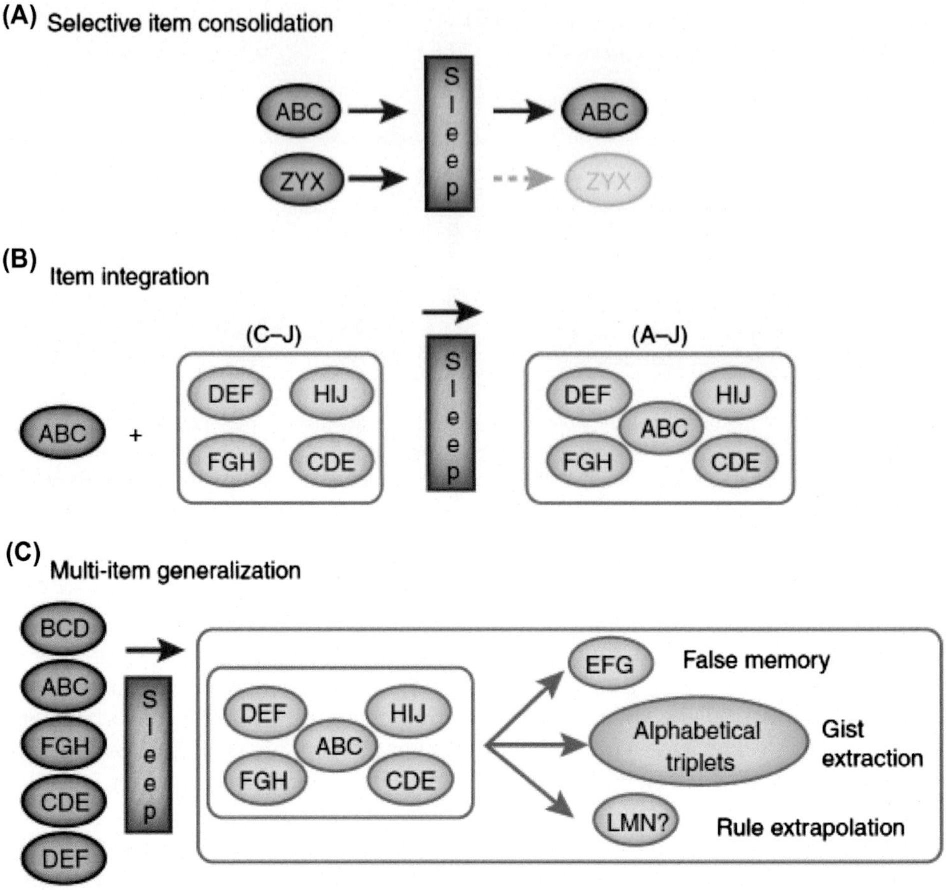

FIGURE 12.21 The concept of *memory evolution through sleep processes* can take many forms. A recent theory put forth by Stickgold and Walker (2013) describes three forms or categories of memory evolution: *item consolidation* (A), *item integration* (B), and *multiitem generalization* (C). In item consolidation (A), individual memories can be stabilized or enhanced (shown in red) or forgotten (shown in blue). In item integration (B), the new item memories (shown in red) can be integrated into existing memory networks (shown in green). In multiitem generalization (C), related new item memories that have been encoded over a brief time interval can create new memory networks and conceptual schema. *Source: Stickgold and Walker (2013).*

items at a similar time—can be integrated into our conceptual schema as well for the extraction of gist information and for rule learning (Fig. 12.21C).

4.3 Encoding and Consolidation: Temporary and Long-Term Memory Stores

A general view of how sleep processes serve to consolidate new memories into a stable, long-term memory system holds that during waking, both a temporary and fast learning storage and a slow learning long-term storage occurs. During sleep, long-term and slow learning processes are activated in concert with the temporary fast store site, transferring and

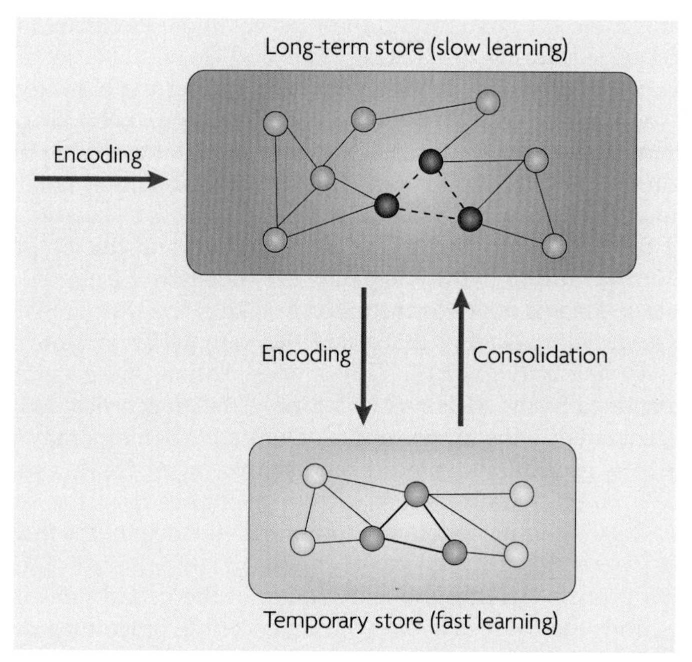

FIGURE 12.22 The two-stage model of memory consolidation. Memory consolation involves the stabilization of new memories so that they become long-lasting memories. To accomplish the integration of these new memories into existing memories, brain areas must be flexible, adaptable, or plastic. A well-established model for describing the stabilization of new memories combined with the plasticity required in existing memory networks is the *two-stage model of memory consolidation*. According to this model, there are two memory stores: a temporary store for fast learning that also serves as an intermediate (and temporary) buffer (shown in orange), and a slow learning store for long-term encoding of new memories (shown in pink). New items or events are initially encoded in both the temporary and long-term stores. Over ensuing memory consolidation occurrences, the new items are repeatedly reactivated until they gradually become a stable part of the long-term store. *Source: Diekelmann and Born (2010).*

integrating the encoded memories from the temporary to the long-term store (Fig. 12.22). This model, termed the Two-Stage Model of Memory Consolidation, has been widely accepted in the field of sleep and memory research (Frankland & Bontempi, 2005; Marr, 1971; McClelland, McNaughton, & O'Reilly, 1995; Rasch & Born, 2007; for a review see Diekelmann & Born, 2010). During each night's sleep, then, there is a reactivation of that day's encoding of new information in both the temporary and the long-term store. Each subsequent reactivation, night after night, serves to both stabilize the new memories in the long-term store and to integrate them into existing long-term memory.

A key next question is to determine just where the temporary and long-term stores are in the brain and what mechanisms are used for consolidating the temporary memory stores into long-lasting memories. The answer, not surprisingly, comes from sleep studies.

4.4 Brain Bases of Two-Stage Model of Memory Consolidation

Groundbreaking studies by Wilson et al. provided important new information that (1) the neural firing pattern in the hippocampus during learning was reactivated in the hippocampus during SWS, (2) importantly, these reactivations occurred in the same order that the learning happened during the waking state, and (3) neural ensembles in the hippocampus and in the cortex were reactivated in parallel during the slow cortical oscillations observed in SWS (Ji & Wilson, 2007; Wilson & McNaughton, 1994). Together, these and many other important studies provided the key groundwork for assessing the activation patterns in the hippocampus and the cortex during SWS to provide information about the brain mechanisms that underlie memory consolidation.

We can think of the temporary store as the hippocampus and the long-term store as the cortex. Memories formed during the waking state are transferred from the hippocampus to the cortex during sleep. Here is how it is theorized to take place: during SWS, there are three characteristic field potential oscillations that are recorded in the brain: slow oscillations, spindles, and sharp wave–ripples (Fig. 12.13). The slow oscillations are generated in the cortex, the spindles are generated in the thalamus and form a thalamocortical loop, and the sharp wave–ripples are generated in the hippocampus. During SWS, temporally coordinated activation between the slow oscillations, spindles, and sharp wave–ripples combine to form the neural basis for the transfer or consolidation of new memories from the temporary store of the hippocampus to the long-term store of the cortex with the help of the thalamocortical synchrony (Fig. 12.23).

The consolidation process includes the reactivation of the neural networks in the hippocampus and in the cortex that were activated during encoding or learning during the waking state. During SWS, this reactivation of networks involved during learning combines with slow oscillations in regions across the cortex to allow *a redistribution* of those new memories

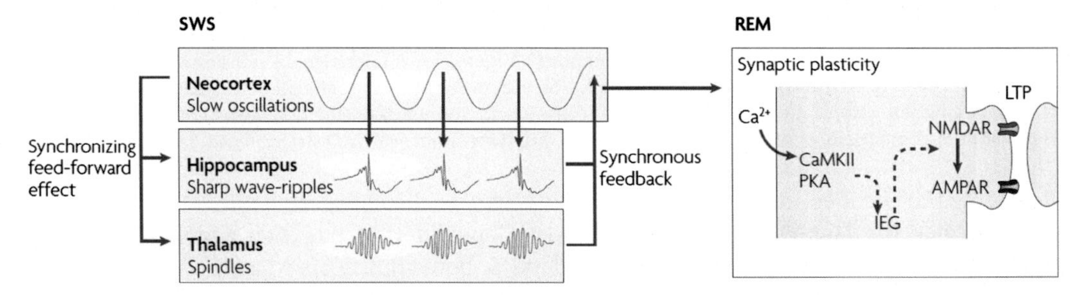

FIGURE 12.23 Slow wave sleep (SWS) and rapid eye movement (REM) sleep processes underlying memory consolidation. Building from the two-stage memory model described above and shown in Fig. 12.22, we can think of the temporary, fast learning store as the hippocampus and the long-term, slow learning store as the cortex. Memories formed during the waking state are transferred from the hippocampus to the cortex during sleep. Here is how it is theorized to take place: during SWS (left panel), there are three characteristic field potential oscillations that are recorded in the brain: slow oscillations, spindles, and sharp wave–ripples (Fig. 12.13). The slow oscillations are generated in the cortex, the spindles are generated in the thalamus and form a thalamocortical loop, and the sharp wave–ripples are generated in the hippocampus. During SWS, temporally coordinated activation between the slow oscillations, spindles, and sharp wave–ripples combine to form the neural basis for the transfer or consolidation of new memories from the temporary store of the hippocampus to the long-term store of the cortex with the help of the thalamocortical synchrony. These processes are typically referred to as *systems consolidation.* During REM sleep (right panel), synaptic level changes provide integrative and stabilizing changes: these processes are typically referred to as *synaptic consolidation. Source: Diekelmann and Born (2010).*

into long-term stores in the cortex. This integration of new information into existing long-term stores necessitates some form of neural plasticity in the cortex so that the new information can be incorporated into existing stores. This plasticity may also allow existing long-term store information to be forgotten.

The sharp wave–ripples of the hippocampus and the spindles of the thalamus are also coupled in time and increase during the up-states of the slow cortical oscillations and decrease during the down-stores (Siapas & Wilson, 1998; Sirota, Csicsvari, Buhl, & Buzsaki, 2003; Wierzynski, Lubenov, Gu, & Siapas, 2009; for a review see Diekelmann & Born, 2010) (Fig. 12.23).

The memory consolidation or evolution processes during SWS are frequently referred to as a "systems consolidation" process because they involve a large-scale transfer and integration of memories from the hippocampus to various locations in the cortex: thus it is occurring on a *brain system level.*

4.5 Memory Consolidation Mechanisms During Rapid Eye Movement Sleep

Far less is known about the role of REM sleep and memory consolidation. While SWS is characterized by slow oscillations across the cortex, brain activation and deactivation patterns during REM sleep are more complex, with many cortical regions active in a similar manner to the waking state (Fig. 12.18). Nevertheless, many studies in humans and in animals have provided evidence that deprivation of REM sleep is detrimental to learning and memory processes, although it is unclear if the stress associated with sleep deprivation may have had a separate effect on learning and memory in addition to REM sleep deprivation.

One current theory holds that during REM sleep, there are consolidations in *synaptic connections* within local networks in the cortex (Fig. 12.23). The basic notion is that local synaptic connections become plastic during REM sleep, allowing for strengthening of recently learned information. This synaptic strengthening may correspond to theories of long-term potentiation, where simultaneously active neurons serve to strengthen their connection in a long-lasting manner. This local plasticity could reflect memory consolidation at the *synaptic level* as opposed to the *systems level* theorized for SWS processes (Diekelmann & Born, 2010; Tononi & Cirelli, 2003).

Another theory about REM sleep memory processes holds that during REM sleep, regions in the limbic system relevant for emotional processing, such as the amygdala, are highly active and thus REM sleep activation may reflect emotional memory processes or consolidation (Genzel, Spoormaker, Konrad, & Dresler, 2015). According to this view, during NREM sharp wave–ripples generated in the hippocampus and slow oscillations from the cortex form the memory reactivation processes key to memory consolidation; however, the brainstem and emotional processing structures such as the amygdala are relatively deactivated (Fig. 12.24, left panel). During REM sleep, however, the amygdala and other regions for emotional processing are highly active (Fig. 12.24, right panel). This activity in limbic/emotional processing regions may correspond both to the emotional sensations experienced during REM dreaming, as well as to the consolidation of the emotional aspects or valence of recently formed memories (Genzel et al., 2015).

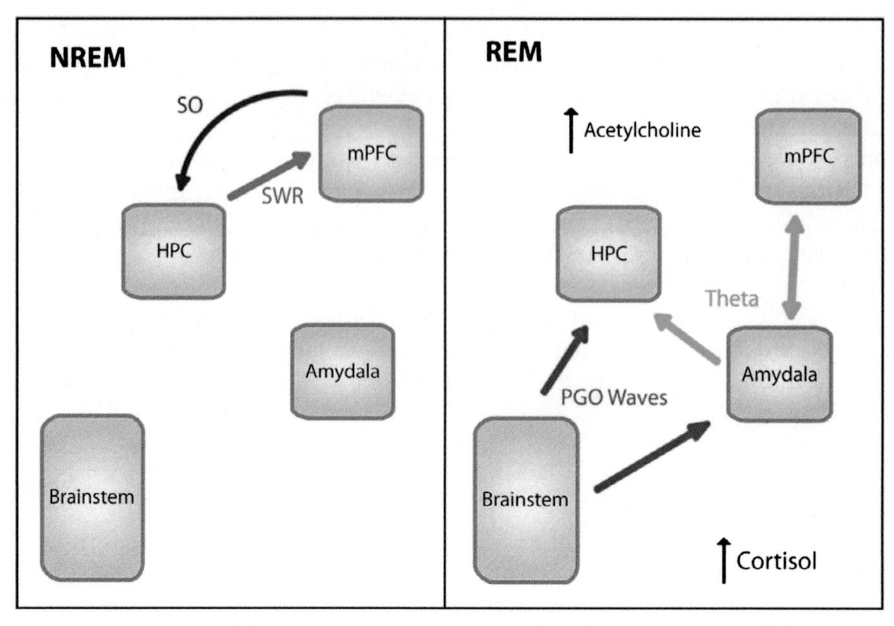

FIGURE 12.24 Genzel et al. (2015) present a theory about rapid eye movement (REM) sleep memory processes: during REM sleep, regions in the limbic system relevant for emotional processing, such as the amygdala, are highly active and thus REM sleep activation may reflect emotional memory processes or consolidation. According to this view, during nonrapid eye movement (NREM) sharp wave–ripples generated in the hippocampus and slow oscillations from the cortex form the memory reactivation processes key to memory consolidation; however, the brainstem and emotional processing structures such as the amygdala are relatively deactivated (left panel). During REM sleep, however, the amygdala and other regions for emotional processing are highly active (right panel). This activity in limbic/emotional processing regions may correspond both to the emotional sensations experienced during REM dreaming, as well as to the consolidation of the emotional aspects or valence of recently formed memories. *Source: Genzel et al. (2015).*

Cumulatively, sleep researchers have yet to determine just what role REM sleep has on memory consolidation. One thing they are certain about, however, is that most of our dreaming takes place during REM sleep!

5. DREAMING

Of the three global states, dreaming is the least understood. In fact, we do not have a single, well-agreed-upon definition of just what a dream is. Does dream content relate to Freud's famous interpretation of our dreams playing out actions that are taboo during waking states? Are dreams predictors of our future? Symbolic? How do dreams relate to human cognition? While those questions are responded to by psychologists and psychiatrists, sleep researchers are typically investigating the brain bases of dreaming.

Although we do not have a shared universal definition of a dream, most of us can do a very fine job of describing our dreams. While most dreams are forgotten as we awaken, some

FIGURE 12.25 Dream images during REM sleep are frequently hallucinatory and bizarre. Flying is frequently a part of dreams, as are bodies of water. *Source: Public domain.*

remain vividly in our memories. When asked to describe their dreams, people from all around the world, from many different cultures, tend to describe them in similar ways. "I am flying, but I do not seem to notice that there is anything strange about being able to fly." "I am trying to escape from a terrible monster, but my legs feel like lead and I can barely move."

When you assess characteristic features of dreams, they are amazingly similar for people across all walks of life and ages: common features experienced in dreams are being chased, flying, being in or near bodies of water...with the state of the water—calm and clear, or turbulent and roiled, possibly reflecting the emotional state of our mind. Many dreams contain otherworldly symbols and creatures that have a hallucinatory feel (Fig. 12.25). Cars, planes, trains, and boats also are frequently reported in dreams. Are these symbols of the direction we are heading in in life? Death figures in many dreams and the feeling of paralysis.

Although we cannot describe the brain bases for most of these frequent dream symbols, the last—paralysis—is something that we can relate to the brain's function during REM sleep. Recall that we described that as we enter REM sleep, our body goes into a state of atonia. This is a reversible state of motor paralysis that corresponds to the onset of REM sleep and that fades just as quickly as we exit REM sleep upon awakening or moving into an NREM sleep stage. The fact that most of us report dreams where we are either paralyzed or have difficulty moving likely reflects that we have some level of awareness of atonia despite being asleep.

5.1 How Is Dreaming Studied?

Dream science has a huge challenge: how can we objectively study human dreaming? One approach has been to awaken sleep-study participants after they have gone through a period of REM sleep as measured using polysomnography (Fig. 12.9). Dream researchers can ask the participants what they just dreamt about and begin to characterize dream topics and characteristics. If this sounds like a difficult process, you are correct. It is difficult indeed to develop objective measurements of the dream states using behavioral or EEG studies. It is even more difficult using modern neuroimaging techniques such as functional magnetic resonance imaging (fMRI): the MRI scanner is incredibly loud. It is also a very cramped and frequently uncomfortable space so falling asleep in a scanner is difficult to do. Because of this, some studies ask participants to be sleep-deprived before they come in for the scanning session. This, however, presents new problems because the exhausted sleep in a novel and noisy scanner may not reflect the brain processes that occur when one is sleeping peacefully in their own bed (De Havas et al., 2012). Perhaps an ideal way to study dreaming is a combination of EEG—to determine which sleep stage the participant is in—and FMRI—to determine which brain areas are activated or deactivated. Future work in this area will shed new light on the brain bases of dreaming.

5.2 When Do We Dream?

Most of our dream state is spent during REM sleep. Fig. 12.26 shows characteristic sleep stages beginning with Stage I at the onset of sleep and moving through the NREM stages II–IV. There are three typical stages/times when people dream: the first is during the initial onset of sleep during Stage I (Fig. 12.26, green circle at hour 23:00). Dreams at this stage tend to be brief, hallucinatory, and fragmented. The next dream state typically happens during the first REM stage, occurring approximately 90 min after sleep onset (shown in grey rectangles

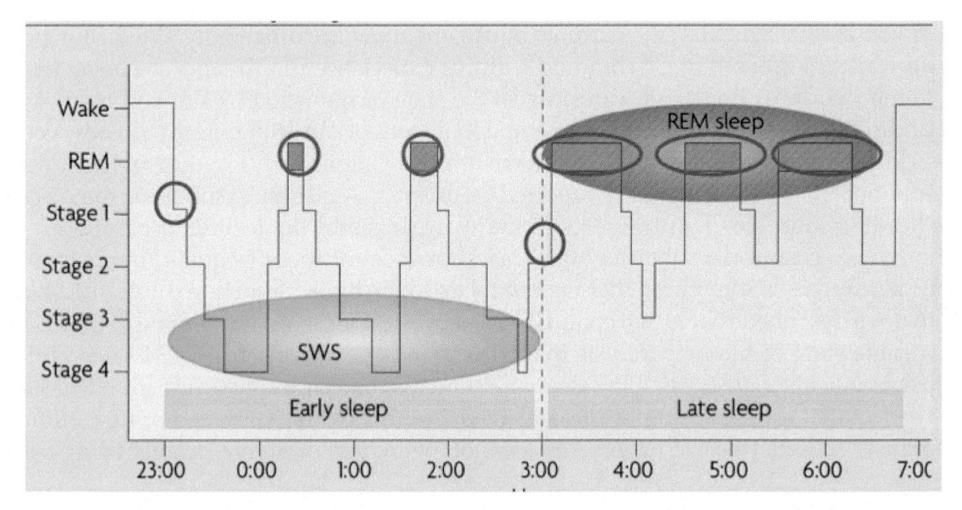

FIGURE 12.26 While most dreams occur during rapid eye movement (REM) sleep stages, two other sleep stage intervals also characteristically have dreams reported. The first is at the onset of sleep in Stage I (shown with a *green circle* at hour 23:00): these dreams are brief and fragmentary and are frequently bizarre and hallucinatory. The second is at the end of the slow wave sleep (SWS) stages in the middle of sleep (~4 h after sleep onset) during the transition to REM sleep (shown with a *green circle* at hour 3:00). REM sleep stages occur several times each night, beginning with brief periods of REM sleep during early SWS and continuing with longer duration and higher frequency during hours 4–8 of sleep time (shown with *pink circles* and *grey-filled shapes*). *Source: Diekelmann and Born (2010).*

with pink circles in Fig. 12.26 beginning after hour 0:00). Note that there are frequently at least five REM stages during a typical night's sleep, thus REM dreams can account for a sizeable portion of a total night's dreaming state. REM dreams tend to be bizarre and hallucinatory. REM dreams tend to be longer than NREM dreams and some are remembered. Another NREM stage for dreaming occurs late in the night in the transition between SWS, up through Stages III and II, as you move toward another REM stage (Fig. 12.26 green circle at hour 3:00) at the end of the SWS portion of the night. Dreams during Stage II middle of the night stage tend to be realistic and more thoughtlike.

5.3 Dreaming in the Brain

Why do these dream stages during the night differ from one another? Look back at Fig. 12.18: during REM sleep—and thus REM dreaming stages—the DLPFC is deactivated. This part of the brain controls "executive" functions such as planning, controlling our actions, solving problems. During REM sleep, the DLPFC is *offline, deactivated*—and these executive functions are not available. This may be why REM dreams tend to be bizarre and unrealistic—we are missing our dorsolateral frontal cortex's executive control! This is not the case for the more thoughtlike dreams that occur in late night Stage II sleep—during this sleep stage, the brain is coming out of SWS and brain areas, including the prefrontal cortex, are becoming more activated.

Despite these many challenges in determining the neural correlates of dreaming, progress has been made in assessing which brain regions are activated during dream sleep. One theory recently put forth is that dreaming involves the default mode network (DMN) of the brain (Domhoff & Fox, 2015). The DMN has been identified as those brain regions active during spontaneous (i.e., not goal-oriented) thought. Key regions included in the DMN are the medial prefrontal cortex (MPFC), the medial temporal lobe, the inferior parietal lobule/temporoparietal junction in both hemispheres, and the PCC (Buckner, 2012). The MPFC and the PCC serve as key connective hubs for this network. In a recent metaanalysis of EEG and fMRI studies of mind wandering during waking and REM sleep (and inferred dreaming), Domhoff and Fox reported similar brain regions activated as during waking DMN, namely the MPFC and the inferior parietal lobule/temporoparietal junction (Fig. 12.27). While intriguing, these results need to be bolstered by future studies that replicate and extend the findings to date. Designing and carrying out a neuroimaging dream study remains a challenge to sleep researchers.

5.4 Nightmares Versus Night Terrors

A very common type of dream is one that is frightening—*a nightmare*. These dreams typically occur during REM sleep and, just as in all REM sleep dreams, they are frequently strange and unrealistic. Children, especially between the ages of 3–6 years, frequently have nightmares: when they wake up from a nightmare, they are typically consolable by their parents. They may not want to go right back to sleep because the dream is still too vivid.

This is *not* the case for night terrors. While almost everyone has a nightmare now and then, night terrors occur in just 3%–5% of the population. Night terrors are a type of NREM sleep disorder in the category of *parasomnia*, discussed in Section 6.6, below.

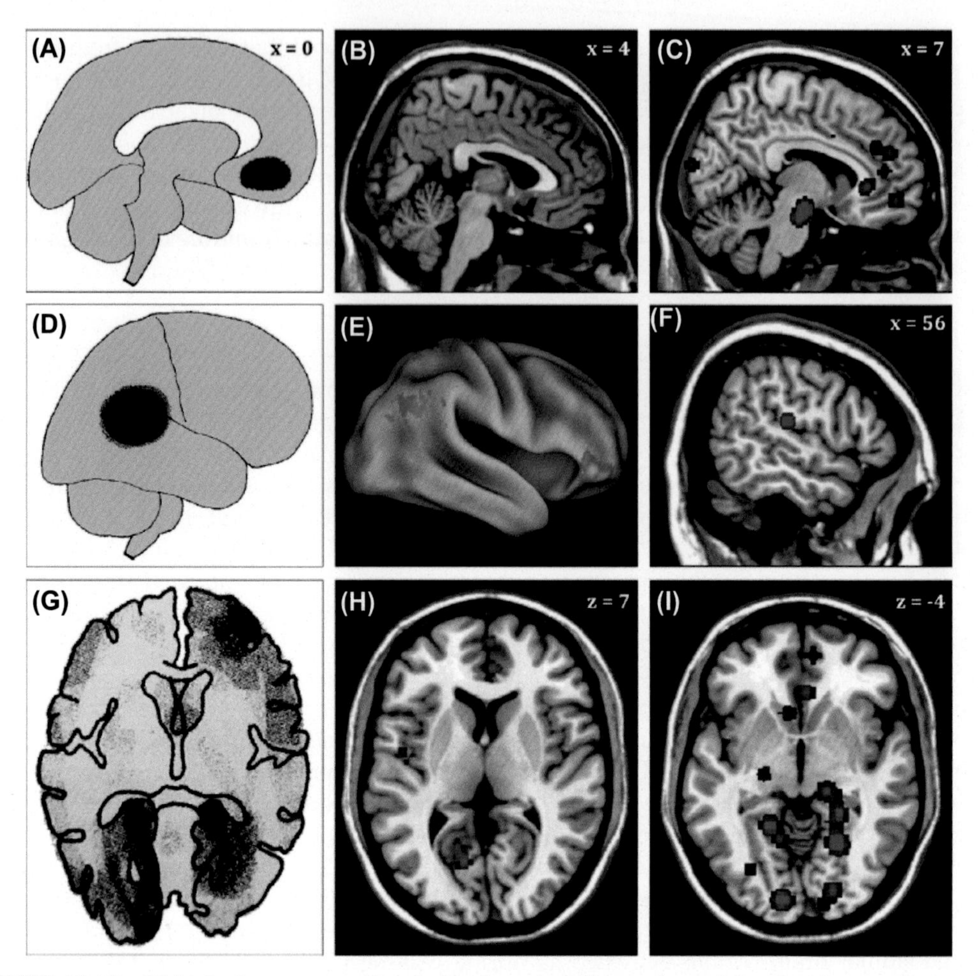

FIGURE 12.27 Which brain areas are active during dreaming? One study used functional magnetic resonance imaging (fMRI) to investigate brain areas active during wakeful daydreaming/mind-wandering and rapid eye movement (REM) sleep and compared them to brain areas shown to be critical for dreaming in studies with patients with brain damage. The left column shows brain areas critical for dreaming based on patient lesion studies. These areas include the ventral medial prefrontal cortex (MPFC) (shown in a sagittal view in (A), or the temporoparietal junction/anterior inferior parietal lobule (shown in a sagittal view in D). These brain areas are shown in a coronal view in (G). Both these regions are part of the proposed default mode network (DMN) of the brain. The center column shows brain areas (in green) active during wakeful daydreaming or mind-wandering. The right column shows brain areas (in blue) active during REM (and presumably, dreaming) sleep. Similar to the lesion studies, the MPFC was activated in both wakeful daydreaming and REM sleep (shown in B and C). The temporoparietal junction/anterior inferior parietal lobule was also activated in both conditions (shown in e and f). The bottom panels h and i show a coronal view of the results shown in a sagittal view in e and f. While these results are compelling, more studies with larger samples are needed to confirm the findings. *Source: Domhoff and Fox (2015).*

6. SLEEP DISORDERS

When sleep disorders do occur, they can be both frightening and life changing. Look back at Fig. 12.1—when sleep is disordered, all of these cognitive and somatic processes and functions are disturbed as well. Let us step through the major types of sleep disorders and their theorized causes and brain bases.

The four most common types of sleep disorders, in order of their prevalence, are *insomnia, sleep apnea, restless legs syndrome, and narcolepsy*. There are many more types of sleep disorders—some estimates range to more than 80—however, these are the core disorders that affect many children and adults. Some are short-lived, some are lifelong: all sleep disorders produce cognitive, emotional, and body-related impairments that can be very destructive to normal health and well-being.

6.1 Diagnosing a Sleep Disorder

How are sleep disorders diagnosed? Typically, a physician who deals with sleep disorders will go through several steps and stages to diagnosis a sleep disorder (Khoury & Doghramji, 2015). It begins with a patient interview to get a sense of the nature of the sleep complaint, instances of daytime sleepiness, and to get a sleep and medical history. Next, the physician typically asks the patient to keep a sleep diary to log their sleeping issues at night and daytime sleepiness. Some basic physical and mental assessments may be done as well. Depending on the results of these interviews and exams, the physician may call for a sleep study where the patient will go to a sleep lab and have a polysomnography test conducted while they sleep. Note: the first night's sleep in a sleep study is almost always a bit disordered. Consider how odd it would feel to be in a strange room with electrodes attached to your head and face! So the first night of a sleep study is typically not used; subsequent nights in the sleep lab tend to provide better data for analyzing the sleep issues.

6.2 Insomnia

By far, the most common sleep disorder is *insomnia*. People with insomnia can have trouble falling asleep, staying asleep, or waking up too early in the morning (https://sleepfoundation.org/insomnia/content/what-is-insomnia). Some instances of insomnia are brief: these are called *acute insomnia* and can happen because of stress, work demands, or emotional issues. However, when insomnia occurs for at least 3 days a week and lasts more than 3 months, it is called *chronic insomnia*. While some instance of insomnia occurs for almost half of all people, chronic insomnia is far less typical, occurring in only about 10% of the population. Chronic insomnia can occur alongside emotional, neurodevelopmental, and neurodegenerative disorders such as posttraumatic stress disorder, autism, and Parkinson's disease.

Insomnia can be due to some medical reasons such as chronic pain, asthma, and heavy allergies (Fig. 12.28). Understandably, all of these issues can cause disrupted sleep. Insomnia

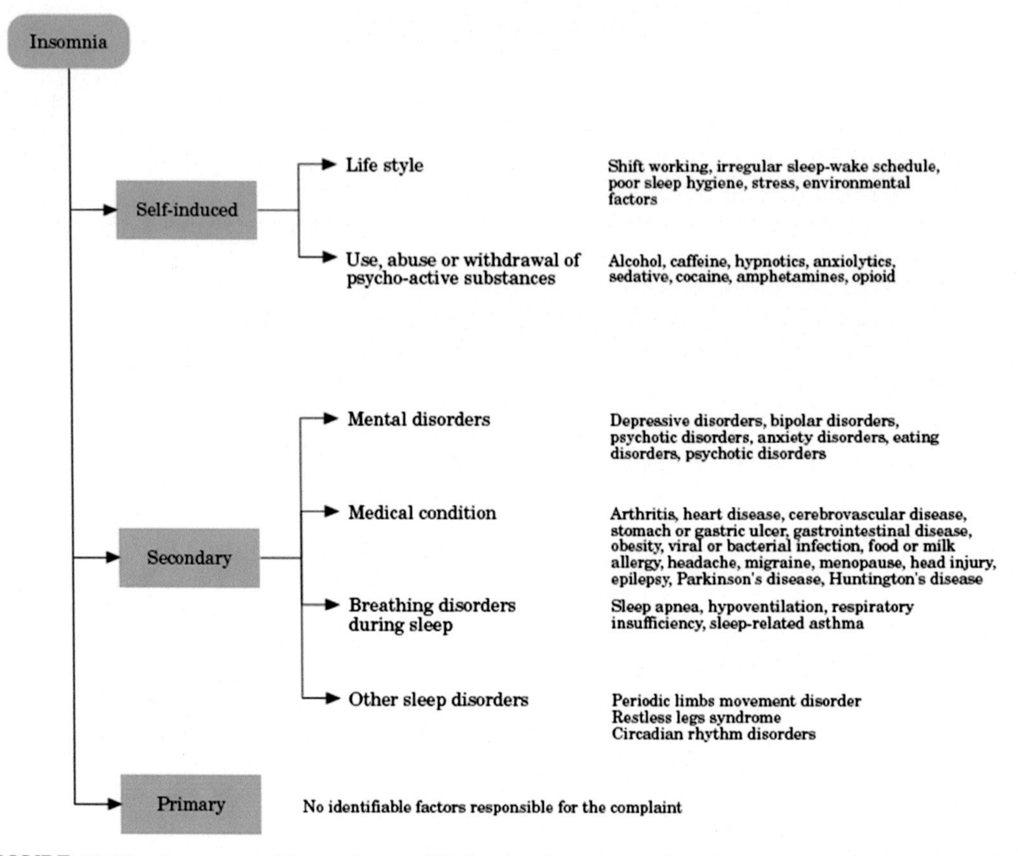

FIGURE 12.28 Some cases of insomnia are *self-induced* and are due to lifestyle situations such as working a late night or overnight shift or abusing alcohol, caffeine, or other drugs. Changing lifestyle and sleep hygiene can improve these cases tremendously. Other cases of insomnia are *secondary*, meaning that they are the result of a primary condition that is leading to a sleep disorder. These primary conditions can range from mental disorders, to medical conditions, to breathing disorders and other sleep disorders. These cases are more difficult to treat but are treatable. Again, a good sleep hygiene generally improves these as well. Still, other cases of insomnia are *primary*, meaning that insomnia is the main diagnosis. Thus in primary insomnia, self-induced and secondary causes of insomnia have been ruled out. *Source: Ohayon (2002).*

can also be the result of diet or medication—we all know that caffeine can help wake us up in the morning, but it can also prevent us from falling asleep due to its blocking of adenosine. Some medications taken for colds and allergies, high blood pressure, and heart disease can lead to insomnia. Insomnias frequently go hand in hand with depression, anxiety, and chronic stress. In these cases, and in cases of individuals with neurodevelopmental disorders, chronic insomnia may be related to a hyperarousal state in the brain (Maski & Owens, 2016). Insomnia may also reflect brain-related abnormalities in the neuromodulation needed for triggering normal sleep and waking, although less is known about these instances.

Many acute and even chronic insomnia cases can be helped by improving *sleep hygiene* (Table 12.2). Sleep hygiene refers to the way you spend your time during the day and how

TABLE 12.2 Sleep Hygiene Measures

THE DOS OF SLEEP HYGIENE
• Increase exposure to bright light during the day
• Establish a daily activity routine
• Exercise regularly in the morning and/or afternoon
• Set aside a worry time
• Establish a comfortable sleep environment
• Do something relaxing before bedtime
• Try a warm bath
THE DON'TS OF SLEEP HYGIENE
• Alcohol
• Caffeine, nicotine, and other stimulants
• Exposure to bright light during the night
• Exercise within 3 h of bedtime
• Heavy meals or drinking within 3 h of bedtime
• Using your bed for things other than sleep (or sexual activity)
• Napping, unless a shiftworker
• Watching the clock
• Trying to sleep
• Noise
• Excessive heat/cold in room

Source: Khoury and Doghramji (2015), Box 3.

you prepare for sleep in the evening. Good sleep hygiene can help establish and maintain healthy sleep even in situations where it is difficult, such as with individuals who work evening or overnight shifts, parents of newborns, and students studying full time and also working.

6.3 Sleep Apnea

The second most common sleep disorder is *sleep apnea*. Sleep apnea occurs when an individual's breathing is periodically blocked or interrupted for several seconds at a time during sleep. *Obstructive sleep apnea* is by far the more common form of sleep apnea. In obstructive sleep apnea, the soft tissue at the back of the throat collapses and blocks—or obstructs—the airway. This blockage can be a partial blockage, which results in snoring, or a full blockage, which causes a complete obstruction of the airways. Obstructive sleep apnea is very common and in fact some individuals do not know they suffer from it until they engage in a sleep study. Obstructive sleep apnea has many potential causes including being overweight, having enlarged tonsils or adenoids, frequent alcohol and tobacco use, and advanced age. In severe cases of obstructive sleep apnea, the individual may stop breathing 30 times in a single hour of sleep.

Obstructive sleep apnea is typically diagnosed through a physical exam by a medical doctor or a polysomnography exam. Treatments include controlling diet and exercise and cutting back on alcohol and tobacco. Therapies include differing types of positive airway pressure (PAP) devices that prevent the collapsing of the soft tissue at the back of the throat.

Central sleep apnea also leads to failure to breathe, but in this case, the autonomic brain functions in the brainstem that signal the automatic breathing during sleep fails. Both obstructive and central sleep apneas are potentially dangerous sleep disorders. If they are occurring frequently throughout the night, as is common, the individual does not proceed through NREM and REM stages in a normal way. This can lead to daytime sleepiness, fatigue, and memory issues. If not treated, apneas can also lead to cardiovascular problems due to the strain on the heart due to the air blockages.

6.4 Restless Legs Syndrome

Restless Legs Syndrome (RLS) is a fairly common sleep disorder occurring in 10% of adults and 2% of children (https://www.nhlbi.nih.gov/health/health-topics/topics/rls/). Symptoms of RLS include a relentless urge to move the legs and uncomfortable feelings in the legs that typically begins in the early evening. RLS can lead to sleep disruption because of the urge to constantly move, which in turn can lead to daytime fatigue, learning and memory impairments, and depression.

What causes RLS? Some causes of RLS are a secondary effect of a primary disease or illness. If RLS is the primary disorder, however, it is hypothesized that it may relate to low levels of iron in the brain, in particular in the substantia nigra in the subcortex, a key brain area for the production of dopamine (Ferre, Earley, Gulyani, & Garcia-Borreguero, 2016). The normal metabolism of iron has a circadian pattern that is similar to the onset of symptoms of RLS: iron serum is at its lowest point between 8 and 12 p.m., a peak time for RLS symptoms to occur. Converging evidence from autopsy and functional imaging studies are providing new information about the role of iron deficiency in RLS. While this research is still in its early stages, it is a promising new approach for understanding the neural bases for RLS.

RLS is frequently treated with dopaminergic drugs, which stimulate the dopamine system in the brain. Approved by the FDA in 2005 for treatment of RLS, dopamine receptor agonists, such as ropinirole and pramipexole, have shown mild side effects for short-term use. In the longer term, however, a major side effect has been shown to develop: *augmentation* (Ferre et al., 2016). Augmentation is an increase in RLS symptoms—with both more or longer-lasting RLS symptoms in the legs and the addition of other parts of the body, such as the arms, having characteristic RLS movement urges and unpleasant sensations (Garcia-Borreguero & Williams, 2010). New research is investigating the use of alternative treatments not involving dopaminergic drugs to avoid the life-changing side effect of augmentation for individuals with RLS. One promising approach is the investigation of iron deficiency in the blood and the use of intravenous iron preparations to modulate adenosine neurotransmitters that may correspond to the symptoms observed in RLS (Ferre et al., 2016).

6.5 Narcolepsy

Narcolepsy is a rare REM-centered sleep disorder of largely unknown origin. It typically occurs in adolescents and young adults, but is sometimes not accurately diagnosed until years later (Maski & Owens, 2016). There are five central symptoms that together form a likely diagnosis of narcolepsy, but the foremost symptom of narcolepsy is that the patient suddenly falls asleep. The core symptoms associated with this are: persistent daytime sleepiness, hallucinations, *sleep paralysis*, *cataplexy*, and restless sleep during the nighttime (Khoury & Doghramji, 2015). Let us go through these symptoms step by step, but first a brief overview: above all, narcolepsy is a disorder of REM sleep. When someone with narcolepsy falls asleep, they immediately enter REM sleep without going through the typical course of NREM sleep that always precedes REM sleep whether one is napping during the daytime or sleeping at night. Think of what you know about REM sleep: the brain is activated but differing regions are active in REM sleep versus waking, in particular the dorsolateral prefrontal lobe. With the onset of REM sleep comes the onset of atonia, paralysis. Dreams during REM sleep are typically bizarre, hallucinatory. Now think about these characteristics of REM suddenly happening to you in the middle of a busy day: you are awake one moment and the next, you are plunged into a dream state where you experience bizarre sights and sounds and you are paralyzed.

Daytime sleepiness is a hallmark of narcolepsy—however, other factors leading to this sleepiness must be ruled out before a diagnosis of narcolepsy makes sense. Many patients experience daytime sleepiness due to insomnia, sleep apnea, and other temporary or chronic sleep disorders. Next, the narcolepsy patient has hallucinations either during the rapid fall into REM sleep or upon awakening. These can be quite frightening and vivid. *Sleep paralysis* is described below in the section on REM parasomnias. Normal REM sleep is associated with the onset of atonia, where the body is weak and paralyzed. Patients with narcolepsy are awake and aware of being paralyzed either when atonia sets in at the onset of REM sleep or upon awakening from REM sleep: this is atypical and is called sleep paralysis. Sleep paralysis can occur in the absence of narcolepsy or another sleep disorder, in which case it is called *isolated sleep paralysis*.

Cataplexy is another rather frightening aspect of narcolepsy: it is a sudden paralysis of the muscles and can result in the person not being able to speak or move, or in some cases, the patient may fall to the ground and stay in a frozen state for minutes at a time. It can occur any time, while the patient is wide awake, and frequently is associated with a strong emotional experience (anger, laughter, surprise). Cataplexy occurs in more than half of narcolepsy cases and can be confused with an epileptic seizure; however, in cataplexy, the patient is conscious (Maski & Owens, 2016). If the patient experiences cataplexy, as well as the other hallmark symptoms of narcolepsy, the diagnosis is narcolepsy type 1. If the patient does not experience cataplexy, the diagnosis is narcolepsy type 2.

What causes narcolepsy? It is not known what causes this rare disorder, but narcolepsy type 1 is thought to be related to the loss of neurons that produce the neuropeptide hypocretin (also called orexin) in the lateral hypothalamus (Fig. 12.29). Hypocretin regulates arousal and wakefulness. The lack of hypocretin neurons is thought to lead to narcolepsy with cataplexy. Patients with narcolepsy type 2—without cataplexy—have normal levels of hypocretin, thus this may not be ruled in as a brain-based cause. Future research is needed to shed light on why narcolepsy occurs and to develop effective treatments for it. Narcolepsy is a lifelong, highly debilitating disorder that is associated with social and emotional stress due to its symptomology.

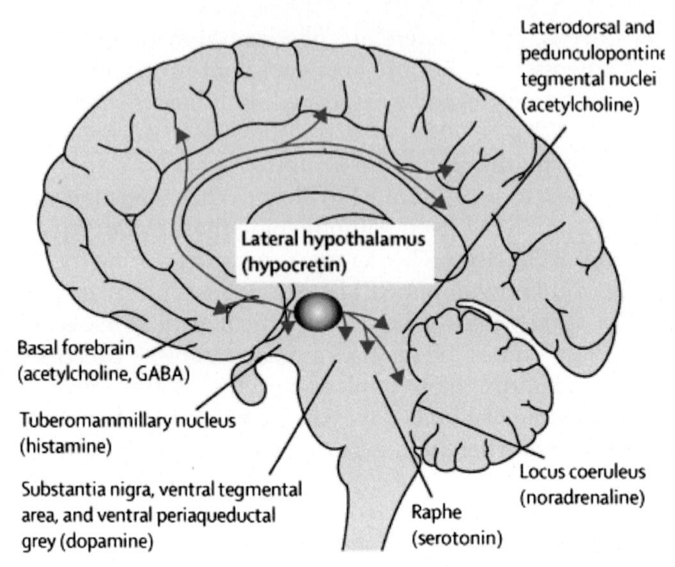

FIGURE 12.29 The brain bases of narcolepsy are largely unknown; however, narcolepsy type 1 is thought to be related to the loss of neurons that produce the neuropeptide hypocretin (also called orexin) in the lateral hypothalamus (shown in purple in center of figure). Hypocretin regulates arousal and wakefulness. The lack of hypocretin neurons is thought to lead to narcolepsy with cataplexy. Patients with narcolepsy type 2—without cataplexy—have normal levels of hypocretin, thus this may not be ruled in as a brain-based cause for narcolepsy in general. *Source: Maski and Owens (2016).*

6.6 Parasomnias: Nonrapid Eye Movement Sleep Disturbances

Parasomnias are sleep disturbances that can occur in either the NREM or REM sleep stages. They have very different characteristics due to the very different states of the brain and body during these sleep stages. Let us begin with NREM parasomnias: confusional arousals, sleepwalking, and night terrors occur in 1%–4% of adults and are even more common in children (Maski & Owens, 2016). These commonly occurring NREM parasomnias share a key feature—they almost always occur during SWS. The observation of more common NREM parasomnias in children is hypothesized to be due to the greater amount of SWS for children as compared to adults. NREM parasomnias can last anywhere from a few minutes to more than half an hour and are typified by the individual experiencing them having no memory of the event in the morning. This is a key aspect of NREM parasomnias: they occur during heavy SWS when the individual is unconscious—even if they are moving about with their eyes open. NREM parasomnias can be caused by stress, but they are not typically related to mental or psychological issues (although some individuals with posttraumatic stress disorder suffer from night terrors). Nevertheless, since they involve a wakeful appearance despite the lack of conscious control, NREM parasomnias can be frightening for individuals and their families and treatment should be sought.

6.6.1 Confusional Arousals

Confusional arousals occur during SWS and not surprisingly, because of the general slowing of brain function during SWS; the individual being aroused at this stage is confused, disoriented, and very groggy. Sometimes the individual will cry out or moan when they have these arousals during SWS. Confusional arousals can last for several minutes, and in some cases, they last for hours (Maski & Owens, 2016). In a way, we all have experience with confusional arousals: if you have ever been suddenly awakened in the middle of the night during your SWS, you will have similar symptoms of disorientation, confusion, and grogginess. The key difference is that the sudden arousal in that case is due to an external event such as an alarm clock, noisy roommate, or environmental noise.

6.6.2 *Night Terrors*

Night terrors occur during late night sleep in the transition from SWS to REM sleep (Fig. 12.26). Night terrors are prevalent in young children of 1–5 years, with almost a third of children in this age range experiencing them at some time (Maski & Owens, 2016). This prevalence declines as the child matures and is reduced to 10% by age 7. When a night terror occurs, the child awakens or partially awakens suddenly and has a strong sense of intense fear that is typically combined with a rapidly beating heart and shallow breathing. There is no dream—just a sudden awakening or partial awakening in a frightened state.

What causes night terrors? It is theorized that in the ascent out of SWS, through NREM sleep and moving toward REM sleep, there is an overreaction of the neuromodulators that are shifting the brain between sleep stages during a child's maturing in the early preschool years. Night terrors can run in families. Night terrors, although frightening for the child and disturbing to the parent, are not a sleep or a neurological disorder. Sleep hygiene issues such as reducing stress levels before sleep onset and maintaining a quiet and peaceful sleep environment can help reduce night terror occurrences.

6.6.3 *Sleepwalking*

Perhaps the most frightening NREM parasomnia is sleepwalking. Occurring as the other NREM parasomnias during SWS, during sleepwalking, the individual is not conscious but has experienced a partial arousal that allows movement, even complex movements, with eyes open and the appearance of wakefulness. Sleepwalking is common in children in the age range of 5–10 years, and is highly prevalent: experienced in more than one in eight children (Maski & Owens, 2016). As in night terrors and confusional arousals, the brain bases of sleepwalking have not been discovered as of yet, although their occurrence during SWS in young children lends support to the notion that they correspond to overreaction of neuromodulators during sleep.

6.7 Parasomnias: Rapid Eye Movement Sleep Disturbances

6.7.1 *Sleep Paralysis*

We have discussed that the body goes into a state of atonia—paralysis—at the onset of REM sleep. This paralysis is perfectly normal, and in fact it is vital for safe sleeping: see Section 6.7.2 for a disorder of this natural atonia. Because we are shifting between sleep stages—from NREM sleep to REM sleep—atonia takes place while we are asleep and not aware of it. It does tend to enter into our dreams, however, as many people report vivid dreams during REM sleep where they have difficulty moving. This is likely due to some level of awareness of the atonia we experience during REM sleep. *Sleep paralysis* differs from the typical atonia in that the patient is awake and aware of the paralysis. This typically occurs either at the onset of REM sleep and atonia or at the offset. It is *the shift to and from a state of atonia* and the addition of consciousness that combine to form sleep paralysis. In addition to the feeling of total paralysis, some patients experience visual hallucinations that can be quite frightening, especially since they are feeling paralyzed (Sawant, Parkar, & Tambe, 2005). A very frequent hallucination or sensation for patients during sleep paralysis is the strong feeling that someone is in the room with them. The prevalence of sleep paralysis is not well known: people experiencing it occasionally may not seek treatment for it and so there is no record of its occurrence.

6.7.2 Rapid Eye Movement Sleep Behavior Disorder

REM sleep behavior disorder (RBD) is a very common REM parasomnia. RBD is a failure for atonia to occur during REM sleep and therefore the RBD patient is not paralyzed and acts out during dreaming. This can result in the patient or sleeping partner being injured, as many times, the patient flails arms and legs around during a dream, even shouting and swearing (https://sleepfoundation.org/sleep-disorders-problems/rem-behavior-disorder). Remember, however, that the patient is asleep and unaware of this behavior. RBD has a clear demographic: 90% of RBD patients are male and over the age of 50. RBD is thought to be associated with impairments in the motor systems in the brain since they do not activate at the onset of REM with typical atonia. In fact, a large number of older male patients with RBD (more than 80%) were reported to develop Parkinson's disease, a degenerative neurological disease that affects the cortical motor systems (Schenck, Boeve, & Mahowald, 2013). Thus RBD may be an early indicator of a predilection to Parkinsonian disease.

7. SUMMARY

Sleep is a core ingredient in our health: lack of sleep causes cognitive, emotional, and body-based problems beginning with just one night of deprived sleep. The benefits of sleep, however, extend far past these short-term effects: sleep is key to learning and memory functions and emotional well-being.

In the brain, there is a complex push–pull between our circadian rhythms and our homeostasis need for sleep. Neuromodulators and hormones combine in a synchronized dance of up- and down-modulation at the onset of sleep and throughout sleep stages during the night until we awaken in the morning.

Sleep is influenced by our psychological selves: when sleep is disordered or disrupted due to sleep disorders or other diseases or conditions, it can lead to mental health issues. Everyday stresses and tensions combine with bright lights during the evening to impede natural sleepiness signals. Good sleep hygiene can help with temporary and even chronic insomnias.

New research is shedding light on the brain bases for sleep disorders, such as narcolepsy; however, some of the key neuronal processes underlying them remain unknown.

8. STUDY QUESTIONS

1. What are some cognitive effects of sleep deprivation?
2. What are some body-related effects of sleep deprivation?
3. What does circadian rhythm refer to? How do they affect our sleep—wake cycles?
4. How is sleep studied? What do these methods reveal about the brain bases of sleep and sleep stages? What does it not reveal?
5. Give an example of a neuromodulator and describe its function in sleep.
6. What is a polysomnogram and how is it used in sleep research?
7. How does SWS differ from REM sleep?
8. What brain areas are deactivated during REM sleep versus waking? What brain areas are activated?
9. How does memory consolidation occur in the brain? During which sleep stage(s) does it occur?
10. How does the concept of "memory evolution" differ from "memory consolidation?"
11. Which brain areas correspond to the "temporary store" and "long-term store" described in this chapter?
12. What are some self-induced causes for insomnia?
13. How does primary insomnia differ from secondary insomnia?
14. During what stage(s) of sleep does sleepwalking typically occur?

CHAPTER

13

Disorders of Consciousness

HUMAN CONSCIOUSNESS

Disorders of consciousness present both clinical challenges of diagnosis and treatment and unique opportunities for fundamental scientific discoveries about the nature of human consciousness. ***Giacino, Fins, Laureys, and Schiff (2014), American and Belgium Scientists.***

Sleeping beauty, 1899, Henry Meynell Rheam, English painter. *Source: Public domain.*

1. INTRODUCTION

While scientists struggle to develop a clear definition of just what *human consciousness* is, we can agree more readily on describing *human unconsciousness*. From a clinical perspective, unconsciousness is characterized by closed eyes and lack of awareness and responsiveness (although a healthy person may be awakened), slowed breathing and heart rate, and some limited spontaneous movement. We may broadly separate unconscious states into two types: reversible and nonreversible. Reversible unconscious states include the slow wave sleep

Fundamentals of Cognitive Neuroscience
https://doi.org/10.1016/B978-0-12-803813-0.00013-1

(SWS) we encounter with every night's sleep and the sedated state experienced with the application of anesthesia. Disorders of consciousness (DOC) are *potentially nonreversible* unconscious states due to brain trauma or damage.

In this chapter, we will step through the states of unconsciousness from SWS to deep coma. Although we will describe separable and definable states of DOS, it is important to note that the states of unconsciousness and consciousness are not discrete and may be better thought of as a continuum of two facets already described: awareness (or contents of consciousness) and wakefulness. The trading levels of awareness and wakefulness that lie below the surface in an unconscious individual contribute massively to that person's experience and, if known, to their clinical treatment.

2. REVERSIBLE UNCONSCIOUSNESS

The state of unconsciousness is an entirely normal state experienced each night during SWS. The push–pull of neural modulators progress us into a deep unconscious sleep and out again on a nightly basis. Anesthetized unconscious states are not normal, however, can be very well controlled and have a low risk factor of the patient failing to regain consciousness. While they may seem similar upon observation of a person in SWS or a patient who is deeply sedated, brain processes and functions differ sharply across these two reversible unconscious states.

2.1 Reversible Unconsciousness – Slow Wave Sleep

As we described in Chapter 12, Sleep and Levels of Consciousness, during SWS, neuromodulators and hormones work together to slow brain processes down until a deep sleep is attained, characterized by the slow waves observed in the EEG (Fig. 13.1). While there is a

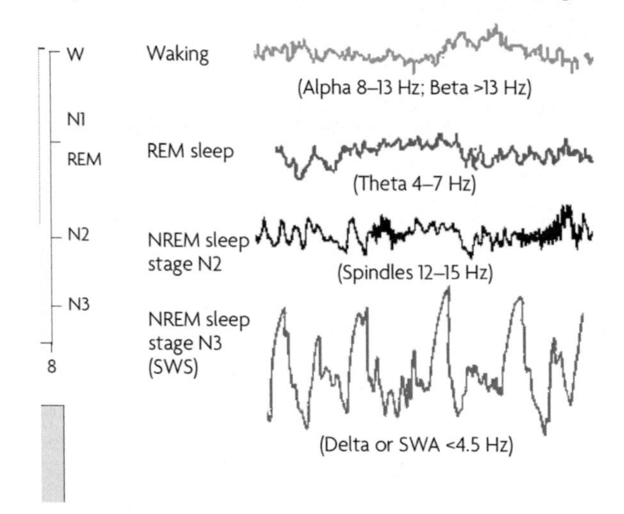

FIGURE 13.1 Electroencephalography (EEG) measures of brain activity across waking, sleeping, and dreaming states. Widespread and mostly uncorrelated neural activity is reflected in relatively high frequency EEG waveforms in the Alpha and Beta bands (Alpha 8–13 cycles per second, Hertz (Hz), Beta >13 Hz). REM sleep has a similar but lower frequency pattern of activity as waking, with waveforms showing Theta band (4–7 Hz) and spindle activity. By sharp contrast, slow wave sleep (SWS) shows low frequency, Delta band (<4.5 Hz) activity. *Source: Cirelli (2009).*

general slowing of brain processes, they differ across brain regions with some brain areas actively engaged in processes that are key to memory and cognitive health while other brain areas are truly "quiet" The differential activity in widely distributed brain areas allows the arousal of the sleeping individual—by a sudden noise, for example. Both awareness and wakefulness are at low levels; however, they are distinguishable from the levels present during anesthesia or coma (Laureys, 2005, Fig. 13.2).

There is a normal progression at the end of SWS stages upward to lighter sleep stages, on to rapid eye movement (REM) stages, and ultimately to awakening.

2.2 Reversible Unconsciousness – Anesthesia

The unconsciousness produced by anesthesia is entirely different than unconsciousness experienced during normal SWS. Anesthesia contains many substances that work together

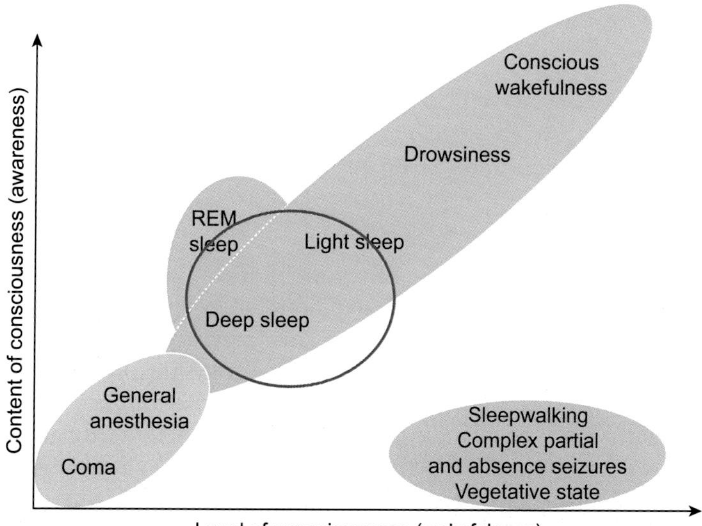

FIGURE 13.2 A simplified version showing levels of the two major components of consciousness: awareness and wakefulness. On the vertical y-axis, levels of awareness are shown beginning at the bottom of the axis with no level of awareness, as in a coma or when under anesthesia (shown in pink). Moving up the y-axis, level of awareness increases as one moves through heavy sleep (*circled in red*), light sleep, and into drowsiness and finally to full conscious wakefulness (shown in blue). Note, however, that during rapid eye movement (REM) sleep, there is a higher level of awareness than during deep or light nonrapid eye movement (NREM) sleep stages. The shape for REM sleep is sitting above the other sleep stages with a higher level of wakefulness, although well below conscious wakefulness. On the horizontal x-axis, levels of wakefulness are shown beginning with coma and moving similarly, through stages of sleep to conscious wakefulness. According to this simplified version, levels of awareness and wakefulness are roughly similar as one moves through the stages from coma to deep sleep to wakefulness, represented by the *pink and blue shapes* forming a diagonal from the bottom left corner, where no awareness or wakefulness are observed in coma, to the upper right, where both levels of awareness and wakefulness are high during conscious wakefulness. Sharply contrasting to this balance of levels of awareness and wakefulness are abnormal states of consciousness such as sleepwalking, epileptic seizures, and vegetative states (VSs) (shown in purple on the x-axis). In these states, the individual shows a high level of wakefulness with an extremely low level of awareness shown by the *purple shape* resting on the bottom of the y-axis but far to the right of the x-axis. *Source: Laureys (2005).*

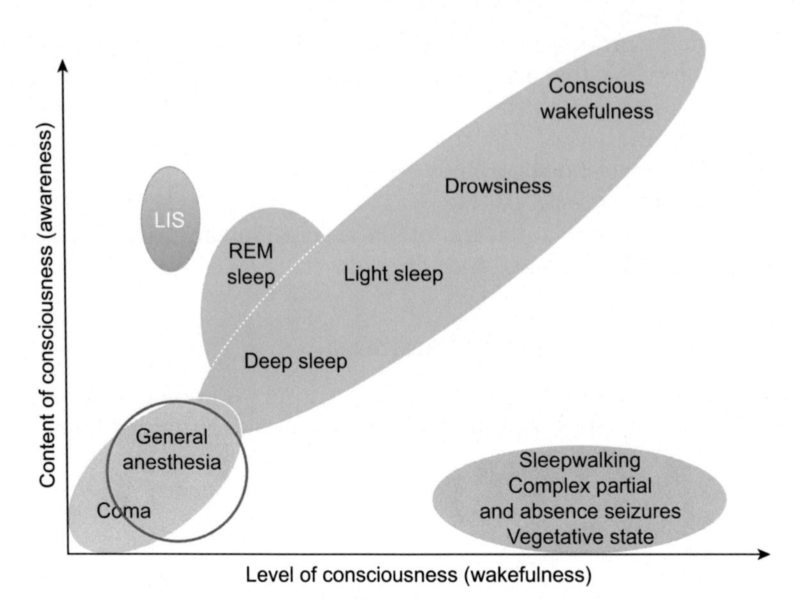

FIGURE 13.3 The same figure as Fig. 13.2, in this case the states of anesthesia-sedated and coma are shown with low levels of awareness and wakefulness (*circled in red*). *Source: Laureys (2005).*

to produce the controlled unconsciousness of a patient, including agents that produce a brief amnesia or memory loss, analgesia that provide pain relief, reflex suppression that cause temporary immobilization or paralysis, and agents that produce a hypnotic state to reduce awareness. How does anesthesia cause the brain to fall into unconsciousness? Simply put, anesthesia alters the flow of sodium molecules into the cell membranes of neurons, blocking the production of action potentials. Under anesthesia, neural activity throughout the brain is reduced, in sharp contrast to the brain activity in SWS. Fig. 13.3 shows this effect on awareness and wakefulness in a schematic form, with the both the *level of awareness* and the *level of wakefulness* reduced as compared to deep sleep.

Unconsciousness caused by anesthesia is reversed by reducing the anesthesia and arousing the patient. As the patient arouses, both the awareness and wakefulness levels increase until full wakefulness occurs. Depending on the type and duration of anesthesia, this waking process can take minutes to hours, again in sharp contrast to moving from SWS into higher sleep stages in normal sleep.

3. DISORDERS OF CONSCIOUSNESS

DOC are typically caused by either focal or more global brain injuries. DOC are clinically categorized by *observable features* in the patient. Consider that you are an emergency room doctor and the paramedics bring in an unconscious patient who was in a car accident. The patient cannot respond to you: you must *infer* their condition through a series of observations and tests.

TABLE 13.1 Characteristic Clinical Features of Disorders of Consciousness

Disorder	Arousal and Attention	Cognition	Receptive Language	Expressive Language	Visuoperception	Motor Function
Coma	No sleep–wake cycles[a]	None	None	None	None	Primitive reflexes only
Vegetative State	Intermittent periods of wakefulness[a]	None	None	None	Inconsistent visual startle	Involuntary movement only
Minimally conscious state	Intermittent periods of wakefulness	Inconsistent but clear-cut behavioral signs of self-awareness or environmental awareness	Inconsistent one-step command following	Aspontaneous and limited to single words or short phrases	Visual pursuit, object recognition	Localization to noxious stimuli, object manipulation, automatic movement sequences
Posttraumatic confusional state	Extended periods of wakefulness	Confused and disoriented	Consistent one-step command following	Sentence-level speech, often confused, perseverative reliable Yes–No responses	Object recognition	Functional use of common objects
Locked-in (LI) syndrome[b]	Normal sleep–wake cycles	Normal to near normal	Normal	Aphonic	Normal	Tetraplegia

[a]*Key distinguishing features.*
[b]*LI syndrome is not a DOC, but is included here for purpose of comparison with syndromes associated with significant disturbance in consciousness.*

Table 13.1 lists the types of observables used to characterize unconscious patients, including EEG tests of the brain to assess neural activity, responsiveness to language, eye movement and gaze, and motor function such as basic reflexes and involuntary movements (Giacino et al., 2014). Keep in mind that although this table—and this chapter—describes DOC according to discrete states, the actual situation varies widely among patients even if they fall into the same category of DOC. This is due to the wide range of brain damage and disruption that can occur in a traumatic injury coupled with the interaction of regions in the brain that are damaged or decoupled. This variability can produce a virtual limitless combination of effects. However, to be effective, the clinical team who are providing treatment to the patient must make some basic assessment as to what state of consciousness/unconsciousness the patient falls into.

As in SWS and anesthesia, DOC result in levels of awareness and wakefulness. Unlike SWS and anesthesia, however, awareness and wakefulness can differ sharply in level.

Let us step through the differing DOC and their clinical features, referring to Table 13.1, beginning with the deepest levels of unconsciousness, coma, and moving to a confusional state.

3.1 Coma

The coma state is usually the initial state caused by severe damage to both hemispheres of the cortex, both of the thalami, and brainstem damage (Fig. 13.4). Coma is defined by the complete loss of spontaneous arousal (Fig. 13.5). No sleep–wake cycles are evident as measured by EEG. Eyes remain closed continuously. There is no speech and no purposeful motor activity. The coma state usually resolves within 2 weeks of the injury and the patient transitions into either a VS or a minimally conscious state (MCS) (Fig. 13.4).

3.2 Vegetative State

The VS differs from the coma state in that there is wakeful unconsciousness. The wakefulness is exhibited by spontaneous eye opening. However, there is no evidence of awareness, language comprehension, visual gaze following despite the eyes opening occasionally, and only sporadic involuntary motor movement (Fig. 13.5). When we open our eyes, observers see this as a sign that we are conscious. This differs in the VS, with levels of awareness that are low while levels of wakefulness relatively higher. Family members may observe the occasional eye-opening and infer that the person is "waking up" but sadly this may not be the case.

Typically, a VS can last for 1–12 months following traumatic brain injury. Following that time window, VS *may* be considered permanent (see Fig. 13.4).

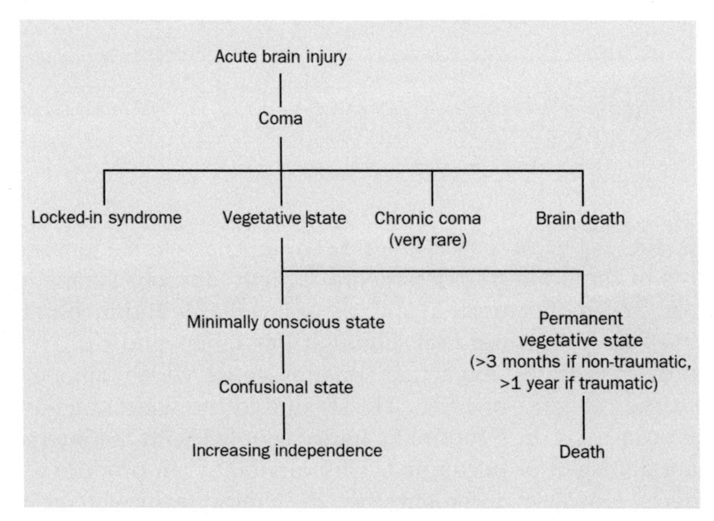

FIGURE 13.4 Transitional pathways following acute brain injury and coma. Following initial coma, patients transition into one of four general states. Rarely, patients with focal bilateral brainstem injury transition to the locked-in syndrome state with an intact cortex coupled with massive paralysis that leaves them with only eye-opening and eye fluttering movement (left of figure). More commonly, patients transition to a vegetative state (VS) and from there onto a minimally conscious state (MCS), to a confusional state, to increasing consciousness (left center of figure). Some patients, however, to not progress to a MCS and remain in a permanent VS until death occurs. Chronic coma (right center of figure) is very rare. Brain death may occur following coma without a transition to a VS (far right of figure). *Source: Laureys et al. (2004).*

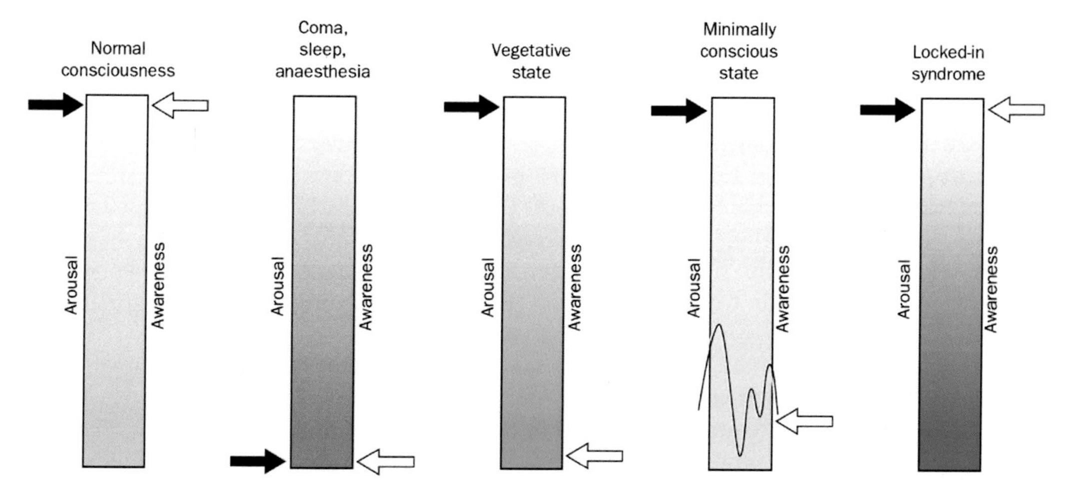

FIGURE 13.5 Levels of arousal/wakefulness and awareness across conscious and unconscious states. In normal consciousness, levels of arousal (*black arrow*) and awareness (*white arrow*) are balanced and high (*light blue shaded bar*, far left of figure). In coma, deep sleep, and anesthesia, these levels are also balanced but very low (*orange shaded bar*). In sharp contrast, in a vegetative state, arousal levels are high coupled with awareness levels that are low (*green shaded bar*). In a minimally conscious state, level of arousal is high and level of awareness is inconsistent (*yellow shaded bar*). In locked-in syndrome, which is not a disorder of consciousness, levels of arousal and awareness are both high and similar to normal conscious individuals (*blue shaded bar*). *Source: Laureys et al. (2004).*

3.3 Minimally Conscious State

A MCS differs sharply from a VS: while an MCS includes eye opening as in VS, that wakefulness sign is coupled with some receptive and expressive language function, some command following behavior, visual pursuit of a moving object, and some automatic motor movement sequences. There are some inconsistent but clear signs of self-awareness in an MCS patient (Fig. 13.5). Some patients in an MCS may respond emotionally to family faces or photographs, they may smile or cry. They may make purposeful movements, for example, reaching for a glass; however, the MCS is characterized by *inconsistency* in these behaviors.

MCS is thought of as a *transitional state*, that is, a state a patient *travels through* during recovery of consciousness with healing or decline with neurodegeneration (Fig. 13.4).

3.4 Posttraumatic Confusional State

Even a nontraumatic brain injury—a sudden fall or a hit on the head—can cause initial confusion. Posttraumatic confusional state is a more extended state than the brief confusion felt when hit on the head. A posttraumatic confusional state may last days, months, or years… or recovery may follow swiftly following transition from MCS, to posttraumatic confusional state, to full consciousness.

Posttraumatic confusional states include extended periods of wakefulness; however, the patient is clearly confused and disoriented. These patients are more consistent than MCS with following one-step commands and they have more functional language abilities. They can recognize objects and can make appropriate movements, such as reaching for a glass.

3.5 Locked-In Syndrome

Locked-in (LI) syndrome is *not* a DOC. It is included here because a LI patient typically has brain damage; however, it is limited to the brainstem region and the cortex is largely unaffected. The brainstem damage is sufficient to produce full paralysis and the only voluntary movement available to the LI patient is eye opening and fluttering. However, unlike the eye-opening observed in VS, eye opening and eye fluttering in LI reflects a high level of *awareness and wakefulness* (Fig. 13.5). The sparing of the cortex and corresponding cognitive function is coupled with the hugely debilitating lack of movement and responsiveness: the patient is *LI* with their brain function intact and no way to communicate this to the treatment providers and family.

If careful notice is not paid, an LI patient may be moved to a facility where insufficient treatment is provided. Their cognitive status may never be known and their existence may consist of being both aware and awake with no way to convey this to others. This situation is even more difficult to diagnose than it appears because frequently an LI patient will *initially* be in a coma or VS; however, they *recover* to an LI status unknown to their caregivers.

Although LI life sounds gruesome, there is a growing awareness of this rare disorder and case studies of the discovery of the cognitive normalness of these patients are available on the internet. In fact, the National Organization for Rare Disorders has a website for LI Syndrome (https://rarediseases.org/rare-diseases/locked-in-syndrome/) that contains helpful advice for treatment and therapeutic interventions.

Together, the several states that combine to form the DOC share many similar and overlapping symptoms that may be in a state of transition at any given point. Understanding the true state of the patient's awareness and wakefulness would lead to better diagnosis, better and more effective treatments, and provide guidance in end-of-life decisions. A central goal, then, of clinicians treating patients in a state of extended unconsciousness is to determine their true brain status.

4. NEUROIMAGING DISORDERS OF CONSCIOUSNESS

Bedside assessments of DOC can be difficult due to the wide variability actually observed in patients coupled with the changes that may be occurring as they transition between DOC states. Many aspects of the patient's actual cognition and state of awareness are opaque to the treating clinicians: they cannot easily discern what a comatose or VS patient is experiencing inwardly.

Laureys et al. (Giacino et al., 2014; Laureys, 2005; Laureys, Owen, & Schiff, 2004) have shed new light on neuroimaging approaches to provide better diagnostic tools for assessing DOC—both when they have newly occurred and when end-of-life decisions are being considered. Much progress has been made in identifying brain metabolic and hemodynamic activity and their correspondence to level of consciousness.

4.1 Metabolic Brain Activity Reflects Conscious State

In normal consciousness, the metabolism of the brain is high, reflecting an active and healthy brain (Fig. 13.6). Brain metabolism shifts as states of unconsciousness are attained,

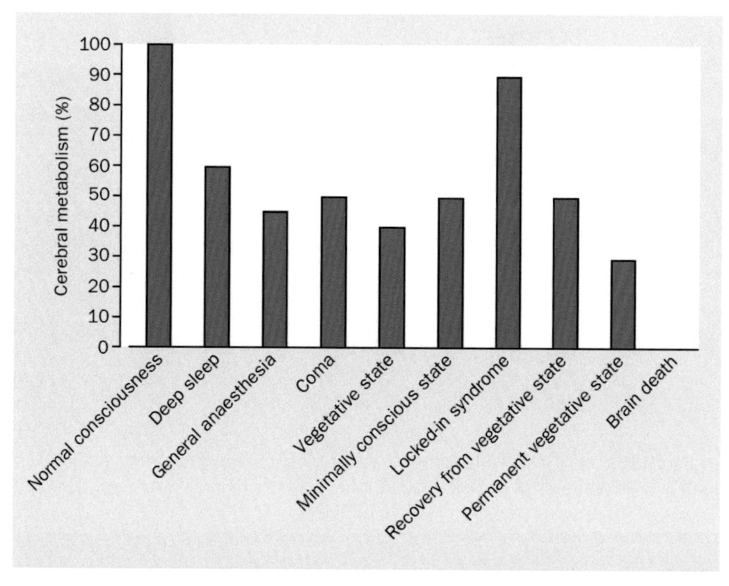

FIGURE 13.6 The percent of the cerebrum that is metabolically active varies across conscious states. The cerebrum shows high (100%) metabolism in normal consciousness (left side of figure). Deep sleep, anesthesia, coma, and vegetative and minimally conscious states show lower (<60%) metabolism, while permanent vegetative state shows very low (<30%) brain metabolism. By contrast, lock-in syndrome shows near normal (~90%) brain metabolism. *Source: Laureys et al. (2004).*

with lower levels of activity seen in deep sleep, anesthesia, and DOC (Laureys et al., 2004; Fig. 13.6). This is *not* the case in LI patients, who show a metabolic rate similar to normal consciousness.

Interestingly, brain metabolism is not uniform across brain regions despite being generally high during normal consciousness. In a healthy brain during normal waking, the *precuneus* and the *posterior cingulate cortex* (PCC) are the most highly active (metabolically) regions in the brain (Laureys et al., 2004, Fig. 13.7). Specifically, the metabolic activity of the precuneus was assessed in a large study of 110 controls and showed consistently higher metabolic activity than any other region. Metabolic activity in the precuneus was also assessed for 33 VS patients, 7 MCS patients in MCS, and 5 LI patients (Fig. 13.8).

Results of the study presented in Fig. 13.8 were dramatic: VS patients showed no precuneus metabolic activity while MCS patients showed just slight activity. In sharp contrast, the LI patients showed near-normal metabolic activity. These results correspond to the cognitive and conscious levels of these patients and thus provide an emergent technique for assessing cognitive health in otherwise unconscious or minimally conscious individuals.

4.2 Hemodynamic Brain Activity Reflects Conscious State

In an fMRI study, hemodynamic activity in the precuneous and the PCC were shown to correlate to state of consciousness ranging from fully conscious Controls, to MCS, VS, and coma patients (Giacino, Fins, Laureys, & Schiff, 2013; Fig. 13.9). That is, *higher activity* in the precuneus and PCC corresponded to *higher level of consciousness* (awareness and

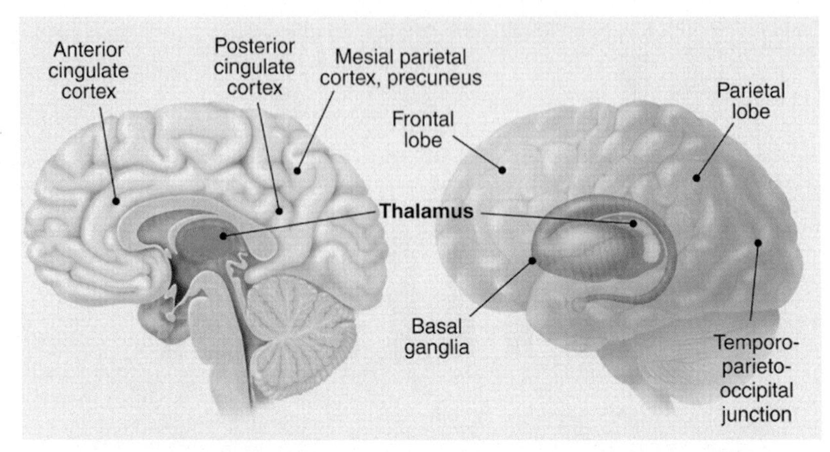

FIGURE 13.7 The precuneus and posterior cingulate cortex (PCC). The precuneus and PCC are tucked into the midline of the brain, on the medial aspect of the parietal lobe. *Source: Alkire, Hudetz, and Tononi (2008).*

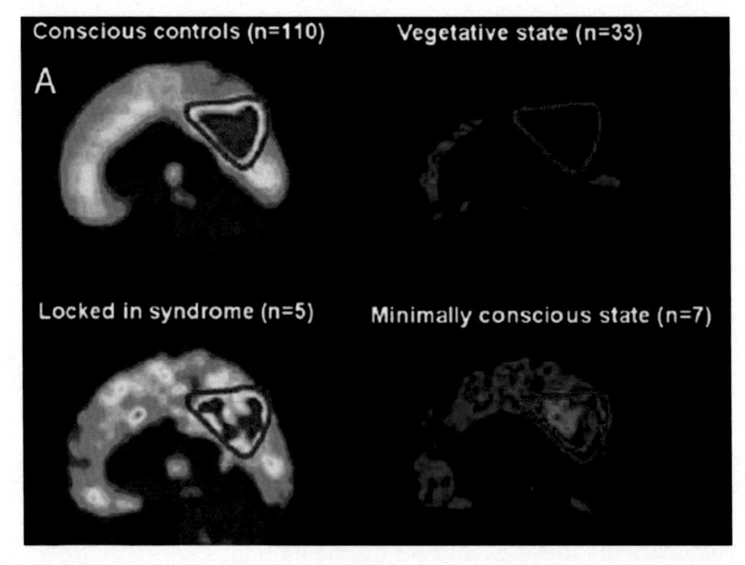

FIGURE 13.8 Metabolic activity (measured using PET) in the precuneus corresponds to level of awareness. In a large (n = 110) sample of healthy controls, the precuneus showed the highest level of metabolic activity (top left panel), high metabolic activity shown in red and yellow colors. In contrast, a sample (n = 33) vegetative state (VS) patients (top right panel) show no activity in the precuneus. A small (n = 5) of locked-in (LI) syndrome patients (bottom left panel) showed near normal metabolic activity (shown in red and yellow colors). A small (n = 7) sample of minimally conscious patients (bottom right panel) showed a much lower level of metabolic activity (shown in yellow colors) than controls or LI patients but higher activity than the VS patients. *Source: Laureys et al. (2004).*

wakefulness), and *lower activity* corresponded to lower *level of consciousness*. This effect was also observed, although less robustly, in the medial prefrontal cortex, the temporoparietal junction, and the parahippocampal gyrus.

Cumulatively, these studies of patients with DOC, and others being conducted around the world, are providing new evidence of how the brain's metabolic and neural activity varies by

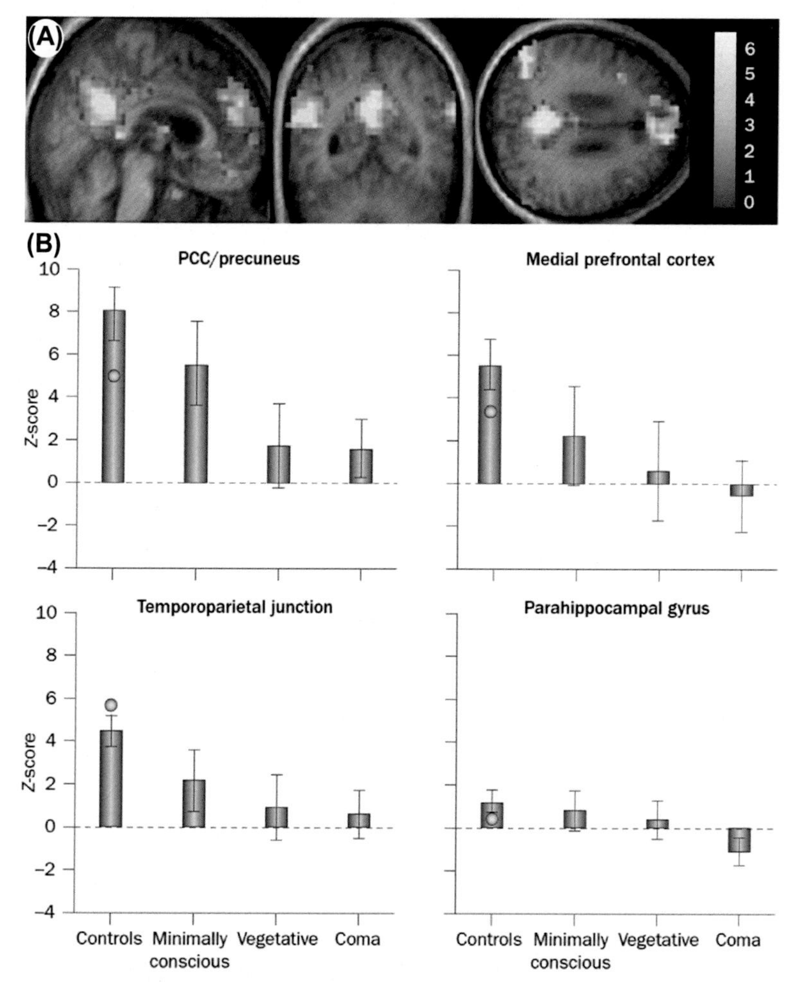

FIGURE 13.9 Neural activity across brain regions corresponds to level of consciousness. (A) Hemodynamic activity measured using functional magnetic resonance imaging (fMRI) reflects level of consciousness in key brain regions. (B) High neural activity for controls was evidenced in the posterior cingulate cortex (PCC) and the precuneus (upper left panel). Z-score activity for locked-in (LI) syndrome patients is shown in a blue circle on the Controls bar. PCC/precuneus activity is significantly lower for minimally conscious patients and even lower for vegetative state and coma patients. A similar pattern holds for the medial prefrontal cortex (upper right panel), the temporoparietal junction (lower left panel) and to a smaller extent to the parahippocampal gyrus (lower right panel). Z-scores for the LI syndrome patients are shown in a blue circle within the bar for the Controls for these brain regions as well. *Source: Giacino et al. (2014).*

conscious/unconscious state. A key goal of these studies is to better understand both the current status of an unconscious or minimally conscious patient and the prognosis for future recovery. End-of-life decisions about withdrawal of life support treatment can be arduous and painful both for healthcare professionals who treat these patients and for the families of the patients. The prolonged survival of patients in a VS or MCS has raised medical, ethical, and public policy concerns around the world. At the core of the medical and ethical issues is the willingness of the healthcare professional to remove treatment for a patient deemed unlikely to ever regain consciousness. A recent survey of almost 2500 European healthcare professionals is presented in Box 13.1.

BOX 13.1

END-OF-LIFE DECISIONS IN DOC

The issue of how to deal with end-of-life situations has been hotly debated across medical, ethical, and public policy grounds. The technological improvements of intensive care facilities have led to the survival of patients who would have died from their injuries previously. These improvements have also allowed pervasively VS or MCS patients to survive for long periods of time. The ethics of whether these artificial methods should be used to extend a life in an unconscious and unresponsive state is a core issue in medical science.

Demertzi et al. (1) conducted a large survey of almost 2500 European healthcare professionals to test the current status of their support for treatment withdrawal (the removal of the artificial life support systems) for VS and MCS patients. The demographics of the survey are presented in Table 13.2. The survey group was almost evenly balanced by gender, with 53% versus 44% proportion of

TABLE 13.2 Demographic Characteristics of the Studied Sample ($n = 2475$)

Age (Years), Mean ± SD (Range)	39 ± 14 (18–88)
GENDER, NO. (%)	
Women	1314 (53%)
Men	1098 (44%)
Missing Data	63 (3%)
RESPONDERS BY EUROPEAN REGION, NO. (%)	
Northern	402 (16%)
Central	1213 (49%)
South	855 (35%)
Missing Data	5 (0%)
PROFESSION, NO. (%)	
Medical professionals	1608 (65%)
Paramedical professionals	651 (26%)
Missing data	216 (9%)
RELIGIOSITY, NO. (%)	
Religious responders	1407 (57%)
Nonreligious responders	1004 (40%)
Missing Data	64 (3%)

Source: Giacino et al. (2014).

BOX 13.1 *(cont'd)*

women versus men. The age of the majority of the survey group was ~35–55 years, although the full range went from 18 to 88 years. Most (49%) survey responders were from Central Europe, many (35%) were from Southern Europe, and a minority (just 16%) were from Northern Europe. Because religiousness may have an influence over end-of-life decision-making, religiosity was also collected, with more than half (57%) of the survey group being religious respondents and less than half (40%) being nonreligious respondents.

Results are presented are presented in Figs. 13.10 and 13.11. For the entire survey group (Fig. 13.10), 66% agreed that it was acceptable to withdraw treatment in chronic VS (>1 year) patients (left side of figure, white-shaded bar) and 28% agreed that it was acceptable for MCS patients (left side of figure, black-shaded bar). Turning the question around to determine of the survey respondent was a patient, would he/she like to be kept alive if in a chronic unconscious state, only 18% responded that they would want to be kept alive in a VS and 33% responded that they would want to be kept alive in an MCS state.

There were regional differences in the respondents' attitudes, with a higher percentage of Northern Europeans responding that it was acceptable to withdraw treatment in chronic VS (Fig. 13.11, top panel, white-shaded bar), fewer Central Europeans responding that it was acceptable to withdraw treatment in chronic VS (Fig. 13.11, top panel, grey-shaded bar), and fewer still Southern Europeans

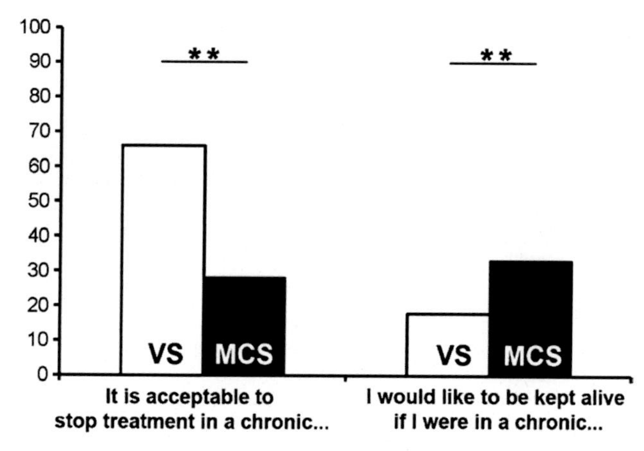

FIGURE 13.10 For the entire survey group, 66% agreed that it was acceptable to withdraw treatment in chronic vegetative state (VS) (>1 year) patients (left side of figure, *white-shaded bar*) and 28% agreed that it was acceptable for a minimally conscious state (MCS) patients (left side of figure, *black-shaded bar*). Turning the question around to determine if the survey respondent was a patient, would he/she like to be kept alive if in a chronic unconscious state, only 18% responded that they would want to be kept alive in a VS and 33% responded that they would want to be kept alive in an MCS state.

Continued

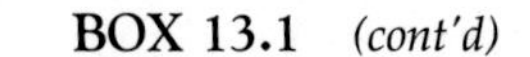

BOX 13.1 *(cont'd)*

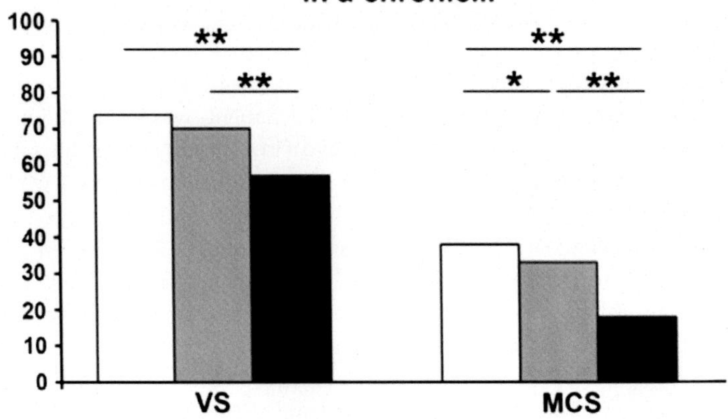

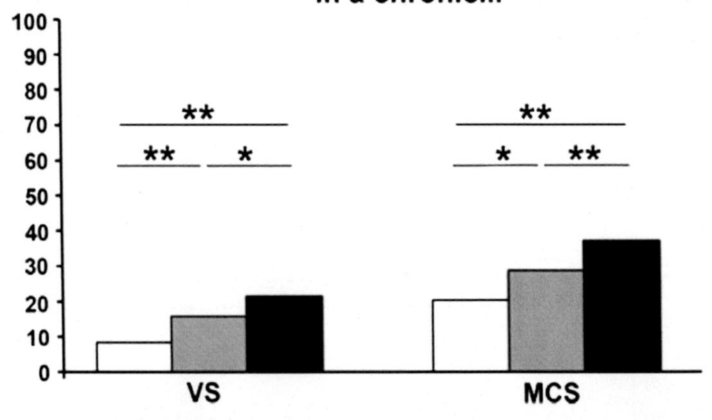

FIGURE 13.11 There were regional differences in the respondents' attitudes, with a higher percentage of Northern Europeans responding that it was acceptable to withdraw treatment in chronic vegetative state (VS) (top panel, *white-shaded bar*), fewer Central Europeans responding that it was acceptable to withdraw treatment in chronic VS (top panel, *grey-shaded bar*), and fewer still Southern Europeans (top panel, *black-shaded bar*). Similarly, fewer Northern European respondents would, themselves, like to be kept alive in chronic VS (bottom panel, *white-shaded bar*), somewhat more Central Europeans responding that they would like to be kept alive (bottom panel, *grey-shaded bar*), and somewhat more Southern Europeans responding that they would like to be kept alive (bottom panel, *black-shaded bar*). A similar pattern was observed for a minimally conscious state (MCS).

BOX 13.1 *(cont'd)*

(Fig. 13.11, top panel, black-shaded bar). Similarly, fewer Northern European respondents would, themselves, like to be kept alive in chronic VS (Fig. 13.11, bottom panel, white-shaded bar), somewhat more Central Europeans responding that they would like to be kept alive (Fig. 13.11, bottom panel, grey-shaded bar), and somewhat more Southern Europeans responding that they would like to be kept alive (Fig. 13.11, bottom panel, black-shaded bar). A similar pattern was observed for MCS.

These survey results provide evidence that, although a majority of healthcare professionals agreed to withdraw treatment for chronic VS and MCS patients, there were still many responders who would not. Clearly the field remains divided on end-of-life issues, both for their patients and for themselves if they were in a chronic unconscious state.

Reference

(1) Demertzi, A., Ledoux, D., Bruno, M.-A., Vanhaudenhuyse, A., Gosseries, O., Soddu, A., Schnakers, C., Moonen, G., & Laureys, S. (2011). Attitudes towards end-of-life issues in disorders of consciousness: a European survey. *Journal of Neurology, 258*, 1058–1065.

5. SUMMARY

The study of human consciousness and unconsciousness continues to challenge scientists, philosophers, and all individuals curious about the human condition. The prolonged lack of consciousness observed in individuals with DOC raises medical, ethical, philosophical, and public policy concerns about the proper treatment of an individual who is not conscious and aware of their surroundings. These issues are very close to a debate of what constitutes human life. Is it being conscious? Or is having a heart that beats and lungs that breathe sufficient to be considered alive?

These debates may not be resolved in any time soon; however, the advent of new and inventive methods to unveil just what a VS patient may be experiencing and to reveal their chances of recovery may serve to set the debate aside due to our advanced knowledge.

6. STUDY QUESTIONS

1. How are brain processes during SWS and anesthesia similar? How do they differ?
2. If a patient is in a coma, what are possible transitions that patient may make in the future?
3. How is a LI syndrome patient different from a minimally conscious patient? How are they alike?
4. What part(s) of the brain are the most highly active from a metabolic perspective?

CHAPTER

14

Growing Up

A CHILD IS BORN

Life is a flame that is always burning itself out, but it catches fire again every time a child is born.
George Bernard Shaw, Irish Playwright.

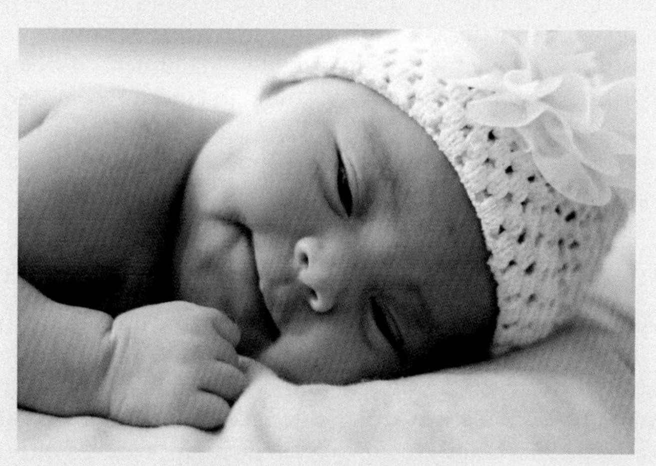

Baby girl. *Source: Public Domain.*

1. INTRODUCTION

In this chapter, we provide an overview of how humans grow and develop across multiple stages of life: from prenatal to infancy, from child to adolescent. Much of our focus will be on early stages of brain and cognitive development because the first years of human life represent a dramatic explosion of neurodevelopmental change as babies learn about their world. We will explore the roles of nature and nurture in the development of the brain and mind, discovering the intricacies of the complex interactions between genetics and experience.

Fundamentals of Cognitive Neuroscience
https://doi.org/10.1016/B978-0-12-803813-0.00014-3

The field of developmental cognitive neuroscience is a relatively young one. The advent of new noninvasive ways to measure brain function in infants and children has literally revolutionized the study of what infants and young children understand about the world surrounding them. These techniques provide a safe way to look at brain changes and development during the early years of life when the brain is changing dynamically.

1.1 New Techniques for Investigating the Developing Brain

Two techniques that have been employed in studies of infants and young children are electroencephalography/event-related potentials (EEG/ERPs) and functional magnetic resonance imaging (fMRI) (Fig. 14.1). While these techniques have revolutionized the field of cognitive neuroscience, nowhere is the effect felt as strongly as in the study of the unfolding of human brain development and its correspondence to behavior. Studies of adult behavior and brain function have informed us about how the typically developing brain functions across domains such as language, emotion, and memory. They also inform us about the effects of brain damage or disease. However, the pattern of deficit found in adults following brain damage differs sharply from the effects when brain damage occurs early in life. Therefore the advent of neuroimaging techniques allows us to understand the brain regions and cognitive capabilities across cognitive domains while it is unfolding in development.

New and sophisticated methods to investigate anatomical developmental changes throughout life have also increased our ability to understand the complex patterns of brain development (Fig. 14.2). These methods allow us to track the development of gray matter (GM) across brain regions, as well as to assess connectivity patterns across and between the cerebral hemispheres.

1.2 The Mystery of the Developing Brain: Old Questions and New

In this chapter, we address some brain questions that have been asked for many years. A central question in human development is the trading roles of nature and nurture. A related issue is to what extent the brain is flexible in adapting to new situations in its environment and to recover from damage. Some new questions can be posed that we were previously unable to address due to the limitation of our experimental approaches or techniques, such as what does a baby know before birth? What are the long-standing effects of very early brain damage? How do dynamic processes in brain development differ across brain regions and hemispheres? We will discuss advances in our knowledge about the developmental pathways of three main areas of cognition that have been a focus in the field: language, executive function, and social cognition.

2. PRENATAL DEVELOPMENT: FROM BLASTOCYST TO BABY

Much of this chapter is devoted to a discussion of brain development and its correspondence to cognition during infancy and childhood. Before we begin that discussion, we provide a brief description of the processes that occur before birth, during *prenatal* development. While little is known about the sensory, perceptual, or cognitive processes of

Inexpensive

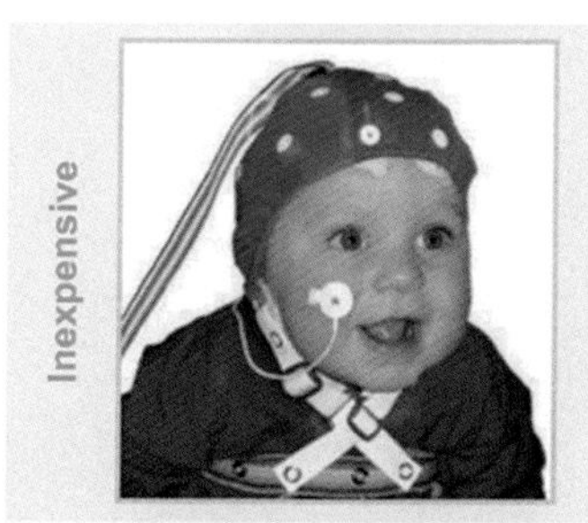

EEG/ERP: Electrical potential changes

- Excellent temporal resolution
- Studies cover the life span
- Sensitive to movement
- Noiseless

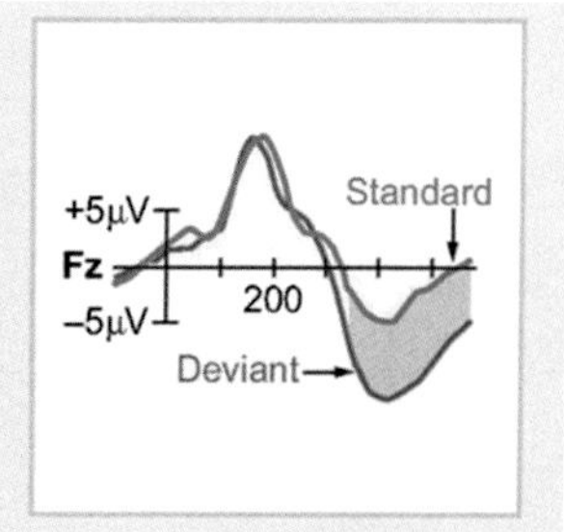

Expensive

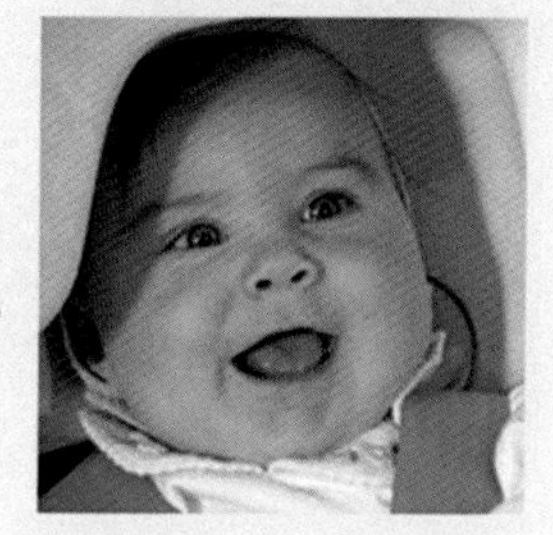

MEG: Magnetic field changes

- Excellent temporal and spatial resolution
- Studies on adults and young children
- Head tracking for movement calibration
- Noiseless

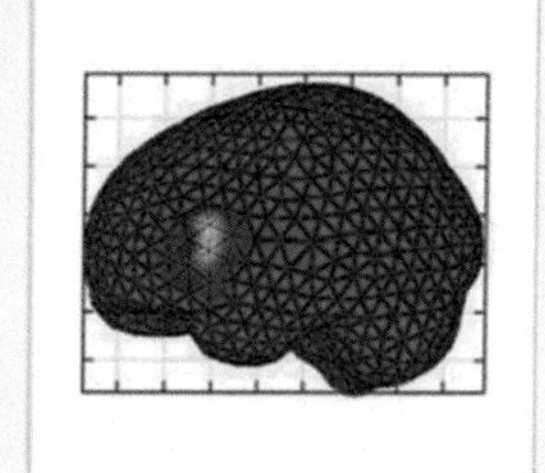

Expensive

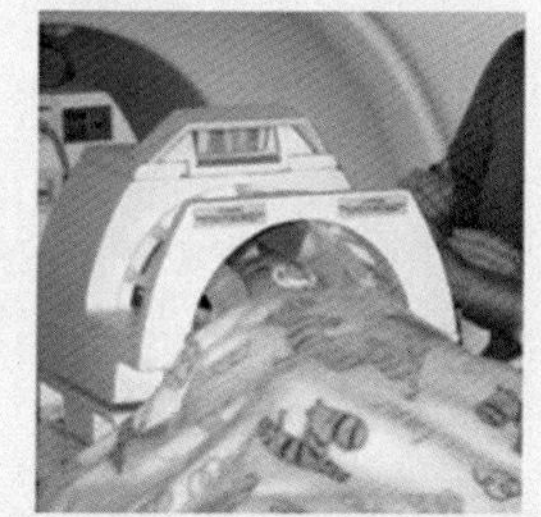

fMRI: Hemodynamic changes

- Excellent spatial resolution
- Studies on adults and a few on infants
- Extremely sensitive to movement
- Noise protectors needed

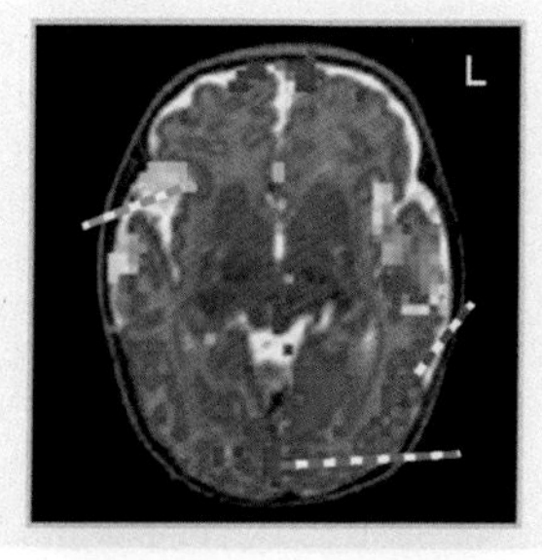

Moderate

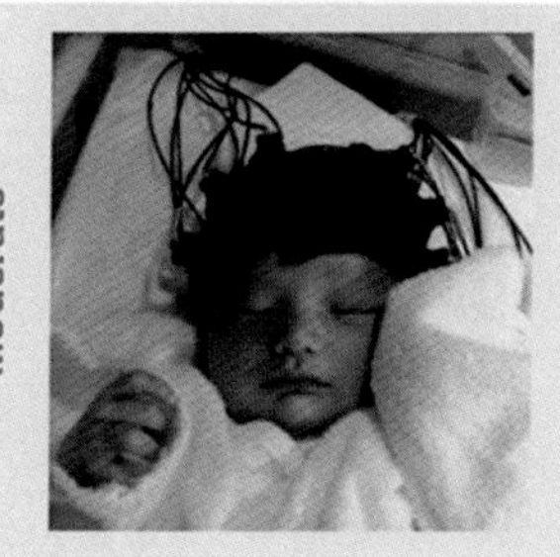

NIRS: Hemodynamic changes

- Good spatial resolution
- Studies on infants in the first 2 years
- Sensitive to movement
- Noiseless

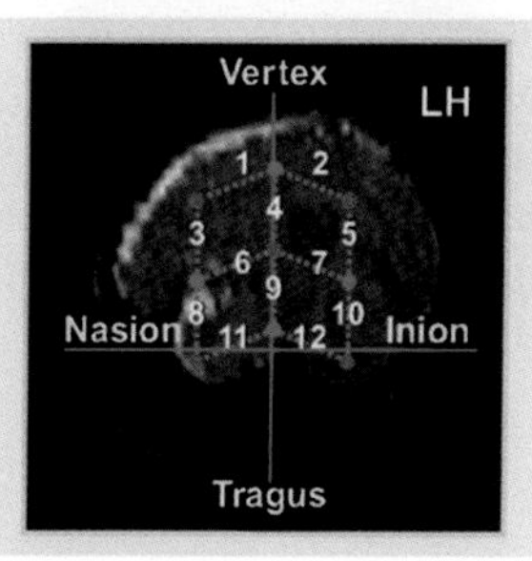

FIGURE 14.1 Shown are four neuroimaging techniques used with infants and young children. Top panel: electroencephalography/event-related potential (EEG/ERP) measures of electrical potentials. Second panel: magnetoencephalography(MEG) measures of magnetic fields. Third panel: functional magnetic resonance imaging (fMRI) measures of hemodynamic changes. Bottom panel: near-infrared spectroscopy (NIRS) measures of near-infrared light changes. *Source: Kuhl and Rivera-Gaxiola (2008), with permission.*

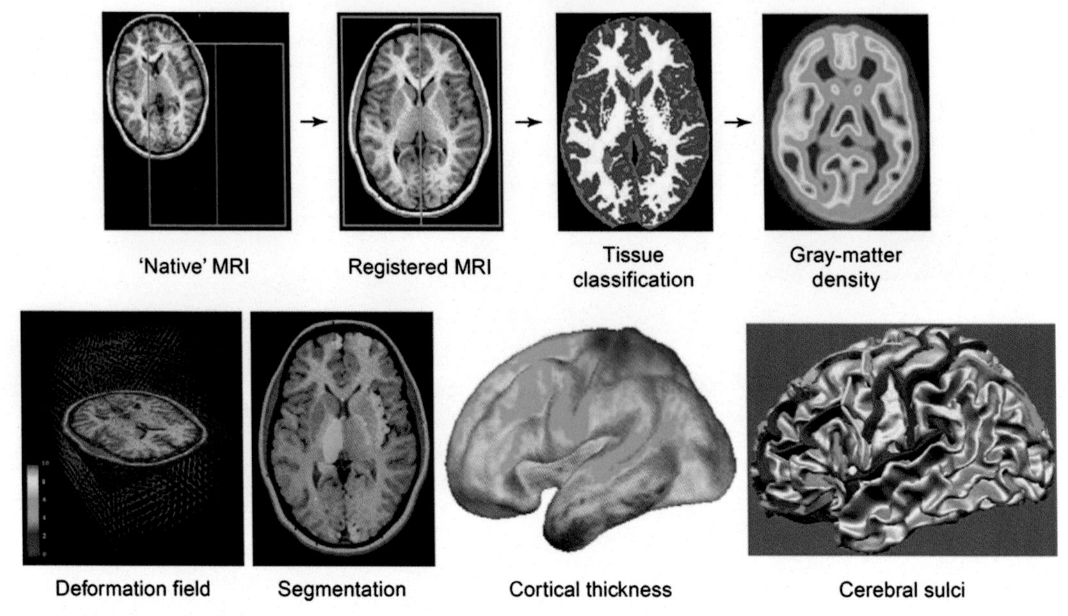

FIGURE 14.2 Image processing pipeline. *Top row*: A typical image processing pipeline begins with a transformation of an MRI image; this process generates an image that is "registered" with the template brain. The next step involves voxel-wise classification of brain tissue into gray matter (GM) (in red), white matter (in white), and cerebrospinal fluid (in green). Each image is filtered to generate "density" images (the "hotter" the color, from blue to red, the higher the GM density). *Bottom row*: By combining nonlinear registration with tissue classification, one can segment various brain structures, such as the frontal lobe or the amygdala. Other techniques produce maps of cortical thickness or identify sulci. *Source: Paus, 2005.*

a fetus in utero, recent investigations have focused on what a baby experiences before birth. These prebirth experiences can be critical for later development. And whether they are positive—hearing a mother's voice or her heartbeat—or negative—experiencing the effects of maternal alcohol abuse—these prenatal experiences can have long-standing effects on later cognitive and social development. Let us begin our prenatal section with a discussion on gene expression and the role of the environment.

2.1 Epigenesis

A central debate in the field of human development is the influence of nature versus nurture. Does our genetic makeup predetermine who we will become? Or does our experience shape who we are? Clearly, both genes and the environment have an impact on the developing human. Does gene expression unfold, followed by the development of brain structures and functions that later are affected by experience? Or does experience—the local environment, whether within a cell, a system, or the brain in toto—have an effect on gene expression? The interplay between genes and the environment is a complex one, with these interactive processes occurring long before birth. Here, we begin the topic of the cognitive neuroscience of human development with a discussion of the nature of epigenesis.

Epigenesis—the unfolding of genetic information within a specific context—is key to modern ideas about development. Different viewpoints on epigenesis underlie different perspectives on developmental cognitive neuroscience. Gottlieb and Halpern (2002) have drawn a useful distinction between "predetermined epigenesis" and "probabilistic epigenesist." *Predetermined epigenesis* assumes that there is a unidirectional causal pathway from genes to brain anatomy to changes in the brain and cognitive function. A hypothetical example of this would be if the endogenous expression of a gene in the brain generated more of a certain neurochemical. Higher levels of this neurochemical might make a particular neural circuit active in the brain, and this additional neural activity allows for more complex computations than were previously possible. This increased cognitive ability will be evident in behavior as the child is able to pass a task that he or she failed at younger ages. In contrast, *probabilistic epigenesis* views the interactions among genes, structural brain changes, and function as bidirectional (Fig. 14.3).

Bidirectional interactions mean that not only can genes trigger a behavioral change but also that sensory input to the child can change patterns of gene expression. For example, we will hear later that newborn infants have primitive brain circuits that bias them to look toward the faces of other humans (Johnson, 1991). This early attention to faces results in some of the neural circuits involved in the visual pathways of the baby becoming shaped or tuned to process faces. The neuroanatomical changes that underlie this shaping process are due to differential gene expression.

2.2 The Anatomy of Brain Development

Much of early brain development occurs in the first weeks following fertilization. Shortly after conception, a fertilized cell undergoes a rapid process of cell division, resulting in a cluster of proliferating cells (called the *blastocyst*) that resembles a tiny bunch of grapes (Fig. 14.4). After a few days, the blastocyst differentiates into a three-layered structure (the embryonic disk). Each of these layers will subsequently differentiate into a major organic system, with the *endoderm* (inner layer) becoming internal organs (digestive, respiratory, etc.), the *mesoderm* (middle layer) becoming skeletal and muscular structures, and the *ectoderm* (outer layer) developing into the skin surface and the nervous system (including the perceptual organs).

The nervous system itself begins with a process known as *neurulation*. A portion of the ectoderm begins to fold in on itself to form a hollow cylinder called the *neural tube* (Fig. 14.5). The neural tube differentiates along three dimensions: length, circumference,

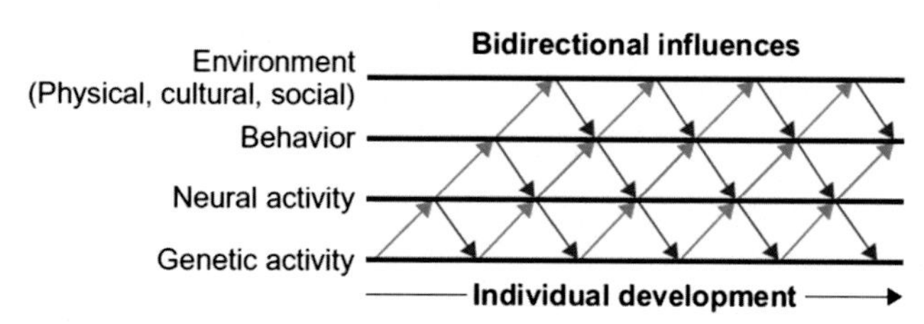

FIGURE 14.3 A systems view of psychobiological development. *Source: Adapted from Gottlieb and Halpern (2002).*

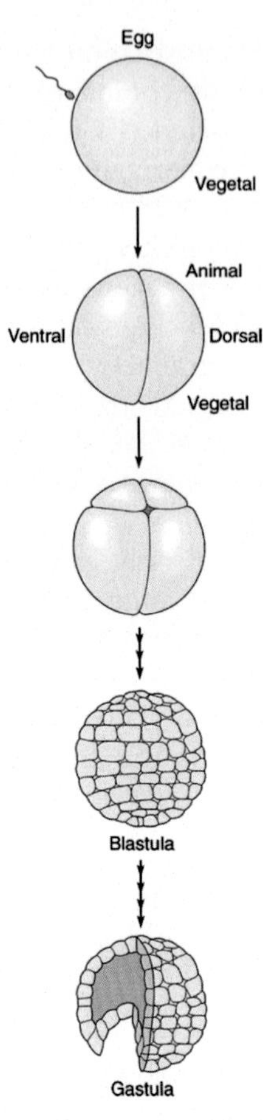

FIGURE 14.4 Blastocyst development. The early processes of animal development follow a conserved pattern; after fertilization, a series of cleavage divisions divide the egg into a multicellular blastula. The process of gastrulation brings some of the cells from the surface of the embryo to the inside and generates the three-layered structure common to most multicellular animals. *Source: Sanes, Reh, and Harris (2006).*

and radius. The length dimension differentiates into components of the central nervous system, with the forebrain and midbrain arising at one end and the spinal cord at the other. The end of the tube that will become the spinal cord differentiates into a series of repeated units or segments, while the other end of the neural tube organizes and forms a series of bulges and convolutions. Five weeks after conception, these bulges become protoforms for parts of the brain. One bulge gives rise to the cortex, a second becomes the thalamus and hypothalamus, a third turns into the midbrain, and others form the cerebellum and medulla.

The distinction between sensory and motor systems develops along the axis tangential to the surface of the neural tube, with the dorsal (top side) becoming mainly sensory cortex, and the ventral (bottom side) developing into motor cortex. The various association cortices and "higher" sensory and motor cortices tend to arise from the tissue between.

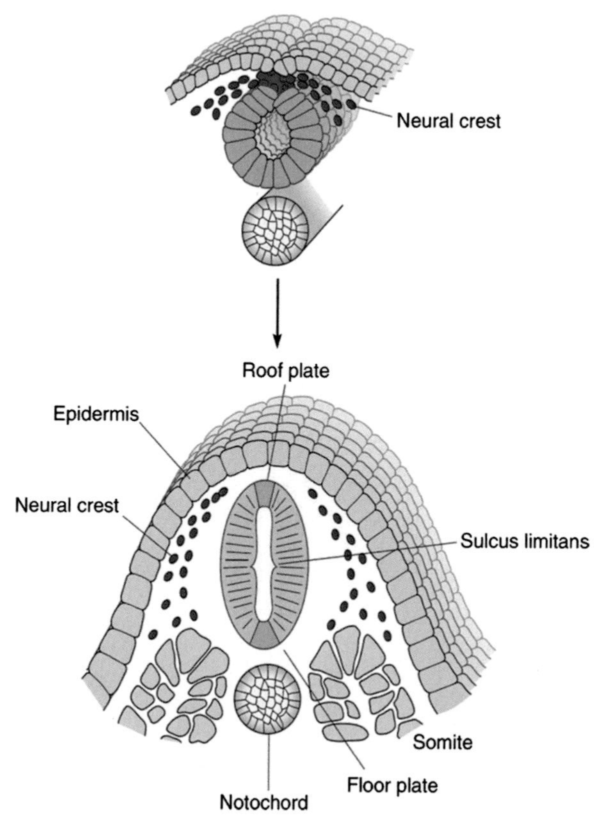

FIGURE 14.5 The neural tube. The overall organization of the neural tube emerges soon after closure. The most ventral (lowest) part of the neural tube becomes flattened into a distinct "floor plate." The most dorsal (highest) aspect develops into a tissue known as the roof plate. *Source: Sanes et al. (2006).*

2.3 Neural Migration

Most cortical neurons in humans are generated outside the cortex itself in a region just underneath what becomes the cortex: the "*proliferative zone*." After young neurons are born, they have to *migrate* from the proliferative zone to the particular region where they will be employed in the mature brain. The most common type of migration involves *passive cell displacement*. In this case, young cells are simply pushed farther away from the proliferative zone by more recently born cells. This gives rise to an outside-to-inside pattern, resulting in the oldest cells ending up toward the surface of the brain, while the youngest cells are toward the inside. This type of migration gives rise to brain structures such as the thalamus and many regions of the brainstem.

The second form of migration is more active and involves the young cell moving past previously generated cells to create an "*inside-out*" *pattern*. This pattern is found in the cerebral cortex and in some subcortical areas that have a laminar structure (divided into parallel layers).

The best-studied example of active migration comes from the prenatal development of the cerebral cortex and the *radial unit model* proposed by Rakic (1988). According to his model, the laminar organization of the cerebral cortex is determined by the fact that each relevant proliferative unit gives rise to about 100 neurons. A radial glial fiber is a long process that stretches from top to bottom of the cortex and originates from a glial cell. Radial glial fibers effectively act like a climbing rope to ensure that neurons produced by one proliferative unit all contribute to one radial column within the cortex. The progeny ("children") from each of the proliferative units all migrate up the same radial glial fiber, with the latest to be born traveling past their older relatives (Fig. 14.6).

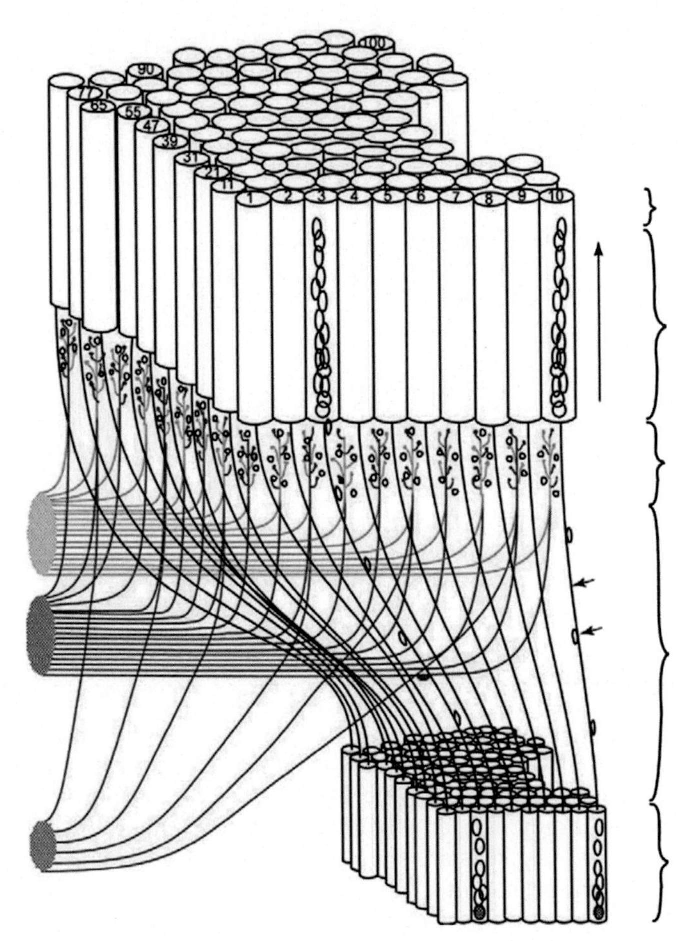

FIGURE 14.6 The radial unit model of Rakic (1988). Note how the radial glial fibers extend from the ventricular zone (VZ, bottom) to the cortical plate (CP) via a number of regions: the intermediate zone (IZ) and the subplate zone (SP). RG indicates a radial glial fiber and MN a migrating neuron. Each MN traverses the IZ and SP zones that contain waiting terminals from the thalamic radiation (TR) and cortico-cortico afferents (CC). After entering the CP, the neurons migrate past their predecessors to the marginal zone (MZ). *Source: Rakic (2009).*

Thus in the early weeks of gestation, the embryo undergoes complex processes that form the basis for the central nervous system. It is important to note that prenatal brain development is not a passive process involving the unfolding of genetic instructions. Rather, from an early stage, *interactions* between cells are critical, including the transmission of electrical signals between neurons. Thus waves of firing intrinsic to the developing organism may play an important role in specifying aspects of brain structure long before sensory inputs from the external world have any effect.

2.4 The Unborn Baby: Prenatal Experience Causes Lifelong Effects

The role of the prenatal environment in brain development can have long-lasting effects, both good and bad. The developing infant is susceptible to events occurring within this environment. One such event is the incursion of a *teratogen*, which we will discuss shortly. However, we are discovering more and more about the positive aspects of prenatal experience and the effect of that experience on the developing infant.

2.4.1 Prenatal Hearing Experience: Voice and Music Perception Before Birth

What do babies know before they are born? Is it important for a mother-to-be to talk to her unborn baby? Read to her baby? Sing to her baby? Is there an impact on later language, music, and cognitive function? In other words, what are the perceptual abilities of an unborn child, and how do they relate to later cognitive development? This question has intrigued developmental psychologists for at least a 100 years (see Kisilevsky & Low, 1998; for a review), but systematic investigations of fetal perception did not get under way until the 1980s. How do you measure a fetal response to sounds? Usually, the investigators measure heart rate changes and sometimes body movements in response to differing types of sounds. These early studies provided evidence that by approximately 30 weeks' gestational age, a fetus hears and responds to simple sounds such as bursts of white noise. By 37–42 weeks, a fetus can discriminate between types of speech sounds (such as vowels, consonant–vowel syllables) (Groome et al., 1999; Lecanuet et al., 1987).

The finding that a fetus can both hear and discriminate between sounds before birth has led to investigations of what a fetus knows about specific sounds—namely, his or her own mother's voice. DeCasper and colleagues studied the listening preferences of newborn infants in a series of investigations in the 1980s. They found that newborns prefer their mother's voice to that of a female stranger (DeCasper & Fifer, 1980). A more recent study used ERPs to compare brain responses in newborn (less than 1 day old) infants to presentations of their mother's voice versus a stranger's voice (Beauchemin et al., 2011). Results showed different brain patterns for the mother's voice, which activated left hemisphere language regions, as compared to the stranger's voice, which activated right-hemisphere voice processing regions (Fig. 14.7). As early as 100 ms (1/10th of a second) after the voice sounds are presented, newborns show a left-hemisphere response for their mother's voice and a differing response—in the right hemisphere—when the voice is of a stranger. These findings provide evidence that a mother's voice is preferentially processed, even immediately after birth, reflecting prenatal experience with the mother's voice.

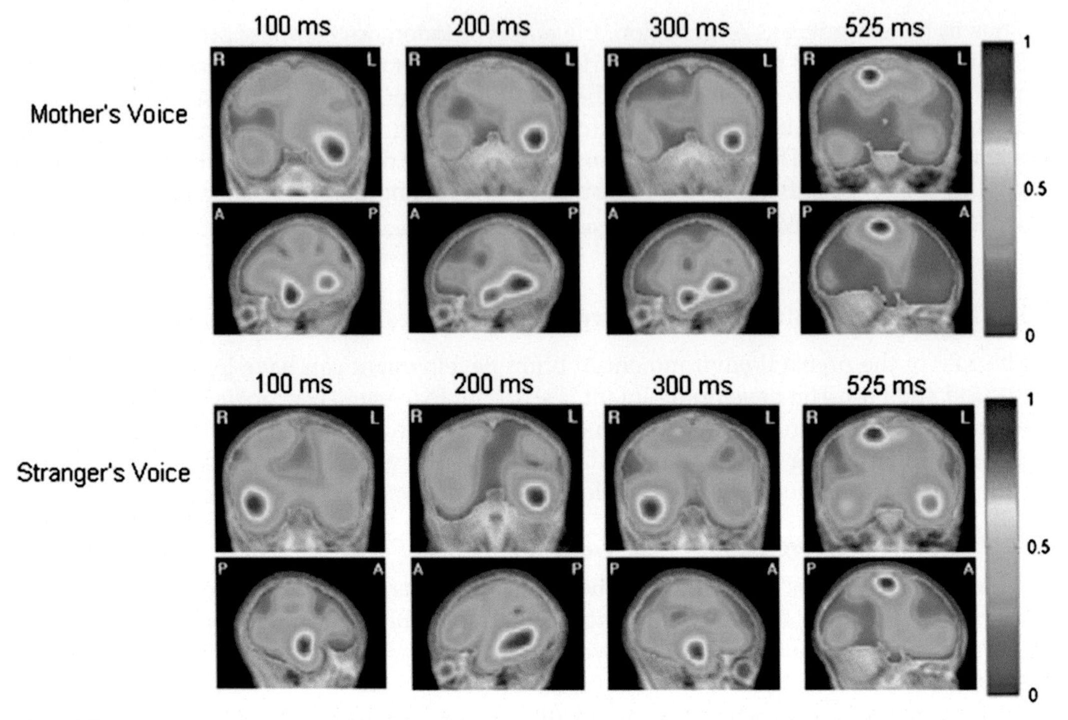

FIGURE 14.7 Brain responses from newborn infants for their mother's voice (top panels) and a stranger's voice (lower panel). Note that as early as 100 ms (1/10th of a second) after the voice sounds are presented, newborns show a left hemisphere response for their mother's voice and a different response—in the right hemisphere—when the voice is of a stranger. These findings provide evidence that a mother's voice is preferentially processed, even immediately after birth, reflecting prenatal experience with the mother's voice. *Source: Beauchemin et al. (2011).*

2.4.2 Teratogens and Their Lasting Effects: Alcohol

The role of the prenatal environment in brain development can have long-lasting effects, both good and bad. The developing infant is susceptible to events occurring within this environment. One such event is the incursion of a *teratogen*. A teratogen is defined as *any environmental agent that causes damage during the prenatal period*. Examples of teratogens are prescription and even nonprescription drugs, caffeine found in coffee and soft drinks, illegal drugs such as cocaine and heroin, tobacco and marijuana products, and alcohol. The effects of the teratogen(s) can be complex depending on the dosage level, the time it occurs during prenatal development, and the genetic makeup of the mother, since some individuals are more susceptible than others.

The prenatal brain is particularly susceptible to the effects of alcohol. Alcohol abuse by the mother during pregnancy results in long-term deficits in cognition, language, and social development called *fetal alcohol syndrome* (FAS) (Jones & Smith, 1973). A recent brain mapping study of children, teenagers, and young adults with severe FAS showed GM density differences in FAS individuals as compared to age- and gender-matched controls. Specific findings

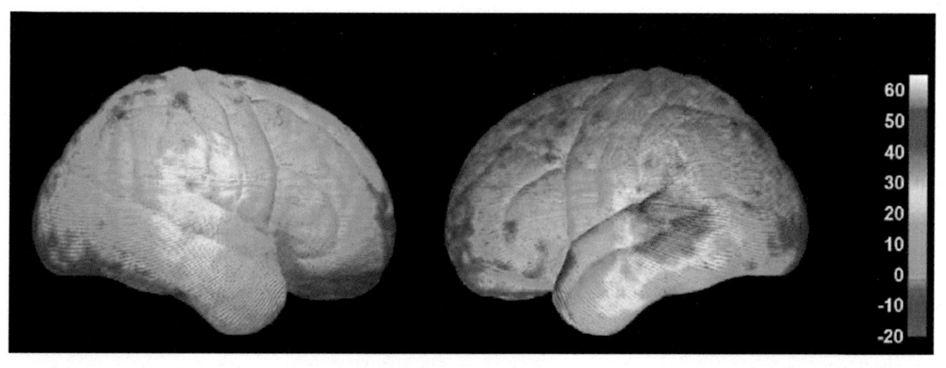

FIGURE 14.8 Differences in gray matter (GM) density between children with fetal alcohol syndrome (FAS) and typically developing controls. Warm (red, yellow) colors represent positive differences, indicating an increase in GM density (and thus a decrease in normal "pruning") in those regions as compared to controls. Note that the children with FAS have much increased densities in temporal lobe regions, particularly in the left hemisphere (Adapted, with permission, from Sowell et al., 2002). *Source: Toga, Thompson, and Sowell (2006).*

were reduced GM density in frontal and parietal areas and increased density in temporal and inferior parietal lobe regions (Fig. 14.8). These brain areas mature throughout childhood and into late adolescence; therefore these GM density differences in FAS individuals indicate that prenatal exposure to alcohol has a resounding and long-lasting impact on brain development and cognitive development throughout the life span.

Another key impact of alcohol use by the mother on the prenatal brain is the *gyrification* of the brain during the early weeks of gestation. Weeks 20–35 of gestation is a critical period of brain development. During this time, cells are proliferating, migration and synaptogenesis occurs, and neuronal connections are blooming. The result of these processes is the gyrification or wrinkling of the cortex: the smooth cortex of the earlier weeks of gestation gradually becomes more and wrinkled due to the addition of cells and connections. In a large sample (175) of children 9–16 years with prenatal alcohol exposure and age-matched controls, cortical gyrification was assessed using MRI methods (Hendrickson et al., 2017). Results showed significantly reduced gyrification in the alcohol-exposed children (Fig. 14.9). Further, the downstream cognitive effects of lower gyrification were observed, with IQ corresponding with amount of smoothness in the cortex. That is, those individuals with less gyrification had lower IQs. Thus the effects of a mother's alcohol use during pregnancy are evidenced early in gestation and will continue throughout the life of the child.

2.4.3 *Teratogens and Their Lasting Effects: Marijuana and Cigarette Smoke Exposure*

Longitudinal studies assessing the long-term effects of smoking cigarettes or marijuana during pregnancy have provided new evidence about their impact on a child's cognitive development. Fried, Watkinson, and Gray (2003) have followed a cohort of children in Canada from birth through young adulthood. Using neuropsychological test batteries to assess cognitive functions such as verbal intelligence, visuospatial processing, language abilities, and attentional function, Fried and colleagues found that there are early occurring (by age 3) and long-lasting cognitive impairments caused by the mother smoking either tobacco or

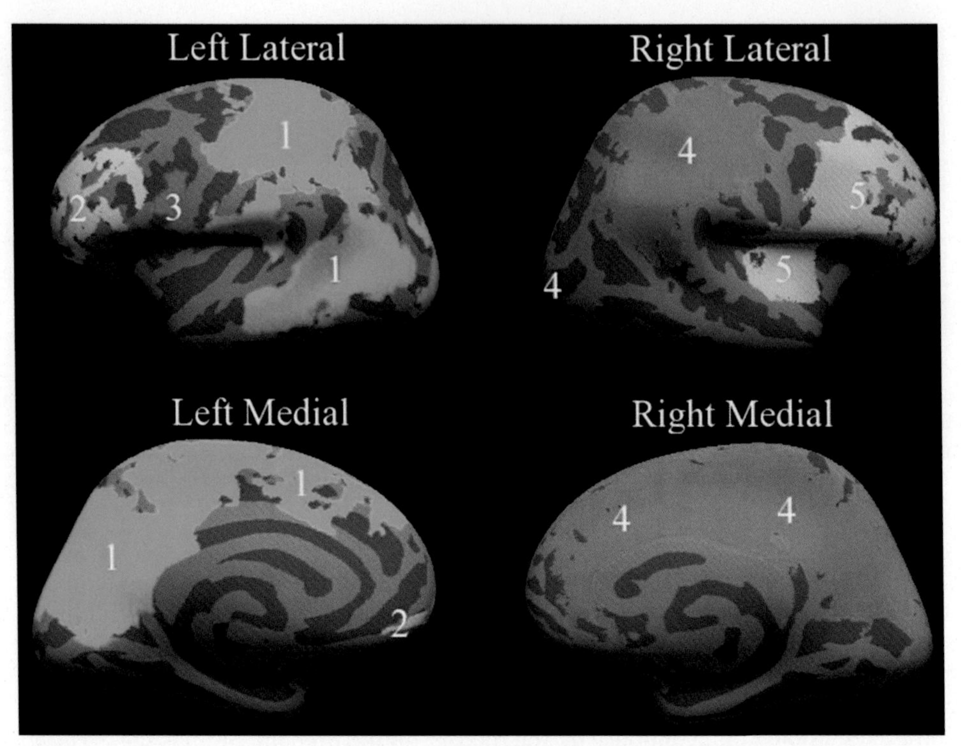

FIGURE 14.9 Cortical gyrification is affected by prenatal alcohol exposure. Using an inflated MRI technique, Hendrickson et al. (2017) evaluated gyral development in control children (top) and children who had been exposed to alcohol prenatally (bottom). The children with alcohol exposure had significantly less gyrification, reflecting abnormalities in early developing brain processes such as cell proliferation, synaptogenesis, myelination, and axon development. The numbers on the figure correspond to the cluster analyses that were used for data mapping. Cluster 1 (shown in green) includes left hemisphere postcentral brain regions, cluster 2 (shown in yellow) includes left hemisphere rostral and middle/frontal regions, cluster 3 (shown in dark green) includes left hemisphere precentral regions, cluster 4 (shown in orange) includes right hemisphere postcentral regions, and cluster 5 (shown in tan) includes right hemisphere rostral and middle/frontal regions. *Source: Hendrickson et al. (2017).*

marijuana during pregnancy. The specific effects of prenatal exposure to cigarette smoke differ sharply from exposure to marijuana, although both cause harm. Exposure to cigarette smoke resulted in lower general intelligence in the children, coupled with deficits in auditory function and verbal working memory that continued from early childhood (age 3) through adolescence (age 16) (Fried et al., 2003). Exposure to marijuana smoke resulted in no general intelligence deficit; however, executive functions, such as attention and working memory, were impaired in these children and, in particular, visual processes such as visual integration, analysis, and reasoning. Again, these impairments were observed early (age 3) and continued through the teenage years.

The cognitive impairments reflected by the neuropsychological test batteries implicated specific brain regions for further study of the effects of prenatal exposure to cigarette and marijuana smoke. Visuospatial integrative processes tap frontal lobe regions in adults.

2.4.3.1 PRENATAL MARIJUANA SMOKE EXPOSURE PRODUCES LONG-LASTING BRAIN CHANGES

Fried and colleagues (Smith, Fried, Hogan, & Cameron, 2006) continued their investigation into the long-term effects of prenatal exposure to marijuana using a neuroimaging technique, fMRI, with a sample of young adults (18–22 years) from the Canadian cohort with prenatal exposure to marijuana smoke. Results of the study revealed a differing pattern of neural activity in frontal lobe regions that are engaged in a visuospatial task. Specific findings were reduced activity in right hemisphere regions (Fig. 14.10, upper panel) and increased activity in left hemisphere regions (Fig. 14.10, lower panel).

The authors interpreted these findings as indicating that right hemisphere neural circuitry engaged in tasks that tap visuospatial short-term memory are less active in children with prenatal marijuana smoke exposure. While it is difficult to know at this stage of the research just why left hemisphere activity was increased, it could be that left hemisphere regions were recruited as a compensatory mechanism due to the decreased neural activation in the right hemisphere. The work of Fried and colleagues provide compelling evidence that the damage appears early in a child's cognitive development and is very long-lasting.

2.4.3.2 PRENATAL NICOTINE EXPOSURE PRODUCES LONG-LASTING BRAIN CHANGES

Longo, Fried, Cameron, and Smith (2013) explored the effects of prenatal exposure to nicotine within this same large cohort of young adults. Fried and colleagues used fMRI

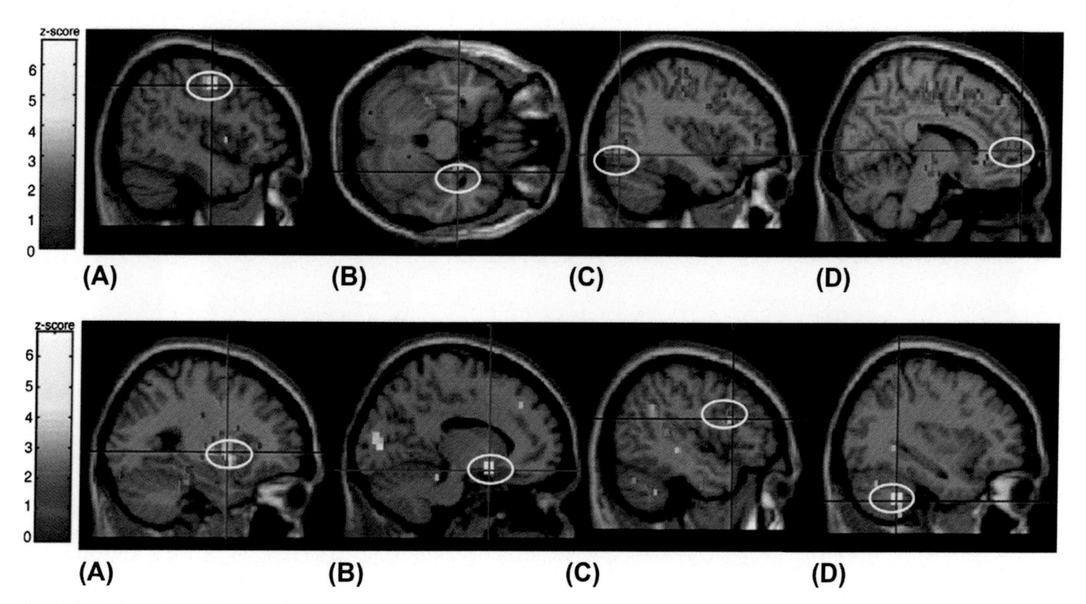

FIGURE 14.10 Effects of prenatal use of marijuana smoke on young adults (18–22 years) measured using functional magnetic resonance imaging. Frontal lobe circuits engaged in a task tapping visuospatial working memory systems, with reduced right hemisphere activity (upper panel) and increased left hemisphere activity (lower panel) as compared to an age-matched control group. *Source: Smith et al. (2006).*

and a *go-no go* task to assess response inhibition in these young adults and age-matched controls. The go-no go task included an array of letters, each containing the letter *x*, and two experimental conditions. In the *go* trials, the subject's task was to "press the letter *x*." In the *no go* trials, the subject's task was to "press all letters except for *x*." This go-no go task is thought to tap brain processes that underlie response inhibition. Although both groups performed similarly on the task, the nicotine-exposed group showed significantly greater activity in regions of the brain that subserve response inhibition. These regions included the inferior frontal gyrus, the inferior parietal lobe, the thalamus, the basal ganglia, and cerebellum (Fig. 14.11). The authors interpreted these findings to indicate that nicotine exposure during the prenatal period affected the developing of these brain regions critical for cognitive control functions, such as response inhibition. The authors further theorized that the abnormal development of these brain regions was compensated later in life by recruitment of greater neural resources within those regions to successfully perform a cognitive control task.

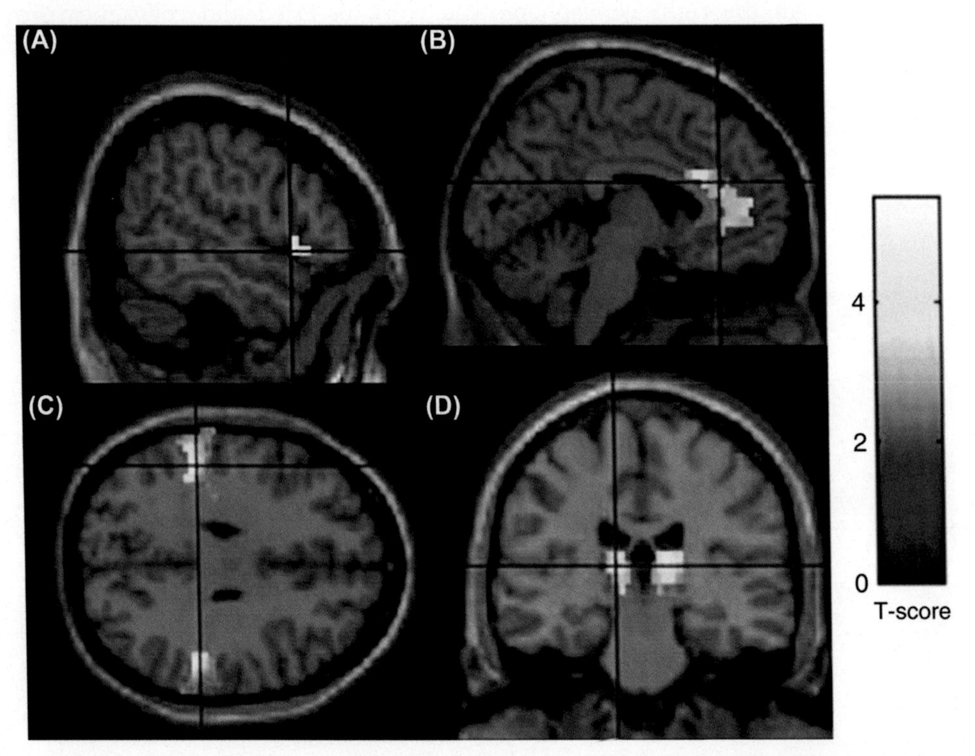

FIGURE 14.11 Fried and colleagues (Longo et al., 2013) explored the effects of prenatal exposure to nicotine using functional magnetic resonance imaging and a *go-no go* task to assess response inhibition in these young adults and age-matched controls. The difference in brain activation for the nicotine-exposed individuals versus the control group is shown in sagittal slices (A and B), an axial slice (C), and a coronal slice (D). Yellow and orange shading indicates high activation levels. Although both groups performed similarly on the task, the nicotine-exposed group showed significantly greater activity in regions of the brain that subserve response inhibition. These regions (shown in blue cross hairs) included (A) the right inferior frontal gyrus, (B) the left anterior cingulate, (C) the right and left parietal lobes, and (D) the right and left thalamus. *Source: Longo et al. (2013).*

Cumulatively, these studies demonstrate that prenatal exposure to marijuana smoke and nicotine causes early brain developmental abnormalities that alter cognitive function throughout life.

3. THE DEVELOPING BRAIN: A LIFETIME OF CHANGE

3.1 The Rise and Fall of Postnatal Brain Development

While the overall appearance of the newborn human brain is rather similar to that in adults, and most neurons have already reached their final locations, a number of substantive *additive* changes occur during postnatal development of the brain. Specifically, brain volume quadruples between birth and adulthood, an increase that comes from a number of sources, but generally not from additional neurons. The generation and migration of neurons takes place almost entirely within the period of prenatal development in the human.

Perhaps the most obvious change during postnatal neural development is the increase in size and complexity of the *dendritic trees* of most neurons. An example of the dramatic increase in dendritic tree extent during human postnatal development is shown in Fig. 14.12. While the extent and reach of a cell's dendritic arbor may increase dramatically, it also often becomes more specific and specialized.

In addition to the more extensive processes involved in the inputs and outputs of cells, there is a steady increase in the density of synapses in most regions of the cerebral cortex in humans (Huttenlocher, 1990, 1994; Huttenlocher, De Courten, Garey, & Van der Loos, 1982). The process of creating new synapses, *synaptogenesis*, begins approximately around the time of birth for all cortical areas studied to date, with the most increases, and the final peak density, occurring at different ages in different areas. For example, in the visual cortex there is rapid synaptogenesis at 3–4 months, and the maximum density of around 150% of that seen in adult humans is reached between 4 and 12 months.

Another additive process is *myelination*. Myelination refers to an increase in the fatty sheath surrounding neuronal processes and fibers that increases the efficiency of electrical

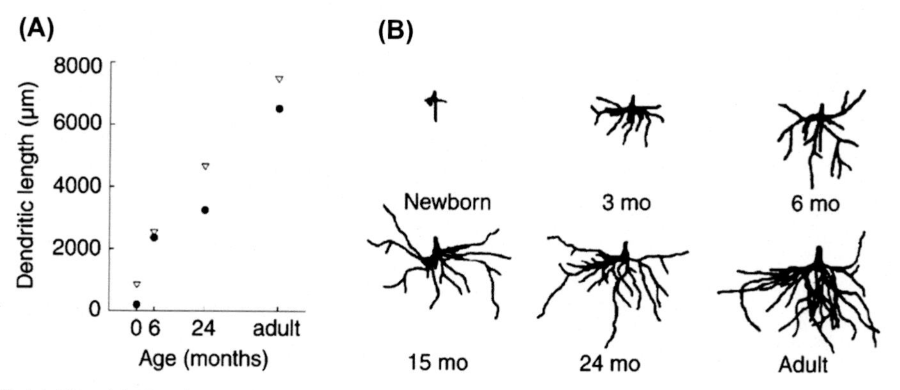

FIGURE 14.12 (A) Dendritic arborization. (B) A drawing of the cellular structure of the human visual cortex based on Golgi stain preparations from Conel (1939–67). *Source: Quartz (1999), from Conel (1951).*

transmission. This sheath is white, so myelinated fibers are referred to as *white matter*. Changes in the extent of white matter are of interest, since they presumably reflect interregional communication in the developing brain. While increases in white matter extend through adolescence into adulthood, particularly in frontal brain regions (Huttenlocher et al., 1982), the most rapid changes occur during the first 2 years. Myelination appears to begin at birth in the pons and cerebellar peduncles and, by 3 months, has extended to the optic radiation and splenium of the corpus callosum. Around 8–12 months, the white matter associated with the frontal, parietal, and occipital lobes becomes apparent (Fig. 14.13).

Surprisingly, human postnatal brain development also involves some significant *regressive events*. One quantitative neuroanatomical measure of a regressive event is the density of synapses, where there is a period of synaptic loss or *pruning* (Huttenlocher, 1990, 1994). Like the timing of bursts of synaptogenesis, and the subsequent peaks of density, the timing of the reduction in synaptic density appears to vary between cortical regions, at least in humans. For example, synaptic density in the visual cortex begins to return to adult levels after about 2 years, while the same point is not reached until adolescence for regions of the prefrontal

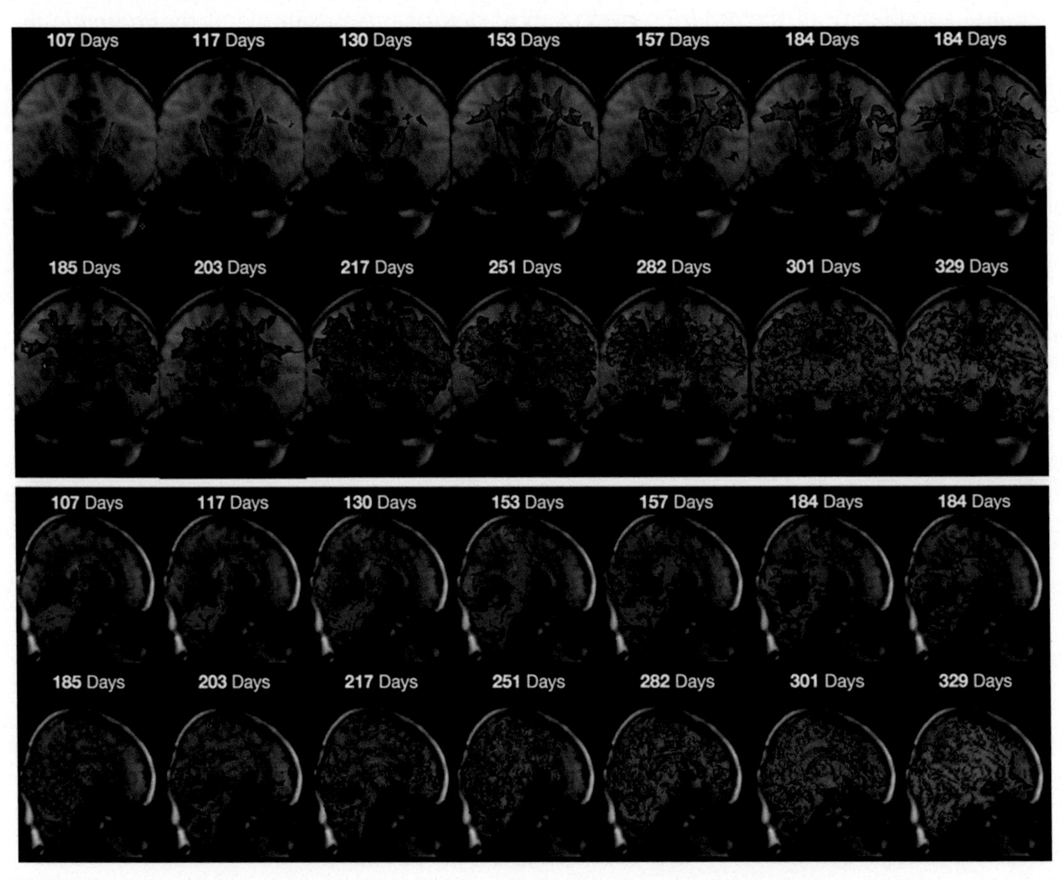

FIGURE 14.13 Myelin maturation in infants. Top panel shows a coronal slice and bottom panel shows a sagittal slice. The age of the infant is shown in days above the images. Note the rapid development of myelination during the first months of life (107 days, blue) to approximately 1 year (329 days, red). *Source: Deoni et al. (2011).*

cortex. Huttenlocher (1990, 1994) suggests that this initial overproduction of synapses may have an important role in the apparent plasticity of the young brain.

Thus the *rise and fall* developmental sequence is seen in a number of different measures of neuroanatomical and physiological development in the human cortex. However, we need to bear in mind that not all measures show this pattern (e.g., myelinization) and that measures of synaptic density are static snapshots of a dynamic process in which both additive and regressive processes are continually in progress.

3.2 Regional Differences in Brain Development

As compared to other species, humans take a very long time to develop into independent creatures. Human postnatal cortical development, for example, is extended roughly four times as long as nonhuman primates. The "down" side of this slow development is that there are many years during which a child is highly dependent on the care provided by family members. The "up" side of this protracted developmental timetable is that the human brain has far more opportunity for experience, and interactions with others, to shape and mold its development.

The rise and fall pattern of additive and regressive events occurs at different time frames in regions and lobes in the human brain. These events are heavily experience driven and reflect synapse formation and dendritic arborization that are due to cognitive and sensory development, learning, and integrative processes that occur throughout infancy and childhood and continue through the teen years. A time course of brain development from conception to late teens is presented in Fig. 14.14 (Casey, Tottenham, Liston, & Durston, 2005). Prenatal

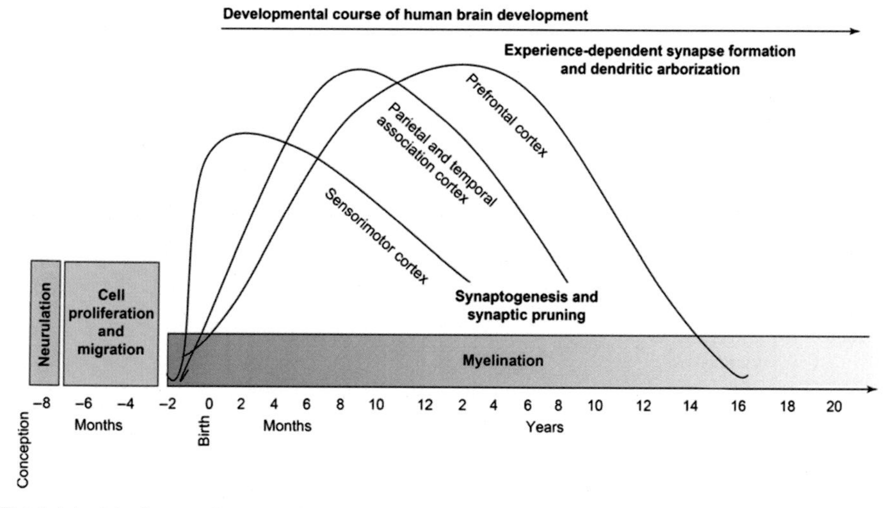

FIGURE 14.14 The human brain undergoes dramatic changes in both its structural architecture and functional organization that reflect a dynamic interplay of simultaneously occurring progressive and regressive events. Although the total brain size is about 90% of adult size by age 6, the brain continues to undergo dynamic changes throughout adolescence and well into adulthood. This figure illustrates some of these developmental changes, including proliferation and migration of cells mostly during fetal development, regional changes in synaptic density during postnatal development, and protracted development well into adulthood (Adapted from Thompson & Nelson, 2001). *Source: Casey et al. (2005).*

changes are shown on the left side of the figure and largely reflect neurulation, cell proliferation, and migration processes. Postnatal changes reflect developmental processes such as synaptogenesis and dendritic arborization. Sensory areas for processing visual and auditory information, for example, develop earlier than frontal lobe regions such as the prefrontal cortex (PFC) for processing executive functions (Casey et al., 2005).

While we are beginning to map out the developmental patterns for brain regions across the life span, a key issue that is being addressed is the notion of individual differences in how the brain matures and the correspondence to cognitive development. In other words, what is the relationship between brain and behavior? Do differing patterns of brain development reflect different levels of intellectual ability? These questions were addressed in a recent neuroimaging investigation of GM density in a large sample of more than 300 children and adolescents (Shaw et al., 2006). The sample was a priori divided into three groups based on their performance on an IQ battery of tests: "Superior" with IQ ranging from 121 to 149; "High" with IQ ranging from 109 to 120; and "Average" with IQ ranging from 83 to 108. The results indicated that there are, indeed, differing patterns of brain change corresponding to overall level of intelligence (Fig. 14.15). The notable finding was that high IQ was associated with thinner cortex, especially in frontal and temporal lobe areas, in early childhood. By late childhood, the opposite pattern was found, with high IQ associated with thicker cortex.

The important finding of this study was that there was a differing pattern of cortical development in frontal lobe regions for children in the "Superior" group as compared to either of the other groups. Specifically, there were differences in the dynamic rate of cortical thickening and thinning throughout early childhood and into adolescence and early adulthood. The authors concluded that differences in GM density in and of itself did not lead to children with superior intellectual abilities, rather, they suggested that the dynamic properties of development of the cortex corresponded to level of intelligence, perhaps enabling the child to extract more information from his environment. Questions have been raised by this study: Are dynamic changes in brain growth patterns due to genetic predispositions in the

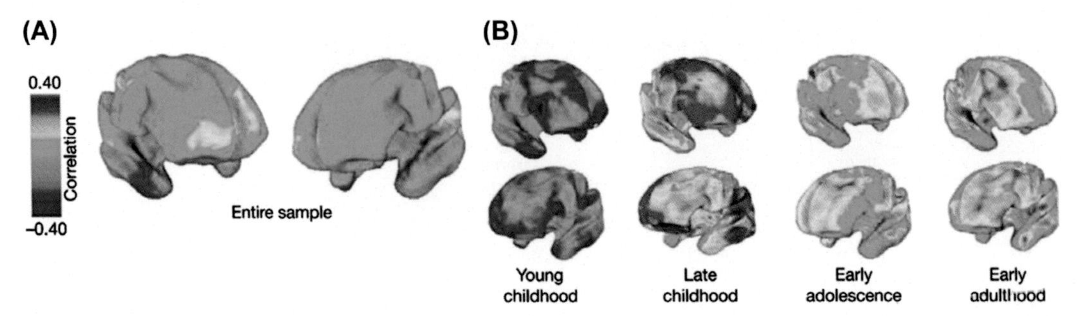

FIGURE 14.15 Correlations between IQ and cortical thickness. (A) correlations for all 307 subjects were generally positive and modest (*P*. 05), with *r* between 0 and 0.10 (green/yellow), except in the anterior temporal cortex (which showed a negative correlation, with *r* between 0 and 21; blue/purple). (B) Correlations in different age groups showed that negative correlations were present in the youngest group, indicating that higher IQ was associated with a thinner cortex, particularly in frontal and temporal regions. The relationship reverses in late childhood, with most of the cerebral cortex correlating positively with IQ. *Source: Shaw et al. (2006).*

"Superior" group? Or do they reflect differences in the environment? Or do they reflect a combination of both genetic influences and experience?

These findings from recent studies of brain development are helping us prepare new models for human cognitive growth and the correspondence to cognition. These studies are still very early work in an ongoing series of investigations of complex brain developmental processes.

4. DEVELOPING MIND AND BRAIN

We have seen that the brain undergoes significant developmental changes throughout childhood. An important question to address in the field of developmental cognitive neuroscience is how these brain changes reflect development of cognitive processes. New techniques are allowing us to begin to map brain development and the correspondence to cognition.

Once again, the question of the role of nature versus nurture is debated in the field. Is the human brain prewired for language? Face perception? Or are these processes based on experience throughout life? Since these topics are of central interest in the study of human development, we will focus on three general areas of developmental cognitive neuroscience: the emergence of language; the development of cognitive control mechanisms; and the development of social cognition, with a specific focus on face perception. In the following section, we provide a brief summary of research to date on these topics in infants during the first year of life. Next, we present findings on these topics for older children and adolescents. Last, we review the effects of early (perinatal) brain damage on these systems.

4.1 The First Year of Life: An Explosion of Growth and Development

The human brain increases fourfold in size from newborn to adult. Many of the dynamic changes that occur in development happen during the first year of life. During this 12-month span, an infant develops from a tiny creature with few voluntary movements to a busy toddler, smiling, reaching for attractive objects, producing many speech sounds, crawling, and even walking as he or she explores the world.

Massive changes are occurring in the brain during these early months of life. New neuroimaging techniques and approaches have provided researchers with new tools to quantify these developmental brain changes. The proliferation of cells and their connections along with maturational processes such as myelination and synaptogenesis causes the initially smooth cortex to become wrinkled—gyrificated—during the early months of life (Fig. 14.16A). GM matter matures rapidly during these early months, reflecting these early brain development processes (Fig. 14.16B). White matter fiber tracts are also exploding in maturity levels, with fiber tracts developing along long cortical pathways (Fig. 14.16C).

While these rapidly changing structural changes are taking place in the brain of infants, a key question is: do they correspond to cognitive function? That question was addressed by a team of researchers who wanted to know "do babies think?" In healthy adults, a *resting state network* has been well defined, showing a set of brain regions that are synchronously, spontaneously activated with absence of a specific stimulation or task (Fox & Raichle.

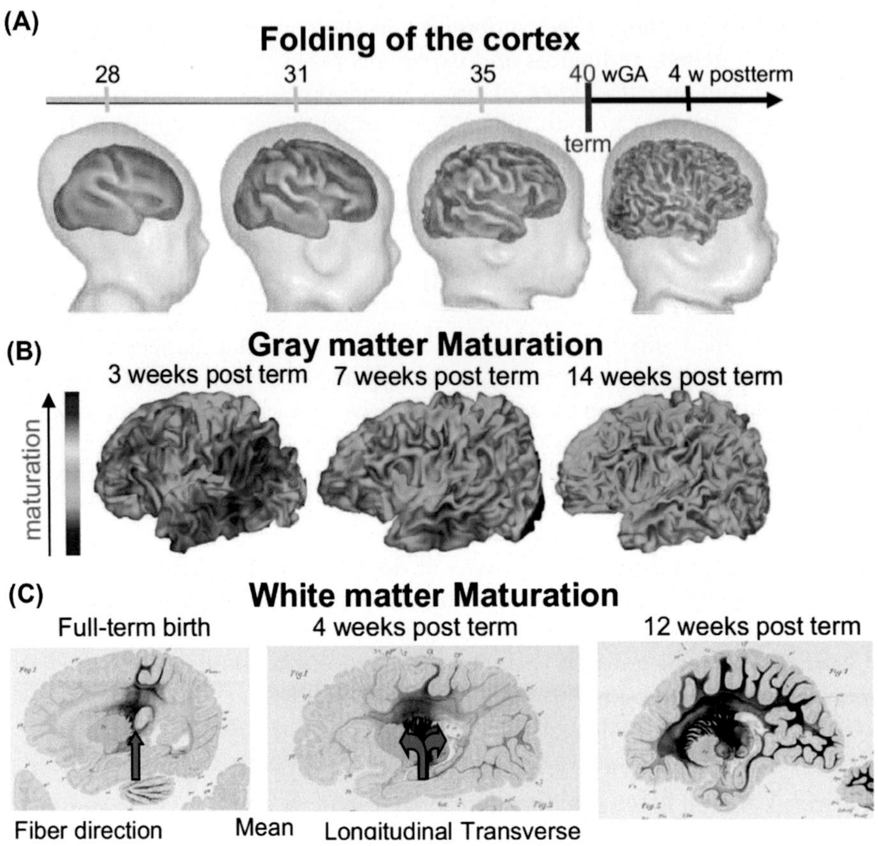

FIGURE 14.16 Massive changes are occurring in the brain during these early months of life. (A) During weeks 28–40 gestation and 4 weeks after birth, the folds of the cortex—the gyri—are enhanced due to brain development processes such as cell proliferation, synaptogenesis, myelination, and axon development. (B) Also due to these processes, the gray matter in the brain matures rapidly during 3–14 weeks post term. (C) White matter matures as well, with fiber tracts developing and rapidly expanding from birth until 12 weeks postterm. These processes continue through infancy and young childhood. *Source: Dehaene-Lambertz and Spelke (2015).*

2007). Interestingly, these canonical resting state networks are also observed in infants (Fig. 14.17), and include a somatomotor network, visual network, and a default mode network (Cusack, Ball, Smyser, & Dehaene-Lambertz, 2016).

4.1.1 Developing the Language Brain: Infant Language Capabilities

Remember the studies showing that babies can hear and discriminate their mother's voice before birth? Studies like these have provided compelling evidence that a newborn infant already has experience with human language. We have discussed the nature versus nurture debate regarding the genetic predisposition for language versus the role of experience. This debate has been influential in the field of developmental cognitive neuroscience, with many studies investigating just what an infant knows about language. Most studies of young

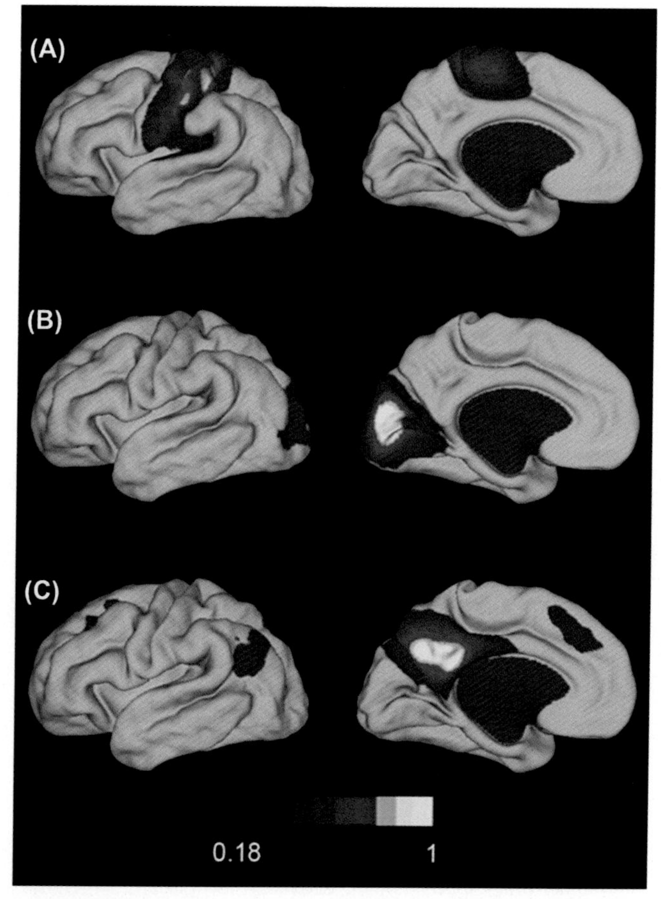

FIGURE 14.17 In healthy adults, a *resting state network* was been well defined, showing a set of brain regions that are synchronously, spontaneously activated with absence of a specific stimulation or task. Interestingly, these canonical restating state networks are also observed in infants, and include (A) a somatomotor network, (B) visual network, and a (C) default mode network. Left column shows a sagittal view of the left hemisphere. Right column shows a midsagittal view of the right hemisphere. *Source: Cusack et al. (2016).*

infants (less than 12 months old) focus on the classes and categories of speech sounds: phonology. Studies with older infants and children also investigate semantic (meaning-based) and syntactic (grammar-based) knowledge.

Is language "biologically special?" One way to address this question is to see whether young babies are specifically sensitive to human speech. If there are specific neural correlates of speech processing observable very early in life, this may indicate language-related neural processing prior to significant experience. Dehaene-Lambertz, Hertz-Pannier, and Dubois (2006) conducted a series of studies investigating this. One finding is that there are hemisphere differences for language processing that are emerging in infants as young as 3 months (Fig. 14.18). Although work with adults has shown increasing evidence for speech processing in both left and right temporal lobes (Hickok & Poeppel, 2007), these data from the lab of Dehaene-Lambertz show a left-lateralized response in infancy.

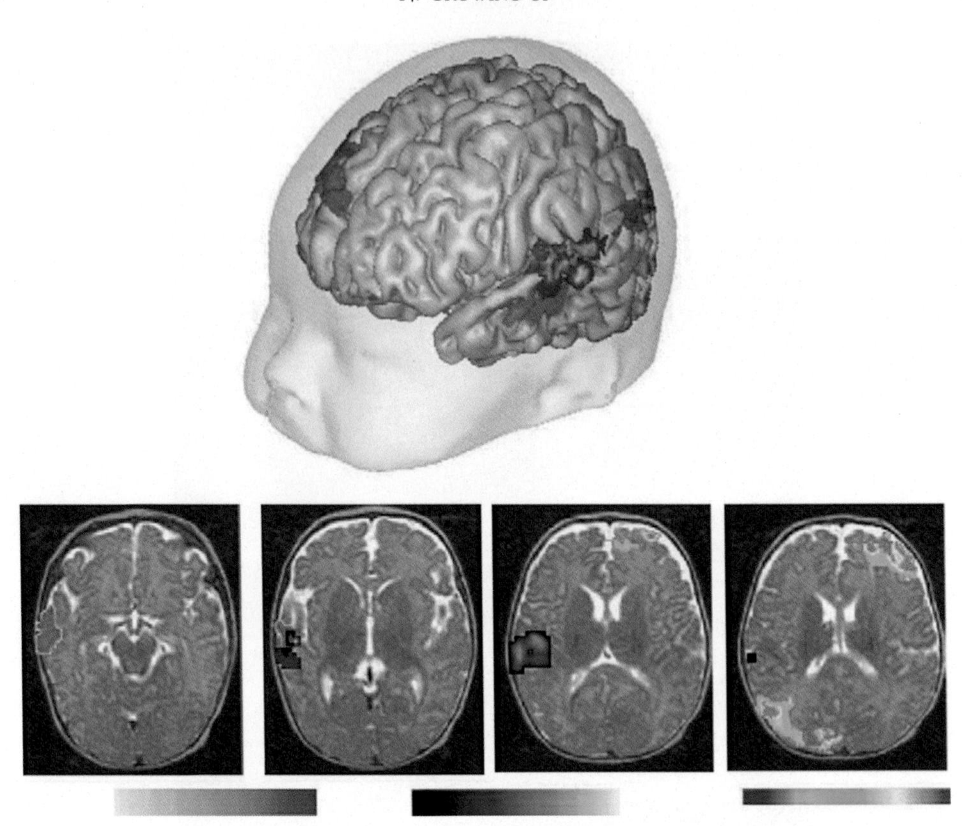

FIGURE 14.18 As early as 3 months, infants show a leftward laterality for human speech. Moving from left to right, the brain figures reflect (i) brain activation in the temporal lobe for backward speech (orange); (ii) brain response in a slightly higher brain slice, showing activation for both backward speech and forward speech (blue); (iii) brain response in a slightly higher brain splice showing activation for forward speech; (iv) the differing in brain activation for forward–backward speech shows activation in the right prefrontal cortex. *Source: Dehaene-Lambertz et al. (2006).*

Intriguingly, and unlike adults, human infants can initially discriminate a very wide range of phonetic constructs, including those not found in their native language (Eimas, Siqueland, Jusczyk, & Vigorito, 1971). For example, Japanese infants, but not Japanese adults, can discriminate between "r" and "l" sounds. However, this ability becomes restricted to the phonetic constructs found in their native language by around 10 months of age. These findings might reflect early speech perceptual processes that take into account the physical or acoustic features in all speech sounds in early infancy, developing later into mechanisms with less reliance on the physical aspects and more on the *abstract representations* of phonemes in their native language. In this way, the role of experience has a strong hand in shaping an infant's language knowledge.

How does infant language learning affect the brain? And how does brain development affect language learning? Take another look at Fig. 14.14, where the time course of brain maturation is shown for different brain regions. Note that the PFC is shown to be the last

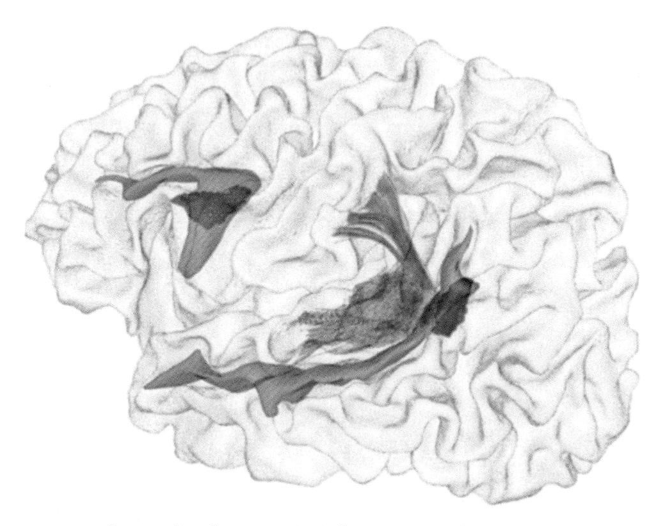

FIGURE 14.19 Language pathway development in infants. Areas shown in red are language pathways in the frontal lobe (left side of figure) and the temporal lobe (right side of figure) that are correlated in terms of their developmental changes. These findings provide evidence that frontal cortex language areas mature earlier than was previously known. *Source: Leroy et al. (2011).*

to mature. There is new evidence, again from the laboratory of Dehaene-Lambertz and colleagues, about the maturation of language pathways in the brain. A key finding was that temporal and frontal lobe language pathways are more developed in infants than was earlier thought (Fig. 14.19). This study provides intriguing new data that the PFC may mature earlier in life than we previously believed.

Language acquisition and speech perception in infancy has been one of the most active areas of developmental cognitive neuroscience. The use of converging methodologies, and frequent comparisons between typical and atypical trajectories of development, make it the domain most likely to see major breakthroughs over the next decade.

4.1.2 *Developing the Executive Brain: What do Babies Know?*

A critical aspect of an infant's cognitive growth during the first year of life is the ability to learn about his or her environment. New items will attract the attention of an infant, and he or she will gaze at these items for longer durations than for items that are accustomed to being seen. While it is important for an infant to gaze at a new item, it is also important for an infant to orient to other aspects of his or her environment. The trading effects of looking at new items and shifting attention to other elements in the world around them provide infants both with learning opportunities for understanding features in new items and a wide range of such experiences by changing the focus of their attention, both of which are critical to cognitive development.

The executive control for directing attention to new items, orienting, maintaining goals, and control over reaching movements are thought to require the most anterior portion of the brain: the PFC. As discussed earlier, the frontal cortex shows the most prolonged period of postnatal development of any region of the human brain, with neuroanatomical changes evident even into the teenage years (Giedd et al., 1999; Huttenlocher, 1990). For this reason, it has been the part of the brain most commonly associated with developments in cognitive abilities during childhood.

One of the most comprehensive attempts to relate a cognitive change to underlying brain developments has concerned the emergence of object permanence in infants. Object permanence is the ability to retain an object in mind after it has been hidden by another object or a cover (Fig. 14.20). Specifically, Piaget observed that infants younger than around 7 months fail accurately to retrieve a hidden object after a short delay period if the object's location is changed from one where it was previously and successfully retrieved. In particular, infants at

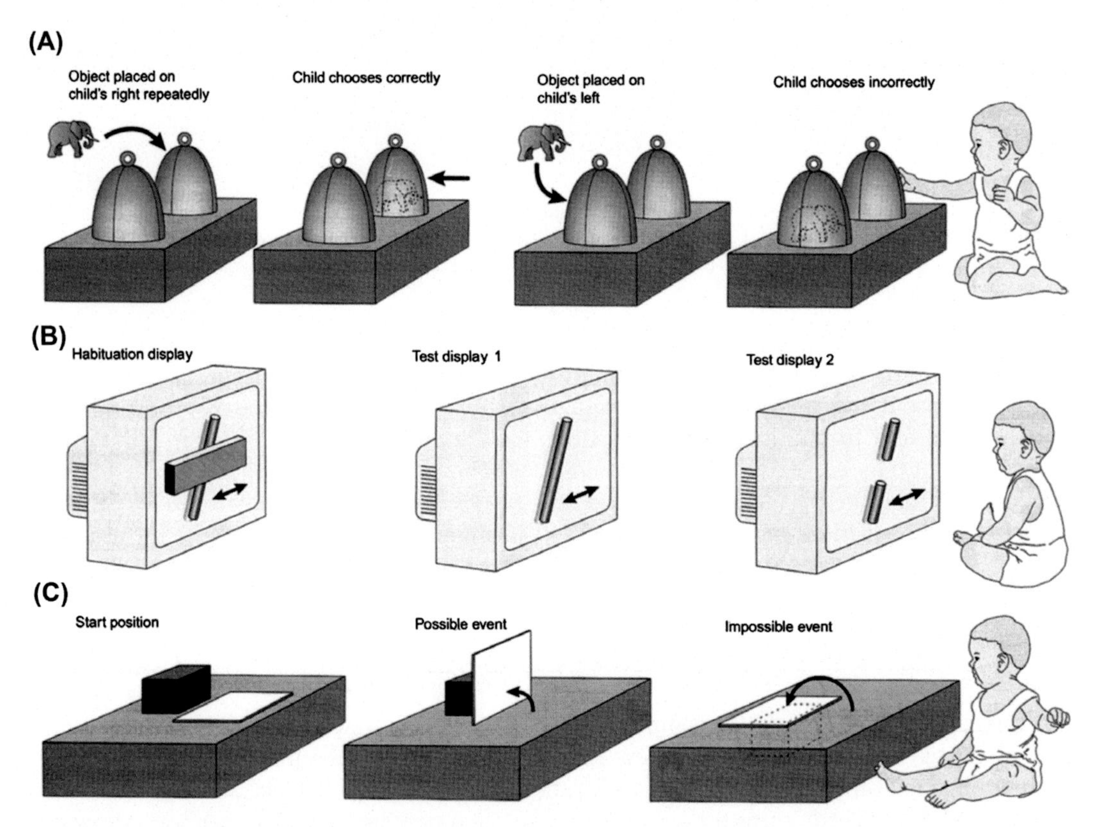

FIGURE 14.20 Behavioral testing in infants. (A) An object retrieval task that infants fail up to 9 months of age. In full view of the infant, the experimenter hides the object in one location and the infant reaches for it successfully. After a few such trials, the experimenter hides the object in a second place, but the infant searches again at the original location (Piaget, 1954). (B) A visual habituation technique can be used to show that infants from as young as 4 months perceive the left-hand figure as a continuous rod moving behind an occluder. Infants dishabituated (found novel) the test display with two short rods, indicating that they perceptually "filled in" the occluded area in the habituation display. Infants under 4 months are only partially successful in such tasks, depending on the complexity of the display. (C) The infant views two event sequences, one possible and one impossible, in which a flap is rotated toward a solid cube. In the "possible" case, the flap stops when it comes into contact with the object. In the impossible case, the flap rotates through the object. Infants as young as 4 months appear surprised (look longer) when viewing the impossible event, showing that they appreciate that objects are solid and (usually) noncompressible. *Source: Johnson (2001).*

this age make a particular perseverative error in which they persistently reach to the hiding location where the object was found on *the immediately preceding trial*. This characteristic pattern of error, called "A not B," was cited by Piaget (1954) as evidence for the failure of infants to understand that objects retain their existence or permanence when moved from view. Beyond about 7 months, infants begin to succeed in the task at successively longer delays of 1–5 s (Diamond, 1985, 2001).

Diamond and Goldman-Rakic (1989) found that infant monkeys also make errors in an adapted version of Piaget's object permanence task. Similar errors were also seen in adult monkeys with damage to the dorsolateral prefrontal cortex (DLPFC). Damage to other parts of the cortex did not have the same effects, indicating a specific role for DLPFC in this task. Evidence linking this change in behavior to brain development also comes from EEG studies with human infants (e.g., Fox & Bell, 1990). In these studies, increases in frontal EEG responses correlate with the ability to respond successfully over longer delays in delayed response tasks.

Thus converging evidence from several sources supports the view that development of PFC allows infants to succeed in the object permanence task. According to Diamond (1991), the critical features of the task carried out by DLPFC are the ability to retain information over spatial delays and to inhibit prepotent (previously reinforced) responses.

More recent evidence has brought in a new question regarding the development of object permanence in infants: Is it the case that the younger (<7 months) infants who fail at this task actually do understand the concept of object permanence but have just not yet developed their reaching skills adequately to do the task? Studies using eye movement rather than reaching as a measure of an infant response have provided evidence that infants as young as 4–5 months of age can perform successfully in an object permanence task (for example, Lecuyer, Abgueguen, & Lemarie, 1992). Thus it is important to understand both the infant behaviors required and the task demands to understand the development of PFC and its role in infant cognition.

Further evidence for the developmental importance of the PFC from early infancy comes from studies of the long-term and widespread effects of perinatal (at birth) damage to the PFC: Damage to the PFC often results in both immediate and long-term difficulties.

The issue raised at the beginning of this section concerned how to reconcile evidence for continuing neuroanatomical development in the frontal cortex until the teenage years on the one hand (see Fig. 14.14), and evidence for some functioning in the region as early as the first few months of age on the other (see Fig. 14.19). One possible resolution to this issue is that representations that emerge within this region of the cortex are initially weak and sufficient only to control some types of output, such as saccades (eye movements), but not others, such as reaching (Munakata, McClelland, Johnson, & Siegler, 1994). Other plausible resolutions of this issue come from Diamond's (1991) proposal that different regions of frontal cortex are differentially delayed in their development, and Thatcher's (1992) suggestion that prefrontal regions may have a continuing role in the cyclical reorganization of the rest of the cortex.

Whether these hypotheses work out or not, there is good reason why some degree of PFC functioning is vital from the first weeks of postnatal life, or even earlier (Fulford et al., 2003). The ability to form and retain goals, albeit for short periods, is essential for generating efforts to perform actions such as reaching for objects. Early and often initially unsuccessful attempts to perform motor actions provide the essential experience necessary for subsequent development (Box 14.1).

BOX 14.1

WHY CAN'T I REMEMBER MY FIRST BIRTHDAY?

We love to celebrate children's birthdays and perhaps the biggest celebration of all is baby's first birthday! Indeed, quite a fuss is made of this milestone...a big party, cake, balloons, presents, with lots of family and friends. But ask a typical college student if they remember this big day and they will have no memory of it. In fact, they will likely have little or no memories of early childhood prior to the ages of 3–5 years. Why is this? Certainly the early birthday parties—and other exciting events like the birth of a sibling—should be important enough that some vague memory would survive. Does this mean that those early memories are not formed as well? Or are they lost?

Infantile Amnesia

Sigmund Freud wrote about what he termed infantile amnesia in the early 1900s based on his observation that his patients did not recall their early childhood years. Infantile amnesia has been well documented in the scientific literature but we are still searching for the reasons behind this phenomenon. An almost complete lack of memory for such a long period of early life—from birth 3 –5 years!—is difficult to understand during this long and intense period of learning.

Are stable memories formed in early childhood?

A first key question is: are stable memories formed during this stage of life? The answer is yes, with some qualifications (see reference 1 for an excellent review of the literature on this topic). It is difficult to test the memory of young children using experimental testing procedures, especially very young preverbal infants and toddlers. However, there are many careful studies that show that infants and toddlers are indeed forming stable memories during the age of birth to 5 years—the very age of infantile amnesia. This makes intuitive sense if you have ever spent time with a three or four year old. Ask him or her about what happened at the zoo yesterday, or what is his or her favorite book is, or who is his or her best friend. This little one will be able to tell you many details about events even far removed from the present. So these memories are being formed to the best of our knowledge. And these memories were stable—at least at the time they were in the midst of being formed in children under the age of 5.

When Are They Lost?

When are these memories lost? Is it a sudden event or is it gradual over middle childhood and adolescence? Most studies of infantile amnesia are conducted with adults. Tustin and Hayne investigated infantile amnesia in groups of children, adolescents, and young adults ranging in age from 5 to 20 years. They found that the younger groups (5, 8–9 year olds) remembered much earlier memories than the older groups (12–13 and 18–20 year olds) (2). In fact, some of these younger children recalled events that occurred before their first birthday! This was not the case for the older groups. While more studies are needed to corroborate these findings, the study by Tustin and Hayne provides evidence that the memory loss of early experiences is a gradual process.

What are the neural mechanisms behind infantile amnesia?

The evidence suggests that stable memories are being formed in the first years of life but nevertheless they are "forgotten" as the child ages and matures. And, this forgetting seems to be a gradual—not sudden—process. How does this happen? Where in the brain is it happening?

A natural place to look for memory loss in the brain is the hippocampus, famously surgically resected in the patient HM, resulting in amnesia (3). An intriguing new study by Sheena Josselyn and Paul Frankland (4) proposes that the growth of new neurons—neurogenesis—in the hippocampus during the early years of life leads to the degradation and thus forgetting of early life memories. Using a mouse model to study the role of neurogenesis in the hippocampus, the scientists evaluated the effects of high rates and lower rates of neurogenesis on forgetting in infant and adult animals. The effects of new neurons added to the hippocampus through the process of neurogenesis aided the formation of new memories and the development of new circuitry for their retrieval; however, conversely, they also served to destabilize existing memories due to the additional or new complexities of those new circuits. See Fig. 14.21 for a schematic of the effects of neurogenesis on existing circuitry.

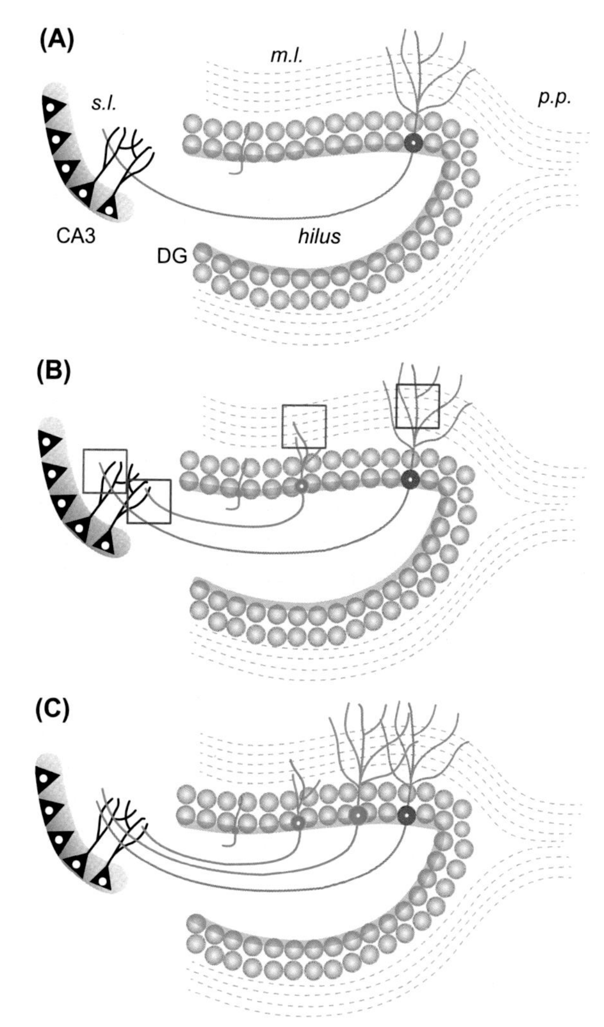

FIGURE 14.21 New neurons are integrated into the circuits in the hippocampus (A) and form new input and output connections (B); however, the new pattern of circuitry can lead to a "forgetting" of older established patterns during periods of high neurogenesis (C). *Source: Frankland, Kohler, and Josselyn (2013).*

Continued

BOX 14.1 *(cont'd)*

If the theory proposed by Frankland, Josselyn, and their colleagues is correct, it would explain why even stable memories become degraded and forgotten during periods of high neurogenesis. While neurogenesis is occurring in the hippocampus throughout life, it is far more prevalent in infancy—thus the larger effect of forgetting during the first years of life.

Future directions

If we can determine just how early childhood memories are established and how they are forgotten in the brain, would that open the door to being able to rekindle those early, lost memories? Could we alter the brain so that we could recapture memories of our first months of life? And would it provide us with a way to "forget" specific memories, such as traumatic memories that cause posttraumatic stress disorder?

References

(1) Mullally, S. L., & Maguire, E. A. (2014). Learning to remember: The early ontogeny of episodic memory. *Developmental Cognitive Neuroscience, 9*, 12–29.
(2) Tustin, K., & Hayne, H. (2010). Defining the boundary of age-related changes in childhood amnesia. *Developmental Psychology, 46*(5), 1049–1061.
(3) Scoville, W. B., & Milner, B. (1957). Loss of recent memory after bilateral hippocampus lesions. *Journal of Neurology, Neurosurgery, and Psychiatry, 20*, 11–21.
(4) Frankland, P. W., Kohler, S., & Josselyn, S. A. (2013). Hippocampal neurogenesis and forgetting. *Trends in Neurosciences, 36*(9), 497–503.

4.1.3 Developing the Social Brain: Faces and Places

One of the major characteristics of the human brain is its social nature. A variety of cortical areas have been implicated in the "social brain," including the superior temporal sulcus, the fusiform "face area" (FFA), and the orbitofrontal cortex. One of the major debates in cognitive neuroscience concerns the origins of the "social brain" in humans, and theoretical arguments abound about the extent to which this is acquired through experience.

One aspect of social brain function in humans that has been the topic of investigation is the perception and processing of faces. There is a long history of research on the development of face recognition in young infants extending back to the studies of Fantz more than 40 years ago (e.g., Fantz, 1964). Over the past decade, numerous papers have addressed the cortical bases of face processing in adults, including identifying areas that may be specifically dedicated to this purpose. Despite these bodies of data, surprisingly little is known about the developmental cognitive neuroscience of face processing.

In a review of the available literature in the late 1980s, Johnsonet al. (1991) revealed two apparently contradictory bodies of evidence: While the prevailing view, and most of the evidence, supported the idea that infants gradually learn about the arrangement of features that compose a face over the first few months of life, the results from at least one study indicated that newborn infants, as young as 10 min old, will track a face-like pattern farther than various "scrambled" face patterns (Goren, Sarty, & Wu, 1975). The key result of this study was that even though face processing is immature at birth, there is a bias or preference for faces over other visual patterns, and this bias is in place at birth. Newborns also show a preference to familiar faces (such as their mother's) versus unfamiliar faces, providing further evidence that face processing is a critical aspect of social development and that it is in place at birth. One hypothesis is that subcortical structures provide the brain mechanisms for the bias for faces observed in newborns, while slower-to-develop regions within the visual cortex then mature to become more adult-like face processors as the infant grows (Fig. 14.22) (Johnson, 2005).

Moving beyond the relatively simple perception of faces, a more complex attribute of the adult social brain is processing information about the eyes of other humans. There are two important aspects of processing information about the eyes. The first of these is being able to detect the direction of another's gaze to direct your own attention to the same object or spatial location. Perception of averted gaze can elicit an automatic shift of attention in the same direction in adults, allowing the establishment of "joint attention." Joint attention to objects is thought to be crucial for a number of aspects of cognitive and social development, including word learning. The second critical aspect of gaze perception is the detection of direct gaze, enabling mutual gaze with the viewer. Mutual gaze (eye contact) provides the main mode of establishing a communicative context between humans and is believed to be important for normal social development. It is commonly agreed that eye gaze perception is important for mother–infant interaction and that it provides a vital foundation for social development. Evidence that mutual eye gaze is important for newborns comes from a recent study where newborns showed a preference to faces with eyes looking directly at them versus averted eyes (Fig. 14.23) (Guellai & Streri, 2011).

FIGURE 14.22 Examples of stimuli that have been used to test newborn's preference for face-related stimuli. Some of the stimuli are designed to test the importance of the spatial arrangement (configuration) of a face (eyes, nose, and mouth) and others to test the importance of particular features. Newborns will preferentially attend to patterns that contain the basic configuration of a face (for example, the second, third, and fourth stimuli from the left) as opposed to the other stimuli. *Source: Johnson (2005).*

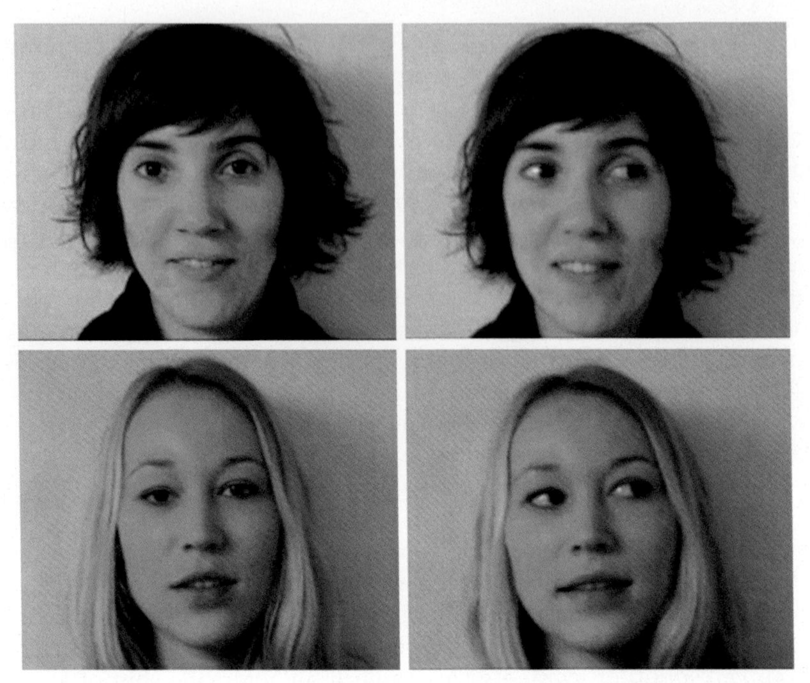

FIGURE 14.23 Examples of the stimuli used in Guellai and Streri's study on newborn face perception. Newborns showed a preference for (looked longer at) faces when the eye gaze was direct (left photos) as opposed to averted (right photos). *Source: Guellai and Streri (2011).*

Beyond face processing and eye gaze detection, there are many more complex aspects of the social brain, such as the coherent perception of human action and the appropriate attribution of intentions and goals to conspecifics (individuals within your own species, such as human–human interpretation of intentions or goals). Investigating the cognitive neuroscience of these abilities in infants and children will be a challenge for the next decade. One way in which these issues have been addressed is through studying genetic developmental disorders in which aspects of social cognition are either apparently selectively impaired (autism) or selectively intact amid otherwise impaired cognition (Williams syndrome).

The first year of life brings dynamic changes in both behavior and the brain. While we provided separate discussions of the emergence of language, executive functions, and social cognition in this busy first year of life, these three domains of human cognition have complex interactions throughout development. An infant's interest in faces, for example, will help him or her to understand speech. The ability to focus on new objects aids the infant in developing knowledge about the world around him or her. While we do not fully understand the explosive growth of brain processes and their relation to behavior in infants, studies such as the ones we presented here are helping to map the complex correspondence between mind and brain.

4.2 Childhood and Adolescence: Dynamic and Staged Growth

While the first year of life represents an unparalleled stage in dynamic human development, many aspects of brain and cognitive growth take years to mature. In this section, we will present results from some studies investigating the development of brain areas subserving language, executive functions, and social cognition in children and adolescents. While these studies are informative, it is important to bear in mind that relating complex brain activation to performance on cognitive tasks is a highly complex process. For example, it may well be the case that the brain activity that we observe in adults—once a cognitive process has been developed—may tap different brain regions while it is being acquired. Thus simply comparing regions of interest in neuroimaging experiments across groups of children versus adults may not provide us with the level of sensitivity that we require to formulate inferences about brain and behavior. Similarly, differing cognitive strategies or coping mechanisms in childhood versus adulthood may also impact the network of brain areas tapped in certain task paradigms.

With those caveats in mind, let us review the evidence regarding the development of neural systems for language, executive control, and social cognition.

4.2.1 The Linguistic Brain: Language Acquisition

Language is not a unitary system. To express our ideas verbally, we need to progress through stages of formulating the concepts, mapping them onto words in our mental lexicon, accessing our mental grammar to form sentences, and mapping this information onto sound-based representations of the articulation of the ideas we want to express. It makes intuitive sense that, early in life, infants develop their knowledge about language based largely on the *sounds of language* that they hear in their environment. Thus it is not surprising that studies with young infants (less than 12 months) typically focus on the phonology of human language. With older infants, children and adolescents, studies typically test other aspects of language, such as lexical-semantic (meaning based) and syntactic (grammar based) knowledge.

Do all aspects of language develop in similar ways, with similar brain developmental processes underlying them? This is a question that has been addressed by developmental cognitive neuroscientists investigating the neural substrates of human language. One way to subdivide the complex neurodevelopmental changes that take place between birth and early adolescence is to think of them in stages.

This approach was presented by Skeide and Friederici (2016), who described early processing capacities in infants and toddlers are largely modulated by data-driven *bottom-up processes.* The events and stimulation the infant is experiences—the data—are "driving" the infant's learning and cognitive development. According to Skeide and Friederici, processing capacities in early childhood through adolescence are largely modulated by cognitively controlled *top-down processes.*

Skeide and Friederici (2016) describe the differential developmental trajectories for these early language bottom-up processes and the later top-down processes and their corresponding brain regions. According to their theory, adult language comprehension involves bottom-up processes for sound, word, phrase, and intonational processing that are distributed in anterior, middle, and posterior temporal regions (shown in red in Fig. 14.24A; arrows

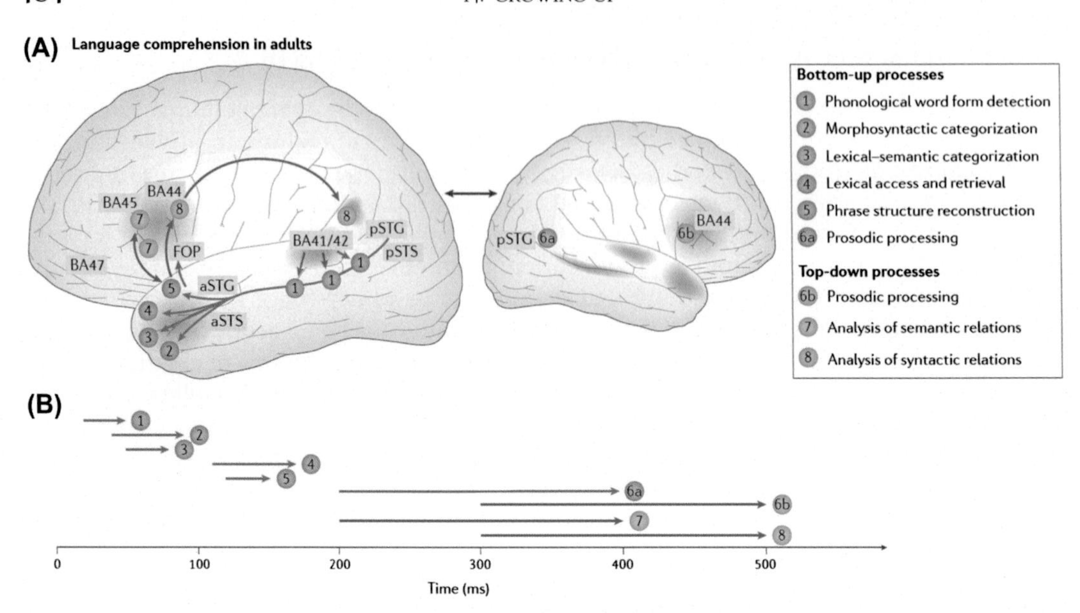

FIGURE 14.24 According to the theory presented by Skeide and Friederici (2016), adult language comprehension involves bottom-up processes for sound, word, phrase, and intonational processing that are distributed in anterior, middle, and posterior temporal regions (shown in red in (A); arrows show the direction of the flow of information). Top-down processes in adults are mediated by regions in the frontal lobe and posterior temporal gyrus (shown in blue in (A); arrows show the direction of the flow of information). The processing time course of the bottom-up and top-down processes are shown in(B), with bottom-up processes occurring largely within ~300 ms after onset of speech and top-down processes occurring largely within ~200 ms after onset. *Source: Skeide and Friederici (2016).*

show the direction of the flow of information). Top-down processes in adults are mediated by regions in the frontal lobe and posterior temporal gyrus (shown in blue in Fig. 14.24A; arrows show the direction of the flow of information). The processing time course of the bottom-up and top-down processes are shown in Fig. 14.24b, with bottom-up processes occurring largely within ~300 ms after onset of speech and top-down processes occurring largely within ~200 ms after onset.

According to Skeide and Friederici (2016), child language comprehension also involves bottom-up processes for sound, word, phrase, and intonational processing that are distributed in anterior, middle, and posterior temporal regions (shown in green in Fig. 14.25A; arrows show the direction of the flow of information). Top-down processes in development are mediated by regions in the frontal lobe and posterior temporal gyrus (shown in orange in Fig. 14.25A; arrows show the direction of the flow of information). The model of the language acquisition timeline for bottom-up and top-down processes is shown in Fig. 14.25B, with bottom-up processes developing before birth and during the first 3 years of life and top-down processes developing after age 3.

4.2.2 The Executive Brain: Taking Cognitive Control

Even young infants must learn what information in their world is important and what is unimportant or irrelevant. These learning mechanisms fall under the general category of

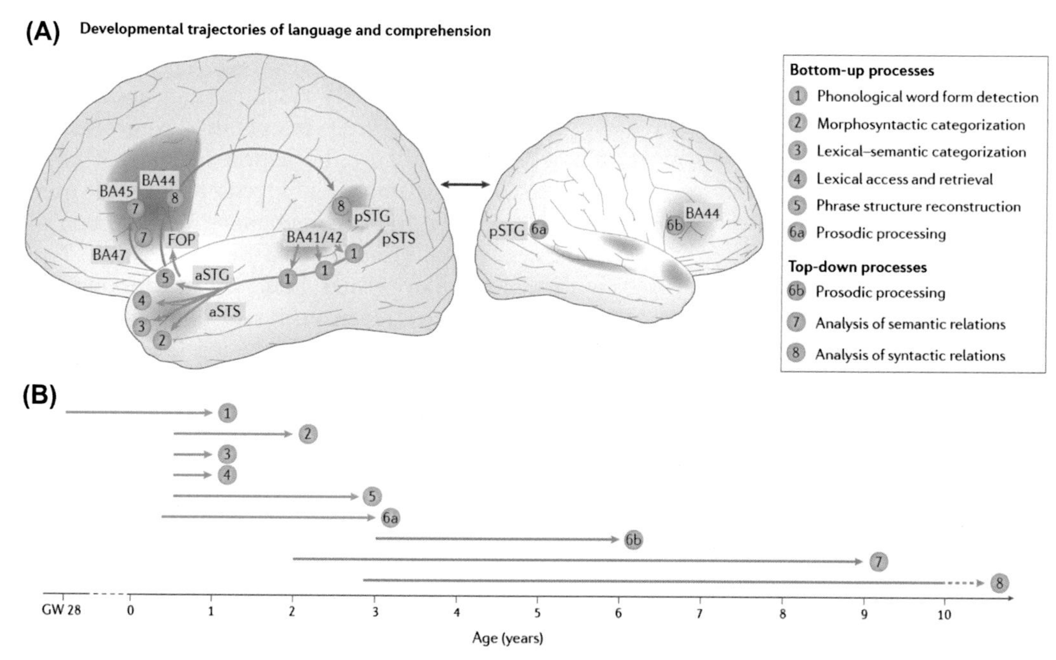

FIGURE 14.25 According to Skeide and Friederici (2016), child language comprehension also involves bottom-up processes for sound, word, phrase, and intonational processing that are distributed in anterior, middle, and posterior temporal regions (shown in green in (A); arrows show the direction of the flow of information). Top-down processes in development are mediated by regions in the frontal lobe and posterior temporal gyrus (shown in orange in (A); arrows show the direction of the flow of information). The model of the language acquisition timeline for bottom-up and top-down processes is shown in (B), with bottom-up processes developing before birth and during the first 3 years of life and top-down processes developing after age 3. *Source: Skeide and Friederici (2016)*

"cognitive control" and have been the focus of much study in infant and child development. Recall that in infants, the A, not B, task has been used to investigate the ability to ignore or inhibit irrelevant information and to inhibit prepotent response (Piaget, 1954; Diamond, 1985). These capabilities become more and more important throughout childhood as a child's environment becomes increasingly complex. Consider a 6-year-old child in a first-grade classroom. This child must be able to pay attention to the teacher or to a task at hand despite the many distractions that surround him or her, such as children talking, books dropping, chairs scraping. The trading of attentional resources toward relevant aspects of the environment and away from less important aspects is a vital element in development.

In adults, the DLPFC is implicated as an important cortical region in tasks that tap cognitive control functions. We know from histological and neuroanatomical studies of developing children that the PFC has a prolonged developmental path that does not reach mature, adult-like stages until midto late adolescence. Behavioral studies of cognitive control function in children and adolescents have provided evidence for a similar time course in the development of cognitive control abilities. An open question in the field of developmental cognitive neuroscience is if there is any correspondence between these late-to-mature brain regions and the late-developing cognitive control abilities.

Studies of the neural substrates of cognitive control have only recently been undertaken with children. Casey et al. (2005) have conducted seminal studies of cognitive control using a combination of fMRI and behavioral methods to investigate the neural patterns of brain activation measured while children perform tasks likely to engage PFC regions. In one experiment, they used event-related fMRI while children and adults were engaged in a "go-no go" task (Durston, Thomas, Worden, Yang, & Casey, 2002). In this task, participants had to suppress their response when presented with a particular visual item within an ongoing sequence of stimulus presentations (e.g., one Pokemon character within a sequence of other Pokemon characters). The difficulty of the task was increased by increasing the number of "go" items that preceded the "no go" character. Successful response inhibition was associated with stronger activation of prefrontal regions for children than for adults.

A general finding of Casey and colleagues has been that younger children exhibit broader, more diffuse brain activation for cognitive control tasks as compared to adults. During development, these brain areas mature and the brain activity that correlates to task performance abilities (such as reaction time and accuracy) becomes more focal and fine-tuned. In Fig. 14.26, we present a figure from a recent article by Casey et al. (2005) reviewing the literature of developmental cognitive control investigations. The general notion of brain activation becoming more focal and defined as a function of a child's age is shown with references to those studies showing increased activation with age and those showing decreased activation with age.

Future work in the investigation of cognitive control and development of PFC in children will need to take into account many other aspects of the prolonged developmental path of the frontal regions and the correspondence to behavior. Some areas of future research may include investigating gender differences in cognitive control functions and related brain activation patterns.

4.2.3 *The Social Brain: Face Perception in Childhood*

Human face perception has been the focus of many neuroimaging and behavioral investigations in adults, in infants, and throughout childhood and adolescence. Why is this area of research important to the field of cognitive neuroscience? Investigating brain regions that may be specialized for perception of our species-specific faces may shed light on the nature versus nurture debate. Are we predisposed to attend to, focus on, and interpret cues in faces? Or does our vast experience with faces provide the information processing abilities that are not specific to faces but rather utilize visual object perception networks? Studying face processing during development may help us to determine how genetic predisposition interacts with experience.

Recall that, in infants, a bias for human faces, and in particular familiar faces, is present in newborns. How do these processes develop during childhood? As in the case of executive functions shown in Fig. 14.26, brain activation for faces appears to be more focal and more face-selective during childhood and adolescence. An fMRI study of brain activation for faces, buildings, and objects with children 5–8 years, 11–14 years, and adults showed an interesting pattern of activations that differed by age (Fig. 14.27). The younger (5–8 years) children

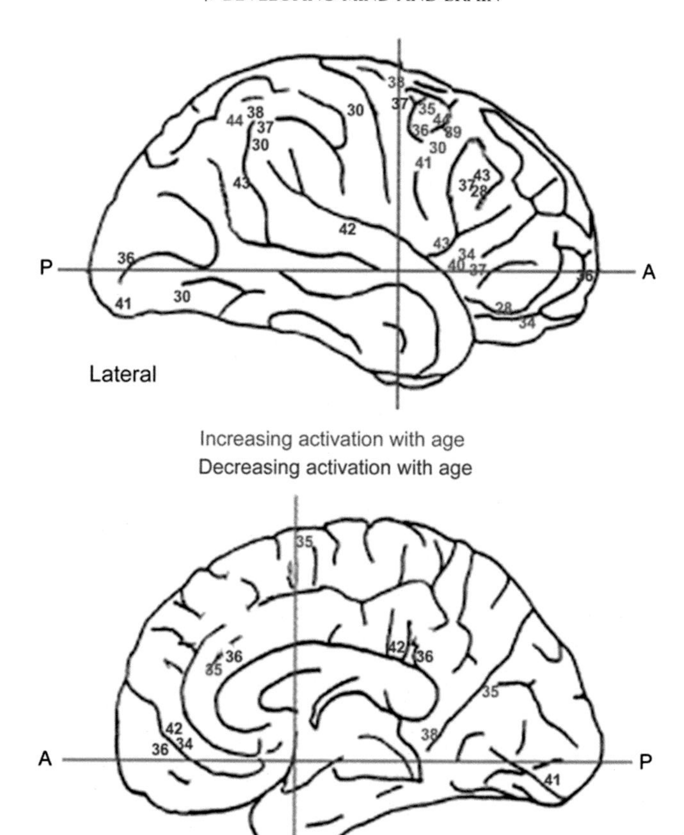

FIGURE 14.26 The development of human cortical function, as measured by contemporary imaging methods, reflects fine-tuning of a diffuse network of neuroanatomical regions. Collectively, developmental neuroimaging studies of cognitive control processes suggest a general pattern of increased recruitment of slow maturing prefrontal cortex (references depicted here in red), especially dorsolateral prefrontal cortex and ventral prefrontal cortex, and decreased recruitment of lower-level sensory regions (references in blue), including extrastriate and fusiform cortex and also posterior parietal areas. This pattern of activity, which has been observed across a variety of paradigms, suggests that higher cognitive abilities supported by association cortex become more focal or fine-tuned with development, whereas other regions not specifically correlated with that specific cognitive ability become attenuated. A = anterior; P = posterior. *Source: Casey et al. (2005).*

showed a less selective response for the three categories of images, while older children and adults showed clear differences for face versus buildings versus objects. Together with the findings of Casey shown in Fig. 14.26, the early results converge to provide evidence for a gradual change in brain responses that reflects maturational changes during childhood.

What have we learned about the development of language, executive function, and social cognition in childhood and adolescence? One central finding from a variety of data sources is that the cortical regions subserving these higher-order cognitive functions have a prolonged

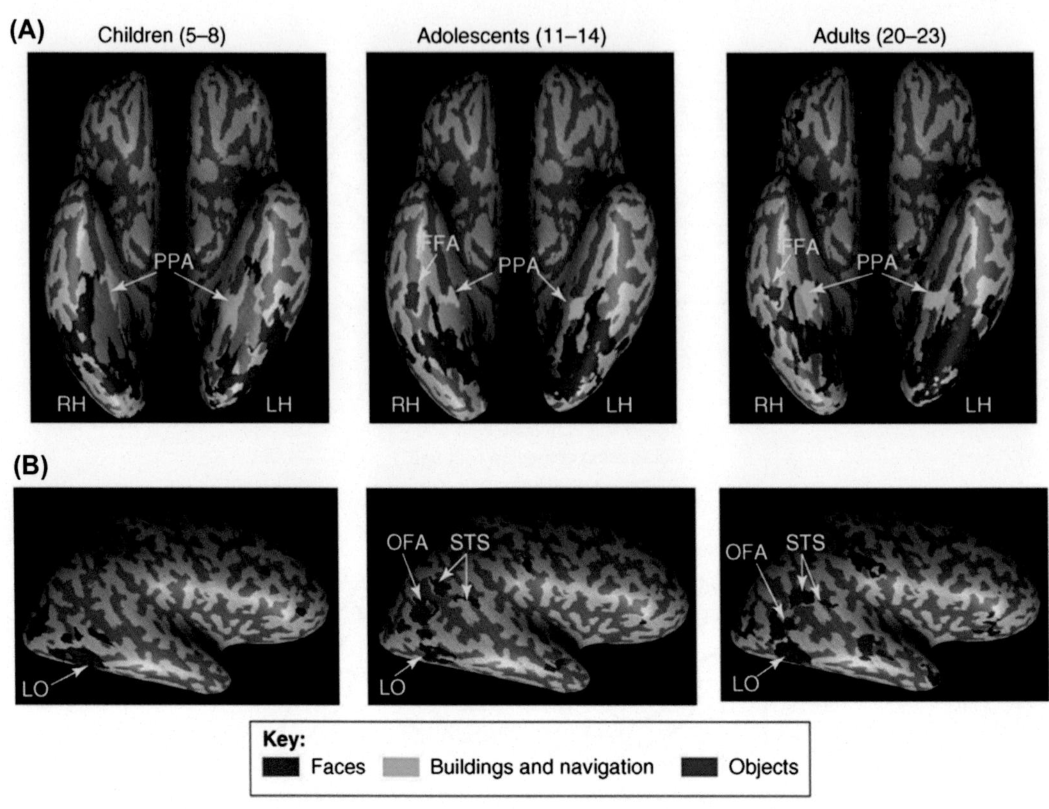

FIGURE 14.27 (A) Brain activation for faces (shown in red), buildings (shown in green), and objects (shown in blue) for children aged 5–8 years (left panel), adolescents aged 11–14 years (center panel), and adults (right panel). (B) Results showed that children aged 5–8 years did not show the face selectivity of older adolescents and adults. *Source: Kadosh and Johnson (2007).*

developmental path extending to mid to late adolescence. We have a wealth of behavioral data showing a similar pattern in tasks that tap more complex and higher-order aspects of these cognitive functions, with task performance not reaching adult-like levels until late adolescence.

An important direction in the field of developmental cognitive neuroscience is to combine methodologies—for example, fMRI with EEG—with behavioral measures to provide converging evidence across methodologies and measures regarding aspects of higher-order cognitive function. Combining fMRI with EEG, for example, can provide high-resolution spatial information regarding brain activations coupled with high resolution of the time course of that activation. Another important direction in the field is to conduct longitudinal studies to track the development over time of individual children. In this way, early "baseline" measures can be taken, and then the development and change in these measures may be assessed at specific intervals. Finally, new experimental design approaches with young infants are demonstrating that babies know and understand a lot more about the world around them than we previously thought. New advances in measuring infant cognition and mapping the relevant brain activity will provide important insights into the developmental changes occurring in the first year of life.

5. EARLY BRAIN DAMAGE AND DEVELOPMENTAL PLASTICITY

We already mentioned the importance of longitudinal studies for tracking individual progress and outcomes throughout development. This type of study is especially important when assessing the long-term outcome of early (perinatal) brain injury. We have seen in earlier chapters that in adults, brain damage due to stroke, disease, or traumatic accident typically leads to deficits in aspects of cognition that are fairly severe, with complete recovery of function unlikely. What happens when brain damage occurs near birth? This question has an important bearing on the nature versus nurture debate. Consider the hypothesis that some brain systems, such as language, have a strong genetic predisposition for their development in specified regions of the brain. If there is early insult to those prespecified regions, will the infant develop language in a typical fashion? Or will language develop in an aberrant fashion due to the early and unrecoverable damage to those brain regions? Alternatively, if experience plays the dominant role in the development of brain regions that become tuned for language function, will the infant develop language in a typical fashion in spite of the early brain damage?

The effects of perinatal brain damage have been extensively investigated in animal studies in the field of neurobiology. The effects of early brain damage and the impact on later cognitive development have been far less studied in humans. One reason for this is that a single, unilateral (in one hemisphere only) pre- or perinatal brain insult is relatively rare. Typically, instances of early brain insult are more global in nature and combine with other neurological complications (Fig. 14.28, left panel). In these cases of larger-scale damage coupled with other

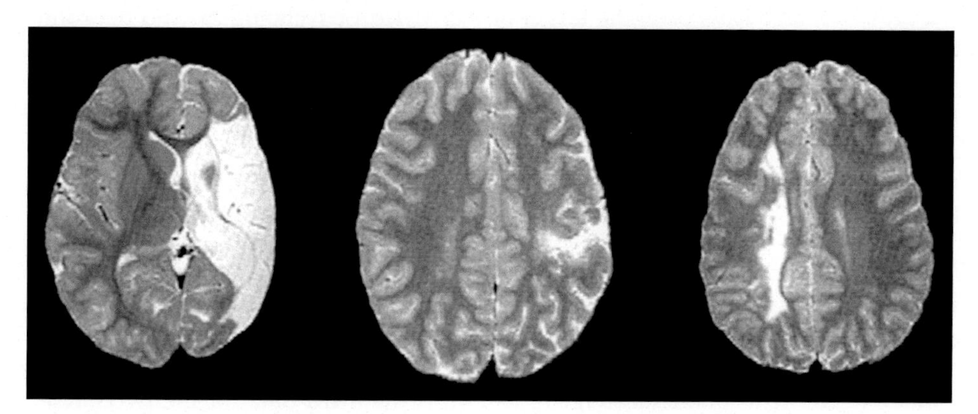

FIGURE 14.28 Large-scale and smaller-scale perinatal brain damage. Structural MRI scans in the axial plane from three children with perinatal brain damage, illustrating different patterns of injury. Left: A large unilateral lesion involving most of one cerebral hemisphere. Middle: A small lesion confined to one cerebral lobe. Right: A deep lesion involving subcortical regions. *Source: Stiles et al. (2005).*

traumatic events, it is difficult to compare cognitive development to children without this early damage and trauma. In cases where the perinatal damage is limited to a circumscribed region (Fig. 14.28, center and right panels), the long-term effects are typically milder. These are the types of cases that we will focus on in this section: early, focal, unilateral brain insult.

While we are just beginning to understand the complexities of early insult on later cognitive growth in humans, a series of longitudinal studies by Stiles, Reilly, Paul, and Moses (2005) shed some light on the long-term effects of perinatal insult, and we highlight some results here. The San Diego Longitudinal Project (Stiles, Bates, Thal, Trauner, & Reilly, 2002) is the largest US investigation of the long-term effects of perinatal brain damage. Stiles and colleagues have followed the cognitive development of several hundred children with perinatal brain damage since 1989.

Much of the focus of the investigations by Stiles and colleagues in the San Diego Longitudinal Project has been on language development in children with perinatal brain damage. One key finding is that whereas focal brain damage in language centers in adults (typically through lesions due to stroke) results in long-lasting deficits, this pattern is quite different with infants who suffer perinatal brain damage. While there is typically a delay in early linguistic milestones, such as onset of word comprehension at 9–12 months and word production at 12–15 months, by the age of 5 years, these children have largely "caught up" in linguistic abilities. The important finding, however, is that when tested carefully, there remain some underlying deficits even at the age of 5, especially in complex sentence structures (Reilly, Losh, Bellugi, & Wulfeck, 2004). These children with early brain damage do ultimately achieve language competence, but the evidence provided by the longitudinal studies by Stiles and colleagues (reviewed in Stiles et al., 2005) indicates that their language proficiency is in the lower than normal range. Thus while the children do acquire many skills and proficiencies with respect to language, there remain throughout childhood, adolescence, and presumably adulthood some key deficits due to the very early damage to important brain regions for language acquisition and processing.

Another aspect of cognition that has been the focus of study by the San Diego Longitudinal Project has been spatial cognition. Spatial cognition and the effects produced by adult-acquired brain damage have been the target of many neuropsychological and neuroimaging investigations. The central findings have been that there is a hemisphere asymmetry in the decoding of visual patterns, with the left hemisphere biased for extracting feature (local) information and the right hemisphere biased for extracting configuration (global) information (Fig. 14.29).

fMRI studies of typically developing adolescents show that they demonstrate a similar hemisphere asymmetry, with greater right hemisphere occipital-temporal activation for global processing and greater left hemisphere occipital-temporal activation for local processing (Fig. 14.30). In stark contrast, a 15-year-old adolescent who had suffered right hemisphere perinatal brain damage showed stronger activation for both global and local processing in the left (undamaged) hemisphere, and a 13-year-old adolescent who had suffered left hemisphere perinatal brain damage showed stronger activation for both global and local processing in the right (undamaged) hemisphere (see Fig. 14.30). Thus the fMRI data for the two adolescents who suffered perinatal brain damage provide evidence for long-lasting damage to spatial

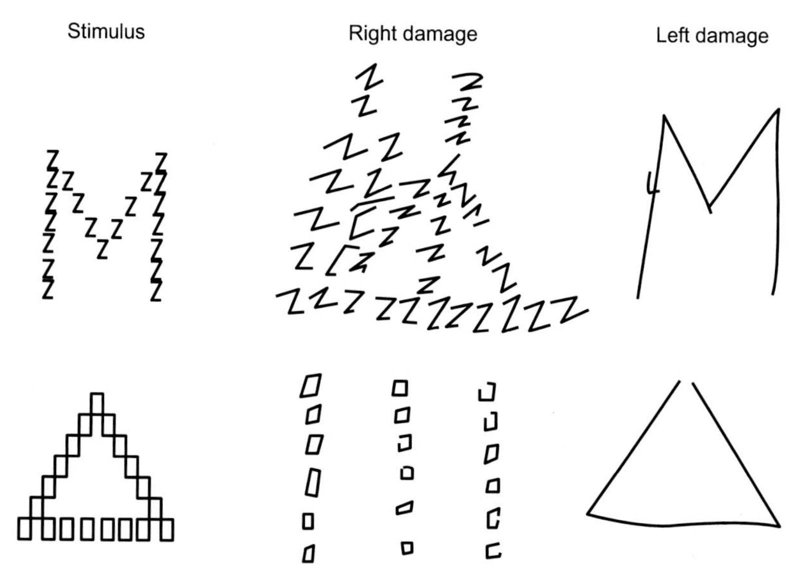

FIGURE 14.29 Global versus local: examples of visuospatial deficits. Examples of memory reproductions of hierarchical form stimuli by adult stroke patients with either right or left hemisphere injury (Adapted, with permission, from Delis et al., 1986). The sample stimulus to be copied is shown on the left side of the figure. Center of the figure: patients with right hemisphere damage typically produce the local (detailed) aspects of the stimulus but omit the global (overall) aspects of the stimulus—in this case, the "M" or triangle shape of the stimulus. Right side of the figure: patients with left hemisphere damage typically produce the global aspects of the stimulus but omit the local aspects of the stimulus. *Source: Stiles et al. (2005).*

cognition mechanisms. However, they also provide intriguing evidence for a brain system that is highly flexible, with recruitment of neural territory in the undamaged hemisphere for spatial cognition functions.

What have we learned about the long-term effects of early brain damage in the longitudinal studies of Stiles and colleagues? And how do they inform us about the complex and highly interactive roles of nature and nurture in human development? While these efforts are still in the early stages, results to date indicate that early brain damage results in long-term, though typically somewhat subtle, deficits. A second important finding is that despite the early insults and the delays that they typically produce in cognitive development, the children mature and acquire higher cognitive function, although sometimes at the lower than normal level. Cumulatively, these findings provide evidence that some brain systems suffer long-term impairments when damaged, even when the damage occurs at or near birth. This provides some support that some systems have a level of genetic predisposition and can suffer long-term harm when disrupted. On a brighter note, these findings provide evidence for significant amounts of early brain plasticity so that the cognitive functions that suffer early damage develop in an alternative manner.

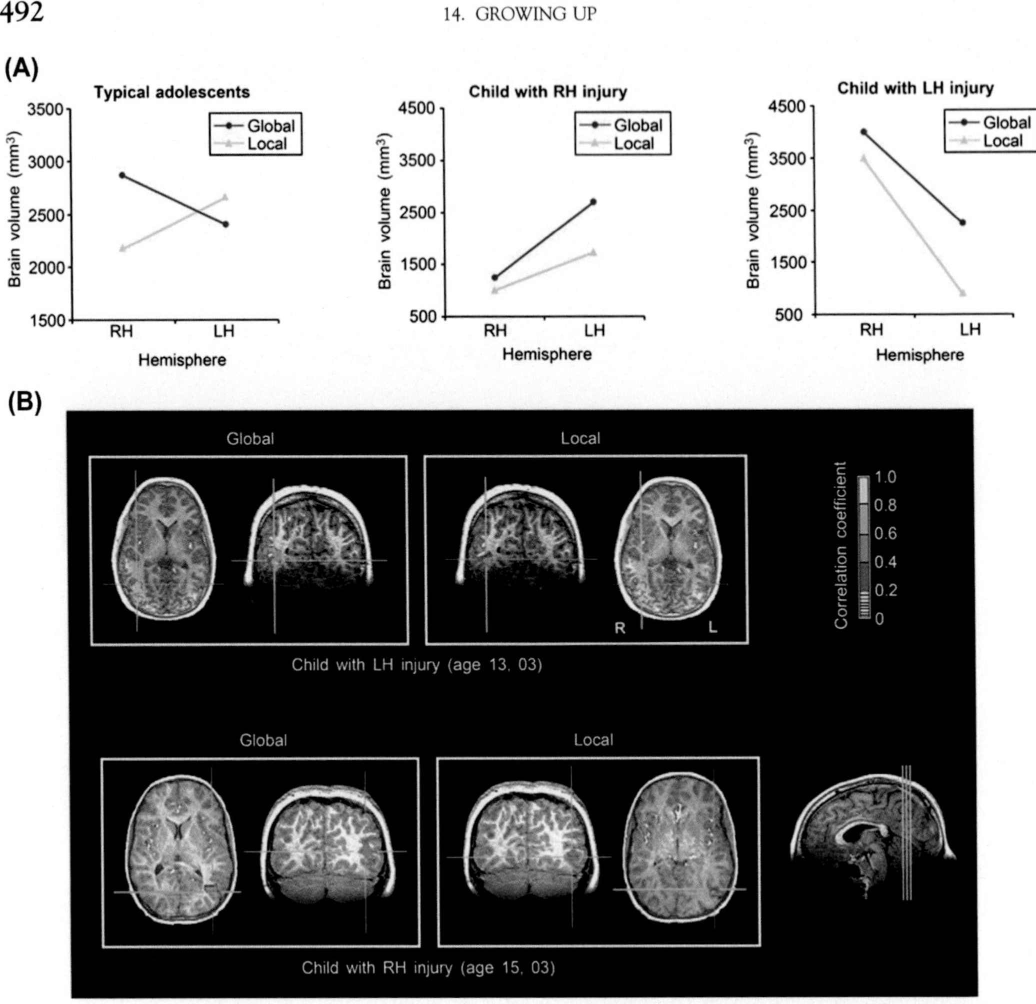

FIGURE 14.30 Global and local processing in the brain. (A) Functional magnetic resonance imaging activation data from two teenagers with prenatal focal brain injury on a hierarchical form-processing task, compared with data from typical adolescents. Each child participated in separate imaging runs, where they were asked to attend to either the global or the local level of the stimulus pattern. Unlike typical controls, who show different patterns of lateralization for global and local processing, the two children with lesions showed activation largely confined to the uninjured hemisphere. (B) Activation brain images for the two children with perinatal brain injury are shown. Top panel shows brain response from a child with left hemisphere injury with activation patterns shown in orange in the right hemisphere. Bottom panel shows brain response from a child with right hemisphere injury with activation patterns shown in orange in the left hemisphere. *Source: Adapted from Stiles et al. (2005), with permission.*

6. SUMMARY

In this chapter, we tracked the stages of human development from early embryo to infant to adolescent. While the field of developmental cognitive neuroscience is still a very young one, nevertheless, the findings presented in this chapter demonstrate the answers to important questions about human brain development and the correspondence to cognition. An overarching topic of much debate in the field of human development is the role of nature versus nurture. From the data presented here, you see that at each stage of human development there are important genetic effects and biological constraints at work in the unfolding of the human brain and mind. Similarly, at each state there are critical effects of the surrounding environment, whether at the level of the cell, the system, or the brain.

The advent of new techniques for noninvasively studying human development has provided the means to address new questions about cognitive development, such as what does a baby know before birth? Does an infant understand the grammar of language? How does the sense of self-development in an infant and a child? What are the long-term effects of focal brain damage? These and other questions will be addressed in future studies investigating the unfolding complex pattern of human brain development and its relation to cognition.

7. STUDY QUESTIONS

1. In what ways have neuroimaging techniques changed the way infant and child development is investigated?
2. What does the term *bidirectional influences* refer to in human development? Why is it an important concept?
3. Provide an example of a nature and nurture interaction that occurs before birth (prenatally).
4. What are some effects of maternal use of alcohol, tobacco, and marijuana on an unborn baby?
5. What brain regions develop and mature early in childhood? What regions develop later in childhood?
6. What language processes are described as bottom-up processes? Top-down processes? How do they differ in type of language knowledge and developmental stage?

Glossary

Acetylcholine (a-SEE-til-kol-en) A neurotransmitter that functions as a modulator of arousal, memory, and attention during waking states. ACh is generated in the brainstem in the pedunculopontine nucleus and laterodorsal tegmental nucleus and in the frontal lobe in the nucleus basalis. See Chapter 12.

Acoustical analysis (uh-KOO-sti-kul uh-NAL-ih-sus; Greek *akoustikos*—of hearing) The process of interpreting physical sound energy, whether linguistic, musical, or sounds in the environment (as in a door slamming or a car starting). See Chapter 5.

Action potential (AK-shun po-TEN-shul) In neurons, an electrochemical signal beginning near the cell body and traveling down the axon to the synaptic terminal. Also called a "spike" or "neuronal firing." See Chapter 3.

Alpha waves (AL-fa WAY-vz; first letter of the Greek alphabet) A regular electromagnetic wave detected in the brain or on the scalp and apparently reflecting the activity of large populations of neurons. Alpha waves have a frequency of 7.5–13 Hz and originate predominantly from the occipital lobe during periods of waking relaxation with the eyes closed. Conversely, alpha waves are decreased when the eyes are open, as well as by drowsiness and sleep. See Chapters 3 and 12.

Amnesia (am-NEE-zhuh; from Greek *a-mn-sia*—not memory) A loss of memory. Two types are anterograde (a loss of memory after the time of the brain injury) and retrograde (a loss of memory before the time of the brain injury). See Chapter 7.

Amygdala (uh-MIG-da-la; from *amygdale*—almond) The amygdalas are two small, almond-shaped masses of neurons located inside the tips of the temporal lobes. They are considered part of the limbic system and play major roles in emotions like fear and trust, as well as in learning. See Chapter 11.

Anterior (ann-TEER-ee-er; from *ante*—in front of) Located in front of something. See Chapter 2.

Anterior cingulate cortex (an-TEER-ee-er SIN-gyu-lut COR-teks; from Latin *ante*—before, in front of; Latin *cingulum*—girdle; Latin *cortex*—bark) The frontal part of the cingulate cortex. The anterior cingulate cortex is involved in executive functioning. See Chapter 2.

Anterior commissure (an-TEER-ee-er KA-mih-shur; from Latin *ante*—before, in front of) A large bundle of nerve fibers connecting the two cerebral hemispheres. See Chapter 2.

Anterograde amnesia (AN-teh-ro-grayd am-NEE-zhuh; from New Latin *antero*—forward; Greek *a-mne-sia*—not memory) A form of amnesia in which events after the brain injury are not encoded in long-term memory, although events may be recalled from the period before the injury. See Chapter 7. See **retrograde amnesia**.

Aphasia (AY-PHAY-zha; from, *a*—without; Latin *phasia*—speech) A loss of language function due to brain injury, such as damage to Broca's area, for speech production, or Wernicke's area, for speech understanding. See Chapter 6.

Arcuate fasciculus (AR-cue-ate fa-SIK-u-lus; Latin for arched bundle) A bundle of axonal fibers, especially the ones connecting Broca's and Wernicke's areas in the left hemisphere. See Chapter 2.

Area MT A part of the visual cortex that represents visual motion. See Chapter 4.

Artificial neural network (ar-ti-FI-shel NOOR-el NET; from Greek *neuron*—nerve) Also known as ANNs or neural models, artificial neural networks are simulated, simplified models of brain functions. Most are relatively small in scale. However, they are important for understanding the principles of neural computation.

Associative process (uh-SO-see-a-tiv PRA-ses; from Latin *ad-* + *sociare*—to join) A process in which one or more sensory and/or response events are linked in the brain.

Attention (a-TEN-shun; from Latin *attende-re*—to stretch out) Selection of some sensory, cognitive, or motor events to the exclusion of others. Attention is often taken to involve a focus on certain conscious events. Also see selective attention. See Chapter 8.

Attention network task (ANT; a-TEN-shun NET-werk TASK) A generalization of the **flanker task**, a tool for studying visual attention. The ANT allows testing of three separate aspects of attention: alerting before an expected signal, orienting to a specific location in space where the target is expected, and executive attention to act against expectations set up by the task. See Chapters 8 and 9.

Auditory cortex (AW-di-tor-ee kor-teks; from Latin *auditorius*—pertaining to one who hears; Latin *cortex*—bark) The parts of the cerebral cortex involved in processing sounds, such as Wernicke's area and Heschl's gyrus. See Chapter 5.

Auditory scene analysis (AW-di-tor-ee SEEN uh-NAL-ih-sus) The process by which the auditory system segments and organizes the listening environment. See Chapter 5.

Autonomic nervous system (ANS; aw-to-NOM-ic NER-vus SIS-tem; from Greek *neuron*—nerve) The division of the peripheral nervous system that acts to maintain homeostasis and to regulate rest and activity. Physiological activities controlled by the ANS, such as blood pressure and sweating, are generally unconscious and nonvoluntary.

Automatic process (au-to-MA-tic PRAH-ses) A highly practiced skill or habit that can be performed with minimal conscious involvement and voluntary effort. See Chapter 7.

Axon (AK-son) A long, slender branch of a nerve cell (neuron) that conducts electrical impulses away from the cell body. See Chapter 2.

Baron-Cohen, Simon (b. 1958) Autism researcher who proposed that young children develop a **theory of mind** capacity composed of four skills: detection of intentions of others, detection of eye direction, shared attention with others, and implicit knowledge about others. See Chapter 10.

Basal ganglia (BAY-zel GAN-glee-uh; from Greek *basis*—step, base; Greek for ganglion 'tumor on or near tendons') A large cluster of subcortical structures just outside of each thalamus, involving motor control, automaticity, cognition, emotions, and learning. See Chapter 2.

Beta waves (BAY-tuh WAYVZ; second letter of Greek alphabet) A band of irregular electromagnetic waveforms detected in the brain or on the scalp, and apparently reflecting the activity of large populations of neurons. The beta band has a frequency above 18–25 Hz and is associated with normal waking consciousness. Low-amplitude beta waves with multiple and varying frequencies often are associated with active, busy, or anxious thinking and active concentration. See Chapter 12.

Binocular disparity (bih-NOC-u-ler dis-PAR-eh-tee; from Latin *bi*—two; Latin *oculus*—eye; Latin *disparare*—to separate) The difference in perceived location of an object seen by the left and right eyes, resulting from the eyes' horizontal separation. The brain uses binocular disparity to obtain depth information from the retinal image in both eyes. See Chapter 4.

Binocular rivalry (bih-NOC-u-ler RYE-vel-ree; from Latin *bi*—two; Latin *oculus*—eye) The alternating perception that occurs when a different pattern is shown to each eye and the brain cannot fuse them into a single, coherent percept. Instead, awareness of each eye's input appears and disappears for a few seconds. See Chapters 4 and 8.

Biological motion (BYE-oh-loj-i-kal MO-shun) Biological motion refers to motions that a body can make as a set of articulated limbs, trunk, and head. Understanding the social implications of body movements is key to detecting any sign of danger or threat—as well as helping understand another person's intentions, emotions, and state of mind. See Chapter 10.

Bistable perception (BYE-STAY-bel per-SEP-shun; from Latin *b-*—two) Sensory events that alternate between two perceptual interpretations. See Chapter 4.

Blindsight (BLIND-site) A type of brain damage in which patients can report some visual events with no subjective sense of seeing them due to impairment of the first cortical area of the visual system, area V1.

Blood-oxygen-level-dependent (BOLD) activity (BLUD OKS-eh-gen LEV-el dee-PEN-dent ak-TI-vi-tee) A magnetically induced physical signal that reflects the flow of oxygen in the bloodstream in specific regions of the brain. The BOLD signal is the physical source for functional magnetic resonance imaging (fMRI). See Chapter 3.

Brainstem (BRAYN-stem) The lower part of the brain connecting to the spinal cord. All major motor and sensory systems pass through it, including the optic and auditory nerves. The brainstem also regulates cardiac and respiratory functions and maintains conscious waking, slow-wave sleep (SWS), and rapid eye movement (REM) dreams. See Chapter 2.

Broca, Pierre Paul (1824–80) A French surgeon who studied a brain-damaged patient with expressive aphasia—the inability to speak, while being able to understand speech. After the patient's death he was able to conduct a postmortem identifying the damaged region as the left inferior frontal gyrus, now called Broca's area. See Chapter 6.

Broca's area (BRO-kas AIR-ee-a) The left inferior frontal gyrus, or its posterior segment, reported by Pierre Paul Broca in 1861 is responsible for the deficit of a patient who could not speak but had preserved speech understanding. Other functions have since been attributed to Broca's area. See Chapter 6.

Brodmann's areas (BROD-mans AIR-ee-uh) About 100 cortical regions defined and numbered by German neurologist **Korbinian Brodmann**, originally based on the microscopic anatomy of **neurons** in different patches of the cortex. They are still widely used for cortical localization, and Brodmann's areas generally have distinctive functions. See Chapter 2.

Cell assemblies (SEL uh-SEM-blees) Also called Hebbian cell assemblies, these are active networks of related **neurons** representing some sensory input or similar event. According to Donald O. Hebb's 1949 hypothesis, "neurons that fire together, wire together," so that simultaneous firing causes the **synaptic** links in a cell assembly to grow stronger. See **Hebbian learning**. See Chapter 7.

Central executive (CEN-trel eks-EK-yoo-tiv; from Latin *centrum*—center) Brain processes for planning, decision-making, abstract thinking, rule acquisition, initiating and inhibiting actions, resolving goal conflicts, and flexible control of attention. These functions relate to **working memory** and tend to involve the **frontal lobes**. See Chapter 9.

Central nervous system (CNS; SEN-trel NER-vus SIS-tem; from Latin *centrum*—center; Latin *nervus*—sinew, nerve) The brain and spinal cord. All neurons outside of the CNS are considered to be the **peripheral nervous system** (PNS). See Chapter 2.

Central sulcus (SEN-tral SUL-cus; from Latin *sulcus*—groove) Also called the central fissure, this fold in the **cerebral** cortex is a prominent landmark of the brain that separates the **parietal lobe** from the **frontal lobe** and the **primary somatosensory cortex** from the **primary motor cortex**. The central sulcus is a clear dividing line between the input- and output-related areas of the cortex. See Chapter 2.

Cerebellum (ser-e-BEL-em; from the Latin word *cerebrum*—brain, *cerebellum* means "little brain") A major region of the brain located just below and to the rear of the occipital lobe of the cerebral cortex. The cerebellum plays an important role in the integration of sensory perception, fine motor control, and sensorimotor coordination. Recent evidence shows cognitive involvement as well. See Chapter 2.

Cerebral cortex (suh-REE-bral KOR-teks; from Latin *cerebrum*—brain; Greek *cortex*—bark) The outer surface of the great cerebrum, the largest part of the human brain, divided into two symmetrical **cerebral hemispheres**. Most of the cortex has six distinctive cellular layers, containing cell bodies with a gray appearance. But its long-distance nerve cells send out axons to other parts of the cortex, to the thalamus, and to other brain regions, which become covered with white supportive cells (**myelin**). As a result, a vertical cut of the cortex appears to the naked eye to have a thin, gray outer layer and a white inner mass, called the "gray matter" and "white matter," respectively. The cerebral cortex plays a key role in sensory analysis, spatial location, speech perception and production, **memory**, **attention**, emotion, motivation, action planning, voluntary control, thought, **executive functions**, and **consciousness**. See Chapters 2 and 3.

Cerebrospinal fluid (CSF; suh-ree-bro-SPEYE-nel floo-id; Latin cerebrum—brain) The internal circulation of the spine and brain. CSF allows for a protected flow of molecules and cells that is not exposed to the bloodstream. See Chapter 2.

Cerebrum (suh-REE-brum; Latin for *brain*) See **cerebral cortex**.

Chunking A way to make efficient use of short-term **memory** limitations by condensing large amounts of knowledge into small symbolic units, rules, or regularities called "chunks." In natural language, nouns can be considered to be chunks because they allow us to refer to large bodies of knowledge by single words. See Chapter 7.

Cingulate cortex (SIN-gyu-lut KOR-teks; from Latin *cingulum*—belt; Latin *cortex*—bark) A part of the **cortex** on the medial (inner) surface of each hemisphere. It is involved in executive functions, the resolution of conflicting goals, and emotion. See Chapters 2 and 9.

Circadian rhythms (ser-KA-dee-an) A daily, 24-h rhythmic cycle that is affected by light and darkness and that guides our daily wakefulness and sleep patterns. See Chapter 12.

Cognitive neuroscience (KOG-ni-tiv NOOR-o–SI–ens; from Latin *cognoscere*—to know) An emerging integration of two previously separate fields of science, cognitive psychology and neuroscience. Most research in cognitive neuroscience makes use of psychological methods simultaneous with brain activity recording.

Coma (KOh-mah) A state of deep and prolonged unconsciousness with a lack of response to any stimuli and loss of normal sleep–wake patterns. See Chapter 12.

Computed tomography (kom-PYOO-ted tom-OG-reh-fee; from Latin *computare*—to consider; Greek *tomos*—slice; Greek *graphein*—writing) Abbreviated as **CT**. Physiological recordings in which a three-dimensional image of a body structure (such as the brain) is constructed by computer from a series of slice images. See Chapter 3.

Confabulation (kon-fab-yoo-LAY-shen; from Latin *fabula*—story) A neurological symptom in which false memories or perceptions are reported with no intention to lie.

Confusional state A state following traumatic head injury. A posttraumatic confusional state may last days, months, or years ... or recovery may follow swiftly following transition from a minimally conscious state, to posttraumatic confusional state, to full consciousness. Posttraumatic confusional states include extended periods of wakefulness; however, the patient is clearly confused and disoriented. See Chapter 13.

Connectionism (keh-NEK-shun-ism; from Latin *com-* + *nectere*—to bind together) The study of artificial or biologically based neural networks.

Connectogram (keh-NEK-oh-gram) The connectogram was developed as a graphical representation of brain connectivity data studied using *connectomics*, the discipline for mapping and interpreting fiber connections in the brain. See Chapter 2.

Consciousness (KON-shes-ness; from Latin *con*—together; *scientia*—knowledge) Awareness, wakefulness. **Consciousness** implies being sensitive and responsive to the environment, in contrast to being asleep or in a coma.

Consolidation hypothesis (kon-SOL-ih-DAY-shun high-POTH-uh-sis) The process by which new memories are transformed into long-term **memory** traces. Memories may be stored in the same areas of the brain that support active moment-to-moment functions such as perception and speech control. Consolidation may involve synaptic changes in such brain regions, which make active neuronal connections more efficient. See Chapter 12.

Contralateral (KON-tra-LAT-er-el; Latin for "against the side") The opposite side of the body or brain. See Chapters 1, 2, and 4. See **ipsilateral**.

Coronal (keh-RONE-el; from Latin *corona*—crown) A crown-shaped vertical slice of the brain that divides it into **anterior** and **posterior** halves. See Chapter 2.

Corpus callosum (KOR-pus kal–OS–um; from Latin *corpus*—body; Latin *callosum*—tough) A massive fiber bridge between the right and left hemispheres, consisting of more than 100 neuronal axons. It appears white when cut because the axons are covered by white **myelin** cells. See Chapter 2.

Cortex (KOR-teks) See **cerebral cortex**.

Cortical column (KOR-ti-kel KAW-lum; from Greek *cort*—skin, husk; Latin *columna*—pillar) A barrel-shaped slice of the six surface layers of the **cortex** that often contain closely related **neurons**. Columns are about 0.5 mm in diameter and 2.5 mm in depth. They may be clustered into **hypercolumns**, which may be part of even larger clusters.

Cortical core (KOR-ti-kel kor) An expression referring to the cerebral cortex and the thalamus in the left and right hemispheres. Together, these structures form a massive hub that serves as a core for brain function. See Chapter 2.

Cortisol (Kor-ih-sal) Cortisol is a hormone key to the circadian rhythms, wake and sleep. Secreted by the adrenal glands, cortisol levels begin to increase in the early morning. They increase by about 50% in the first half hour after awakening, triggering signals throughout the brain and body for arousal. This is termed the "cortisol awakening response." Cortisol levels decrease throughout the day into the evening, reaching their lowest levels between midnight and 4 a.m. during NREM and SWS stages. Cortisol levels increase again in the early morning during REM sleep and as dawn approaches, they ramp up for another day. See Chapter 12.

Creole (KRAY-ole) A true language that children spontaneously evolve in multilingual communities. Creoles often are encountered in island communities where language communities overlap. "Creoles" are contrasted with "pidgins," which are dialects typically spoken by adults as a second language consisting of simplified phrases rather than full grammatical languages: for example, "me go now" as opposed to "I am going now." Creoles are remarkable because they exhibit a full-fledged grammar, unlike pidgins. The spontaneity with which they arise suggests that human infants and children may be equipped with a biological language capacity with universal features.

Crowding *Crowding* is the difficulty in recognizing individual objects in clusters of other objects or natural scenes. Note that we stated "recognizing"—despite being in a crowded part of your visual field, individual features of that crowded scene are detected. The breakdown comes when they are to be identified as individual objects. See Chapter 4.

Declarative memory (deh-KLAR-a-tiv MEM-ree; from Latin *declarare*—Latin *declarare*, from *de- 'thoroughly'* + *clarare 'make clear.'*) The capacity to recall facts and beliefs. A kind of **explicit memory**. See Chapter 7.

Delta waves (DEL-tuh WAYVZ; fourth letter of Greek alphabet) A band of slow, high-amplitude electromagnetic waveforms associated with deep sleep and recorded in the brain or on the scalp, apparently reflecting large populations of neurons. Delta generally is considered to be less than 2.5 Hz. It coexists with waking EEG as well but becomes visible in the raw (unprocessed) EEG only when delta predominates in sleep and drowsy states. See Chapter 3.

Dendrite (DEN-drite; from Greek *dendritēs 'treelike'*) One of numerous thin, branched micron-level tubes extending from the cell body of a **neuron**. Dendrites typically receive synaptic stimulation from other neurons, and therefore serve as the input branches of the neuron. See Chapter 2.

Descartes, René (1596–1650) A French philosopher, mathematician, scientist, and writer who spent most of his adult life in the Dutch Republic. Descartes has been dubbed the "Father of Modern Philosophy" and was also a careful student of the brain. He often is considered the originator of modern mind/body philosophy.

Developmental cognitive neuroscience (deh-vel-op-MEN-tel COG-ni-tiv NUR-o-si-ens) The study of the normal growth of the brain and its mental capacities. See Chapter 14.

Diencephalons (die-en-SEF-a-lon; from Greek *dia*-—through; *enkephalos*—brain) The part of the brain that contains the **thalamus, hypothalamus**, and the posterior half of the **pituitary gland**.

Diffusion tractography (di-FYOO-zhen trak-TOH-greh-fee; from Latin *diffusus*—scatter; *trahere*—to pull; Greek *graphein*—writing) A brain imaging technique that tracks the diffusion of water molecules to trace the major neuronal pathways of the brain. See Chapter 3.

Discourse (DIS-kors) A connected series of utterances, a conversation. Used by linguists to reflect more than a series of sentences, rather a conversation with a theme and with intents. See Chapter 6.

Disorder of consciousness Disorders of consciousness (DOC) are *potentially nonreversible* unconscious states due to brain trauma or damage. See Chapter 13.

Domain specificity (do-MANE spes-i-FIS-ih-tee) Functional specificity of brain regions or mechanisms. The idea that each cognitive function may have its own region or network of brain regions, rather than general-purpose brain mechanisms with multiple cognitive functions.

Dorsal (DOR-sel; from *dorsum*—back) The upper part of a brain structure, also called **superior**.

Dorsolateral prefrontal cortex (DOR-so-LAT-er-el pree-FRON-tal KOR-teks; from Latin *dorsum*—back; *latus*—side; *pre*—in front of; *frons*—the forehead; Greek *cortex*—bark) Prefrontal region involved in motor planning, executive control, self-regulation, emotion, and **working memory**. See Chapters 2 and 9.

Dynamic causal modeling (die-NAM-ic KOS-el MO-del-ing; from Greek *dynamikos*—powerful; Latin *causa*—cause; Latin *modulus*—small measure) A method for interpreting brain data, such as fMRI, that helps to interpret causal relationships among brain activities during a specified task.

Edelman, Gerald M. (b. 1929, d. 2014) American immunologist and neurobiologist who won the Nobel Prize for his work on the structure of antibody molecules. Edelman developed the theoretical framework of Neural Darwinism, which applies Darwinian selectionist principles to the brain, in contrast to the instructionist principles of conventional computers.

Electroencephalography (EEG;eh-LEK-tro-en-sef-eh- LOG-reh-fee; from Greek *e-lektron*—sunlight; *en-* + *kephale*—in the head; *graphein*—writing) Electrical activity that typically is recorded on the scalp and sometimes on the surface of the **cortex**, reflecting the electromagnetic field of large numbers of active neurons. See Chapter 3.

Empathy (EM-path-ee; from Greek *empatheia*—passion) The capability to share one's feelings and understand another person's. See Chapter 11.

Epigenesis (ep-ih-GEN-eh-sis; from Greek *epi*—after; Greek *genesis*—birth, origin) Non-DNA factors that shape cells during gestation (pregnancy) and after birth. Contrasted with the classical central dogma of molecular biology in which DNA is recoded into transfer RNA, which ends in the production of proteins for the structure and functions of all cells. Epigenesis implies a flow of causality in the opposite direction. For example, numerous physiological and environmental factors can influence whether specific genes (DNA) are expressed or not. DNA is the primary molecule that encodes phenotypes, passing the plan for a species from one generation to the next. But non-DNA factors can influence the activation and silencing of DNA, the on/off switches. See Chapter 14.

Episodic memory (ep-i-SOD-ic MEM-ree; from Greek *episeidos*—coming in besides) Memory for conscious experiences, especially those that can be explicitly recalled, such as times, places, events, associated emotions, and other contextual knowledge. The formation of new episodic memories requires the **medial temporal lobe**, especially the **hippocampal region** in combination with the **cerebral cortex**. See Chapter 7.

Evoked potential (EP; ee-VOKD puh-TEN-shul; from Latin *evocare*—to call forth; *potentia*—power) Also called *event-related potential* (ERP). A quite stereotypical electrical voltage pattern obtained from the brain, after averaging a time-locked voltage to a stimulus or other known event. Traditionally, the **EP** was obtained by averaging the stimulus-locked **EEG** over numerous trials. Although the exact brain sources of **EPs** are still debated, they are highly sensitive to cognitive and emotional variables. See Chapter 3.

Executive attention (ek-ZEK-u-tiv a-TEN-shun) Also called **voluntary, goal-directed**, or **top-down attention**. The act of voluntarily focusing on one stream of conscious events while ignoring others. Also see **selective attention, stimulus-driven attention**. See Chapters 8 and 9.

Executive function (ek-ZEK-u-tiv FUNK-shun) Also called **executive control** or **frontal lobe function**. Capacities such as planning, cognitive flexibility, voluntary action, abstract thinking, rule acquisition, initiating correct actions and inhibiting incorrect ones, impulse control, and emotional regulation. See Chapter 9.

Explicit memory (eks-PLI-sit MEM-ree; from Latin *explicitus*—clear) A type of **memory** involving conscious, intentional recollection of stored experiences and knowledge. See **implicit memory**, **implicit learning**. See Chapter 7.

Feedback (FEED-bak) 1. In goal-guided systems, a signal from the environment indicating the degree of error in achieving the goal. 2. In neuroscience and psychology, an environmental signal reflecting some neuronal event. This kind of neurofeedback often allows people to learn to control otherwise spontaneous neuronal activities. 3. In neural networks, a flow of information returning an output signal to the input layer of the network. Some theorists make a strong distinction between **feedback** and **reentrant** signaling in the **thalamo-cortical** system of the brain. See **Neural Darwinism**. Chapter 7.

Feedforward (feed-FOR-werd) 1. Signal passing from a simpler to a more complex stage of processing. 2. In sensorimotor guidance, preparing an internal action trajectory to obtain more precise feedback when the action is executed. This strategy is used in fast movement control in birds and humans, and even in machines like aerodynamically unstable jet planes. 3. In neural networks, passing information from earlier to later layers of the network.

Fetal alcohol syndrome (FEE-tel AL-ko-hol SIN-drum) (FAS) Brain damage in a fetus due to the mother's alcohol consumption, a major health risk. See Chapter 14.

Flanker task (FLANK-er TASK) A tool for studying visual attention in which the subject is asked to respond as quickly as possible to a target at the center of gaze or one located off-center by a known distance. The target is flanked by distracting stimuli, such as arrows or letters. The flanker task permits quantitative assessment of the subject's speed and accuracy in shifting attention to an expected or unexpected position of the target. Moving attention in an unexpected direction is taken to require **executive attention**, since it must override the prepared, expected shift. Flanker tasks generally require subjects to avoid voluntary eye movements, so that any change in accuracy or speed in response to a target can be attributed to **implicit** shifts of attention.

Fourier analysis (FOOR-ee-ay uh-NAL-a-sis) Named after French mathematician and physicist Joseph Fourier, who showed that any complex function can be decomposed into a finite set of sine and cosine functions. In music, for example, this implies that any complex sound can be decomposed into a set of pure tones (sine waves). Fourier analysis is routinely applied to decompose EEG and other complex brain signals into frequency bands.

Frontal lobe (FRON-tal lobe) A large region of the cortex located at the front of each cerebral hemisphere and positioned forward of the **parietal lobes** and above and in front of the **temporal lobes**. The executive functions of the frontal lobes include the ability to anticipate the consequences of actions, to plan and make decisions, to speak, to override inappropriate impulses and resolve conflicting goals, to understand the mental states of others, and to hold information in working memory. See Chapters 2 and 9.

Functional fixedness (FUNK-shun-el FIKS-ed-ness) A cognitive set that tends to block a person from novel ways of acting, perceiving, or solving problems. See Chapter 7.

Functional magnetic resonance imaging (fMRI; FUNK-shun-el mag-NET-ic REZ-nence IH-ma-jing) fMRI uses a combination of the MRI signal—which provides images of the brain with high anatomical accuracy—with measures to assess sensory, motor, and cognitive processes. fMRI experiments typically use a subtraction method where brain activation for one experimental condition is literally subtracted from another. Early fMRI studies employed a task + resting state method where, for example, brain activation for a particular task would be compared for activation during a rest (no task) condition. fMRI methods have continued to change and evolve and currently there are many other methods employed that are more sensitive to providing information about complex cognitive processes. fMRI helped to make cognitive neuroscience possible. See Chapter 3.

Functional neuroanatomy (FUNK-shun-el NUR-el-an-at-oh-mee) The study of how the many brain areas contribute to our functional selves: ranging from emotional processing to social cognition, to language and thought, and to attentional processes and future planning. See Chapters 2 and 3.

Functional redundancy (FUNK-shun-el ree-DUN-den-see) Built-in backup functions in a system to prevent the complete failure of critical functions. For example, mammals have two lungs so that if one fails, the organism still has one lung to breathe. The brain has multiple redundant capabilities.

Fusiform face area (FFA; FYOO-ze-form; from Latin *fusus*—spindle, after its shape) A specialized region in the **medial temporal lobe** that responds selectively to visual faces compared with other objects. See Chapters 4 and 10.

Gage, Phineas (1823–60) A historic brain damage patient, whose railroad accident demonstrated remarkable spared cognitive capacities in spite of severe damage to the frontal lobes. Gage was a railroad foreman who had a two-foot long thin tamping iron shot through the upper orbit of the left eye and out through the medial scalp, when an unstable dynamite charge exploded unexpectedly. Although Gage appeared to have no loss of perception, motor control, or speech, his personality changed in ways that have come to typify frontal lobe damage, especially a major loss of impulse control and long-term motivation.

Gamma waves (GAM-a WAYVZ; third letter of Greek alphabet) A band of fast, low-amplitude electromagnetic waveforms associated with wakefulness and active thinking, and recorded in the brain or on the scalp, apparently reflecting the activities of large populations of neurons. The gamma band is thought to be centered near 40 Hz, ranging from 25 to 60 Hz. However, higher frequency waves are sometimes labeled gamma as well. Gamma is thought to reflect regional connectivity in the service of current tasks. See Chapter 3.

Ganglion (GAN-glee-on; Greek for ganglion 'tumor on or near tendons') A large cluster of **neurons**. The major subcortical organs may be considered to be ganglia, such as the **basal ganglia**. They are often very large structures and have multiple functions. They are typically composed of subdivisions, which themselves are often layered and folded arrays of nerve cells.

Gestalt (gesh-TALT; German for form) 1. A perceptual stimulus that cannot be reduced to simple subcomponents. 2. A branch of psychology based on the German concept of *Gestalt*, often summed up with the slogan that "The whole is more than the sum of its parts." Gestalt psychology has profoundly influenced the study of perception. See Chapter 6.

Glial cell (GLEE-el SEL; from Greek *glia*—glue) Nonneuronal cells in the brain that support neurons, maintain neurochemical homeostasis, form a protective **myelin sheath** around neurons, and process information. See Chapter 14.

Global states Most scientists agree to the existence of three global brain states for humans: *waking, sleeping, and dreaming*. See Chapter 12.

Gray matter The outer layers of the **cerebral cortex**, as seen with the naked eye. Gray matter contains the cell bodies of tens of billions of **neurons** that send white-covered **axons** in many directions below the cortical mantle. See **white matter**. See Chapter 2.

Hebbian learning (HEB-ee-en LUR-ning) According to Donald O. Hebb, "neurons that fire together, wire together." That is, two neurons strengthen their **synaptic** links if they are active at the same moment. This process forms **cell assemblies**. Introduced by Donald Hebb in 1949, it is also called Hebb's rule. See Chapter 7.

Hemispheric lateralization (hem-is-FEER-ik lat-er-al-ih-ZAY-shun; from Greek *hēmi-* (half) + *sphairion*—sphere; and Latin *lateralis*—side) The degree to which certain brain functions are performed primarily by one *cerebral* hemisphere, the most prominent being speech production on the left side for most people.

Hemodynamics (HEE-mo-dye-NAM-ics; from Greek *hema*—blood; *dynamos*—force or power) The study of blood flow changes, particularly in the brain, as an index of local neural activity. See Chapter 2.

Hippocampus (hip-o-KAM-pes; Greek seahorse, from *hippos*—horse, *kampos*—sea monster) In the human brain, the hippocampi are looped structures in each of the two **medial temporal lobes**. The hippocampi are part of the **limbic system** and play basic roles in encoding and retrieving **episodic** and **semantic memories** and in spatial navigation. See Chapters 2 and 7.

Homunculus (ho-MUN-cue-lus; Latin—little man) The distorted human body maps in the **primary somatosensory cortex** (the sensory homunculus) and in the **primary motor cortex** (the motor homunculus). The lips, hands, feet, and sex organs have more sensory neurons than other parts of the body, so the homunculus has correspondingly distorted large lips, hands, feet, and genitals. Each hemisphere contains a sensory and motor homunculus of the opposite side of the body. These body maps were discovered by neurosurgeon Wilder Penfield at the University of Montreal in the 1950s and 1960s.

Hypothalamus (hie-po-THAL-a-mus; from Latin *hypo*—below; Greek *thalamos*—chamber) The major neuroendocrine organ of each side the brain, with vital roles in the regulation of blood nutrients, motivation, appetite control, and other major life functions. The hypothalamus is located below each **thalamus** just above the **brainstem**. It is closely related to the **pituitary** and **pineal glands**. See Chapter 2.

Immediate memory (ih-MEE-dee-et MEM-er-ee) Also called short-term **memory**. The ability to recall something for 10–30 s without rehearsal. **Working memory** and **sensory memories** can be seen as specific kinds of immediate memory. See Chapter 7.

Implicit memory (im-PLI-sit MEM-er-ee; from Latin *implicitus*—obscure) Unconscious **memory**, which may arise from conscious or unconscious events. See Chapter 7.

Inattentional blindness (in-uh-TEN-shun-el BLIND-ness) A reliable experimental phenomenon in which one is not able to see things that are normally clearly visible. See Chapter 8.

Inference (IN-fer-ens) Drawing a conclusion based on knowledge rather than direct observation.

Inferior (in-FEER-ee-er; from Latin *inferus*—lower) Below. See Chapter 2.

Insomnia Insomnia is by far the most common sleep disorder. People with insomnia can have trouble falling asleep, staying asleep, or waking up too early in the morning. Some instances of insomnia are brief: these are called *acute insomnia* and can happen because of stress, work demands, or emotional issues. However, when insomnia occurs for at least 3 days a week and lasts more than 3 months, it is called *chronic insomnia*. While some instance of insomnia occurs for almost half of all people, chronic insomnia is far less typical, occurring in only about 10% of the population. See Chapter 12.

Insula (IN-soo-la; Latin for island) A structure that is hidden in and underneath the **lateral sulcus**, covered up by the **temporal** and **parietal lobes**, and therefore appears as an island when the covering tissues are gently pulled away. The insula is involved in "gut feelings," such as the sense of nausea and disgust, and possibly in emotional feelings and cravings. See Chapter 2.

Intentionality (in-ten-shen-AL-ih-tee) The "aboutness" of mental events, their ability to represent aspects of the world. Distinguished from *intention* as a mental goal. See Chapter 10.

Interaural level difference (in-ter-OR-el; from Latin *inter-*—between; *auris*—ear) A method of sound localization in which the brain detects the small difference in loudness between the two ears that occurs when a sound travels toward the head from an angle. See Chapter 5.

Interaural time difference (in-ter-OR-el) A method of **sound localization** in which the brain detects the split-second delay between the time when sound from a lateral source reaches the near ear and when it reaches the far ear. See Chapter 5.

Intonation contour (in-toh-NAY-shun kon-TOOR) The "melody" or singsong of normal speech. In English and other languages, questions typically are given a different intonation contour (a rising tone) compared with affirmative statements (a falling tone).

Ipsilateral (IP-si-LAT-er-al; from Latin *ipse*—self; *latus*—side) On the same side of the body. See **contralateral**. See Chapter 2.

Lateral (LAT-er-al; from Latin *lateralis*—side) On the side(s) of the brain. See Chapter 2.

Lateral geniculate nucleus (LGN; LAT-er-el jen-IK-yoo-let NOO-klee-us; from Latin *latus*—side; *genu*—knee-shaped; Latin *nux*—nut) A nucleus consisting of "knee-shaped" layers of cells in the thalamus. It is the primary relay center between the **retina** of the eye and the **primary visual cortex** (Area V1). See Chapter 2.

Lateral inhibition (LAT-er-el in-hi-BI-shun; from Latin *latus*—side; Latin *inhibitus*—restrain) The capacity of a **neuron** to reduce the activity of its neighboring cells in the same layer of neurons. See Chapter 4.

Lateral occipital complex (LOC; LAT-er-el ox-SIP-it-al KOM-pleks; from Latin *latus*—side; Latin *occiput*—rearmost part of the skull) A region on the side of the **occipital lobe** that has a general role in visual object recognition. See Chapter 4.

Lateral sulcus (LAT-er-al SUL-cus; from Latin *latus*—side; *sulcus*—groove) Also called Sylvian fissure or lateral fissure. This prominent "valley" divides the **temporal lobe** from the **frontal** and **parietal lobes**. See Chapter 2.

Lexical identification (LEKS-ih-kul eye-den-tih-fih-KAY-shun; from Greek *lexis*—word) The process of assigning words to speech sounds. See Chapter 6.

Lexicon (LEKS-ih-con; from Greek *lexis*—word) The vocabulary of a language.

Limbic system (LIM-bik sis-tem; from Latin *limbus*—border) A set of brain structures involved in emotion, memory, olfaction, and action control, including the **hippocampus**, **amygdala**, **thalamus**, **hypothalamus**, and **cingulate gyrus**. The limbic system is interwoven with the endocrine system and **autonomic nervous system**. See Chapter 11.

Locked-in syndrome Although discussed in depth in Chapter 13, Disorders of Consciousness (DOC), locked-in (LI) syndrome is *not* a DOC. An LI patient typically has brain damage; however, it is limited to the brainstem region and the cortex is largely unaffected. So, although they may present as similar to a patient in a vegetative state (VS) without the ability to move or speak, the LI patient actually has an intact cortex and preserved cognition. See Chapter 13.

Long-term depression (LTD; LONG TERM de-PRE-shun; from Latin *deprimere*—to press down) A lasting decrease in the strength of a **synapse**. Along with **long-term potentiation** (**LTP**), **LTD** is thought to be a synaptic basis for **learning** and long-term **memory**.

Long-term potentiation (LTP; LONG TERM puh-ten-shee-AY-shun; from Latin *potentia*—power) A long-lasting strengthening of a synaptic link. Along with **LTD, LTP** is thought to be the synaptic basis of **learning** and long-term **memory**.

Longitudinal fissure (lon-gi-TOOD-in-al FISH-er; from Latin *fissus*—crack, opening) The deep valley that divides the right and left hemispheres of the vertebrate brain. See Chapter 2.

Magnetic resonance imaging (MRI; mag-NET-ik REZ-nence IH-ma-jing; Latin *resonare*—to sound; *imago*—imitation) Based on the spin resonance of atomic nuclei, MRI is a technique used to visualize the internal structures of the body, including the brain. **Functional MRI (fMRI)** records brain activity and is often superimposed on the structural brain image obtained via **MRI**. See Chapter 3.

Magnetoencephalography (MEG; mag-NET-o-en-sef-eh-LOG-gra-fee; Greek *en-* + *kephale-*—in the head; *graphein*—writing) An imaging technique based on the magnetic fields produced by brain activity. MEG is silent and noninvasive and has good temporal and spatial resolution. See Chapter 3.

Medial (MEE-dee-al) Toward the midline of the body or the brain. **midsagittal**. See Chapter 2.

Medial temporal lobe (MTL; MEE-dee-el TEM-per-el LOBE) The bottom aspect of the **temporal** lobes, which are arranged symmetrically around the midline and contain evolutionarily ancient olfactory structures, memory encoding and recall, and emotional functions. See Chapters 2 and 7.

Memory (MEM-ree, MEM-eh-ree; from Latin *memor*—mindful) A lasting brain representation that is reflected in thinking, experience, or behavior. See Chapter 7.

Mental flexibility (MEN-tel fleks-ih-BIL-ih-tee; from Latin *mens*—mind; Latin *flexus*—bent; Latin *-ibilis*, from *-bilis*—capable or worthy of) Also called ability to shift **cognitive set**. The capacity to respond rapidly to unanticipated environmental contingencies. See Chapter 7.

Mentalize (MEN-tel-ize; from Latin *mens*—mind) The ability to understand the self and others, not just as sensory objects but also as subjective beings with mental states. See Chapter 11.

Metacognition (MET-a-cog-NI-shen; from Greek *meta*—above; Latin *cognere*—to know) Knowing about cognition **awareness and understanding of one's own thought processes**.

Midsagittal (mid-SAJ-i-tal; from Latin *sagitta*—arrow) medial. The midline plane of section, going from the nose to the middle of the back of the head. In the brain, a midsagittal view shows a single hemisphere with a view of the brainstem, the corpus callosum, and the four lobes. See Chapter 2.

Mind (from Greek *menos*—spirit) Those aspects of intellect and **consciousness** manifested in thought, perception, **memory,** emotion, will, and imagination, including all of the brain's conscious and unconscious cognitive processes.

Minimally conscious state A minimally conscious state (MCS) differs sharply from a VS patient (see **vegetative state)**. While an MCS includes eye opening as in VS, that wakefulness sign is coupled with some receptive and expressive language function, some command-following behavior, visual pursuit of a moving object, and some automatic motor movement sequences. There are some inconsistent but clear signs of self-awareness in an MCS patient. Some patients in an MCS may respond emotionally to family faces or photographs; they may smile or cry. They may make purposeful movements, for example, reaching for a glass; however, the MCS is characterized by *inconsistency* in these behaviors. MCS is thought of as a *transitional state*, that is, a state a patient *travels through* during recovery of consciousness with healing or decline with neurodegeneration. See Chapter 13.

Morpheme (MOR-feem; from Greek *morphe*—form) The smallest linguistic unit that can convey meaningful information by itself. In English, prefixes and suffixes are considered to be morphemes (e.g., "pre-" and "post-") as well as the single phoneme/s/, which can signify the plural. See Chapter 5.

Motion blindness (MO-shun BLIND-nes) A symptom caused by injury to brain regions needed for motion perception, such as **area MT** of the visual cortex, resulting in an inability to perceive visual motion. See Chapter 4.

Multistable perception (MUL-tee-STAY-bel per-SEP-shun) from Latin *stabilis*, from the base of ***stare 'to stand.'*** Alternating visual perceptions of an ambiguous stimulus. See bistable perception and Chapter 4.

Myelin (MY-e-lin; from Latin *myel*—marrow) A sheath of glial cells, called the myelin sheath, surrounding the **axons** of many **neurons**. Myelinated axons appear **white**, hence the "**white matter**" of the visible brain. See Chapters 2 and 3.

Narcolepsy (NAR-ko-lep-see) Narcolepsy is a rare REM-centered sleep disorder of largely unknown origin. It typically occurs in adolescents and young adults but is sometimes not accurately diagnosed until years later. There are five central symptoms that together form a likely diagnosis of narcolepsy, but the foremost symptom of narcolepsy is that the patient suddenly falls asleep. The core symptoms associated with this are persistent daytime sleepiness, hallucinations, *sleep paralysis*, *cataplexy*, and restless sleep during the nighttime. See Chapter 12.

Neocortex (NEE-o-COR-tex; from Latin *neo*—new; Greek *cort*—bark) The largest and most visible part of the human cerebral cortex. It is the "new" cortex from an evolutionary point of view, as contrasted with the "old" cortex of the **medial temporal lobe**, **hippocampus**, and olfactory brain. See Chapter 2.

Neon color spreading (NEE-on CAW-ler SPRED-ing) A perceptual illusion in which white space appears to be tinted by proximity to colored and black lines. See Chapter 4.

Neural Darwinism (NUR-el DAR-win-izm; from Greek *neuron*—nerve) A theory proposed by physician and neuroscientist Gerald Edelman that suggests that neurons develop and make connections following Darwinian principles. In biological evolution, species adapt by reproduction, mutations leading to diverse forms, and selection among the resulting repertoire of slightly different organisms. Neural Darwinism suggests that brains develop in similar fashion, both in the reproduction, variation, and selection of developing neurons, and in a later stage, in the Darwinian selection of synaptic connections. Brains are said to be *selectionist* rather than *instructionist*, unlike the program of a digital computer.

Neural migration (NUR-el my-GRAY-shun; from Greek *neuron*—nerve) Movement of nerve cells from their place of origin toward their final location in the developing brain. See Chapter 14.

Neural net model (NUR-el NET MO-del; from Greek *neuron*—nerve) Also known as **artificial neural networks (ANNs)**, neural models are simulated, simplified models of selected brain functions. Most are relatively small in scale and do not represent the great complexity of the brain. However, they are important for a better understanding of how neural computation might work.

Neuromodulator (NOOR-o-MOD-u-lay-ter; from Greek *neuron*—nerve; *modulate* is used in the sense of "influence" or "regulate") Certain neurochemicals have very widespread effects in large regions of the brain. These are called **neuromodulators**, whereas **neurotransmitters** are molecules with very local effects in specific synapses. See Chapter 12.

Neuron (NOOR-on or NYOO-ron; from Greek *neuron*—nerve) Nerve cells that transmit information by electrochemical signaling. They are the core components of the human brain, spinal cord, and peripheral nerves. Many different types of neurons exist, from sensory receptors and motor units and neuroendocrine cells to pyramidal neurons, which have long-distance axons; interneurons, which form bushy local connections; and a wide variety of specialized cells. See Chapter 2.

Neuron doctrine (NOOR-on or NYOO-ron DOK-trin; from Greek *neuron*—nerve; *doctor*—teacher) The central theory of the neuron doctrine was a theory credited to the Spanish histologist **Santiago Ramon y Cajal**, stating that "the nervous system consists of numerous nerve units (**neurons**), anatomically and genetically independent." This has been one of the basic assumptions of brain science for the past century. However, the discovery of large numbers of electrical synapses (gap junctions) may challenge some aspects of the neuron doctrine.

Neuroanatomy (NOOR-oh-an-at-oh-mee) The study of the structures and regions of the brain. See Chapter 2.

Neuroconnectivity (NOOR-oh-con-ekt-iv-ih-tee) The study of the connected pathways of the brain. See Chapter 2.

Neurodynamics (NOOR-oh-die-nam-ikz) The study of the neural transmission and oscillations in the brain. See Chapter 2.

Neurophysiology (NOOR-oh-fiz-ee-ol-oh-gee) The study of the cellular properties in the brain. See Chapter 2.

Neurotransmission (NOOT-oh-trans-MISH-en) Electrochemical signaling between nerve cells. See Chapter 3.

Neurotransmitter (NOOR-o-TRANS-mit-er; from Greek *neuron*—nerve; Latin *trans*—moving through) Chemicals that act to relay a signal from one neuron to the next across a synaptic cleft. Some neurotransmitters are packaged into **vesicles** that cluster beneath the membrane on the presynaptic side of a **synapse** and are released into the **synaptic cleft**, where they bind to receptors located on the postsynaptic membrane. Release of neurotransmitters often is driven by **action potentials** in the presynaptic axon. There is a low level of baseline release even in the absence of an action potential. See Chapter 2.

Nonrapid eye movement sleep Sleep falls into two general patterns: nonrapid eye movement (NREM) and rapid eye movement (REM) sleep. NREM sleep is described as occurring during sleep stages I–IV. It is characterized by slowed brain functions and slow waves as measured by the electroencephalogram. See Chapter 12.

Object permanence (OB-jekt PER-ma-nens) The knowledge that perceptual objects continue to exist even when they cannot be seen or touched. Object permanence begins in infants around 7 months. See Chapter 14.

Occipital lobe (ox-SIP-it-al lowb; from Latin *occiput*—a back bone of the skull) The occipital lobes, which contain the earliest visual region of the cortex, are the smallest of four lobes in the human **cerebral cortex**. See Chapter 2.

Orbitofrontal cortex (or-bit-oh-FRON-tel COR-teks; from Latin *orbis*—circle, orb, orbit, world) Refers to the part of the brain immediately above the sockets or *orbits* of the two eyes. See Chapters 2 and 11.

Output functions (OWT-put FUNK-shuns) Brain processes controlled by the **frontal lobes** that include the **central executive**, action planning, and motor output. See Chapter 2.

Parahippocampal place area (PAIR-a-HIP-o-KAMP-el [PPA]; from Greek *para*—before; see hippocampus [PPA]) A region near the hippocampus that responds more strongly to landscapes and visual scenes than to isolated objects such as houses or faces. See Chapter 4.

Parietal lobe (puh-REYE-uh-tl lowb; from Latin *parietalis*—relating to walls) A large cortical region located above the **occipital lobe** and behind the **frontal lobe**. The parietal lobe integrates sensory information from different modalities and contains constantly updated maps of the position of the body and nearby objects. See Chapter 2.

Penfield, Wilder (1891–1976) A neurosurgeon and researcher in Montreal who performed pioneering work in epileptic surgery. Before operating, he performed exploratory brain stimulation in awake patients (who were free of pain using only a local anesthetic in the surgical opening). Thus patients could report their experiences upon electrical brain stimulation. Penfield and coworkers were able to determine functions of the human brain that were previously only approachable via postmortem studies of brain-damaged patients.

Perceptual filling in (per-SEP-choo-el FIL-ing in) A general feature of sensory perception in which the brain fills in missing parts of a visual object or scene, often far beyond the direct sensory input. See Chapter 4.

Perceptual memory (per-SEP-choo-el MEM-ree, MEM-eh-ree; from Latin *memor*—mindful) Long-lasting changes in one's ability to perceive the world—for example, the ability to perceive the sounds of speech and to recognize visual objects under changes in orientation and lighting.

Peripheral nervous system (PNS; per-IF-er-el NUHR-vus SIS-tem) The extensive network of neurons outside of the brain and spinal cord. The PNS includes sensorimotor neurons below the neck and autonomic neurons that innervate the smooth musculature of the digestive tract, heart, and circulatory system. See Chapter 2.

Perseveration (per-sev-er-AY-shun; from Latin *perseverare*—persist) A symptom involving the inappropriate and uncontrollable repetition of a specific word, phrase, or gesture.

Phoneme (FO-neem; from Greek *phōnē*—sound) In human languages, the smallest lexically distinctive category of sound, such as consonants and vowels. See Chapter 5.

Planum temporale (PLAH-num tem-por-AHL-eh; from Latin *planum*—a flat surface; Latin *temporalis*—of the temple) A part of the **auditory cortex** involved in sound analysis and particularly speech perception. Recently, the posterior portion of the planum temporale has been implicated in sensorimotor processing. See Chapter 5.

Plasticity (plas-TI-SI-tee; from Greek *plastikos*, from *plassein*—to mold, form) The ability of the brain to adapt and reorganize to new environmental inputs or demands or following brain damage. See Chapter 14.

Polysomnography (POL-ee-som-nog-raf-ee) Electroencephalography (EEG) is combined with an assortment of eye, muscle, and heart measures using a technique termed *polysomnography*. The recordings produced using polysomnography—the polysomnogram—are used to diagnose sleep disorders such as sleep apnea, narcolepsy, and insomnia. See Chapter 12.

Pons (PONZ; Latin, *pons*—bridge) A prominent anterior bulge in the **brainstem**. The pons relays sensory information between the **cerebellum** and the forebrain and spinal cord, helps to control sleep and wakefulness, and regulates respiration among other functions. It also generates **REM sleep** signals that are interpreted by the cortex as visually vivid, narrative dreams. See Chapter 2.

Positron emission tomography (PET; POH-zi-tron ee-MISH-en tom-OG-reh-fee; Latin *emitterre*—to send out; Greek *tomos*—section; Greek *graphein*—writing) Positrons are positively charged subnuclear particles, typically produced by a particle accelerator. PET is a low-level radioactive imaging technique that allows the computational extraction of brain or body slice maps, from which a three-dimensional image can be constructed. Typically, PET reflects metabolic activity. See Chapter 3.

Postcentral gyrus (post-SEN-tral JEYE-res; from Latin *post*—behind; Latin *gyrus*—ridge) A protruding fold in the **parietal lobe** of the human brain immediately behind the **central sulcus**. It includes the **primary somatosensory cortex**, the first cortical map of the body senses, also called the **sensory homunculus**, which represents the opposite or **contralateral** side of the body. See Chapter 2.

Posterior (pos-TEER-ee-er; from Latin *post*—after) Behind. In brain anatomy, **posterior** is synonymous with **caudal**.

Precuneus (pree-KUN-ee-us) A region at the superior aspect of the parietal lobe. In a healthy brain during normal waking, the *precuneus* is the most highly active (metabolically) regions in the brain. See Chapter 13.

Prefrontal cortex (pree-FRON-tal KOR-teks; from Latin *prae*—in front of; *frons*—the forehead; Greek *cort*—bark) The large, forward portion of the **frontal lobes**, not including the motor cortex. Prefrontal cortex includes **executive** functions and Broca's area, and is sometimes called "the organ of civilization." See Chapters 2 and 9.

Primary motor cortex (PRIE-mar-ee MO-ter KOR-teks; from Latin *primus*—first; Greek *cort*—bark) The brain region that directly controls skeletal (voluntary) muscles. It corresponds to the **motor homunculus** and works in close association with other sensory body and motor maps, such as the premotor cortex. See Chapter 2.

Primary somato Sensory cortex (PRIE-mar-ee so-MAT-o-SENS-ery KOR-tex; from Latin *primus*—first, most important; *soma*—body; *sensus*—sense; Greek *cort*—bark) The sensory homunculus (body map), located on the **postcentral gyrus** of the cortex, is the first cortical area for the body senses of touch, pressure, and pain. See Chapter 2.

Primary visual cortex (PRIE-mar-ee VIZH-oo-el KOR-teks; from Latin *primus*—first, most important; Latin *visus*—sight; Greek *cort*—bark) (also called V1) The first cortical map of the visual system, located within the calcarine sulcus in the **occipital lobe**. See Chapter 4.

Problem space (PROB-lem SPAYS) A graph of the decision points in problem-solving, often in the form of a tree structure. See Chapter 7.

Procedural memory (pruh-SEE-der-el MEM-ree, MEM-eh-ree; from Latin *procedere*—a way of doing things; Latin *memor*—mindful) A form of **implicit memory** equivalent to skill memory (such as riding a bicycle) or knowing how to do a task. It appears to be largely unconscious. This type of **memory** is often very durable. See Chapter 7.

Radial unit model (RAY-dee-el YOO-nit MAH-del) A model of **neural migration** proposed by neuroscientist **Pasko Rakic** that asserts that in the developing cerebral cortex, the cells are created at the base of each cortical column and each new cell migrates past its predecessors. See Chapter 14.

Rakic, Pasko (b. 1933) (rah-KEECH) A neuroanatomist who showed that neural migration occurs radially as well as rostrally, like the outflowing spokes of a forward-moving wheel. See Chapter 14.

Rapid eye movement sleep Rapid eye movement (REM) sleep involves a major shifting of brain processes from nonrapid eye movement sleep, as can be seen in the changing EEG to a pattern that is similar to an awake state. Most dreaming occurs during REM sleep. Another major change occurs in the body during REM sleep: the body enters a state of *atonia* or paralysis. Atonia is perfectly normal, and it is thought to be a protection from a person acting out a dream during REM sleep. You may try to move—and in fact many dreams involve a feeling of slow motion or actual paralysis, reflecting the true state of the body even though in our REM dreams we move about freely. See Chapter 12.

Receptive field (ree-SEP-tiv FEELD; from Latin *recipere*—to take) The receptive field of a nerve cell in the visual system, for example, is the region of the visual field that can activate or inhibit the firing of the cell. The receptive field of a retinal receptor is therefore different from the receptive field of a higher level cell tuned to detect motion or visual object identity. Analogous receptive fields have been found for visual attention in the parietal lobe. Receptive fields are found in other sensory systems as well, such as the auditory and somatosensory systems. See Chapter 2, and see Chapter 4 for more discussion on receptive fields in the visual system.

Reentrant connectivity (ree-EN-trent con-ec-TIV-e-tee) Most brain connections are bidirectional, in that activity at point A triggers activity at point B and vice versa.

Reentry (ree-entry) In Neural Darwinism, the resonant looping between two neurons or arrays of neurons so that neuron A activates neuron B and vice versa. Reentry can also take place between neuronal populations. It is believed to be the primary signaling mechanism among brain regions and therefore closely related to brain rhythms.

Reflex circuit (REE-fleks SIR-kut) Also called a reflex arc, this is the relatively simple pathway that mediates a reflex action. The most common example is the knee-jerk (or patellar tendon) spinal reflex, which occurs even when the spinal cord is isolated from the brain. However, spinal reflexes can be quite fast, complex, and coordinated, and may interact with the brainstem and the vestibular (balance) system, as in the case of a cat reorienting its body during a fall. Normally reflexes work in close coordination with voluntary control via the frontal lobes, cerebellum, and basal ganglia. Cranial reflexes like the pupillary reflex are under the joint control of autonomic, visual, and emotional regions of the brain.

Reticular formation (reh-TIC-u-ler for-MAY-shun; from Latin *reticulum*—network) A part of the brainstem that is involved in the sleep–waking cycle and many other functions. It receives collateral input from all sensory and motor systems, as well as from higher-level brain structures. It is evolutionarily one of the oldest parts of the brain. See Chapter 2.

Retina (REH-tin-a) The array of light receptors lining the inner surface of the eye. Light striking retinal receptors (rods or cones) trigger a chemical reaction that evokes a change in electrical potential across the cell membrane. This may trigger activity in retinal ganglion cells that project their axons to make up the optic nerve, which terminates in the visual relay nucleus of the thalamus. See Chapter 4.

Retrograde amnesia (RET-ro-grayd am-NEE-zhuh; from Latin *retrogradus*—going back; Greek *a-mne-sia*—without memory) A form of memory loss extending before the time of brain injury. Contrasted with **anterograde amnesia**. See Chapter 7.

Sagittal (SAJ-i-tal; from Latin *sagitta*—arrow) Any section of the brain that runs parallel to the **medial** or **midline** cut. See Chapter 2.

Selective attention (suh-LEC-tiv a-TEN-shun) The ability to pay attention to one aspect of the environment while ignoring competing stimuli. This may occur voluntarily, as in choosing to read an interesting book while sitting on a noisy bus, or when one sensory experience is biologically or personally significant. See Chapters 8 and 9.

Semantic memory (seh-MAN-tic MEM-ree or MEM-er-ee) A type of **declarative memory** that involves meanings, factual beliefs, categories, and other general knowledge going beyond specific experiences. See Chapter 7.

Semantics (seh-MAN-tiks) The study of meaning in language. See Chapter 7.

Sensory system (SEN-suh-ree SIS-tem) Part of the nervous system responsible for processing sensory information (the primary five senses being visual, auditory, tactile, taste, and olfaction [smell]). A sensory system consists of sensory receptors, neural pathways, and mostly posterior cortex involved in sensory perception. The classical senses have many subsenses, such as pain and even tickle sensations, light receptors in the eye that trigger melatonin as a sleep hormone, the balance sense, and the like. Not all sensory systems yield conscious experiences; blood pressure, for example, which is sensed by hypothalamic neurons, is rarely conscious. The classical senses begin with receptor surfaces containing many millions of receptors, such as the retina and the basilar membrane.

Sequential grouping (seh-KWEN-shul GROOP-ing) One way in which the human auditory system organizes sound into perceptually meaningful elements. If sound properties are repeated in the same sequence, they may be grouped together. For example, the sound properties of your friend's voice may help you hear him speak in a noisy environment. See Chapter 5.

Simultaneous grouping (SEYE-mul-TAY-nee-us GROOP-ing) Latin *simul 'at the same time'* If two sounds have common onsets (beginnings) and offsets (endings), they may be grouped together. One way in which the human auditory system organizes sound into meaningful elements. See Chapter 5.

Sleep disorder The four most common types of sleep disorders, in order of their prevalence, are *insomnia, sleep apnea, restless legs syndrome, and narcolepsy*. There are many more types of sleep disorders—some estimates range to more than 80—however, these are the core disorders that affect many children and adults. Some are short lived, some are life long. All sleep disorders produce cognitive, emotional, and body-related impairments that can be very destructive to normal health and well-being. See Chapter 12.

Sleep stages Sleep falls into two general patterns: nonrapid eye movement (NREM) and rapid eye movement (REM) sleep. Within NREM sleep, there are I–III or I–IV stages depending on whose theories you follow. Let us take the assumption that there are four NREM sleep stages. Sleep stages are sometimes labeled as stages 1–4 rather than I–IV. See Chapter 12.

Slow-wave sleep (SWS) is a nonrapid eye movement sleep stage in which brain functions and activity slow down markedly versus awake states. The EEG measures of brain waves during SWS show characteristic slow waves, thus the name of these sleep stages. See Chapter 12.

Sound localization (SOUND lo-cal-ih-ZAY-shun) Identifying the location of a sound, often based on **binaural disparities** of timing and loudness between the two ears. See interaural level difference, interaural time difference, and Chapter 5.

Source memory (SORS MEM-ree or MEM-er-ee) Memory for the specific time, place, and circumstances when an event was experienced. For example, you may remember not only when you learned about the theory of gravity (the memory) but also who first told you about it (the source). See Chapter 7.

Spiking code (SPI-king CODE) The rate and pattern of action potentials, which may transmit useful information in the brain. See Chapter 3.

Stimulus-driven attention (STIM-u-lus DRI-vn a-TEN-shun) The capture of **attention** by salient stimuli—such as the sudden honking of a car horn or the crash of a glass breaking. See Chapter 8.

Stroop test (STROOP test) Named after American psychologist John Ridley Stroop, who first wrote about this phenomenon in English in 1935. When the name of a color, such as *blue*, *green*, or *red*, is printed in a color differing from that expressed by the word's meaning (e.g., the word *red* is printed in blue ink), a subject has more difficulty naming the color of the word and is slower and more prone to errors than when the meaning of the word is congruent with its color. This phenomenon is known as the Stroop effect. To correctly name the color of the ink (e.g., 'red') rather than the word (e.g., 'green'), the subject has to suppress the near automatic reading response to respond "red." Variations of the Stroop task have been used to investigate many aspects of automatic processing. The Stroop effect is useful in activating conflict-related regions of the brain and generalizes well to related tasks, like the "emotional Stroop."

Superior (soo-PEER-ee-er) Latin ***superior***, comparative of ***superus 'that is above,'*** from ***super 'above.'*** Above. In the human brain, it is synonymous with **dorsal**.

Supratemporal plane (SOO-pra-tem-per-el PLANE) A flat region of **cortex** in the Sylvian fissure, where primary and secondary auditory cortex and parts of **Wernicke's area** are located. See Chapters 5 and 6.

Sylvian fissure (SIL-vee-en FISH-er) Also called the **lateral sulcus** or lateral fissure. This prominent "valley" of the cortex divides the **frontal lobe** and parietal lobe above from the **temporal lobe** below. See Chapter 2.

Synapse (SIN-aps) Synapses are tiny gaps between neurons that communicate by way of chemical neurotransmitters. Synapses are a basic computational element of the brain, a kind of traffic control point for the flow of information. The brain has tens of billions of neurons, but it has many trillions of synapses. See Chapter 3.

Synaptic cleft (sin-AP-tic CLEFT) The space between two **neurons** that can communicate with each other via neurotransmitters. See Chapter 3.

Synaptic pruning (sin-AP-tik PROO-ning) The selective loss of **synapses** in the brain when some potential connections are not utilized. See **Hebbian learning, neural Darwinism**. See Chapter 14.

Synaptogenesis (sin-AP-toe-GEN-eh-sis) The birth of **synapses** in the brain. See Chapter 14.

Syntactic analysis (sin-TAK-tik uh-NAL-ih-sus) The identification of grammatical structures from words, phonemes, and morphemes. See Chapter 6.

Syntax (sin-TAKS) The rules and regularities of sentences in natural languages. See Chapter 6.

Talairach coordinates (tal-AY-rahk co-ORE-din-etz) A precise three-dimensional coordinate system for the human brain that can localize any point in the brain with millimeter precision. See Chapter 3.

Temporal lobe (TEM-por-al lobe) from Latin *temporalis*, from *tempus*, *tempor- 'time.'* The temporal lobes are parts of the cerebral **cortex** that are involved in visual perception, hearing and speech perception, and **memory encoding** and **recall**. They emerge from the sides of the cortex, beneath the **lateral sulcus**. In profile, if the human brain resembles a boxing glove, the temporal lobes would be the thumb of each side. The temporal lobe envelops the **hippocampus** and **amygdala** and is therefore involved in emotion and memory formation as well. The **medial temporal** lobe (most easily seen from the bottom perspective of the brain) is ancient paleocortex, including the olfactory cortex. See Chapter 2.

Teratogen (ter-AT-e-jen) A chemical or other factor (such as prescription or nonprescription drugs or cigarette smoke) that causes developmental malformations. See Chapter 14.

Terminal (TER-mi-nul) The distal end of an **axon**. See Chapter 2.

Thalamocortical system (THAL-a-mo COR-ti-kel SIS-tem; from Greek *thalamos*—chamber; Greek *cort*—bark) A central hub in the brain involving the **cortex** and **thalamus**, allowing signal traffic to flow flexibly back and forth in both directions. See Chapter 2.

Thalamus (THAL-a-mus; from Greek *thalamos*—room, chamber) A pair of symmetric egg-shaped structures in the brain that provide the main cortical input hub and cortico-cortical traffic hub. Plural: thalami.

Theory of mind (THEE-eh-ree or THIR-ee of MIND) The ability to attribute mental states—beliefs, desires, intentions—to others. See Chapter 10.

Transcranial magnetic stimulation (TMS; trans-CRAY-nee-el mag-NET-ic stim-yoo-LAY-shun) A relatively noninvasive method using powerful electromagnets outside of the head to stimulate or inhibit cortical neurons. TMS shows good temporal and spatial resolution. See Chapter 3.

Unconscious perception (un-CON-shus per-SEP-shun) From Latin un (not) + ***conscious 'knowing with others or in'*** Sensory stimulus processing without awareness of the stimulus—such as the sudden honing of a car horn or the crash of a glass breaking. See Chapter 8.

Vegetative state The vegetative state (VS) differs from the coma state in that there is wakeful unconsciousness. The wakefulness is exhibited by spontaneous eye opening. However, there is no evidence of awareness, language comprehension, visual gaze following despite the eyes opening occasionally, and only sporadic involuntary motor movement. When we open our eyes, observers see this as a sign that we are conscious. This differs in the VS, with levels of awareness that are low while levels of wakefulness relatively higher. Typically, a VS can last for 1–12 months following traumatic brain injury. Following that time window, VS *may* be considered permanent. See Chapter 13.

Ventral (VEN-trel; from Latin ***venter*****—the belly)** The lower part of a brain structure, inferior.

Ventricles (VEN-trik-lz) Four small cavities in the brain containing circulating cerebrospinal fluid. The ventricular walls have been found to be sites for neural stem cells. See Chapter 2.

Ventromedial prefrontal cortex (ven-tro-MEE-dee-el pree-FRON-tal KOR-teks; Latin ***venter*****—the belly;** ***medialis*****—in the middle)** The bottom midline structures of the frontal lobe, especially in humans and other primates. This region, extending backward from the top of the nose, is involved in emotions, infant–mother bonding, fear, and risk in decision-making. See Chapter 9.

Verbal rehearsal (VER-bel ree-HER-sel) Mental repetition of words to be remembered, using the "inner speech" component of *working memory*. Inner speech involves a spontaneous commentary on current concerns, goals, and emotions.

Vesicle (VES-i-cl; from Latin ***vesicula*****—small bladder)** The small bubbles filled with neurotransmitter molecules that travel through the axon to the synaptic terminals, where they fuse with the synaptic membrane to release neuromolecules into the cleft when an action potential occurs. Neurotransmitters then diffuse across the synapse to trigger depolarization of the postsynaptic membrane, ultimately leading to another axonal spike. Vesicles are essential for the propagation of signals between neurons and are constantly recreated by the cell.

Visual agnosia (VI-zhoo-el ag-NO-zhe; from Greek ***agno-sia*****—lacking knowledge)** A condition in which a person has difficulty recognizing objects because of damage to object-recognition regions of the cortex, such as the inferior temporal lobe. See Chapter 4.

Visual backward masking (VI-zhoo-el BAK-werd MAS-king) A conscious visual image can be "erased" by a subsequent visual event, such as a cross-hatch display, even though the conscious event is not physically blocked from reaching the retina.

Visuospatial sketchpad (vizh-oo-oh-SPAY-shul SKECH-pad) The ability to hold visual and spatial information momentarily in **working memory**.

Volition (vuh-LI-shun) From Latin ***volitio(n-)*****, from** ***volo 'I wish.'*** Voluntary control of actions, as contrasted with automatic control, as in the case of highly practiced habits. Many brain disorders involve a loss of voluntary control.

Voxel (VAHX-ul) A voxel is the smallest unit imaged using magnetic resonance imaging (MRI). The actual size of a voxel varies depending upon factors such as the resolution of the MRI scanner, the size of the brain being scanned, and the brain region being scanned. A typical voxel for a T1-weighted scan is about one cubic millimeter (mm^3). If it is from the cortex, a single voxel may contain tens of thousands of neurons. See Chapter 3.

Wearing, Clive (b. 1938) A prominent British classical musician who suffered a viral brain infection in his forties that destroyed both hippocampi and some frontal lobe regions. Wearing's case has become well known due to the efforts of his wife, Deborah Wearing, to raise public awareness of such medical conditions. Wearing lives in a single, blindered moment, without the ability to store information for later recall. Despite his **memory** problems, he is still able to play the piano and conduct musical pieces he knew well before the brain injury. See **anterograde amnesia**.

Wernicke, Carl (1848–1905) German physician and discoverer of a selective cortical region for speech comprehension. This region is now referred to as **Wernicke's area**, and the associated deficit is known as **Wernicke's** or **receptive aphasia**. Patients with this deficit cannot understand speech, including their own, but produce fluent-sounding (but not usually meaningful) speech.

Wernicke's area (WER-nik-ees AIR-ee-a) An area of the upper posterior temporal lobe that is needed for language comprehension.

White matter In the brain, white matter consists of dense bundles of myelinated **axons**, which connect various **gray matter** areas of the brain to each other. White matter is named for the appearance or massive numbers of **myelinated** nerve axons, which appear to form the visible core of brain structure. See Chapter 2.

Working memory (WUR-king MEM-ree or MEM-er-ee) A cognitive capacity for storing and manipulating novel information over 10–30 s. Working memory includes **central executive, working storage, verbal rehearsal**, and the **visuospatial sketchpad**. See Chapter 7.

References

Adolphs, R. (2002). Recognizing emotion from facial expressions: Psychological and neurological mechanisms. *Behavioral and Cognitive Neuroscience Reviews, 1*(1), 21–62.

Adolphs, R. (March 2003). Cognitive neuroscience of human social behaviour. *Nature Reviews Neuroscience, 4*(3), 165–178. Review. PubMed PMID:12612630.

Albright, T. D. (1984). Direction and orientation selectivity of neurons in visual area MT of the macaque. *Journal of Neurophysiology, 52*(6), 1106–1130.

Albright, T. D. (1992). Form-cue invariant motion processing in primate visual cortex. *Science, 255*(5048), 1141–1143.

Alkire, M. T., Hudetz, A. G., & Tononi, G. (2008). Consciousness and anesthesia. *Science, 322*(5903), 876–880.

Allison, T., Puce, A., & McCarthy, G. (2000). Social perception from visual cues: Role of the STS region. *Trends in Cognitive Sciences, 4*(7), 267–278.

Aminoff, M., & Daroff, R. (Eds.). (2003). *Encyclopedia of the neurological sciences*. San Diego: Academic Press.

Amodio, D. M. (2014). The neuroscience of prejudice and stereotyping. *Nature Reviews Neuroscience, 15*, 670–682.

Anderson, S. W., Bechara, A., Damasio, H., Tranel, D., & Damasio, A. R. (1999). Impairment of social and moral behavior related to early damage in human prefrontal cortex. *Nature Neuroscience, 2*(11), 1032–1037.

Antony, J. W., Ferreira, C. S., Norman, K. A., & Wimber, M. (2017). Retrieval as a fast route to memory consolidation. *Trends in Cognitive Sciences, 21*(8), 573–576.

Arnsten, A. F., Raskind, M. A., Taylor, F. B., & Connor, D. F. (2015). The effects of stress exposure on prefrontal cortex: Translating basic research into successful treatments for post-traumatic stress disorder. *Neurobiology of Stress, 1*, 89–99.

Awh, E., Jonides, J., Smith, E. E., Schumacher, E. H., Koeppe, R. A., & Katz, S. (1996). Dissociation of storage and rehearsal in verbal working memory: Evidence from positron emission tomography. *Psychological Science, 7*(1), 25–31.

Baars, B. J. (2002). The conscious access hypothesis: Origins and recent evidence. *Trends in Cognitive Sciences, 6*(1), 47–52.

Badre, D., & D'Esposito, M. (2007). Functional magnetic resonance imaging evidence for a hierarchical organization of the prefrontal cortex. *Journal of Cognitive Neuroscience, 19*, 2082–2099.

Badre, D., & D'Esposito, M. (2009). Is the rostro-caudal axis of the frontal lobe hierarchical? *Nature Reviews Neuroscience, 10*, 659–669.

Badre, D. (2008). Cognitive control, hierarchy, and the rostro-caudal axis of the frontal lobes. *Trends in Cognitive Sciences, 12*, 193–200.

Banaji, M. R., & Greenwald, A. G. (1995). Implicit gender stereo-typing in judgments of fame. *Journal of Personality and Social Psychology, 68*(2), 181–198.

Bandettini, P. A. (2012). Functional MRI: A confluance of fortunate circumstances. *Neuroimage, 61*, A3–A11.

Barkley, R. A. (1997). *ADHD and the nature of self-control*. New York: The Guilford Press.

Barlow, H. B., Blakemore, C., & Pettigrew, J. D. (1967). The neural mechanism of binocular depth discrimination. *Journal of Physiology, Paris, 193*(2), 327–342.

Baron-Cohen, S. (1995). *Mindblindness: An essay on autism and theory of mind*. Boston: MIT Press/Bradford Books.

Baumeister, R. F., Sparks, E. A., Stillman, T. F., & Vohs, K. D. (2008). Free will in consumer behavior: Self-control, ego depletion, and choice. *Journal of Consumer Psychology, 18*, 4–13.

Beauchamp, M. S. (2015). The social mysteries of the superior temporal sulcus. *Trends in Cognitive Science, 19*(9), 489–490.

Beauchemin, M., González-Frankenberger, B., Tremblay, J., Vannasing, P., Martínez-Montes, E., Belin, P., et al. (2011). Mother and stranger: An electrophysiological study of voice processing in newborns. *Cerebral Cortex, 21*(8), 1705–1711.

Bechara, A., Damasio, A. R., Damasio, H., & Anderson, S. W. (1994). Insensitivity to future consequences following damage to human prefrontal cortex. *Cognition, 50*, 7–15.

Bell, A. H., Pessoa, L., Tootell, R. B. H., & Ungerleider, L. G. (2013). Visual perception of objects. In L. R. Squire, D. Berg, F. E. Bloom, S. du Lac, A. Ghosh, & N. C. Spitzer (Eds.), *Fundamental neuroscience* (4th ed., pp. 947–968). San Diego: Academic Press.

Benson, H., Alexander, S., & Feldman, C. L. (1975). Decreased premature ventricular contractions through use of the relaxation response in patients with stable ischaemic heart-disease. *Lancet, 2*(7931), 380–382.

Berwick, R. C., Friederici, A. D., Chomsky, N., & Bolhuis, J. J. (2013). Evolution, brain, and the nature of language. *Trends in Cognitive Sciences, 17*(2), 89–98.

Binder, J. R., Frost, J. A., Hammeke, T. A., Bellgowan, P. S., Rao, S. M., & Cox, R. W. (1999). Conceptual processing during the conscious resting state. A functional MRI study. *Journal of Cognitive Neuroscience, 11*, 80–95.

Bizley, J. K., & Cohen, Y. E. (2013). The what, where, and how of auditory-object perception. *Nature Reviews Neuroscience, 14*, 693–707.

Blanche, Spacek, Hetke, & Swindale. (2005). Polytrodes: High-density silicon electrode arrays for large-scale multiunit recording. *Journal of Neurophysiology, 93*, 2987–3000.

Bodamer, J. L. (1947). Die prosopagnosie. *European Archives of Psychiatry and Clinical Neuroscience, 179*(1), 6–53.

Borbély, A. A. (2009). Refining sleep homeostasis in the two-process model. *Journal of Sleep Research, 18*(1), 1–2.

Bouvier, S. E., & Engel, S. A. (2006). Behavioral deficits and cortical damage loci in cerebral achromatopsia. *Cerebral Cortex, 16*(2), 183–191.

Bowden, E. M., & Jung-Beeman, M. (2007). Methods for investigating the neural components of insight. *Methods, 42*, 87–99.

Bregman, A. S. (1990). *Auditory scene analysis: The perceptual organization of sound*. Cambridge: MIT Press.

Broadbent, D. E. (1982). Task combination and selective intake of information. *Acta Psychologica (Amst), 50*(3), 253–290.

Broca, P. (1861). Remarques sur le siege de la faculte du langage articule, suives d'une observation d'aphemie (pert de la parole). *Bulletin de la Societe Anatomica, 6*(330–357), 398–407.

Brodmann, K. (1909). *Vergleichende lakalisationslehre der grosshirnrinde: In ihren prinzipien dargestellt anf grund des zellenbaues*. Leipzig: Verlag von Johann Ambrosisu Barth.

Brooks, R., & Meltzoff, A. (2003). Gaze following at 9 and 12 months: A developmental shift from global head direction to gaze. In *Poster presented at SRCD conference*. Tampa: Society for Research in Child Development.

Brown, M. C., & Santos-Sacchi, J. (2013). Audition. In L. R. Squire, D. Berg, F. E. Bloom, S. du Lac, A. Ghosh, & N. C. Spitzer (Eds.), *Fundamental neuroscience* (4th ed., pp. 553–576). San Diego: Academic Press.

Buchbinder, B. R. (2016). Functional magnetic resonance imaging. In J. C. Masdeu, & R. G. Gonzalez (Eds.), *Handbook of clinical neurology (3rd series)* (Vol. 135, pp. 61–92). Amsterdam: Elsevier.

Buckley, M. J., & Gaffan, D. (2006). Perirhinal cortical contributions to objet perception. *Trends in Cognitive Sciences, 10*(3), 100–107.

Buckner, R. L., Goodman, J., Burock, M., Rotte, M., Koutstaal, W., Schacter, D., et al. (1998). Functional-anatomic correlates of object priming in humans revealed by rapid presentation event-related fMRI. *Neuron, 20*(2), 285–296.

Buckner, R., Andrews-Hanna, J. R., & Schacter, D. (2008). The brain's default network: Anatomy, function, and relevance to disease. *Annals of the New York Academy of Sciences, 1124*, 1–38.

Buckner, R. (2012). The serendipitous discovery of the brain's default network. *Neuroimage, 62*, 1137–1145.

Bunge, S. A. (2004). How we use rules to select actions: A review of evidence from cognitive neuroscience. *Cognitive, Affective, and Behavioral Neuroscience, 4*(4), 564–579.

Bunzeck, N., Wuestenberg, T., Lutz, K., Heinze, H. J., & Jancke, L. (2005). Scanning silence: Mental imagery of complex sounds. *Neuroimage, 26*(4), 1119–1127.

Bush, G., Luu, P., & Posner, M. I. (2000). Cognitive and emotional influences in anterior cingulate cortex. *Trends in Cognitive Sciences, 4*(6), 215–222.

Bushnell, M. C., Ceko, M., & Low, L. A. (2013). Cognitive and emotional control of pain and its disruption in chronic pain. *Nature Reviews Neuroscience, 14*, 502–511.

Cahill, L., Prins, B., Weber, M., & McGaugh, J. L. (1994). Beta-adrenergic activation and memory for emotional events. *Nature, 371*(6499), 702–704.

Cahill, L., Uncapher, M., Kilpatrick, L., Alkire, M. T., & Turner, J. (2004). Sex-related hemispheric lateralization of amygdala function in emotionally influenced memory: An fMRI investigation. *Learning & Memory, 11*, 261–266.

Cahn, B. R., & Polich, J. (2006). Meditation states and traits: EEG, ERP, and neuroimaging studies. *Psychological Bulletin, 132*(2), 180–211.

Cahn, B. R., & Polich, J. (2009). Meditation (Vipassana) and the P3a event-related brain potential. *International Journal of Psychophysiology, 72*(1), 51–60.

Caldara, R., Segher, M., Rossion, B., Lazeyras, F., Michel, C., & Hauert, C. (2006). The fusiform face area is tuned for curvilinear patterns with more high contrasted elements in the upper part. *Neuroimage, 31*, 313–319.

Carlin, J. D., Calder, A. J., Kriegeskorte, N., Nilli, H., & Rowe, J. B. (2011). A head view-invariant representation of gaze direction in anterior superior temporal sulcus. *Current Biology, 21*, 1817–1821.

Casey, B. J., Tottenham, N., Liston, C., & Durston, S. (2005). Imaging the developing brain: What have we learned about cognitive development? *Trends in Cognitive Science, 9*(3), 104–110.

Catani, M., & Thiebaut de Schotten, M. (2008). A diffusion tensor imaging tractography atlas for virtual in vivo dissections. *Cortex; A Journal Devoted To the Study of the Nervous System and Behavior, 44*, 1105–1132.

Chanes, L., & Barrett, L. F. (2016). *Trends in Cognitive Sciences, 20*(2), 96–106.

Chase, W. G., & Simon, H. A. (1973). The mind's eye in chess. In W. G. Chase (Ed.), *Visual information processing* (pp. 215–281). New York: Academic Press.

Cherry, E. C. (1953). Some experiments on the recognition of speech, with one and with two ears. *The Journal of the Acoustical Society of America, 25*, 975.

Chomsky, N. (1995). *The minimalist program*. Cambridge: MIT Press.

Chun, M. M., & Phelps, E. A. (1999). Memory deficits for implicit contextual information in amnesic subjects with hippocampal damage. *Nature Neuroscience, 2*(9), 844–847.

Cirelli, C. (2009). The genetic and molecular regulation of sleep: From fruit flies to humans. *Nature Reviews Neuroscience, 10*, 549–560.

Clayton, M. S., Yeung, N., & Kadosh, R. C. (2015). The roles of cortical oscillations in sustained attention. *Trends in Cognitive Sciences, 19*(4), 188–195.

Cohen, J., Perlstein, W., Braver, T., Nystrom, L. E., Noll, D. C., Jonides, J., et al. (1997). Temporal dynamics of brain activation during a working memory task. *Nature, 386*(6625), 604–608.

Conel, J. L. (1951). The postnatal development of the human cerebral cortex. In *The cortex of the six-month infant* (Vol. IV). Cambridge, Mass: Harvard University Press.

Corbetta, M., Kincade, J. M., & Shulman, G. L. (2002). Neural systems for visual orienting and their relationships to spatial working memory. *Journal of Cognitive Neuroscience, 14*(3), 508–523.

Corkin, S., Amaral, D., Gonzalez, R. G., Johnson, K. A., & Hyman, B. T. (1997). H.M's medial temporal lobe lesion: Findings from magnetic resonance imaging. *The Journal of Neuroscience, 17*(10), 3964–3979.

Corkin, S. (1965). Tactually-guided maze learning in man: Effects of unilateral cortical excisions and bilateral hippocampal lesions. *Neuropsychologia, 3*, 339–351.

Cowan, N., Izawa, C., & Ohta, N. (2005). Working memory capacity limits in a theoretical context. In *Human learning and memory: Advances in theory and application. The 4th Tsukuba international conference on memory* (p. 155). Mahwah: Lawrence Erlbaum Associates, Publishers.

Cowey, A., & Stoerig, P. (1995). Blindsight in monkeys. *Nature, 373*(6511), 247–249.

Cowey, A., & Walsh, V. (2001). Tickling the brain: Studying visual sensation, perception and cognition by transcranial magnetic stimulation. *Progress in Brain Research, 134*, 411–423.

Culpepper, L. (2010). Why do you need to move beyond first-line therapy for major depression? *Journal of Clinical Psychiatry, 71*(Suppl. 1), 4–9.

Cumming, B. G. (2002). An unexpected specialization for horizontal disparity in primate primary visual cortex. *Nature, 418*(6898), 633–636.

Curtis, C. E., & D'Esposito, M. (2003). Persistent activity in the prefrontal cortex during working memory. *Trends in Cognitive Sciences, 7*(9), 415–423.

Cusack, R., Ball, G., Smyser, C. D., & Dehaene-Lambertz, G. (2016). A neural window on the emergence of cognition. *Annals of the New York Academy of Science, 1369*, 7–23.

Cusack, R. (2005). The intraparietal sulcus and perceptual organization. *Journal of Cognitive Neuroscience, 17*(4), 641–651.

D'Esposito, M., & Postle, B. R. (1999). The dependence of span and delayed-response performance on prefrontal cortex. *Neuropsychologia, 37*(11), 1303–1315.

Damasio, A. R. (1995). On some functions of the human prefrontal cortex. *Annals of the New York Academy of Sciences, 769*, 241–251.

Danek, A. H., Fraps, T., von Muller, A., Grothe, B., & Ollinger, M. (2014). *Frontiers in Psychology, 5*, 1–11.

Daselaar, S. M., Fleck, M. S., Prince, S. E., & Cabeza, R. (2006). The medial temporal lobe distinguishes old from new independently of consciousness. *The Journal of Neuroscience, 26*(21), 5835–5839.

Davidson, R. J., & Irwin, W. (1999). The functional neuroanatomy of emotion and affective style. *Trends in Cognitive Sciences, 3*(1), 11–21.

de Gelder, B., Tamietto, M., van Boxtel, G., Goebel, R., Sahraie, A., van den Stock, J., et al. (2008). Intact navigation skills after bilateral loss of striate cortex. *Current Biology: CB, 18*(24), 1128–1129.

de Groot, A. D., Gobet, F., & Jongman, R. W. (1966). *Perception and memory in chess: Heuristics of the professional eye*. Assen: Van Gorcum.

De Havas, J., Parimal, S., Soon, C., & Chee, M. (2012). Sleep deprivation reduces default mode network connectivity and anti-correlation during rest and task performance. *Neuroimage, 59*, 1745–1751.

DeCasper, A. J., & Fifer, W. P. (1980). Of human bonding: Newborns prefer their mothers' voices. *Science, 208*(4448), 1174–1176.

Deen, B., Koldewyn, K., Kanwisher, N., & Saxe, R. (2015). Functional organization of social perception and cognition in the superior temporal sulcus. *Cerebral Cortex*. https://doi.org/10.1093/cercor/bhv111. Published online June 5, 2015.

Degonda, N., Mondadori, C. R., Bosshardt, S., Schmidt, C. F., Boesiger, P., Nitsch, R. M., et al. (2005). Implicit associative learning engages the hippocampus and interacts with explicit associative learning. *Neuron, 46*(3), 505–520.

Dehaene, S., & Changeux, J.-P. (1997). A hierarchical neuronal network for planning behavior. *Proceedings of the National Academy of Sciences of the United States of America, 94*, 13292–13298.

Dehaene, S. (2014). *Consciousness and the brain*. New York: Penguin.

Dehaene-Lambertz, G., & Spelke, E. S. (2015). The infancy of the human brain. *Neuron, 88*, 93–109.

Dehaene-Lambertz, G., Hertz-Pannier, L., & Dubois, J. (2006). Nature and nurture in language acquisition: Anatomical and functional brain-imaging studies in infants. *Trends in Neurosciene, 29*(7), 367–373.

Del Cul, A., Baillet, S., & Dehaene, S. (2007). Brain dynamics underlying the nonlinear threshold for access to consciousness. *PLoS Biology, 5*(10), 2408–2423.

Delis, D. C., et al. (1986). Hemispheric specialization of memory for visual hierarchical stimuli. *Neuropsychologia, 24*, 205–214.

Deoni, S. C. L., Mercure, E., Blasi, A., Gasston, D., Thomson, A., Johnson, M., et al. (2011). Mapping infant brain myelination with magnetic resonance imaging. *The Journal of Neuroscience, 31*(12), 784–791.

D'Esposito, M., & Chen, A. J. (2006). Neural mechanisms of prefrontal cortical function: Implications for cognitive rehabilitation. *Progress in Brain Research, 157*, 123–139.

D'Esposito, M., Detre, J. A., Alsop, D. C., Shin, R. K., Atlas, S., & Grossman, M. (1995). The neural basis of the central executive system of working memory. *Nature, 378*(16), 279–281.

Deutsch, G., & Eisenberg, H. M. (1987). Frontal blood flow changes in recovery from coma. *Journal of Cerebral Blood Flow and Metabolism, 7*(1), 29–34.

Diamond, A., & Goldman-Rakic, P. S. (1989). Comparison of human infants and rhesus monkeys on Piaget's AB task: Evidence for dependence on dorsolateral prefrontal cortex. *Experimental Brain Research, 74*(1), 24–40.

Diamond, A. (1985). Development of the ability to use recall to guide action, as indicated by infants' performance on AB. *Child Development, 56*(4), 868–883.

Diamond, A. (1991). Neuropsychological insights into the meaning of object concept development. In S. G. R. Carey (Ed.), *The epigenesis of mind: Essays on biology and cognition* (pp. 67–110). Hillsdale: Lawrence Erlbaum Associates.

Diamond, A. (2001). A model system for studying the role of dopamine in the prefrontal cortex during early development in humans: Early and continuously treated phenylketonuria. In C. A. L. Nelson (Ed.), *Handbook of developmental cognitive neuroscience* (pp. 433–472). Cambridge: MIT Press.

Diana, R. A., Yonelinas, A. P., & Ranganath, C. (2007). Imaging recollection and familiarity in the medial temporal lobe: A three-component model. *Trends in Cognitive Sciences, 11*(9), 379–386.

Diekelmann, S., & Born, J. (2010). The memory function of sleep. *Nature Reviews Neuroscience, 11*, 114–126.

Domhoff, G. W., & Fox, K. C. (2015). Dreaming and the default network: A review, synthesis, and counterintuitive research proposal. *Conscious Cognition, 33*, 342–353.

Driver, J., & Mattingley, J. B. (1998). Parietal neglect and visual awareness. *Nature Neuroscience, 1*(1), 17–22.

Dudai, Y. (2004). The neurobiology of consolidations, or, how stable is the engram? *Annual Review of Psychology, 55*, 51–86.

Duncker, K. (1945). On problem-solving. *Psychological Monographs, 58*(5).

Durston, S., Thomas, K. M., Worden, M. S., Yang, Y., & Casey, B. J. (2002). The effect of preceding context on inhibition: An event-related fMRI study. *Neuroimage, 16*(2), 449–453.

Dusek, J. A., Chang, B. H., Zaki, J., et al. (2006). Association between oxygen consumption and nitric oxide production during the relaxation response. *Medical Science Monitor: International Medical Journal of Experimental and Clinical Research, 12*(1), CR1–CR10.

Edelman, G. M. (1989). *The remembered present: A biological theory of consciousness*. New York: Basic Books Inc. [Author Note: Incorrectly cited in chapter as Edelman, 1992].

Egner, T., Jamieson, G., & Gruzelier, J. (2005). Hypnosis decouples cognitive control from conflict monitoring processes of the frontal lobe. *NeuroImage, 27*(4), 969–978.

Eimas, P. D., Siqueland, E. R., Jusczyk, P., & Vigorito, J. (1971). Speech perception in early infancy. *Science, 171*, 304–306.

Ekman, P. (1992). An argument for basic emotions. *Cognition & Emotion, 6*(3/4), 169–200.

Eldridge, L. L., Sarfatti, S., & Knowlton, B. J. (2002). The effect of testing procedure on remember know judgments. *Psychonomic Bulletin & Review, 9*(1), 139–145.

Epstein, R., & Kanwisher, N. (1998). A cortical representation of the local visual environment. *Nature, 392*(6676), 598–601.

Etkin, A., Büchel, C., & Gross, J. J. (2015). The neural bases of emotion regulation. *Nature Reviews Neurosciences, 16*(11), 693–700.

Fan, J., McCandliss, B. D., Fossella, J., Flombaum, J. I., & Posner, M. I. (2005). The activation of attentional networks. *NeuroImage, 26*(2), 471–479.

Fantz, R. L. (1964). Visual experience in infants: Decreased attention to familiar patterns relative to novel ones. *Science, 146*, 668–670.

Felleman, D. J., & Van Essen, D. C. (1991). Distributed hierarchical processing in the primate cerebral cortex. *Cerebral Cortex, 1*(1), 1–47.

Ferre, S., Earley, C., Gulyani, S., & Garcia-Borreguero, D. (2016). In search of alternatives to dopaminergic ligands for the treatment of restless legs syndrome: Iron, glutamate, and adenosine. *Sleep Medicine, 31*, 86–92.

Fishman, Y. I., Reser, D. H., Arezzo, J. C., & Steinschneider, M. (2001). Neural correlates of auditory stream segregation in primary auditory cortex of the awake monkey. *Hearing Research, 151*(1–2), 167–187.

Foster, D. H. (2011). Color constancy. *Vision Research, 51*(7), 674–700.

Fourier, J. (1822). *The analytical theory of heat. (English Transl. Freeman, 1878). Republished 1955*. New York: Dover.

Fox, N. A., & Bell, M. A. (1990). Electrophysiological indices of frontal lobe development. In A. Diamond (Ed.), *The development and neural bases of higher cognitive functions* (Vol. 608, pp. 677–698). New York: New York Academy of Sciences.

Fox, M. D., & Raichle, M. E. (2007). Spontaneous fluctuations in brain activity observed with functional magnetic resonance imaging. *Nature Reviews Neuroscience, 8*(9), 700–711.

Frankland, P. W., & Bontempi, B. (2005). The organization of recent and remote memories. *Nature Reviews Neuroscience, 6*, 119–130.

Frankland, P. W., Kohler, S., & Josselyn, S. A. (2013). Hippocampal neurogenesis and forgetting. *Trends in Neurosciences, 36*(9), 497–503.

Franz, E. A., & Gillett, G. (2011). John Hughlings Jackson's evolutionary neurology: A unifying framework for cognitive neuroscience. *Brain: A Journal of Neurology, 134*(10), 3114–3120.

Freeman, J. B., & Johnson, K. L. (2016). More than meets the eye: Split-second social perception. *Trends in Cognitive Sciences, 20*(5), 362–374.

Freud, S. (1961). The ego and the id (J. Strachey, Trans.). In *The standard edition of the complete psychological works of Sigmund Freud* (Volume 19, pp. 3–66). London: Hogarth Press. Original work published 1923.

Fried, P. A., Watkinson, B., & Gray, R. (2003). Differential effects on cognitive functioning in 13- to 16-year-olds prenatally exposed to cigarettes and marihuana. *Neurotoxicology and Teratology, 25*(4), 427–436.

Friederici, A. (2015). White-matter pathways for speech and language processing. In G. G. Celesia, & G. Hickok (Eds.), *Handbook of clinical neurology* (Vol. 129, pp. 177–186). San Diego: Academic Press, 3rd series.

Frith, C. D., & Frith, U. (1999). Interacting minds—a biological basis. *Science, 286*(5445), 1692–1695.

Fulford, J., Vadeyar, S. H., Dodampahala, S. H., Moore, R. J., Young, P., Baker, P. N., et al. (2003). Fetal brain activity in response to a visual stimulus. *Human Brain Mapping, 20*(4), 239–245.

Funahashi, S., Bruce, C. J., & Goldman-Rakic, P. S. (1993). Dorsolateral prefrontal lesions and oculomotor delayed-response performance: Evidence for mnemonic scotomas. *The Journal of Neuroscience, 13*(4), 1479–1497.

Fuster, J. M., & Alexander, G. E. (1971). Neuron activity related to short-term memory. *Science, 173*(997), 652–654.

Fuster, J. M. (1997). Network memory. *Trends in Neurosciences, 20*(10), 451–459.

Fuster, J. M. (2008). *The prefrontal cortex* (4th ed.). London: Academic Press.

Gabrieli, J. D., Keane, M. M., Stanger, B. Z., Kjelgaard, M. M., Corkin, S., & Growdon, J. H. (1994). Dissociations among structural-perceptual, lexical-semantic, and event-fact memory systems in Alzheimer, amnesic, and normal subjects. *Cortex, 30*(1), 75–103.

Gage, N. M., & Roberts, T. P. (2000). Temporal integration: Reflections in the M100 of the auditory evoked field. *Neuroreport, 11*(12), 2723–2726.

Gage, N., Poeppel, D., Roberts, T. P., & Hickok, G. (1998). Auditory evoked m100 reflects onset acoustics of speech sounds. *Brain Research, 814*(1–2), 236–239.

Gage, N. M., Roberts, T. P., & Hickok, G. (2002). Hemispheric asymmetries in auditory evoked neuromagnetic fields in response to place of articulation contrasts. *Brain Research. Cognitive Brain Research, 14*(2), 303–306.

Gage, N., Roberts, T. P., & Hickok, G. (2006). Temporal resolution properties of human auditory cortex: Reflections in the neuromagnetic auditory evoked m100 component. *Brain Research, 1069*(1), 166–171.

Galaburda, A. M., & Pandya, D. N. (1983). The intrinsic architectonic and connectional organization of the superior temporal region of the rhesus monkey. *The Journal of Comparative Neurology, 221*(2), 169–184.

Galvan, H., Hare, T. A., Parra, C. E., Penn, J., Voss, H., Glover, G., et al. (2006). Earlier development of the accumbens relative to orbitofrontal cortex might underlie risk-taking behavior in adolescents. *Journal of Neuroscience, 26*(25), 6885–6892.

Garcia-Borreguero, D., & Williams, A. M. (2010). Dopaminergic augmentation of restless legs syndrome. *Sleep Medicine Reviews, 14*, 339–346.

Gauthier, I., Skudlarski, P., Gore, J. C., & Anderson, A. W. (2000). Expertise for cars and birds recruits brain areas involved in face recognition. *Nature Neuroscience, 3*(2), 191–197.

Genzel, L., Spoormaker, V. I., Konrad, B. N., & Dresler, M. (2015). The role of rapid eye movement sleep for amygdala-related memory processing. *Neurobiology of Learning and Memory, 122*, 110–121.

Geschwind, N., & Galaburda, A. M. (1985a). Cerebral lateralization. Biological mechanisms, associations, and pathology. I. A hypothesis and a program for research. *Archives of Neurology, 42*(6), 428–459.

Geschwind, N., & Galaburda, A. M. (1985b). Cerebral lateralization. Biological mechanisms, associations, and pathology. II. A hypothesis and a program for research. *Archives of Neurology, 42*(6), 521–552.

Geschwind, N., & Galaburda, A. M. (1985c). Cerebral lateralization. Biological mechanisms, associations, and pathology. III. A hypothesis and a program for research. *Archives of Neurology, 42*(6), 634–654.

Giacino, J. T., Fins, J. J., Laureys, S., & Schiff, N. D. (2014). Disorders of consciousness after acquired brain injury: The state of the science. *Nature Reviews Neuroscience, 10*, 99–114.

Gick, M. L., & Lockhart, R. S. (1995). Cognitive and affective components of insight. In R. J. Sternberg, & J. E. Davidson (Eds.), *The Nature of insight* (pp. 197–228). Cambridge, MA: MIT Press.

Giedd, J. N., Blumenthal, J., Jeffries, N. O., Castellanos, F. X., Liu, H., Zijdenbos, A., et al. (1999). Brain development during childhood and adolescence: A longitudinal MRI study. *Nature Neuroscience, 2*(10), 861–863.

Gilboa, A., & Marlatte, H. (2017). Neurobiology of schemas and schema-mediated memory. *Trends in Cognitive Sciences, 21*(8), 618–631.

Gluck, M. A., Meeter, M., & Myers, C. E. (2003). Computational models of the hippocampal region: Linking incremental learning and episodic memory. *Trends in Cognitive Sciences, 7*(6), 269–276.

Gobet, F., & Simon, H. A. (1997). Recall of random and distorted chess positions: Implications for the theory of expertise. *Memory & Cognition, 24*, 493–503.

Goldberg, E., & Bougakov, D. (2000). Novel approaches to the diagnosis and treatment of frontal lobe dysfunction. In *International handbook of neuropsychological rehabilitation. Critical issues in neuropsychology* (pp. 93–112).

Goldberg, E., & Costa, L. D. (1985). Qualitative indices in neuropsychological assessment: An extension of Luria's approach to executive deficit following prefrontal lesion. In I. Grant, & K. M. Adams (Eds.), *Neuropsychological assessment of neuropsychiatric disorders* (pp. 48–64). New York: Oxford University Press.

Goldberg, E., Bilder, R. M., Hughes, J. E., Antin, S. P., & Mattis, S. (1989). A reticulo-frontal disconnection syndrome. *Cortex: A Journal Devoted to the Study of the Nervous System and Behavior, 25*(4), 687–695.

Goldberg, E. (1992). Introduction: The frontal lobes in neurological and psychiatric conditions. *Neuropsychology Neuropsychiatry and Behavioral Neurology, 5*(4), 231–232.

Goldberg, E. (2001). *The executive brain*. New York: Oxford University Press.

Golden, H. L., et al. (2015). Functional neuroanatomy of auditory scene analysis in Alzheimer's disease. *Neuroimage: Clinical, 7*, 699–708.

Goldman-Rakic, P. S. (1987). Circuitry of primate prefrontal cortex and regulation of behavior by representational memory. *Handbook Physiologica, 5*, 373–417.

Goldstein, R. Z., & Volkow, N. D. (2011). Dysfunction of the prefrontal cortex in addiction: Neuroimaging findings and clinical implications. *Nature Reviews Neurosciences, 12*(11), 652–669.

Goodale, M. A., Milner, A. D., Jakobson, L. S., & Carey, D. P. (1991). A neurological dissociation between perceiving objects and grasping them. *Nature, 349*(6305), 154–156.

Goodale, M. A., & Humphrey, G. K. (1998). The objects of action and perception. *Cognition, 67*(1–2), 181–207.

Goren, C. C., Sarty, M., & Wu, P. Y. (1975). Visual following and pattern discrimination of face-like stimuli by newborn infants. *Pediatrics, 56*(4), 544–549.

Gottlieb, G., & Halpern, C. T. (2002). A relational view of causality in normal and abnormal development. *Development and Psychopathology, 14*(3), 421–435.

Graham, N. L., Patterson, K., & Hodges, J. R. (2000). The impact of semantic memory impairment on spelling: Evidence from semantic dementia. *Neuropsychologia, 38*(2), 143–163.

Green, M. F., Horan, W. P., & Lee, J. (2015). Social cognition in schizophrenia. *Nature Review Neurosciences, 16*, 620–631.

Grillner, S. (2013). Fundamentals of motor systems. In L. R. Squire, D. Berg, F. E. Bloom, S. du Lac, A. Ghosh, & N. C. Spitzer (Eds.), *Fundamental neuroscience* (4th ed., pp. 499–511). San Diego: Academic Press.

Groome, L. J., Mooney, D. M., Holland, S. B., Smith, L. A., Atterbury, J. L., & Dykman, R. A. (1999). Behavioral state affects heart rate response to low-intensity sound in human fetuses. *Early Human Development, 54*(1), 39–54.

Grossman, E. D., Jardine, N. L., & Pyles, J. A. (2010). fMRI-adaptation reveals invariant coding of biological motion on the human STS. *Frontiers in Human Neuroscience, 4*(15), 1–18.

Guellai, B., & Streri, A. (2011). Cues for early social skills: Direct gaze modulates newborns' recognition of talking faces. *PLoS One, 6*(4).

Hall, D. A., & Susi, K. (2015). Hemodynamic imaging of auditory cortex. In G. G. Celesia, & G. Hickok (Eds.), *Handbook of clinical neurology (3rd series)* (Vol. 129, pp. 257–275).

Hall, A. (2014). Sleep physiology and the perioperative care of patients with sleep disorders. *Continuing Education of Anaesthesia Critical Care Pain, 15*(4), 167–172.

Hamann, S. (2012). Mapping discrete and dimensional emotions onto the brain: Controversies and consensus. *Trends in Cognitive Sciences, 16*(9), 458–466.

Hari, R., & Salmelin, R. (2012). Magnetoencephalography: From SQUIDs to neuroscience. Neuroimage 20th anniversary special edition. *Neuroimage, 61*, 386–396.

Hartley, A. A., & Speer, N. K. (2000). Locating and fractionating working memory using functional neuroimaging: Storage, maintenance, and executive functions. *Microscopy Research and Technique, 51*(1), 45–53.

Hatfield, E., Cacioppo, J. T., & Rapson, R. L. (1993). Emotional contagion. *Current Directions in Psychological Science, 2*(3), 96–99.

Haxby, J. V., Gobbini, M. I., Furey, M. L., Ishai, A., Schouten, J. L., & Pietrini, P. (2001). Distributed and overlapping representations of faces and objects in ventral temporal cortex. *Science, 293*(5539), 2425–2430.

Haxby, J. V., Hoffman, E. A., & Gobbini, M. I. (January 1, 2002). Human neural systems for face recognition and social communication. *Biological Psychiatry, 51*(1), 59–67. Review. PubMed PMID:11801231.

Heeger, D. J., & Ress, D. (2002). What does fMRI tell us about neuronal activity? *Nature Reviews Neuroscience, 3*, 142–151.

Hendrickson, T. J., Mueller, B. A., Sowell, E. R., Mattson, S. N., Coles, C. D., Kable, J. A., et al. (2017). Cortical gyrification is abnormal in children with prenatal alcohol exposure. *Neuro Image: Clinical, 15*, 391–400.

Hendry, S. H., & Hsiao, S. S. (2014). Fundamentals of sensory systems. In L. R. Squire, D. Berg, F. E. Bloom, S. du Lac, A. Ghosh, & N. C. Spitzer (Eds.), *Fundamental neuroscience* (4th ed., pp. 499–511). San Diego: Academic Press.

Henke, K., Mondadori, C. R., Treyer, V., Nitsch, R. M., Buck, A., & Hock, C. (2003). Nonconscious formation and reactivation of semantic associations by way of the medial temporal lobe. *Neuropsychologia, 41*(8), 863–876.

Henke, K. (2010). A model for memory systems based on processing modes rather than consciousness. *Nature Reviews Neuroscience, 11*, 523–529.

Henry, J. D., von Hippel, W., Molenberghs, P., Lee, T., & Sachdev, P. S. (2015). Clinical assessment of social cognitive function in neurological disorders. *Nature Reviews, 12*, 28–39.

Henson, R. N. (2001). Repetition effects for words and non-words as indexed by event-related fMRI: A preliminary study. *Scandinavian Journal of Psychology, 42*(3), 179–186.

Hermann, L. (1870). Eine Erscheinung simultanen Contrastes. *Pflügers Archiv für die gesamte Physiologie, 3*, 13–15.

Hickok, G., & Poeppel, D. (2007). The cortical organization of speech processing. *Nature Reviews Neuroscience, 8*(5), 393–402.

Hickok, G. (2009). The functional neuroanatomy of language. *Physics of Life Reviews, 6*(3), 121–143.

Hilgard, E. (1977). *Divided Consciousness: Multiple controls in human thought and action* [Wiley series in behavior]. New York: John Wiley & Sons Inc.

Hirshkowitz, M., Whiton, K., Albert, S. M., Alessi, C., Bruni, O., Don Carlos, L., et al. (2015). National sleep Foundation's sleep time duration recommendations: Methodology and results summary. *Sleep Health, 1*, 40–43.

Hobson, J. A., & Pace-Schott, E. F. (2002). The cognitive neuroscience of sleep: Neuronal systems, consciousness, and learning. *Nature Reviews Neuroscience, 3*, 679–693.

Hobson, J. A. (2009). REM sleep and dreaming: Towards a theory of protoconsciousness. *Nature Reviews Neuroscience, 10*, 803–814.

Hofer, S., & Frahm, J. (2006). Topography of the human corpus callosum revisited: Comprehensive fiber tractography using magnetic resonance imaging. *NeuroImage, 32*(3), 989–994.

Hubel, D. H., & Wiesel, T. N. (1962). Receptive fields, binocular interaction and functional architecture in the cat's visual cortex. *Journal of Physiology, Paris, 160*, 106–154.

Hubel, D. H., & Wiesel, T. N. (1968). Receptive fields and functional architecture of monkey striate cortex. *Journal of Physiology, Paris, 195*(1), 215–243.

Huttenlocher, P. R., De Courten, C., Garey, L. J., & Van der Loos, H. (1982). Synaptic development in human cerebral cortex. *International Journal of Neurology, 16–17*, 144–154.

Huttenlocher, P. R. (1990). Morphometric study of human cerebral cortex development. *Neuropsychologia, 28*(6), 517–527.

Huttenlocher, P. R. (1994). Synaptogenesis, synapse elimination, and neural plasticity in human cerebral cortex: Threats to optimal development. In C. A. Nelson (Ed.), *The Minnesota symposia on child psychology* (Vol. 27, pp. 35–54). Hillsdale: Lawrence Erlbaum Associates.

Ingvar, D. H., & Franzen, G. (1974). Abnormalities of cerebral blood flow distribution in patients with chronic schizophrenia. *Acta Psychiatrica Scandinavica, 50*(4), 425–462.

Ingvar, D. H. (1985). Memory of the future: An essay on the temporal organization of conscious awareness. *Human Neurobiology, 4*(3), 127–136.

Irimia, A., Chambers, M. C., Torgerson, C. M., & Van Horn, J. D. (2012). Circular representation of human cortical networks for subject and population-level connectomic visualization. *Neuroimage, 60*, 1340–1351.

Izhikevich, E. M., & Edelman, G. M. (2008). Large-scale model of mammalian thalamocortical systems. *Proceedings of the National Academy of Sciences of the United States of America, 105*(9), 3593–3598.

Jacobs, G. D., Benson, H., & Friedman, R. (1996). Topographic EEG mapping of the relaxation response. *Biofeedback and Self-regulation, 21*(2), 121–129.

James, T. W., Culham, J., Humphrey, G. K., Milner, A. D., & Goodale, M. A. (2003). Ventral occipital lesions impair object recognition but not object-directed grasping: An fMRI study. *Brain, 126*(Pt 11), 2463–2475.

Jancke, D., Chavane, F., Naaman, S., & Grinvald, A. (2004). Imaging cortical correlates of illusion in early visual cortex. *Nature, 428*(6981), 423–426.

Javitt, D. C., & Sweet, R. A. (2015). Auditory dysfunction in schizophrenia: Integrating clinical and basic features. *Nature Reviews Neuroscience, 16*, 535–550.

Ji, D., & Wilson, M. A. (2007). Coordinated memory replay in the visual cortex and hippocampus during sleep. *Nature Neuroscience, 10*, 100–107.

Johnson, M. H. (1991). *Biology and cognitive development: The case of face recognition*. Oxford: Blackwell.

Johnson, M. H. (2001). Functional brain development in humans. *Nature Reviews. Neuroscience, 2*(7), 475–483.

Johnson, M. H. (2005). Subcortical face processing. *Nature Reviews. Neuroscience, 6*(10), 766–774.

Jonathan Coe. (2001). *The Rotter's Club.*

Jones, K. L., & Smith, D. W. (1973). Recognition of the fetal alcohol syndrome in early infancy. *Lancet, 2*(7836), 999–1001.

Jung-Beeman, M., Bowden, E. M., Haberman, J., et al. (2004). Neural activity when people solve verbal problems with insight. *PLoS Biology, 2*(4), 500–510.

Kadosh, K. C., & Johnson, M. H. (2007). Developing a cortex specialized for face perception. *Trends in Cognitive Sciences, 11*(9), 367–369.

Kanske, P., Bockler, A., Trautwein, R.-M., & Singer, T. (2015). Dissecting the social brain: Introducing the EmpaTOM to reveal distinct neural networks and brain-behavioral relations for empathy and Theory of Mind. *Neuroimage, 122*, 6–19.

Kanwisher, N., & Yovel, G. (2006). The fusiform face area: A cortical region specialized for the perception of faces. *Philosophical Transactions of the Royal Society B, 361*, 2109–2128.

Kanwisher, N., McDermott, J., & Chun, M. M. (1997). The fusiform face area: A module in human extrastriate cortex specialized for face perception. *The Journal of Neuroscience, 17*(11), 4302–4311.

Kaszniak, A. W. (1990). Psychological assessment of the aging individual. In J. E. Birren, & K. W. Schaie (Eds.), *Handbook of the psychology of aging* (3rd ed., Vol. xvii, pp. 427–445). San Diego: Academic Press.

Kelly, O. E., Johnson, D. H., Delgutte, B., & Cariani, P. (1996). Fractal noise strength in auditory-nerve fiber recordings. *The Journal of the Acoustical Society of America, 99*(4 Pt 1), 2210–2220.

Kessler, R. C., Demler, O., Frank, R. G., Olfson, M., Pincus, H. A., Walters, E. E., et al. (2005). Prevalence and treatment of mental disorders, 1990 to 2003. *The New England Journal of Medicine, 352*(24), 2515–2523.

Khoury, J., & Doghramji, K. (2015). Primary sleep disorders. *Psychiatric Clinics of North America, 38*, 683–704.

Kisilevsky, B. S., & Low, J. A. (1998). Human fetal behavior: 100 years of study. *Development Review, 18*, 1–29.

Koch, C., Massimini, M., Boly, M., & Tononi, G. (2016). Neural correlates of consciousness: Progress and problems. *Nature Reviews Neuroscience, 17*, 307–321.

Koechlin, E., et al. (2003). The architecture of cognitive control in the human prefrontal cortex. *Science, 302*, 1181–1185.

Koelsch, S. (2005). Neural substrates of processing syntax and semantics in music. *Current Opinion in Neurobiology, 15*(2), 207–212.

Koelsch, S. (2014). Brain correlates of music-evoked emotions. *Nature Review Neuroscience, 15*(3), 170–180.

Koenigs, M. (2012). The role of the prefrontal cortex in psychopathy. *Reviews Neuroscience, 23*(3), 253–262.

Koffka, K. (1935). *Principles of gestalt psychology*. Harcourt, Brace and World, Jovanovic.

Köhler, W. (1921). *Intelligenzprüfungen am Menschenaffen*. Berlin: Springer. https://doi.org/10.1007/978-3-642-47574-0.

Kounios, J., & Beeman, M. (2014). The cognitive neuroscience of insight. *Annual Review of Psychology, 65*, 71–93.

Kounios, J., Frymiare, J. L., Bowden, E. M., Fleck, J. I., Subramaniam, K., Parrish, T. B., et al. (2006). The prepared mind. *Psychological Science, 17*, 882–890.

Kramer, A. D., Guillory, J. E., & Hancock, J. T. (2014). Experimental evidence of massive-scale emotional contagion through social networks. *Proceedings of the National Academy of Sciences, 111*(24), 8788–8790.

Kreiman, G., Koch, C., & Fried, I. (2000). Category-specific visual responses of single neurons in the human medial temporal lobe. *Nature Neuroscience, 3*, 946–953.

Kringelbach, M. (2005). The human orbitofrontal cortex: Linking reward to hedonic experience. *Nature Reviews. Neuroscience, 6*, 691–702.

Kuffler, S. W. (1953). Discharge patterns and functional organization of mammalian retina. *Journal of Neurophysiology, 16*(1), 37–68.

Kuhl, P. K., & Rivera-Gaxiola, M. (2008). Neural substrates of early language acquisition. *Annual Review of Neuroscience, 31*, 511–534.

Kumar, S., Stephan, K. E., Warren, J. D., Friston, K. J., & Griffiths, T. D. (2007). Hierarchical processing of auditory objects in humans. *PLoS Computational Biology, 3*(6), 977–985.

Lameka, K., Farwell, M. D., & Ichise, M. (2016). Positron emission tomography. In J. C. Masdeu, & R. G. Gonzalez (Eds.), *Handbook of clinical neurology (3rd series)* (Vol. 135, pp. 209–227). Amsterdam: Elsevier.

Lamme, V. A. F., & Roelfsema, P. R. (2000). *Trends in Neuroscience, 23*, 571–579.

Lancaster, J. L., Laird, A. R., Eickhoff, S. B., Martinez, M. J., Fox, P. M., & Fox, P. T. (2012). Automated regional behavioral analysis for human brain images. *Frontiers in Neuroinformatics, 6*(23), 1–12.

Landolt, H.-P. (2008). Sleep homeostasis: A role for adenosine in humans? *Biochemical Pharmacology, 75*(11), 2070–2079.

Laureys, S., Owen, A. M., & Schiff, N. D. (2004). Brain function in coma, vegetative state, and related disorders. *Lancet, 3*, 537–546.

Laureys, S. (2005). The neural correlates of (un)awareness: Lessons from the vegetative state. *Trends in Cognitive Sciences, 9*(12), 556–559.

Laviolette, S. R., & van der Kooy, D. (2004). The neurobiology of nicotine addiction: Bridging the gap from molecules to behavior. *Nature Reviews Neuroscience, 5*, 55–65.

Lazar, S. W., Bush, G., Gollub, R. L., Fricchione, G. L., Khalsa, G., & Benson, H. (2000). Functional brain mapping of the relaxation response and meditation. *NeuroReport, 11*(7), 1581–1585.

Lecanuet, J. P., Granier-Deferre, C., DeCasper, A. J., Maugeais, R., Andrieu, A. J., & Busnel, M. C. (1987). Fetal perception and discrimination of speech stimuli; demonstration by cardiac reactivity; preliminary results. *Comptes Rendus De L'Academie Des Sciences. Serie III, Sciences De La Vie, 305*(5), 161–164.

Lecuyer, R., Abgueguen, I., & Lemarie, C. (1992). 9-and-5 month olds do not make the AB error if not required to manipulate objects. In *Proceedings of the VIIth international conference on infant studies. Miami.*

LeDoux, J. E. (1996). *The emotional brain*. New York: Simon & Schuster.

Lees, G. V., Jones, E. G., & Kandel, E. R. (2000). Expressive genes record memories. *Neurobiology of Disease, 7*(5), 533–536.

Leisman, G., Machado, C., Melillo, R., & Mualem, R. (2012). Intentionality and "free-will" from a neurodevelopmental perspective. *Frontiers in Integrative Neuroscience, 6*(36), 1–12.

Leroy, F., Glasel, H., Dubois, J., Hertz-Pannier, L., Thirion, B., Mangin, J.-F., et al. (2011). Early maturation of the linguistic dorsal pathway in human infants. *The Journal of Neuroscience, 31*(4), 1500–1506.

Levy, R., & Goldman-Rakic, P. S. (2000). Segregation of working memory functions within the dorsolateral prefrontal cortex. *Experimental Brain Research, 133*(1), 23–32.

Lim, S.-J., Fiez, J. A., & Holt, L. L. (2014). How may the basal ganglia contribute to auditory categorization and speech perception? *Frontiers in Neuroscience, 8*(230), 1–18.

Longo, C. A., Fried, P. A., Cameron, I., & Smith, A. M. (2013). The long-term effects of prenatal nicotine exposure on response inhibition: An fMRI study of young adults. *Neurotoxicity and Teratology, 39*, 9–18.

Luria, A. R. (1966). *Higher cortical functions in man* (B. Haigh, Trans.). London: Tavistock Morris et al., 1993.

Lutz, A., Greischar, L. L., Rawlings, N. B., Ricard, M., & Davidson, R. J. (2004). Long-term meditators self-induce high-amplitude gamma synchrony during mental practice. *Proceedings of the National Academy of Sciences of the United States of America, 101*(46), 16369–16373.

Maffei, C., Soria, G., Prats-Galino, A., & Catani, M. (2015). Imaging white-matter pathways of the auditory system with diffusion tensor tractography. In G. G. Celesia, & G. Hickok (Eds.), *Handbook of clinical neurology (3rd series)* (Vol. 129, pp. 277–288).

Maguire, E. A., Gadian, D. G., Johnsrude, I. S., Good, C. D., Ashburner, J., Frackowiak, R. S., et al. (2000). Navigation-related structural change in the hippocampi of taxi drivers. *Proceedings of the National Academy of Sciences of the United States of America, 97*(8), 4398–4403.

Malach, R., Reppas, J. B., Benson, R. R., et al. (1995). Object-related activity revealed by functional magnetic resonance imaging in human occipital cortex. *Proceedings of the National Academy of Sciences of the United States of America, 92*(18), 8135–8139.

Manji, H. K., & Duman, R. S. (2001). Impairments of neuroplasticity and cellular resilience in severe mood disorders: Implications for the development of novel therapeutics. *Psychopharmacol Bulletin, 35*(2), 5–49.

Marcus, D. S., Wang, T. H., Parker, J., Csernansky, J. G., Morris, J. C., & Buckner, R. L. (2007). Open Access Series of Imaging Studies (OASIS): Cross-sectional MRI data in young, middle aged, nondemented, and demented older adults. *Journal of Cognitive Neuroscience, 19*(9), 1498–1507.

Marr, D. (1971). Simple memory: A theory for archicortex. *Philosophical Transactions of the Royal Society B: Biological Sciences, 262*, 23–81.

Maski, K., & Owens, J. A. (2016). Insomnia, parasomnias, and narcolepsy in children: Clinical features, diagnosis, and management. *Lancet, 15*, 1170–1181.

McCarthy, G., Puce, A., Gore, J. C., & Allison, T. (1997). Face-specific processing in the human fusiform gyrus. *Journal of Cognitive Neuroscience, 9*(5), 605–610.

McClelland, J. L., McNaughton, B. L., & O'Reilly, R. C. (1995). Why there are complementary learning systems in the hippocampus and neocortex: Insights from the successes and failures of connectionist models of learning and memory. *Psychological Review, 102*, 419–457.

McDougall, G. J. (1990). A review of screening instruments for assessing cognition and mental status in older adults. *The Nurse Practitioner, 15*(11), 18–28.

McGaugh, J. L. (2000). Memory—A century of consolidation. *Science, 287*(5451), 248–251.

McGaugh, J. L. (2002). Memory consolidation and the amygdala: A systems perspective. *Trends in Neurosciences, 25*(9), 456.

McIntosh, A. R., Lobaught, N. J., Cabeza, R., Bookstein, F. L., & Houle, S. (1998). Convergence of neural systems processing stimulus associations and coordinating motor responses. *Cerebral Cortex, 8*(7), 648–659.

Meadows, J. C. (1974a). Disturbed perception of colours associated with localized cerebral lesions. *Brain, 97*(4), 615–632.

Meadows, J. C. (1974b). The anatomical basis of prosopagnosia. *Journal of Neurology, Neurosurgery, and Psychiatry, 37*(5), 489–501.

Meng, M., Remus, D. A., & Tong, F. (2005). Filling-in of visual phantoms in the human brain. *Nature Neuroscience, 8*(9), 1248–1254.

Metcalfe, J., & Wiebe, D. (1987). Intuition in insight and noninsight problem solving. *Memory & Cognition, 15*, 238–246.

Metcalfe, J. (1986). Feeling of knowing in memory and problem solving. *Journal of Experimental Psychology, 12*, 288–294.

Michel, C. M., & Murray, M. M. (2012). Towards the utilization of EEG as a brain imaging tool. *Neuroimage, 61*, 371–385.

Miller, E. K., & Wallis, J. D. (2013). Theories of prefrontal cortex function. In L. R. Squire, D. Berg, F. E. Bloom, S. du Lac, A. Ghosh, & N. C. Spitzer (Eds.), *Fundamental neuroscience* (4th ed., pp. 553–576). San Diego: Academic Press.

Milner, B., Corkin, S., & Teuber, H. L. (1968). Further analysis of the hippocampal amnesic syndrome: 14-year follow-up study of H.M. *Neuropsychologia, 6*(3), 215–234.

Mirskey, A. F., & Duncan, C. C. (2005). Pathophysiology of mental illness: A view from the fourth ventricle. *International Journal of Psychophysiology, 58*, 162–178.

Mitz, A. R., Bartolog, R., Saunders, R. C., Browning, P. G., Talbot, T., & Averbeck, B. B. (2017). High channel count single-unit recordings from nonhuman primate frontal cortex. *Journal of Neuroscience Methods, 289*, 39–47.

Moerel, M., de Martino, F., & Formisano, E. (2014). An anatomical and functional topography of human auditory cortical areas. *Frontiers of Neuroscience, 8*, 1–14.

Morris, R. G., Ahmed, S., Syed, G. M., & Toone, B. K. (1993). Neural correlates of planning ability: Frontal lobe activation during the tower of London test. *Neuropsychologia, 31*, 1367–1378.

Moscovitch, M., & McAndrews, M. P. (2002). Material specific deficits in 'remembering' in patients with unilateral temporal lobe epilepsy and excisions. *Neuropsychologia, 40*(8), 1335–1342.

Moscovitch, M., & Nadel, L. (1998). Consolidation and the hippocampal complex revisited: In defense of the multiple-trace model. *Current Opinion in Neurobiology, 8*(2), 297–300.

Moscovitch, M., Rosenbaum, R. S., Gilboa, A., Addis, D. R., Westmacott, R., Grady, C., et al. (2005). Functional neuroanatomy of remote episodic, semantic and spatial memory: A unified account based on multiple trace theory. *Journal of Anatomy, 207*(1), 35–66.

Moscovitch, M. (1992). Memory and working-with memory: A component process model based on modules and central systems. *Journal of Cognitive Neuroscience, 4*(3), 257–267.

Moscovitch, M. (1995). Recovered consciousness: A hypothesis concerning modularity and episodic memory. *Journal of Clinical and Experimental Neuropsychology, 17*(2), 276–290.

Muckli, L., Kohler, A., Kriegeskorte, N., & Singer, W. (2005). Primary visual cortex activity along the apparent-motion trace reflects illusory perception. *PLoS Biology, 3*(8), e265.

Munakata, Y., McClelland, J. L., Johnson, M. H., & Siegler, R. S. (1994). *Now you see it, now you don't: A gradualistic framework for understanding infants' successes and failures in object permanence tasks.* Carnegie Mellon University. Technical Report PDP. CNS.94.2.

Murray, M. M., & Herrmann, C. S. (2013). Illusory contours: A window onto the neurophysiology of constructing perception. *Trends in Cognitive Neuroscience, 17*(9), 471–481.

Nadel, L., & Hardt, O. (2011). Update on memory systems and processes. *Neuropsychopharmacology, 36*, 251–273.

Nadel, L., & Moscovitch, M. (1997). Memory consolidation, retrograde amnesia and the hippocampal complex. *Current Opinion in Neurobiology, 7*(2), 217–227.

Nadel, L., & Moscovitch, M. (1998). Hippocampal contributions to cortical plasticity. *Neuropharmacology, 37*(4–5), 431–439.

Nadel, L., Samsonovich, A., Ryan, L., & Moscovitch, M. (2000). Multiple trace theory of human memory: Computational, neuroimaging, and neuropsychological results. *Hippocampus, 10*(4), 352–368.

Nader, K. (2003). Memory traces unbound. *Trends in Neurosciences, 26*(2), 65–72.

Naghavi, H. R., & Nyberg, L. (2005). Common fronto-parietal activity in attention, memory, and consciousness: Shared demands on integration? *Cognition and Consciousness, 14*, 390–425.

Nair, J., Klaassen, A.-L., Poirot, J., Vyssotski, A., & Rasch, B. (2016). Gamma band directional interactions between basal forebrain and visual cortex during wake and sleep states. *Journal of Physiology – Paris, 110*, 19–28.

Nauta, W. J. (1972). Neural associations of the frontal cortex. *Acta Neurobiologiae Experimentalis (Wars), 32*(2), 125–140.

Navratilova, E., & Porreca, F. (2014). Reward and motivation in pain and pain relief. *Nature Neuroscience, 17*(10), 1304–1312.

Nilsson, L. G., & Markowitsch, H. J. (1999). *Cognitive neuroscience of memory*. Seattle: Hogrefe & Huber.

Northoff, G. (2012). From emotions to consciousness – a neuro-phenomenal and neuro-relational approach. *Frontiers in Psychology, 3*, 1–17.

Nourski, K. V., & Howard, M. A., III (2015). Invasive recordings in the human auditory cortex. In G. G. Celesia, & G. Hickok (Eds.), *Handbook of clinical neurology (3rd series)* (Vol. 129, pp. 225–244). Amsterdam: Elsevier.

Oatley, K., & Johnson-Laird, P. N. (2014). Cognitive approaches to emotion. *Trends in Cognitive Sciences, 18*(3), 134–140.

Ohayon, M. M. (2002). Epidemiology of insomnia: What we know and what we still need to learn. *Sleep Medicine Reviews, 6*(2), 97–111.

Öllinger, M., & Knoblich, G. (2009). Psychological research on insight problem solving. In H. Atmanspacher, & H. Primas (Eds.), *Recasting Reality: Wolfgang Pauli's philosophical ideas and contemporary science* (pp. 275–300). Berlin: Springer.

Ostergaard, A. L. (1987). Episodic, semantic and procedural memory in a case of amnesia at an early age. *Neuropsychologia, 25*(2), 341–357.

Panksepp, J. (1998). *Affective neuroscience: The foundations of human and animal emotions*. New York: Oxford University Press.

Panksepp, J. (2005). Affective consciousness: Core emotional feelings in animals and humans. *Consciousness and Cognition, 14*(1), 30–80.

Panksepp, J. (2006). Emotional endophenotypes in evolutionary psychiatry. *Progress in Neuro-psychopharmacology & Biological Psychiatry, 30*(5), 774–784.

Panksepp, J. (2011). The basic emotional circuits of mammalian brains: Do animals have affective lives? *Neuroscience and Biobehavioral Reviews, 35*, 1791–1804.

Pardo, J. V., Pardo, P. J., Janer, K. W., & Raichle, M. E. (1990). The anterior cingulate cortex mediates processing selection in the Stroop attentional conflict paradigm. *Proceedings of the National Academy of Sciences of the United States of America, 87*, 256–259.

Pardo, J. V., Fox, P. T., & Raichle, M. E. (1991). Localization of a human system for sustained attention by positron emission tomography. *Science, 349*, 61–64.

Pasternak, T., & Merigan, W. H. (1994). Motion perception following lesions of the superior temporal sulcus in the monkey. *Cerebral Cortex, 4*(3), 247–259.

Pasupathy, A., & Connor, C. E. (2002). Population coding of shape in area V4. *Nature Neuroscience, 5*(12), 1332–1338.

Paulesu, E., Frith, C. D., & Frackowiak, R. S. (1993). The neural correlates of the verbal component of working memory. *Nature, 362*(6418), 342–345.

Paus, T. (2005). Mapping brain maturation and cognitive development during adolescence. *Trends in Cognitive Sciences, 9*(2), 60–68.

Pelli, D. G., Palomares, M., & Majaj, N. J. (2004). Crowding is unlike ordinary masking: Distinguishing feature integration from detection. *Journal of Vision, 4*(12), 1136–1169.

Peretz, I., & Zatorre, R. J. (2005). Brain organization for music processing. *Annual Review of Psychology, 56*, 89–114.

Petersen, S. E., & Posner, M. I. (2012). The attention system of the human brain: 20 years after. *Annual Review of Neuroscience, 35*, 73–89.

Petrides, M., & Pandya, D. N. (1994). Comparative architectonic analysis of the human and the macaque frontal cortex. In F. Boller, & J. Grafman (Eds.), *Handbook of neuropsychology* (vol. 9, pp. 17–58). Amsterdam: Elsevier.

Pfeiffer, U. J., Vogeley, K., & Schilbach, L. (2013). From gaze cuing to dual eye-tracking: Novel approaches to investigate the neural correlates of gaze in social interaction. Fig 1. *Neuroscience and Biobehavioral Reviews, 37*, 2516–2528. https://doi.org/10.1016/j.neubiorev.2013.07.017. Epub 2013 Aug 5. Review. PubMed PMID:23928088.

Piaget, J. (1954). *The construction of reality in the child* (M. Cook, Trans.). New York: Basic Books.

Plant, G. T., Laxer, K. D., Barbaro, N. M., Schiffman, J. S., & Nakayama, K. (1993). Impaired visual motion perception in the contralateral hemifield following unilateral posterior cerebral lesions in humans. *Brain, 116*(Pt 6), 1303–1335.

Poeppel, D., & Hickok, G. (2015). Electromagnetic recording of the auditory system. In G. G. Celesia, & G. Hickok (Eds.), *Handbook of clinical neurology (3rd series)* (Vol. 129, pp. 245–255).

Poldrack, R. A., Sabb, F., Foerde, K., Tom, S. M., Asarnow, R. F., Bookheimer, S. Y., et al. (2005). The neural correlates of motor skill automaticity. *Journal of Neuroscience, 25*, 5356–5364.

Porkka-Heiskanen, T., & Kalinchuk, A. V. (2011). Adenosine, energy metabolism, and sleep homeostasis. *Sleep Medicine Reviews, 15*, 123–135.

Portas, C. M., Krakow, K., Allen, P., Josephs, O., Armony, J. L., & Frith, C. D. (2000). Auditory processing across the sleep-wake cycle: Simultaneous EEG and fMRI monitoring in humans. *Neuron, 28*(3), 991–999.

Posner, M. I., & Petersen, S. E. (1990). The attention system of the human brain. *Annual Review of Neuroscience, 13*, 25–42.

Posner, M. I., & Raichle, M. E. (1997). *Images of mind*. New York: Henry Holt and Company.

Posner, M. I., & Rothbart, M. K. (1998). Attention, self regulation and consciousness. *Philosophical Transactions of the Royal Society of London. Series B, Biological Sciences, 353*(1377), 1915–1927.

Posner, M. I., Petersen, S. E., Fox, P. T., & Raichle, M. E. (1988). Localization of cognitive operations in the human brain. *Science, 240*(4859), 1627–1631.

Posner, M. I. (1980). Orienting of attention. *The Quarterly Journal of Experimental Psychology, 32*(1), 3–25.

Price, C. J. (2012). A review and synthesis of the first 20 years of PET and fMRI studies of heard speech, spoken language, and reading. *NeuroImage, 62*, 816–847.

Quartz, S. R. (1999). The constructivist brain. *Trends in Cognitive Sciences, 3*(2), 48–57.

Raichle, M. E., Fiez, J. A., Videen, T. O., MacLeod, A. M., Pardo, J. V., Fox, P. T., et al. (1994). Practice-related changes in human brain functional anatomy during nonmotor learning. *Cerebral Cortex, 4*(1), 8–26.

Raichle, M. E., MacLeod, A. M., Snyder, A. Z., Powers, W. J., Gusnard, D. A., & Shulman, G. L. (2001). A default mode of brain function. *Proceedings of the National Academy of Science of the United States of America, 98*, 676–682.

Raine, A., Buchsbaum, M., & LaCasse, L. (1997). Brain abnormalities in murderers indicated by positron emission tomography. *Biological Psychiatry, 42*(6), 495–508.

Rakic, P. (1988). Specification of cerebral cortical areas. *Science, 241*(4862), 170–176.

Rakic, P. (2009). Evolution of the neocortex: A perspective from developmental biology. *Nature Reviews. Neuroscience, 10*(10), 724–735.

Ranganath, C. (2006). Working memory for visual objects: Complementary roles of inferior temporal, medial temporal, and prefrontal cortex. *Neuroscience, 139*(1), 277–289.

Rasch, B., & Born, J. (2007). Maintaining memories by reactivation. *Current Opinions in Neurobiology, 17*, 698–703.

Rauschecker, J. P., & Scott, S. K. (2009). Maps and streams in auditory cortex: Nonhuman primates illuminate human speech processing. *Nature Neuroscience, 12*(6), 718–724.

Redcay, E., Kleiner, M., & Saxe, R. (2012). Look at this: The neural correlates of initiating and responding to bids for joint attention. *Frontiers in Human Neuroscience, 6*(169), 1–14.

Reid, R. C., & Usrey, W. M. (2013). Vision. In L. R. Squire, D. Berg, F. E. Bloom, S. du Lac, A. Ghosh, & N. C. Spitzer (Eds.), *Fundamental neuroscience* (4th ed., pp. 577–595). San Diego: Academic Press.

Reilly, J., Losh, M., Bellugi, U., & Wulfeck, B. (2004). Frog, where are you?' narratives in children with specific language impairment, early focal brain injury, and Williams syndrome. *Brain and Language, 88*(2), 229–247.

Ress, D., & Heeger, D. J. (2003). Neuronal correlates of perception in early visual cortex. *Nature Neuroscience, 6*(4), 414–420.

Roediger, H. L., & McDermott, K. B. (1993). Implicit memory in normal human subjects. *Handbook Neuropsychology, 8*, 63–131.

Roy, M., Shohamy, D., & Wager, T. D. (2012). Ventromedial prefrontal-subcortical systems and the generation of affective meaning. *Trends in Cognitive Sciences, 16*(3), 147–156.

Ryan, J. D., Althoff, R. R., Whitlow, S., & Cohen, N. J. (2000). Amnesia is a deficit in relational memory. *Psychological Science, 11*(6), 454–461.

Rypma, B., Berger, J. S., & D'Esposito, M. (2002). The influence of working-memory demand and subject performance on prefrontal cortical activity. *Journal of Cognitive Science, 14*(5), 721–731.

Sacks, O. (1985). *The man who mistook his wife for a hat: And other clinical tales*. New York: Simon & Schuster.

Sacks, O. (1992). Tourette's syndrome and creativity. *BMJ: British Medical Journal, 305*(6868), 1151–1515.

Sanes, D. H., Reh, T. A., & Harris, W. A. (2006). *Development of the nervous system* (2nd ed.). Amsterdam, Boston: Elsevier Academic Press.

Saper, C. B., Scammell, T. E., & Lu, J. (2005). Hypothalamic regulation of sleep and circadian rhythms. *Nature, 437*, 1257–1263.

Sasaki, Y., & Watanabe, T. (2004). The primary visual cortex fills in color. *Proceedings of the National Academy of Sciences of the United States of America, 101*(52), 18251–18256.

Sawant, N. S., Parkar, S. R., & Tambe, R. (2005). Isolated sleep paralysis. *Indian Journal of Psychiatry, 47*(4), 238–240.

Schaafsma, S. M., Pfaff, D. W., Spunt, R. P., & Adolphs, R. (2015). Deconstructing and reconstructing theory of mind. *Trends in Cognitive Sciences, 19*(2), 65–72. https://doi.org/10.1016/j.tics.2014.11.007. Epub 2014 Dec 11. Review. PubMed PMID:25496670; PubMed Central PMCID:PMC4314437.

Schacter, D. L., Dobbins, I. G., & Schnyer, D. M. (2004). Specificity of priming: A cognitive neuroscience perspective. *Nature Reviews Neuroscience, 5*(11), 853–862.

Schenck, C. H., Boeve, B. F., & Mahowald, M. W. (2013). Delayed emergence of a parkinsonian disorder or dementia in 81% of older men initially diagnosed with idiopathic rapid eye movement sleep behavior disorder: A 16-year update on a previously reported series. *Sleep Medicine, 14*(8), 744–748.

Schendan, H. E., Searl, M. M., Melrose, R. J., & Stern, C. E. (2003). An FMRI study of the role of the medial temporal lobe in implicit and explicit sequence learning. *Neuron, 37*(6), 1013–1025.

Scherf, K. S., Luna, B., Minshew, N., & Behrmann, M. (2010). Location, location, location: Alterations in the functional topography of face- but not object- or place-related cortex in adolescents with autism. *Frontiers in Human Neuroscience, 4*(26), 1–16.

Schloesser, R. J., Martinowich, K., Manji, H. K., et al. (2011). Mood-stabilizing drugs: Mechanisms of action. *Trends in Neurosciences, 35*(1), 36–46.

Schmid, M. C., Mrowka, S. W., Turchi, J., Saunders, R. C., Wilke, M., Peters, A. J., et al. (2010). Blindsight depends on the lateral geniculate nucleus. *Nature, 466*, 373–377.

Schore, A. (1999). *Affect regulation and the origin of the self: The neurobiology of emotional development*. Hillsdale: Lawrence Erlbaum Associates.

Schreiner, C. E., & Winer, J. A. (2007). Auditory cortex mapmaking: Princples, projections, and plasticity. *Neuron, 56*(2), 356–365.

Schumann, C. M., & Nordahl, C. W. (2011). Bridging the gap between MRI and postmortem research in autism. *Brain Research*, 175–186.

Schurz, M., Radua, J., Aichhorn, M., Richlan, F., & Perner, J. (2014). Fractionating theory of mind: A meta-analysis of functional brain imaging studies. *Neuroscience and Biobehavioral Reviews, 42*, 9–34.

Schwarzlose, R. F., Baker, C. I., & Kanwisher, N. (2005). Separate face and body selectivity on the fusiform gyrus. *Journal of Neuroscience, 25*(47), 11055–11059.

Scoville, W. B., & Milner, B. (1957). Loss of recent memory after bilateral hippocampal lesions. *Journal of Neurology, Neurosurgery, and Psychiatry, 20*(1), 11–21.

Senju, A., & Johnson, M. (2009). The eye contact effect: Mechanisms and development. *Trends Cogn Sci, 13*, 127–134.

Seth, A. K., & Baars, B. J. (2005). Neural darwinism and consciousness. *Consciousness and Cognition, 14*(1), 140–168.

Shackman, A. J., Salomons, T. V., Slagter, H. A., Fox, A. S., Winter, J. J., & Davidson, R. J. (March 2011). The integration of negative affect, pain and cognitive control in the cingulate cortex. Fig 1. *Nature Reviews Neuroscience, 12*(3), 154–167. https://doi.org/10.1038/nrn2994. Review. PubMed PMID:21331082; PubMed Central PMCID:PMC3044650.

Shallice, T., & Warrington, E. K. (1970). Independent functioning of verbal memory stores: A neuropsychological study. *The Quarterly Journal of Experimental Psychology, 22*(2), 261–273.

Shallice, T. (1982). Specific impairments of planning. *Philosophical Transactions of the Royal Society of London. Series B, Biological Sciences, 298*(1089), 199–209.

Shapiro, A. K., & Shapiro, E. (1974). Gilles de la Tourette's syndrome. *American Family Physician, 9*(6), 94–96.

Shaw, P., Greenstein, D., Lerch, J., Clasen, L., Lenroot, R., Gogtay, N., et al. (2006). Intellectual ability and cortical development in children and adolescents. *Nature, 440*(7084), 676–679.

Siapas, A. G., & Wilson, M. A. (1998). Coordinated interactions between hippocampal ripples and cortical spindles during slow-wave sleep. *Neuron, 21*, 1123–1128.

Simon, H. A. (1948). *Administrative behaviour: A study of the decision making processes in administrative organisation* (4th ed.). New York: The Macmillan Co., Free Press, 1997.

Simons, D. J., & Chabris, C. F. (1999). Gorillas in our midst: Sustained inattentional blindness for dynamic events. *Perception, 28*(9), 1059–1074.

Simons, J. S., & Spiers, H. J. (2003). Prefrontal and medial temporal lobe interactions in long-term memory, Nature Reviews. *Neuroscience, 4*(8), 637–648.

Simons, D. J. (2010). Monkeying around with the gorillas in our midst: Familiarity with an inattentional-blindness task does not improve the detection of unexpected events. *I-perception, 1*, 3–6.

Sirota, A., Csicsvari, J., Buhl, D., & Buzsaki, G. (2003). Communication between neocortex and hippocampus during sleep in rodents. *Proceedings of the National Academy of Sciences of the United States of America, 100*, 2065–2069.

Skeide, M. A., & Friederici, A. D. (2016). The ontogeny of the cortical language network. *Nature Reviews Neuroscience, 17*, 323–332.

Skerry, A. E., & Saxe, R. (2015). Neural representations of emotion are organized around abstract event features. *Current Biology*.

Smith, E. E., & Jonides, J. (1998). Neuroimaging analysis of human working memory. *Proceedings of the National Academy of Sciences of the United States of America, 95*(20), 12061–12068.

Smith, E. E., Jonides, J., & Koeppe, R. A. (1996). Dissociating verbal and spatial working memory using PET. *Cerebral Cortex, 6*(1), 11–20.

Smith, A. M., Fried, P. A., Hogan, M. J., & Cameron, I. (2006). Effects of prenatal marijuana on visuospatial working memory: An fMRI study in young adults. *Neurotoxicology and Teratology, 28*(2), 286–295.

Sowell, E. R., Thompson, P. M., Mattson, S. N., Tessner, K. D., Jernigan, T. L., Riley, E. P., et al. (2002). Regional brain shape abnormalities persist into adolescence after heavy prenatal alcohol exposure. *Cerebral Cortex, 12*, 856–865.

Spiegel, D. (2003). Negative and positive visual hypnotic hallucinations: Attending inside and out. *The International Journal of Clinical and Experimental Hypnosis, 51*(2), 130–146.

Squire, L. R., & Alvarez, P. (1995). Retrograde amnesia and memory consolidation: A neurobiological perspective. *Current Opinion in Neurobiology, 5*(2), 169–177.

Squire, L. R. (1992). Declarative and nondeclarative memory: Multiple brain systems supporting learning and memory. *Journal of Cognitive Neuroscience, 4*(3), 232–243.

Squire, L. R. (2004). Memory systems of the brain: A brief history and current perspective. *Neurobiology of Learning and Memory, 82*, 171–177.

Standing, L. (1973). Learning 10,000 pictures. *Quarterly Journal of Experimental Psychology, 25*(2), 207–222.

Steriade, M. (1997). Synchronized activities of coupled oscillators in the cerebral cortex and thalamus at different levels of vigilance. *Cerebral Cortex, 7*(6), 583–604.

Steriade, M. (2005). Sleep, epilepsy and thalamic reticular inhibitory neurons. *Trends in Neuroscience, 28*(6), 317–324.

Sterman, M. B., & Egner, T. (2006). Foundation and practice of neurofeedback for the treatment of epilepsy. *Applied Psychophysiology and Biofeedback, 31*(1), 21–35.

Sternberg, R. J. (1996). Striving for creativity. *Science, 272*(5270), 1857–1858.

Stickgold, R., & Walker, M. P. (2013). Sleep-dependent memory triage: Evolving generalization through selective processing. *Nature Neuroscience, 16*(2), 139–145.

Stiles, J., Bates, E. A., Thal, D., Trauner, D. A., & Reilly, J. (2002). Linguistic and spatial cognitive development in children with pre- and perinatal focal, brain injury: A ten-year overview from the San Diego longitudinal Project. In M. H. Johnson, Y. Munakata, & R. O. Gilmore (Eds.), *Brain development and cognition: A reader* (2nd ed., pp. 272–291). Oxford: Blackwell Publishing.

Stiles, J., Reilly, J., Paul, B., & Moses, P. (2005). Cognitive development following early brain injury: Evidence for neural adaptation. *Trends in Cognitive Sciences, 9*(3), 136–143.

Stoerig, P., & Cowey, A. (1997). Blindsight in man and monkey. *Brain, 120*, 535–559.

Stoerig, P., Zontanou, A., & Cowey, A. (2002). Aware or unaware: Assessment of cortical blindness in four men and a monkey. *Cerebral Cortex, 12*(6), 565–574.

Striano, T., & Reid, V. M. (2006). Social cognition in the first year. *Trends in Cognitive Sciences, 10*(10), 471–476.

Stuss, D. T., & Benson, D. F. (1986). *The frontal lobes*. New York: Raven Press.

Sussman, E. S. (2005). Integration and segregation in auditory scene analysis. *The Journal of the Acoustical Society of America, 117*(3 Pt 1), 1285–1298.

Sweller, J. (1988). Cognitive load during problem solving: Effects of learning. *Cognitive Science, 12*, 257–285.

Talairach, J., & Tournoux, P. (1988). Co-planar stereotaxic atlas of the human brain. In *3-dimensional proportional system: An approach to cerebral imaging*. Stuttgart: Thieme Medical Publishers.

Tamietto, M., & de Gelder, B. (2010). Neural bases of the non-conscious perception of emotional signals. *Nature Reviews Neuroscience, 11*, 697–709.

Tang, Y. Y., Ma, Y., Wang, J., Fan, Y., Feng, S., Lu, Q., et al. (2007). Short-term meditation training improves attention and self-regulation. *Proceedings of the National Academy of Sciences of the United States of America, 104*(43), 17152–17156.

Taubert, M., Draganski, B., & Anwander, A. (2010). Dynamic properties of human brain structure: Learning-related changes in cortical areas and associated fiber connections. *The Journal of Neuroscience, 30*(35), 11670–11677.

Teyler, T. J., & DiScenna, P. (1986). The hippocampal memory indexing theory. *Behavioral Neuroscience, 100*(2), 147–154.

Thatcher, R. W. (1992). Cyclic cortical reorganization in early childhood. *Brain and Cognition, 20*, 24–50.

Toga, A. W., Thompson, P. M., & Sowell, E. R. (2006). Mapping brain maturation. *Trends in Neurosciences, 29*(3), 148–159.

Tong, F., Nakayama, K., Vaughan, J. T., & Kanwisher, N. (1998). Binocular rivalry and visual awareness in human extrastriate cortex. *Neuron, 21*, 753–759.

Tong, F., Nakayama, K., Moscovitch, M., Weinrib, O., & Kanwisher, N. (2000). Response properties of the human fusiform face area. *Cognitive Neuropsychology, 17*, 257–279.

Tong, F. (2003). Out-of-body experience: From Penfield to present. *Trends in Cognitive Sciences, 7*(3), 104–106.

Tononi, G., & Cirelli, C. (2003). Sleep and synaptic homeostasis: A hypothesis. *Brain Research Bulletin, 62*(2), 143–150.

Treisman, A. M., & Gelade, G. (1980). A feature-integration theory of attention. *Cognitive Psyhcology, 12*(1), 97–136.

Tsao, D. Y., Freiwald, W. A., Tootell, R. B., & Livingstone, M. S. (2006). A cortical region consisting entirely of face-selective cells. *Science, 311*(5761), 670–674.

Tsuchiya, N., & Adolphs, R. (2007). Emotion and consciousness. *Trends in Cognitive Sciences, 11*(4), 158–167.

Tulving, E. (1972). Episodic and semantic memory. In E. Tulving, W. Donaldson, & G. H. Bower (Eds.), *Organization of memory* (pp. 381–403). New York: Academic Press.

Tulving, E. (1985). How many memory systems are there. *The American Psychologist, 40*(4), 385–398.

Uhlhaas, P. J., Pipa, G., Lima, B., Melloni, L., Neuenschwander, S., Nikolic, D., et al. (2009). Neural synchrony in cortical networks: History, concept and current status. *Frontiers in Integrative Neuroscience, 3*(17), 1–19.

Ungerleider, L. G., & Mishkin, M. (1982). Two cortical visual systems. In M. A. Ingle, M. A. Goodale, & J. W. Mansfield (Eds.), *Analysis of visual behavior*. Cambridge: The MIT Press.

van Horn, J. D., Irimia, A., Torgerson, C. M., Chambers, M. C., Kikinis, R., & Toga, A. W. (2012). Mapping connectivity damage in the case of Phineas Gage. *PLoS One, 7*(5), 1–23.

Volavka, J., Mohammad, Y., Vitrai, J., Connolly, M., Stefanovic, M., & Ford, M. (1995). Characteristics of state hospital patients arrested offenses committed during hospitalization. *Psychiatric Services, 46*(8), 796–800.

Vuilleumier, P., & Pourtois, G. (2007). Distributed and interactive brain mechanisms during emotion face perception: Evidence from functional neuroimaging. *Neuropsychologia, 45*, 174–194.

Wager, T. D., & Smith, E. E. (2003). Neuroimaging studies of working memory: A meta-analysis. *Cognitive, Affective & Behavioral Neuroscience, 3*(4), 255–274.

Warrier, C., Wong, P., Penhune, V., Zatorre, R., Parrish, T., Abrams, D., et al. (2009). Relating structure to function: Heschl's gyrus and acoustic processing. *The Journal of Neuroscience, 29*(1), 61–69.

Wearing, D. (2005). *Forever today: A true story of lost memory and never-ending love*. London: Corgi Books.

Weiler, I. J., Hawrylak, N., & Greenough, W. T. (1995). Morphogenesis in memory formation: Synaptic and cellular mechanisms. *Behavioural Brain Research, 66*(1–2), 1–6.

Weiskrantz, L., Warrington, E. K., Sanders, M. D., & Marshall, J. (1974). Visual capacity in the hemianopic field following a restricted occipital ablation. *Brain, 97*(4), 709–728.

Weiskrantz, L. (1986). *Blindsight: A case study and implications*. Oxford: Oxford University Press.

Wernicke, C. (1874). *Der Aphasische Symptomencomplex: Eine Psychologische Studie auf Anatomischer Basis*. Breslau: Cohn and Weigert.

Wertheimer, M. (1912). Experimentelle Studien ü ber Sehen von Bewegung. *Zeits Psychology, 61*, 161–265.

Wessinger, C. M., VanMeter, J., Tian, B., Van Lare, J., Pekar, J., & Rauschecker, J. P. (2001). Hierarchical organization of the human auditory cortex revealed by functional magnetic resonance imaging. *Journal of Cognitive Neuroscience, 13*(1), 1–7.

Westmacott, R., Black, S. E., Freedman, M., & Moscovitch, M. (2004). The contribution of autobiographical significance to semantic memory: Evidence from Alzheimer's disease, semantic dementia, and amnesia. *Neuropsychologia, 42*(1), 25–48.

Whitney, D., & Levi, D. M. (2011). Visual crowding: A fundamental limit on conscious perception and object recognition. *Trends in Cognitive Neuroscience, 15*(4), 160–168.

Wierzynski, C. M., Lubenov, E. V., Gu, M., & Siapas, A. G. (2009). State-dependent spike-timing relationships between hippocampal and prefrontal circuits during sleep. *Neuron, 61*, 587–596.

Wigg, C. L., & Martin, A. (1998). Properties and mechanisms of perceptual priming. *Current Opinion in Neurobiology, 8*, 227–233.

Wilson, M. A., & McNaughton, B. L. (1994). Reactivation of hippocampal ensemble memories during sleep. *Science, 265*, 676–679.

Woodward, S. H., Kaloupek, D. G., Schaer, M., Martinez, C., & Eliez, S. (2008). Right anterior cingulate cortical volume covaries with respiratory sinus arrhythmia magnitude in combat veterans. *Journal of Rehabilitation Research and Development, 45*(3), 451–463.

Wulff, K., Gatti, S., Wettstein, J. G., & Foster, R. G. (2010). Sleep and circadian rhythm disruption in psychiatric and neurodegenerative disease. *Nature Reviews Neuroscience, 11*, 1–11.

Xuan, B., Mackie, M.-A., Spagna, A., Wu, T., Tian, Y., Hof, P. R., et al. (2016). The activation of interactive attentional networks. *NeuroImage, 129*, 308–319.

Yakovlev, P. I., & Lecours, A. R. (1967). The myelogenetic cycles of regional maturation of the brain. In A. Minokowski (Ed.), *Regional development of the brain in early life* (pp. 3–70). Philadelphia: Davis.

Yan, F., Duncan, N. W., de Greck, M., & Northoff, G. (January 2011). Is there a core neural network in empathy? An fMRI based quantitative meta-analysis Fig 2. *Neuroscience and Biobehavioral Reviews, 35*, 903–911. https://doi.org/10.1016/j.neubiorev.2010.10.009. ISSN:0149-7634.

Yang, D. Y.-L., Rosenblau, G., Keifer, C., & Pelphrey, K. A. (2015). An intergrative neural model of social perception, action observation, and theory of mind. *Neuroscience and Biobehavioral Reviews, 51*, 264–275.

Yantis, S. (2008). The neural basis of selective attention. *Current Directions in Psychological Science, 17*(2), 86–90.

Yonelinas, A. P., Otten, L. J., Shaw, K. N., & Rugg, M. D. (2005). Separating the brain regions involved in recollection and familiarity in recognition memory. *The Journal of Neuroscience, 25*(11), 3002–3008.

Yonelinas, A. P. (2002). The nature of recollection and familiarity: A review of 30 years of research. *Journal of Memory and Language, 46*(3), 441–517.

Zahr, N. M., & Sullivan, E. V. (2008). Translational studies of alcoholism: Bridging the gap. Fig 2. Alcohol Res Health. *2008, 31*(3), 215–230. Review. PubMed PMID:20041042; PubMed Central PMCID:PMC2798743.

Zatorre, R. J., & Halpern, A. R. (2005). Mental concerts: Musical imagery and auditory cortex. *Neuron, 47*(1), 9–12.

Zatorre, R. J., Chen, J. L., & Penhune, V. B. (2007). When the brain plays music: Auditory-motor interactions in music perception and production. *Nature Reviews Neuroscience, 8*(7), 547–558.

Zeki, S. M. (1974). Functional organization of a visual area in the posterior bank of the superior temporal sulcus of the rhesus monkey. *Journal of Physiology, Paris, 236*(3), 549–573.

Zeki, S. M. (1977). Colour coding in the superior temporal sulcus of rhesus monkey visual cortex. Proceedings of the Royal Society of London. Series B. *Philosophical Transactions of the Royal Society of London Series B, Biological Sciences, 197*(1127), 195–223.

Zeki, S. (1983). Colour coding in the cerebral cortex: The reaction of cells in monkey visual cortex to wavelengths and colours. *Neuroscience, 9*, 741–765.

Index

'*Note*: Page numbers followed by "f" indicate figures, "t" indicate tables and "b" indicate boxes.'

M

R

T

W

Y